Food Systems and Biodiversity in the Context of Environmental and Climate Risks

Mohamed Behnassi · Mirza Barjees Baig ·
Himangana Gupta · Rachid Sabbahi ·
Gitanjali Nain Gill · Mahjoub El Haiba
Editors

Food Systems and Biodiversity in the Context of Environmental and Climate Risks

Dynamics and Evolving Solutions

Springer

Editors
Mohamed Behnassi
Economic Social and Environmental Council (CESE), Center for Environment Human Security & Governance (CERES)
Ministry of Higher Education and Innovation
Rabat, Morocco

Himangana Gupta
United Nations University Institute for the Advanced Study of Sustainability (UNU-IAS)
Tokyo, Japan

Gitanjali Nain Gill
Faculty of Business and Law
Northumbria Law School
Northumbria University
London, UK

Mirza Barjees Baig
Prince Sultan Institute for Environmental Water and Desert Research
King Saud University
Riyadh, Saudi Arabia

Rachid Sabbahi
Research Team in Science and Technology
Higher School of Technology of Laayoune
Ibn Zohr University
Laayoune, Morocco

Mahjoub El Haiba
Faculty of Law
Economics and Social Sciences
University of Hassan II Casablanca
Casablanca, Morocco

ISBN 978-3-031-89166-3 ISBN 978-3-031-89167-0 (eBook)
https://doi.org/10.1007/978-3-031-89167-0

This Springer imprint is published by the registered company Springer Nature Switzerland AG
The registered company address is: Gewerbestrasse 11, 6330 Cham, Switzerland

Foreword by Dr. Ismahane Elouafi

In an era marked by escalating climate pressures and a critical need to build sustainable food systems, biodiversity stands at the core of our capacity to adapt and thrive. This volume presents both a reality check and a roadmap for transformative solutions in our food and agricultural systems.

Humanity has already transgressed six out of the nine global environmental boundaries: (i) bio-geochemical flows with reference to nitrogen and phosphorus, (ii) freshwater change, (iii) land system change, (iv) biosphere integrity, (v) climate change, and (vi) novel entities. Unless drastic actions are taken to halt business-as-usual, the next and seventh boundary to be transgressed will be "ocean acidification."

We simply cannot take the risks. Immediate, bold, and sustainable actions are required to safeguard the planet and the people on it. This book sets critical perspectives for the future we want. It examines pathways for sustainable food, land, and water systems, as we balance various crises including climate change, population pressure, and geopolitical tensions among others.

The themes in this book echo CGIAR's call to sustainably transform food, land, and water systems in a climate crisis. It also reflects the critical importance of integrated implementation of the three Rio Conventions (biodiversity, climate, and desertification) to deliver on the United Nations Sustainable Development Goals (SDGs). Importantly, it very-much resonates with my priorities as the CGIAR Executive Managing Director, and those of our close partners. Similarly, it responds to global challenges faced during my tenure as the Chief Scientist at the Food and Agriculture Organization of the United Nations (FAO).

The book examines how climate change—driven by large-scale emissions, unsustainable agricultural practices, and land-use changes, along with pollution and the spread of invasive pests and plant diseases—accelerates biodiversity loss. These forces disrupt essential ecosystem functions and services, such as water provision, soil fertility, and pollination, which are vital for food security and human well-being. (Behnassi et al.; Gurbanov et al.). It highlights transformative and nature-based approaches required to halt the transgression of the planetary environmental boundaries—solutions like organic farming, agroecology, agroforestry, biofortification, sustainable intensification, and land restoration. Each of these can bolster ecosystem resilience and safeguard food systems from challenges such as climate change, population pressures, and geopolitical tensions (Mrabet and Behnassi; Ampitiyawatta et al.; Dagar and Gupta; and Alemagi et al., Sabbahi). Using regional case studies to reflect worldwide contexts, these chapters stress the importance of holistic solutions, integrating all sources of knowledge—from traditional wisdom to cutting-edge science—to guide policy decisions, support capacity building, and drive technological, technical, and social innovations.

As land-use changes, pollution, and invasive species continue to drive forest and biodiversity losses, these solutions become even more critical for maintaining climate resilience and safeguarding essential ecosystem functions and services. The costs of inaction would be high—further habitat losses, drastic reduction to extinction of selected species, and forced migration. Crop biofortification (Krishnan et al.), plant improvement (Tibanda et al.), land use planning and governance (Rabi et al.), and agroecological transition (Chari and Aziz) are some of the book's promising solutions.

The book presents options for integrating biodiversity and ecosystem gains into current food, land, and water systems to enhance climate adaptation and mitigation, achieve land degradation neutrality, and improve food security, nutrition, and livelihoods for all. This would be particularly instrumental for farming communities, indigenous peoples and local communities—particularly women, youth, and other resource-constrained social groups—to boost socio-ecological resilience. There is an overdue need to bridge economic growth, agricultural productivity, environmental sustainability, and social inclusion, and the authors share options that can be integrated to effectively deliver on the three Rio Conventions. A "one health approach" is now more critical than ever, to unify environmental health, animal health, and peoples' health.

Overall, *Food Systems and Biodiversity in the Context of Environmental and Climate Risks* is a significant contribution to the growing scientific debate on the food-biodiversity-climate nexus. It offers a comprehensive, multi-regional, and cross-sectoral analysis of the various natural and anthropogenic factors that affect ecosystem integrity and resilience. The chapters are grounded in empirical research and provide actionable insights for future-proof policy innovations, making this volume a vital resource for researchers, policymakers, and practitioners seeking to build sustainable, climate-resilient food, land, and water systems.

The authors call for the whole-of-society to take participatory actions to transition our food, land and water systems into solutions to biodiversity loss, climate change, and ecosystem degradation. As the issues and challenges reflected in this volume evolve with time, the book represents an inspiration for further research based on various megatrends like climate change, population pressure, and geopolitical tensions among others.

Having had the privilege of working with most of the authors during my career, it is my great honor to encourage all actors to explore the technological, technical and social innovations in this important book.

Dr. Ismahane Elouafi
CGIAR Executive Managing Director
Montpellier, France

Dr. Ismahane Elouafi is the Executive Managing Director of CGIAR, effective December 1, 2023. She previously held the position of Chief Scientist at the Food and Agriculture Organization (FAO) of the United Nations. Prior to that, she was the Director General at the International Center for Biosaline Agriculture (ICBA) based in the United Arab Emirates. She has held several senior scientific and leadership positions, including: Senior Adviser to the Assistant Deputy Minister, Agriculture and Agri-Food Canada Research Branch in Ottawa (AAFC), Canada (2006–2007); the National Manager of Plant Research Section (2007–2010); and Director of Research Management and Partnerships Division at the Canadian Food Inspection Agency (CFIA) (2010–2012). She has worked as a scientist with several international research organizations and has been a member of various strategy expert panels and advisory groups, including the Global Commission on Adaptation (GCA), and HarvestPlus. She has been a member of several governance boards, including: the boards of the International Food Policy Research Institute (IFPRI) in the United States; the Centre for Agriculture and Bioscience International (CABI) in the United Kingdom; and the Professional Development Institute in Canada. She was also a member of the Scientific Group for the 2021 UN Food Systems Summit and of the CGIAR System Board.

Foreword by Juan Lucas Restrepo

Today, the challenges we face at the intersection of food systems, biodiversity, and climate risks are unprecedented in both scale and urgency. This volume, Food Systems and Biodiversity in the Context of Environmental and Climate Risks, brings these issues to the forefront, offering an in-depth exploration of the complex dynamics that drive biodiversity loss, climate vulnerability, and food insecurity. Authored by an esteemed group of researchers and practitioners, it provides a rigorous analysis of the current threats to ecosystems and highlights transformative solutions that can mitigate these risks.

This book underscores that food systems are at once contributors to and victims of climate change. Unsustainable agricultural practices, land use, and pollution are not only driving biodiversity decline but also intensifying climate risks. Biodiversity, in turn, is a fundamental resource for resilient food systems, playing a crucial role in ecosystem services such as soil fertility, biological control, water purification, and pollination—services essential for food security and human well-being. With the alarming prospect of one million species at risk of extinction in coming decades, the urgent need for sustainable and biodiversity-friendly food systems has never been clearer.

This volume highlights several elements and approaches to drive sustainable, climate resilient, and biodiverse food systems. It signals how agroecological practices, organic farming, agroforestry, and sustainable intensification can reduce greenhouse gas emissions, promote soil health, and support ecosystem restoration, all while securing the food base. By examining case studies across diverse regions, the authors make a compelling case for adopting biodiversity-centered approaches to agriculture.

Second, the book emphasizes the importance of blending traditional ecological knowledge with modern scientific progress to generate climate-smart, biodiversity-rich food systems. By fostering cross-cultural and interdisciplinary collaboration, solutions can be developed that respect local knowledge systems and are tailored to specific environmental and cultural contexts. Effective policy reforms, responsible investment, and nature-based solutions are also highlighted as pathways to ensure that agriculture supports both food security and ecosystem health.

Finally, the book calls for horizontal partnering, capacity-sharing networks, and co-innovation among stakeholders. Solutions to biodiversity and climate challenges require the active participation of local communities, policymakers, scientists, academic and research organizations and the private sector. By creating platforms for co-learning and capacity building, communities are better enabled to adopt sustainable practices and improved resilience.

Through empirical research and actionable insights, Food Systems and Biodiversity in the Context of Environmental and Climate Risks serves as a vital resource for those working at the intersection of biodiversity, climate resilience, and food security. Its message is both urgent and inspiring: the time for transformative action is now, and by working together across disciplines and regions, we can safeguard our ecosystems and secure a sustainable future.

Juan Lucas Restrepo
Global Director
Partnerships and Advocacy, CGIAR
Director General
Alliance of Bioversity International and CIAT
Rome, Italy

Juan Lucas Restrepo is Global Director of Partnerships and Advocacy, CGIAR, and Director General of the Alliance of Bioversity International and the International Center for Tropical Agriculture (CIAT). A Colombian and French national, he worked in the agricultural domain, both in the public and private sectors, for the past 25 years. He has significant experience in policy, value chains, markets, and leading agricultural research. He has supported the work of CGIAR through the various governance roles he held over the years in the Committee of Genetic Resources Policy, Oversight and Executive Committees, and as the representative of the Colombian Government on the CIAT Board of Directors as an ex-officio member. Among other positions, in Colombia, he served as the Vice Minister of the Ministry of Agriculture and Rural Development, as the Executive Director of AGROSAVIA, the largest agricultural research organization in the country, and as Chief Commercial Officer of the National Federation of Coffee Growers. As its Chair, he also led the Global Forum for Agricultural Research (GFAR), hosted by the Food and Agriculture Organization of the United Nations (FAO), and its efforts to better integrate the agricultural research for the development community, with change and innovation by farmers. He holds a degree in civil engineering from Universidad de los Andes and an MS in agricultural economics from Cornell University.

Acknowledgements

The completion of this volume would not have been possible without the contributions and support of numerous individuals and institutions, to whom we extend our deepest gratitude.

First and foremost, we wish to express our sincere thanks to the distinguished chapter authors, whose expertise and intellectual contributions form the core of this work. Their research, insights, and scholarly rigor have significantly enriched the quality and scope of this volume. We are particularly appreciative of their responsiveness and professionalism throughout the entire editorial process.

We extend our deepest appreciation to the co-editors of this volume: *Dr. Mirza Barjees Baig* of the Prince Sultan Institute for Environmental, Water, and Desert Research, King Saud University, Saudi Arabia; *Dr. Himangana Gupta* of the United Nations University—Institute for the Advanced Study of Sustainability, Japan; *Dr. Rachid Sabbahi* of the Higher School of Technology of Laayoune, Ibn Zohr University, Morocco; *Dr. Gitanjali Nain Gill* of the Faculty of Business and Law, Northumbria University, United Kingdom; and *Dr. Mahjoub El Haiba* of Hassan II University of Casablanca, Morocco. Their collaborative spirit, expertise, and commitment were invaluable in shaping the direction and content of this work.

We would like to express our sincere appreciation to the institutions that provided critical support throughout this project: *The Center for Environment, Human Security, and Governance (CERES), the Prince Sultan Institute for Environmental, Water, and Desert Research*, and *King Saud University*. Their resources, encouragement, and institutional backing were pivotal in facilitating the collaboration and research necessary for the completion of this volume.

Additionally, we wish to recognize the invaluable assistance of our administrative and research teams, whose dedication and efficiency in handling the logistics and editorial coordination have been critical to the success of this project. Their efforts behind the scenes have ensured a smooth editorial process, for which we are deeply thankful.

While every effort has been made to ensure the accuracy and scholarly integrity of this volume, any oversights or errors remain solely the responsibility of the editors.

Lastly, we are profoundly grateful to the reviewers, whose critical assessments and constructive feedback were indispensable in refining the chapters. Their thorough evaluations have greatly enhanced the academic quality of the volume, ensuring its coherence and scholarly rigor. We acknowledge the time and effort they devoted to this endeavor and are deeply appreciative of their expertise. On behalf of my co-editors, I would specifically like to acknowledge, with sincere and deepest thanks, the following reviewers:

- Prof. Dr. Muhammad Farooq, Sultan Qaboos University, Oman
- Dr. Muhammad Siddiq, Food Sciences, Michigan State University, USA
- Dr. AbdulMalek A. Al-Shaikh, Director, Prince Sultan Institute for Environmental, Water and Desert Research (PSIEWDR), King Saud University, Riyadh, Kingdom of Saudi Arabia
- Prof. Dr. Loutfy El-Juhany, King Saud University, Saudi Arabia
- Prof. Dr. Muhammad Ismail Kumbhar, Dean of the Faculty of Agriculture, Sindh Agriculture University, Tandojam, Pakistan
- Azizullah Shah, Director, Pakistan Agricultural Research Council, Islamabad, Pakistan
- Imtiaz Ahmad, Director, Pakistan Agricultural Research Council, Islamabad, Pakistan
- Prof. Dr. Inayatullah Rajpar, Department of Soil Sciences, Sindh Agriculture University, Tandojam, Pakistan
- Eng. Ali Wafa A. Abu-Risheh, Prince Sultan Institute for Environmental, Water and Desert Research (PSIEWDR), King Saud University, Riyadh, Kingdom of Saudi Arabia
- Dr. Abed Alataway, Prince Sultan Institute for Environmental, Water and Desert Research (PSIEWDR), King Saud University, Riyadh, Kingdom of Saudi Arabia
- Prof. Dr. Mohamed Esham, Faculty of Agricultural Sciences, Sabaragamuwa University of Sri Lanka, Sri Lanka
- Dr. Ghulam M. Shah, International Centre for Integrated Mountain Development (ICIMOD), Kathmandu, Nepal
- Dr. Abid Hussain, Senior Economist, Food Systems, Mountain Economy, and Sustainable Development, International Centre for Integrated Mountain Development (ICIMOD), Kathmandu, Nepal
- Dr. Faisal M. Qamer, Lead, Resilient River Basins Intervention, International Centre for Integrated Mountain Development (ICIMOD), Kathmandu, Nepal
- Dr. Farid Ahmad, International Centre for Integrated Mountain Development (ICIMOD), Kathmandu, Nepal
- Prof. Dr. Ata Ullah Shah, Vice Chancellor, International Karakorum University, Gilgit Baltistan, Pakistan
- Prof. Mohamed Habib Ben Jamâa, General Director of Plant Health and Control of Agricultural Inputs, Tunis, Tunisia
- Dr. Muazamma Batool, Senior Scientific Officer, Regional Agricultural Research Institute, Bhawalpur, Pakistan

- Dr. Turki K. F. Asiri, Prince Sultan Institute for Environmental, Water and Desert Research (PSIEWDR), King Saud University, Riyadh, Kingdom of Saudi Arabia
- Dr. Zulfqar Ali Gurmani, National Agricultural Research Centre, Islamabad, Pakistan
- Dr. M. I. Sultani, Director, National Agricultural Research Centre, Islamabad, Pakistan
- Prof. Dr. Ghulam Jilani, Department of Soil Sciences, Pir Mahar Ali Shah Arid Agriculture University, Rawalpindi
- Dr. Ali Raza Gurmani, Department of Soil and Climate Science, The University of Haripur, Pakistan
- Prof. Dr. Ayoub Rashid, Government College University, Lahore, Pakistan
- Dr. Rajendra Uprety, Division Chief, Food Security and Agribusiness Promotion Division, Ministry of Land Management Agriculture and Cooperative, Biratnagar, Nepal
- Dr. Muazamma Batool, Senior Scientific Officer, Regional Agricultural Research Institute, Bhawalpur, Pakistan
- Dr. Rodolfo Dirzo, Department of Biology, Stanford University, USA
- Prof. Dr. Gottlieb Basch, University of Evora, Portugal
- Dr. Rurik List, Universidad Autónoma Metropolitana, Mexico
- Dr. José Luis Riveros, Pontificia Universidad Católica de Chile, Santiago, Chile
- Dr. Salih Gucel, Yakın Doğu, Environmental Research Center Üniversitesi, Near East University, Turkey
- Dr. Mohamed Zakaria Hussin, Department of Forest Science and Biodiversity, Serdang, Selangor, Malaysia
- Dr. Faiza Shaikh, Plant Ecologist/Environmentalist, Milton, Ontario, Canada
- Dr. Humera Nazneen, Department of Botany, Nizam College, Osmania University, Hyderabad, Telangana, India
- Dr. Brent L. Nielsen, Department of Microbiology and Molecular Biology, Brigham Young University, Provo, Utah, USA
- Dr. Moulay Chrif Smaili, National Institute for Agricultural Research (INRA) and CRRA-Kénitra, Morocco
- Dr. Yinglong Chen, The UWA Institute of Agriculture and School of Agriculture and Environment, the University of Western Australia, Perth, Australia
- Dr. Md. Nur Alam, Department of Pharmacy, Jahangirnagar University, Savar, Dhaka, Bangladesh
- Dr. Fazleen Abdul Fatah, Faculty of Plantation and Agrotechnology, Universiti Teknologi MARA, Jasin Campus, Melaka, Malaysia
- Dr. Karim-Aly S. Kassam, College of Agriculture and Life Sciences, Cornell University, Ithaca, New York, USA
- Dr. Aziz Ali Khan, University of Central Asia (UCA), Khorog Campus, Tajikistan
- Prof. Dr. Michael Keusgen, University of Marburg, Marburg, Germany
- Prof. Dr. Cristián Bonacic, Director Fauna Australis, Pontificia Universidad Católica de Chile

- Dr. Asif Sajjad, Department of Entomology, Islamia University, Bahawalpur, Pakistan
- Dr. nat. techn. Demet Cekin, RWTH Aachen University, Academic and Research Department Engineering Hydrology UNESCO Chair in Hydrological Changes and Water Resources Management Mies-van-der-Rohe-Straße, Aachen, Germany
- Prof. Dr. A. D. Ampitiyawatta, Faculty of Agriculture, Sabaragamuwa University of Sri Lanka
- Prof. Dr. Azeem Muhammad Khan, Agriculture University, Peshawar, KPK, Pakistan
- Dr. Muhammad Adnan, Department of Botany, KUST, Kohat
- Dr. Asif Khan, Department of Technology, Universidade Estadual de Maringá, Umuarama, Brazil
- Prof. Dr. Waheed Murad, Department of Botany, Abdul Wali Khan University Mardan, Mardan, KPK Pakistan
- Dr. Muhammad Jawad, Center for Geographical Information System, University of Punjab, Lahore, Pakistan
- Dr. Sajid Ali, Department of Horticulture and Life Science, Yeungnam University, Republic of Korea
- Prof. Dr. Asad Ali, Department of Agriculture, Abdul Wali Khan University Mardan, Mardan, KPK Pakistan
- Prof. Dr. Zaheer-ud-din Khan, Distinguished Professor, Department of Botany, Government College University (GCU), Lahore, Pakistan
- Prof. E. P. N. Udaya Kumara, Sabaragamuwa University of Sri Lanka, Sri Lanka
- Dr. Murodbek Laldjebaev, School of Arts and Sciences, University of Central Asia, Republic of Tajikistan
- Prof. Dr. Paul Dijkwel, School of Fundamental Sciences, Massey University, Palmerston North, New Zealand
- Dr. Midhula Gireesh, Gulf Coast Research and Education Centre, University of Florida, Wimauma, FL USA
- Prof. Wathugala Gamage Deepthi Lakmini, Faculty of Agriculture, Ruhuna University, Sri Lanka
- Prof. Nimal Abesinghe, Faculty of Agriculture, Rajarata University of Sri Lanka, Sri Lanka
- Prof. Dr. Monier M. Abd El-Ghani, Faculty of Science, Cairo University, Giza, Egypt
- Dr. Amanullah Mahar, Centre for Environmental Science, University of Sindh, Jamshoro, Pakistan
- Dr. Hadi Bux, Institute of Plant Sciences, University of Sindh, Jamshoro, Pakistan
- Prof. Dr. Shaukat Ali, Department of Zoology, GCU Lahore, Pakistan
- Dr. Mazharul Aziz, Agricultural Meteorology Expert, Agrometeorological Advisory Services, Climate Change Adaptation and Mitigation in Crop Sector, Dhaka, Bangladesh
- Dr. Sumera Gul Bhatti, Department of Environmental Science, Lahore College Women University, Lahore

- Dr. Habib Rehman, Troy University, Troy, Alabama, USA
- Prof. Dr. Karim-Aly S. Kassam, College of Agriculture and Life Sciences, Cornell University, Ithaca, New York, USA
- Prof. Dr. Prasajith Kapila Dissanayake, Sabaragamuwa University of Sri Lanka, Sri Lanka
- Dr. Patrick Lhomme, Research Institute for Biosciences, University of Mons, Mons, Belgium
- Dr. Ijaz Ahmad, National Agriculture Research Center, Islamabad, Pakistan
- Prof. Dr. Xiangkai Li, School of Life Sciences, Lanzhou University, Gansu, P.R. China
- Professor Rodolfo Dirzo, Department of Biology, Stanford University, USA
- Professor Rurik List, Universidad Autónoma Metropolitana, Mexico
- Dr. José Luis Riveros, Pontifical Catholic University, Chile
- Dr Juan David Amaya, Javeriana University, Colombia
- Dr. Salih Gucel, Yakın Doğu Üniversitesi, Çevre Araştırmaları Merkezi, Turkey
- Prof. Dr. Mohamed Zakaria Hussin, Department of Forest Science and Biodiversity Faculty of Forestry and Environment, Universiti Putra Malaysia, Selangor, Malaysia
- Prof. Dr. Ibrahim Ozdemir, Department of Wildlife Ecology and Management, Faculty of Forestry, Isparta University of Applied Sciences, Cunur Isparta, Turkey
- Dr. Humera Nazneen, Assistant professor, Department of Botany, Nizam College Osmanai University, Hyderabad, India
- Dr. Faiza Shaikh, Plant Ecologist/Environmentalist, Milton, Ontario, Canada
- Prof. Dr. John Fike, School of Plant and Environmental Sciences, Virginia Tech University, Blacksburg, VA, USA
- Prof. Dr. Brent L. Nielsen, Department of Microbiology and Molecular Biology, Brigham Young University, Provo, Utah, USA
- Dr. Syed Amir Manzoor, Bahauddin Zakariya University, Multan, Pakistan
- Dr. Fawad Z. A. Khan, University of Georgia, USA
- Dr. Shimat V. Joseph, University of Georgia, USA
- Prof. Dr. Faiza Sharif, Sustainable Development Study Centre, GC University, Lahore, Pakistan
- Dr. Syed Aziz, Department of Environmental science, University of Veterinary and Animal Sciences (UVAS), Lahore, Pakistan
- Dr. Syed Ajijur Rahman, College of Urban and Environmental Science, Peking University, Beijing, China
- Ms. Rimsha Javed, Sustainable Development Study Centre, GC University Lahore, Pakistan
- Prof. Dr. Weldesemayat G. Sileshi, University of Kwazulu-Natal, Pietermaritzburg, South Africa
- Prof. Dr. Amtul Bari Tabinda, Sustainable Development Study Centre, GC University Lahore, Pakistan
- Dr. Sumera Aslam, GC University, Faisalabad, Pakistan
- Dr. Um-e- Kalsoom Naqvi, GC University, Lahore, Pakistan

- Dr. Nauman Khalid, School of Food and Agricultural Sciences, University of Management and Technology (UMT), Lahore, Pakistan
- Dr. Sumera Gul Bhatti, Assistant Professor, Department of Environmental Science, Lahore College Women University, Lahore, Pakistan
- Ms. Javairia Ansar, Sustainable Development Study Centre, GC University Lahore, Pakistan
- Prof. Jameel M. Al-Khayri, Department of Agricultural Biotechnology, College of Agriculture and Food Sciences, King Faisal University, Kingdom of Saudi Arabia
- Prof. Dr. Abdullah Yasar, Sustainable Development Study Centre, GC University Lahore, Pakistan
- Prof. Dr. João Paulo Jorge, School of Tourism and Maritime Technology (ESTM), Portugal
- Prof. Dr. Anastasia Pantera, Agricultural University of Athens, Greece
- Prof. Dr. Saman Herath, Uwa Wellassa University, Sri Lanka
- Prof. Dr. Sohaib Muhammad, Government College University, Lahore, Pakistan
- Dr. Anderson Kehbila, Research fellow, Stockholm Institute of Environment, Nairobi, Kenya
- Dr. Nandita Baruah, Biswanath College of Agriculture, Assam, India
- Dr. Renu Gupta, Sher-e-Kashmir University of Agricultural Sciences and Technology of Jammu, Jammu (J&K), India
- Prof. Dr. Anshuman Kohli, Bihar Agricultural University, Bhagalpur, Bihar, India
- Dr. Jaime de Pablo Valenciano, University of Almeria, Spain
- Prof. Dr. J. Bimali Koongolla, Faculty of Fisheries and Marine Sciences and Technology, University of Ruhuna, Matara, Sri Lanka
- Prof. Dr. Mohan Kumar B., Ex-Vice Chancellor, Arunachal University of Studies, Thrissur, Kerala, India
- Dr. Rajeswari Das, School of Agriculture, GIET University, India
- Dr. Vinayak U, Department of Forest Products and Utilization, College of Forestry, Karnataka, India
- J. Hunter Adams, Cypress Environmental Laboratory, Wichita Falls, Texas, USA
- Prof. Dr. Riaz Hussain Qureshi, Ex-Vice Chancellor, University of Agriculture, Faisalabad, Pakistan
- Prof. Dr. Mohammad Suhail Director, Centre of Applied Remote Sensing and GIS Applications, Samarkand State University, Uzbekistan
- Prof. Dr. Mohamed Najim M. Mujithaba, Faculty of Agriculture, University Islam Sultan Sharif Ali—Brunei, Darussalam
- Dr. Tefide Kizildeniz, Biosystem, Faculty of Agricultural Sciences and Technologies, Niğde Ömer Halisdemir University, Turkey
- Prof. Dr. Hemanthi Ranasinghe, Forestry and Environmental Science, University of Sri Jayewardenepura Nugegoda, Sri Lanka
- Prof. Dr. Ayad Al-Quraishi, Tishk International University, Kurdistan Region, Iraq

- Dr. Solh Mahmoud, Former Director General International Center for Agricultural Research in the Dry Areas (ICARDA)
- Dr. Mina Devkota, Agronomist, International Center for Agricultural Research in the Dry Areas (ICARDA), Nepal
- Prof. Dr. Rahmatullah Qureshi, Dean of the Faculty of Sciences, Pir Mehr Ali Shah Arid Agriculture University, Rawalpindi, Pakistan
- Dr. Muneeb M. Musthafa, Department of Biosystems Technology, Faculty of Technology, South Eastern University of Sri Lanka, Sri Lanka
- Prof. Dr. Hasan Hüseyin Atar, Dean of the Faculty of Agriculture, Ankara University, Turkey
- Prof. Dr. Mohammed Bahir, Faculty of Science, Cadi Ayyad University, Marrakech, Morocco
- Prof. Dr. Asad Ullah, Centre of Plant Biodiversity, University of Peshawar, Khyber Pakhtunkhwa, Peshawar, Pakistan
- Dr. Ahmet Nuri Karabaş, Expert on Soils, Ankara, Turkey
- Prof. Dr. Ahmet Çolak, Faculty of Agriculture, Ankara University, Turkey
- Dr. Akif Özkaldi, Former Vice Minister of Agriculture and Forestry, Turkey
- Dr. Muhammad Munir Goryia, Pakistan Agricultural Research Council, Islamabad, Pakistan
- Dr. Nizar Haddad, Director General, National Agriculture Research Centre (NARC), Jordan
- Prof. Dr. Imran Khan, Department of Agronomy, University of Agriculture, Faisalabad, Pakistan
- Prof. Dr. John Eriba Ochi, Abubakar Tafawa Balewa University (ATBU), Bauchi State, Nigeria
- Ms. Saraswati Aryal, Ministry of Forest and Environment, Kathmandu, Nepal
- Dr. Techn Vesela Tanaskovic Gassner, Founder of Afforest4Future, Vienna, Austria
- Dr. Hemu (Kharel) Kafle, Kathmandu Institute of Applied Sciences, Weizmann Institute of Science, Kathmandu, Nepal
- Prof. Dr. Muhammad Abdullah, The Islamia University of Bahawalpur, Bahawalpur, Pakistan
- Prof. Dr. Mohamed Neffati, Director of Research and Training, Arid Zone Research Institute (IRA), Médenine, Tunisia
- Dr. Susma Giri, Scientist/Director of Research, Kathmandu Institute of Applied Sciences Kathmandu, Nepal
- Prof. Dr. Javed Iqbal, Plant Protection Department, King Saud University, Riyadh, Saudi Arabia
- Prof. Dr. Mirza B. Baig, Prince Sultan Institute for Environmental, Water and Desert Research (PSIEWDR), King Saud University, Riyadh, Kingdom of Saudi Arabia
- Prof. Dr. Uzma Hanif, Department of Botany, GC University Lahore, Pakistan
- Dr. Memuna Ghafoor, Department of Botany, GC University Lahore, Pakistan

- Dr. Ranjan Prakash Shrestha, European Union Delegation to Nepal, Kathmandu, Nepal
- Dr. Yasin Wazir, Director General Soil and Water Conservation, KPK, Pakistan
- Prof. Dr. Gary S. Straquadine, Utah State University, UT, USA
- Dr. Muhammad Waheed, Lecturer in Botany, GC University, Lahore, Pakistan
- Dr. Sohaib Muhammad, Assistant Professor of Botany, GC University, Lahore, Pakistan

Mohamed Behnassi

About the Publishing Institutions

Center for Environment, Human Security and Governance (CERES), Morocco

CERES, formerly the North-South Center for Social Sciences (NRCS), emerged in 2015 as an independent, non-profit research institute built upon the expertise of Moroccan researchers. Its global ambition is to be a leading think tank and a reference point for positive change, especially in the Global South. Since its founding, CERES has fostered a robust network encompassing researchers, policymakers, students, practitioners, and journalists across diverse fields. This network has propelled CERES members to contribute to prestigious assessments like Ipbes, Medec, and EuroMeSco. The Center's expertise is further recognized by its membership in the MedThink 5+5, a group shaping research and policy agendas in the Mediterranean. CERES not only provides expert knowledge to various institutions, but also organizes international conferences and capacity-building workshops. Their extensive publications, including books, scientific papers, and policy studies, enjoy global recognition and distribution. The Center focuses on emerging research areas within the human-environment-climate nexus, employing multidimensional, multiscale, interdisciplinary, and policy-making perspectives. Through these initiatives, CERES aims to deliver cutting-edge expertise, advance scientific understanding and application, and ultimately bridge the gap between science and policy for impactful solutions.

The Prince Sultan Institute for Environmental, Water and Desert Research (PSIEWDR), King Saud University, Riyadh, Kingdom of Saudi Arabia

PSIEWDR was established in 1986 to conduct scientific research related to environmental issues and water resources. It also engages with vital issues related to the problem of aridity and the desert environment. It conducts development initiatives for the country's desert areas, particularly programs for combating desertification in the Arabian Peninsula. PSIEWDR designed and carried out two major water harvesting and storage programs, including the construction of purpose-built infrastructure, throughout the Kingdom of Saudi Arabia using novel techniques and equipment. The Institute actively applies remote sensing technologies using advanced satellite image processing systems and GIS to study the country's environment and natural resources. In 2007, the Institute published *The Space Image Atlas of the Kingdom of Saudi Arabia*, and it is currently developing *The Environmental Atlas of the Kingdom of Saudi Arabia*. The Institute has been the primary sponsor of the biennial International Conference on Water Resources and Arid Environments (ICWRAE) held in Riyadh, Saudi Arabia since 2004. The institute hosts the General Secretariat of the Prince Sultan Bin Abdulaziz International Prize for Water (PSIPW) which honors scientists all over the world for their innovative water-related research. PSIPW, in turn, has many agreements with various international water associations as well as a close partnership with the United Nations. PSIPW and the United Nations Office of Outer Space Affairs (UNOOSA) jointly produce and maintain the International Space4Water Portal, an online hub for all stakeholders involved in utilizing space technologies for water resources applications.

Contents

Editors and Contributors

About the Editors

Dr. Mohamed Behnassi is currently a Senior Environmentalist Expert at the Economic Social and Environmental Council (CESE), Rabat, Morocco. From 2004 to 2022, he served as a Professor, Head of Public Law Department, and Director of the Research Laboratory for Territorial Governance, Human Security and Sustainability (LAGOS), Faculty of Law, Economics, and Social Sciences of Agadir, Ibn Zohr University. He holds a Ph.D. in International Environmental Law and Governance (2003), a M.Sc. in Political Sciences and International Relations (1997), and a B.A. of Administration (1995) from the Faculty of Law, Economics, and Social Sciences, Hassan II University of Casablanca. He obtained a Diploma in International Environmental Law and Diplomacy from the University of Eastern Finland and UNEP, 2015 and is also Alumnus of both the International Visitors Leadership Program of the Department of State (2011, USA), and the Near East South Asia Center for Strategic Studies (NESA), National Defense University, Washington, DC (2022–24, USA).

Dr. Behnassi is currently the Founding Director of the Center for Environment, Human Security and Governance (CERES). He was appointed in 2019 as Expert Evaluator for the National Center for Scientific and Technical Research (CNRST/Morocco), and selected twice (2019–2024) as an Assessment Scoping Expert and a Review Editor by the Intergovernmental Science-Policy Platform on Biodiversity and Ecosystem

Services (IPBES—Thematic assessment of the interlinkages among biodiversity, water, food and health). He was a Lead Author (LA) of the Human Security Chapter of 1st MEDECC Assessment Report (MAR1): Climate and Environmental Change in the *Mediterranean Basin - Current Situation and Risks for the Future* (2021) and Coordinating Lead Author (CLA) of the MEDECC Special Report on "the Nexus between Water, Energy, Food and Ecosystems". In 2024, he was nominated and certified as the UNFCCC Expert Evaluator (Roster of Experts), and as a Member of the PRIMA Foundation Scientific Evaluation Committee (SEC) to evaluate proposals submitted to the Thematic Area "Leveraging urban and local food systems for sustainable food systems transformation".

Dr. Behnassi is a Senior Researcher in the areas of International Law and Politics of Environment and Human Security. In addition to his participation in more than 70 international and national conferences, he has published 66 scientific papers and 22 books in these areas, including recent ones on: *The Water, Climate, and Food Nexus: Linkages, Challenges and Emerging Solutions* (Springer, 2024); *Food Security and Climate-Smart Food Systems—Building Resilience for the Global South* (Springer, 2022); and *The Climate-Conflict-Migration Nexus from a Human Security Perspective* (Springer, 2022). Dr. Behnassi serves as a reviewer for many global publishers (such as Routledge and Springer) and scientific journals with high impact factor. He has organized many international conferences covering the above research areas, managed many research and expertise projects for national and international institutions, and is regularly requested to provide scientific expertise nationally and internationally. Other professional activities include social compliance auditing and consultancy by monitoring human rights at work and the sustainability of the global supply chain.

Dr. Mirza Barjees Baig is a professor at the Prince Sultan Institute for Environmental, Water, and Desert Research, King Saud University, Saudi Arabia. He earned his MS degree in International Agricultural Extension in 1992 from Utah State University, Logan, Utah, USA, and was placed on the "Roll of Honor". He completed his Ph.D. in Extension for Natural Resource Management from the University of Idaho, USA, and was honored with the "1995 Outstanding Graduate Student Award". He has published extensively on the issues associated with natural resources in national and international journals. He has also made oral presentations about agriculture, natural resources, and the role of extension education at various international conferences. He serves on the editorial boards of many international journals and is a member of many international professional organizations. He has produced six books on climate change, food security, and climate-smart agriculture. Food waste, water management, degradation of natural resources, the deteriorating environment, and their relationship with society and the community are his areas of interest. He has attempted to develop strategies to conserve natural resources, promote the environment, and develop sustainable communities. He started his scientific career in 1983 as a researcher at the Pakistan Agricultural Research Council, Islamabad, Pakistan. He served at the University of Guelph, Ontario, Canada, as a Special Graduate Faculty member from 2000 to 2005. He served as a foreign professor at the Allama Iqbal Open University (AIOU), Pakistan, through the Higher Education Commission from 2005 to 2009.

Dr. Himangana Gupta is Research Fellow and Academic Associate at United Nations University—Institute for the Advanced Study of Sustainability (UNU-IAS). She is an expert in climate change and biodiversity policy and has worked on climate adaptation, forestry, women and climate change, energy efficiency, and national GHG inventories. She was also a part of the National Communication Cell of the Indian Ministry of Environment, Forest and Climate Change from 2016–2019, and contributed to India's Second and Third Biennial Update Reports to the UNFCCC. She is a certified expert in climate adaptation finance. She has seven edited books to her credit and more than 40 research publications, including book chapters. Currently, she is also serving as Lead Author for the Nexus Assessment Report of the Intergovernmental Platform on Biodiversity and Ecosystem Services.

Dr. Rachid Sabbahi has been an associate professor at the Higher School of Technology—Laayoune, Ibn Zohr University, Morocco, since 2019. He holds a bachelor's degree (2000) in animal biology from Hassan II University, Casablanca, Morocco, and a master's degree (2003) and doctorate (2008) in biological sciences from Quebec University, Montreal, Canada. Dr. Sabbahi specializes in the biological control of insect pests and plant diseases, plant-microbe interaction, environmental microbial biotechnology, the biological activity of medicinal plants, biocomposite materials for the removal of organic pollutants and heavy metals, insect pollination, and crop production. In addition, he has co-managed many national and international research projects, published several scientific papers, and participated in many international conferences.

Dr. Gitanjali Nain Gill is a tenured full professor of environmental law at the Faculty of Law, Northumbria University, Newcastle, UK. Her research interests and published work form a coherent thematic body that includes climate change, access to environmental justice with a focus on India's National Green Tribunal, land rights, SDGs and sustainability. Professor Gill acts as an international expert member to prestigious international organizations including Regional Leader for Asia at the Sabin Centre Peer Review Network on Climate Change Litigation, Columbia University; British Institute of International and Comparative Law, London; and World Commission on Environmental Law Special Task Force on Rights of Nature, IUCN. She has also provided intellectual and disciplinary expertise by identifying and contributing specialized environmental courts and tribunals best practices, these being particularly instructive for the global south, to the United Nations Environment Programme (2016 and 2022), and to the United Nations Expert Consultation Group OHCHR Accountability and Remedy Project

II. Professor Gill is conferred the title of the prestigious Fellow of the Academy of Social Sciences, UK. She is also a recipient of the Global Initiative of Academic Networks in Higher Education Award (GIAN) by the Ministry of Human Resources and Development, Government of India. She was awarded prestigious research grants including British Academy Grants on environmental justice in India (2013–16; 2020–2023—Principal Investigator), National Environment Research Council Grant on climate change and decarbonization agenda (2022—Principal Investigator), and Responsible AI and Law Enforcement (AHRC and UK Keystone- Co- Investigator). Through awareness and capacity-building empowerment programmes, Prof. Gill's action research helps build coalitions, trust and confidence between Indian NGOs and the marginalized communities to support victims of environmental degradation.

Dr. Mahjoub El Haiba is a highly respected expert in human rights, international law, and environmental law, currently serving as a member of the United Nations' Human Rights Committee. He has been a professor at Hassan II University, Casablanca, since 1983 and a part-time lecturer at Mohamed V University in Rabat since 1997. Dr. El Haiba holds a State Doctorate in Public Law and Political Science from Hassan II University, alongside postgraduate degrees in international relations and political science. Over the course of his career, he has played a pivotal role in advancing Morocco's human rights and environmental policies, particularly during his tenure as the First Inter-ministerial Delegate for Human Rights from 2011 to 2018. In this capacity, he coordinated the country's reports for the Universal Periodic Review (UPR) and other United Nations human rights treaty bodies, leading delegations in interactive dialogues with UN bodies. A former Secretary General of the Advisory Council on Human Rights (CCDH) and a member of the Equity and Reconciliation Instance (IER), Dr. El Haiba was deeply involved in Morocco's transitional justice process. He has been a founding member of the Moroccan Human Rights Organization (OMDH) and played a key role in shaping Morocco's human rights landscape, advocating for governance reforms and justice mechanisms. His work in environmental law is equally impressive, as he is a founding member of the Moroccan Society for Environmental Law (SOMADE) and the Arab Network for Environment and Development (RAED). He has consulted for the United Nations Environment Programme (UNEP) and the Food and Agriculture Organization (FAO), contributing to environmental protection projects and policies in Morocco and other regions. Dr. El Haiba has published extensively on human rights, environmental law, and sustainable development, and he has translated and coordinated key studies on topics such as biodiversity and women's rights in environmental contexts. His academic roles include teaching public international law, environmental law, and human rights mechanisms, while supervising graduate and doctoral research. He has been an active participant in international forums, representing Morocco at human rights conferences in Geneva, New York, The Hague, and other global centers, and played a key role in organizing the World Human Rights Forum in Marrakech. Additionally, he has served as President of the jury for various environmental awards, including the League of Arab States' Environment Award and the Hassan II Environment Award. Dr. El Haiba's contributions to international human rights and environmental governance extend to his advisory roles, including being a member of scientific and ethics committees at universities and research centers. His leadership in both fields underscores his commitment to advancing Morocco's legal frameworks and its international presence in human rights and environmental protection. Throughout his career, Dr. El Haiba has made substantial contributions to policy development, academic research, and international diplomacy, solidifying his reputation as a leading figure in both human rights and environmental law.

Contributors

Monier M. Abd El-Ghani Cairo University Herbarium, Department of Botany and Microbiology, Faculty of Science, Cairo University, Giza, Egypt

Muhammad Awais Ahmad Institute of Plant Protection and Department of Outreach and Continuing Education, Muhammad Nawaz Shareef University of Agriculture, Multan, Pakistan

Mukhtar Ahmad Department of Agronomy, Pir Mehr Ali Shah-Arid Agriculture University, Rawalpindi, Pakistan

Mushtaq Ahmad Department of Plant Sciences, Quaid-i-Azam University, Islamabad, Pakistan

Shabir Ahmad Department of Plant Sciences, Quaid-i-Azam University, Islamabad, Pakistan

Abdulmalek A. Al Shaikh Prince Sultan Institute for Environmental, Water and Desert Research, King Saud University, Riyadh, Kingdom of Saudi Arabia

Jameel M. Al-Khayri Department of Agricultural Biotechnology, College of Agriculture and Food Sciences, King Faisal University, Al-Ahsa, Saudi Arabia

Dieudonne Alemagi FOKABS INC., Ottawa, ON, Canada

Asad Ali Department of Agriculture, Abdul Wali Khan University Mardan, Mardan, Khyber Pakhtunkhwa, Pakistan

Mudssar Ali Institute of Plant Protection, MNS University of Agriculture, Multan, Pakistan

Sajid Ali Department of Agronomy, University of the Punjab, Lahore, Pakistan; Department of Horticulture and Life Science, Yeungnam University, Gyeongsan, Republic of Korea

A. D. Ampitiyawatta Department of Export Agriculture, Faculty of Agricultural Sciences, Sabaragamuwa University of Sri Lanka, Belihuloya, Sri Lanka

Javairia Ansar Sustainable Development Study Centre, Government College University Lahore, Lahore, Pakistan

Alavudeen Asmiya Department of Biosystems Technology, Faculty of Technology, South Eastern University of Sri Lanka, University Park, Oluvil, Sri Lanka

Hichem Asses General Directorate of Plant Health and Control of Agricultural Inputs, Ministry of Agriculture, Water Resources and Fisheries, Tunis, Tunisia

Larbi Aziz Department of Rural Development Engineering, National School of Agriculture of Meknes, Meknes, Morocco

Mirza Barjees Baig Prince Sultan Institute for Environmental, Water and Desert Research, King Saud University, Riyadh, Kingdom of Saudi Arabia

Mohamed Behnassi Economic Social & Environmental Council (ESEC-Rabat), Center for Environment, Human Security & Governance (CERES), Ministry of Higher Education and Innovation, Rabat, Morocco

Ahlem Belhaj General Directorate of Plant Health and Control of Agricultural Inputs, Ministry of Agriculture, Water Resources and Fisheries, Tunis, Tunisia

Sivakumar Bhadrapriya Department of Life Sciences, School of Sciences, CHRIST (Deemed to be University), Bengaluru, India

Mohamed Chaouachi General Directorate of Plant Health and Control of Agricultural Inputs, Ministry of Agriculture, Water Resources and Fisheries, Tunis, Tunisia

Mouad Chari Department of Rural Development Engineering, National School of Agriculture of Meknes, Meknes, Morocco

Elhoucine Chougrani International Law, University Cadi Ayyad, Marrakesh, Morocco

Jagdish Chander Dagar Natural Resource Management Division, Krishi Anusandhan Bhavan—II, Indian Council of Agricultural Research, Pusa, New Delhi, India

Asim Chandra Dash Department of Agronomy, Odisha University of Agriculture and Technology, Bhubaneswar, Odisha, India

Nadia Debbabi General Directorate of Plant Health and Control of Agricultural Inputs, Ministry of Agriculture, Water Resources and Fisheries, Tunis, Tunisia

P. K. Dissanayake Department of Export Agriculture, Faculty of Agricultural Sciences, Sabaragamuwa University of Sri Lanka, Belihuloya, Sri Lanka

Lalisa Duguma World Agroforestry Centre, Nairobi, Kenya

Mahjoub El Haiba Faculty of Law, Economics and Social Sciences, UN Human Rights Committee, Hassan II University of Casablanca, Casablanca, Morocco

Ammara Fatima Department of Environmental Science, Lahore College for Women, University, Lahore, Pakistan

Faten Zubair Filimban Plant Taxonomy of the Arabian Peninsula, Division of Botany, Department of Biological Sciences, College of Sciences, King Abdulaziz University, Jeddah, Saudi Arabia

Kalame Fobissie FOKABS INC., Ottawa, ON, Canada

Gitanjali Nain Gill Faculty of Business and Law, Northumbria Law School, Northumbria University, Newcastle Upon Tyne, UK

Syed Waseem Gillani Department of Plant Sciences, Quaid-i-Azam University, Islamabad, Pakistan

Himangana Gupta United Nations University—Institute for the Advanced of Sustainability, Tokyo, Japan

Sharda Rani Gupta Faculty of Life Sciences, Kurukshetra University, Kurukshetra, Haryana, India

Sadraddin Gurbanov Institute of Genetic Resources of the National Academy of Sciences of Azerbaijan, Baku, Azerbaijan

Javaid Iqbal Department of Plant Protection, College of Food and Agriculture Sciences, King Saud University, Riyadh, Saudi Arabia

Imed Jaouadi General Directorate of Plant Health and Control of Agricultural Inputs, Ministry of Agriculture, Water Resources and Fisheries, Tunis, Tunisia

Rimsha Javed Sustainable Development Study Centre, Government College University Lahore, Lahore, Pakistan

Muhammad Jawad Center for Geographical Information System, University of Punjab Lahore, Lahore, Pakistan

Asif Khan Department of Technology, Universidade Estadual de Maringá, Umuarama, Brazil

Danyal Haider Khan Institute of Plant Protection, MNS University of Agriculture, Multan, Pakistan

Fawad Zafar Ahmad Khan Institute of Plant Protection and Department of Outreach and Continuing Education, Muhammad Nawaz Shareef University of Agriculture, Multan, Pakistan

Zaheer-ud-din Khan Department of Botany, Government College University, Lahore, Pakistan

Ben Jamâa Mohamed Lahbib General Directorate of Plant Health and Control of Agriculture Inputs; Ministry of Agriculture, Water Resources and Fisheries, Eaux et Forêts (INRGREF), University of Carthage, Institut National de Recherches en Génie Rural, Tunis, Tunisia

Essia Limem General Directorate of Plant Health and Control of Agricultural Inputs, Ministry of Agriculture, Water Resources and Fisheries, Tunis, Tunisia

Adeel Mahmood Department of Environmental Sciences, GC Women University Sialkot, Sialkot, Pakistan

Muhammad Manzoor Department of Plant Sciences, Quaid-i-Azam University, Islamabad, Pakistan

Faiz Mohideen Mohamed Thassim Marikar General Sir John Kotelawala Defence University, Ratmalana, Sri Lanka

Rosa María Martínez-Vázquez Faculty of Economics and Business, University of Almería, Almería, Spain

Lassaad Medhioub General Directorate of Plant Health and Control of Agricultural Inputs, Ministry of Agriculture, Water Resources and Fisheries, Tunis, Tunisia

Mouna Mhafdhi General Directorate of Plant Health and Control of Agricultural Inputs, Ministry of Agriculture, Water Resources and Fisheries, Tunis, Tunisia

Peter A. Minang World Agroforestry Centre, Nairobi, Kenya

Besma Mrabet General Directorate of Plant Health and Control of Agricultural Inputs, Ministry of Agriculture, Water Resources and Fisheries, Tunis, Tunisia

Rachid Mrabet National Institute for Agricultural Research (INRA), Rabat, Morocco

Sadia Muneer University of Agriculture, Faisalabad, Pakistan

Mumna Munir Department of Plant Sciences, Quaid-i-Azam University, Islamabad, Pakistan

Waheed Murad Department of Botany, Abdul Wali Khan University Mardan, Mardan, Khyber Pakhtunkhwa, Pakistan

Muneeb M. Musthafa Department of Biosystems Technology, Faculty of Technology, South Eastern University of Sri Lanka, University Park, Oluvil, Sri Lanka

Praveen Nagella Department of Life Sciences, School of Sciences, CHRIST (Deemed to be University), Bengaluru, India

Mohamed Mujithaba Mohamed Najim Faculty of Agriculture, Sultan Sharif Ali Islamic University, Jln Tutong, Kampong Sinaut, Brunei

J. Nayana Department of Life Sciences, School of Sciences, CHRIST (Deemed to be University), Bengaluru, India

S. Nithya Department of Life Sciences, School of Sciences, CHRIST (Deemed to be University), Bengaluru, India

Meenakhi Prusty AICRP on LTFE, College of Agriculture, Odisha University of Agriculture and Technology, Bhubaneswar, Odisha, India

Aziz Radi English for Economics, University Cadi Ayyad, Marrakesh, Morocco

Rohan Rama Krishnan Department of Life Sciences, School of Sciences, CHRIST (Deemed to be University), Bengaluru, India

Nachiketha R. Rao Department of Life Sciences, School of Sciences, CHRIST (Deemed to be University), Bengaluru, India

Rachid Sabbahi Research Team in Science and Technology, Higher School of Technology of Laayoune, Ibn Zohr University, Laayoune, Morocco

Gyanaranjan Sahoo Krishi Vigyan Kendra, Odisha University of Agriculture and Technology, Nayagarh, Odisha, India

Amita Sharma Krishi Vigyan Kendra, Rajmata Vijayaraje Scindia Krishi Vishwa Vidyalaya, Gwalior, Madhya Pradesh, India

Kashaf Siddiqi Department of Agronomy, University of the Punjab, Lahore, Pakistan

Amir Sohail Department of Plant Sciences, Quaid-i-Azam University, Islamabad, Pakistan

Amtul Bari Tabinda Sustainable Development Study Centre, Government College University Lahore, Lahore, Pakistan

Noura Touati General Directorate of Plant Health and Control of Agricultural Inputs, Ministry of Agriculture, Water Resources and Fisheries, Tunis, Tunisia

Eranga M. Wimalasiri Department of Export Agriculture, Faculty of Agricultural Sciences, Sabaragamuwa University of Sri Lanka, Belihuloya, Sri Lanka

Abdullah Yasar Sustainable Development Study Centre, Government College University Lahore, Lahore, Pakistan

Iqra Zaheer Livestock Production Research Institute, Okara, Pakistan

Adnan Zahid Department of Agronomy, University of the Punjab, Lahore, Pakistan

Aisha Zahra Department of Agronomy, University of the Punjab, Lahore, Pakistan

Abbreviations and Acronyms

ADB	Asian Development Bank
AZN	New Azerbaijani Manat—monetary unit of Azerbaijan
BYMV	Bean Yellow Mosaic Virus
CBD	Convention on Biological Diversity
CCAFS	Climate Change, Agriculture and Food Security
CEC	Central Election Commission
CGIAR	Consultative Group on International Agricultural Research
CIIFL	Centre Inter-Professional des Fruits et Légumes
CIS	Commonwealth of Independent States
CITES	Convention on International Trade in Endangered Species of Wild Fauna and Flora
CRI	Climate Risk Index
CSA	Climate-Smart Agriculture
CSCCC	Civil Society Coalition for Climate Change
DGPA	Direction Générale de la Production Agricole
DGSVCIA	Direction Générale de la Santé Végétale et du Contrôle des Intrants Agricoles (Directorate-General of Plant Health and Agricultural Inputs Control)
DPMV	Deforming Pea Mosaic Virus
EPPO	European and Mediterranean Plant Protection Organization
ESF	Energy Smart Food for People and Climate
FAO	Food and Agriculture Organization
GBF	Global Biodiversity Framework
GCM	Global Climate Model
GDP	Gross Domestic Product
GHG	Greenhouse Gas
GHI	Global Hunger Index
GID	Groupement Interprofessionnel des Dattes
GIFRUIT/GIF	Groupement Interprofessionnel des Fruits
GIL	Groupement Interprofessionnel des Légumes
GMST	Global Mean Surface Temperature

HRC	Human Rights Council
HUBR	Harrat Uwayrid Biosphere Reserve
IFAD	International Fund for Agricultural Development
INRA	Institut National de Recherche Agronomique
IPPM	Integrated Pest and Pollinator Management
IRESA	Institution de la Recherche et de l'Enseignement Supérieur Agricole
NPPO	National Plant Protection Organization
OJSC	Open Joint-Stock Company
OsNAS	Nicotianamine Synthase (Rice Biofortification)
OsYSL	Rice Metal Nicotianamine Transporter
PEMV	Pea Enation Mosaic Virus
PLRV	Potato Leaf Roll Virus
PPMV	Plain Pea Mosaic Virus
PSTVd	Potato Spindle Tuber Viroid
PVX	Potato Virus X
PVY	Potato Virus Y
RCU	Royal Commission for Al-Ula
REDD+	Reducing Emissions from Deforestation and Forest Degradation
RNA	Ribonucleic Acid
RPW	Red Palm Weevil
TM/CAM	Traditional Medicines/Complementary and Alternative Medicine
UNCTAD	United Nations Conference on Trade and Development
UNEP	United Nations Environment Programme
UNFCCC	United Nations Framework Convention on Climate Change
UNGA	United Nations General Assembly
WFPs	Wild Food Products
WHO	World Health Organization
WRI	World Resources Institute

List of Figures

List of Tables

Chapter 1
Food Systems and Biodiversity in the Context of Environmental and Climate Risks-Dynamics and Evolving Solutions: An Introduction

Mohamed Behnassi, Mirza Barjees Baig, Himangana Gupta, Rachid Sabbahi, Gitanjali Nain Gill, and Mahjoub El Haiba

Abstract His introductory chapter to the book *Food Systems and Biodiversity in the Context of Environmental and Climate Risks* places biodiversity at the core of the food-climate nexus, emphasizing its essential role in sustaining resilient food systems. The chapter examines how climate change—driven by large-scale emissions, unsustainable agricultural practices, and land-use changes—pollution, and the spread of invasive pests and plant diseases accelerate biodiversity loss and disrupts

M. Behnassi (✉)
Senior Environmentalist Expert, Economic Social & Environmental Council (CESE), Center for Environment, Human Security & Governance (CERES), Ministry of Higher Education and Innovation, Rabat, Morocco
e-mail: behnassi@gmail.com

M. B. Baig
Prince Sultan Institute for Environmental, Water and Desert Research, King Saud University, Riyadh, Kingdom of Saudi Arabia
e-mail: mbbaig@ksu.edu.sa

H. Gupta
United Nations University—Institute for the Advanced of Sustainability, Tokyo, Japan
e-mail: himangana@gmail.com

R. Sabbahi
Research Team in Science and Technology, Higher School of Technology of Laayoune, Ibn Zohr University, Laayoune, Morocco
e-mail: r.sabbahi@uiz.ac.ma

G. N. Gill
Faculty of Business and Law, Northumbria Law School, Northumbria University, Newcastle Upon Tyne, UK
e-mail: gireeshctripathi@gmail.com

M. El Haiba
Faculty of Law, Economics and Social Sciences, UN Human Rights Committee, Hassan II University of Casablanca, Casablanca, Morocco
e-mail: elhaiba.elmahjoub@gmail.com

M. Behnassi et al. (eds.), *Food Systems and Biodiversity in the Context of Environmental and Climate Risks*, https://doi.org/10.1007/978-3-031-89167-0_1

crucial ecosystem services—such as water provision, soil fertility, and pollination—that are vital for food security and human well-being. It highlights the dual role of food systems as contributors to and victims of climate change, advocating for transformative, nature-centered approaches. These mainly include organic farming, agroforestry, biofortification, sustainable intensification, and land restoration, which can bolster ecosystem resilience and safeguard food systems from climate risks. Drawing on case studies from many regions covered in this book, the chapter stresses the importance of integrating traditional knowledge with scientific innovation and policy reforms to build climate-smart, biodiversity-rich food systems. It calls for coordinated global action to align agricultural practices with the goals of the Paris Agreement to the United Nations Framework Convention on Climate Change (UNFCCC), the 2030 Agenda for Sustainable Development, and the Kunming-Montreal Global Biodiversity Framework (GBF) under the Convention on Biological Diversity (CBD). This volume serves as a comprehensive guide for policymakers, researchers, and practitioners working at the intersection of biodiversity, climate resilience, and food security.

Keywords Biodiversity conservation · Climate resilience · Sustainable agriculture · Food security · SDGs · Climate-smart practices

1.1 Introduction

The rapid pace of climate change has intensified existing challenges to the sustainability and resilience of social-ecological systems globally. This crisis is compounded by land-use changes, increased pollution, species invasions, and declining biodiversity (Ukhurebor and Aidonojie 2021). In critical areas such as agriculture, water management, ecosystems, and public health, the dynamics of climate and environmental change present complex and growing risks. Global biodiversity is in peril, with up to one million species facing extinction in the coming decades (Tollefson 2019). Climate change exacerbates these trends by altering habitats, causing sensitive species to die off or forcing them to migrate, disrupting the delicate ecological balance. These shifting dynamics weaken the resilience of ecosystems and disrupt the services they provide, such as clean water, fertile soil, and pollination, which are essential for food production and human survival (Behnassi and Gupta 2022; Pecl et al. 2017).

Agricultural expansion and habitat conversion for food production have been the principal drivers of biodiversity decline over the past five decades. Current food systems rely heavily on monocropping, excessive water use, pesticide application, and energy-intensive practices, leading to ecosystem degradation and increased greenhouse gas emissions. The focus on maximizing yields at minimal costs has resulted in unsustainable practices that harm soils, reduce land productivity, overuse water resources, and disrupt biodiversity. This intensification of agricultural activities contributes to climate change, as food production—including livestock—and loss is now one of the largest sources of greenhouse gas emissions. The global food system thus plays a dual role: it not only suffers from the impacts of climate change but is also a significant driver of environmental degradation and global warming.

As ecosystems deteriorate, species are forced to relocate or face extinction, creating new interactions between species and fostering conflicts over shrinking habitats. In some cases, these interactions create opportunities for the emergence of new infectious diseases as human-animal contact increases in disrupted environments. The ongoing spread of zoonotic diseases is a stark reminder of the interconnectedness of human health, biodiversity, and ecosystem health. Such a complex web of interactions calls for transformative changes in food systems that can simultaneously address climate change, biodiversity loss, and food security.

Therefore, without a major shift in food systems, both climate change, ecosystem degradation, and biodiversity loss will keep going up. Continued environmental and climate disruptions and the resulting habitat loss will jeopardize our capacity to maintain the human security and resilience of social-ecological systems. Significant, transformative changes in food systems are necessary to create a world free of hunger and malnutrition by protecting the natural resource base needed to feed current and future generations, while also reducing environmental and climate risks (Caron et al. 2018). A coordinated effort to transform food systems is critical to implementing the Paris Agreement, the 2030 Agenda for Sustainable Development, and the Post-2020 Global Biodiversity Framework (GBF).

Building sustainable, climate-friendly, and resilient food systems has emerged as a dominant endeavor aimed at redirecting our food systems and policies toward more appropriate goals and improved societal well-being (Allen and Prosperi 2016). The challenge is to enhance agricultural production on current agricultural lands while maintaining biodiversity, ecosystem integrity and services. The most efficient strategy to conserve biodiversity is to safeguard land from conversion or other developmental purposes (such as mining); hence, avoiding further land-use change. For example, land-sharing (promoting ecosystem services in agricultural settings), biodiversity-friendly farming systems, and land-sparing (i.e., separation of high-yielding agriculture and natural habitats) can help support multifunctionality of agricultural landscapes (Grass et al. 2019).

Restoring native ecosystems on abandoned agricultural land, minimizing the use of chemical inputs, and replacing monoculture with polyculture farming practices, can help boost biodiversity in agricultural areas. Global guidelines in policy areas like dietary modification, responsible investment, and nature-based climate change mitigation and adaptation solutions will be required to lead national-level action plans that may collectively lead to transformative change in the global food system.

The volume *Food Systems and Biodiversity in the Context of Environmental and Climate Risks* contributes to the growing scientific debate on the food-biodiversity-climate nexus, offering a comprehensive, multi-regional, and cross-sectoral analysis of the various natural and anthropogenic factors that affect ecosystem integrity and resilience. The volume explores the complex linkages between biodiversity, food systems, and climate change while assessing the effectiveness of response mechanisms aimed at mitigating these challenges. The chapters are grounded in empirical research and provide actionable insights for future-proof policy innovations, making this volume a vital resource for researchers, policymakers, and practitioners seeking to build sustainable, climate-resilient food systems.

1.2 Synthesis of Contributions

As climate change accelerates and exerts increasing pressure on ecosystems and food systems, it becomes clear that food security and biodiversity conservation must be addressed simultaneously. In their exploration of organic farming systems, Rachid Mrabet and Mohamed Behnassi argue that organic agriculture can play a pivotal role in achieving this balance. Organic farming prioritizes biodiversity by fostering healthier ecosystems through practices such as crop rotation, polyculture, and the avoidance of synthetic pesticides and fertilizers. These practices not only enhance soil health but also contribute to the conservation of beneficial species such as pollinators and natural enemies, which are critical for maintaining ecosystem services. Despite its potential, organic farming faces significant challenges, particularly regarding lower short-term yields compared to conventional farming. However, the authors stress that the long-term benefits of organic farming–such as improved soil fertility, reduced environmental contamination, and enhanced climate resilience–make it a crucial component of sustainable agriculture. The chapter by Mrabet and Behnassi sets the stage for a broader discussion on how agricultural practices can be reoriented toward environmental stewardship without compromising food security.

Agroforestry, as discussed by Jagdish Chander Dagar and Sharda Rani Gupta, offers another solution to the challenge of creating climate-smart, biodiversity-friendly agricultural systems. By integrating trees into farming landscapes, agroforestry systems enhance carbon sequestration, improve soil structure, and increase biodiversity. These systems also diversify farmers' income streams by providing additional products such as fruits, nuts, timber, and medicinal plants. The authors emphasize that agroforestry is particularly well-suited to regions facing climate-induced stress, as it enhances resilience to extreme weather events such as droughts and floods. In this way, agroforestry serves as a holistic approach that addresses both climate mitigation and adaptation while also contributing to the conservation of biodiversity. Their chapter builds on the idea that agricultural intensification can occur in harmony with nature, provided that sustainable practices are adopted.

Ampitiyawatta and his colleagues further contribute to this discourse by presenting case studies from Sri Lanka, where climate change is already severely impacting crop production. They highlight how the introduction of drought-resistant crop varieties and the diversification of cropping systems have helped farmers adapt to increasingly unpredictable rainfall patterns and rising temperatures. Agroecological diversity, they argue, is essential for building resilience in agricultural systems. Their work underscores the importance of locally tailored adaptation strategies that draw on traditional knowledge, innovative practices, and scientific research to safeguard food security in the face of climate variability.

Asif Khan, Muhammad Jawad, Sajid Ali, and other authors explore how climate change in Khyber Pakhtunkhwa (KP), Pakistan, exacerbates biodiversity loss and food insecurity through extreme weather patterns like droughts and floods. These climatic disruptions worsen deforestation, soil erosion, and crop losses, particularly in wheat and maize, which destabilizes food systems and threatens livelihoods in an agriculturally dependent region. The authors emphasize the interconnectedness of climate change, biodiversity, and food security, advocating for integrated solutions such as climate-resilient crop varieties, sustainable agricultural practices, reforestation, and improved water management. They also call for stronger environmental law enforcement and alternative livelihoods to mitigate overgrazing and illegal logging, offering practical and policy recommendations to build resilience and address the broader global issues of biodiversity loss and food insecurity.

Dieudonne Alemagi and co-authors focus on cocoa agroforestry in Cameroon, examining how incentive mechanisms can promote sustainable intensification without expanding agricultural land at the expense of forests. Through well-designed financial incentives, farmers are encouraged to adopt agroforestry practices that not only increase cocoa yields but also protect biodiversity and enhance ecosystem services such as carbon sequestration. The authors emphasize that agroforestry represents a win–win strategy that balances economic development with environmental sustainability. By linking economic incentives to environmental conservation, their chapter highlights the potential for policy interventions to drive sustainable development in tropical regions.

The chapter by Munir et al. examines the potential of Saudi Arabia's wild edible plants to bolster food security in the face of climate change and environmental degradation. It highlights the rich biodiversity of the region, home to over 2200 plant species, many of which are undervalued in modern agriculture despite their significant nutritional, medicinal, and ecological benefits. Wild edibles, which thrive in arid conditions with minimal resource use, offer a sustainable alternative to conventional crops by enhancing dietary diversity, reducing reliance on food imports, and supporting local economies. These plants are rich in antioxidants, anti-inflammatory compounds, essential vitamins, and trace elements, making them valuable for both nutrition and disease prevention. Leveraging qualitative and quantitative research methods, including field studies, dietary assessments, and stakeholder engagement, the study demonstrates the adaptability of wild plants to extreme weather and their role in mitigating food scarcity and malnutrition. It also underscores their socioeconomic value, particularly in empowering marginalized communities such as women and promoting circular economy practices. By advocating for climate-smart agricultural practices, policy interventions, and collaborative research, this chapter positions wild edibles as a critical resource for building resilient, equitable food systems that align with global sustainability goals and address the multifaceted challenges of food security in Saudi Arabia amidst climate change.

The critical role of tropical forests in carbon sequestration and biodiversity conservation is explored by Gyanaranjan Sahoo and co-authors. Tropical forests act as carbon sinks, absorbing large amounts of carbon dioxide and thus mitigating the effects of climate change. However, deforestation for agricultural expansion, logging, and other human activities has significantly reduced these forests' capacity to sequester carbon. The authors call for stronger governance frameworks and international cooperation to protect tropical forests, arguing that forest conservation is directly linked to the health of food systems and global climate stability. Without urgent action to halt deforestation, the authors warn that the loss of these vital ecosystems will have far-reaching consequences for biodiversity, climate regulation, and food security.

Alavudeen Asmiya and co-authors talk about the marine environment, where the sustainable use of marine resources is integral to food security and economic development, particularly for coastal and island nations. Their exploration of the blue economy emphasizes that oceans provide not only food but also livelihoods for millions of people. The authors argue that integrating marine and terrestrial food systems is essential for creating resilient economies capable of withstanding environmental shocks. Sustainable marine resource management, they contend, is crucial for ensuring that ocean ecosystems continue to provide essential services such as food production, climate regulation, and biodiversity conservation.

The role of pollinators in food security is critically examined by Mudssar Ali, Fawad Zafar Ahmad Khan, and Muhammad Awais Ahmad, who discuss how climate change threatens pollinator populations. As insect pollinators are responsible for the reproduction of over 75% of the world's food crops, their decline presents a direct threat to global food production. The authors emphasize the importance of integrated pest and pollinator management (IPPM) systems that protect pollinators while controlling pests in a sustainable manner. By highlighting successful IPPM practices from various regions, they demonstrate how these systems can enhance agricultural productivity while maintaining the health of insect pollinator populations, which are essential for ecosystem stability.

Expanding on this theme, Fawad Zafar Ahmad Khan and co-authors delve deeper into the concept of IPPM, providing case studies that illustrate how pest control strategies can be adapted to changing climatic conditions while simultaneously promoting pollinator health. They argue that IPPM is not only an effective approach for managing agricultural pests but also a crucial strategy for enhancing ecosystem resilience. Their work underscores the interconnectedness of pest control, pollinator conservation, and food security, reinforcing the need for holistic approaches to agricultural management.

In a stark warning about the growing threats posed by climate change, Rachid Sabbahi examines the increasing frequency and severity of desert locust swarms, which devastate crops and threaten food security across North Africa, the Middle East, and South-West Asia. His chapter highlights how changing weather patterns, particularly increased rainfall in arid regions, have created ideal breeding conditions for locusts, worsening the risk to agricultural systems. Sabbahi calls for stronger regional cooperation in monitoring locust activity and improving early warning systems to mitigate the agricultural and socio-economic impacts of locust outbreaks, emphasizing the need for proactive and coordinated responses to climate-induced agricultural challenges.

Ben Jamâa Mohamed Lahbib and co-authors focus on the threat posed by invasive species in Tunisia, which are increasingly disrupting local ecosystems and agricultural productivity. The chapter explores how climate change exacerbates the spread of invasive species, leading to significant economic and ecological damage. The authors advocate for integrated management strategies that combine scientific research with community engagement to protect native biodiversity and support sustainable agriculture.

The complex relationship between climate change and plant diseases, or phytopathogens, is examined by Sadraddin Gurbanov and co-authors. Climate change is altering the behavior and distribution of phytopathogens, leading to the spread of new diseases that threaten crop yields and further endanger food security. The authors highlight that rising temperatures, increased humidity, and unpredictable weather patterns create favorable conditions for pathogens to thrive, causing outbreaks that devastate crops. To mitigate these threats, they stress the importance of breeding disease-resistant crop varieties and improving disease management practices, drawing on both traditional knowledge and modern agricultural science. Their chapter underscores the critical need for continued research and innovation to safeguard food systems against the evolving challenges posed by climate change.

Rohan R. Krishnan and co-authors focus on the role of biofortification as a promising strategy to improve both food security and public health, particularly in regions where micronutrient deficiencies are widespread. They discuss the biofortification of rice, one of the world's most important staple crops, as a way to enhance the nutritional value of food while maintaining or increasing yields. By integrating biofortification with sustainable intensification practices–such as improving soil health and water efficiency–the authors argue that we can address both the quantity and quality of food produced, helping to combat malnutrition while supporting agricultural resilience in the face of climate change.

Amtul Bari Tabinda and co-authors further explore crop resilience practices by mapping various mitigation strategies that can improve crop yields under increasingly challenging climatic conditions. They highlight the importance of developing climate-resilient crop varieties through advanced breeding techniques, which can help crops withstand extreme weather events such as droughts, floods, and heatwaves. In addition to genetic improvements, they advocate for the adoption of advanced irrigation methods, such as drip irrigation and rainwater harvesting, to maximize water efficiency and protect crops from water stress. The chapter emphasizes the need for a multifaceted approach that combines scientific innovation with practical solutions to ensure that food systems remain productive and resilient in the context of climate change.

The chapter by Abd El-Ghani et al. delves into the floristic diversity and habitat composition of the Al-Ula oasis in northwestern Saudi Arabia, a region renowned for its rich cultural heritage, ecological significance, and distinctive biodiversity. Situated along the historic Incense Road, Al-Ula has served as a hub for trade and human settlement for millennia, showcasing the ingenuity of ancient civilizations in adapting to arid conditions through sustainable agricultural practices, such as date palm cultivation and qanat irrigation systems. The chapter identifies a number of plant species across numerous families, thriving in diverse ecosystems ranging from sandstone mountains to volcanic formations and lush oases. Efforts by the Royal Commission for Al-Ula (RCU) aim to balance conservation with sustainable development, particularly through initiatives tied to Saudi Vision 2030, including heritage preservation, biodiversity protection, and the promotion of eco-tourism. While highlighting Al-Ula's importance as a biodiversity hotspot, the chapter emphasizes the necessity of sustainable land use and conservation measures to safeguard rare and endangered species and maintain ecological balance. The chapter also explores Al-Ula's role in global efforts to combat desertification, as outlined in the Saudi Green Initiative, and its potential for sustainable agriculture, renewable energy, and research-driven development. By drawing parallels with other arid regions, it underscores the need for continued botanical exploration and integrated strategies to protect Al-Ula's unique natural and cultural assets while fostering resilience against environmental challenges.

The environmental and health risks associated with pesticide pollution are analyzed by Javaid Iqbal and co-authors, who provide a critical examination of how the overuse of pesticides not only harms ecosystems but also exacerbates the impacts of climate change. Pesticides can disrupt soil health, contaminate water supplies, and harm non-target species, including beneficial insects. Moreover, the production and application of pesticides contribute to greenhouse gas emissions, further intensifying climate-related risks. The authors call for a shift toward more sustainable pest management practices, including the use of biopesticides and integrated pest management (IPM), to reduce reliance on harmful chemicals and promote biodiversity conservation. Their chapter ties into the broader theme of rethinking agricultural practices to minimize environmental harm while maintaining food production.

In a fascinating exploration of traditional knowledge systems, Aziz Radi, Mohamed Behnassi, and Elhoucine Chougrani examine the Agdal management system in Morocco's High Atlas. The Agdal system, a communal land management practice, has long served as a model for sustainable resource use, balancing the needs of local communities with the conservation of biodiversity. However, modern pressures such as climate change, population growth, and economic development are threatening the sustainability of this system. The authors argue that revitalizing traditional governance mechanisms, such as the Agdal system, could play a crucial role in enhancing the resilience of rural communities and protecting local ecosystems. They call for policy frameworks that integrate traditional knowledge with modern conservation strategies to promote sustainable resource management in the face of environmental change.

Complementing this, Mouad Chari and Larbi Aziz provide an insightful analysis of how grassroots movements and local exchange collectives are driving agroecological transitions in Morocco. They highlight the importance of community-based initiatives in fostering sustainable agricultural practices that are both resilient to climate change and supportive of local economies. By creating networks of exchange and cooperation among farmers, these collectives help spread agroecological knowledge and practices that improve soil health, conserve water, and enhance biodiversity. The authors' case study offers a powerful example of how local governance and community engagement can lead the way in building more sustainable and resilient food systems.

1.3 Conclusion

This volume presents a comprehensive and nuanced view of the critical intersections between food systems, biodiversity, and climate resilience, offering a wide array of regional perspectives and sectoral approaches. The chapters collectively underscore the urgent need for transformational changes in agricultural practices, governance structures, and policy frameworks to address the multifaceted challenges posed by climate change, biodiversity loss, and food insecurity. The contributors make a compelling case that sustainable food systems are not only possible but essential for a resilient future—one that balances human well-being with environmental integrity.

At the core of this volume is the recognition that food security cannot be achieved in isolation from environmental sustainability and biodiversity conservation. The global scope of the volume reflects the interconnectedness of food systems, ecosystems, and climate, demonstrating that solutions must be both local and global, drawing on the strengths of traditional knowledge, scientific innovation, and policy integration. From the pollinator-dependent crops of South Asia to the deforestation-threatened cocoa plantations of Cameroon, the chapters illustrate the complex challenges facing food systems around the world and offer practical solutions for overcoming them.

Ultimately, this book serves as a call to action for researchers, policymakers, and practitioners to collaborate in developing and implementing innovative solutions that integrate ecological principles into food production, ensuring that climate resilience, biodiversity, and human security remain central pillars of future agricultural strategies. By fostering cross-disciplinary and regional cooperation, we can build food systems that are more inclusive, sustainable, and resilient—capable of feeding the world's growing population while protecting the planet's invaluable natural resources. The path forward lies in embracing holistic approaches that promote both human well-being and the health of the ecosystems upon which we all depend, ensuring a future where food security and environmental sustainability are no longer competing goals but shared priorities.

References

Allen T, Prosperi P (2016) Modeling sustainable food systems. Environ Manage 57(5):956–975. https://doi.org/10.1007/s00267-016-0664-8

Behnassi M, Gupta H (2022) Managing the food security, biodiversity, and climate nexus: transformative change as a pathway, In Behnassi M, Gupta H, Baig MB, Noorka IR (eds) The food security, biodiversity, and climate nexus. Berlin, Springer Cham. https://doi.org/10.1007/978-3-031-12586-7

Caron P, Ferrero y de Loma-Osorio G, Nabarro D, Hainzelin E, Guillou M, Andersen I, Arnold T, Astralaga M, Beukeboom M, Bickersteth S, Bwalya M, Caballero P, Campbell BM, Divine N, Fan S, Frick M, Friis A, Gallagher M, Halkin J-P, Verburg G (2018) Food systems for sustainable development: proposals for a profound four-part transformation. Agron Sustain Dev 38(4):41. https://doi.org/10.1007/s13593-018-0519-1

Grass I, Loos J, Baensch S et al (2019) Land-sharing/-sparing connectivity landscapes for ecosystem services and biodiversity conservation. People Nat 1:262–272. https://doi.org/10.1002/pan3.21

Pecl GT, Araújo MB, Bell JD, Blanchard J, Bonebrake TC, Chen I-C, Clark TD, Colwell RK, Danielsen F, Evengård B, Falconi L, Ferrier S, Frusher S, Garcia RA, Griffis RB, Hobday AJ, Janion-Scheepers C, Jarzyna MA, Jennings S, Williams SE (2017) Biodiversity redistribution under climate change: impacts on ecosystems and human well-being. Science, 355(6332):eaai9214. https://doi.org/10.1126/science.aai9214

Tollefson J (2019) Humans are driving one million species to extinction. Nature 569(7755):171–171. https://doi.org/10.1038/d41586-019-01448-4

Ukhurebor KE, Aidonojie PA (2021) The influence of climate change on food innovation technology: review on topical developments and legal framework. Agric Food Secur 10(1):50. https://doi.org/10.1186/s40066-021-00327-4

Dr. Mohamed Behnassi is currently a Senior Environmentalist Expert at the Economic Social and Environmental Council (CESE), Rabat, Morocco. From 2004 to 2022, he served as a Professor, Head of Public Law Department, and Director of the Research Laboratory for Territorial Governance, Human Security and Sustainability (LAGOS), Faculty of Law, Economics, and Social Sciences of Agadir, Ibn Zohr University. He holds a Ph.D. in International Environmental Law and Governance (2003), a M.Sc. in Political Sciences and International Relations (1997), and a B.A. of Administration (1995) from the Faculty of Law, Economics, and Social Sciences, Hassan II University of Casablanca. He obtained a Diploma in International Environmental Law and Diplomacy from the University of Eastern Finland and UNEP, 2015 and is also Alumnus of both the International Visitors Leadership Program of the Department of State (2011, USA), and the Near East South Asia Center for Strategic Studies (NESA), National Defense University, Washington, DC (2022–24, USA).

Dr. Behnassi is currently the Founding Director of the Center for Environment, Human Security and Governance (CERES). He was appointed in 2019 as Expert Evaluator for the National Center for Scientific and Technical Research (CNRST/Morocco), and selected twice (2019–2024) as an Assessment Scoping Expert and a Review Editor by the Intergovernmental Science-Policy Platform on Biodiversity and Ecosystem Services (IPBES—Thematic assessment of the interlinkages among biodiversity, water, food and health). He was a Lead Author (LA) of the Human Security Chapter of 1st MEDECC Assessment Report (MAR1): Climate and Environmental Change in the *Mediterranean Basin - Current Situation and Risks for the Future* (2021) and Coordinating Lead Author (CLA) of the MEDECC Special Report on "the Nexus between Water, Energy, Food and Ecosystems". In 2024, he was nominated and certified as the UNFCCC Expert Evaluator (Roster of Experts), and as a Member of the PRIMA Foundation Scientific Evaluation Committee (SEC) to evaluate proposals submitted to the Thematic Area "Leveraging urban and local food systems for sustainable food systems transformation".

Dr. Behnassi is a Senior Researcher in the areas of International Law and Politics of Environment and Human Security. In addition to his participation in more than 70 international and national conferences, he has published 66 scientific papers and 22 books in these areas, including recent ones on: *The Water, Climate, and Food Nexus: Linkages, Challenges and Emerging Solutions* (Springer, 2024); *Food Security and Climate-Smart Food Systems—Building Resilience for the Global South* (Springer, 2022); and *The Climate-Conflict-Migration Nexus from a Human Security Perspective* (Springer, 2022). Dr. Behnassi serves as a reviewer for many global publishers (such as Routledge and Springer) and scientific journals with high impact factor. He has organized many international conferences covering the above research areas, managed many research and expertise projects for national and international institutions, and is regularly requested to provide scientific expertise nationally and internationally. Other professional activities include social compliance auditing and consultancy by monitoring human rights at work and the sustainability of the global supply chain.

Dr. Mirza Barjees Baig is a professor at the Prince Sultan Institute for Environmental, Water, and Desert Research, King Saud University, Saudi Arabia. He earned his MS degree in International Agricultural Extension in 1992 from Utah State University, Logan, Utah, USA, and was placed on the "Roll of Honor". He completed his Ph.D. in Extension for Natural Resource Management from the University of Idaho, USA, and was honored with the "1995 Outstanding Graduate Student Award". He has published extensively on the issues associated with natural resources in national and international journals. He has also made oral presentations about agriculture, natural resources, and the role of extension education at various international conferences. He serves on the editorial boards of many international journals and is a member of many international professional organizations. He has produced six books on climate change, food security, and climate-smart agriculture. Food waste, water management, degradation of natural resources, the deteriorating environment, and their relationship with society and the community are his areas of interest. He has attempted to develop strategies to conserve natural resources, promote the environment, and develop sustainable communities. He started his scientific career in 1983 as a researcher at the Pakistan Agricultural Research Council, Islamabad, Pakistan. He served at the University of Guelph, Ontario, Canada, as a Special Graduate Faculty member from 2000 to 2005. He served as a foreign professor at the Allama Iqbal Open University (AIOU), Pakistan, through the Higher Education Commission from 2005 to 2009.

Dr. Himangana Gupta is Research Fellow and Academic Associate at United Nations University—Institute for the Advanced Study of Sustainability (UNU-IAS). She is an expert in climate change and biodiversity policy and has worked on climate adaptation, forestry, women and climate change, energy efficiency, and national GHG inventories. She was also a part of the National Communication Cell of the Indian Ministry of Environment, Forest and Climate Change from 2016–2019, and contributed to India's Second and Third Biennial Update Reports to the UNFCCC. She is a certified expert in climate adaptation finance. She has seven edited books to her credit and more than 40 research publications, including book chapters. Currently, she is also serving as Lead Author for the Nexus Assessment Report of the Intergovernmental Platform on Biodiversity and Ecosystem Services.

Dr. Rachid Sabbahi has been an associate professor at the Higher School of Technology—Laayoune, Ibn Zohr University, Morocco, since 2019. He holds a bachelor's degree (2000) in animal biology from Hassan II University, Casablanca, Morocco, and a master's degree (2003) and doctorate (2008) in biological sciences from Quebec University, Montreal, Canada. Dr. Sabbahi specializes in the biological control of insect pests and plant diseases, plant-microbe interaction, environmental microbial biotechnology, the biological activity of medicinal plants, biocomposite materials for the removal of organic pollutants and heavy metals, insect pollination, and crop production. In addition, he has co-managed many national and international research projects, published several scientific papers, and participated in many international conferences.

Dr. Gitanjali Nain Gill is a tenured full professor of environmental law at the Faculty of Law, Northumbria University, Newcastle, UK. Her research interests and published work form a coherent thematic body that includes climate change, access to environmental justice with a focus on India's National Green Tribunal, land rights, SDGs and sustainability. Professor Gill acts as an international expert member to prestigious international organizations including Regional Leader for Asia at the Sabin Centre Peer Review Network on Climate Change Litigation, Columbia

University; British Institute of International and Comparative Law, London; and World Commission on Environmental Law Special Task Force on Rights of Nature, IUCN. She has also provided intellectual and disciplinary expertise by identifying and contributing specialized environmental courts and tribunals best practices, these being particularly instructive for the global south, to the United Nations Environment Programme (2016 and 2022), and to the United Nations Expert Consultation Group OHCHR Accountability and Remedy Project II. Professor Gill is conferred the title of the prestigious Fellow of the Academy of Social Sciences, UK. She is also a recipient of the Global Initiative of Academic Networks in Higher Education Award (GIAN) by the Ministry of Human Resources and Development, Government of India. She was awarded prestigious research grants including British Academy Grants on environmental justice in India (2013–16; 2020–2023—Principal Investigator), National Environment Research Council Grant on climate change and decarbonization agenda (2022—Principal Investigator), and Responsible AI and Law Enforcement (AHRC and UK Keystone- Co- Investigator). Through awareness and capacity-building empowerment programmes, Prof. Gill's action research helps build coalitions, trust and confidence between Indian NGOs and the marginalized communities to support victims of environmental degradation.

Dr. Mahjoub El Haiba is a highly respected expert in human rights, international law, and environmental law, currently serving as a member of the United Nations' Human Rights Committee. He has been a professor at Hassan II University, Casablanca, since 1983 and a part-time lecturer at Mohamed V University in Rabat since 1997. Dr. El Haiba holds a State Doctorate in Public Law and Political Science from Hassan II University, alongside postgraduate degrees in international relations and political science. Over the course of his career, he has played a pivotal role in advancing Morocco's human rights and environmental policies, particularly during his tenure as the First Inter-ministerial Delegate for Human Rights from 2011 to 2018. In this capacity, he coordinated the country's reports for the Universal Periodic Review (UPR) and other United Nations human rights treaty bodies, leading delegations in interactive dialogues with UN bodies. A former Secretary General of the Advisory Council on Human Rights (CCDH) and a member of the Equity and Reconciliation Instance (IER), Dr. El Haiba was deeply involved in Morocco's transitional justice process. He has been a founding member of the Moroccan Human Rights Organization (OMDH) and played a key role in shaping Morocco's human rights landscape, advocating for governance reforms and justice mechanisms. His work in environmental law is equally impressive, as he is a founding member of the Moroccan Society for Environmental Law (SOMADE) and the Arab Network for Environment and Development (RAED). He has consulted for the United Nations Environment Programme (UNEP) and the Food and Agriculture Organization (FAO), contributing to environmental protection projects and policies in Morocco and other regions. Dr. El Haiba has published extensively on human rights, environmental law, and sustainable development, and he has translated and coordinated key studies on topics such as biodiversity and women's rights in environmental contexts. His academic roles include teaching public international law, environmental law, and human rights mechanisms, while supervising graduate and doctoral research. He has been an active participant in international forums, representing Morocco at human rights conferences in Geneva, New York, The Hague, and other global centers, and played a key role in organizing the World Human Rights Forum in Marrakech. Additionally, he has served as President of the jury for various environmental awards, including the League of Arab States' Environment Award and the Hassan II Environment Award. Dr. El Haiba's contributions to international human rights and environmental governance extend to his advisory roles, including being a member of scientific and ethics committees at universities and research centers. His leadership in both fields underscores his commitment to advancing Morocco's legal frameworks and its international presence in human rights and environmental protection. Throughout his career, Dr. El Haiba has made substantial contributions to policy development, academic research, and international diplomacy, solidifying his reputation as a leading figure in both human rights and environmental law.

Chapter 2
Organic Farming Systems: Navigating the Path to Resilience and Food Security

Rachid Mrabet and Mohamed Behnassi

Abstract Researchers, policymakers, and civil society are focused on producing healthy and nutritious food to meet the evolving needs of the growing population while lessening environmental and climate impact. However, we are heading on a trajectory that is far from ideal. In 2023, between 713 and 757 million people, one out of 11 people in the world, and one out of every five in Africa, still suffer from hunger or malnutrition. Hence, innovative solutions are crucial for advancing the food security as they address both environmental and social challenges. In this regard, this chapter lays the groundwork for understanding the complex interplay between organic farming and food security while paving the way for sustainability and resilience. Organic farming (OF) that eschews synthetic fertilizers and pesticides has gained popularity (4.5 million producers and 96 million hectares). The organic food market has experienced significant growth and diversity in the recent years due to health-conscious consumers and environmental benefits. Global sales of organic foods and drink reached almost 135 billion euros in 2022. Scientific literature indicates that organic food is more nutrient-dense and promotes biodiversity, climate resilience, and profitability. Nonetheless, organic sites generally had 20% lower yields than conventional sites, posing a challenge to ensuring financial stability for producers. OF faces challenges with nutrient addition and nitrogen availability, limiting crop productivity. However, it offers health benefits by reducing exposure to harmful chemicals for farmers and consumers. Organic food is rich in antioxidants, vitamins, and minerals, and its consumption is associated with lower obesity and cancer risk. It has economic benefits for farmers, consumers, and environmental organizations, but these benefits vary depending on farm size, market access, and local economic conditions. The chapter can be summarized in three main points: (i) While OF may not always result in the highest short-term yields, it offers significant benefits for long-term sustainability, soil health, and biodiversity; (ii) by considering the factors that influence crop yields and adopting appropriate organic practices,

R. Mrabet (✉)
National Institute for Agricultural Research (INRA), Rabat, Morocco
e-mail: rachid.mrabet@inra.ma

M. Behnassi
Economic Social & Environmental Council (ESEC), Center for Environment, Human Security & Governance (CERES), Ministry of Higher Education and Innovation, Rabat, Morocco

M. Behnassi et al. (eds.), *Food Systems and Biodiversity in the Context of Environmental and Climate Risks*, https://doi.org/10.1007/978-3-031-89167-0_2

farmers can achieve high yields and deliver sustainable and healthy food systems—in other terms, OF offers a promising path toward a more sustainable, resilient, and food-secure future—(iii) to reach its full potential, OF requires continued research, innovation, and supportive policies for expanding market access, growing consumer demands, product categories and farmer's profitability and ensuring compliance with standards and certifications.

Keywords Food security · OF · Organic foods · Nutrition security · Crop and livestock integration · Crop and food quality · Soil health · Environmental stewardship · Sustainability

2.1 Introduction

As we advance into the twenty-first century, climate disasters and extreme events are sounding an ever-louder alarm (Fletcher et al. 2024). Scientific evidence for global warming is irrevocable and witnessed by record rates of increase in atmospheric and sea temperatures and is correlated to rapid rises in atmospheric carbon dioxide (CO_2). Since the signing of the Paris Agreement in 2015, the concentration of CO_2 in the atmosphere has increased from 399.4 parts per million (ppm) to 419.1 ppm today. According to the 6th IPCC assessment report, with the continued rise in Greenhouse Gases (GHG) emissions, the probability of irreversible climate change and subsequent ecological breakdown grows exponentially. In other claim, the urgency of radical actions cannot be overstated (IPCC 2021, 2022a). In 2023, the world has experienced 12 consecutive months with temperatures exceeding 1.5 °C above the pre-industrial average. Such heat anomalies wreak havoc on the world's water resources, energy needs, and ecosystems, with the most disenfranchised communities hit hardest. At the same time, poverty levels in low-income countries remain above pre-covid-19 levels and food inflation remains unacceptably high, pushing up the number of food- and water-insecure populations, while the biodiversity that our agri-food systems depend on is at greater risk than ever. Given these immense, complex, and interlinked challenges, new policies, institutions, and solutions are urgently required to drive the transformation of land, water, and food systems, enhance the lives and livelihoods of the poorest and most food- and water-insecure communities and peoples. These calls were repeatedly stated in all forums and high-level meetings both at global and regional levels.

The world agricultural paradigm is progressively symbolized by specialized-intensive corporate farming systems, characterized by their high intensity of production, large scale, and mainly specialization. With the advancement of machinery, fertilizers, agrochemicals, breeding programs, precision tools and globalized supply chains, production systems are producing food and feed in record quantities leading to the gradual diminishing of the threat of global hunger and malnourishment. This increased food demand has led to significant agricultural expansion globally and provided opportunities to support economic growth in developing countries through exports of high-value agricultural commodities however with high negative impacts on resources and the environment including biodiversity. Much of the fruits and vegetables are grown in arid or semi-arid climates and water-stressed river basins where irrigation is used to increase yields and improve levels of crop quality for export markets. This has been linked to rapidly rising water abstraction, which has negatively impacted surface and groundwater resources, contributed to higher environmental and carbon footprints, increased energy, and land use change (Kanter et al. 2018; Zabel et al. 2019; Correa-Cano et al. 2022).

The issue of food security has become increasingly critical with the expanding global population, the diminishing natural resources, and the increasing allocation of food production to alternative uses, such as bio-energy production (Gibbs and Salmon 2015). As the population grows, there is mounting pressure to enhance the efficiency of food production within the constraints of limited land resources (Bestelmeyer et al. 2015; Müller et al. 2018). Sustainable agriculture is widely acknowledged as essential to address the new socio-economic and ecological challenges confronting agriculture and livestock (Nabuurs et al. 2022). A variety of low-impact, clean and green strategies aimed at improving crop productivity, water availability and use efficiency, and mitigating the impacts of climate change are available and being adapted and adopted to varying extents worldwide (HLPE 2019; Epule 2019; Hatfield and Dold 2019; Mrabet 2023; Grigorieva et al. 2023; Panday et al. 2024).

OF has gained significant importance among producers, consumers, and policymakers (Rahman et al. 2021). The groundwork for organic agriculture was laid in the mid-1950s by Howard (1947) and Rodale (1948), emphasizing the importance of soil health, natural pest control, and the potential hazards of synthetic chemicals. Their primary objective was to develop environmentally friendly and sustainable organic farms. Over time, organic products have gained broader recognition and acceptance due to their health and ecological benefits following certification. This has transformed OF from a niche movement into a global industry, driven by the increasing global demand for organic food and the conversion of conventional agriculture to organic methods worldwide. It is often cited as a model for sustainable agriculture. However, Giampieri et al. (2022) have argued that OF may not be a sustainable strategy for optimizing land use efficiency, supporting the observations of Tuomisto et al. (2012) that OF requires approximately 84% more land compared to conventional farming due to lower yields and the need to build soil fertility. This suggests that lower environmental impact is not always guaranteed per product unit.

In this chapter, through an analysis of the academic and gray literature, including reviews and meta-analyses, the aim is to have a sound understanding of the benefits and effects of OF systems in terms of food and nutrition security. Wherever the relevant information is available, the present analysis contrasts the impacts of OF systems with other conventional or traditional farming in terms of productivity and nutrition quality, as pillars of food security, but also sustainability indicators. Recently, these questions have been at the center of sprightly debates among farmers, the civil society, policymakers, and in academic and research. Calls for more funding for ecologically based cropping systems research and extension (including OF systems) are echoed by Shennan et al. (2017) and many other scholars and institutions.

2.2 Statistics and Emerging Trends in OF

The International Federation of Organic Agriculture Movements (IFOAM) defines OF (OF) as the agricultural process that utilizes ecologically based biofertilizers and pest controls abstracted substantially from plant and animal waste and organic manure (IFOAM 2022). However, OF is an agricultural approach or system that advocates healthy products free from components that may harm humans and nature. They include but are not limited to industrial pesticides, insecticides, fertilizers, GMOs, chemical medications, hormones, growth-boosters, etc. OF provides food and feed with optimum nutritional value and minimum dangerous ingredients, with only permitted substances used. The principle also requires 100% natural forage for livestock and its further processing without synthetics. It is also qualified as an environment-friendly and nutritive-sensitive agriculture system. In simplest way, the cultivation of arable and permanent crops without the use of artificial pesticides, growth hormones, antibiotics, genetically modified organisms, or fertilizers is known as OF or agriculture.

Organic agriculture or farming (OF) occupies 2% of agricultural land. The growth of organic agriculture has been accelerating worldwide, and is practiced in 188 countries, with 4.5 million farmers and producers operating over 96 million hectares of farmland organically. OF is revolutionary as it has moved from local and niche initiative to being an industry and expanding market (Willer et al. 2024; Pandey et al. 2024).

The top three countries with the highest number of organic producers were India (2,48 million), Uganda (0.40 million), and Thailand (0.12 million) (Willer et al. 2024). There has been an increase in the number of producers of nearly 919,000 and 2,590,000, or 25.6 and 134.7%, compared to 2021 and 2011, respectively. In Asia, Europe and Oceania, the number of producers increased; however, in Africa, North and Latin America there was a decrease in 2022 (Willer et al. 2024). Hopefully, the Ecological Organic Agriculture Initiative[1] will be able to rescale up organic strategies and practices through sound institutional capacity development, scientific innovations, market participation, public policies and programs, outreach and communication, efficient coordination, networking, and partnerships in Africa.

In 2000, the OF acreage and number of farmers were 14.9 million hectares and 200 000, respectively. In 2022, organic agricultural land increased on all continents. The top three countries are Australia (53.0 million hectares or a share of 14.7%), India (4.7 million hectares or a share of 2.6%) and Argentina (4.1 million hectares or a share of 3%) and they comprise 64% of the total OF area (Willer et al. 2021, 2024) (Fig. 2.1). In 2000, the share of OF was only 0.3% while in 2022, it has increased 7 folds (Fig. 2.2).

The Unites States, in contrast, has 2.06 million hectares in 2022, which means Australia has 25 times the amount. In fact, Australia has more organic farmland than the rest of the countries of the world combined. That said, it is possible

[1] https://eoai-africa.org.

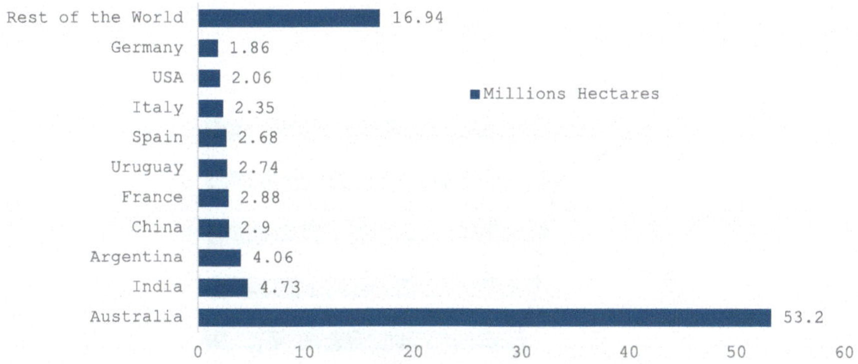

Fig. 2.1 Ten countries with the largest areas of organic agricultural land in 2022. *Source* Willer et al. (2024)

some countries have farms that use an organic style or method but do not report or certify it as being organic due to the costs.

The good signs for the future are: (i) the continuous increase in the number of organic certifications in most countries; and (ii) the global and rapid increase in the past two decades of organic food consumption, with the biggest consumers coming from the United States, Germany, France, and Australia.

India has witnessed a phenomenal growth in organic agriculture and has emerged as the single largest country having the highest number of organic producers in the world. Indian government has set up zones for OF, giving farmers financial aid for using organic inputs, and implementing systems for organic certification (Bhujel and Joshi 2023). It is worth noting that the country had only 41,000 hectares of organic land in 2002 (Das et al. 2020).

The cropland area (arable land with 15.1 million hectares and permanent crops with 6.4 million hectares) constituted 21.5 million hectares, which is less than a quarter of the organic agricultural lands. Hence, more than two-thirds of the 96.4 million hectares of organic agricultural land in 2022 were from grassland/grazing areas (over 67.6 million hectares) mostly in Oceania (Willer et al. 2024). From the same report, the key arable crops were cereals, green fodder from arable land and oilseeds. For permanent crops, nuts, olives, coffee, grapes, and citrus were the most relevant (Willer et al. 2024). Organic arable crops represent 1.1% of world area of arable crops while organic permanent crops form 7% of the permanent crops grown.

It is important to note that arable and permanent crops increased from 3.7 and 1.1 million hectares in 2000, while grassland/grazing areas at the same year were estimated 21.7 million hectares (Willer et al. 2024). According to these authors, from 2013 and 2022, the change in area of arable and permanent crops was 82.2%.

Organic agriculture offers immense trading opportunities to the producers in 188 countries, mainly those with certification regulations. The organic market evolved, diversifying its offerings, which now span from staple produce to processed foods, textiles, cosmetics, and beyond. The global sales of organic food and drink reached

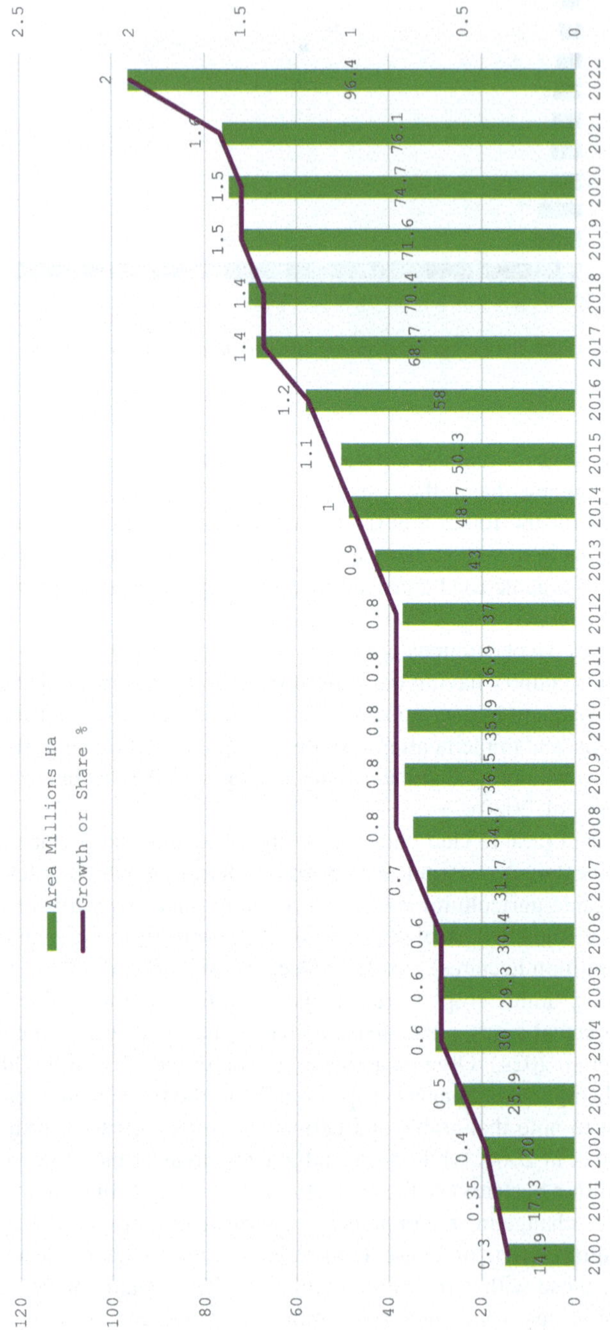

Fig. 2.2 Growth of the organic agricultural land and organic share 2000–2022. *Source* Willer et al. (2024)

more than 134.8 billion euros in 2022 (15.1 billion euros in 2000) (Willer et al. 2024) and USD 231.52 billion in 2023 (Grand View Research 2023). In 2024, the market size value is expected to reach USD 258.94 billion. Overall, organic food continues to rise in quality, choice, and affordability. It is expected to grow at a compound annual growth rate of 13.9% from 2024 to 2030 (Grand View Research 2023). However, still the demand for organic products mainly remained concentrated in North America and Europe. Unites States is having the single largest market for organic products (40.6 billion euros, 42 percent of the global market) followed by European Union (37.3 billion euros, 38.5 percent) and China (8.1 billion euros, 8.3 percent). Some countries in Northern Europe are currently witnessing a boom in sales of organic foods. The highest per capita consumption was found in Switzerland and Denmark (312 euros) followed by Sweden (231 euros) and Luxembourg (221 euros).

According to Eurostat (2022),[2] in 2020, in EU 27, areas under organic farming amounted to 14.7 million hectares and accounted for 9.1% of the Utilized Agricultural Area. Countries with the largest extension of organic farming were France (2.5 million hectares); Spain (2.4 million hectares); Italy (2.0 million hectares), and Germany (1.6 million hectares). Out of 57.8% of areas under organic farming in EU was concentrated in these four countries. Among these four countries, Italy had the highest incidence of areas under organic farming (15.9%); while Spain (10.0%), Germany (9.6%) and France (8.7%) were closer to the average value of EU 27 (Calabro and Vieri 2024). These values have slightly increased in 2022 as shown in Fig. 2.1. The growth and acceptance of OF practices and food by EU farmers and consumers make OF a good point of departure for agroecological transitions. The annual compound growth rate between 2012 and 2021 was 6%. Meeting the 25% target by 2030 would require a nearly doubled annual compound growth rate of 10.8% for the 2021–2030 period. This would require the conversion of 27 thousand km^2 per year in 2021–2030. EU Farm-to-Fork Strategy actions should disable barriers such as complexity of certification processes, market saturation for organic products, uncertainty in sales and a lack of premium prices (Moschitz et al. 2021; EC 2020).

A well-structured regulatory framework is crucial for the development of the OF sector (Brito et al. 2021). Effective regulations ensure that organic certification processes are transparent, accessible, and affordable, particularly for smallholder farmers. However, certification often remains a significant barrier, as costs and bureaucratic hurdles can discourage small-scale farmers from transitioning to organic methods. Governments must therefore simplify the organic certification process, ensuring that certification bodies are well-regulated and accessible. This can be achieved by offering government-backed subsidies or grants to cover certification costs for smallholders. Furthermore, regulations should ensure that certified organic products meet international standards, making them competitive in global markets. In Morocco, the regulatory framework must also address the alignment of organic standards with those of key export markets, such as the European Union, which has

[2] EUROSTAT (2020), Available at: https://ec.europa.eu/eurostat/web/agriculture/data/database (accessed 30 January 2023).

stringent organic certification requirements. Ensuring compliance with such standards will facilitate the entry of Moroccan organic products into these lucrative markets, thereby supporting the country's economic growth. In addition to certification, the regulatory framework should include robust mechanisms for consumer protection, ensuring that organic labeling is clear and accurate. This builds consumer trust in organic products, further driving demand and encouraging more farmers to transition to organic practices.

Liechtenstein has a staggering 43.0% share of organic agriculture, followed by Austria (27.5%) and Estonia (23.4%). Similarly, 22 countries exhibited 10 percent share or more of the total organic agricultural land. The total organic agricultural land and organic share have increased beyond sixfold in the last decade (1999–2022). Organic producers have increased significantly worldwide because of the popularization of OF in the last decade (Willer et al. 2024).

Over the past decade, the OF sector has experienced significant growth, yet it has not achieved full maturity (Willer et al. 2021). This growth analysis indicates that the suitability of OF, along with its production and environmental effects, is highly dependent on specific contexts (Seufert and Ramankutty 2017). Furthermore, several critics argue that organic agriculture is inefficient, particularly regarding land utilization (Kirchmann et al. 2008; Connor and Mínguez 2012; Röös et al. 2018). Additionally, some researchers question whether there is an adequate supply of organic feed materials—such as animal manures, off-farm organic wastes, crop residues, green manures, and biofertilizers—to support extensive OF and fulfill global food demands (Manna et al. 2021).

Malek et al. (2019) found that organic crop farmers are mostly present in areas with favorable socio-economic and climatic conditions, both globally but also within countries. They also reported that within developed countries, the locations of organic crop farmers often do not differ significantly from the locations of conventional crop farmers. In developing countries, there are, however, larger differences and organic crop farmers concentrate in the more accessible and developed regions. Such situation may prevent poor organic farmers from certification and access to markets. According to UNCAD and UNEP (2008), OF systems and practices are ideally suited for many poor, marginalized smallholder farmers in Africa, as they require minimal or no products, use locally available natural materials to produce high-quality and safe products, and encourage a whole systemic approach to farming that is more diverse and resilient to stress. However, Falconnier et al. (2023) denoted agricultural production in Sub-Saharan Africa as 'agroecological' by default leading to poor soil fertility because of widespread soil nutrient mining. The authors concluded that for first stage, crops need mineral fertilizers for their growth and development and in the long-term OF or other agroecological systems should be the model. In other terms, it not possible in the current situations of soils to close crop nutrient gaps and increase crop productivity by relying solely on N fixation and transfers of biomass using manure under OF.

Another significant question frequently posed by stakeholders in both the public and private sectors is the degree to which OF systems can improve food and nutrition security while promoting environmental sustainability (Röös et al. 2018). Additionally, there is concern regarding the overall impact on biodiversity, particularly if more land is transitioned to organic practices. One more important consideration is that the willingness to adopt and the level of adoption are influenced by the perceived welfare and economic benefits that farmers and producers, regardless of their size, experience when shifting to organic production (Xu et al. 2018; Aihounton and Henningsen 2023). According to Bouttes et al. (2019), farmers view OF to enhance their adaptive capacity through four key mechanisms: mitigating risks, fostering learning, boosting professional satisfaction, and allowing them to sustain their family farms.

2.3 Agricultural Production and the Guarantee of Food Security and Availability

Crop yield or productivity is the key to global food security. Studies on future global food production and global food security generally take as a premise that global demand for food will increase by 60% between 2005/2007 (the base period) and 2050 and that the global crop demand may increase by 100–110% over the same period (Giller et al. 2021a). At the same time, food systems are facing major complex challenges of increased food insecurity and malnutrition (FAO et al. 2020; FAO 2021).

Soil degradation caused by both intensive agriculture and smallholder farming is threatening global food security and continues to negatively affect natural ecosystems under increasingly dry climatic conditions and high resource pressure (Gibbs and Salmon 2015; Bestelmeyer et al. 2015; Jung et al. 2024; IPCC 2022a, b). Humankind is seriously challenged by poverty, loss of (agro-) biodiversity, and soil and water insecurity, which oblige for comprehensive and integrated solutions mainly of low-impacts on environment and society (Giller et al. 2021b).

Literature is brimming with studies that attempt to link OF to food security, although the feasibility of OF systems for achieving food and nutrition security is still contested and revealed mixed evidence (Wezel et al. 2014; Kirchmann et al. 2016; Nyantakyi-Frimpong et al. 2017; Muller et al. 2017a, b; Röös et al. 2018; Steufert 2019; Das et al. 2020; Barbieri et al. 2021; Yadav et al. 2021). Available studies show a wide range of results, depending on the context (Table 2.1).

In some situations, organic yields were found to be higher than conventional yields, whereas in other situations they were considerably lower (Liu et al. 2010; Venkat 2012; Abeliotis et al. 2013; Rodale Institute 2022; Meemken and Quaim 2018; Das et al. 2020; Yadav et al. 2021).

Olle and Williams (2012) reported about 59% of studies observed positive effects on vegetable yields from organic cultivation, while 29% reported a negative impact, and 12% found no significant influence. Such diversity in results can be explained by a narrower spectrum of choices in organic nutrient sourcing and applications under organic agriculture (Panday et al. 2024).

Table 2.1 Overview of organic crop yield gaps

Meta-analysis/field trials	Crops	Regions	Gap (%)
Fjelkner-Modig et al. (2000)	Vegetables (cabbage, potato, peas and carrot)		– 12 to – 30
Pimentel et al. (2005)	Soybean Corn	USA	– 2 0
Badgley et al. (2007)	Oilseed crops Cereals Vegetables Legumes/pulses	Worldwide	+ 8 + 31 + 6 + 52
Kirchmann et al. (2008)	Corn Soybean Wheat Oat Tomato	USA	– 2 to – 66 4 to – 20 – 15 to – 42 – 5 + 11
Seufert et al. (2012)	Oilseed crops Fruit crops Cereals Vegetables Legumes/pulses	Worldwide	– 11 – 3 – 26 – 33 – 10
de Ponti et al. (2012)	Oilseed crops Fruit crops Cereals Vegetables Roots and tubers Legumes/pulses	Worldwide	– 26 – 28 – 21 – 20 – 26 – 12
Ponisio et al. (2015)	Oilseed crops Fruit crops Cereals Vegetables Roots and tubers Legumes/pulses	Worldwide	– 11 – 8 – 22 – 13 – 29 – 15
Meier et al. (2015)	Arable crops Fruits and vegetables	Worldwide	– 68 to + 32 – 65 to + 76
Kniss et al. (2016)	Wheat Soybean Barley Corn Oat Hay and alfalfa Vegetables Fruits	USA	– 34 – 32 – 24 – 35 – 20 + 80 – 23 to – 62 – 40 to – 60
Ranganathan and Selvaseelan (1997)	Rice	India	+ 20
Chan et al. (2008)			– 6 to – 45
Gao et al. (2023)	Tomato	Worldwide	+ 42.2

The work of Badgley et al. (2007) initiated a significant debate regarding the capacity of organic agriculture to sustain the global population. Their analysis, which incorporated findings from multiple studies, revealed that organic agriculture achieved an average crop yield that was 33% higher than that of conventional methods on a global scale. Specifically, their examination of 46 paired organic and conventional systems indicated that while organic yields in developed nations were 9% lower than those of conventional agriculture, in developing countries, organic practices could enhance crop yields by as much as 74%. Similar conclusions were drawn by Clark and Tilman (2017). Nevertheless, some researchers have questioned the reliability and significance of the findings presented by Badgley et al. (2007), leading to critiques from authors such as Cassman (2007), Connor (2008), and Meemken and Quaim (2018).

A review by Kirchmann et al. (2008) found organic crop yields being 25 to 50% lower than conventional ones and explained the gaps by lower nutrient availability, poorer weed control, and limited possibilities to improve the nutrient status of soils. They also linked the yield gaps to whether the organic system has access to animal manure. They also found that the amounts of manure available on organic farms are usually not sufficient to produce comparable crop yields as in conventional systems, and therefore farmers are commonly using green manures. They also assume saying that high crop yields are not proof OF the productivity of an organic system if it uses large amounts of manure transferred from other systems and not produced by the farming system itself.

Out of the 12 crops analyzed by Venkat (2012), higher productivities in organic systems were only reported for alfalfa, blueberry, raisin, and wine grape and apple. The author found that converting cropland to organic production over the next few decades may offer significant GHG reduction opportunities by way of increasing the SOC stocks during the transition.

In their global analysis of organic agriculture, Seufert et al. (2012) found that the yields of organic fruits and oilseed crops exhibit a minor gap of − 3% and − 11%, respectively, compared to conventional crops, although these differences are not statistically significant. In contrast, organic cereals and vegetables demonstrate significantly lower yields, with reductions of -26% and -33%, respectively, when compared to their conventional counterparts. The study also indicated that OF yields are 35% lower than conventional agriculture under irrigated conditions, while the difference narrows to 17% under rain-fed conditions. The authors emphasized that yield disparities are highly context-dependent, influenced by specific system and site characteristics, with organic yields ranging from 5% lower for rain-fed legumes and perennials on weak-acidic to weak-alkaline soils, to 13% lower when optimal organic practices are implemented, and up to 34% lower when comparing the most similar conventional and organic systems. Furthermore, Seufert et al. (2012) noted that perennial crops and legumes tend to yield more comparably to conventional crops than other organic varieties, a finding supported by Kniss et al. (2016), which highlighted the superior performance of organic perennial hay crops over annual silage crops in the USA.

Crop protection is a key activity for the sustainability and feasibility of agriculture in a current context of climate change. Weed and pest management in OF can be labor-intensive and increase costs (Panday et al. 2024). In 12-year field trials representing India, Bolivia and Kenya, Bhullar et al. (2020) reported that substitution of conventional pesticides with those allowed in organic production is not sufficient to solve pest and disease problems. They recommended a system approach to pest and disease management in organic farming. Hence, weed and pest control can be a major source of concern for organic farmers, but with the help of intercropping and mixed cropping, these enemies can be managed, and their infestation reduced.

Integrated pest and disease management approach and enhanced uses of biopesticides are intimately linked with OF to limit or suppress pest and disease development and guaranty good crop productivity. In fact, a plethora of biocontrol agents have been successfully used to manage the effect of pests and diseases, which also can enhance the quality of crops (Oyetunji et al. 2024). Among such agents, plant growth-promoting bacteria (PGPB) are sustainable alternatives that positively impact plant growth, yield, and abiotic stress mitigation. They also aid in nutrient transportation, enhancing the nutrient bioavailability, mineral solubilization, nitrogen fixation, and production of phytohormones that ultimately enhance crop productivity and contribute to food safety and food nutrition (Saleem et al. 2024).

Compared to conventional farms, organic farms typically have longer crop rotations, which increase the diversity of crops grown on the farm and potentially enhance crop yields and quality. Organic farmers also diversify their productivity through intercropping and mixed cropping systems (Rani et al. 2023). Such diversification and integration of practices boost soil structure and nutrient levels alongside hindering soil-borne pests and diseases (Howard 1947; Soni et al. 2022; Bhujel and Joshi 2023) but also minimize risks by reducing the economic dependence on a single crop (Seufert and Ramankutty 2017; Panday et al. 2024). However, the logistical challenges of sourcing organic seeds and fertilizers and managing extended crop rotations add hurdles for farmers.

In irrigated OF, generally, water use is high due to pre-irrigation for weed germination and control. In addition, climate change is expected to increase the geographical areas where certain insects and diseases impact crops. For mitigating such stresses, crop varieties have yet to be developed to produce high yields and quality and nutritious products under organic conditions (Lammerts van Bueren et al. 2011; Murphy et al. 2007).

Life-cycle assessment by van der Werf et al. (2020) favored high-input intensive agricultural systems over OF systems and did not fully distinguish the beneficial impacts of practices on ecosystem services. Using the same assessment, Meier et al. (2015) also found that organic crops have 34% lower yields than conventional crops. Further, as OF approaches can have lower yields, there is concern that shifting to lower-yielding organic agricultural systems implies land use expansion potentially causing higher GHG emissions and reduced biodiversity due to the conversion of natural ecosystems to agricultural lands (Muller et al. 2017a, b; Smith et al. 2019a; Seufert and Ramankutty 2017). Several authors reported that OF provides lower yields compared to intensive and tillage-based cropping systems (Stanhill 1990; de

Ponti et al. 2012; Seufert et al. 2012; Seufert and Ramankutty 2017; Ponisio et al. 2015). For instance, Cooper et al. (2016) found that, compared to deep inversion tillage, reduced tillage in organic systems resulted in yield losses of 8% in the humid continental zone, but that yield losses were negligible in the humid oceanic zone (< − 1%). However, Van Balen et al. (2023) showed from a ten-year experiment (2009–2018) that integrating OF and reduced tillage systems permitted improved soil health and similar or even higher marketable yields of root crops in comparison to conventional farming.

Ponisio et al. (2015) also recommended crop diversification and use of biological tools to both increase resilience of organic crops to biotic and abiotic stresses while enhancing soil biological dynamics to boost productivities. Owing to its diverse mixed farming approach, OF can abate risks by reducing the economic reliance on a single crop or produce as modern corporate farming.

Organic fertilizer is a blend of several nutrients available in organic form, which need to be mineralized to release the nutrients for plant uptake. In this respect, authors linked reduced organic crop yields to the fact the release of plant-available nitrogen from organic sources is slow and can often not keep up with the nitrogen demand during peak crop growth periods and especially for high nutrient-demanding crops. They also reported that amount of phosphorus provided in organic systems is sometimes insufficient to replenish the quantities lost due to harvest (Seufert and Ramankutty 2017). This would be more critical when growers produce longer-growing season greenhouse crops like cucumber, tomato, and sweet pepper. These crops need a continuous and regular supply of nutrients throughout the growth and development stages (Burnett et al. 2016; Bergstrand 2022). Other authors also gave critics showing that under conventional systems, synthetic fertilizers deliver excess quantities of nutrients to crops, often past what is needed and cause low system's efficiency and soil and water pollution (i.e., Worthington 2001; West et al. 2014). The study by West et al. (2014) reported that less than half of the supplied fertilizer is used by crops, and the remainder is squandered to the environment. The remaining nutrients are washed out of the soil and discharged into waterways (e.g., streams, rivers, reservoirs, and seas), causing various environmental problems.

In high pest pressure environments, and where pests and diseases that are difficult to control with biological methods are found, and due to the ban of chemical pesticides, yield gaps of organic agriculture are higher than in low pest pressure environments. Leifeld (2016) contended that OF is unable to cope with high-fecundity and rapidly dispersing pests, which could result in yield losses more severe than those observed in isolated, small-plot experiments or under conventional agriculture. However, Ponisio and Kremen (2016) and Parizad and Bera (2023) defied with evidence of the positive effects of organic farmland on pest suppression at the landscape scale. According to Seufert and Ramankutty (2017), biological pest control tools and measures (mainly bio-pesticides) available to farmers for effectively controlling weeds, insect pests, and plant diseases are limited and hence not able to keep pest impacts below the economic injury level. Porcel et al. (2018) found that local natural enemy populations—particularly those of predatory bugs—are preserved by organic management in apple orchards, which is crucial for the early suppression of aphid colonies.

A meta-analysis by Smith et al. (2019b) concluded that efforts to reliably curb environmental impacts through organic certification may come at a cost of lower and more variable yields. They also pointed out that despite the lower yields and greater yield variability on organic farms, such farms were more profitable, and had similar production costs compared to conventional farms. They reported that while organic farms provided a 'win-win' (high means and low variability) for environmental sustainability, conventional farms provided a 'win-win' for production by promoting high crop yields with low variability. This raises two crucial and important issues: (i) OF crop high yield variability does not allow producers to achieve consistent production and avoid unprofitable years and not ensuring that consumers have reliable access to nutritious and sufficient food; and (ii) OF systems, due their low variability in ecosystem services, are more resilient to changing climatic conditions, including an increased frequency of drought and extreme weather events.

Pretty et al. (2003) concluded from an analysis of 208 projects among 8.98 million farmers and on 28.92 million hectares of arable and permanent crops in Africa, Asia, and Latin America that OF led to improvement in food security and nutrition among farm households. While Bhullar et al. (2020) observed no significant difference in farm productivity between organic and non-organic farmers for selected crops.

Aihounton, and Henningsen (2023), for the case of small farmers in Benin, indicated that adopting OF substantially reduces their food security, while it tends to slightly reduce the nutritional quality of their diets. Such results could be explained by the conclusions of Kirchmann et al. (2008) reporting that organic manure is a severely limited resource, unavailable in quantities sufficient for sustaining high crop yields; legumes are not a free to replace inorganic fertilizers; and low native soil fertility cannot be overcome with local inputs alone.

In India, Manibharathi and Senthamil (2024) concluded their review stating that OF increases food security by increasing yields and incomes, improving food availability through mixed farming and diversification, and lowering the risk of crop failure in the event of adverse weather conditions. Sow et al. (2023) confirmed the positive yield impact of OF using varying forms of organic fertilizers for rice, chickpea, potato tuber, pea, turmeric, and wheat as compared to conventional systems. These two examples explain the increase in adoption of OF in India and its reduced uptake in Sub-Saharan Africa during the past years.

In a review of a 22-year farming trial study by Pimentel et al. (2005), OF produced the same yields of corn and soybeans as does conventional farming, but used 30 percent less energy, less water, and no pesticides and fertilizers. The authors also found that OF conserved more water in the soil, induced less erosion, maintained soil quality, and conserved more biological resources than conventional agriculture. Pimentel et al. (2005) found that crop yields were improving over time due to soil health improvement and crop produced higher yields under drought conditions. The meta-analysis carried by Lori et al. (2017) revealed that organic systems had 32%

to 84% greater microbial biomass carbon, microbial biomass nitrogen, total phospholipid fatty acids, and dehydrogenase, urease and protease activities than conventional systems.

Analyses by Meemken and Quaim (2018) indicated that the mean yield reduction of crops in organic agriculture is 19–25%. The authors analyzed several studies and observed considerable differences across different crop species, with legumes and fruits showing smaller yield gaps than cereals and root and tuber crops. They summarized showing that nutrient limitations are an important factor to explain the observed yield gaps in organic agriculture. The lower-than-average yield gaps observed for legumes and fruits are acceptable since legumes can fix atmospheric nitrogen and are hence less dependent than other crops on externally added nitrogen and fruit trees have longer-growing seasons and extensive root systems and are hence better able to absorb nutrients in synchrony with crop demand. According to another analysis by Leifeld (2016), organic agriculture makes use of biological N fixation via legumes as a primary N source; however, owing to the many non-fixing crops in rotation, N often limits productivity. In other terms, restricted use of synthetic fertilizers under certified OF downregulates productivity of crops compared to conventional systems. However, according to the meta-analysis of Fan et al. (2023), synthetic nitrogen fertilizer substitution practice by organic amendments is considered as a possible strategy to decrease the on-field GHG emissions and carbon footprints in upland cropping systems and simultaneously increase the soil organic carbon and yields in cropping systems. Biernat et al. (2020) found that arable OF showed the ability to produce agricultural commodities with low N_2O emissions per unit area, and no differences in product-related emissions compared with conventional farming. These effects offset increased emissions due to higher land use and deforestation needed for conversion to OF to feed growing population (Muller et al. 2017a, b).

With the yield decline by immediate shifting from conventional systems to organic systems, many food products would not be accessible to a larger community due to less availability, and less affordability due to higher prices. In addition, many poor households in developing countries spend over 50% of their income on food. In such situations, high organic food price increases are associated with higher levels of food insecurity and undernutrition, especially in urban areas as shown by Meemken and Quaim (2018). However, the authors also mentioned that yield gaps may be due to higher skills and technicity of farmers and producers of conventional crops than organic farmers. This conclusion was also supported by Muluneh et al. (2022). In contrast, OF promotes a closer connection between farmers and consumers, fostering a sense of community and shared responsibility for the environment and food quality. With time, such connections and responsibilities will make OF more accessible and viable for a broader range of farmers in addition to increased access to high-value markets.

Crop production and animal husbandry are closely linked on organic farms (Karlsson and Röös 2019). However, for smallholder farmers and peasant agriculture, who depend extensively upon animal manure and cover crops to generate soil organic matter, OF systems may not build or restore degraded soils as rapidly as the world food needs will require. Several studies show that organic systems are limited in nitrogen and phosphorus (Kirchmann et al. 2008; Meemken and Quaim 2018; Panday et al. 2024). To guarantee high yields and farm profitability, soil fertility must be strengthened through balanced crop rotation and organic fertilization using biofertilizers, recycling organic wastes, composts, mulching, and green manure (UNCAD and UNEP 2008; Diacono et al. 2019; Krauss et al. 2020; Manna et al. 2023; Panday et al. 2024). Crop yields can also be boosted using diversified rotations, incorporation of legumes, cover crops and inter/rely cropping (Sharma et al. 2018; Baritz et al. 2018). Authors also showed the importance of gypsum to replace chemical fertilizers for soil fertility enhancement (Tavakkoli et al. 2022) and combining biofortification with OF to increase crop production (Dhaliwal et al. 2022).

The efficiency of nitrogenous fertilizers is on a declining trajectory due to increased losses (denitrification, nitrate leaching, and ammonia volatilization) in conventional systems while in OF, nutrients—including nitrogen—are less available to crops to satisfy their needs due to their immobilization in soils (Barbieri et al. 2021). To address the low-nutrient availability and slow-release nature of organic fertilizers (manure and compost) and to increase nutrient-use efficiency, the market offers a variety of organic substrates and commercial fast-releasing organic fertilizers provide higher nutrient levels than traditional compost. Vermicomposting, for instance, was found to positively correlate with income and crop production in India (Atiyeh et al. 2000; Samawat et al. 2001; Kumar et al. 2018; Bhujel and Joshi 2023; Cidón et al. 2021; Yadav et al. 2021). However, municipal solid waste compost and sewage water are being increasingly employed in organic agriculture and very large amounts of organic residues and pollutants are added to the soil. If not appropriately used, these practices may limit the environment stewardship related to OF (Manna et al. 2021). Use of these practices could eventually lead to higher cadmium levels in organic foods as in conventional ones in addition to greater microbiological risk.

From the various studies, OF systems are knowledge and skill intensive, and achieving yields depend on timely and spatially management interventions. It is interesting that most of the meta-analysis come from experimental trials and research studies (Krupnik et al. 2019). In other terms, experimental yields are often higher than farmer's yields due to difficulties or uncertainty to replicate recommended management practices. Hence, while the reported yield differences between organic and conventional agriculture may be true under experimental conditions (Kravchenko et al. 2017), they may possibly underestimate the yield gaps that occur in real world farming situations (Seufert et al. 2012; Taheri et al. 2017; Meemken and Quaim 2018). From the analysis by Kirchmann et al. (2016), it is difficult or biased to compare paired side-by-side systems and the authors pointed out to specific pitfalls and shortcomings in the comparative analysis, which are due to differences in initial soil fertility, legacy in crop management and reliance on external resources and inputs.

According to Krupnik et al. (2019), in meta-analyses based on farm-and plot-scale data, smaller organic yield gaps were reported than in agricultural statistics from large-scale farming. In other terms, OF at large scales may lead to a global reduction rather than an increase of agricultural production as reported by several authors.

Pimentel et al. (2005) reported organic cereal yields to be 50% lower than conventional yields under practical conditions in some parts of Europe. These yield gaps under OF mean additional land requirements, which were estimated to 23–67%, are needed to feed a growing population (Meemken and Quaim 2018).

Kniss et al. (2016) stated that the analyses documenting the yield deficiency in organic production have relied mostly on data generated under experimental conditions. They assumed that these studies do not necessarily reflect the full range of innovation or practical limitations that are part of commercial agriculture. Kniss et al. (2016) compared OF and conventional systems in terms of crop yields for commercial farms and found the following: (i) Averaged across all crops, organic yield averaged 80% of conventional yield; (ii) for cereals, yields on organic farms were approximately 20% lower than on conventional farms, which is like the yield gaps reported in the meta-analyses; (iii) for certain vegetables (tomato, potato, bell pepper, and onion) and fruits (watermelon, grape, blueberry, and apple) reported yield gaps were 50% and more, which is significantly higher than what the yield reported by meta-analyses; (iv) organic yields surpassed conventional yields for some hay crops; and (v) Organic squash, snap bean (Phaseolus vulgaris), sweet maize, and peach (Prunus persica) yields were not statistically different from conventional. The lower organic cereal yields are likely associated with the challenges of balancing soil quality and weed management in organic grain production as given in Table 2.2.

This expansion of agricultural land will be a major contributor to biodiversity loss and climate change due increased livestock dependence of OF systems for manure production. However, the attention-grabbing simulation studies by Erb et al. (2016) and Muller et al. (2017a, b) showed that the predicted world population in 2050 could be fed even with lower yields and without the loss of additional natural habitat if only vegan diets were consumed.

Corn productivity was found lower during the initial year in an organically managed field than in subsequent years as soil fertility levels increase over time with the addition of organic materials in the organic management system (Crowder and Reganold 2015). Similarly, Surekha (2007) revealed that a gradual increase in rice yield with the use of organic fertilizers over a period was observed. According to Reddy et al. (2023b), the lower yields are seen during the initial stages of OF, where stabilization of the yields may see in later stages. Remarkably, organic agriculture outperforms conventional farming in times of drought or flooding. Therefore, the adoption of OF will act as a sustainable way to combat yield losses due to such abiotic stresses. Sharma et al. (2021) requested a need to estimate the breakeven point of profitability to farmers vis a vis food security over time and space to upscale the organic agriculture in view of changing climate. Consequently, increasing investment in the research, capacity development, and technology transfer efforts could possibly contribute to reducing the yield gaps between OF and conventional systems over time in both developed and developing regions. In addition, the adoption of a whole-system approach is needed to assess and address complex issues and aspects of OF transitions (Musche, et al. 2019).

Table 2.2 Environmental advantages and ecosystem services offered by OF systems

Pollution/contamination control of soil and ecosystems	Achieved by prohibiting synthetic pesticides and fertilizers in OF, resulting in better health outcomes and lower healthcare expenses related to chemical exposure
Soil health and structure improvement	The implementation of practices such as crop rotation, intercropping, and the use of biofertilizers has significantly improved soil health, fertility, and resilience by fostering a balanced ecosystem of soil microorganisms
Biodiversity conservation	OF promotes biodiversity and habitat conservation by minimizing harmful chemical use and monoculture, enhancing pollination, and supporting beneficial wildlife, which collectively enriches ecosystem services and maintains ecological balance
Water quality and conservation	Reduced risk of water contamination from runoff or leaching contributes to cleaner aquatic ecosystems, including rivers, lakes, and groundwater sources
Sustainable use of water resources	Enhanced soil quality and increased water retention capabilities lead to more efficient water usage, decreasing reliance on irrigation and alleviating drought effects
Climate change mitigation (lower carbon footprint)	OF practices are associated with lower greenhouse gas emissions per area, improved carbon sequestration, and reduced energy consumption by minimizing the use of synthetic inputs and fossil fuels
Soil conservation and erosion mitigation	Preservation of soil health and reduction of erosion through the upkeep of plant life and varied agricultural practices
Sustainable and circular approach	Adoption of sustainable and circular methodologies that promote resource efficiency by recycling nutrients and materials while improving ecosystem functions. Implementation of strategies that ensure the longevity of agricultural systems and the resilience of natural habitats

Commercial organic fertilizers are available in dry (e.g., bone meal, blood meal) and liquid (e.g., fish emulsion, seaweed extract) forms, each providing specific nutrients essential for plant growth and nutrient cycling (Durán-Lara et al. 2020). These products are mainly byproducts of different industries (i.e., fisheries, livestock, and food processing (Panday et al. 2024). However, these uses are knowledge intensive and noticeable gap exist in farmers' understanding of the products in terms of selection, cost, availability, scalability, and, to a certain level, the anticipated impacts on crop yields. Practical experiences of farmers have a pivotal role in constructing new

OF knowledge to develop further the perception and adoption of OF systems in their diversity in effects and trade-offs. So, for successful adoption of OF in complex social contexts like in developing countries, knowledge transmission should be strongly combined with knowledge co-production (Charatsari and Lioutas 2019).

According to Manna et al. (2023), for positive benefits from composting and other organic fertilizers, their management and optimal applications are crucial for synchronizing nutrient delivery and crop needs, improving crop productivity while minimizing environmental and health risks. The authors reported that blending/integrating the organic manures with holistic agronomic practices would be necessary to double fertilizer-use efficiency from ~ 30 to ~ 60% in India as realized in most of the agriculturally advanced countries (Koninger et al. 2021).

Maltais-Landry et al. (2023) reported an increase in the marketable yield of cabbage when using processed fertilizers—including blood meal, bone meal, feather meal, and meat meal—compared to composted manure and control in a two-year experiment. A meta-analysis by Gao et al. (2023), including 107 research papers, found that animal-based organic fertilizers originating improved the yield of tomato production by 42.2%.

2.4 Food Quality, Nutritional Value, and Health Security

The well-being and health of individuals are closely linked to the quality of their diet, which is fundamentally connected to nutrition security. Food quality is influenced by various factors that encompass both physiological and genetic aspects of crops. Variations in nutrient levels, vitamins, and other compounds can be traced back to changes in plant metabolism that occur due to the differing types of fertilizers—organic versus conventional—employed for crop nutrition or soil fertility enhancement.

Research conducted by multiple authors (Panday et al. 2024; Rahman et al. 2024; Rani et al. 2023) indicates that organically farmed crops tend to be more nutritious and beneficial for health compared to their conventional counterparts. A study by Olle and Williams (2012) revealed that in most cases (65%), vegetables grown organically exhibited superior nutritional values compared to those grown conventionally; 20% showed no significant difference, while only 15% demonstrated a decline in nutritional quality.

Numerous meta-analyses and studies, including those by Baker et al. (2002), Benbrook et al. (2008), Lairon (2010), Crinnion (2010), Brandt et al. (2011), Barański et al. (2014), Yu et al. (2018), Rahman et al. (2021), have highlighted notable differences between organic and conventional produce that may lead to enhanced nutrition and health benefits for consumers. These studies reported higher levels of essential minerals and vitamins, increased antioxidants such as vitamin C and polyphenols, reduced cadmium and pesticide residues, a lower occurrence of antibiotic-resistant bacteria, and a decreased water content in organic produce, resulting in greater dry matter per unit of fresh weight.

The vitamin levels in crops are notably affected by the specific agronomic practices linked to different farming systems. For instance, the application of nitrogen fertilizers in conventional agriculture can enhance beta-carotene concentrations while simultaneously lowering vitamin C levels. Research conducted by Worthington (2001) indicated that organically cultivated lettuce, spinach, potatoes, and cabbage exhibited vitamin C levels that were 27% higher than their conventionally grown counterparts. A review by Lairon (2010) further supported these findings, showing elevated vitamin C levels in organic potatoes, tomatoes, kale, and celeriac, along with increased vitamin E in organic olive oils. These observations were corroborated by subsequent studies on leeks by Golubkina et al. (2018) and onions by Hallmann and Rembiałkowska (2006). However, a more recent investigation by Reddy et al. (2023a) found no significant differences in water-soluble vitamins between organic and conventional crops, except for Moringa leaves and carrots. Earlier studies have aligned with these results (Woese et al. 1997; Warman and Havard 1998; Kumpulainen 2001; Bourn and Prescott 2002). In contrast, the research by Fjelkner-Modig et al. (2000) indicated that the farming system did not affect vitamin C content.

Several authors (Reddy et al. 2023a; Czech et al. 2022) observed a reduction in protein levels in conventional crops when compared to their organic counterparts. In contrast, research by Dos Santos et al. (2019) in Brazil indicated no significant differences in the centesimal composition, including protein, lipids, and carbohydrates, between the two varieties of sweet potato. Other studies (Giampieri et al. 2022; Yu et al. 2018; Golijan and Secanski 2021; Rembiałkowska 2007; Worthington 2001) suggest that the high nitrogen availability from mineral fertilizers in conventional agricultural practices may result in lower protein content in organic foods. Nevertheless, conventional crops exposed to elevated nitrogen levels tend to enhance protein production while simultaneously reducing carbohydrate levels, which can ultimately lead to a decrease in vitamin C synthesis.

For healthy life, humans need a variety of minerals in their diets, encompassing both major and trace elements. Research by Yu et al. (2018) indicated that organic corn contains 20% more magnesium (Mg) and 30% more phosphorus and potassium than its conventionally grown counterparts. This observation was supported by Rembialkowska (2016), whose experiments consistently showed elevated levels of iron, phosphorus, and magnesium in organically cultivated products compared to those grown conventionally. Additionally, Reddy et al. (2023a) noted a trend of increased mineral content, including iron, zinc, potassium, sodium, magnesium, and manganese, in organic Indian foods relative to conventional options. These findings align with the earlier work of de Souza Araújo et al. (2014), which reported higher levels of magnesium, manganese, potassium, and sodium in organic vegetables compared to conventional ones. Numerous prior studies have corroborated the trend of enhanced mineral composition in organic foods (Williams 2002; Magkos et al. 2003; Hunter et al. 2011; Asaolu et al. 2012; Głodowska and Krawczyk 2017). Furthermore, a study by Dos Santos et al. (2019) from Brazil highlighted that certain sweet potato cultivars exhibited higher levels of calcium, copper, iron, potassium, magnesium, manganese, and phosphorus in organic produce compared to conventional samples.

High-profile meta-analyses conducted by Smith-Spangler et al. (2012) have raised doubts about the widely held belief that organic foods are inherently healthier. In their review of 223 studies that compared organic and non-organic fruits, vegetables, and grains, focusing on nutrients such as ascorbic acid, phosphorus, calcium, magnesium, iron, and various vitamins, Smith-Spangler et al. (2012) found no significant differences between the two product categories.

In a recent publication, Munné-Bosch and Bermejo (2024) concluded that while OF may not match the productivity levels of conventional agriculture, it can yield higher nutrition quality in certain fruit crops. This quality enhancement is attributed to the absence of synthetic fertilizers and pesticides, improved pollination, and reduced protective treatments, which collectively contribute to increased production of antioxidant compounds.

Supporting this view, Reeve et al. (2016) provided evidence indicating that organically grown fruits and vegetables may possess elevated levels of health-promoting phytochemicals, potentially due to factors such as greater plant stress, diverse rhizosphere microbial communities, and lower nitrogen availability. Additionally, a report by Matt et al. (2011) highlighted that organic food products generally contain fewer pesticides and lower nitrate levels, along with enhanced nutritional value, as evidenced by higher concentrations of polyphenols (Flavanol), vitamin C, essential amino acids, and sugars compared to those produced through conventional farming methods.

It is well studied that flavonoids exhibit potent antioxidant properties and possess the ability to chelate metals, which plays a significant role in neutralizing free radicals. This activity is associated with various health benefits, including the inhibition of cancer progression, the mitigation of atherosclerosis, a reduction in obesity risk, and the fortification of blood vessel walls, thereby preventing micro-bleeding and strokes. Additionally, flavonoids contribute to a decrease in blood clot formation, further lowering stroke risk. Research has demonstrated that organic produce tends to have higher concentrations of various flavonoids. In a decade-long study conducted by Mitchell et al. (2007), it was observed that flavonoid levels in samples from OF practices increased over time, while those in conventional farming showed little variation. Specifically, the average concentrations of quercetin and kaempferol in organic tomatoes were found to be 79% and 97% higher, respectively, compared to their conventional counterparts, with values of 115.5 and 63.3 mg g^{-1} of dry matter (DM) versus 64.6 and 32.06 mg g^{-1} of DM. These variations were attributed to the enhanced organic matter present in soils managed under OF practices.

Research indicates that organic plants may possess twice the amount of phenolic compounds compared to their conventional counterparts (Rembiałkowska 2007). A subsequent study by Rembiałkowska (2016) found that the concentration of these compounds in organic products is approximately 20% higher than in conventional ones. Similarly, Faller and Fialho (2009) noted that vegetables and fruits cultivated organically exhibit elevated levels of polyphenols. Das et al. (2020) further highlighted that organic products tend to have greater amounts of dry matter, minerals, and antioxidants, including polyphenols and salicylic acids. Additionally, research by Talhaoui et al. (2015) revealed that vegetables grown organically possess various beneficial properties, such as antimicrobial, antioxidant, immune-modulating, analgesic, anti-inflammatory, cardioprotective, and vasodilatory effects. Furthermore, a study conducted by Barański et al. (2017) demonstrated that organic crops can show an increase in antioxidant activity ranging from 18 to 69% compared to conventional crops.

The concentration of salicylic acid was observed to be greater in organic foods. This organic compound is known for its significant anti-inflammatory properties, and its consumption is associated with various health benefits, including a reduced risk of cardiovascular diseases and colorectal cancers, as noted by Baxter et al. (2001), Rossi et al. (2008), Lairon (2010), and Ferreira et al. (2020). Carotenoids, such as lycopene, β-carotene, and lutein, contribute to lowering blood cholesterol levels, thereby promoting heart health. They also play a role in bolstering the immune system, particularly through beta-carotene, which encourages an increase in lymphocyte count. Additionally, these compounds exhibit anti-tumor effects, primarily attributed to their antioxidant capabilities and their ability to neutralize free radicals.

A study by Rani et al. (2023), comparing the nutritional content of crops and vegetables—including apples, tomatoes, eggplants, cauliflowers, carrots, and cereals—from both organic and conventional farming, revealed that organic sources are richer in nutrients such as vitamin C, carotenoids, amino acids, and polyphenols. Specifically, organic tomatoes showed increases in soluble solids, soluble sugars, lycopene, vitamin C, and nitrate by 11.86%, 42.18%, 23.95%, 18.97%, and 8.36%, respectively, compared to their conventional counterparts, as reported by Gao et al. (2023). Furthermore, Gomiero (2021) affirmed the superior nutritional quality of organic foods over conventional ones, highlighting the absence of pesticide residues and mycotoxins. The review by Rahman et al. (2024) also supported the notion that organic foods contain elevated levels of iron, magnesium, and vitamin C.

In a research conducted by Kumar et al. (2022) in India, it was demonstrated that biofertilizers play a crucial role in the decomposition of complex nutrients, enabling plants to efficiently absorb these nutrients, which is vital for the optimal growth and development of nutrient-rich crops. Additionally, Basalingappa et al. (2018) emphasized that biofertilizers help to inhibit pathogens that could compromise product quality, thereby ensuring that the crops remain safe and healthy for both consumption and re-cultivation. Mehata et al. (2023) concluded from a review that biofertilizers could be a practical and long-term answer for organic farmers who seek to improve soil health and crop productivity.

The protein concentration in crops, particularly in cereals such as wheat and rice, is a significant characteristic that influences both market and nutritional value (Geyer et al. 2022). It has been projected that rising levels of atmospheric CO_2 could lead to a 4.1% decrease in the global availability of dietary protein, which may disproportionately impact nations already facing high nutritional deficiencies (Loladze 2014). This author's meta-analysis indicated a widespread decline in plant mineral levels globally, revealing that increased CO_2 concentrations diminish the nutritional quality of plants, resulting in an average reduction of 8% in the concentrations of 25 essential minerals, including calcium, potassium, zinc, and iron. Such declines in crop nutritional quality can adversely affect diets and exacerbate the existing challenges of micronutrient malnutrition, ultimately impacting human health. These findings were corroborated by Ben Mariem et al. (2021), who noted that elevated CO_2 levels adversely affect cereal quality regardless of geographic location. OF systems may mitigate these effects through the strategic application of organic amendments or fertilizers, thereby enhancing the availability of nitrogen and other vital nutrients for crops. Furthermore, employing metagenomics and metabolomics research can deepen our understanding of the interactions between the soil microbiome, biofertilizers, and crops, while breeding organic food crops for improved nutritional content presents a promising alternative for addressing global food quality challenges.

It is suggested that organic produce possesses a higher sugar content, enhanced metabolic integrity, and a superior cellular structure, which contribute to its improved taste and longer shelf life (Bourn and Prescott 2002). Pawar et al. (2022) further assert that organic foods are not only more flavorful but also emit a distinctive aroma.

Gopalakrishnann (2019) notes that organic fruits and vegetables exhibit a significantly richer flavor profile compared to their non-organic counterparts, potentially leading to a decreased reliance on quality and flavor enhancers, such as preservatives and synthetic flavors, that are often utilized in the food industry. Moreover, the author highlights that organic crops demonstrate increased antioxidant activity, with concentrations of various antioxidants being 18 to 69% higher. This enhanced antioxidant presence is associated with a reduced risk of several health issues, including cardiovascular and neurodegenerative diseases, as well as certain types of cancer.

A meta-analysis conducted by Barański et al. (2014) indicated that organic crops may offer various nutritional and health advantages, including higher levels of antioxidants and polyphenolics, reduced risks from harmful pesticide residues, and lower cadmium (Cd) concentrations due to limited use of mineral fertilizers. The findings showed that organic crops have, on average, 48% less Cd compared to their conventional counterparts. Additionally, the study noted that organic crops exhibited significantly higher total carbohydrate levels while containing lower amounts of proteins, amino acids, and fiber. Although most macro and micro-elements remained consistent between organic and conventional farming, there was a slight increase in magnesium, molybdenum, and zinc in organic produce.

The substantial concentrations of vitamins and minerals present in organic foods and products contribute significantly to enhancing the human immune system. Additionally, antioxidants play a beneficial role in mitigating risks associated with heart disease, cancer, visual impairments, premature aging, and cognitive decline (Pawar et al. 2022). According to the review conducted by Rahman et al. (2024), individuals who consume organic products experience various health advantages, including lower rates of obesity and body mass index (BMI), improved blood nutrient profiles, and decreased risks of maternal obesity and pregnancy-related preeclampsia.

Research by Muller et al. (2017a, b) indicates that OF can be viable on a global scale, provided it is paired with strategies aimed at reducing demand, such as dietary changes and minimizing food waste. Their findings suggest that food supply can be adequately met by implementing these strategies, and that a sustainable nitrogen supply can be achieved through the promotion of integrated crop-livestock systems. Furthermore, Clark and Tilman (2017) highlighted that since more than 30% of food production is lost, addressing food waste could yield significant environmental benefits without necessitating changes in production methods or dietary habits. They also advocated for a transition toward low-input and low-impact food systems.

2.5 Connecting Environmental Sustainability and SDGs Through OF Large-Scale Adoption

We are currently navigating a poly-crisis, characterized by a series of interconnected challenges that intensify one another due to various converging and interdependent issues, including rising poverty and social as well as economic inequalities (UNEP 2024). Climate change, pollution, and desertification remain persistent global threats that contribute to increasing food insecurity and hunger in developing regions, while also affecting agricultural productivity in more affluent areas, especially within Euro-Mediterranean countries. Land use challenges have a profound impact on food availability and supply, subsequently influencing public health. A multitude of prominent institutions and numerous governments are advocating for regenerative and ecologically intensive strategies to tackle this complex crisis, including the promotion of OF as outlined by various reports and reviews (HLPE 2019; FAO 2018; IPCC 2019; EC 2020; Giller et al. 2021a). Additionally, Kanter et al. (2018) emphasized the integral role of agriculture in the realization of all Sustainable Development Goals (SDGs), underscoring its unifying nature across the 17 SDGs.

As OF continues to gain popularity, understanding its potential environmental benefits, significant sustainability challenges, and broader implications for sustainable agriculture and SDGs has become increasingly important (Eyhorn et al. 2019; Šeremešic et al. 2021; Manna et al. 2021). In this regard, extensive research has been conducted by the scientific community to evaluate the environmental effects of both organic and conventional farming methods (Tuomisto et al. 2012; Meier et al. 2015; Chiriacò et al. 2017; Smith et al. 2019a, b), revealing that organic systems tend to perform better on a per-unit-area basis. It has been noted that OF often leads to lower crop yields and less yield stability when compared to conventional methods, which presents challenges in the fight against hunger. However, the certification of OF systems for export markets can positively influence Sustainable Development Goals (SDG) 1 and the premium prices of organic products can significantly reduce poverty, a key factor contributing to food insecurity (UNCAD and UNEP 2008). These higher prices can help compensate for the lower yields typically associated with OF (Panday et al. 2024). Studies indicate that consuming organic foods can alleviate micronutrient deficiencies in various regions (Reddy et al. 2023a), thus aiding in the achievement of SDG 2. This issue is particularly pressing for the least developed countries striving to overcome poverty and hunger. Consequently, it is crucial to approach OF policies with care, particularly in sub-Saharan African nations that have not yet met SDG 1 and SDG 2 (Eyhorn et al. 2019; Zhang et al. 2024).

Sustainable Development Goal 12 emphasizes the importance of promoting sustainable consumption and production practices, which are vital for the well-being of both present and future generations. Achieving this goal involves reducing consumption levels, choosing products with a lower environmental footprint, such as organic goods, and minimizing the carbon emissions associated with daily activities and consumption, facilitated by OF supply chains (Eyhorn et al. 2019).

Additionally, the certification of OF practices plays a significant role in decreasing contaminant levels in food, animal feed, and the environment, especially in developed nations where regulatory measures are more rigorous and effectively implemented. Organic agriculture represents a sustainable and ecologically sound approach to production, offering a wide range of economic, environmental, social, and cultural advantages over different time periods. Organic farmers primarily rely on ecosystem services to achieve optimal crop yields, in contrast to conventional farmers who often depend on external inputs. Studies indicate that organic farms can generate profits that are 22–35% higher than those of their conventional counterparts, largely due to the higher prices that organic products command in the market.

In his 2021 study, Singh highlighted the necessity of ensuring that the anthropogenic impact of OF systems is in harmony with the adaptive resources available in agricultural ecosystems or basins, while preserving the morphological and functional integrity of the environment (Singh 2021). A substantial body of research supports the notion that OF can reduce the agricultural sector's effects on climate change (Aguilera et al. 2015; Niemiec et al. 2020; Goel et al. 2020; Biernat et al. 2020; Bellassen et al. 2021; Nabuurs et al. 2022; Singh et al. 2024). According to these studies, OF is associated with a 45% reduction in greenhouse gas (GHG) emissions compared to conventional farming, thus enhancing environmental sustainability.

Conventional agricultural methods are a major source of GHG emissions, primarily due to the reliance on synthetic fertilizers and pesticides, which organic farmers consciously avoid. The implementation of OF practices can lead to improved carbon sequestration in soils and permanent crops, thus enhancing climate resilience (Gattinger et al. 2012; Skinner et al. 2014; Madhavi et al. 2021; Hansen et al. 2019; Gross and Glaser 2021; Gamage et al. 2023; Kishore et al. 2023). According to a review by Fess and Benedito (2018) and Sharma et al. (2021), numerous studies have indicated that organic management results in higher levels of soil organic matter compared to conventional management. Gross and Glaser (2021) found that the application of farmyard-, cattle-, and pig-manure showed the highest soil organic carbon (SOC) increases (50%, 32%, and 41%, respectively), while green manure and straw showed only minor effects. They also reported that if manure applications were combined with additional mineral fertilizer, the SOC increases were higher (+ 1.7 Mg ha^{-1}) compared to manure alone. In a 9-year rice–wheat cropping system-based field experiment, Zhao et al. (2023) found that organic manure treatment accelerated the synergistic process between organic C sequestration and soil aggregation and showed great potential to increase soil organic C accumulation compared to mineral fertilizers.

OF systems are implemented and adopted across all significant climatic zones, ranging from the warm, humid tropics to cooler temperate and arid regions, and are integral to the primary agricultural practices worldwide, including rain-fed and irrigated systems, annual and perennial crops, mixed crop-animal systems. Furthermore, when integrated with advanced technologies such as precision agriculture, plant breeding, biotechnology, microbial innovations, smart biofertilizers, and biochar applications, organic agriculture can enhance carbon sequestration in the soil, bolster soil resilience, ensure food security, mitigate climate change, and support higher and more consistent yields even under drought or extreme conditions (Venkat 2012; Paustian et al. 2016, 2019; Kassam 2022; Yang et al. 2024).

To tackle the nutritional challenges for a growing world population in a context of changing climatic conditions, several studies and projects aimed to enhance the productivity of crops in organic farming through breeding and biotechnological technologies. In other terms, targeting plant breeding for OF can contribute to reduce its yield gaps vis-à-vis conventional farming (Crespo-Herrera and Ortiz 2015). According to Mmbando (2024), omics technologies are game changer strategies for improving yields and stress resilience of strategic crops (especially in Africa).

Additionally, research by Das et al. (2020) and Pimentel et al. (2005) indicates that organic soils exhibit superior texture and water retention capabilities, leading to increased crop yields, particularly during periods of drought. However, for improved yields and qualities, climate-resilient and efficient crops are of high needs and breeding programs are the sole pathway.

Several authors explored the motivations behind adopting agroecological practices, including OF, from various disciplinary viewpoints in the Global South. Findings from Ume et al. (2022) indicated a strong connection between values and environmentalism. Jouzi et al. (2017) determined that while OF poses notable challenges for small-scale farmers, it should still be regarded as a viable solution for addressing climate change and enhancing food security, as well as improving farmers' livelihoods. Their research highlighted that the primary benefits of OF include environmental protection, increased resilience to environmental fluctuations, higher farmer incomes, reduced reliance on external inputs, strengthened social capacity, and greater employment opportunities. The price premium associated with organic products can offset the potential drawbacks of lower yields and higher production costs, ultimately contributing to better financial stability for farmers (Crowder and Reganold 2015; Reganold and Wachter 2016; Seufert and Ramankutty 2017; Panday et al. 2024). Furthermore, OF emphasizes the use of natural processes to enhance yields, in contrast to the synthetic fertilizers used in conventional agriculture. This approach often requires more labor, which can lead to increased job opportunities.

Organic foods often command higher prices in the market, resulting in net economic returns per hectare that can equal or exceed those of conventionally grown crops (Pimentel et al. 2005). In fact, OF approach integrates traditional practices, innovative methods, and scientific principles to enhance the shared environment while fostering equitable relationships and improving the quality of life for all stakeholders. Key practices of OF include the use of cover crops, crop rotation with legumes, and the application of manure and compost, rather than a heavy reliance on synthetic fertilizers and pesticides for higher economic benefits (Sharma et al. 2018; Martin et al. 2018).

The significance of organic agriculture in preserving the natural resource base and supporting biodiversity and ecosystems is clear from substantial and sizeable amount of literature (Altieri et al. 2012; Seitz et al. 2018; Oteros-Rozas et al. 2019; Brito et al. 2021; Tscharntke et al. 2021; Koninger et al. 2021). As an example, research conducted by Tscharntke et al. (2021) revealed that local species richness and abundance can rise by about 34% and 50%, respectively, across different crops globally when compared to traditional farming methods.

The separation of crop and livestock production in traditional agricultural systems significantly contributes to environmental pollution, primarily due to excessive nitrogen loss and soil erosion, particularly in nations experiencing rapid growth in specialized livestock farming (Garrett et al. 2020; de Faccio Carvalho et al. 2021; Delandmeter et al. 2024; Gu 2022; Yang et al. 2022). In contrast, OF promotes a more integrated approach that combines trees, livestock, and crops, thereby aiming to minimize environmental harm. Furthermore, OF principles emphasize the importance of animal welfare. This holistic integration enhances nutrient cycling and energy efficiency, leading to improved animal productivity and higher quality outputs, whether in meat or milk (Badgley et al. 2007; van Wagenberg et al. 2017; Röös et al. 2018). The latter studies have also proposed methods to enhance or mitigate the performance of organic livestock in comparison to conventional systems. Perrin et al. (2020) found that in addition to higher autonomy in investments and economic efficiency, diversification within the organic farm via crop rotations, herd composition, and farm products improve farm resilience.

Reviews conducted by Pimentel et al. (2005) and Reganold and Wachter (2016) indicate that OF systems typically exhibit higher soil carbon content, better soil quality, reduced erosion, and lower nitrogen inputs when compared to conventional agricultural practices. Additionally, Pimentel et al. (2005) and Seitz et al. (2018) have estimated that the potential for soil erosion can be decreased by more than 61% through the implementation of OF techniques. Earlier, Niggli et al. (2008, 2009) insisted that organic agriculture substantially reduces the risk of soil loss through erosion by wind and water as well as by overgrazing, which is considered a crucial precondition for future food security.

A meta-analysis conducted by Smith et al. (2019b) indicated that organic sites outperform conventional ones in terms of mean biotic abundance and richness, soil organic carbon, carbon stocks, and profitability. Additionally, Schneider et al. (2014) found that, when comparing twelve countries across and Africa, species richness in organic production fields was, on average, 10.5% higher than in non-organic fields, with the most significant increases observed in intensive arable areas, reaching approximately 0.45%.

Tuck et al. (2014) highlighted a meta-analysis revealing that OF typically enhances species richness by around 30%. This finding has remained consistent over three decades of research and shows no indication of decline. Further validation of these results has been done by Crowder et al. (2010, 2012), Gattinger et al. (2012), Kennedy et al. (2013), Crowder and Reganold (2015), Reganold and Wachter (2016), and Lichtenberg et al. (2017).

Shrivas et al. (2019) emphasized the critical role of organic amendments and fertilizers in fostering microbial communities and enhancing their functions, which are essential for developing healthy and resilient soils capable of withstanding climatic and anthropogenic stresses. Nonetheless, a significant challenge in biological or OF remains the effective regeneration and reproduction of soil fertility, as noted by Puyu et al. (2023). Additionally, many regions have experienced soil degradation, leading to the depletion of organic matter that is often utilized to provide nitrogen. Therefore, increasing the adoption of OF practices could be a vital strategy for restoring and revitalizing soil health (Barbieri et al. 2021).

Despite the existing yield gaps of approximately 20% between conventional and OF, the integration of reduced feed production—characterized by lower outputs of forage maize, cereals, and soy for concentrated feed—along with a decrease in livestock numbers, can contribute to a more sustainable and climate-friendly agricultural system that is capable of meeting food demands (Reganold and Wachter 2016; Muller et al. 2017a, b).

The increasing demand for organic foods can be attributed, in part, to consumers' beliefs that these products offer superior nutrition and health benefits as well safety (Hoefkens et al. 2009; Smith-Spangler et al. 2012; Palupi et al. 2012; Rahman et al. 2021; Panday et al. 2024). By minimizing the use of synthetic chemicals, OF aims to lower the health risks associated with pesticide and metal residues (Hurtado-Barroso et al. 2017; Giampieri et al. 2022; Rahman et al. 2024). Smith-Spangler et al. (2012) reported that the risk for contamination with detectable pesticide residues was lower among organic than conventional produces, but differences in risk for exceeding maximum allowed limits were small. Nevertheless, research by Hoefkens et al. (2009) and Munné-Bosch and Bermejo (2024) indicates that OF is not without challenges, including the potential for microbial and heavy metal contamination.

Reganold and Wachter (2016) found soybeans and oilseeds to be the highest yielding globally but yield gaps of 37% are present in wheats and 33% in vegetables. The authors highlight that while conventional farming may yield higher crop outputs, OF systems provide healthier products along with enhanced ecosystem services and social advantages. Their findings suggest that OF improves water retention and soil infiltration, which can help mitigate the impact of reduced crop yields when transitioning from conventional methods. Furthermore, the reliance on ecosystem services, such as natural pest control, plays a crucial role in maintaining sufficient yields and minimizing production gaps (Karp et al. 2018).

In OF systems, practices that enhance soil health, such as mulching, the use of green manures, nitrogen-fixing crops and trees, and the incorporation of animal manures, play a crucial role in fostering below-ground biodiversity. In organic systems, reliance on N-fixing legume cover crop rotations with cereal grains can improve on N synchrony as the rate of N mineralization from decomposing cover crops more closely matches N uptake demands by the cereal (Crews and Peoples 2005).

The promotion of essential mycorrhizal fungi and various soil microorganisms is vital for numerous ecosystem functions and soil fertility. These practices are particularly significant for alleviating the impacts of drought and enhancing the water-use efficiency of crops. Research indicates that the densities of bacteria, protozoa, nematodes, and arthropods are greater in soils managed under OF compared to those under conventional methods. Furthermore, OF positively influences the micro-climate within the rhizosphere, contributing to improved crop productivity and resilience against drought and water scarcity (Brito et al. 2021; Durán-Lara et al. 2020).

Organic certification programs and schemes are instrumental in mitigating the adverse environmental impacts of agricultural practices on local communities, thereby supporting the populations of beneficial natural enemies and pollinators. This increase in beneficial biodiversity is essential for boosting crop yields, contrasting sharply with the outcomes observed in pesticide-dependent agricultural systems (Brito et al. 2021).

Cassman (2007) suggests that OF can enhance product quality, is beneficial for the environment, and reduces disease risks, which may offset some declines in crop yields. These factors underscore the significance of OF systems and the necessity for a large-scale transition to these practices. A long-term Farming Systems Trial (FST) conducted by the Rodale Institute over 40 years has demonstrated that regenerative organic agriculture can surpass conventional grain cropping systems, particularly during extreme weather events, with corn yields being 31% higher in OF during drought conditions (Rodale Institute 2022). Nonetheless, there remains a contentious debate regarding the comparison of food quality between organic and conventional farming methods (Brandt and Mølgaard 2001).

Based on the findings from various studies and the overview presented in Table 2.2, OF emerges as a vital strategy for land sharing, fostering biodiversity while simultaneously promoting food production on the same land without significantly harming the environment or society. Table 2.2 clearly illustrates the numerous positive outcomes associated with organic agriculture when evaluated against various environmental indicators. However, due to the lower yields produced by organic systems, they necessitate larger land areas to achieve equivalent outputs compared to conventional systems, particularly when assessing environmental benefits on a per unit of product basis rather than per unit of area (Nabuurs et al. 2022). Research by Verdi et al. (2022) indicates that approximately 192 million hectares of OF would be required to sustain current wheat production levels in the EU, in contrast to just 99 million hectares needed for conventional farming. Furthermore, Muller et al. (2017a, b) projected that a complete shift to organic agriculture would demand more land than conventional methods, although it would lead to reductions in nitrogen surplus and pesticide usage. The adoption of OF systems is complex and promotes a community-oriented approach that connects all stakeholders in the value chain. The advantages cited in Table 2.2 can be translated in Table 2.3 showing that OF systems offer a promising avenue for achieving several SDGs. In other terms, this Table explains that by promoting environmental sustainability, social equity, and economic resilience, large-scale adoption of OF can contribute significantly to navigating sustainability pathways. In fact, the correlative analyses by Smith et al. (2022) suggested that countries with greater progress toward the SDGs of no poverty, zero hunger, good health and well-being, quality education, decent work, economic growth, and reduced inequalities had the highest production of organic agriculture and, to a lesser extent, intensive use of inputs.

Table 2.3 Key SDGs addressed by large-scale adoption of OF systems

SDG 1: no poverty	OF often requires more labor-intensive practices than conventional agriculture, leading to increased job opportunities in rural areas Organic products often command higher prices than conventional ones, providing farmers with higher incomes and improving their livelihoods OF can foster community development by supporting local economies and creating opportunities for collaboration among farmers which lessen poverty and vulnerability with the communities
SDG 2: zero hunger	OF can enhance food security by promoting sustainable and resilient food production systems Organic practices can lead to crops with higher nutritional content, improving food quality and value
SDG 3: good health and well-being	Reduced exposure to harmful chemicals limits the risk of human health issues Improved food safety by reducing the presence of harmful contaminants
SDG 5: gender equality	Empowerment of women and youth through large-scale use of OF practices and products can empower human capital in rural areas by providing them with economic opportunities, green jobs and wider decision-making power OF can contribute to rural development by revitalizing rural communities and improving the quality of life for rural residents
SDG 6: clean water and sanitation	OF practices can help reduce water pollution by minimizing the use of synthetic fertilizers and pesticides
SDG 12: responsible consumption and production	OF promotes sustainable production practices by minimizing environmental impact OF can reduce food waste through practices like composting and integrated pest management
SDG 13: climate action	OF can help sequester carbon in the soil, contributing to climate change mitigation Organic practices promote biodiversity, which is essential for climate resilience and adaptation
SDG 15: life on land	OF supports biodiversity conservation by fostering natural ecosystems Organic practices improve soil health, which is essential for sustainable land management

2.6 Integrating OF with Regenerative Practices: The Agricultural Frameworks of the Future Must Prioritize Both Sustainability and High Productivity

The interconnected land challenges necessitate comprehensive and sustainable strategies to enhance resilience and livability in both rural and urban settings. To ensure food security and sustainability, agricultural systems must produce high-quality food, enhance environmental health and ecosystem services, be economically viable, and foster social well-being. In other terms, there is a need to reconnect agriculture with nature for long-term sustainability as proposed by de Rhodes (2012) and Faccio Carvalho et al. (2024). In this context, the OF model is widely recognized for its alignment with various sustainability principles (Smith et al. 2019a). The growing demand for safe and clean food production has led to an expansion of land dedicated to OF globally (Lockeretz 2007; Willer et al. 2024). Essentially, organic agriculture represents an ethical approach that fosters a more balanced relationship between humanity and the environment (Sumberg 2022). Scholars generally agree that OF yields beneficial environmental, economic, and social outcomes. Nevertheless, if OF systems continue to rely on tillage for seedbed preparation and weed management, soil erosion could pose a significant threat, potentially diminishing crop yields (Cooper et al. 2016; Moos et al. 2017). Furthermore, inadequate planning and rapid transitions within OF can also lead to food insecurity. Table 2.4 addresses the key barriers for OF development and progress and present solving options.

Low organic crop yields have been linked to restricted and unpredictable availability of essential nutrients, particularly nitrogen and phosphorus, in OF systems (Panday et al. 2024). Muller et al. (2017a, b) suggest that the discontinuation of synthetic nitrogen fertilizers may result in nutrient deficiencies, even with a higher presence of legumes in organic agricultural practices. Consequently, various studies have explored the integration of diverse regenerative or resilient systems, such as OF, conservation agriculture, and agroforestry, to support sustainability objectives and promote more sustainable food systems and multifunctional agriculture (Peigné et al. 2015; Cooper et al. 2016; Knapp and Heijden 2018; Ali and Dahlhaus 2022; Mrabet et al. 2024; Aboutayeb et al. 2024; Puyu et al. 2023; van Balen et al. 2023). Migliorini and Wezel (2017) examined the similarities and differences between OF and agroecology, concluding that both methodologies, beyond their regulatory frameworks, provide valuable contributions to the advancement of sustainable agricultural production and food systems. They emphasize that these approaches are grounded in holistic principles, advocate for the sustainable utilization of natural resources and inputs and prioritize biodiversity conservation along with the enhancement of ecosystem services. Beaury et al. (2024) recognized the need to shift from business-as-usual agricultural and restoration models to integration of land-based climate mitigation strategies to address both climate change and food security challenges.

In low-input cropping systems, a recent review by Dreyfus et al. (2023) based on published literature has provided compelling evidence on the technical maturity, financial viability, and broad applicability of site-specific nutrient management

Table 2.4 Identified constraints hindering OF progress for effective problem-solving

Constraints hindering OF progress	Detailed clarification and analysis	Targeted interventions to narrow the gap
Insufficient nutrient application and reduced effectiveness	• The available materials do not meet all the nutritional needs of crops effectively • Organic fertilizers often fall short in supplying sufficient nitrogen and essential nutrients for optimal crop production • The timing of nitrogen release does not align with crop requirements, leading to lower efficiency in the use of organic inputs such as animal manure, green manure, and compost, while stable yields rely on the timely provision of nutrients like phosphorus and potassium • Multiple nutrient deficiencies	• Enhancing productivity through selective breeding of livestock and crops • Incorporating food legumes and pasture crops into agricultural rotations • Implementing nutrient recycling practices, including the use of manure, biofertilizers, and high-quality compost, while utilizing local resources for sustainable food systems • Implementing eco-friendly practices for waste management and resource recovery
Inadequate management of weed populations	• Organic agriculture faces challenges in effective weed management due to the limited application of herbicides, leading to significant weed infestations often exacerbated by improper practices and manure usage • The reduction of the weed seed bank through organic methods remains minimal, contributing to ongoing weed issues • Additionally, there are prevalent nutrient deficiencies that further complicate the management of weeds in OF systems	• Effective design and management of crop rotation practices • Advanced strategies for weed management and control including techniques such as cover cropping, soil seed bank management, false seedbed application, and precision farming

(continued)

Table 2.4 (continued)

Constraints hindering OF progress	Detailed clarification and analysis	Targeted interventions to narrow the gap
Prevalence of pests and diseases	• Insufficient management of diseases and pests resulting from a lack of viable solutions • Presence of multiple symptoms alongside a significant level of infestation • Decreased quality of produce, leading to lower premium prices and potentially depriving organic farmers of additional benefits	• Certified crop varieties that exhibit tolerance or resistance • Effective design and management strategies for crop rotation • Utilization of intercropping, cover crops, and biological control methods, including biocontrol organisms
Limited natural soil fertility, particularly in sub-Saharan Africa	• Relying solely on organic practices for large-scale soil restoration may fall short in improving agricultural productivity and produce standards	• Effective planning and administration of crop rotation practices • Incorporation of food legumes within crop cycles • Implementation of green manure crops and diverse cropping systems • Minimal or no-tillage practices—integrated crop and livestock systems • Enhancement of microbial activity in soil • Application of soil amendments and restoration techniques • Adherence to conservation regulations

(SSNM) across geographies. Further, in a meta-analysis covering 11 countries in Africa and Asia, Chivenge et al. (2021) demonstrated that SSNM increased yields by 12% for maize, rice, and wheat and led to a reduction in nitrogen application by 10% and an increase in profitability of 15% at farm level.

Using modeling and scenario analysis, Muller et al. (2017a, b) showed that OF can contribute to providing sufficient food and improving environmental impacts, only if adequately high proportions of legumes are produced and with significant reductions of food-competing feed use, livestock product quantities, and food wastage. The development of organic agriculture in the future should take up these challenges on the consumption side, and not only focus on sustainable production. This scenario will also reduce the need for increased lands under OF and yield gaps over time.

2.7 Facilitating Factors that Promote OF to Improve Food Security

According to the latest report on food security (FAO et al. 2024), the world is off track to meet SDG Targets 2.1 and 2.2—to end hunger, food insecurity and all forms of malnutrition by 2030. In other terms, the distribution of challenges and opportunities related to enhancing global food security through the adoption of OF systems merits high considerations as they vary significantly depending on the social and economic contexts and policy and institutional commitments. Therefore, it is essential to promote and facilitate the uptake of OF through robust policies and incentives at both national and regional levels. The collaboration between farmers, their organizations, and public authorities should create synergistic effects that contribute to public goods, such as climate and environmental protection. Achieving these synergistic effects necessitates appropriate incentives and motivations, which can help prevent political compromises and support evidence-based approaches for scaling up OF practices. The successful and sustainable implementation of OF requires a combination of managerial, financial, and technological strategies, alongside institutional support aimed at enhancing the productivity of crops, trees, pastures, and livestock.

As indicated above, OF has the potential to play a key role in achieving national and global sustainability goals, but only if supported by coherent and integrated policy frameworks. The role of government policies in scaling up OF practices is crucial for addressing issues such as food security, environmental degradation, and climate resilience (El Ghmari et al. 2021). For example, the alignment of OF policies with broader sustainability objectives can help countries meet their commitments under the SDGs, particularly SDGs 2 (Zero Hunger), 12 (Responsible Consumption and Production), and 13 (Climate Action). Such alignment ensures that OF contributes to food security while also addressing environmental concerns like biodiversity loss, water scarcity, and carbon emissions (FAO 2019).

Policy integration should transcend traditional agricultural policies and involve cross-sector collaboration between ministries responsible for agriculture, water, energy, and environment. For instance, promoting OF as part of a Nexus approach that integrates water, energy, and food policies is essential in water-scarce regions (Ciccarese and Silli 2016). In Morocco, the current Agricultural policy—Generation Green 2020–2013—could be expanded to more explicitly promote OF as a sustainable agricultural practice that reduces dependence on imported chemical inputs and conserves rare resources. Countries like Germany provide strong examples of how integrated policies can ensure that policy fragmentation does not hinder progress and create favorable environments for OF. These countries have embedded OF in their national sustainability strategies, offering financial incentives, subsidies, and technical assistance to farmers transitioning to organic practices (Federal Ministry of Food and Agriculture 2022).

Stakeholder engagement is a crucial component in any policymaking process. Governments, research institutions, and development organizations must work together to provide the necessary technical assistance, capacity-building programs, and financial incentives to help farmers transition to organic agriculture. In addition, stakeholders must navigate between advancing global long-term goals and addressing short-term local socio-economic, cultural, and political realities in shifting to organic or sustainable farming or food systems.

Institutional support is critical for the success of OF, particularly for smallholder farmers who often lack the resources to adopt sustainable practices. In Morocco, where the agricultural strategy (Green Generation) is seeking 100,000 Ha per 2030, institutions like the National Institute for Agricultural Research (INRA) play a vital role in advancing OF through research and innovation. By developing sustainable farming practices that are adaptable to the local context, INRA and similar institutions can help farmers enhance their productivity while preserving the environment. Additionally, educational institutions and agricultural cooperatives must be involved in disseminating knowledge about OF practices, enabling farmers to access cutting-edge techniques and technologies and developing curricula related to OF systems for new generations of scientists, technicians and extension agents (Mrabet et al. 2024).

Moreover, institutional support extends beyond research and knowledge dissemination to include financial mechanisms that facilitate farmers' access to markets. Small-scale farmers often struggle to cover the high initial costs associated with organic fertilizers and bio-pesticides, particularly when they face credit limitations (Cafer and Rikoon 2018). As a result, these financial constraints represent a significant barrier to the adoption of OF practices (Sieber et al. 2018). One of the most effective policy measures to address this issue is the establishment of multi-annual incentive programs aimed at encouraging farmers and value chains to adopt OF methods. Indeed, governments should establish public-sector programs that offer subsidies for organic inputs, facilitate access to certification processes, and develop infrastructure for organic food distribution. Such support ensures that smallholder farmers can participate in organic value chains, increasing their profitability while contributing to national food security.

Additionally, governance frameworks should encourage investment in OF by creating incentives for private companies to invest in sustainable agriculture. Tax breaks, subsidies, and streamlined regulatory processes can make OF more attractive to investors, thereby facilitating the growth of the sector.

In this regard, public-private partnerships (PPPs) can also play a transformative role, as they bring together the resources and expertise of governments and private-sector entities to foster innovation in organic agriculture. By developing infrastructure, expanding market access, and fostering local entrepreneurship, institutional support can lay the foundation for the growth of OF across, especially in developing regions.

Information asymmetry has created unnecessary barriers to farmers' decisions regarding production inputs and has adversely affected agricultural productivity (Ullah et al. 2020). This lack of information is a primary reason for the reluctance of farmers to embrace OF systems. However, digital technology has the potential to significantly reduce this asymmetry in agricultural production and marketing, allowing farmers to optimize fertilizer use, adopt new technologies, and manage resources with greater precision (Harou et al. 2022). Hence, vital enablers for OF enactment and functioning are access to up-to-date information on innovative technologies and organic quality standards, simplified certification and labeling of products, investment opportunities, PPPs, communication strategy to raise awareness of farmers and continuous improvement of stimulating innovation and research. In addition, developing a robust supply chain and infrastructure for OF is crucial to support farmers and make their products more accessible to consumers both domestically and abroad. Table 2.5 detailed enabling factors that ease adoption of OF for ensuring food security pillars.

PPPs are essential for addressing the challenges of market access and scaling innovations in OF. OF, particularly in developing countries, often faces barriers related to market access, certification, and the high costs of transitioning from conventional to organic practices. PPPs can help overcome these barriers by facilitating investments in the necessary infrastructure and supply chains for organic products. For example, partnerships between governments and private food companies can support farmers in accessing both domestic and international markets. The role of retail chains and global organic brands in sourcing organic products from local farmers is critical for scaling organic production. These partnerships ensure that smallholder farmers gain access to certified organic markets, where they can sell their produce at premium prices. In countries like India and Uganda, PPPs have been successfully used to promote OF, with private companies offering technical training, input supplies, and marketing support, while governments ensure regulatory frameworks and market access. By establishing such partnerships, concerned countries may become key players in the global organic market, while simultaneously promoting sustainability and rural development.

Proximity to markets in larger urban centers has been identified as an important spatial driver of OF in several countries (Europeans countries, USA, Korea etc.). Most certified OF systems are globally found in areas with lower poverty levels and good soils, climate conditions and infrastructure (Malek et al. 2019). Willer and Lernoud (2018) found that in developing countries, certified organic farmers mostly produce commodity crops for export to more developed markets with only a minor portion being produced for domestic consumption. However, to increase OF adoption and acreage for domestic use of OF produces, poor farmers should be given support to access certification and provided privileged market access or market-production nexus (Ume 2023).

Table 2.5 Enabling factors for OF and food security

Enabling factor	Ways for achievements and defying the status quo
Government and institutional Support	• Subsidies and incentives: providing financial support to organic farmers can help offset the initial costs of transitioning to organic practices • Research and development: investing in research to develop OF techniques and technologies can improve yields and efficiency • Certification programs: establishing and supporting certification programs can ensure the integrity of organic products and build consumer trust
Consumer awareness and demand	• Education and awareness campaigns: raising public awareness about the benefits of OF and the importance of sustainable food systems • Accessible markets/improved market-production nexus: ensuring that organic products are available and affordable to a wide range of consumers • Labeling and standards: implementing clear and consistent labeling standards to help consumers make informed choices • Workshops and training: organizing workshops and training sessions to educate the community about the benefits of OF and sustainable food practices • Community events: hosting events and festivals to raise awareness about organic agriculture and local food production
Technological advancements	• Precision agriculture: utilizing technology to optimize resource use, reduce waste, and improve yields in OF • Organic inputs: developing sustainable and effective organic inputs, such as biofertilizers and natural pest control methods • Equipment and machinery: designing and adapting equipment and machinery specifically for OF practices • Composting: encouraging composting to recycle food waste and create nutrient-rich soil for OF
International cooperation	• Knowledge sharing: facilitating the exchange of best practices and technologies among countries to advance OF • Trade agreements: promoting fair trade practices and reducing barriers to international trade in organic products • Global standards: developing and harmonizing international standards for organic agriculture to ensure consistency and quality

(continued)

Table 2.5 (continued)

Enabling factor	Ways for achievements and defying the status quo
Community involvement	• Local food systems: promoting local food production and consumption to reduce food miles and support local economies • Reducing food waste: promoting practices like meal planning and proper food storage to minimize food waste • Ownership and self-sufficiency: encouraging community members to grow their own food, fostering a sense of ownership and self-sufficiency • Cooperatives and associations: establishing cooperatives and associations to provide support, resources, and market access to organic farmers • Pooling resources: cooperatives and associations can help organic farmers share knowledge, resources, and market access • Advocacy: cooperatives and associations can advocate for policies that support OF and promote sustainable food systems
Private sector engagement, investment and funding	• Venture capital: investing in OF startups and businesses can provide the necessary capital for growth and expansion • Corporate social responsibility (CSR): companies can allocate resources to support OF initiatives, such as partnerships with organic farmers or investments in sustainable agriculture research • Market development through retailers and distributors and export promotion • Marketing and branding: promoting organic products through effective marketing campaigns and branding strategies • Public relations and communication: engaging in public relations activities to raise awareness about the benefits of OF and the importance of sustainable food systems • Certification and Standards as well as research and innovation

2.8 Conclusion and Final Remarks: The Upheavals Will Be Important for Farmers and Consumers

Gomiero et al. (2011) claimed that preserving or increasing soil health and reducing environmental impacts must not only be linked with a farm long-term sustainability (and benefit), but, and maybe most importantly, with preserving a country's long-term food security.

The goals of SDG2 are to promote healthy diets and to ensure the productivity and sustainability of agricultural systems. However, if we project ourselves to 2050, the cross-reference of climate data from the IPCC and demographic data reveals a worrying picture. In addition, massive urbanization and increasing disposable incomes favor a rapid transition in diets and lifestyle mainly featured by unhealthy consumption of processed, animal-protein based and energy intensive foods.

From the landmark reports on food security (FAO 2021; FAO et al. 2023, 2024; FAO and IFAD 2021) both growing hunger and obesity rates are harming health and well-being, and adversely affecting health and social systems, governmental budgets, and economic productivity and growth. Arable lands will be reduced by desertification due to climate change, and by the progression of metropolises since it will be necessary to house and feed in a healthy and sustainable way between 9 and 10 billion human beings. It is estimated that around 10% of arable land should disappear. The challenges facing farmers on a global scale seem dizzying. The ultimatum is immense, and to be met, it will require mobilizing a wide range of tools and strategies and requires bold commitments from policy makers.

Currently, 86% of the world's food is produced by conventional agricultural systems. Food supply per capita has increased more than 30%, accompanied by greater use of nitrogen fertilizers (increase of about 800%) and water resources for irrigation (increase of more than 100%) (Mbow et al. 2019). Furthermore, despite advances in plant genetics and crop management, yields under these systems are no longer or sufficiently increasing due to climate change and decreasing levels of soil quality. To this difficult food equation, we must add another parameter: tomorrow's agriculture will have to, like the rest of the economy, use much less fossil fuels, and itself contribute to storing carbon and producing renewable energy from biomass. In the other side, 85% of farmers worldwide still operate on 20 hectares or less of land, which is known as smallholder farming using low-input and mostly managed as uncertified OF.

A critical aspect of building global climate resilience and ensuring food security is the simultaneous promotion of land restoration, sustainable land management, and nature-positive agricultural practices such as OF. Organic agriculture is gaining increased scientific and consumer recognition worldwide. While the largest share of the global organic area and the highest consumption are found in developed countries, 89% of the organic farmers (4 million out of 4.5 million) live in developing countries in Asia, Africa, and Latin America.

Despite this, given that OF today only accounts for 2% of the agricultural land, a total conversion to organic agriculture does not seem to be a realistic scenario in the foreseeable future. In addition, lower yields under OF would mean additional land requirements, which was estimated to 23–67% (Meemken and Quaim 2018). According to Wani et al. (2017), it is also difficult to meet the nutrient requirement from organic source, if entire cultivable land is converted into OF.

Organic agriculture and related approaches also build overall resilience in agricultural systems by enhancing soil health and fertility, increasing soil water-holding potential, and increasing the diversity of soil microflora and fauna, keeping in mind that these soil qualities will be critical in dealing with the varied impacts of drought and flooding due to climate change. Such soil health improvements offer innovative solutions to the challenges posed by conventional agricultural methods and support the swift transition to sustainable agriculture and in achieving several SDG as shown by Šeremešic et al. (2021), Sharma et al. (2024) and Bhullar et al. (2020).

Studies have reviewed the emissions footprint of organic systems compared to conventional systems and evidence suggests organic production to typically generate lower emissions per unit of area, while emissions per unit of product vary and depend on the product. Even with relatively high environmental benefits and economic profits of OF systems, a critical hurdle for these systems, is to identify ways to improve and support consistently high crop yields. Nabuurs et al. (2022) summarized that large-scale conversion, without fundamental changes in food systems and diets may lead to increases in absolute emissions from land use change, driven by greater land requirements to maintain production at level to feed growing population.

It is not enough for farmers to opt for OF, diversify their rotations, use green manures or diversified forms of composting, plant more legumes, which capture nitrogen from the soil and replace fertilizers, it is still necessary to create outlets for these practices and crops including legumes. Since farmers are just one link in the chain, the success of the transition relies on the citizen engagement. Consumers will have to agree to pay a little more for their food.

The primary reported motivations are health concerns, including nutritional quality, freshness, lack of pesticides, and food safety. In other words, most research works reported that organic food is free from residues of pesticides, it does not present a higher level of heavy metals and tends to contain significantly more valuable nutritional compounds than the conventional equivalent. In addition, organic food is not more contaminated by mycotoxins than conventional food.

Fertilization is the most important and controllable factor affecting nutritional value of crops. If not well management, fertilizers, especially N fertilizers, are detrimental to ecosystem services and impeding SDGs achievement as explained by Zhang et al. (2024). In fact, induced micronutrient deficiencies are a significant public health concern. For two nutrients, vitamin C and protein, there is enough evidence to suggest that organic crops are superior to conventional ones. From this study, the key issues that emerged include yield reduction under OF mainly due to nutrient availability and the slow-release nature of organic fertilizers, lack of raw materials for organic fertilizer production. Other factors for yield gaps are lack of technical skills among farmers and development agents, short-term soil fertility enhancement, certification constraints, and marketing, policy regulation, and support.

Biocontrol or organic fertilization solutions are more complex to apply than a traditional phytosanitary product or chemical fertilizer. Introducing new mechanical tools or new crops requires investments and capital as well as the development of capacities and skills (machinery, seeds, processing, etc.). Hence, organic systems are more information and skill intensive that conventional systems. OF requires additional learning and experiences from farmers and farm labor in managing crops rotations and intercropping systems, growing different and new crops, and develop new cultural and mechanical strategies for battling pest, weeds, and disease. However, Bouttes et al. (2019) reported that OF was perceived by farmers as stimulating their learning, especially through a collective dynamic and an open exchange of experiences. Farmers also have their skills improved and are aware of technical risks linked to new production practices but were confident in their ability to manage them. These changes in perception associated with higher satisfaction and skills enable OF farmers to better face current turbulences and future changes and risks in the broader context.

OF faces criticism for its elevated food prices, which are often higher than those of conventionally produced food. This is primarily attributed to lower crop yields, coupled with the significant costs associated with certification and labor-intensive practices. Therefore, it is essential to explore and enhance management techniques that can mitigate the yield limitations inherent in organic systems. Without such advancements, the adoption of OF may remain constrained, as further expansion is necessary to bridge the yield gaps and meet the increasing food demands of a growing population.

Organic agriculture is often termed as knowledge-based rather than input-based agriculture. Experts suggest that the integration of big data and Artificial Intelligence into the rural economy is essential for improving global food and nutrition security. This integration can help reduce crop losses, enhance resource efficiency, and boost crop yields (Lassoued et al. 2021; Mondejar et al. 2021; Mesıas-Ruiz et al. 2023; Yang et al. 2024), while also impacting the entire food supply chain (Wolfert et al. 2017). Genomics, AI-driven and precision agriculture tools are becoming key contributors to sustainable development and should be increasingly adopted in OF. This adoption can lead to greater production efficiency, stimulate economic growth, and lower costs associated with energy, water usage, and waste recycling (Dong et al. 2020; Timsina et al. 2021; De Marchi et al. 2022; Meena et al. 2023; Panday et al. 2024; Yang et al. 2024). It is worth noticing that the use of such high-level technologies should not favor the concept of yield maximization valued by conventional agriculture. In addition, integration of local knowledge should not be disregarded.

Numerous studies are believed to concentrate primarily on direct economic impacts, such as yields, while overlooking a broader spectrum of potential societal benefits (Garibaldi et al. 2017). Frequently, products cultivated organically achieve higher market prices—whether local, national, or international—due to the increasing consumer preference for organic goods. Additionally, it is crucial to highlight that the variations, commonalities, and gaps identified in the literature can be linked to significant differences in methodological approaches and management practices between the agricultural systems (conventional vs OF). These discrepancies may complicate or undermine the validity of comparisons.

Mäder et al. (2002) noted that OF may need more land to produce the same yield, but it maintains fertility and biodiversity of the cropped land. In other terms, there is significant progress in integrating the principles of OF with those of agroecology, regenerative agriculture, agroforestry, and conservation agriculture, particularly in addressing trade-offs associated with tillage, synthetic fertilizer and pesticide applications and uses (Newton et al. 2020; Mrabet et al. 2024). The potential for additive benefits is substantial and warrants thorough evaluation and comprehension (Cooper et al. 2016). By implementing nature-positive agricultural practices, such as cultivating drought-resistant varieties, utilizing efficient irrigation techniques, and adopting no-till OF along with other soil conservation strategies, farmers can improve both the yield and quality of their crops while mitigating the adverse effects of drought and soil degradation. Nevertheless, the topic of genetically modified organisms (GMOs) remains a contentious issue within the OF community. It is essential for crop breeders and biotechnologists to identify specific plant traits that can enhance the nutritional quality of organic crops in response to rising atmospheric CO_2 levels, thereby integrating these traits into value chains as viable and compelling alternatives for sustainable and healthy food systems.

As the international community confronts the challenges of environmental degradation and the imperative for sustainable development, OF emerges as a vital element in the shift toward more sustainable and resilient agricultural practices. Ongoing investment in research, innovation, and policy formulation related to OF is essential for unlocking its full potential and overcoming existing obstacles. By integrating scientific research, community-driven innovation, and traditional dietary practices, OF can serve as a cornerstone for agricultural sustainability. Nevertheless, there is a pressing need for more comprehensive research that encompasses all levels of the value chain and the ecosystem to fully appreciate and quantify the benefits of organic foods. This will provide consumers, farmers, researchers, industries, and policymakers with the evidence-based information necessary for making informed food choices and dietary decisions. To foster sustainable development in OF, it is crucial to combine knowledge creation with social learning, bringing together diverse stakeholders to share insights and practices, thereby enhancing the exchange of both tangible and intangible resources (Charatsari et al. 2022).

In summary, the path to implementing effective policies and institutional frameworks for transitioning to organic agriculture is complex and requires navigating numerous challenges. However, through strategic, inclusive, and context-sensitive policy-making, it is possible to overcome these barriers as outlined by IFOAM's Organic 3.0 initiative.[3] In other words, while the challenges facing organic farmers are daunting, the opportunities are also vast and need a jumpstart or a fresh look. The organic movement through this initiative should enhance its relevance and effectiveness in addressing the pressing challenges faced by global organic agriculture. As consumer demand continues to grow and diversify and technological advancements support sustainable and fair behavioral practices, Organic 3.0 is poised to play a significant role in shaping the future of agriculture and food systems worldwide. Science and research institutions have a clear potential to play a crucial role in this transition. Leaders in the scientific community must use advocacy, diplomacy, and modern communication tools to improve public understanding of the benefits of OF and influence policy responses to ensure that OF is prioritized on the sustainability agenda of every country. Also, there is a strong need upon the recognition of voices and contributions to organic farming that come from indigenous knowledge of farmers.

[3] www.ifoam.bio/organic3zero.

References

Abeliotis K, Detsis V, Pappia C (2013) Life cycle assessment of bean production in the Prespa National Park. Greece J Clean Prod 41:89e96

Aboutayeb R, Boutarfa A, Mrabet R (2024) Carbon sequestration strategies in agricultural lands: towards a sustainable agriculture. In Karmaoui A (ed) Water-soil-plant-animal nexus in the era of climate change. IGI Global, pp 28–48. https://doi.org/10.4018/978-1-6684-9838-5.ch002

Aguilera E, Guzmán G, Alonso A (2015) Greenhouse gas emissions from conventional and organic cropping systems in Spain. II. Fruit tree orchards. Agron Sustain Dev 35(2):725–737

Aihounton GBD, Henningsen A (2023) Does organic farming jeopardize food and nutrition security? IFRO Working Paper, No. 2023/02, University of Copenhagen, Department of Food and Resource Economics (IFRO), Copenhagen

Ali B, Dahlhaus P (2022) Roles of selective agriculture practices in sustainable agricultural performance: a systematic review. Sustainability 14:3185. https://doi.org/10.3390/su14063185

Altieri MA, Funes-Monzote FR, Petersen P (2012) Agroecologically efficient agricultural systems for smallholder farmers: contributions to food sovereignty. Agron Sustain Dev 32:1–13. https://doi.org/10.1007/s13593-011-0065-6

Asaolu SS, Adefemi OS, Oyakilome IG, Ajibulu KE, Asaolu MF (2012) Proximate and mineral composition of Nigerian leafy vegetables. J Food Res 1(3):214–218

Atiyeh RM, Arancon N, Edwards CA, Metzger JD (2000) Influence of earthworm-processed pig manure on the growth and yield of greenhouse tomatoes. Biores Technol 75(3):175–180

Badgley C, Moghtader J, Quintero E, Zakem E, Chappell MJ et al (2007) Organic agriculture and the global food supply. Renew Agric Food Syst 22:86–108

Baker BP, Benbrook CM, Groth E, Lutz Benbrook K (2002) Pesticide residues in conventional, integrated pest management (IPM)-grown and organic foods: insights from three US data sets. Food Addit Contam 19(5):427–446

Barański M, Rempelos L, Iversen PO, Leifert C (2017) Effects of organic food consumption on human health; the jury is still out! Food Nutr Res 61:1287333

Barański M, Srednicka-Tober D, Volakakis N, Seal C, Sanderson R et al (2014) Higher antioxidant and lower cadmium concentrations and lower incidence of pesticide residues in organically grown crops: a systematic literature review and meta-analyses. Br J Nutr 112:794–811

Barbieri P, Pellerin S, Seufert V et al (2021) Global option space for organic agriculture is delimited by nitrogen availability. Nat Food 2:363–372. https://doi.org/10.1038/s43016-021-00276-y

Baritz R, Wiese L, Verbeke I, Vargas R (2018) Voluntary guidelines for sustainable soil management: global action for healthy soils. International yearbook of soil law and policy 2017. Springer, Cham, Switzerland, pp 17–36

Basalingappa KM, Nataraj R, Thangaraj G (2018) Biofertilizer for crop production and soil fertility. Acad J Agric Res 6(8):299–306. https://doi.org/10.15413/ajar.201

Baxter GJ, Graham AB, Lawrence JR, Wiles D, Paterson JR (2001) Salicylic acid in soups prepared from organically and non-organically grown vegetables. Eur J Nutr 40:289–292

Beaury EM, Smith J, Levine JM (2024) Global suitability and spatial overlap of land-based climate mitigation strategies. Global Change Biology 30:e17515. https://doi.org/10.1111/gcb.17515

Bellassen V et al (2021) The carbon and land footprint of certified food products. J Agric Food Ind Organ 19(2):113–126. https://doi.org/10.1515/jafio-2019-0037

Benbrook C, Zhao X, Davies N et al (2008) New evidence confirms the nutritional superiority of plant-based organic foods. http://organic-center.org/reportfiles/NutrientContent-Report.pdf

Ben Mariem S, Soba D, Zhou B, Loladze I, Morales F, Aranjuelo I (2021) Climate change, crop yields, and grain quality of C_3 cereals: a meta-analysis of CO_2, temperature, and drought effects. Plants 10(6):1052. https://doi.org/10.3390/plants10061052

Bergstrand K-J (2022) Organic fertilizers in greenhouse production systems—a review. Sci Hortic 2022(295):110855

Bestelmeyer BT, Okin GS, Duniway MC et al (2015) Desertification, land use, and the transformation of global drylands. Front Ecol Environ 13:28–36

Bhujel RR, Joshi HG (2023) Organic agriculture in India: a review of current status, challenges, and future prospects. Univers J Agric Res 11(2):306–313. https://doi.org/10.13189/ujar.2023.110208

Bhullar GS, Bautze D, Adamtey N, Armengot L, Cicek H, Goldmann E, Riar A, Rüegg J, Schneider M, Huber B (2020) What is the contribution of organic agriculture to sustainable development? A synthesis of twelve years (2007–2019) of the "long-term farming systems comparisons in the tropics (SysCom)". https://orgprints.org/id/eprint/39536/1/Syscom_Synthesis_Report.pdf

Biernat L, Taube F, Loges R, Kluß C, Reinsch T (2020) Nitrous oxide emissions and methane uptake from organic and conventionally managed arable crop rotations on farms in Northwest Germany. Sustainability 12(8):3240. https://doi.org/10.3390/su12083240

Bourn D, Prescott J (2002) A comparison of the nutritional value, sensory qualities, and food safety of organically and conventionally produced foods. Crit Rev Food Sci Nutr 42:1–34

Bouttes M, Darnhofer I, Martin G (2019) Converting to organic farming as a way to enhance adaptive capacity. Org Agric 9:235–247. https://doi.org/10.1007/s13165-018-0225-y

Brandt K, Molgaard JP (2001) Organic agriculture: does it enhance or reduce the nutritional value of plant foods? J Sci Food Agric. 81:924–931. https://doi.org/10.1002/jsfa.903

Brandt K, Leifert C, Sanderson R, Seal CJ (2011) Agroecosystem management and nutritional quality of plant foods: the case of organic fruits and vegetables. Crit Rev Plant Sci 30:177–197

Brito TP, de Souza-Esquerdo VF, Borsatto RS (2021) State of the art on research about organic certification: a systematic literature review. Org Agric 12(2):177–190

Burnett SE, Mattson NS, Williams KA (2016) Substrates and fertilizers for organic container production of herbs, vegetables, and herbaceous ornamental plants grown in greenhouses in the United States. Sci Hortic 208:111–119

Cafer AM, Rikoon JS (2018) Adoption of new technologies by smallholder farmers: the contributions of extension, research institutes, cooperatives, and access to cash for improving tef production in Ethiopia. Agric Hum Values 35(3):685–699. https://doi.org/10.1007/s10460-018-9865-5

Calabro G, Vieri S (2024) Limits and potential of organic farming towards a more sustainable European agri-food system. Br Food J 126(1):223–236. https://doi.org/10.1108/BFJ-12-2022-1067

Cassman KG (2007) Editorial response by Kenneth Cassman: can organic agriculture feed the world—science to the rescue? Renew Agric Food Syst 22:83–84

Chan KY, Dorahy C, Wells T et al (2008) Use of garden organic compost in vegetable production under contrasting soil P status. Aust J Agric Res 59(4):374–382

Charatsari C, Lioutas ED, Papadaki-Klavdianou A, Koutsouris A, Michailidis A (2022) Experiential, social, connectivist, or transformative learning? Farm advisors and the construction of agroecological knowledge. Sustainability 14:2426. https://doi.org/10.3390/su14042426

Charatsari C, Lioutas ED (2019) Is current agronomy ready to promote sustainable agriculture? Identifying key skills and competencies needed. Int J Sustain Dev World Ecol 26:232–241

Chiriacò MV, Grossi G, Castaldi S, Valentini R (2017) The contribution to climate change of the organic versus conventional wheat farming: a case study on the carbon footprint of wholemeal bread production in Italy. J Clean Prod 153:309–319. https://doi.org/10.1016/j.jclepro.2017.03.111

Chivenge P, Saito K, Bunquin MA, Sharma S, Dobermann A (2021) Co-benefits of nutrient management tailored to smallholder agriculture. Glob Food Secur 30(2):100570. https://doi.org/10.1016/j.gfs.2021.100570

Ciccarese L, Silli V (2016) The role of organic farming for food security: local nexus with a global view. Futur Food J Food Agric Soc 4(1):56–67

Cidón CF, Figueiró PS, Schreiber D (2021) Benefits of organic agriculture under the perspective of the bioeconomy: a systematic review. Sustainability 13(12):1–19. https://doi.org/10.3390/su13126852

Clark M, Tilman D (2017) Comparative analysis of environmental impacts of agricultural production systems, agricultural input efficiency, and food choice. Environ Res Lett 2:064016

Connor DJ (2008) Organic agriculture cannot feed the world. Field Crops Res 106:187–190
Connor DJ, Mínguez MI (2012) Evolution not revolution of farming systems will best feed and green the world. Glob Food Sec 1:106–113. https://doi.org/10.1016/j.gfs.2012.10.004
Cooper J, Baranski M, Stewart G, Nobel-de Lange M, Bàrberi P et al (2016) Shallow non-inversion tillage in organic farming maintains crop yields and increases soil C stocks: a meta-analysis. Agron Sustain Dev 36(1):1–20. https://doi.org/10.1007/s13593-016-0354-1
Correa-Cano ME, Salmoral G, Rey D, Knox JW, Graves A, Melo O, Foster W, Naranjo L, Zegarra E, Johnson C, Viteri-Salazar O, Yan X (2022) A novel modelling toolkit for unpacking the water-energy-food-environment (WEFE) nexus of agricultural development. Renew Sustain Energy Rev 159:112182. https://doi.org/10.1016/j.rser.2022.112182
Crespo-Herrera LA, Ortiz R (2015) Plant breeding for organic agriculture: something new? Agric Food Secur 4:25. https://doi.org/10.1186/s40066-015-0045-1
Crews TE, Peoples MB (2005) Can the synchrony of nitrogen supply and crop demand be improved in legume and fertilizer-based agroecosystems? A review. Nutr Cycl Agroecosyst 72(2):101–120
Crinnion WJ (2010) Organic foods contain higher levels of certain nutrients, lower levels of pesticides, and may provide health benefits for the consumer. Altern Med Rev J Clin Ther 15:4–12
Crowder DW, Northfield TD, Strand MR, Snyder WE (2010) Organic agriculture promotes evenness and natural pest control. Nature 466:109–112. https://doi.org/10.1038/nature09183
Crowder DW, Northfield TD, Gomulkiewicz R, Snyder WE (2012) Conserving and promoting evenness: organic farming and fire-based wildland management as case studies. Ecology 93:2001–2007. https://doi.org/10.1890/12-0110.1
Crowder DW, Reganold JP (2015) Financial competitiveness of organic agriculture on a global scale. Proc Natl Acad Sci USA 112:7611–7616. https://doi.org/10.1073/pnas.1423674112
Czech A, Szmigielski M, Sembratowicz I (2022) Nutritional value and antioxidant capacity of organic and conventional vegetables of the genus Allium. Sci Rep 12(1):18713
Das S, Chatterjee A, Pal TK (2020) Organic farming in India: a vision towards a healthy nation. Food Qual Saf 4(2):69–76. https://doi.org/10.1093/fqsafe/fyaa018
de Faccio Carvalho PC, da Silveira Pontes L, Barro RS et al (2024) Integrated crop-livestock-forestry systems as a nature-based solution for sustainable agriculture. Agroforest Syst 98(7):2309–2323. https://doi.org/10.1007/s10457-024-01057-9
de Faccio Carvalho PC, de Albuquerque Nunes PA, Pontes-Prates A, Szymczak LS, de Souza Filho W, Moojen FG, Lemaire G (2021) Reconnecting grazing livestock to crop landscapes: reversing specialization trends to restore landscape multifunctionality. Front Sustain Food Syst 5:750765. https://doi.org/10.3389/fsufs.2021.750765
Delandmeter M, de Faccio Carvalho PC, Bremm C, dos Santos Cargnelutti C, Bindelle J, Dumont B (2024) Integrated crop and livestock systems increase both climate change adaptation and mitigation capacities. Sci Total Environ 912:169061. https://doi.org/10.1016/j.scitotenv.2023.169061
De Marchi M, Pappalardo SE, Diantini A (2022) Agroecology and sustainable food systems: inquiring technological approaches. Drones and geographical information technologies in agroecology and organic farming contributions to technological sovereignty. CRC Press, Boca Raton, pp 1–21
de Ponti T, Rijk B, van Ittersum MK (2012) The crop yield gap between organic and conventional agriculture. Agric Syst 108:1–9
de Souza Araújo DF, da Silva AMRB, de Andrade Lima LL, da Silva Vasconcelos MA, Andrade SAC, Sarubbo LA (2014) The concentration of minerals and physicochemical contaminants in conventional and organic vegetables. Food Control 44:242–248
Dhaliwal SS, Sharma V, Shukla AK, Verma V, Kaur M, Shivay YS, Nisar S, Gaber A, Brestic M, Barek V et al (2022) Biofortification—a frontier novel approach to enrich micronutrients in field crops to encounter the nutritional security. Molecules 27:1340. https://doi.org/10.3390/molecules27041340

Diacono M, Persiani A, Testani E, Montemurro F, Ciaccia C (2019) Recycling agricultural wastes and by-products in organic farming: biofertilizer production, yield performance and carbon footprint analysis. Sustainability 11(14):3824. https://doi.org/10.3390/su11143824

Dong Y, Xu F, Liu L, Du X, Ren B, Guo A, Geng Y, Ruan C, Ye H, Huang W et al (2020) Automatic system for crop pest and disease dynamic monitoring and early forecasting. IEEE J Sel Top Appl Earth Obs Remote Sens 13:4410–4418

Dos Santos AM, Lima JS, Dos Santos IF, Silva EF, de Santana FA, de Araujo DG, Dos Santos LO (2019) Mineral and centesimal composition evaluation of conventional and organic cultivars sweet potato (Ipomoea batatas (L.) Lam) using chemometric tools. Food Chem 273:166–171

Dreyfus G, Frederick C, Larkin E, Powers Y, Chatterjee J, Sharma S (2023) Reducing nitrous oxide emissions from smallholder farmer agriculture through site specific nutrient management. https://precisiondev.org/wp-content/uploads/2024/12/IGSDPxD-Reducing-nitrous-oxide-emissions-02-06-23.pdf

Durán-Lara E, Valderrama A, Marican A (2020) Natural organic compounds for application in organic farming. Agriculture 10(2):41. https://doi.org/10.3390/agriculture10020041

European Commission (EC) (2020) Farm to fork strategy—for a fair, healthy, and environmentally friendly food system. European Commission. https://ec.europa.eu/food/farm2fork_en

El Ghmari H, Harbouze R, El Bilali H (2021) Transition to organic agriculture in Morocco. AGROFOR Int J 6(3):5–14

Epule TE (2019) Contribution of organic farming towards global food security: an overview. In: Chandran S, Unni MR, Thomas S (eds) Woodhead publishing series in food science, technology and nutrition, organic farming. Woodhead Publishing, pp 1–16. https://doi.org/10.1016/B978-0-12-813272-2.00001-X.

Erb KH, Lauk C, Kastner T et al (2016) Exploring the biophysical option space for feeding the world without deforestation. Nat Commun 7:11382. https://doi.org/10.1038/ncomms11382

Eyhorn F, Muller A, Reganold J, Frison E, Herren HR, Luttikholt L, Mueller A, Sanders J, Scialabba N, Seufert V, Smith P (2019) Sustainability in global agriculture driven by organic farming. Nat Sustain 2:253–255. https://doi.org/10.1038/s41893-019-0266-6

Falconnier GN, Cardinael R, Corbeels M, Baudron F, Chivenge P, Couëdel A, Ripoche A, Affholder F, Naudin K, Benaillon E, Rusinamhodzi L, Leroux L, Vanlauwe B, Giller KE (2023) The input reduction principle of agroecology is wrong when it comes to mineral fertilizer use in sub-Saharan Africa. Outlook Agric 52(3):311–326. https://doi.org/10.1177/00307270231199795

FAO (2018) The future of food and agriculture—alternative pathways to 2050. Food and Agriculture Organization of the United Nations, Rome. http://www.fao.org/3/I8429EN/i8429en.pdf

FAO (2019) Organic agriculture and the sustainable development goals: part of the solution. https://archive.ifoam.bio/sites/default/files/nm19_329_report_sdg_lr.pdf

FAO (2021) The state of the world's land and water resources for food and agriculture—systems at breaking point. Synthesis Report. Rome. https://doi.org/10.4060/cb7654en

FAO and IFAD (2021) The state of food security and nutrition in the world 2021: transforming food systems for food security, improved nutrition and affordable healthy diets for all. FAO, Rome, Italy. https://doi.org/10.4060/cb4474en

FAO, IFAD, UNICEF, WFP, and WHO (2020) The state of food security and nutrition in the world 2020. Transforming food systems for affordable healthy diets. FAO, Rome. https://doi.org/10.4060/ca9692en

FAO, IFAD, UNICEF, WFP, WHO (2023) The state of food security and nutrition in the world 2023. Urbanization, agrifood systems transformation and healthy diets across the rural–urban continuum. FAO, Rome. https://doi.org/10.4060/cc3017en

FAO, IFAD, UNICEF, WFP, WHO (2024) The state of food security and nutrition in the world 2024—financing to end hunger, food insecurity and malnutrition in all its forms. Rome. https://doi.org/10.4060/cd1254en

Faller ALK, Fialho E (2009) The antioxidant capacity and polyphenol content of organic and conventional retail vegetables after domestic cooking. Food Res Int 42(1):210–215

Fan H, Zhang Y, Li J, Jiang J, Waheed A, Wang S, Rasheed SM, Zhang L, Zhang R (2023) Effects of organic fertilizer supply on soil properties, tomato yield, and fruit quality: a global meta-analysis. Sustainability 2023(15):2556. https://doi.org/10.3390/su15032556
Federal Ministry of Food and Agriculture (2022) Organic farming in Germany. https://www.ovgu.de/unimagdeburg_media/Forschung+_+Wirtschaft/Forschungsförderung/WKZ/Organic+farming+in+Germany.pdf
Ferreira F, Mendes-Moreira P, Botelho G (2020) Is organic agriculture a potential public health indicator? Open Agric 5(1):914–929. https://doi.org/10.1515/opag-2020-0088
Fess TL, Benedito VA (2018) Organic versus conventional cropping sustainability: a comparative system analysis. Sustainability 10(1):272. https://doi.org/10.3390/su10010272
Fjelkner-Modig S, Bengtsson H, Stegmark R, Nystrom S (2000) The influence of organic and integrated production on nutritional, sensory and agricultural aspects of vegetable raw materials for food production. Acta Agric Scand Sect B Soil Plant Sci 50:102–113
Fletcher C, Ripple WJ, Newsome T et al (2024) Earth at risk: an urgent call to end the age of destruction and forge a just and sustainable future. PNAS Nexus 3:1–20. https://doi.org/10.1093/pnasnexus/pgae106
Gamage A, Gangahagedara R, Gamage J, Jayasinghe N, Kodikara N, Suraweera P, Merah O (2023) Role of organic farming for achieving sustainability in agriculture. Farming Syst 1(1):100005. https://doi.org/10.1016/j.farsys.2023.100005
Gao F, Li H, Mu X, Gao H, Zhang Y, Li R, Cao K, Ye L (2023) Effects of organic fertilizer application on tomato yield and quality: a meta-analysis. Appl Sci 13:2184
Garrett RD, Ryschawy J, Bell LW, Cortner O, Ferreira J et al (2020) Drivers of decoupling and recoupling of crop and livestock systems at farm and territorial scales. Ecol Soc 25(1):24. https://doi.org/10.5751/ES-11412-250124
Garibaldi LA, Gemmill-Herren B, D'Annolfo R, Graeub BE, Cunningham SA, Breeze TD (2017) Farming approaches for greater biodiversity, livelihoods, and food security. Trends Ecol Evol 32:68–80. https://doi.org/10.1016/j.tree.2016.10.001
Gattinger A et al (2012) Enhanced topsoil carbon stocks under organic farming. Proc Natl Acad Sci 109:18226–18231
Geyer M, Mohler V, Hartl L (2022) Genetics of the inverse relationship between grain yield and grain protein content in common wheat. Plan Theory 11:2146. https://doi.org/10.3390/plants11162146
Giampieri F, Mazzoni L, Cianciosi D, Alvarez-Suarez JM, Regolo L, Sanchez-Gonzalez C, Battino M (2022) Organic vs conventional plant-based foods: a review. Food Chem 383:132352
Gibbs HK, Salmon JM (2015) Mapping the world's degraded lands. Appl Geogr 57:12–21
Giller KE, Hijbeek R, Andersson JA, Sumberg J (2021) Regenerative agriculture: an agronomic perspective. Outlook Agric 50(1):13–25. https://doi.org/10.1177/0030727021998063
Giller KE, Delaune T, Silva JV et al (2021) The future of farming: who will produce our food? Food Secur 13:1073–1099. https://doi.org/10.1007/s12571-021-01184-6
Głodowska M, Krawczyk J (2017) Heavy metals concentration in conventionally and organically grown vegetables. Qual Assur Saf Crop Foods 9(4):497–503. https://doi.org/10.3920/QAS2017.1089
Goel R, Debbarma P, Kumari P, Suyal DC, Kumar S, Mahapatra BS (2020) Assessment of soil chemical quality, soil microbial population and plant growth parameters under organic and conventional rice–wheat cropping system. Agric Res 10(2):1–12
Golijan J, Secanski M (2021) The development of organic agriculture in Serbia and worldwide. Contemp Agric 70(3–4):85–94. https://doi.org/10.2478/contagri-2021-0013
Golubkina NA, Seredin TM, Antoshkina MS, Kosheleva OV, Teliban GC, Caruso G (2018) Yield, quality, antioxidants and elemental composition of new leek cultivars under organic or conventional systems in a greenhouse. Horticulturae 4(4):39
Gomiero T, Pimentel D, Paoletti MG (2011) Environmental impact of different agricultural management practices: conventional vs. organic agriculture. Crit Rev Plant Sci 30(1–2):95–124

Gomiero T (2021) Organic agriculture: impact on the environment and food quality. In Galanakis CM (ed) Environmental impact of agro-food industry and food consumption. Academic Press, pp 31–58. https://doi.org/10.1016/B978-0-12-821363-6.00002-3. ISBN 9780128213636

Gopalakrishnann R (2019) Advantages and nutritional value of organic food on human health. Int J Trend Sci Res Dev 3:242–245

Grand View Research (2023) Organic food and beverages market size report, 2030. https://www.grandviewresearch.com/industry-analysis/organic-foods-beverages-market

Grigorieva E, Livenets A, Stelmakh E (2023) Adaptation of agriculture to climate change: a scoping review. Climate 11:202. https://doi.org/10.3390/cli11100202

Gross A, Glaser B (2021) Meta-analysis on how manure application changes soil organic carbon storage. Sci Rep 11(1):5516. https://doi.org/10.1038/s41598-021-82739-7

Gu B (2022) Recoupling livestock and crops. Nat Food 3:102–103. https://doi.org/10.1038/s43016-022-00466-2.

Hallmann E, Rembiałkowska E (2006) Antioxidant compounds content in selected onion bulbs from organic and conventional cultivation. J Res Appl Agric Eng 51(2):42–46

Hansen S, Froseth RB, Stenberg M, Stalenga J, Olesen JE et al (2019) Reviews and syntheses: review of causes and sources of N_2O emissions and NO_3 leaching from organic arable crop rotations. Biogeosciences 16(14):2795–2819

Harou AP, Madajewicz M, Michelson H, Palm CA, Amuri N, Magomba C, Semoka JM, Tschirhart K, Weil R (2022) The joint effects of information and financing constraints on technology adoption: evidence from a field experiment in rural Tanzania. J Dev Econ 155:102707. https://doi.org/10.1016/j.jdeveco.2021.102707

Hatfield JL, Dold C (2019) Water-use efficiency: advances and challenges in a changing climate. Front Plant Sci 10:103. https://doi.org/10.3389/fpls.2019.00103

High Level Panel of Experts (HLPE) (2019) Agroecological and other innovative approaches for sustainable agriculture and food systems that enhance food security and nutrition. A report by the high-level panel of experts on food security and nutrition of the committee on world food security. HLPE report 14. Rome, 163 pp. http://www.fao.org/3/ca5602en/ca5602en.Pdf

Hoefkens C, Verbeke W, Aertsens J, Mondelaers K, Van Camp J (2009) The nutritional and toxicological value of organic vegetables: consumer perception versus scientific evidence. Br Food J 111(10):1062–1077. https://doi.org/10.1108/00070700910992916

Howard A (1947) The soil and health: a study of organic agriculture. Lexington, KY, USA, University Press of Kentucky. ISBN 978-0-8131-9171-3

Hunter D, Foster M, McArthur JO, Ojha R, Petocz P, Samman S (2011) Evaluation of the micronutrient composition of plant foods produced by organic and conventional agricultural methods. Crit Rev Food Sci Nutr 51:571–582

Hurtado-Barroso S, Tresserra-Rimbau A, Vallverdú-Queralt A, Lamuela-Raventós RM (2017) Organic food and the impact on human health. Crit Rev Food Sci Nutr. https://doi.org/10.1080/10408398.2017.1394815

IFOAM (2022) Definition of organic agriculture. https://www.ifoam.bio/why-organic/organic-landmarks/definition-organic

IPCC (2019) Summary for policymakers. In: Shukla PR, Skea J, Calvo Buendia E, Masson-Delmotte V, Pörtner H-O, Roberts DC, Zhai P, Slade R, Connors S, van Diemen R, Ferrat M, Haughey E, Luz S, Neogi S, Pathak M, Petzold J, Portugal Pereira J, Vyas P, Huntley E, Kissick K, Belkacemi M, Malley J (eds) Climate change and land: an IPCC special report on climate change, desertification, land degradation, sustainable land management, food security, and greenhouse gas fluxes in terrestrial ecosystems (in press)

IPCC (2021) Summary for policymakers. In: Masson-Delmotte V, Zhai P, Pirani A, Connors SL, Péan C, Berger S, Caud N, Chen Y, Goldfarb L, Gomis MI, Huang M, Leitzell K, Lonnoy E, Matthews JBR, Maycock TK, Waterfield T, Yelekçi O, Yu R, Zhou B (eds) Climate change 2021: the physical science basis. Contribution of working group I to the sixth assessment report of the intergovernmental panel on climate change

IPCC (2022a) Summary for policymakers. In: Shukla PR, Skea PR, Slade R, Al Khourdajie A, van Diemen R, McCollum D, Pathak M, Some S, Vyas P, Fradera R, Belkacemi M, Hasija A, Lisboa G, Luz S, Malley J (eds) Climate change 2022: mitigation of climate change. Contribution of working group III to the sixth assessment report of the intergovernmental panel on climate change. Cambridge University Press, Cambridge, UK and New York, NY. https://doi.org/10.1017/9781009157926.001

IPCC (2022b) Summary for policymakers. In: Pörtner H-O, Roberts DC, Poloczanska ES, Mintenbeck K, Tignor M, Alegría A, Craig M, Langsdorf S, Löschke S, Möller V, Okem A (eds) Climate change 2022: impacts, adaptation and vulnerability. Contribution of working group II to the sixth assessment report of the intergovernmental panel on climate change. Cambridge University Press, Cambridge, UK and New York, NY, pp 3–33. https://doi.org/10.1017/9781009325844.001

Jouzi Z, Azadi H, Taheri F, Zarafshani K, Gebrehiwot K, Van Passel S, Lebailly P (2017) Organic farming and small-scale farmers: Main opportunities and challenges. Ecol Econ 132:144–154

Jung M, Boucher TM, Wood SA, Folberth C, Wironen M, Thornton P et al (2024) A global clustering of terrestrial food production systems. PLoS ONE 19(2):e0296846. https://doi.org/10.1371/journal.pone.0296846

Kanter DR, Musumba M, Wood SLR, Palm C, Antle J, Balvanera P, Dale VH, Havlik P, Kline KL, Scholes RJ, Thornton P, Tittonell P, Andelman S (2018) Evaluating agricultural trade-offs in the age of sustainable development. Agric Syst 163:73–88

Kassam A (ed) (2022) Advances in conservation agriculture: volume 3: adoption and spread. https://doi.org/10.1201/9781003180678

Karlsson JO, Röös E (2019) Resource-efficient use of land and animals—environmental impacts of food systems based on organic cropping and avoided food-feed competition. Land Use Policy 85:63–72. https://doi.org/10.1016/j.landusepol.2019.03.035

Karp DS, Chaplin-Kramer R, Meehan TD, Martin EA, De Clerk F, Grab H et al (2018) Crop pests and predators exhibit inconsistent responses to surrounding landscape composition. Proc Natl Acad Sci USA 115:E7863–E7870. https://doi.org/10.1073/pnas.1800042115

Kennedy CM, Lonsdorf E, Neel MC, Williams NM, Ricketts TH, Winfree R et al (2013) A. global quantitative synthesis of local and landscape effects on wild bee pollinators in agroecosystems. Ecol Lett 16:584–599. https://doi.org/10.1111/ele.12082

Kirchmann H, Bergström L, Kätterer T, Andrén O, Andersson R (2008) Can organic crop production feed the world? In: Kirchmann H, Bergström L (eds) Organic crop production—ambitions and limitations. Springer, Dordrecht, The Netherlands, pp 39–72. http://pub.epsilon.slu.se/3514/

Kirchmann H, Kätterer T, Bergström L, Börjesson G, Bolinder MA (2016) Flaws and criteria for design and evaluation of comparative organic and conventional cropping systems. Field Crop Res 186:99–106

Kishore AR, Niveditha K, Uriti A, Anilkumar C, Sarabu A, Prasanna KVL (2023) An in-depth analysis of the elements shaping organic farmers: a systematic review. In: International conference on energy, materials and communication engineering (ICEMCE), Madurai, India, pp 1–5. https://doi.org/10.1109/ICEMCE57940.2023.10434263

Knapp S, van der Heijden MGA (2018) A global meta-analysis of yield stability in organic and conservation agriculture. Nat Commun 9:3632. https://doi.org/10.1038/s41467-018-05956-1

Kniss AR, Savage SD, Jabbour R (2016) Commercial crop yields reveal strengths and weaknesses of organic agriculture in the United States. PLoS ONE 11:e0161673. https://doi.org/10.1371/journal.pone.0161673

Koninger J, Lugato E, Panagos P, Kochupillai M, Orgiazzi A, Brionesa MJI (2021) Manure management and soil biodiversity: towards more sustainable food systems in the EU. Agric Syst 194:103251. https://doi.org/10.1038/ncomms11382

Krauss M, Berner A, Perrochet F, Frei R, Niggli U, Mäder P (2020) Enhanced soil quality with reduced tillage and solid manures in organic farming—a synthesis of 15 years. Sci Rep 10:4403. https://doi.org/10.1038/s41598-020-61320-8

Kravchenko AN, Snapp SS, Robertson GP (2017) Field-scale experiments reveal persistent yield gaps in low-input and organic cropping systems. PNAS 114:926–931

Krupnik TJ, Andersson JA, Rusinamhodzi L, Corbeels M, Shennan C, Gérard B (2019) Does size matter? A critical review of meta-analysis in agronomy. Exp Agric 55(2):200–229. https://doi.org/10.1017/S0014479719000012

Kumar A, Prakash CHB, Brar NS, Kumar B (2018) Potential of vermicompost for sustainable crop production and soil health improvement in different cropping systems. Int J Curr Microbiol Appl Sci 7(10):1042–1055. https://doi.org/10.20546/ijcmas.2018.710.116

Kumar S, Diksha, Sindhu SS, Kumar R (2022) Biofertilizers: an ecofriendly technology for nutrient recycling and environmental sustainability. Curr Res Microb Sci 3:100094. https://doi.org/10.1016/j.crmicr.2021.100094

Kumpulainen J (2001) Organic and conventional grown foodstuffs: nutritional and toxicological quality comparisons. Proc Int Fertil Soc 472(120):6779

Lairon D (2010) Nutritional quality and safety of organic food. A review. Agron Sustain Dev 30:33–41

Lammerts Van Bueren E, Jones SS, Tamm L, Murphy KM, Myers JR, Leifert C, Messmer MM (2011) The need to breed crop varieties suitable for organic farming, using wheat, tomato and broccoli as examples: a review. NJAS Wagening J Life Sci 58(3–4):193–205. https://doi.org/10.1016/j.njas.2010.04.001

Lassoued R, Macall DM, Smyth SJ, Phillips PW, Hesseln H (2021) Expert insights on the impacts of, and potential for, agricultural big data. Sustainability 13(5):2521

Leifeld, J (2016) Current approaches neglect possible agricultural cutback under large-scale organic farming. A comment to Ponisio et al. Proc R Soc B 283:1623. https://doi.org/10.1098/rspb.2015.1623.

Liao X, Zhao J, Yi Q, Li J, Li Z, Wu S, Zhang W, Wang K (2023) Metagenomic insights into the effects of organic and inorganic agricultural managements on soil phosphorus cycling. Agric Ecosyst Environ 343:108281. https://doi.org/10.1016/j.agee.2022.108281

Lichtenberg EM, Kennedy CM, Kremen C, Batáry P, Berendse F, Bommarco R et al (2017) A global synthesis of the effects of diversified farming systems on arthropod diversity within fields and across agricultural landscapes. Glob Chang Biol 23:4946–4957. https://doi.org/10.1111/gcb.13714

Liu Y, Langer V, Høgh-Jensen H, Egelyng H (2010) Life cycle assessment of fossil energy use and greenhouse gas emissions in Chinese pear production. J Clean Prod 18:1423e1430

Lockeretz W (2007) Organic farming: an international history. CABI, Wallingford, UK

Loladze I (2014) Hidden shift of the ionome of plants exposed to elevated CO_2 depletes minerals at the base of human nutrition. eLife 3:e02245. https://doi.org/10.7554/eLife.02245

Lori M, Symnaczik S, Mäder P, De Deyn G, Gattinger A (2017) Organic farming enhances soil microbial abundance and activity—A meta-analysis and meta-regression. PLoS ONE 12(7):e0180442. https://doi.org/10.1371/journal.pone.0180442

Mäder P, Fliessbach A, Dubois D, Gunst L, Fried P, Niggli U (2002) Soil fertility and biodiversity in organic farming. Science 296(5573):1694–1697. https://doi.org/10.1126/science.1071148

Madhavi R, Vijaya GS, Malathi H (2021) Sustainable organic farming in India. Int J Mod Agric 10(2):1080–1088. http://www.modernjournals.com/index.php/ijma/article/view/827

Magkos F, Arvaniti F, Zampelas A (2003) Putting the safety of organic food into perspective. Nutr Res Rev 16:211–221. https://doi.org/10.1079/NRR200361

Malek Ž, Tieskens KF, Verburg PH (2019) Explaining the global spatial distribution of organic crop producers. Agric Syst 176:102680. https://doi.org/10.1016/j.agsy.2019.102680

Maltais-Landry G, Buchanan C, Longanecker J (2023) Using processed fertilizers or composted poultry manure results in similar yields but contrasting nutrient budgets in organic cabbage production. J Plant Nutr 46:2462–2472

Manna MC, Rahman MM, Naidu R, Bari AF, Singh AB, Thakur JK, Rao AS (2021) Organic farming: a prospect for food, environment and livelihood security in Indian agriculture. Adv Agron 170:101–153

Manna MC, Ghosh A, Rao AS (2023) Composting: a green technology for soil health management. Indian J Fertil 19(4):352–367

Manibharathi S, Senthamil E (2024) Organic farming practices. In: Waghmare et al (eds) Sustainable crop management in modern agronomy, pp 38–51. ISBN 978-81-969162-7-5

Martin G, Allain S, Bergez J-E, Burger-Leenhardt D, Constantin J, Duru M, Hazard L, Lacombe C, Magda D, Magne M-A, Ryschawy J, Thénard V, Tribouillois H, Willaume M (2018) How to address the sustainability transition of farming systems? A conceptual framework to organize research. Sustainability 10(6):2083. https://doi.org/10.3390/su10062083

Matt D, Rembialkowska E, Luik A, Peetsmann E, Pehme S (2011) Quality of organic vs. conventional food and effects on health: report. Estonian University of Life Sciences, 104p. ISBN 978-9949-484-06-5

Mbow C, Rosenzweig C, Barioni LG, Benton TG, Herrero M, Krishnapillai M, Liwenga E, Pradhan P, Rivera-Ferre MG, Sapkota T, Tubiello FN, Xu Y (2019) Food security. In: Shukla PR, Skea J, Calvo Buendia E, Masson-Delmotte V, Pörtner H-O, Roberts DC, Zhai P, Slade R, Connors S, van Diemen R, Ferrat M, Haughey E, Luz S, Neogi S, Pathak M, Petzold J, Portugal Pereira J, Vyas P, Huntley E, Kissick K, Belkacemi M, Malley J (eds) Climate change and land: an IPCC special report on climate change, desertification, land degradation, sustainable land management, food security, and greenhouse gas fluxes in terrestrial ecosystems. https://doi.org/10.1017/9781009157988.007

Meena RH, Meena K, Choudhary M (2023) Organic farming—a key to food security and agricultural sustainability. In: Sarathchandran UMR, Thomas S, Meena DK (eds) Woodhead publishing series in food science, technology and nutrition, organic farming, 2nd edn. Woodhead Publishing, pp 1–30. https://doi.org/10.1016/B978-0-323-99145-2.00007-0

Meemken EM, Quaim M (2018) Organic Agriculture, food Security and the environment. Annu Rev Resour Econ 10:39–63

Mehata DK, Kattel I, Sapkota P, Ghimire NP, Kumar Mehta RK (2023) Biofertilizers: a sustainable strategy for organic farming that would increase crop production and soil health. Plant Physiol Soil Chem 3(2):49–53

Meier S, Stoessel F, Jungbluth N, Juraske R, Schader C, Stolze M (2015) Environmental impacts of organic and conventional agricultural products—are the differences captured by life cycle assessment? J Environ Manag 149:193–208

Mesıas-Ruiz GA, Perez-Ortiz M, Dorado J, de Castro AI, Peña JM (2023) Boosting precision crop protection towards agriculture 5.0 via machine learning and emerging technologies: a contextual review. Front Plant Sci 14:1143326. https://doi.org/10.3389/fpls.2023.1143326

Migliorini P, Wezel A (2017) Converging and diverging principles and practices of organic agriculture regulations and agroecology. A review. Agron Sustain Dev 37:1–18

Mitchell AE, Hong Y-J, Koh E, Barrett DM, Bryant DE, Denison RF, Kaffka S (2007) Ten-year comparison of the influence of organic and conventional crop management practices on the content of flavonoids in tomatoes. J Agric Food Chem 55:6154–6159

Mmbando GS (2024) Omics: a new, promising technologies for boosting crop yield and stress resilience in African agriculture. Plant Stress 11:100366. https://doi.org/10.1016/j.stress.2024.100366

Mondejar ME, Avtar R, Diaz HLB, Dubey RK, Esteban J, Gomez-Morales A, Hallam B, Mbungu NT, Okolo CC, Prasad KA, She QH, Garcia-Segura S (2021) Digitalization to achieve sustainable development goals: steps towards a smart green planet. Sci Total Environ 794:148539. https://doi.org/10.1016/j.scitotenv.2021.148539

Moos JH, Schrader S, Paulsen HM (2017) Reduced tillage enhances earthworm abundance and biomass in organic farming: a meta-analysis. Landbauforschung 67(3–4):123–128. https://doi.org/10.3220/LBF1512114926000

Moschitz H, Muller A, Kretzschmar U, Haller L, de Porras M, Pfeifer C, Oehen B, Willer H, Stolz H (2021) How can the EU farm to fork strategy deliver on its organic promises? Some critical reflections. EuroChoices 20:30–36. https://doi.org/10.1111/1746-692X.12294

Mrabet R (2023) Sustainable agriculture for food and nutritional security. In: Farooq M, Gogoi N, Pisante M (eds) Sustainable agriculture and the environment. Academic Press, pp 25–90. https://doi.org/10.1016/B978-0-323-90500-8.00013-0. ISBN 9780323905008

Mrabet R, Alaoui SB, Daoui K (2024) Bundling climate smart agriculture systems in Morocco: insights, advances and perspectives in a changing climate. In: Behnassi M, Al-Shaikh AA, Hussain Qureshi R, Barjees Baig M, Faraj TKA (eds) Climate-smart and resilient food systems and security. Springer, Cham. https://doi.org/10.1007/978-3-031-65968-3_5

Müller C, Elliott J, Pugh TAM, Ruane AC, Ciais P, Balkovic J et al (2018) Global patterns of crop yield stability under additional nutrient and water inputs. PLoS ONE 13:e0198748. https://doi.org/10.1371/journal.pone.0198748

Muller A, Schader C, El-Hage Scialabba N et al (2017a) Can organic agriculture feed the world in 2050: a critical assessment of the challenges and opportunities. Nat Commun 8:1290. https://doi.org/10.1038/s41467-017-01410-w

Muller A, Schader C, El-Hage Scialabba N et al (2017) Strategies for feeding the world more sustainably with organic agriculture. Nat Commun 8(1):1290. http://www.nature.com/articles/s41467-017-01410-w

Muluneh MW, Talema GA, Abebe KB, Dejen TB, Kassaw MA, Teka MA (2022) Determinants of organic fertilizers utilization among smallholder farmers in south Gondar zone, Ethiopia. Environ Health Insights 16:11786302221075448. https://doi.org/10.1177/11786302221075448

Munné-Bosch S, Bermejo NF (2024) Fruit quality in organic and conventional farming: advantages and limitations. Trends Plant Sci 29(2). https://doi.org/10.1016/j.tplants.2024.01.011

Murphy KM, Garland-Campbell KA, Lyon SR, Jones SS (2007) Evidence of varietal adaptation to organic farming systems. Field Crop Res 102:172–177

Musche M, Adamescu M, Angelstam P, Bacher S, Bäck J, Buss HL, Duffy C, Flaim G, Gaillardet J, Giannakis GV, Haase P, Halada L, Kissling WD, Lundin L, Matteucci G, Meesenburg H, Monteith D, Nikolaidis NP, Pipan T, Pyšek P, Rowe EC, Roy DB, Andrew Sier U, Tappeiner M, Vilà T, White MZ, Klotz S (2019) Research questions to facilitate the future development of European long-term ecosystem research infrastructures: a horizon scanning exercise. J Environ Manage 250:109479. https://doi.org/10.1016/j.jenvman.2019.109479

Nabuurs G-J, Mrabet R, Abu Hatab A, Bustamante M, Clark H, Havlík P, House J, Mbow C, Ninan KN, Popp A, Roe S, Sohngen B, Towprayoon S (2022) Agriculture, forestry and other land uses (AFOLU). In: Shukla PR, Skea J, Slade R, Al Khourdajie A, van Diemen R, McCollum D, Pathak M, Some S, Vyas P, Fradera R, Belkacemi M, Hasija A, Lisboa G, Luz S, Malley J (eds) IPCC, 2022: climate change 2022: mitigation of climate change. Contribution of working group III to the sixth assessment report of the intergovernmental panel on climate change. Cambridge University Press, Cambridge, UK and New York, NY. https://doi.org/10.1017/9781009157926.009

Newton P, Civita N, Frankel-Goldwater L, Bartel K, Johns C (2020) What is regenerative agriculture? A review of scholar and practitioner definitions based on processes and outcomes. Front Sustain Food Syst 4:577723. https://doi.org/10.3389/fsufs.2020.577723

Niemiec M, Chowaniak M, Sikora J, Szeląg-Sikora A, Gródek-Szostak Z, Komorowska M (2020) Selected properties of soils for long-term use in organic farming. Sustainability 12(6):2509

Niggli U, Fließbach A, Hepperly P, Scialabba NE (2009) Low greenhouse gas agriculture. Mitigation and adaptation potential of sustainable farming systems. Edited by FAO. http://www.fao.org/tempref/docrep/fao/010/ai781e/ai781e00.pdf. Accessed on 28 Aug 2020

Niggli U, Slabe A, Schmid O, Halberg N, Schlüter M (2008) Vision for an organic food and farming research agenda to 2025. Organic knowledge for the future. Edited by IFOAM EU, FiBL. https://orgprints.org/13439/1/niggli-etal-2008-technology-platform-organics.pdf. Accessed on 28 Aug 2020

Nyantakyi-Frimpong H, Kangmennaang J, Bezner Kerr R, Luginaah I, Dakishoni L, Lupafya E, Shumba L, Katundu M (2017) Agroecology and healthy food systems in semi-humid tropical Africa: participatory research with vulnerable farming households in Malawi. Acta Trop 175:42–49. https://doi.org/10.1016/j.actatropica.2016.10.022

Olle M, Williams IH (2012) Organic farming of vegetables. In: Lichtfouse E (ed) Sustainable agriculture reviews, vol 11. Springer, Dordrecht, The Netherlands, pp 63–76. ISBN 978-94-007-5449-2

Oteros-Rozas E, Ravera F, García-Llorente M (2019) How does agroecology contribute to the transitions towards social-ecological sustainability? Sustainability 11(16):4372. https://doi.org/10.3390/su11164372

Oyetunji OE, Ojuederie OB, Thonda OA, Kotun B, Glick BR, Babalola OO (2024) The expediency of fungi as biocontrol agents for the enhancement of food security. In: Kumar A, Santoyo G, Singh J (eds) Plant and soil microbiome, biocontrol agents for improved agriculture. Academic Press, pp 1–28. https://doi.org/10.1016/B978-0-443-15199-6.00005-1. ISBN 9780443151996

Palupi E, Jayanegara A, Ploeger A, Kahl J (2012) Comparison of nutritional quality between conventional and organic dairy products: a meta-analysis. J Sci Food Agric 92(14):2774–2781. https://doi.org/10.1002/jsfa.5639

Panday D, Bhusal N, Das S, Ghalehgolabbehbahani A (2024) Rooted in nature: the rise, challenges, and potential of organic farming and fertilizers in agroecosystems. Sustainability 16:1530. https://doi.org/10.3390/su16041530

Parizad S, Bera S (2023) The effect of organic farming on water reusability, sustainable ecosystem, and food toxicity. Environ Sci Pollut Res 30:71665–71676. https://doi.org/10.1007/s11356-021-15258-7

Paustian K, Lehmann J, Ogle S, Reay D, Robertson GP, Smith P (2016) Climate-smart soils. Nature 532:49–57

Paustian K, Collier S, Baldock J, Burgess R, Creque J, DeLonge M et al (2019) Quantifying carbon for agricultural soil management: from the current status toward a global soil information system. Carbon Management 10(6):567–587. https://doi.org/10.1080/17583004.2019.1633231

Pawar J, Choudhari V, Choudhari G, Wagh P, More G, Ramanathan V (2022) Organic food: the importance from public health perspective. J Pharm Res Int 34(29B):28–37. https://doi.org/10.9734/JPRI/2022/v34i29B36055

Peigné J, Casagrande M, Payet V, David C, Sans FX, Blanco-Moreno JM, Cooper J, Gascoyne K, Antichi D, Bàrberi P, Bigongiali F, Surböck A, Kranzler A, Beeckman A, Willekens K, Luik A, Matt D, Grosse M, Heß J, Maurice C, Dierauer H, Mäder P (2015) How organic farmers practice conservation agriculture in Europe. Renewable Agric Food Syst. https://doi.org/10.1017/S1742170514000477

Perrin A, Milestad R, Martin G (2020) Resilience applied to farming: organic farmers' perspectives. Ecol Soc 25(4):5. https://doi.org/10.5751/ES-11897-250405

Pimentel D, Hepperly P, Hanson J, Douds D, Siedel R (2005) Environmental, energetic, and economic comparisons of organic and conventional farming systems. Bioscience 55(7):573–582

Ponisio CP, M'Gonigle LK, Mace KC, Palomino J, de Valpine P, Kremen C (2015) Diversification practices reduce organic to conventional yield gap. Proc R Soc B 282:20141396

Ponisio LC, Kremen C (2016) System-level approach needed to evaluate the transition to more sustainable agriculture. Proc R Soc B Biol Sci 283(1824):20152913

Porcel M, Andersson GKS, Pålsson J, Tasin M (2018) Organic management in apple orchards: higher impacts on biological control than on pollination. J Appl Ecology 55:2779–2789. https://doi.org/10.1111/1365-2664.13247

Pretty JN, Morison JIL, Hine RE (2003) Reducing food poverty by increasing agricultural sustainability in developing countries. Agr Ecosyst Environ 95(1):217–234. https://doi.org/10.1016/S0167-8809(02)00087-7

Puyu V, Vakhnyak V, Lapchynskyi V, Koberniuk O, Tarasyuk V (2023) Comparative study of organic farming systems in different geographical regions. Sci Horiz 26(12):124–134. https://doi.org/10.48077/scihor12.2023.124

Rahman SME, Mele MA, Lee YT, Islam MZ (2021) Consumer preference, quality, and safety of organic and conventional fresh fruits, vegetables, and cereals. Foods 10(1):105. https://doi.org/10.3390/foods10010105

Rahman A, Baharlouei P, Koh EHY, Pirvu DG, Rehmani R, Arcos M, Puri SA (2024) Comprehensive analysis of organic food: evaluating nutritional value and impact on human health. Foods 13:208. https://doi.org/10.3390/foods13020208

Ranganathan DS, Selvaseelan DA (1997) Mushroom spent rice straw compost and composted coir pith as organic manures for rice. J Indian Soc Soil Sci 45(3):510–514

Rani R, Kaushik P, Bhayana S, Kapoor S (2023) Impact of organic farming on soil health and nutritional quality of crops. J Saudi Soc Agric Sci 22:560–569. https://doi.org/10.1016/j.jssas.2023.07.002

Reddy AA, Vimala Y, Goudar G, Mergu N, Rao JS (2023a) Nutritional compilation of commonly consumed organic and conventional fruits and vegetables from India. Food Hum 1:1652–1658. https://doi.org/10.1016/j.foohum.2023.11.006

Reddy MVM, Sujatha HT, Korav S, Kumari G, Meher A (2023b) A review on scope and challenges of organic farming in India. Eco Env Cons 29(3):1328–1336. https://doi.org/10.53550/EEC.2023.v29i03.050

Reganold J, Wachter J (2016) Organic agriculture in the twenty-first century. Nat Plants 2:15221. https://doi.org/10.1038/nplants.2015.221

Reeve JR, Hoagland LA, Villalba JJ, Carr PM, Atucha A et al (2016) Organic farming, soil health, and food quality: considering possible links. Adv Agron 137:319–367

Rembiałkowska E (2016) Production system and composition of plant foods. In: Human health implications of organic food and organic agriculture, pp 33–40

Rembiałkowska E (2007) Quality of plant products from organic agriculture. J Sci Food Agric 87:2757–2762. https://doi.org/10.1002/jsfa.3000

Rhodes CJ (2012) Feeding and healing the world: through regenerative agriculture and permaculture. Sci Prog 95:345–446

Rodale JI (1948) The organic front, 1st edn. Rodale Press, New York, NY

Rodale Institute (2022) Farming systems trial: 40-year report. Rodale Institute, Kutztown, PA

Röös E, Mie A, Wivstad M, Salomon E, Johansson B, Gunnarsson S, Wallenbeck A et al (2018) Risks and opportunities of increasing yields in organic farming. A review. Agron Sustain Dev 38:1–21

Rossi F, Godani F, Bertuzzi T, Trevisan M, Ferrari F, Gatti S (2008) Health-promoting substances and heavy metal content in tomatoes grown with different farming techniques. Eur J Nutr 46:266–272. https://doi.org/10.1007/s00394-008-0721-z

Saleem S, Mushtaq NUI, Shah WH, Rasool A, Rehman RUI (2024) Plant growth-promoting bacteria in food security and plant disease management. In: Kumar A, Gustavo Santoyo G, Singh J (eds) Plant and soil microbiome, biocontrol agents for improved agriculture. Academic Press, pp 275–292. https://doi.org/10.1016/B978-0-443-15199-6.00017-8. ISBN 9780443151996

Samawat S, Lakzian A, Zamirpour A (2001) The effect of vermicompost on growth characteristics of tomato. Agric Sci Technol 15(2):83–89

Schneider M, Lüscher G, Jeanneret P et al (2014) Gains to species diversity in organically farmed fields are not propagated at the farm level. Nat Commun 5:4151. https://doi.org/10.1038/ncomms5151

Seitz S, Goebes P, Puerta VL, Pereira EIP, Wittwer R, Six J, van der Heijden MGA, Scholten T (2018) Conservation tillage and organic farming reduce soil erosion. Agron Sustain Dev 2018(39):4

Šeremešic S, Dolijanovic Ž, Tomaš-Simin M, Vojnov B, Glavaš-Trbic D (2021) The future we want: sustainable development goals accomplishment with organic agriculture. Probl Ekorzwoju 16:171–180

Seufert V, Ramankutty N, Foley JA (2012) Comparing the yields of organic and conventional agriculture. Nature 485:229–232. https://doi.org/10.1038/nature11069

Seufert V (2019) Comparing yields: organic versus conventional agriculture. In: Ferranti P, Berry E, Jock A (eds) Encyclopedia of food security and sustainability. Elsevier, San Diego, pp 196–208. https://doi.org/10.1016/B978-0-08-100596-5.22027-1

Seufert V, Ramankutty N (2017) Many shades of gray—the context dependent performance of organic agriculture. Sci Adv 3(3):1602638

Sharma P, Sharma P, Thakur N (2024) Sustainable farming practices and soil health: a pathway to achieving SDGs and future prospects. Discov Sustainability 5:250. https://doi.org/10.1007/s43621-024-00447-4

Sharma SK, Ravishankar N, Chander M, Jat G (2021) Organic agriculture under changing climate scenario. Indian J Agron 66(5th IAC Special issue):S82__S94

Sharma SK, Choudhary R, Jat G, Ravishankar N (2018) Strategies for achieving sustainable food systems through organic agriculture lead paper: symposium on doubling farmers income through agronomic interventions under changing scenario organized by Indian Society of Agronomy, New Delhi and Maharana Pratap University of Agricultural and Technology during 24–26 October 2018 at Udaipur, p 97

Shennan C, Krupnik TJ, Baird G, Cohen H, Forbush K, Lovell RJ, Olimpi EM (2017) Organic and conventional agriculture: a useful framing? Annu Rev Environ Resour 42(1):317–346

Shrivas VL, Singh U, Weisskopf L, Hariprasad P, Sharma S (2019) Effect of organic farming on structural and functional diversity of soil microbiome: benefits and risks. In: Varma A, Tripathi S, Prasad R (eds) Plant biotic interactions. Springer, Cham. https://doi.org/10.1007/978-3-030-26657-8_9

Sieber S, Graef F, Amjath-Babu TS, Mutabazi KD, Tumbo SD, Faße A, Gomez y Paloma S, Rybak C, Lana MA, Ndah HT, Uckert G, Schuler J, Grote U (2018) Trans-SEC's food security research in Tanzania: from constraints to adoption for out-and upscaling of agricultural innovations. Food Secur 10(4):775–783. https://doi.org/10.1007/s12571-018-0822-3

Singh M (2021) Organic farming for sustainable agriculture. Indian J Org Farming 1(1):1–8

Singh M, Jaswal Y, Manavi, Mishra S, Kushwaha A, Ekka SK, Kujur R (2024) Organic farming in agricultural practices: a review. Asian J Biol 20(3):34–39. https://doi.org/10.9734/AJOB/2024/v20i3395

Skinner C, Gattinger A, Muller A, Mader P, Fliebbach A, Stolze M, Ruser R, Niggli U (2014) Greenhouse gas fluxes from agricultural soils under organic and non-organic management—a global meta-analysis. Sci Total Environ 468–469:553–563. https://doi.org/10.1016/j.scitotenv.2013.08.098

Smith LG, Kirk GJD, Jones PJ et al (2019) The greenhouse gas impacts of converting food production in England and Wales to organic methods. Nat Commun 10:4641. https://doi.org/10.1038/s41467-019-12622-7

Smith OM, Cohen AL, Rieser CJ, Davis AG, Taylor JM, Adesanya AW et al (2019) Organic farming provides reliable environmental benefits but increases variability in crop yields: a global meta-analysis. Front Sustain Food Syst 3:82. https://doi.org/10.3389/fsufs.2019.00082

Smith OM, Jocson DMI, Lee BW, Orpet RJ, Taylor JM, Davis AG, Rieser CJ, Clarke AE, Cohen AL, Hayes AM, Auth CA, Bergeron PE, Marshall AT, Reganold JP, Crowder DW, Northfield TD (2022) Identifying farming strategies associated with achieving global agricultural sustainability. Front Sustain Food Syst 6:882503. https://doi.org/10.3389/fsufs.2022.882503

Smith-Spangler C, Brandeau ML, Hunter GE, Bavinger JC, Pearson M, Eschbach PJ, Sundaram V, Liu H, Schirmer P, Stave C, Olkin I, Bravata DM (2012) Are organic foods safer or healthier than conventional alternatives? A systematic review. Ann Intern Med 157(5):348–366. https://doi.org/10.7326/0003-4819-157-5-201209040-00007. Erratum in: Ann Intern Med. 2012 Nov 6;157(9):680. Erratum in: Ann Intern Med. 2012 Oct 2;157(7):532. PMID: 22944875

Soni R, Gupta R, Agarwal P, Mishra R (2022) Organic farming: a sustainable agricultural practice. Vantage J Themat Anal 3(1):21–44. https://doi.org/10.52253/vjta.2022.v03i01.03

Sow S, Ranjan S, Kumari J, Kumar A, Kumar N, Ghosh M, Kumar S, Roy DK (2023) Organic farming: a step towards sustainable agriculture and its challenges. In: Kumar S, Kumar V, Singh DV, Patel JN (eds) Novel technologies in modern agriculture. Vital Biotech Publication Kota, Rajasthan, India, pp 256–268

Stanhill G (1990) The comparative productivity of organic agriculture. Agr Ecosyst Environ 30:1–26. https://doi.org/10.1016/0167-8809(90)90179-H

Sumberg J (2022) Future agricultures: the promise and pitfalls of a (re)turn to nature. Outlook Agric 51:3–10

Surekha K (2007) Nitrogen-release pattern from organic sources of different C: N ratios and lignin content, and their contribution to irrigated rice (Oryza sativa). Indian J Agron 52(3):220–224

Taheri F, Azadi H, D'Haese M (2017) A world without hunger: organic or GM crops? Sustainability 9:580

Talhaoui N, Taamalli A, Gómez-Caravaca AM, Fernández-Gutiérrez A, Segura- Carretero A (2015) Phenolic compounds in olive leaves: analytical determination, biotic and abiotic influence, and health benefits. Food Res Int 77:92–108

Tavakkoli E, Uddin S, Rengasamy P, McDonald GK (2022) Field applications of gypsum reduce pH and improve soil C in highly alkaline soils in southern Australia's dryland cropping region. Soil Use Manag 38:466–477

Timsina J, Dutta S, Devkota KP, Chakraborty S, Neupane RK, Bishta S, Amgain LP, Singh VK, Islam S, Majumdar K (2021) Improved nutrient management in cereals using nutrient expert and machine learning tools: productivity, profitability and nutrient use efficiency. Agric Syst 192:103181

Tscharntke T, Grass I, Wanger TC, Westphal C, Batáry P (2021) Beyond organic farming-harnessing biodiversity-friendly landscapes. Trends Ecol Evol 36(10):919–930. https://doi.org/10.1016/j.tree.2021.06.010

Tuck SL, Winqvist C, Mota F, Ahnström J, Turnbull LA, Bengtsson J (2014) Land-use intensity and the effects of organic farming on biodiversity: a hierarchical meta-analysis. J Appl Ecol 51:746–755. https://doi.org/10.1111/1365-2664.12219

Tuomisto HL, Hodge ID, Riordan P, Macdonald DW (2012) Does organic farming reduce environmental impacts?—a meta-analysis of European research. J Environ Manage 112:309–320. https://doi.org/10.1016/j.jenvman.2012.08.018

Ullah A, Arshad M, Kachele H, Khan A, Mahmood N, Muller K (2020) Information asymmetry, input markets, adoption of innovations and agricultural land use in Khyber Pakhtunkhwa, Pakistan. Land Use Policy 90:104261. https://doi.org/10.1016/j.landusepol.2019.104261

Ume C, Nuppenau E-A, Domptail SE (2022) A feminist economics perspective on the agroecology-food and nutrition security nexus. Environ Sustain Indic 16:100212. https://doi.org/10.1016/j.indic.2022.100212

Ume C (2023) The role of improved market access for small-scale organic farming transition: implications for food security. J Clean Prod 387:135889. https://doi.org/10.1016/j.jclepro.2023.135889

UNCAD and UNEP (2008) Organic agriculture and food security in Africa. https://unctad.org/system/files/official-document/ditcted200715_en.pdf

United Nations Environment Programme (UNEP) (2024) Global resources outlook 2024: bend the trend—pathways to a liveable planet as resource use spikes. International Resource Panel. Nairobi. https://wedocs.unep.org/20.500.11822/44901

Van Balen D, Cuperus F, Haagsma W, De Haan J, Van Den Berg W, Sukkel W (2023) Crop yield response to long-term reduced tillage in a conventional and organic farming system on a sandy loam soil. Soil Tillage Res 225:105553

Van der Werf H, Knudsen MT, Cederberg C (2020) Towards better representation of organic agriculture in life cycle assessment. Nat Sustain 3(6):419–425. https://doi.org/10.1038/s41893-020-0489-6.hal-02983631

Van Wagenberg C, De Haas Y, Hogeveen H et al (2017) Animal board invited review: comparing conventional and organic livestock production systems on different aspects of sustainability. Animal 11(10):1839–1851. https://doi.org/10.1017/S175173111700115

Venkat K (2012) Comparison of twelve organic and conventional farming systems: a life cycle greenhouse gas emissions perspective. J Sustain Agric 36:620–649. https://doi.org/10.1080/10440046.2012.672378

Verdi L, Dalla Marta A, Falconi F, Orlandini S, Mancini M (2022) Comparison between organic and conventional farming systems using Life Cycle Assessment (LCA): a case study with an ancient wheat variety. Eur J Agron 141:126638. https://doi.org/10.1016/j.eja.2022.126638

Wani SA, Wani MA, Mehraj S, Padder BA, Chand S (2017) Organic farming: present status, scope and prospects in northern India. J Appl Nat Sci 9(4):2272–2279

Warman PR, Havard KA (1998) Yield, vitamin and mineral contents of organically and conventionally grown potatoes and sweet corn. Agr Ecosyst Environ 68(3):207–216

West PC, Gerber JS, Engstrom PM, Mueller ND, Brauman KA, Carlson KM et al (2014) Leverage points for improving global food security and the environment. Science 345:325–328. https://doi.org/10.1126/science.1246067. PMID: 25035492

Wezel A, Casagrande M, Celette F, Vian J-F, Ferrer A et al (2014) Agroecological practices for sustainable agriculture: a review. Agron Sustain Dev 34(1):1–20. https://doi.org/10.1007/s13593-013-0180-7

Willer H, Lernoud J (2018) The world of organic agriculture. Statistics and emerging trends 2018. FiBL—Research Institute of Organic Agriculture. IFOAM—Organics International, Frick, Switzerland

Willer H, Travnícek J, Meier C, Schlatter B (eds) (2021) The world of organic agriculture. Statistics and emerging trends 2021. Research Institute of Organic Agriculture FiBL, and IFOAM-Organic International, Bonn, v20210301

Willer H, Trávníček J, Schlatter B (eds) (2024) The world of organic agriculture. Statistics and emerging trends. Research Institute of Organic Agriculture FiBL, Frick, and IFOAM—Organics International, Bonn

Williams CM (2002) Nutritional quality of organic food: shades of grey or shades of green? Proc Nutr Soc 61:19–24. https://doi.org/10.1079/PNS2001126

Woese K, Lange D, Boess C, Bogl KW (1997) A comparison of organically and conventionally grown foods: results of a review of the relevant literature. J Sci Food Agric 74(3):281–293

Worthington V (2001) Nutritional quality of organic versus conventional fruits, vegetables, and grains. J Altern Complement Med 7:161–173

Wolfert S, Ge L, Verdouw C, Bogaardt M-J (2017) Big data in smart farming—a review. Agric Systems 153:69–80

Xu, Q., Huet, S., Poix, C., Boisdon, I., & Deffuant, G. (2018). Why do farmers not convert to organic farming? Modeling conversion to organic farming as a major change. *Nat. Resour. Model.* https://doi.org/10.1111/nrm.12171

Yadav A, Bhuj BD, Ram S, Singh CP, Dhar S, Yadav RK, Mahindra Singh M, Ashok K (2021) Organic farming for sustainable agriculture: a review. Annals of R.S.C.B. 25(6):8088–8123

Yang C, Ji X, Cheng C, Liao S, Obuobi B, Zhang Y (2024) Digital economy empowers sustainable agriculture: implications for farmers' adoption of ecological agricultural technologies. Ecol Ind 159:111723. https://doi.org/10.1016/j.ecolind.2024.111723

Yang G, Li J, Liu Z, Zhang Y, Xu X, Zhang H, Xu Y (2022) Research trends in crop-livestock systems: a bibliometric review. Int J Environ Res Public Health 19(14):8563. https://doi.org/10.3390/ijerph19148563

Yu X, Guo L, Jiang G, Song Y, Muminov MA (2018) Advances of organic products over conventional productions with respect to nutritional quality and food security. Acta Ecol Sin 38:53–60

Zabel F, Delzeit R, Schneider JM, Seppelt R, Mauser W, Vaclavík T (2019) Global impacts of future cropland expansion and intensification on agricultural markets and biodiversity. Nat Commun 10(1):1–10

Zhang C, Gu B, Liang X, Lam SK, Zhou Y, Chen D (2024) The role of nitrogen management in achieving global sustainable development goals. Resour Conserv Recycl 201:107304. https://doi.org/10.1016/j.resconrec.2023.107304

Zhao Z, Mao Y, Gao S, Lu C, Pan C, Li X (2023) Organic carbon accumulation and aggregate formation in soils under organic and inorganic fertilizer management practices in a rice–wheat cropping system. Sci Rep 13:3665. https://doi.org/10.1038/s41598-023-30541-y

Dr. Rachid Mrabet is a Research Director at INRA Morocco with a Ph.D. in Agronomy from Colorado State University (1997) and an M.Sc. in Soil Physics from the Joint Project between Hassan II Veterinary and Agronomic Sciences Institute (Rabat/Morocco) and University of Minnesota (1989). With over 35 years of experience in agronomy, soil science, conservation agriculture, and climate change research, Dr. Mrabet has authored 19 books, 82 book and conference chapters, and 98 peer-reviewed papers. He has served as an IPCC review editor and Coordinating Lead Author for the MEDECC, IPCC and UNEP/Geo_7 reports. Dr. Mrabet has received several prestigious awards, including Distinguished Research Award from WASWAC, the 2020 North-South Prize of the Council of Europe and the 2022 Gulbenkian Prize for Humanity. He is also involved in more than 60 national and international research projects.Member several evaluation panels including the PRIMA Foundation Scientific Evaluation Committee (SEC).

Dr. Mohamed Behnassi is currently a Senior Environmentalist Expert at the Economic Social & Environmental Council (CESE), Rabat, Morocco. From 2004 to 2022, he served as a Professor, Head of Public Law Department, and Director of the Research Laboratory for Territorial Governance, Human Security and Sustainability (LAGOS), Faculty of Law, Economics, and Social Sciences of Agadir, Ibn Zohr University. He holds a Ph.D. in International Environmental Law and Governance (2003), a M.Sc. in Political Sciences and International Relations (1997), and a B.A. of Administration (1995) from the Faculty of Law, Economics, and Social Sciences, Hassan II University of Casablanca. He obtained a Diploma in International Environmental Law and Diplomacy from the University of Eastern Finland and UNEP, 2015 and is also Alumnus of both the International Visitors Leadership Program of the Department of State (2011, USA), and the Near East South Asia Center for Strategic Studies (NESA), National Defense University, Washington, DC (2022–24, USA).

Dr. Behnassi is currently the Founding Director of the Center for Environment, Human Security and Governance (CERES). He was appointed in 2019 as Expert Evaluator for the National Center for Scientific and Technical Research (CNRST/Morocco), and selected twice (2019-2024) as an Assessment Scoping Expert and a Review Editor by the Intergovernmental Science-Policy Platform on Biodiversity and Ecosystem Services (IPBES—Thematic assessment of the interlinkages among biodiversity, water, food and health). He was a Lead Author (LA) of the Human Security Chapter of 1st MEDECC Assessment Report (MAR1): Climate and Environmental Change in the *Mediterranean Basin—Current Situation and Risks for the Future* (2021) and Coordinating Lead Author (CLA) of the MEDECC Special Report on "the Nexus between Water, Energy, Food and Ecosystems". In 2024, he was nominated and certified as the UNFCCC Expert Evaluator (Roster of Experts), and as a Member of the PRIMA Foundation Scientific Evaluation Committee (SEC) to evaluate proposals submitted to the Thematic Area "Leveraging urban and local food systems for sustainable food systems transformation".

Dr. Behnassi is a Senior Researcher in the areas of International Law and Politics of Environment and Human Security. In addition to his participation in more than 70 international and national conferences, he has published 66 scientific papers and 22 books in these areas, including recent ones on: *The Water, Climate, and Food Nexus: Linkages, Challenges and Emerging Solutions* (Springer, 2024); *Food Security and Climate-Smart Food Systems—Building Resilience for the Global South* (Springer, 2022); and *The Climate-Conflict-Migration Nexus from a Human Security Perspective* (Springer, 2022). Dr. Behnassi serves as a reviewer for many global publishers (such as Routledge and Springer) and scientific journals with high impact factor. He has organized many international conferences covering the above research areas, managed many research and expertise projects for national and international institutions, and is regularly requested to provide scientific expertise nationally and internationally. Other professional activities include social compliance auditing and consultancy by monitoring human rights at work and the sustainability of the global supply chain.

Chapter 3
Climate Change, Adaptation Measures Concerning Crops, and Agroecological Diversity to Meet the Food Security Challenges in Sri Lanka

A. D. Ampitiyawatta, P. K. Dissanayake, and Eranga M. Wimalasiri

Abstract In parallel with the global climate change, the climate of Sri Lanka is also changing. High intensity rainfall, extreme warming and cooling, flash floods, and droughts are more common during the past couple of decades. It is obvious that these sudden changes in the environment may greatly affect food production, creating a tremendous pressure to the food security. To cope with these changes and minimize their impact, farmers are practicing various adaptation strategies. Researchers and scientists are also looking for possible adaptation measures in the long-term perspective. Further, by nature, Sri Lanka is blessed with a diverse spectrum of climatic conditions, as well as a great variability in soil and topographical characteristics. All these variations favorably affected the availability of a wide diversity of food crops, which range from temperate to tropical fruits and vegetables, while maintaining the nation's food security in a wider perspective. Further, the availability of a wide array of underutilized fruits and other food crops, which have not been paid public or research attention yet, has a huge potential to increase the nation's food security. The ability to withstand harsh environmental conditions of these underutilized, naturally grown food crops has to be extracted in future food security programs. This chapter's goal is to establish a relationship between the nation's food security and the consequences of climate change on agriculture and to discuss various adaptation and mitigation measures practiced by farmers and the climatic, soil, and topographic potential to ensure the food security. Further, possibilities of utilizing underutilized food crops in this regard was also discussed. The chapter concludes with a case study based on the crop model of the Agricultural Production Systems Simulator (APSIM),

A. D. Ampitiyawatta · P. K. Dissanayake (✉) · E. M. Wimalasiri
Department of Export Agriculture, Faculty of Agricultural Sciences, Sabaragamuwa University of Sri Lanka, Belihuloya, Sri Lanka
e-mail: kapila@agri.sab.ac.lk

A. D. Ampitiyawatta
e-mail: ada@agri.sab.ac.lk

E. M. Wimalasiri
e-mail: eranga@agri.sab.ac.lk

M. Behnassi et al. (eds.), *Food Systems and Biodiversity in the Context of Environmental and Climate Risks*, https://doi.org/10.1007/978-3-031-89167-0_3

which may adapt a number of crops as a response to the scenario of future climate change while maintaining food security in tandem with population increase.

Keywords Adaptation measures · Climate change · Food crop diversity · Food security · Soil diversity · Underutilized crops · Weather patterns

3.1 Introduction

The world's climate has changed dramatically since the onset of the Planet Earth, 4.5 billion years ago. Regardless of the significant climate change that has occurred on Earth in its early phases, the last 8000 years, which cover the majority of documented human history, have been rather steady near the warmer end of this temperature range. Agriculture, permanent settlements, and population growth were all enabled by this stability (Australian Academy of Science 2021). Every operation in today's world is based on these climatic conditions in this small time period. Apart from the natural events affecting the changes in climatic conditions, attention to climate change aspects has been increasing in the recent past, mainly due to anthropogenic interventions, which tend to change the climate, causing global warming, at an unusual rate. Since 1880, the overall land and ocean temperature has increased at an average rate of 0.08 °C every decade (NOAA 2021); but since 1981, the average rate of the atmospheric temperature increase (0.18 °C) was more than doubled. The Intergovernmental Panel on Climate Change (IPCC) predicts that average air temperatures would rise by 1.4–4.4 °C by the end of the century (2081–2100), as compared to the 1850–1900 period (IPCC 2021). Other studies expect that rising air temperatures will result in more frequent instances of catastrophic drought, flooding, or heat waves than in the past (Assad et al. 2004).

Major concern about climate change is mainly due to the fact that it makes human life more threatened by unfavorable environmental conditions (EBoDE 2010; GA2LEN 2010:2; WHO 2006). Agriculture is the section most vulnerable, as crop species' responses mainly depend on climatic conditions (Adams et al. 1990). According to Ahmed and Suphachalasai (2014), climate change is expected to cost Sri Lanka 1.2% of its GDP by the year 2050. As climate change is in progress, agriculture policies and research should focus on crop species and cultivation practices that are more adaptive to the changing climate. Sri Lanka, as a tropical country has many advantages, which consist of a wide diversity in ecosystems and food crops. In our opinion, Sri Lanka has the potential to cope with climatic disasters predicted to happen. In this chapter, we intend to explore the country's attempts at climate change adaptation strategies and the natural strength of crop, climatic, and ecological diversity on a broader scale toward food security in the nation. Prevailing climatic conditions and changes that have occurred for a few decades are also discussed. Further, adaptation measures of common agricultural crops in Sri Lanka are explained as a case study using the Agricultural Production Systems Simulator (APSIM) crop model (Holzworth et al. 2014).

3.2 Climate Change and Its Impacts

3.2.1 Climate and Cultivating Seasons in Sri Lanka

Sri Lanka is a small country with a size of 65,610 km^2 placed in the Indian ocean, off the coast of India's southern point, between 5° 55′ and 9° 55′ N, and between 79° 41′ and 81° 53′ E. The central highlands, which rise to more than 2500 m in the middle of the country, and the majority of the rest of the country are generally flat with scattered rocks. Being located in the tropical zone, the country basically has a tropical monsoon climate which comprises tropical monsoon (Am), tropical rainforest (Af), and tropical savanna-wet (Aw), according to the Köppen-Geiger climatic zones. In addition, some parts of the country have temperate oceanic climate (Cfb) due to the topographical variations. The tropical monsoon system and its associated winds, known as the Southwest and Northeast monsoons (SWM and NEM), dominate the country's rainfall climate (Domroes 1974), resulting in two principle monsoon rainfall seasons (SWM: from May to September and NEM: from December to February) and two inter-monsoon rainfall seasons (First inter-monsoon; FIM: from March to April and Second inter-monsoon; SIM: from October to November) (Ranatunge et al. 2003; Suppiah 1996; Thambyahpillay 1954). Conventional types of rainfall are dominant during inter-monsoon seasons due to calm atmospheric conditions.

Besides, considerable amount of rainfall brings to the country (especially during the SIM) by the tropical depressions emerging in the Bay of Bengal. The type of rainfall and the seasonality, combined with the local and regional topographic variability, results in diverse rainfall climates over the island, both spatially and temporally, ranging from 900 to 5000 mm, with the annual mean of around 1850 mm. Accordingly, the country is classified into three main climatic zones (Panabokke 1996) as wet (annual average rainfall is greater than 2500 mm), intermediate (1750–2500 mm), and dry (annual average rainfall is less than 1750 mm) (Fig. 3.1). Based on the unique climate, soil, and topographic conditions (Panabokke 1996), the country is divided further into 46 agroecological zones. Regional elevation differences are the key determinant of the temperature variation across the country, whereas regional geographical characteristics also influence the temperature climate. As a result, the country's annual average temperature ranges between 26.5 and 28.5 °C, with the central highlands experiencing temperatures ranging between 14° and 16 °C (Abhayasinghe 2007; Basnayake 2007; Chandrapala 2007b).

The crop cultivation seasons of the country predominantly depend on the monsoon rainfall pattern, which has two different seasons, known as the major (*Maha*) and the minor (*Yala*) (Zubair 2002). The *Maha* season (October–March) follows the NEM and SIM rainfalls, while the *Yala* season (April to September) follows the SWM and FIM rainfalls. During SWM and NEM seasons, the central highlands divide the rainfall climate into windward and leeward sides and mainly govern the distribution of rainfall over wet and dry zones. During the SWM rainfall season, wet zone areas act as windward side and receive more rainfall, while dry zone areas with less rainfall are on the leeward side. Therefore, the *Yala* season for the wet zone is

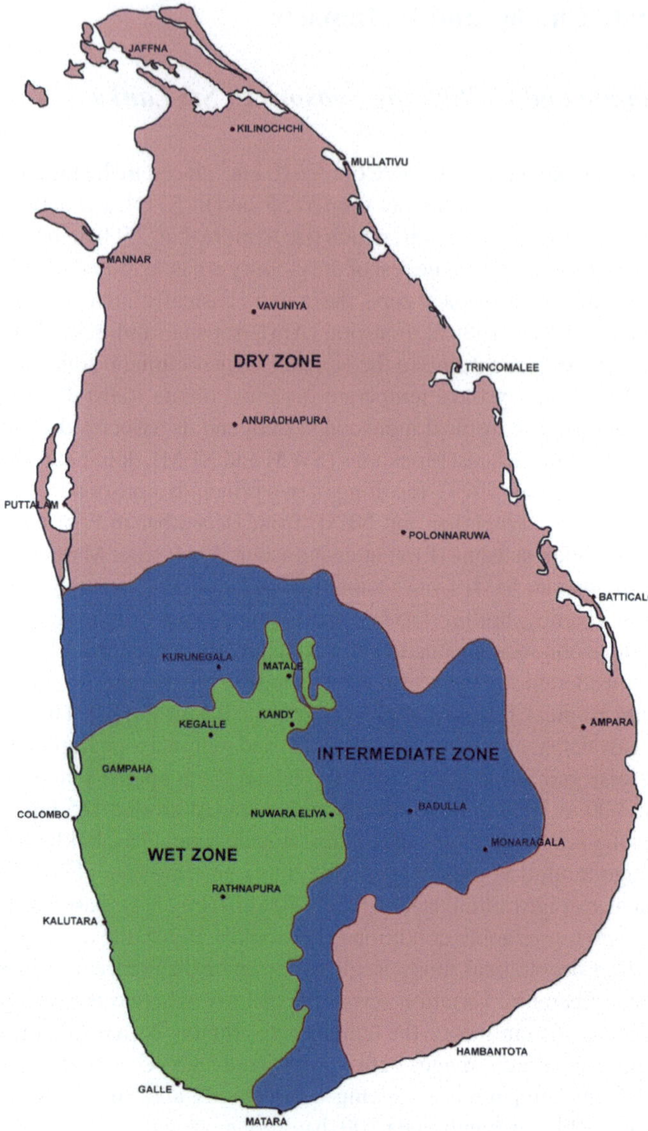

Fig. 3.1 Map of Sri Lanka showing wet, intermediate, and dry zones. *Source* Department of Agriculture, Sri Lanka (https://doa.gov.lk/)

rich with moisture, whereas the dry zone is moisture deficient. The dry zone receives a substantial amount of rainfall during *Yala* only from the FIM, but it is limited to mid-March to early-May. The *Maha* season of the dry zone receives plentiful rainfall from both SIM and NEM rainfalls (Punyawardena et al. 2013a, b), from October to February, being the major cultivation season nationwide.

3.2.2 Climate Change and Its Impact on Weather Patterns

Temperature and rainfall are the most influential climatic parameters, and their changes, variability, and extreme events greatly affect food production and food security in Sri Lanka. According to the Fifth Assessment Report (AR5) of the Intergovernmental Panel on Climate Change (IPCC 2014) the temperature of the South-Asian region may rise by more than 2 °C by the mid twenty-first century, the frequency of the occurrence of warm days will increase, and the rainfall may display more erratic patterns with increasing extreme events. However, the warming trend in Sri Lanka is more pronounced than in the rest of South Asia (Sheikh et al. 2015).

Many researchers have found that there is a gradual increasing trend in the atmospheric temperature almost all over the country (Chandrapala 2007b; De Costa 2008; Eriyagama et al. 2010a; Nissanka et al. 2011; Sathischandra et al. 2014). However, the rates of increase vary by place, and trends have accelerated in recent years (Chandrapala 2007b; De Costa 2008; Sathischandra et al. 2014). Annual mean air temperature anomalies have been increasing significantly in all meteorological stations across Sri Lanka in recent years (Basnayake 2007). The mean daytime maximum and mean nighttime minimum temperatures have been increased, and the contribution from the nighttime minimum temperature increase is more significant in the case of the annual average temperature (Basnayake 2007; Zubair et al. 2005). It has been found that the average air temperature has increased by 0.64 °C over the past 40 years and by 0.97 °C over the last 72 years, with a trend of 0.14 °C per decade. However, a recent study reveals that the increase in the air temperature was 0.45 °C over 22 years at a rate of 0.2 °C per decade (Punyawardena et al. 2013a, b). Further, the number of warm days and warm nights has increased while the number of cold days and cold nights has decreased throughout Sri Lanka.

In contrast to the temperature, a wide range of variations are found for rainfall. According to the previous studies, a distinct pattern or trend could not be observed in temporal and spatial dimensions. Based on mean annual rainfall, some researchers have identified a significant decreasing trend in rainfall in certain locations of Sri Lanka during last two decades (Basnayake 2007; Chandrapala 2007a; De Costa 2008; Jayatilake et al. 2005). It was accounted for as a drop of 144 mm of annual average rainfall during 1961—with reference to the period from 1931 to 1960 (Jayatilake et al. 2005). Though this is the overall scenario of the rainfall trend, some opposing trends could also be observed in different locations. For example, heavy rainstorms have become increasingly common in the central highlands in recent years (Punyawardena, Mehmood et al. 2013a, b). However, all accept that there is a variability in rainfall and it has increased over time (Chandrapala 2007a; Eriyagama et al. 2010a; Punyawardena et al. 2013a, b). Studies revealed that the number of consecutive dry days has grown while the number of consecutive wet days has dropped, accordingly (Premalal 2009; Ratnayake and Herath 2005). The spatial distribution of the rainfall seems to be changing, but the changing pattern is contradictory (Basnayake 2007; Marambe et al. 2013; Nissanka et al. 2011; Sathischandra et al. 2014). Some researchers argue that the variation in rainfall distribution may result in modifying the agro-ecological boundaries as well (Ampitiyawatta and Wijeratne 2015; Eriyagama et al. 2010a; Mutuwatte and Liyanage 2013).

Temporal and spatial variations as well as pattern changes in rainfall cause frequent floods, landslides, and drought conditions (Aheeyar 2012; Imbulana et al. 2010; Ratnayake and Herath 2005), posing a high threat to crop cultivation. For example, during the period 2010–2015, Sri Lanka suffered five large floods and four droughts, which had a significant influence on crop production (Perera 2015). The drought that occurred in 2012 was the worst during the past 20 years, and the Central Bank of Sri Lanka (2020) highlighted in its annual report this year as one of the worst agricultural years in the country. Many of the 2012 drought-affected districts have subsequently been affected by floods, further putting the affected communities and agriculture in more trouble. The widespread drought within 14 districts in 2014 badly affected 1.8 million people, and it has created a food insecurity situation among 750,000 people (Central Bank of Sri Lanka 2020). The drought conditions are further aggravated by the increase in temperature. The occurrence of landslides is very much correlated with high-intensity rainfall events (Ratnayake and Herath 2005). With the heavy rainfall, landslides have been more common in the last two decades than the previous ones.

3.2.3 Impacts of Climate Change on Agriculture

Climate change has had a significant impact on agriculture, particularly agricultural productivity. Both short-term and long-term changes in the climate affect various aspects of crop production. Apart from crop production, other fields of agriculture, such as livestock production and fisheries are also extremely vulnerable to the changes in the climate. The variability of rainfall pattern and amount, the increase in temperature, and the increase in atmospheric CO_2 concentration are the major consequences of climate change, which directly affect agriculture. Recent estimates of food production predict that world agricultural production will decline by 16% due to climate change impacts by 2080, and this scenario will be more severe in the Sri Lankan context, where up to 20% production loss is predicted (Cline 2007).

Since the staple food of the Sri Lankan community is rice, it is cultivated all over the country except for very high mountain peaks, and that is highly influenced by both rainfall and temperature variations (Fernando 2000; Peiris et al. 2004). Short-term variations such as high intensity rainfall and flash floods cause physical damage to the crop, leading to reduced yield. However, it may totally destroy the harvest in certain instances. During the past decade, droughts have occurred frequently, being the main issue with rice farming. Rice cultivation needs between 1500 and 2000 mm of water per growing season, depending on soil and meteorological conditions (Bouman 2001). Any water deficit condition during the growing stage, especially during the milking stage, severely affects the quantity and quality of the harvest. Rice is extremely temperature sensitive, especially the higher temperatures, which affect anthesis (flowering) and pollen fertility. Though paddy can be grown in temperatures ranging from 12 to 35 °C, the optimum temperature is around

25–30 °C (Yoshida 1973). Grain sterility commences at 30 °C and complete sterility is achieved at 36 °C (Abeysiriwardena et al. 2002; Weerakoon et al. 2008). Grain sterility conditions have been reported during the *Yala* season in many districts of the dry zone in Sri Lanka in which the warm dry weather conditions are available, exceeding the upper threshold of 34 °C. Even though the elevated atmospheric CO_2 will increase the yield of rice due to enhancing the photosynthetic process, this will be negated by the increasing temperature by the greenhouse effect (Ainsworth 2008). This has been proved by De Costa (2000) in his studies relevant to Sri Lankan conditions.

Climate change has a negative influence on other major crop types such as fruits, vegetables, tuber crops, grains, and legumes (Titumil and Basak 2010). Maize, being a widely cultivated crop throughout the dry zone of Sri Lanka during the *Yala* seasons when the water is limited for rice cultivation, has shown yield reductions with the increasing temperature (Amarasingha et al. 2018; Malaviarachchi et al. 2015). Research evidence is available for the yield reduction of green gram, also similar to maize with the increasing temperature (Malaviarachchi et al. 2015; Malaviarachchi et al. 2016). However, opposite results are found with regard to chili yield in response to the increasing temperature (Abhayapala et al. 2018). Impacts of rainfall variations such as flash floods and droughts affect similarly all the above crops by physically damaging or drying the crop totally or to a considerable extent. Fruits and vegetables grown mainly in the dry zone are also significantly affected by droughts and floods. The dry conditions and the increasing temperature may result in increased evapotranspiration rates, creating a soil moisture deficit in the crop root zone, crusting the topsoil layer and developing salinity conditions in upland crops. Apart from these direct impacts, increasing temperature causes the respiration rates of perishable horticultural crops to increase resulting post-harvest losses and storage expenses rise. The rise in nighttime temperatures has been more noticeable in recent decades in Sri Lanka, causing it to reduce the diurnal temperature variation. This is negatively affected by dry matter assimilation in tuber crops and finally reduces the tuber yield (Marambe et al. 2015).

Plantation crops, including tea, rubber, coconut, sugarcane, etc., are the main sources of the national economy of Sri Lanka. All of these plantation crops are also highly sensitive to climate change. Tea is widely grown in the low, intermediate, and high elevations of Sri Lanka's wet zone, and it requires well-distributed rainfall of at least 200 mm per month (Watson 1986). Short and intermediate dry spells (7-day, 10-day) have become more common in the wet zone in recent years as a result of climate change, resulting in a higher yield drop. High intensity rainfalls are causing physical damage to tender shoots, while continuous rainfalls are causing fungal diseases. Since the process of initiation and expansion of young shoots, which are harvested as yield, is also highly influenced by temperature variations (De Costa et al. 2007). Frequent rainfall reduces the amount of rubber tapping days while also harboring a number of fungal infections, resulting in a drop-in yield. On the other hand, long dry spells are affecting the drying of the tapping panel, and consequently totally collapsing the cultivation. The production capacity of coconut is also highly sensitive to rainfall and dry spells (Peiris et al. 2008). Changes in the monsoon rainfall pattern, as well as an increase in maximum air temperature, are two important factors influencing the variability of coconut production (Peiris et al. 2004).

Although the plantation and non-plantation sectors of agriculture are both affected by climate change, the implications for the non-plantation sector are more severe because the majority of farmers are smallholders whose livelihood is primarily based on farming operations. Therefore, the effects of climate change have a direct influence on the economy and food security of rural populations.

3.3 Adaptation and Mitigation Measures to Cope with Climate Change Impacts

As mentioned previously, Sri Lanka is extremely sensitive to the effects of climate change. Agriculture, tourism, and fisheries are among the key sectors of the economy that are being impacted by climate change. Apart from these economic sectors, climate change has a significant impact on ecosystems, including the water cycle and biodiversity (Eriyagama et al. 2010b; Marambe et al. 2012; Ministry of Environment 2012; Weerahewa et al. 2012). Agriculture is the most affected among these threatening sectors, which results in highly sensitive social issues in the country. A considerable population (27.1%) is dependent on agricultural-related livelihoods, affecting their living conditions directly. Further, it causes food insecurity conditions in the country (De Costa 2008; De Silva 2006; Marambe et al. 2013).

Adaptation and mitigation are the primary measures that can be used to mitigate the negative effects of climate change. Mitigation does not aim to reduce or minimize the direct impact of climate change, but rather to address the causes of climate change, such as reducing greenhouse gas emissions from anthropogenic activities, reducing fossil fuel consumption by increasing the use of renewable energy sources, reducing deforestation, and so on. Therefore, mitigation is a long-term strategy for minimizing the effects of climate change. On the other hand, adaptation refers to short-term modifications that can be implemented to reduce the effects of climate change. Hence, adaptation can be defined, more precisely, as 'adjustments in behavior of natural or social systems and their members in response to actual expected variability of change in climate in order to moderate and cope with harmful impacts or to take advantage of opportunities' (Adger et al. 2009; Smit et al. 2000).

A wide range of adaptation strategies can be summarized into four categories: anticipatory, reactive, planned, and autonomous (Fankhauser et al. 1999; Smith 1997; Tol 2005). Adaptive strategies practiced in advance to the climate change events are known as anticipatory measures where reactive adaptation strategies are practiced at the time of event in response to the climate change (Smith 1997). Planned adaptation refers to decisions made by various government authorities to mitigate the effects of climate change, whereas autonomous adaptation refers to adaption measures that are not decided by government agencies (Stage 2010). Smit and Skinner (2002) classified agricultural adaptation strategies into four major groups: (1) technological,

(2) government programs and insurance, (3) farm production practices, and (4) farm financial management. Categories 1 and 2 address government policy actions, while categories 3 and 4 describe farm-level adaptive methods used by farmers. The main focus of all these agricultural adaptation measures is to establish food and social security status among the poor and marginalized communities at the national level. Possible agricultural adaptation measures that could be practiced or promoted toward food security as a whole could be summarized as in Table 3.1.

3.4 Food Security of the Nation

Food security has many dimensions, such as availability, accessibility, utility, and stability. Changes in climatic conditions have high impact on all the above. Sri Lanka is primarily an agriculture-oriented country. The report from Central Bank of Sri Lanka (2020) shows that the agricultural sector contributes 7% of the Gross Domestic Production (GDP) of the country. Nevertheless, 27.1% of the Sri Lankans are directly or indirectly depending on the agriculture sector, supporting food security as well as the development of other industries related to agriculture (Department of Census and Statistics 2020). As the staple food, rice production exceeds the total rice requirement of the country, which leads to over 100% of self-sufficiency in rice in Sri Lanka. By 2015, Sri Lanka had over a 120% rate of self-sufficiency in rice with 4.8 million metric tons of rice production (Department of Census and Statistics 2021) (Fig. 3.2). As a consequence, rice importation was reduced to the level of 0.024 million metric tons per year by 2019, confirming food security in the country with concern for the staple food (Fig. 3.2). However, breaking the consistency of self-sufficiency over a period of time, especially in 2014 and 2017, rice importation rose again to 0.6 million metric tons and 0.75 million metric tons, respectively (Fig. 3.3). Domestic rice production in 2017 alone dropped by 49% compared to the previous year. This is mainly due to the prolonged drought conditions in the country, which directly affected the cultivation of rice during those years. This alerted scientists and make them more aware about the possible effects of climate change disasters on food production. Studies highlight that adverse climatic conditions will limit rice yield and production in the future too, especially the potential negative effect of high temperatures on rice yields (Stuecker et al. 2018). According to Delpitiya et al. (2014), the 2 °C increase of temperature would result in 12 percent drop of rice yield.

Prolonged dry conditions due to reducing rainfall as well as unexpected heavy rainfall seriously affect rice production in different areas of Sri Lanka. The major issue is that low rainfall in the catchments of the respective reservoirs cause negative impact on irrigated rice fields resulting in low yields. Nevertheless, flooding is the most prominent barrier to rice production, especially in eastern and western provinces of Sri Lanka's (Nguyen 2002). Rise of sea level would be a reason for increasing salinity in paddies and could cause potential damage to rice production in coastal region (De Costa 2010). There are potential rice varieties available in Sri Lanka, especially with salinity tolerance, such as Pokkali, At 401, At 95-26-4, and Bw 400

Table 3.1 National climate change adaptation plan for food security

Adaptation needs	Adaptation options	Action
1. Enhance the resilience of crops	A. Germplasm improvement	• Examine existing cultivars for water and heat stress • Produce tolerable varieties (Paddy, other field crops, fruit, and vegetables) – Heat tolerant – Drought tolerant – Short age (early maturity)
	B. Water management on farms	• Reduce irrigation water losses at the field level • Encourage the use of micro-irrigation technology • Develop water-saving farming techniques • Encourage rainwater collecting on farms • Encourage water reuse
	C. Encourage the use of resource-efficient farming methods	• Enhance cropping systems and conservation farming techniques • Enhance nursery safety • Employ flower-inducing strategies in fruits • Improve soil quality by increasing the utilization of organic materials (Integrated nutrient management) • Encourage low-water-demanding crops and types, as well as agricultural diversification (For Dry and intermediate zones)
	D. Sectoral capacity development	• Enhance the capacity of research institutes to perform research on tolerant cultivars and water-saving farming methods
2. Reduce the danger of crop and health impact from biological agents	A. Germplasm enhancement	• Examine current cultivars for pest and disease resistance • Produce pest-resistant cultivars • For paddy, other field crops, horticulture; insect pests and diseases)

(continued)

Table 3.1 (continued)

Adaptation needs	Adaptation options	Action
	B. Improvements to supporting infrastructure	• Boost vaccination campaigns • Create a mechanism for pest forecasting • Perform researches on parasites and diseases • Encourage crop clinics
	C. Promotion of best practices	• Encourage integrated pest management
3. Reduce the impact of irregular precipitation changes on agriculture	A. Creating an effective climate information management and communication system	• Create a system for issuing seasonal and intermediate-term weather forecasts on schedule (mobile and internet alert systems) • Adjust cropping schedules based on seasonal weather projections
4. Improve crop, animal, fish, and agroecosystem resistance to harsh weather events	A. Germplasm improvement	• Examine existing types for resistance to extreme events • Produce tolerable varieties (paddy) • Flood resistant • Drought resistant
	B. Creating an effective climate information management and communication system	• Create a system for issuing seasonal and intermediate-term weather forecasts on schedule • Improve early warning systems • Create communication networks based on mobile phones • Create safety plans and encourage the use of safety equipment
	C. Improvements in catastrophe risk management and preparedness	• Identify and collect data on flood and drought-prone areas • Determine the capacity of food storage in vulnerable areas • Create and manage buffer stocks on a regular basis

(continued)

Table 3.1 (continued)

Adaptation needs	Adaptation options	Action
5. Reduce the effects of sea-level rise on agriculture in coastal zones	A. Germplasm improvement	• Examine existing types for resistance to salinity/alkalinity • Produce salt/alkalinity tolerant cultivars. (paddy)
	B. Improving climate effect monitoring	• Regularly monitor the growth of salinity/alkalinity levels • Build saltwater defense structures to control seawater intrusions into coastal paddy lands
	C. Exploring alternatives	• Transform severely damaged paddy fields to alternative uses (e.g. brackish water aquaculture)
6. Examine the effects of increased CO_2 levels in the atmosphere on crop productivity and weed populations	A. Starting research studies to determine the effects of climate change	• Conduct study on the effects of rising CO_2 on agriculture • Crop Productivity – Weed populations (Invasive alien species)

Source National Adaptation Plan for Climate Change Impacts (2016)

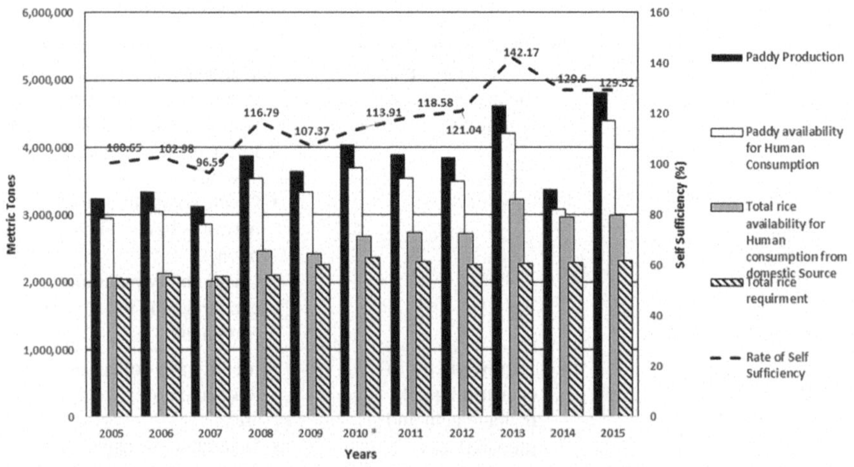

Fig. 3.2 Paddy production (metric ton), total rice requirement (metric ton), and rice self-sufficiency in Sri Lanka, from 2005 to 2015. *Source* Graph was developed from the data from the Department of Census and Statistics (2021)

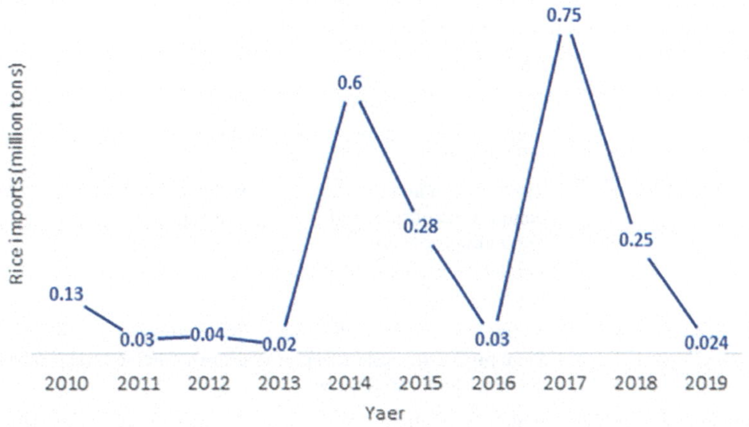

Fig. 3.3 Rice imports (million tons) from 2010 to 2009. *Source* Graph was developed using data from IRRI, Bathalagoda, Sri Lanka

(Dissanayake and Wijeratne 2006). Improving genetic diversity by breeding would possibly help increase rice production amidst future climatic changes (Yoshida 1983).

Horticultural crops such as vegetables and fruits are also having a significant contribution to the country's food security. They are available in a variety of forms such as fresh, cooked, value added, and partially processed. Fruits and vegetables are the main sources of vitamin A, vitamin C, potassium, folic acid, and dilatory fiber. They can thus be viewed as an affordable and easily accessible healthy food for the consumer who is very concerned about their health as well as for communities that are at risk of malnourishment. Major contributions of vegetables and fruits are from the home garden cultivation in Sri Lanka, which occupies 18.18% land area (Agstat 2020). The volume index of production of vegetables in Sri Lanka showed an increasing trend from a value of 130.2 in 2007 to 177.14 in 2017 (Agstat 2018). Fruit production is also increasing in the production area. However, the production and market details of underutilized fruit crops are not available at a negligible level. Other field crops such as chili, onion, root, and tuber crops occupy 0.15 million hectares of land.

Future pressure on food production and agriculture is projected to come from a growing world population and a wide range of dietary needs (Castañeda-Álvarez et al. 2016). The Sri Lankan population increased annually by 1.1% (Central Bank of Sri Lanka 2020) demanding more food with diverse requirements. As a result, more varied agricultural and food systems are required to meet these expanding demands as well (Padulosi et al. 2013). According to recent studies, underutilized fruit species have a great deal of potential to help rural residents in drought-prone regions secure their food, nutrition, and income (Chivenge et al. 2015). Consequently, boosting these underutilized neglected fruit crops to make them more 'commercial' and 'consumer friendly' might be seen as a potent way to achieve sustainable development goals like lowering poverty and malnutrition. Because these plants are hardy and adaptable

Table 3.2 Types and characteristics of food systems in Sri Lanka

Type	Characteristics*	Example
Traditional food system	Subsistence, unorganized Supply chains, limited market infrastructure	Home garden, chena (shifting) cultivation
Structured food system	Traditional market actors, complex set of rules and regulations applied to marketplaces, higher degree of market infrastructure	Paddy cultivation, vegetable cultivation
Industrialized food system	A high degree of coordination, large processing sector, export market, organized retailers	Tea cultivation (plantation crops), organic food production
Alternative food system	Semi-closed circuits of exchange	Alternative organic market

Source Kledal (2009)

to such changes, they have the ability to increase the diversity of food systems and, therefore, make agricultural production systems less vulnerable to climate change (Bandula et al. 2016; Malkanthi 2017; Padulosi et al. 2013). They can survive in arid climates and ecosystems that lack nutrients (Singh et al. 2019). These species have gained more recognition for the potential to adjust food security to climate change in the future (Padulosi et al. 2011).

3.5 Food Production Systems in Sri Lanka

In Sri Lanka, a variety of food systems are present, which can be broadly categorized under traditional, structured, industrialized, and alternative food systems. Table 3.2 summarizes the characteristics of different food systems found in the country.

3.6 Diversity of Sri Lankan Soils

Six major soil groups are found in Sri Lanka, namely: Alfisols, Entisols, Histosols, Inceptisols, Ultisols, and Vertisols (Mapa et al. 1999). These soil types came from a variety of parent substances, such as local alluvium produced from Precambrian metamorphic rocks, colluvium and alluvium derived from micaceous schists, and many more. The Entisols, Histosols, Inceptisols, and Ultisols can be found in wet and semi-wet intermediate zones. Alfisols, Inceptisols, and Ultisols are present in the intermediate zone. Alfisols, Inceptisols, Entisols, and Vertisols are found in the semi-dry intermediate and dry zones of the country (Mapa 2020). In Sri Lanka's dry zone, there are 11 agroecological areas with a total of 54 known soil series. In the wet zone, 28 soil series were described, while 36 soil profiles were identified in the intermediate zone of the country (Mapa 2020).

Due to the variation of other environmental factors such as rainfall, a diverse range of soil types that vary in morphological characteristics and chemical, physical, and biological properties can be observed in Sri Lanka (Figs. 3.4 and 3.5). Due to their natural variability along with the climate, these soils support different types of crops, which are supportive for the country's food security. Soil characteristics have a key impact on the crop selection and performance. Table 3.3 summarizes the average soil properties of Sri Lanka.

However, poor or problematic soils are also present in different parts of Sri Lanka, which is a challenge to climate adaptation and national food security. Therefore, proper attention should be given to crop production in the problematic soils. Some of the problematic soils in Sri Lanka are shown in Table 3.4 (Vidhanarachchi 2011).

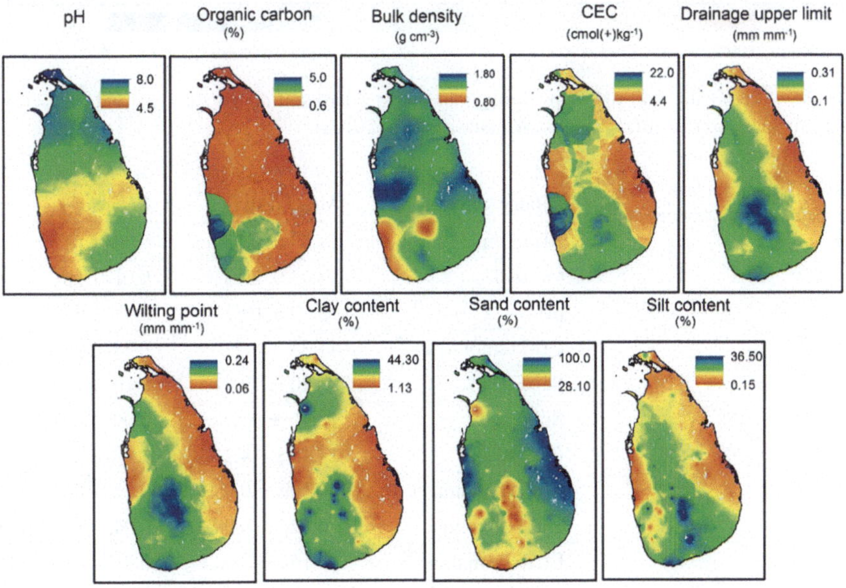

Fig. 3.4 Interpolated soil properties in Sri Lanka for the 0–5 cm layer. *Source* Adapted from Wimalasiri et al. (2020b)

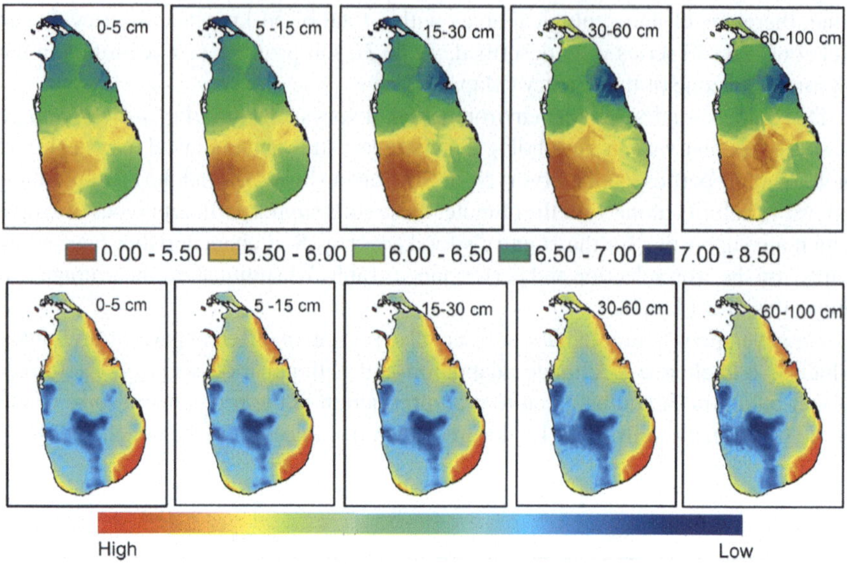

Fig. 3.5 Interpolated soil pH maps throughout five standard soil depths (above), and their error maps (below). *Source* Adapted from Wimalasiri et al. (2020a)

Table 3.3 Summary statistics of soil properties in Sri Lanka

Parameter	Mean with SD	Range
BD (g cm^{-3})	1.43 ± 0.23	0.30–1.82
OC (%)	1.13 ± 2.13	0.01–16.81
pH	5.79 ± 0.97	2.37–8.27
CEC (cmol + kg^{-1})	11.33 ± 8.21	0.78–51.49
Sand (%)	61.20 ± 19.54	3.70–99.00
Silt (%)	13.66 ± 9.59	0.00–62.07
Clay (%)	25.13 ± 14.26	1.00–67.35
SAT (mm mm^{-1})	0.409 ± 0.089	0.263–0.836
DUL (mm mm^{-1})	0.220 ± 0.095	0.045–0.423
LL15 (mm mm^{-1})	0.162 ± 0.083	0.017–0.350
AD (mm mm^{-1})	0.129 ± 0.076	0.009–0.350

where, BD = bulk density, OC = organic carbon, CEC = cation exchange capacity, SAT = saturation, DUL = drainage upper limit, LL15 = wilting point and AD = soil moisture limit to which soil can dry by evaporation

Source Adapted from Wimalasiri et al. (2020b)

Table 3.4 Problem soils in Sri Lanka

Problem	Location	Description
Acid sulfate soil	Palatuwa series (Nilwala river flood plain)	Strong acidity, currently under paddy cultivation
Acid soil	Malaboda series	pH values less than 5.5 with low base saturation
Calcareous soil	Hunugala series	Reaction of soil is mild to strong alkaline
Histosols	Wagura series, Muthurajawela	High organic matter content, poorly drained,
Saline soils	Puttalam soil series	High salt content and electrical conductivity
Vertisols	Palatuwa soil series, Wagura series	High clay content and high moisture retention capacity, swelling and shrinking characteristics

3.7 Diversity of Food Crops

Biodiversity in Sri Lanka is at a high level as a as a tropical country with a favorable climate, available solar radiation, and a rich diversity of crop species, which leads to the country's world-renowned unique agricultural diversity (Biodiversity conservation in Sri Lanka 1999; Galhena et al. 2013; Gaston 2000; Stevens 1989). Despite the fact that the climate is typically tropical, variations exist due to variations in rainfall patterns and elevation around the nation. (MoFE 1999). The Sri Lankan economy, in a variety of sectors, including agriculture, greatly benefits from benefits gained from biodiversity.

When considering the agricultural diversity in Sri Lanka, it includes *Oryza sativa* (rice) with over 2,000 varieties and five wild species *Zea mays* (maize) and *Sorghum bicolor* (sorghum) include seven species and their traditional cultivars; 14 grain legume species; eight cucurbitaceous, two solanaceous and four other vegetables species; and 17 root and tuber crop species. The economically useful spices are eight species of cinnamon, *Elettaria cardamomum* (L.) Maton, three Piper species (with seven wild relatives), clove, nutmeg, betel nut, vanilla, chili, and ginger. Others of importance include three major beverage species (tea, coffee, and cocoa), sugarcane, kitul plam, arecanut, tamarind, citronella, three species of oil crops, and two fiber crops (Wijesinghe et al. 1993). About 196 species from 46 plant families make up Sri Lanka's fruit species diversity. This species diversity is composed of 18% endemic, 41% indigenous, and 41% exotic fruit species. Of the 196 species, 56 species are considered wild relatives of fruit species (Pushpakumara et al. 2016a; b). Crop Wild Relatives (CWRs) have not yet been properly utilized for varietal improvement in Sri Lanka. A new project to conserve this valuable biological resource is being implemented by the Department of Agriculture in collaboration with the Ministry of Environment and Natural Resources. The project is funded by UNEP/GEF and globally implemented by the International Plant Genetic Resources Institute.[1]

[1] https://www.doa.gov.lk.

The dependency of a small number of food species would not be confident enough to cope with predicting climate change scenarios which the world would be adversely affected by (IPCC 2014; NRC 2010). Future global food and nutrition security is at risk due to the world economy's significant reliance on a small number of variations in crops. Many recognized plant genetic resources have been lost during the previous century, and by 2050, one-third of the current variety might be gone (FAO 2011). Additionally, despite the fact that the world is currently being severely affected by global climate change, the diversity of tree species found in nature may help to strengthen food security. To preserve and comprehend the range and diversity of the crop in nature as genetic resources, thorough study on neglected trees as a food crop is crucial.

3.7.1 Rice Plant Diversity

The traditional rice of Sri Lanka comes in about 500 varieties, and most of them have more than one accession from the country's 46 distinct agro-ecological regions. Variation of days to flowering (DF), plant morphology, yield, and photoperiodic response are most recognized genetic diversity of Sri Lankan traditional rice (Irangani and Shiratake 2013). Rice flowering time and yield are mainly affected by important environmental factors such as photoperiod and temperature (Luan et al. 2009; Vergara and Chang 1985). However, there are few studies on understanding the temperature responses of traditionally grown rice in Sri Lanka. Most of the rice being cultivated is currently in regions where temperatures are above the optimal for growth. Any further rise in mean temperatures or the frequency of high temperature events during vulnerable growth phases might significantly lower rice harvests (Krishnan et al. 2011). A study reported that 22 accessions of traditional rice varieties out of 40 accessions are temperature responsive and others are not (Rathnathunga et al. 2019). This implies that rice germplasm is characterized by high genetic variation, which could potentially be important to investigate for future use under harsh climatic conditions.

As climatic advantages due to elevation, vegetable cultivation progresses in Sri Lanka into two main groups, such as upcountry vegetables and low country vegetables. These climatic differences help to increase the diversity of commercially grown vegetables in Sri Lanka, supporting food security. Up country vegetables are mainly adapted to temperate climates and were mainly introduced during the British colonial period. Mainly, potato, bean, beetroot, cabbage, carrot, knolkhol, leek, radish, and tomato are the major vegetable crops cultivated in the area (Agstat 2020; Padmajani et al. 2014). In addition to those major up-country vegetables, many other temperate vegetables, to a lesser extent, can be found in the highest elevations of the country, such as broccoli, cauliflower, and asparagus (Personal experience).

3.7.2 Fruit Crop Diversity

Besides the natural gift of diversity, commonly available fruit and vegetable crops are not very diverse when considering the tropical potential. The entire nation is largely dependent on a small number of crop species, neglecting a significant number of vegetables, fruits, and other crop species, which has led to these valuable crop species becoming critically endangered. This is primarily due to the commercialization and modernization of agricultural practices. They are especially facing the state of speedy genetic erosion of their gene pools and are often regarded as lost crops (William and Haq 2002). This situation is not confined to Sri Lanka but to the whole world. Only about 30 crop species provide 95% of the entire world's food energy, but over 7000 crop species have been known to be used for food and are either partly or fully domesticated. This large range of crop plant are recognized to be underutilized crops or minor crops (William and Haq 2002). Banana (*Musa* spp.), pineapple (*Ananas comosus* (L.) Merr.), papaya (*Carica papaya* (L.)), mango (*Mangifera indica* (L.)), avocado (*Persea americana* Mill.), and rambutan (*Nephelium lappaceum* (L.)) are identified as the major fruit crop species grown in Sri Lanka (Heenkenda 2014). In addition to that, there are numerous minor, underused fruit species (at least 50–60 species) cultivated around the island that have just become popular island wide (Dahanayake 2015; Pushpakumara et al. 2007, 2011).

Different underutilized fruit species bear flowers at different periods of the year, which will help to ensure fruit availability widely spread throughout the year. In weera tree (*Drypetes sepiaria* (Wight & Arn.) Pax & K. Hoffm), flowering can be seen in January–February, and hence during May we can expect fruits, in palu (*Manilkara Hexandra* (Roxb.) Dubard), and himbutu (*Salacia chinensis* (L.)), flowering can be seen in February–March which will produce fruit in May–June. In mora (*Nephelium longana* (Lam.) Cambess), kirikon (*Walsura piscidia* Roxb.), divul (*Feronia limonia* (L.)), and kon (*Schleichera oleosa* (Lour.) Oken), flowering can be seen in March–April, and hence fruiting in June–July. In syambala (*Tamarandus indica* (L.)), madang (*Syzygium cumini* (L.) Skeels.), flowering can be seen in April–May. Therefore, fruiting can be expected from July to August. This showed the potential of climatic resilience species and might be photosensitivity. Further, due to less attention they are mostly confined to wild and hence show high variability and diversity in same species (Dissanayake et al. 2022). This diversity could potentially be helpful for avoiding some food unavailability under different climatic conditions. However, attention to possible improvements is scarce in the country.

Another reason for not being concerned about other small crops is the absence of broad understanding among development organizations of linkages between agriculture and biodiversity (William and Haq 2002; Srivastava et al. 1996). We do need an understanding of the value of our underutilized fruits in nature. There are a few groups, and the Department of Agriculture is currently working on the identification of wild fruits for commercial use. These attempts should be accelerated to face any circumstances of climatic change that severely affect commonly grown fruits.

3.7.3 Home Gardens and Climate Change

Home gardens are significant source of food, energy, and nutritional security as well as a source of income and has been a long-standing practice in Sri Lanka. A home garden is defined as a small plot of ground that is typically close to or accessible by foot from the family house, with a mixed cropping system which consists of food crops, tree crops, spices, ornamentals, medicinal plants, and livestock, that primarily provides food for domestic use and can be used as a secondary source of income (Jacob and Alles 1987; Thamilini et al. 2019; Johnson-Welch et al. 2000; Krishna 2006; Odebode and Odebode 2006). In Sri Lanka, there are over 1.42 million home gardens, covering a total area of 76,483 ha. These gardens are regarded as the source of the nation's agricultural biodiversity and are widespread throughout the country including dry zones as well as wetlands.

In addition to supplying rural and peri-urban households with a consistent supply of nutritious foods like rice, yams, underutilized fruit species, indigenous vegetables, leafy vegetables, and spices, home gardens play a significant role in the conservation of hundreds of useful plant species and their genetic diversity. About 452 crop wild relative species were recorded during the conduct of the Crop Wild Relatives Conservation Project. High crop genetic variation in backyard gardens creates a wealth of agro-biodiversity across the nation Weerahewa et al. (2012) noted that the composition of home gardens in Sri Lanka has not altered significantly over the past 20 years, despite signs of climate change. Over the past twenty years, home gardeners have developed strategies that have helped them adapt to climatic changes. Furthermore, Weerahewa et al. (2012) proposed that development programs to promote climate change adaptation should be designed with four main strategies in mind: (i) changes in technology; (ii) changes in planting date; (iii) changes in agronomic practices; and (iv) adopting soil and water conservation strategies. The job situation, the variety of home gardens, the age, gender, and education of the household head, awareness of climate change, and the size of the home garden all have a substantial impact on the decision to implement new adaptation techniques (Weerahewa et al. 2012).

3.7.4 Kandyan Home Gardens

Kandyan home gardens (KHGs) are Sri Lankan home gardens that are located in Kandy and nearby districts including Badulla, Kegalle, Kurunegala, Matale, Nuwara Eliya, and Rathnapura. According to Pushpakumara et al. (2016a; b) KHGs represent as a scattered but important human-made land use system that increases the tree cover of this area of Sri Lanka.

Approximately half of Sri Lanka's fruit crop species variety is preserved through KHGs, of which 6% are endemic, 34% are indigenous, and 60% are of exotic. The average density of trees in the KHGs is around 200 trees/ha, of which 40–50% are fruit trees (Ariyadasa 2002). Consequently, KHGs are a beneficial practice for diversity and offer a crucial choice for the on-farm conservation of species diversity of naturalized exotic fruit crops (Pushpakumara et al. 2016a, b). This diversity in KHGs can be taken into consideration as a source of providing wider support to get into suitable adaptive and climatic resilience crop species or varieties to cope with adverse climate changes.

3.8 Case Study

A case study was conducted to examine the effects of change adaptation on crop yield using the below-described crop modeling approach.

The test crop was selected as rice, which is the major cereal of Sri Lanka and is cultivated throughout the country. The Oryza module of the Agricultural Production Systems Simulator (APSIM–Version 7.10) (Holzworth et al. 2014) was used in this study. The APSIM is a process-based crop model that simulates several physiological and growth parameters of crops. The study site was chosen to be Thanamalwila Lanka (6.428° N, 81.090° E) in the dry zone of Sri Lanka.

In simulations, Zubair et al. (2014)'s publication of genetic coefficients for paddy cultivars was used, but all other APSIM model parameters were left at their default values. November 10 was set as the planting date while other required crop management data for simulations were from Wimalasiri et al. 2020a, b. Daily rainfall, minimum and maximum temperatures, and solar radiation with bias correction for the years 2040–2069 (mid-century) using three General Circulation Models (GCMs) under the Representative Concentration Pathway (RCP) 4.5 scenario were downscaled, namely; BNU-ESM of the College of Global Change and Earth System Science, Beijing Normal University, China (GCM1); CSIRO-Mk3-6-0 of the Commonwealth Scientific and Industrial Research Organization and the Queensland Climate Change Centre of Excellence, Australia (GCM2); and GFDL-ESM2M of the Geophysical Fluid Dynamics Laboratory, USA (GCM3). The data were downscaled from the Consultative Group for International Agricultural Research (CGIAR) Program on Climate Change, Agriculture and Food Security (CCAFS) data portal (http://ccafs-climate.org/). The 1980–2009 period was used as the baseline period.

The following adaptation measures were tested using the APSIM model:

- Variety: short duration (BG300) and long duration (BG358)
- Late planting: 2 weeks delayed planting
- Early planting: 2 weeks earlier planting.

Figure 3.6 shows the variation of simulated rice yield for baseline and future climates. The rice yield of BG300 under the baseline climate was 3711 and 4524 kg ha^{-1} in rice variety BG358. Except for GCM1 and GCM3 in BG358, the average yield decreased in all other scenarios.

The yield change relative to the baseline yield under different adaptation measures (variety and late/early planting) is shown in Fig. 3.7. Increased the yield of short-duration rice variety (BG300) by delayed planting by two weeks the decreased that of

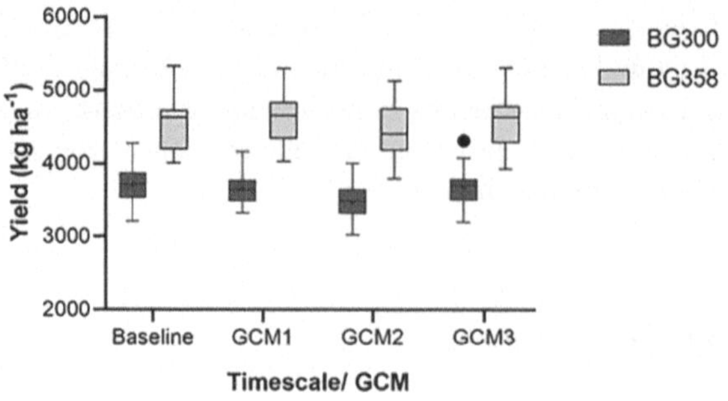

Fig. 3.6 Variation of the yield of short-duration (BG300) and long-duration (BG358) rice varieties under baseline and future climates

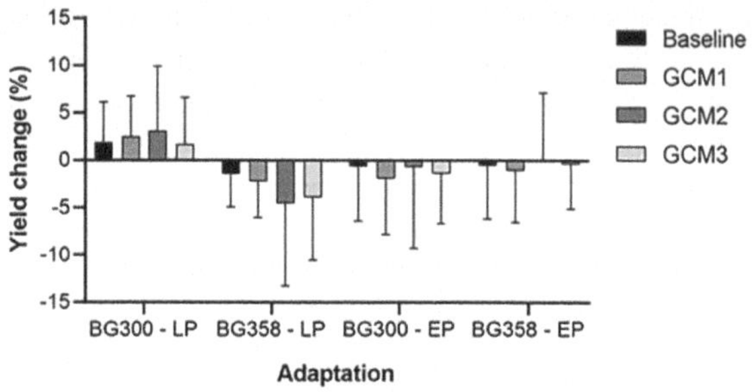

Fig. 3.7 Variation of the yield change for different adaptation measures under baseline and future climates. LP = Late/delayed plating by two weeks and EP = Early planting by two weeks

the long-duration variety (BG358). In BG300, a comparatively higher yield increment from delayed planting was reported from both GCM 1 and 2. In long-duration rice variety, the yield increased in GCM2 when the planting was started 2 weeks earlier than the regular date.

According to the simple simulation described above, both early and late planting of paddy yield changed under both current and future climates. Further detailed simulations and field experiments are needed to validate and/or strengthen the findings.

3.9 Conclusion

As with tropical potentials, Sri Lanka is comparatively in a favorable position against any climatic change issues for the long-term with respect to crop diversity and agroecological diversity concerns. However, sudden changes in weather patterns for a short period have a huge impact on the agriculture sector, which hampers food security. The main three climatic zones are helpful for a wider distribution of crop cultivation throughout the year. This variation is mainly due to monsoon winds which bring the main two seasons of rainfall, named as *Yala* and *Maha*, to two different regions. Further, topographical variation makes more food crop diversity, ranging from temperate to tropical, which is helpful for food security. However, these diversities need more attention to explode some underutilized food crops' availability to cope with the climatic change that could happen on any scale in the future. The soil factors also give high support to having diverse cropping patterns and diverse food crops to be available. National Adaptation Plan for Climate Change on Food Security is also very helpful in dealing with food security in the face of climatic changes. The case study we presented showed that adaptation measures in terms of selection of short-term and long-term varieties and planting dates are important to compensate for the yield losses due to climate change.

References

Abeysiriwardena DSZ, Ohba K, Maruyama A (2002) Influence of temperature and relative humidity on grain sterility in Rice. J Natl Sci Found Sri Lanka 30(1–2):33–41

Abhayapala R, De Costa J, Malaviarachchi W, Kumara A, Suriyagoda L, Fonseka R (2018) Exploitation of differential temperature-sensitivities of crops for improved resilience of tropical smallholder cropping systems to climate change: a case study with temperature responses of tomato and chilli. Agr Ecosyst Environ 261:103–114. https://doi.org/10.1016/j.agee.2017.10.027

Abhayasinghe KR (2007) Climate. In: The national atlas of Sri Lanka. Survey Department of Sri Lanka, pp 52–53

Adams RM, Rosenzweig C, Peart RM, Ritchie JT, McCarl BA, Glyer JD, Curry RB, Jones JW, Boote KJ, Allen LH (1990) Global climate change and US agriculture. Nature 345(6272):219–224. https://doi.org/10.1038/345219a0

Adger WN, Dessai S, Goulden M, Hulme M, Lorenzoni I, Nelson DR, Naess LO, Wolf J, Wreford A (2009) Are there social limits to adaptation to climate change? Clim Change 93(3):335–354. https://doi.org/10.1007/s10584-008-9520-z

Agstat (2018) Agricultural Statistics (XV). Socio Economics and Planning Centre Department of Agriculture

Agstat (2020) Agricultural Statistics (XVII). Socio Economics and Planning Centre Department of Agriculture

Aheeyar M. MM (2012) Climate change adaptation in water management for food security: Recent developments in Sri Lanka-A review of Existing Knowledge and Information. Sri Lanka Water Partnership & Global Water Partnership–South Asia. http://lankajalani.org/wp-content/uploads/2015/03/APAN-report.pdf

Ahmed, M., & Suphachalasai, S. (2014). *Assessing the* costs of climate change and adaptation in South Asia. Asian Development Bank. . http://www.adb.org/sites/default/files/publication/42811/assessing-costsclimate-change-and-adaptation-south-asia.pdf

Ainsworth EA (2008) Rice production in a changing climate: a meta-analysis of responses to elevated carbon dioxide and elevated ozone concentration. Glob Change Biol 14(7):1642–1650. https://doi.org/10.1111/j.1365-2486.2008.01594.x

Amarasingha RK, Suriyagoda LDB, Marambe B, Galagedara LW, Punyawardena R (2018) Impact of climate change on rice yield in Sri Lanka: a crop modelling approach using agriculture production system simulator (APSIM). Sri Lanka J Food Agric 4(1):21–26. https://doi.org/10.4038/sljfa.v4i1.54

Ampitiyawatta AD, Wijeratne AW (2015) Impact of temporal changes of rainfall on major climatic zones of Sri Lanka. Trop Agric Res Ext 18(4):148–152

Ariyadasa KP (2002) Assessment of tree resources in the home gardens of Sri Lanka. FAO Regional Office for Asia and the Pacific. http://www.fao.org/3/AC629E/AC629E00.htm#TOC

Assad ED, Pinto HS, Zullo Junior J, Ávila AMH (2004) Impacto das mudanças climáticas no zoneamento agroclimático do café no Brasil. Pesq Agrop Brasileira 39:1057–1064. https://doi.org/10.1590/S0100-204X2004001100001

Australian Academy of Science (2021) How has climate changed? https://www.science.org.au/learning/general-audience/science-climate-change/2-how-has-climate-changed

Bandula A, Jayaweera C, De Silva A, Oreiley P, Karunarathne A, Malkanthi SHP (2016) Role of underutilized crop value chains in rural food and income security in Sri Lanka. Procedia Food Sci 6:267–270. https://doi.org/10.1016/j.profoo.2016.02.049

Basnayake BRSB (2007) Climate change. In: The national atlas of Sri Lanka. Survey Department of Sri Lanka, pp 54–55

Biodiversity Conservation in Sri Lanka (1999) A framework for action. Ministry of Forestry and Environment

Bouman BAM (2001) Water-efficient management strategies in rice production. Int Rice Res Notes 26(2):1–1

Castañeda-Álvarez NP, Khoury CK, Achicanoy HA, Bernau V, Dempewolf H, Eastwood RJ, Guarino L, Harker RH, Jarvis A, Maxted N, Müller JV, Ramirez-Villegas J, Sosa CC, Struik PC, Vincent H, Toll J (2016) Global conservation priorities for crop wild relatives. Nature Plants 2(4):1–6. https://doi.org/10.1038/nplants.2016.22

Central Bank of Sri Lanka (2020) Annual Report, vol 1. Central bank of Sri Lanka

Chandrapala L (2007a) Rainfall. In: The national atlas of Sri Lanka. Survey Department of Sri Lanka, pp 58–59

Chandrapala L (2007b) Temperature. In: The national atlas of Sri Lanka. Survey Department of Sri Lanka, pp 56–57

Chivenge P, Mabhaudhi T, Modi AT, Mafongoya P (2015) The potential role of neglected and underutilised crop species as future crops under water scarce conditions in sub-Saharan Africa. Int J Environ Res Public Health 12(6):5685–5711. https://doi.org/10.3390/ijerph120605685

Cline WR (2007) Global warming and agriculture: impact estimates by country. Center for Global Development, Peterson Institute for International Economics

Dahanayake N (2015) Some neglected and underutilized fruit crops in Sri Lanka. Int J Sci Res Publ 5(2):165–171

De Costa WAJM (2000) Prediction of the effects of elevated CO2 and temperature on irrigated rice yields in the lowcountry dry zone of Sri Lanka using a process-based simulation model. J Natl Sci Found 28:165–184. https://doi.org/10.4038/jnsfsr.v28i3.2659

De Costa WAJM (2008) Climate change in Sri Lanka: Myth or reality? Evidence from long-term meteorological data. J Natl Sci Found 36:63. https://doi.org/10.4038/jnsfsr.v36i0.8048

De Costa WAJM (2010) Adaptation of agricultural crop production to climate change: a policy framework for Sri Lanka. J Natl Sci Found 38(2):79. https://doi.org/10.4038/jnsfsr.v38i2.2032

De Costa WAJM, Mohotti AJ, Wijeratne MA (2007) Ecophysiology of tea. Braz J Plant Physiol 19:299–332. https://doi.org/10.1590/S1677-04202007000400005

De Silva CS (2006) Impacts of climate change on water resources in Sri Lanka. In: 32nd WEDC international conference, Colombo, Sri Lanka

Delpitiya DPGS, Weerakoon WMW, Nissanka SP, Punyawardena BVR, Karunarathne MA, Herath RM (2014) Impacts of climate change on rice yields in Sri Lanka: a crop modelling approach using decision support system for agro technology transfer (DSSAT4.5). Ann Sri Lanka Dep Agric 16:71–81

Department of Census and Statistics (2020) Sri Lanka Labour force statistics quarterly Bulletin, Department of Census and statistics, Ministry of finance. http://www.statistics.gov.lk/Resource/en/LabourForce/Bulletins/LFS_Q1_Bulletin_2020

Department of Census and Statistics (2021) Agriculture and environment statistics division, department of census and statistics, Ministry of Finance. http://www.statistics.gov.lk/Agriculture/StaticalInformation/rubpaddy

Dissanayake PK, Wekumbura WGC, Wijeratne AW, Wijesundara DSA (2022) Morphological characterization, antioxidant capacity and diversity of *Syzygium cumini* trees from Sri Lanka. Hortic Plant J 8(1):53–67. https://doi.org/10.1016/j.hpj.2021.09.002
Dissanayake PK, Wijeratne AW (2006) Development of a varietial screening procedure for salt tolerance of rice (*Oryza sativa* L.) varieties at germination stage. J. Agric. Sci. **2**(1):63–72
Domroes M (1974) The agroclimate of ceylon. franz steiner verlag
EBoDE (2010) Environmental Burden of Disease in Europe (EBoDE) pilot project. http://en.opasnet.org/w/Ebode
Eriyagama N, Smakhtin V, Chandrapala C, Fernando F (2010a) Impacts of climate change on water resources and agriculture in Sri Lanka: a review and preliminary vulnerability mapping. IWMI
Eriyagama N, Smakhtin V, Chandrapala L, Fernando K (2010b) Impacts of climate change on water resources and agriculture in Sri Lanka: a review and preliminary vulnerability mapping. IWMI
Fankhauser S, Smith JB, Tol RSJ (1999) Weathering climate change: some simple rules to guide adaptation decisions. Ecol Econ 30(1):67–78. https://doi.org/10.1016/S0921-8009(98)00117-7
FAO (2011) Save and grow: a policymaker's guide to the sustainable identification of smallholder crop production (p. Rome). Food and Agriculture Organization of the United Nations
Fernando TK (2000) Impact of climate change on paddy production in Sri Lanka. Glob Environ Res 2:169–176
GA2LEN (2010) Global allergy and asthma European Network. www.ga2len.net
Galhena DH, Freed R, Maredia KM (2013) Home gardens: a promising approach to enhance household food security and wellbeing. Agric Food Secur 2(1):8. https://doi.org/10.1186/2048-7010-2-8
Gaston KJ (2000) Global patterns in biodiversity. Nature 405(6783):220–227. https://doi.org/10.1038/35012228
Heenkenda HMS (2014) Better future through fruit crop research. Fruit Research and Development Institute
Holzworth DP, Huth NI, deVoil PG, Zurcher EJ, Herrmann NI, McLean G, Chenu K, van Oosterom EJ, Snow V, Murphy C, Moore AD, Brown H, Whish JPM, Verrall S, Fainges J, Bell LW, Peake AS, Poulton PL, Hochman Z, Keating BA (2014) APSIM—Evolution towards a new generation of agricultural systems simulation. Environ Model Softw 62:327–350. https://doi.org/10.1016/j.envsoft.2014.07.009
Imbulana KAUS, Wijesekara NTS, Neupane BR (2010) Sri Lanka water development report. UNESCO and Ministry of Irrigation and Water Resources Mangement
IPCC (2014) Climate change 2014: synthesis report. contribution of working Groups I, II and III to the fifth assessment report of the intergovernmental panel on climate change. intergovernmental panel on climate change
IPCC (2021) Climate change 2021: the physical science basis. Contribution of working group i to the sixth assessment report of the intergovernmental panel on climate change. Cambridge University Press
Irangani MKL, Shiratake Y (2013) Indigenous techniques used in rice cultivation in Sri Lanka: an analysis from an agricultural history perspective. IJTK **12**(4) (October 2013). http://nopr.niscair.res.in/handle/123456789/22194
Jacob VJ, Alles WS (1987) The Kandyan gardens of Sri Lanka'. Agrofor Syst 5:123–137
Jayatilake HM, Chandrapala L, Basnayake BRSB, Dharmaratne GHP (2005) Water resources and climate change. In: Proceedings of the preparatory workshop on Sri Lanka national water development report. Ministry of Irrigation, Mahaweli and Rajarata Development, pp 62–87
Johnson-Welch C, Alemu B, Msaki TP, Sengendo M, Kigutha H, Wolff A (2000) Improving household food security: institutions, gender, and integrated approaches. University of Wisconsin. https://www.africabib.org/rec.php?RID=W00079330
Kledal PR (2009) The four food systems in developing countries and the challenges of modern supply chain inclusion for organic small-holders (Conference paper, poster, etc.). https://orgprints.org/id/eprint/18553/

Krishna GC (2006) Home gardening as a household nutrient garden. In: Enhancing the contribution of home garden to on-farm management of plant genetic resources and to improve the livelihoods of nepalese farmers: lessons learned and policy implications

Krishnan P, Ramakrishnan B, Reddy KR, Reddy VR (2011) Chapter three—high-temperature effects on rice growth, yield, and grain quality. In: Sparks DL (ed) Advances in agronomy, vol 111. Academic Press, pp 87–206. https://doi.org/10.1016/B978-0-12-387689-8.00004-7

Luan W, Chen H, Fu Y, Si H, Peng W, Song S, Liu W, Hu G, Sun Z, Xie D, Sun C (2009) The effect of the crosstalk between photoperiod and temperature on the heading-date in rice. PLoS ONE 4(6):e5891. https://doi.org/10.1371/journal.pone.0005891

Malaviarachchi MAPWK, Costa WAJMD, Fonseka RM, Kumara JBDAP, Abhayapala KMRD, Suriyagoda LDB (2015) Response of Maize (*Zea mays* L.) to a temperature gradient representing long-term climate change under different soil management systems. Trop Agric Res **25**(3):327. https://doi.org/10.4038/tar.v25i3.8043

Malaviarachchi MAPWK, De Costa WAJM, Kumara JBDAP, Suriyagoda LDB, Fonseka RM (2016) Response of Mung Bean (*Vigna radiata* (L.) R. Wilczek) to an increasing natural temperature gradient under different crop management systems. J Agron Crop Sci **202**(1):51–68. https://doi.org/10.1111/jac.12131

Malkanthi SHP (2017) Importance of underutilized crops in Thanamalwila divisional secretariat division in monaragala district in Sri Lanka. http://repo.lib.sab.ac.lk:8080/xmlui/handle/123456789/1058

Mapa RB (2020) Soil research and soil mapping history. In: Mapa RB (ed) The soils of Sri Lanka. Springer International Publishing, pp 1–11. https://doi.org/10.1007/978-3-030-44144-9_1

Mapa RB, Somasiri S, Magarajah S (1999) Soils of the wet zone of Sri Lanka: morphology, characterization and classification: special publication No. 1. Soil Science Society of Sri Lanka, Sarvodaya Wishva Lekha

Marambe B, Punyawardena R, Silva P, Premalal S, Rathnabharathie V, Kekulandala B, Nidumolu U, Howden S (2015) Climate, climate risk, and food security in sri lanka: need for strengthening adaptation strategies. In: Handbook of climate change adaptation. Springer, pp 1759–1789. https://doi.org/10.1007/978-3-642-38670-1_120

Marambe B, Pushpakumara G, Silva P (2012) Biodiversity and agro-biodiversity in Sri Lanka: a case in village tank systems. In: Biodiversity observation network in the asia-pacific region: towards further development of monitoring. Springer, pp 403–430

Marambe B, Pushpakumara G, Silva P, Weerahewa J, Punyawardena R (2013) Climate change and housdehold food security in homegardens of Sri Lanka. In: Proceedings of the international conference on climate change impacts and adaptation for food and environmental security. World Agroforestry, pp 87–100. https://www.worldagroforestry.org/publication/climate-change-and-housdehold-food-security-homegardens-sri-lanka

Ministry of Environment (2012) The national climate change policy of Sri Lanka. Ministry of Environment

MoFE (1999) Biodiversity Conservation in Sri Lanka: a framework for action. Ministry of Forestry and Environment

Mutuwatte L, Liyanage PKNC (2013) Impacts of climate change on agro-ecological regions of Sri Lanka. In: Proceedings of the international conference on climate change impacts and adaptations for food and environmental security. Coconut Research Institute

National Adaptation Plan for Climate Change Impacts (2016) National adaptation plan for climate change impacts in Sri Lanka (2016–2025). Ministry of Mahaweli development and environment

Nguyen NV (2002) Global climate changes and rice food security. Food Agric Organ. http://www.fao.org/forestry/15526-03ecb62366f779d1ed45287e698a44d2e.pdf

Nissanka S, Punyawardena R, Premalal K, Thattil R (2011) Recent trends in annual and growing seasons rainfall of Sri Lanka. In: Proceedings of the international conference on the impact of climate change in agriculture. Faculty of Agriculture,University of Ruhuna, pp 249–263

NOAA (2021) State of the climate: global climate report for 2020. https://www.ncdc.noaa.gov/sotc/global/202013/supplemental/page-3

NRC (2010) Advancing the science of climate change. The National Academies Press

Odebode SO, Odebode SO (2006) Assessment of home gardening as a potential source of household income in Adinyele local government are of Oyo state. Niger J Hortic Sci 11(1):47–55. https://doi.org/10.4314/njhs.v11i1.3398

Padmajani MT, Aheeyar MMM, Bandara MMM (2014) Assessment of pesticide usage in up-country vegetable farming in Sri Lanka (HARTI Research Report No. 164). Hector Kobbekaduwa Agrarian Research and Training Institute

Padulosi S, Heywood V, Hunter D, Jarvis A (2011) Underutilized species and climate change: current status and outlook. In: Yadav SS, Redden RJ, Hatfield JL, Lotze-Campen H, Hall AE (eds) Crop adaptation to climate change. Wiley-Blackwell, pp 507–521. https://doi.org/10.1002/9780470960929.ch35

Padulosi S, International B, Thompson J, Rudebjer PG (2013) Fighting poverty, hunger and malnutrition with neglected and underutilized species: Needs, challenges and the way forward. Bioversity International. https://cgspace.cgiar.org/handle/10568/68927

Panabokke C (1996) Soils and agro-ecological environments of Sri Lanka (Natural Resources Series—No.2). Natural Resources, Energy and Science Authority of Sri Lanka

Peiris TSG, Hansen JW, Zubair L (2008) Use of seasonal climate information to predict coconut production in Sri Lanka. Int J Climatol 28(1):103–110. https://doi.org/10.1002/joc.1517

Peiris TSG, Wijeratne M, Ranasinghe CS, Aanadacumaraswamy A, Fernando MTN, Jayakody A, Ratnasiri J (2004) Impact of climate change on coconut and tea industry in Sri Lanka. In: 2nd AIACC regional workshop for Asia and the Pacific, Manila, the Philippines

Perera A (2015) El Nino creates topsy turvy weather in Sri Lanka. Inter Press News Agency

Premalal KHMS (2009) Weather and climate trends, climate controls and risks in Sri Lanka. Sri Lanka Monsoon Forum, Department of Meteorology, Sri Lanka, Colombo, Sri Lanka

Punyawardena BVR, Dissanaike T, Mallawathanthri A (2013a) Vulnerability of Sri Lanka to climate change. Monograph. Department of Agriculture, Sri Lanka

Punyawardena BVR, Mehmood S, Hettiarachchi AK, Iqbal M, Silva SHSA, Goheer A (2013) Future climate of Sri Lanka: an approach through dynamic downscaling of ECHAM4 General Circulation Model (GCM). Trop Agric 161:35–52

Pushpakumara DKNG, Heenkenda HMS, Marambe B, Ranil RHG, Punyawardena BVR, Weerahewa J, Silva GLLP (2016a) Kandyan home gardens. A time-tested good practice from Sri Lanka for conserving tropical fruit tree diversity. In: Sthapit B, Lamers HAH, Ramanatha Rao V, Bailey A (eds) Tropical fruit tree diversity, good practices for in situ and on-farm conservation. Routledge, pp 127–146

Pushpakumara G, Heenkenda HMS, Marambe B, Ranil RHG, Punyawardena BVR, Weerahewa J, Silva GLLP, Hunter D, Rizvi J (2016) A time-tested good practice from Sri Lanka for conserving tropical fruit tree diversity. In: Tropical fruit tree diversity; conservation and use. Earthscan from Routledge, pp 127–146

Pushpakumara V, Gunasena HMP, Singh VP (2007) Underutilized fruit trees in Sri Lanka, vol 1. World Agroforestry Centre, South Asia Office

Pushpakumara V, Gunasena HMP, Singh VP (2011) Underutilized fruit trees in Sri Lanka, vol 2. World Agroforestry Centre, South Asia Office

Ranatunge E, Malmgren BA, Hayashi Y, Mikami T, Morishima W, Yokozawa M, Nishimori M (2003) Changes in the Southwest Monsoon mean daily rainfall intensity in Sri Lanka: relationship to the El Niño-Southern Oscillation. Palaeogeogr Palaeoclimatol Palaeoecol 197(1):1–14. https://doi.org/10.1016/S0031-0182(03)00383-3

Rathnathunga EUU, Senanayake G, Seneweera S, Geekiyanage S (2019) Varietal diversity of Sri Lankan traditional rice based on sensitivity to temperature and photoperiod at vegetative stage. J Natl Sci Found 47(1):51–68

Ratnayake U, Herath G (2005) Changes in Water cycle: effect on natural disasters and ecosystems. In: Proceedings of the preparatory workshop on Sri Lanka national water development report. World Water Assessment Programme, pp 192–205

Sathischandra H, Marambe B, Punyawardena R (2014) Seasonal changes in temperature and rainfall and its relationship with the incidence of weeds and insect pests in rice (*Oryza sativa* L) cultivation in Sri Lanka. Clim Chang Environ Sustain 2:105–115. https://doi.org/10.5958/2320-642X.2014.00002.7
Sheikh MM, Manzoor N, Ashraf J, Adnan M, Collins D, Hameed S, et al. (2015). Trends in extremedaily rainfall and temperature indices over South Asia. Int J Climatol 35(7):1625–1637. https://doi.org/10.1002/joc.4081
Singh A, Dubey PK, Chaurasia R, Dubey RK, Pandey KK, Singh GS, Abhilash PC (2019) Domesticating the undomesticated for global food and nutritional security: four steps. Agronomy 9(9):491. https://doi.org/10.3390/agronomy9090491
Smit B, Burton I, Klein RJT, Wandel J (2000) An anatomy of adaptation to climate change and variability. In: Kane SM, Yohe GW (eds) Societal adaptation to climate variability and change. Springer Netherlands, pp 223–251. https://doi.org/10.1007/978-94-017-3010-5_12
Smit B, Skinner MW (2002) Adaptation options in agriculture to climate change: a typology. Mitig Adapt Strat Glob Change 7(1):85–114. https://doi.org/10.1023/A:1015862228270
Smith JB (1997) Setting priorities for adapting to climate change. Glob Environ Chang 7(3):251–264. https://doi.org/10.1016/S0959-3780(97)00001-0
Srivastava J, Smith NJH, Forno D (1996) Biodiversity and agriculture—implications for conservation and development (Technical Paper 321). World Bank
Stage J (2010) Economic valuation of climate change adaptation in developing countries. Ann N Y Acad Sci 1185(1):150–163
Stevens GC (1989) The latitudinal gradient in geographical range: how so many species coexist in the tropics. Am Nat 133(2):240–256. https://doi.org/10.1086/284913
Stuecker MF, Tigchelaar M, Kantar MB (2018) Climate variability impacts on rice production in the Philippines. PLoS ONE 13(8):e0201426. https://doi.org/10.1371/journal.pone.0201426
Suppiah R (1996) Spatial and temporal variations in the relationships between the southern oscillation phenomenon and the rainfall of Sri Lanka. Int J Climatol 16(12):1391–1407. https://doi.org/10.1002/(SICI)1097-0088(199612)16:12%3c1391::AID-JOC94%3e3.0.CO;2-X
Thambyahpillay G (1954) The rainfall rhythm in Ceylon. University Ceylon Rev 12:224–273
Thamilini J, Wekumbura C, Mohotti AJ, Kumara AP, Kudagammana ST, Silva KDRR, Frossard E (2019) Organized homegardens contribute to micronutrient intakes and dietary diversity of rural households in Sri Lanka. Front Sustain Food Syst 3:94. https://doi.org/10.3389/fsufs.2019.00094
Titumil RAM, Basak JK (2010) Agriculture and food security in South Asia. A historical analysis and a long run perspective. The Innovators
Tol RSJ (2005) Adaptation and mitigation: trade-offs in substance and methods. Environ Sci Policy 8(6):572–578. https://doi.org/10.1016/j.envsci.2005.06.011
Vergara BS, Chang TT (1985) The flowering response of the rice plant to photoperiod—a review of the literature. The International Rice Research Institute
Vidhanarachchi LP (2011) Reclamation of degraded lands in Sri Lanka. Godage International Publishers (Pvt) Ltd
Watson M (1986) Soil and climatic requirements. In: Handbook on Tea. Tea Research Institute of Sri Lanka
Weerahewa J, Pushpakumara G, Silva P, Daulagala C, Punyawardena R, Premalal S, Miah G, Roy J, Jana S, Marambe B (2012) Are homegarden ecosystems resilient to climate change? An analysis of the adaptation strategies of homegardeners. APN Sci Bull 2:22–27
Weerakoon WMW, Maruyama A, Ohba K (2008) Impact of humidity on temperature-induced grain sterility in rice (*Oryza sativa* L). J Agron Crop Sci 194(2):135–140. https://doi.org/10.1111/j.1439-037X.2008.00293.x
WHO (2006) Preventing disease through healthy environments. World Health Organisation
Wijesinghe LCA, Gunatilleke IAUN, Jayawardana SDG, Kotagama SW, Gunatilleke CVS (1993) Biological conservation in Sri Lanka: a national status report. The World Conservation Union

William JT, Haq N (2002) Global research on underutilized crops—an assessment of current activities and proposals for enhanced cooperation. Int Cent Underutil Crop

Wimalasiri EM, Jahanshiri E, Suhairi TASTM, Mapa RB, Karunaratne AS, Vidhanarachchi LP, Udayangani H, Nizar NMM, Azam-Ali SN (2020) The first version of nation-wide open 3D soil database for Sri Lanka. Data in Brief 33:106342. https://doi.org/10.1016/j.dib.2020.10634242

Wimalasiri EM, Jahanshiri E, Suhairi TASTM, Udayangani H, Mapa RB, Karunaratne AS, Vidhanarachchi LP, Azam-Ali SN (2020) Basic soil data requirements for process-based crop models as a basis for crop diversification. Sustainability 12(18):7781. https://doi.org/10.3390/su12187781

Yoshida S (1973) Effects of temperature on growth of the rice plant (Oryza sativa L.) in a controlled environment. Soil Sci Plant Nutr 19(4): 299–310

Yoshida S (1983) Rice. In: Smith WH (ed) Symposium on potential productivity of field crops under different environments. The International Rice Research Institute

Zubair L (2002) El Niño–southern oscillation influences on rice production in Sri Lanka. Int J Climatol 22(2):249–260. https://doi.org/10.1002/joc.714

Zubair L, Hansen J, Chandimala J, Siraj M, Siriwardhana M, Tennakoon U, Yahiya Z, Ariyarathne K, Bandara I, Bulathsinhala H, Rajakaruna S, Abeyratne T, Samuel T (2005) Current climate and climate change assessments for coconut and tea plantations in Sri Lanka. IRI, FECT, NRMS and UoP Contribution to AS12 Project Report to be submitted to START

Zubair L, Nissanka S P, Weerasinghe KDN, Punyawardhene BVR, Weerakoon WMW, Wickramagamage P, Agalawatte P, Chandrasekera SC, Delpitiya P, Navaratne CM, Gunaratna J, Herath DI, Herath RM, Karunaratne A,Ratnayake S, Samaratunga P, Sanmuganathan K, Vishwanathan J, Wijekoon, E, Weerasinghe, A K, Yahiya Z,Perera TMRS, Gunesekara, AAI, Ratnasekera H, Ralapanawe V, Seneviratne G, Wallach D (2014) Modeling the impacts of a variable and changing climate on rice agricultural systems in Sri Lanka for AgMIP. Sri Lanka Project Final Report for AgMIP, Technical Report, Foundation for Environment, Climate and Technology, Kandy, Sri Lanka.

A. D. Ampitiyawatta Dr. A.D. Ampitiyawatta is a professor in the Department of Export Agriculture at the Faculty of Agricultural Sciences, Sabaragamuwa University of Sri Lanka. He holds degrees in Agriculture and Hydrology and Water Resources Engineering from Wuhan University, China. His research focuses on the impact of climate change on water resources and agriculture, with a particular interest in climate extremes, adaptation, and mitigation strategies for farming communities.

P. K. Dissanayake Dr. P.K. Dissanayake is a professor in the Department of Export Agriculture at Sabaragamuwa University of Sri Lanka. His research focuses on chlorophyll degradation in horticultural crops, biodiversity, and post-harvest physiology. Collaborating with Yamaguchi University in Japan, Dr. Dissanayake has contributed to breakthroughs in understanding alternative chlorophyll degradation pathways. His interests also extend to insect behavior and biotechnological applications in agriculture.

Eranga M. Wimalasiri Dr. Eranga M. Wimalasiri is a senior lecturer at the Department of Export Agriculture, Sabaragamuwa University of Sri Lanka. He earned his PhD in crop modeling from the University of Nottingham, Malaysia campus. His research focuses on decision support systems for neglected crops, climate change adaptation, and crop modeling. He has been recognized as the best researcher at the Faculty of Agricultural Sciences, Sabaragamuwa University in 2021.

Chapter 4
Impact of Climate Change on Biodiversity and Food Security in Khyber Pakhtunkhwa: Challenges, Vulnerabilities, and Strategies for Sustainable Development

Asif Khan, Muhammad Jawad, Sajid Ali, Asad Ali, Zaheer-ud-din Khan, Waheed Murad, and Mirza Barjees Baig

Abstract Khyber Pakhtunkhwa (KP), located in northern Pakistan, is characterized by its diverse landscapes, historical significance, and rich cultural heritage. The province's economy heavily relies on agriculture, horticulture, livestock, and forestry, making it particularly vulnerable to the impacts of climate change. The region's varied topography, encompassing alpine meadows, dense forests, and arid zones, supports unique ecosystems but also faces significant environmental challenges.

A. Khan
Department of Technology, Universidade Estadual de Maringá, Umuarama, Brazil

M. Jawad
Center for Geographical Information System, University of Punjab Lahore, Lahore, Pakistan

S. Ali
Department of Horticulture and Life Science, Yeungnam University, Gyeongsan, Republic of Korea

A. Ali
Department of Agriculture, Abdul Wali Khan University Mardan, Mardan, Khyber Pakhtunkhwa, Pakistan
e-mail: drasadali@awkum.edu.pk

Z. Khan
Department of Botany, Government College University, Lahore, Pakistan
e-mail: zaheeruddin@gcu.edu.pk

W. Murad (✉)
Department of Botany, Abdul Wali Khan University Mardan, Mardan, Khyber Pakhtunkhwa, Pakistan
e-mail: waheedmurad@awkum.edu.pk

M. B. Baig
Prince Sultan Institute for Environmental, Water and Desert Research, King Saud University, Riyadh, Kingdom of Saudi Arabia
e-mail: mbbaig@ksu.edu.sa

M. Behnassi et al. (eds.), *Food Systems and Biodiversity in the Context of Environmental and Climate Risks*, https://doi.org/10.1007/978-3-031-89167-0_4

Climate change is increasingly affecting KP, influencing crop yields, soil health, and water availability, thus threatening food security and biodiversity. This chapter aims to analyze the threats and causes of climate change in KP, its impact on biodiversity loss, and the implications for food security, while evaluating the interconnections among these issues. It proposes integrated strategies to enhance resilience and sustainable development in the face of climate challenges. The chapter also highlights the province's susceptibility to extreme weather events such as droughts, heat waves, and floods, which endanger ecosystems, agricultural productivity, and community well-being. The loss of biodiversity and habitat degradation further exacerbates these challenges by undermining essential ecosystem services. To address these issues, the chapter advocates for collaborative strategies involving experts across various fields to develop comprehensive plans focused on ecosystem protection and climate change adaptation, ensuring a sustainable future for the region.

Keywords Khyber Pakhtunkhwa · Pakistan · Climate change · Biodiversity · Habitat degradation · Ecosystem services

4.1 Introduction

Khyber Pakhtunkhwa is well known for its diverse landscape, significant historical importance, and richer cultural heritage (Ullah et al. 2023). Furthermore, the province's economy is mostly driven by agriculture, horticulture, livestock, and forest on which most residents are directly and indirectly dependent. In addition, KP is richer in natural resources like precious and semi-precious minerals, hydroelectric potential, and dense forests. However, with its peculiar location, precious natural resources, and agriculture-based economy, KP is one of the most susceptible provinces to the impacts of climate change (Akbar 2020; Saleem and Shrestha 2019). The province's dependence on agriculture makes it particularly sensitive to weather change patterns, while its mountainous terrain makes it susceptible to natural disasters such as floods and landslides (Hussain et al. 2019). Unfortunately, these challenges intensify socioeconomic factors like a high population growth rate, poverty, and a limited framework that makes the region more vulnerable to environmental changes (Zada et al. 2022).

Food security (FS) in KP is closely associated with its agriculture, a sector significantly immune to varied climatic fluctuations (Javed et al. 2023). Since a large part of the population depends directly on agriculture for their livelihood, ensuring reliable food production and access is crucial. However, environmental change influences net crop yields, soil richness, and water accessibility, and may also decrease the nutritional quality of food, posing an additional burden on public health (Imran 2013).

Biodiversity in KP is an important feature of natural resources and is vital for ecosystem services such as soil fertility, water purification, and pollination. However, the region's diverse ecosystem also supports several precious species, many of which are endemic and under serious threat due to climate change, anthropogenic activities, and habitat loss (Khan et al. 2021a, b, c). However, it is quite important to consider that this decline in biodiversity threatens resilience and stability, further intensifying the challenges created by food insecurity and climate change (Rehman et al. 2022; Zada et al. 2022).

Addressing the interconnections among climate change, biodiversity, and food security is essential for the sustainable development of KP. Implementing various integrated approaches that prefer to consider the synergies and trade-offs between these complications would help enhance adaptive capacity and awareness, build resilience, and enhance the long-term well-being of the KP population and ecosystem, accordingly. The objectives of this chapter are to: (1) gain comprehensive insights into the causes and threats of climate change and its profound effects on biodiversity; (2) assess the vulnerabilities of food security within this context; (3) explore the intricate interconnections between climate change, biodiversity loss, and food security in KP; and (4) propose integrated strategies to build resilience and promote sustainable development, while protecting both biodiversity and mitigating food losses.

4.1.1 Khyber Pakhtunkhwa: An Overview

Khyber Pakhtunkhwa is located in the North West of Pakistan, at 34.9526° N latitude and 72.3311° E longitude, covering an area of approximately 101, 741 km^2 (Fig. 4.1). The province has a population of around 40.86 million as per Gallup Pakistan Analysis of Census reports, 2023 (Nasir 2024) with quite diverse ethnic groups. Each of these groups has its interesting languages and cultural practices (Khan et al. 2021a, b, c). The strategic location underscores its importance for trade, for example, on the west and north sides, it shares borders with Afghanistan, followed by Baluchistan toward the south, Azad Jammu and Kashmir at the east, and Baluchistan and Punjab on the south sides, respectively (Farooq et al. 2021).

A few important pieces of information are listed as,

- Total land area of KP is 101,741 km^2;
- Major revisers in KP are Swat, Indus, Kabul, and Chitral;
- Elevation ranges from 300 m in the Southern plains to over 7000 m in the Northern mountainous regions;
- A total of nine languages are spoken (Pashto is major among all, 75%);
- Climate: Varies from semi-arid in the south to alpine in the north.

Interestingly, the region is characterized by mountain ranges, rolling submontane areas, and plains, all surrounded by hills, with the northern Hindu Kush area renowned for its exotic beauty. The highest peak of the northern Hindu Kush is Tirich Mir famous for its unique and natural attractiveness. Moreover, from the extraordinary peaks of the Hindu Kush and Karakoram Mountain ranges in the north to the

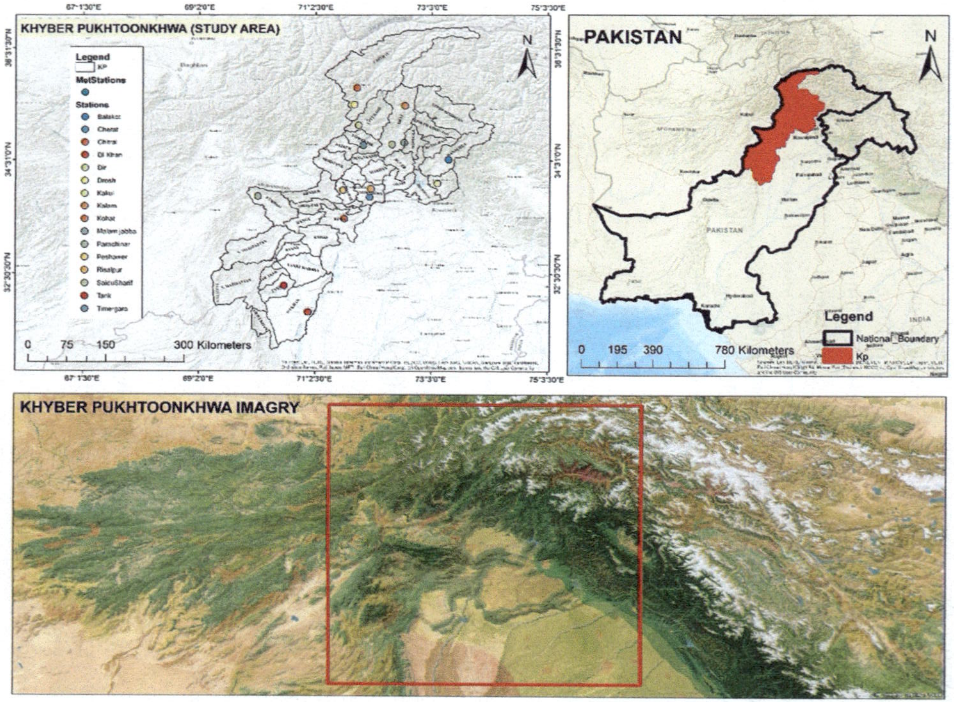

Fig. 4.1 Map of Khyber Pakhtunkhwa (study area), along with a map of Pakistan and general imagery of KP, is presented. The location of KP is depicted by districts, with district borders represented by black lines and legends provided for clarity. In the Pakistan map, KP is highlighted in red, while the study area within KP is marked by a red rectangle. The maps were designed using ARC-GIS software, version 10.8

productive field of the Peshawar Valley in the south, the variety of KP is quite critical. This varied topography is home to a wide range of ecosystems, including alpine meadows, dense forests, and arid zones, each supporting a unique flora and fauna, respectively (Zada et al. 2021).

4.2 Climate Change

Climate change refers to long-term changes in temperature, atmospheric conditions, and precipitation patterns on the surface of the Earth, driven by anthropogenic activities (Dore 2005). Examples of these activities include deforestation, rapid industrialization, and the burning of fossil fuels that significantly increase the concentration of greenhouse gases (GHGs) in the atmosphere, leading to global warming and following changes in weather patterns. The manifestations of climate change include rising sea levels, global temperature increases, melting of ice caps, and severe extreme weather patterns such as droughts and storms that directly disrupt ecosystems and biodiversity (Hoffmann and Spekat 2021). South Asian countries, including Pakistan, are among the areas severely affected by the climate risks (Ali and Erenstein 2017).

Unfortunately, Pakistan's climate vulnerability has been observed to increase over time, despite its negligible contribution to global GHG emissions. For example, Pakistan ranked 12th in 2012, 8th in 2015 and recently ranked 7th, among the top countries most directly exposed to the vagaries of climate change and global warming (Ahmed et al. 2020; Salik et al. 2015). Recent climate-related disasters, including droughts, floods, and storms, have increasingly impacted Pakistan, causing widespread damage and affecting millions of people, animals, plants, and beneficial microorganisms. Despite the rising frequency and severity of these events, the country's response has remained inadequate, leaving the population and its resources vulnerable to such natural hazards (Mumtaz et al. 2019).

4.2.1 Climate of Khyber Pakhtunkhwa

KP is a topographically diverse province of Pakistan and is located in the north-west region of the country (Fig. 4.2). KP is home to Hindukush, Karakoram, and Himalayan ranges predominantly in the northern, northern-west, and eastern regions of the province (Ashraf et al. 2021). On the other hand, southern regions are dominated by central plain valleys that encompass agriculture and rangelands. The southern region has a wide spectrum of extremely hot weather conditions, during summer, with less harsh winters and moderate weather (Munawar and Udelhoven 2020). In contrast, the northern region of the province experiences harsh winters with heavy snowfall, abundant rainfall, and mild, pleasant summers. Since Chitral, the highest district in KP, has the lowest winter temperatures, there are many glaciers,

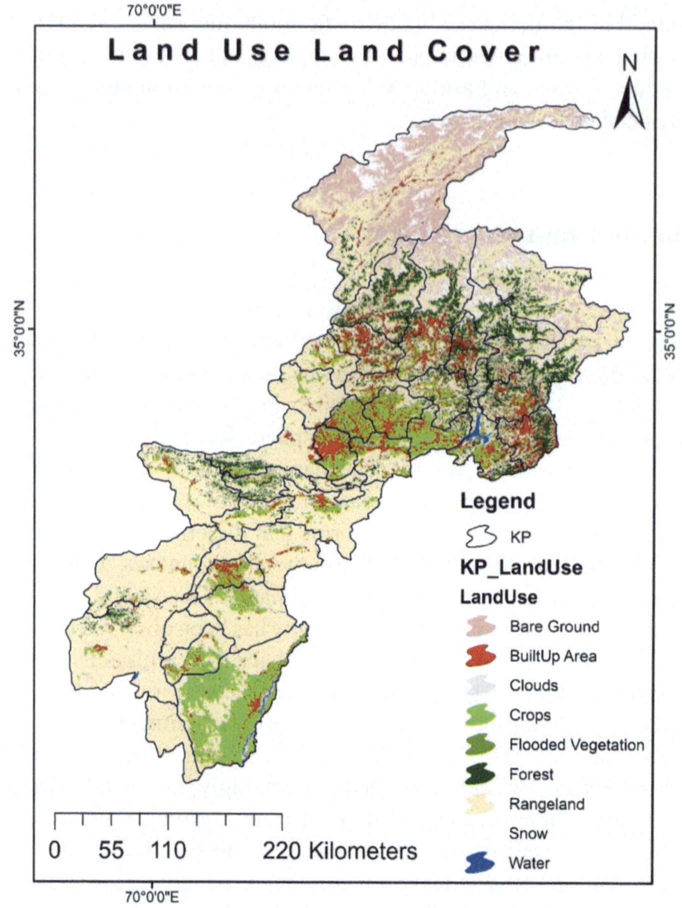

Fig. 4.2 Land use and land cover (LULC) model of Khyber Pakhtunkhwa was developed using ARC-GIS software (version 10.8). The map visually represents different land uses such as forests, agriculture, water bodies, and urban areas, with clear boundaries marked for each district. Color coding and legends are used to differentiate between land types, providing an overview of the region's spatial land use patterns

while Dera Ismael Khan is the most southern region with moderate winter (Gul et al. 2019; Khan and Baig 2020).

4.2.2 Climatic Zones of Khyber Pakhtunkhwa

The comprehensive analysis of climate, i.e., rainfall, temperature, altitude, and topography in the environmental profile of KP has been developed by the Environmental Protection Agency (Nazeer 2012). These regional classifications serve as a critical tool in understanding and predicting the region's potential vulnerability to other external factors such as climate change. However, by this categorization of KP to different zones, becomes easier and more interesting to assess a particular impact of environmental changes on each zone, allowing more effective and targeted adaptation strategies (Khan et al. 2019a, b). Understanding the type of impact each zone is prone to requires knowledge of land cover and land use of each area, accordingly. The land cover in KP varies significantly from north to south, leading to diverse patterns of land use. KP contains 40% of Pakistan's forests, with 17% of its area covered with forests, primarily in the Hindukush Himalayan region (Hussain and Bangash 2017). Additionally, 28% of KP's land is rangelands, 8% is committed to agricultural practices, and 13% is covered with snow and ice glaciers. KP is divided into four agro-natural zones, each zone can have overlapping areas, such as agriculture, forests, water, and biodiversity (Fig. 4.2 and Table 4.1) (POLICY 2016).

4.2.3 *Effect of Climate Change on Khyber Pakhtunkhwa*

The effects of climate change are already evident in various parts of the world. The Intergovernmental Panel on Climate Change (IPCC) acknowledged the significant link between climate change and its impacts on natural and human systems in their latest Assessment Report 5 (AR5) published in 2014 (Yousaf et al. 2023). Furthermore, there is increasing evidence that various natural disasters are linked to climate change, including altered hydrological systems, shifts in species diversity due to rerouted migration paths, changes in crop yields affecting food production, and weather patterns that disrupt and accelerate glacier melting. The data indicate that heat waves will become more common and last longer in Asia; there will be more days and nights with warmth instead of cold (Ikram et al. 2016; Yousaf et al. 2023).

On the globe, KP is located in the mid-latitude, where the possibilities of changes toward severe weather patterns are expected (Woodward et al. 2014). It is also speculated that as temperatures rise, monsoon rains will rise and move further to the north side, accordingly. Moreover, increased precipitation is expected in wet areas (like Zone B's subhumid wet mountains), and decreased precipitation is expected in dry areas (like Zone C and D's Central Valley and Piedmont Plains). In dry places, less precipitation will lead to droughts, while these high precipitation patterns can accelerate glacier melting and cause flash floods (Iqbal et al. 2022).

KP is in a sensitive situation due to its climate patterns and diverse topology, including low-fertile lands and high-altitude mountain ranges such as the Hindu Kush (Fig. 4.3). The region is increasingly vulnerable to extreme weather patterns such as drought, heat waves, and heavy floods that could endanger millions of people's lives, biodiversity, natural resources, and agricultural lands and products (Ali 2014). The loss of biodiversity and habitat degradation due to climate change further undermines, directly and indirectly, ecosystem services, impacting agriculture and community well-being. Addressing these challenges requires a collaborative strategy involving

Table 4.1 Different agroecological zones of KP with districts and other important factors of each zone

Zones	Districts	Description	Elevation range (m)	Average annual rainfall (mm)	Average temperature range (°C)	Dominant vegetation type	Key environmental issues	Natural disaster vulnerability
A	Buner, Shangla, Dir (lower and upper), Swat, Chitral	Higher northern mountains and northern mountains	800–4500	1000–2000	May-20	Coniferous forests	Deforestation and soil erosion	High flood and landslide risks
B	Haripur, Batagram, Mansehra, Abbottabad, Kohistan, and Torghar	Subhumid eastern mountains and wet mountain	500–3000	1200–2500	Oct-25	Mixed forests and alpine meadows	Habitat fragmentation, deforestation, and soil erosion	Prone to landslides and moderate flood risk
C	Peshawar, Mardan, Charsadda, Nowshera, Swabi, Kohat, and Hangu	Central Valley Plain	300–1000	500–1000	15–30	Scrubland and cultivated land	Water scarcity and urban sprawl	Moderate flood risk and urbanization
D	Bannu, Karak, Lakki Marwat, Tank and D.I. Khan	Piedmont plain and Suleiman piedmont	200–800	300–600	20–35	Arid zones and sparse vegetation	Desertification and water scarcity	High drought risk and urbanization

experts in various fields, including ecology, politics, geology, agriculture, climatology, botany, forest, and zoology, to develop comprehensive plans focusing on ecosystem protection and risk management for climate change (Ullah et al. 2024).

It was considered that the average temperature of different Met stations varied in time intervals, i.e., 1991, 2001, 2011, and 2018 revealing an increasing trend (Fig. 3c).

4.2.4 Causes of Climate Change in Khyber Pakhtunkhwa

4.2.4.1 Deforestation

According to Pakistan's government officials, around 4.2 million hectares (4.8%) of the country's total land area are covered by forests (Tahir and Khaliq 2018). However, KP is 1.21 million hectares higher than other provinces (i.e., Baluchistan 0.33, Sindh 0.92, and Punjab 0.69%, respectively) of Pakistan. Interestingly, KP hosts a significant portion of Pakistan's forests (40%) (Khan et al. 2019a, b). The forests found in the KP province are mostly confined to the Hindukush, Karakorum, and Himalayan mountains. In addition, Pakistani forests exhibit various animal and plant species due to their adaptable physiographic and unique climatic conditions. Numerous forest types can be distinguished in Pakistan, including 'tropical thorn forest', swamp and littoral forests, subtropical broad leaves forests (evergreen), subtropical pine forests, tropical dry deciduous forests, Himalayan moist temperate forests, Himalayan dry temperate forests, subalpine forests, etc. However, among all, conifer forests are the dominant overall (Rauf et al. 2019).

Unfortunately, every year, more than 39,000 hectares of forests in Pakistan disappear; for example, in 1990 and 2000, the annual deforestation rate in Pakistan was 1.5%. Notwithstanding Pakistan's limited forest resources, the country's forests contribute significantly to economic development and are essential for maintaining ecological balance, controlling stream flow, protecting water supplies, and conserving soil. In KP's hilly regions, most people, directly and indirectly, depend on forests, i.e., fuel, lumber, medicines, and food (Forward 2018; Siddiqui et al. 2018). The KP's forests are under threat due to climate change impacts. As climate changes, plants and animal species are forced to adapt their habitats by altering their environmental niches (Schwitzer et al. 2011). The species found in forests have been shown to adapt to the severe occurrences caused by climate change by changing their habitat and ascending to higher heights. It has been considered that a temperature increase of 3 °C could result in a 500 mm shift in the elevation of forest zones (Tang et al. 2018). Therefore, the KP forests, which are mainly coniferous and range in elevation from 1000 to 4000 m, will continue to develop at higher altitudes. Additionally, the researcher found that a temperature increase of 1 °C may result in a 30 times greater chance of wildfires (Cruz et al. 2022).

Additionally, excessive wood harvesting and overgrazing worsen the effects of climate change on KP forests. According to estimates, the consumption of fuel food

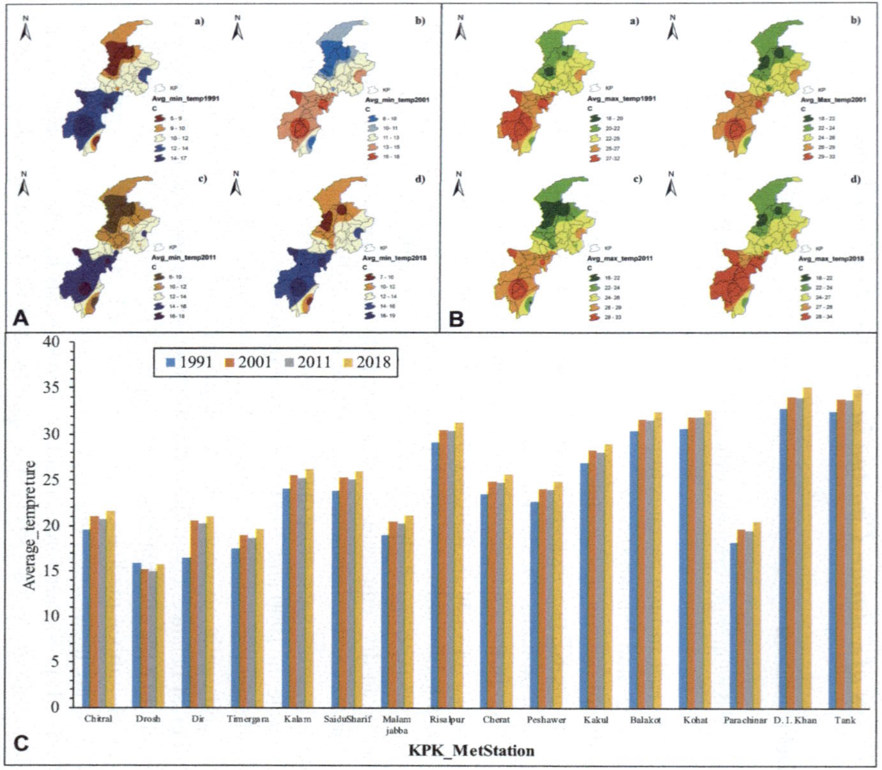

Fig. 4.3 Climatic scenarios in KP are illustrated through two maps: **A** shows the average minimum temperature for the years **a** 1991, **b** 2001, **c** 2011, and **d** 2018. Similarly, **B** depicts the average maximum temperature for the same years. The comparison highlights changes in temperature across different meteorological stations in KP. The maps were generated using ARC-GIS software (version 10.8) on 20/08/2024

is three times higher than reforestation, while the intensity of grazing is 6.5 times higher than the forest's ability to produce fodder. KP's forests would suffer severely from both overexploitation and climate change (Khattack 1992; Munir and Munir 2023).

4.2.4.2 Flood

A complex network of rivers drains water in the Induce River that flows through KP. Similarly, the Kabul River, Swat River, Kurram River, Gomal River, and Zhob River are the main rivers that cross in the KP area (Kakakhel 2018). Among all, the Indus River has the potential of 154 million metric tons of water and its tributaries can hold yearly and are made up of 9 million metric tons from eastern rivers (mostly Ravi and Sutlej) and 145 million metric tons from northwestern rivers (the Indus, Jhelum, Chenab, Kabul, and their neighboring minor rivers) (Rahman et al. 2022). These river networks make the area susceptible to active flash floods during the intense monsoon seasons. However, in recent years, the frequency of heavy flooding has increased, especially in the KP region (Rahman et al. 2021). Similarly, the area is flooded every year as a result of heavy rain; from July to August the glaciers start melting and in heavy rain, the Kohistan (northern side of the KP) faces severe flooding. For example, in 2007 the flood was intense; however, a more severe situation was seen in 2010 in KP (Khan 2013).

According to OCHA statistics, 4,725,695 individuals in KP were affected by the 2010 flood and migrated to other districts and their homes, crops, and other agricultural and horticulture resources were run off with the flood. Interestingly, more intense patterns were taken into consideration in the next years, i.e., 2011, followed by 2012, 2013, 2014, and 2015, accordingly (Sawa 2023). Due to these events, the conditions became much more difficult, challenging and costly, as a result of years of flooding, with efforts to recover from earlier floods still ongoing (Shehzad 2023). According to the (Lee and Markham 2016) the PDMA, one of the worst floods ever recorded is the 2015 flood, which damaged 1200 acres of standing crops and affected 321,644 people in Chitral (Maan and Jameel 2021).

The Himalayan wet temperature forests, located between 1525 and 3660 m, are expected to remain the primary flood-prone areas in KP. However, that area is likely to get a lot of precipitation, especially in the southern sections of Pakistan—for example, Galiat and the lower Kaghan Valley. As a result, flooding is a major concern in Zones A and B of KP; therefore, it is crucial to take precautions and adjust to severe weather events (Shaheen et al. 2012).

4.2.4.3 Drought

Drought can be defined as a 'prolonged period of unusually low rainfall'. This condition causes several problems such as water scarcity, low agricultural and horticultural production, high migration of humans and animals from one area to another, and negative impacts on ecosystems (Shen et al. 2007). Additionally, drought disrupts the natural water cycle, depletes soil moisture, and places significant stress on wildlife habitats. Thus, substantial financial losses, food insecurity, and land degradation are the main consequences of drought. Moreover, understanding these consequences is essential to maintaining the ecological balance and advancing sustainable development (Clark et al. 2016).

Droughts are more prevalent in the southern regions (Zones C and D) than in Zones A and B in the north. The areas primarily used for agriculture include the Suleiman Piedmont, Piedmont Plain, and Central Plain Valley, in Zones C and D. In general, most of the area is used for agriculture, however, Zones C and D combined have a cultivated area of 1,115,000 ha, whereas Zones A and B have a cultivable area of 626,000 ha. Zones C and D are expected to also experience water stress in the future due to rising temperatures and decreasing rainfall in arid areas. For example, Zone C experiences a warm and humid environment, whereas Zone D experiences a hot and dry climate, accordingly. The region that would be most vulnerable to droughts is Dera Ismail Khan (Fahim et al. 2016; Hina and Saleem 2019; Mahmood et al. 2023).

The province's Zones C and D have drier winters, warmer summers, and less rainfall. According to climate predictions, with time, the rainy regions will become wetter and the dry parts will become dryer. There will likely be an increase in average summer temperatures because these zones are hot in the summer and dry in the winter. This extreme weather may adapt to these warmer conditions; however, high temperatures also increase evapotranspiration, increasing the need for water from crops. Thus, drought in the region will be largely determined by the amount and accessibility of irrigation and groundwater (Ali 2014; Ayub et al. 2021).

4.2.4.4 Agriculture

Pakistan generates just 0.43% of GHG emissions, while it is among the countries most vulnerable to the consequences of climate change. With rising temperatures, longer summers, droughts, and an excess of precipitation, climate change is predicted to pose serious problems to the KP's agriculture sector (Michel Devadoss et al. 2021). Agriculture production in KP and throughout Pakistan would also decline as a result of climate change. According to Ali and Iqbal, around 2/3 of the 11% of people employed in agriculture would be impacted by the fall in the productivity of the industry. For example, major crops including food grains, sugar cane, and fruits would suffer from the rise in temperatures caused by climate change. However, in addition to the challenges posed by seasonal changes in crop development, variations in the weathering process can also alter the characteristics and attributes of the soil (Iqbal and Ahmad 2005). According to (Sivakumar and Stefanski's (2011) prediction, there would be a 5.7% decrease in wheat productivity in Pakistan for every 1-degree Celsius increase in temperature (Sivakumar and Stefanski 2011). Although it is expected that, by 2080, the growing season for crops will be shortened due to early plant maturity caused by high temperatures in other crops, production will also be negatively affected 2017.

Since KP also produces corn and other cereals (Shah et al. 2020), however, its productivity will decrease as crop yields increase at moderate temperatures; higher temperatures cause yields to decrease (Hashim 2020; Sonia 2019). Therefore, as a result of climate change, KP is likely to see a shortfall of grain crops in the future (Abid et al. 2014). As a result of climate change, plants generally experience an increase in temperature that affects their development of roots and shoots, nutrient absorption, and the ratio of carbon (CO_2) to nitrogen (N_2) in their tissues (Shah and Khalil 2017). Consequently, a negligible shift in temperature will cause plants to lessen their yield and later on, will start to disappear and might result in desertification (Rashid and Chung 2017). In addition, rising temperatures will attract more pests and accelerate the development of crop diseases, both of which would have an impact on Pakistan's food security (Miller et al. 2021; Shah and Khalil 2017).

4.2.4.5 Soil Erosion

Soil erosion is one of the key factors that leads to global soil degradation (Panagos et al. 2016). According to IPCC research, soil erosion is exacerbated and made worse by the effects of land use and climate change (Ma et al. 2021). Due to different soil features, soil erosion, biodiversity degradation, and the disruption of biogeochemical cycles pose a main risk to various ecological processes, soil biochemistry, and physiological conditions. Furthermore, the IPBES 2018 global assessment report on land degradation and restoration (IPBES) revealed that preventing land degradation is a top priority in fulfilling Sustainable Development Goal 15 (Life on Land) (Bertrand et al. 2015; Fisher et al. 2018).

Due to the strong association between climate change and the hydrological cycle, any type of changes in the climate will have direct and indirect impacts on water resources and water-related issues by modifying the hydrological cycle both temporally and geographically. Similarly, the water cycle will also be disrupted by rising temperatures, which will have an impact on evapotranspiration (Kundzewicz 2008). Variations in temperature and precipitation dramatically impact the production and movement of sediments. Numerous studies show that the rate of discharge (Menzel and Bürger 2002), sediment flux (Jiongxin 2003), and soil erosion may be significantly impacted by climate change (Williams et al. 2002).

Globally, Asia holds the highest land degradation rate, with 663 million hectares affected primarily by soil erosion. Additionally, 76% of Pakistan's land is susceptible to erosion by wind and water in South Asia. However, more than 11 Mha of the 23 Mha of agricultural land is subjected to soil erosion (Barman et al. 2020). For example, the Mangla and Terbela dams have been swollen due to the yearly erosion of the Indus River System of about 40 million tons of soil in the upper catchment areas of the Hindu Kush, Himalayas, and Karakoram, respectively (Khan et al. 2014).

The storage capacity of most dams is expected to decrease by 50% by 2050 (Chuenchum et al. 2023). At first, the main Asian dams had a total storage capacity of 1450 × 109 m^3, but siltation in their reservoirs has caused this to decrease by 0.8% per year. The initial storage capacity of most Asian dams will be reduced by more than 40% due to siltation. The availability of agricultural water and the production of hydropower could be negatively impacted by this problem. Almost all of the storage capacity of the Warsak dam (KP, Pakistan) has been depleted due to silting within the first three years of operation (Sabir et al. 2013).

4.3 Biodiversity

The term biodiversity refers to 'the variety of all living things and their interactions on the Earth' (Jordano 2016). According to the 'United Nations Convention on Biological Diversity (UNCBD)', the diversity of genes, ecosystems, and the ecological processes that sustain them are included (Dobson 2005). However, according to UNEP (2004), biodiversity may alternatively be defined as the average abundance of the original species at a particular period as opposed to its original abundance when humans had little influence over ecosystems. Currently, the world is experiencing a very serious threat to biodiversity due to an increase in anthropogenic activities, including climate change and habitat fragmentation. Many species have lost large-scale populations due to climate change; habitat loss and habitat fragmentation can even cause a greater drop in abundance than would be predicted from habitat loss alone (Koellner et al. 2013; Navarro-Perez and Tidball 2012).

4.3.1 Impact of Climate Change on Biodiversity

2006 Climate change can be defined as 'the variability in climate properties like changes in its average state, standard deviations and the occurrence of extreme events that go beyond individual climate events on spatial and temporal scales and persist for a long period, typically decades or more (IPCC 2006; Stocker et al. 2013).

The IPCC (2006) identified key indicators of climate change, including a 32% increase in atmospheric CO_2 concentrations and a 151% rise in CH4 concentrations between 1750 and 2000 (IPCC 2006). Additionally, global average temperatures have risen by approximately 0.6 °C over the past 100 years. Precipitation patterns have also shifted, with a 5–10% increase in most mid- and high-latitude regions of the Northern Hemisphere and up to a 3% decrease in many subtropical areas. Furthermore, there has been a 10% decline in snow and ice cover across the Northern Hemisphere, particularly in North America, Europe, and Asia (Solomon 2007; Zhao et al. 2012).

Climate change projections for the twenty-first century demonstrate that existing climate-related risks will amplify and create new threats to natural systems (Fig. 4.4). Some of these risks are limited to a certain sector and/or region and will have cascading effects. Examples include the significant reduction in surface and underground water resources and the increased threat of extinction of terrestrial and water

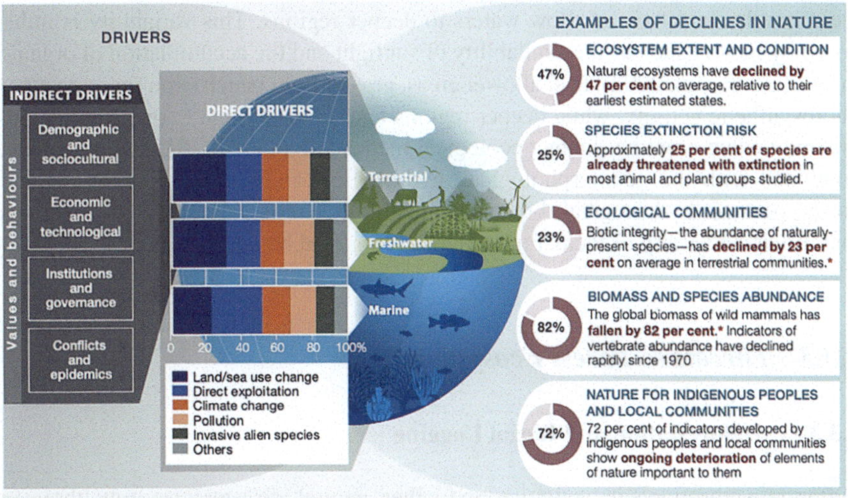

Fig. 4.4 Global loss of biodiversity and key drivers. *This figure is sourced from the IPBES global assessment report on biodiversity and ecosystem services. Source* Kuiper (2023)

species. In the case of wetlands, a high risk of change on a regional scale is expected, with an abrupt and irreversible impact on the composition, structure and function of ecosystems (Field et al. 2014).

Climate change has enormous effects on the distribution and interaction of species (Carpenter et al. 2007). Similarly, in humid areas, its direct effects can be distinguished, such as an increase in temperature and CO_2 concentration in the atmosphere and indirect effects, such as regional and global changes in precipitation regimes and their impacts on evaporation, radiation, wind speed, and changes in hydrological regimes (Acreman et al. 2009). There is interference at different scales of biological organization (genetic, population, species, communities, and ecosystems) and spatial location (habitat, location, regional, and continental) as a result of its highly variable character (Van der Putten et al. 2010). The disturbance occurs in almost all aspects of people's lives. Organisms are associated with the climatic niche, an aspect of the environmental niche, defined by limits of climatic variation that guarantee the positive growth rate of organisms as long as they are contained therein. The link takes place within complex ecological relationships where climate elements (temperature and precipitation) are involved (Pearman et al. 2014).

4.3.2 Biodiversity of Khyber Pakhtunkhwa

The transitional nature and diverse location of floodable regions across the landscape creates a wide range of environments within a relatively narrow surface strip (Cantonati et al. 2020). These habitats vary significantly, from the headwaters of rivers to their mouths, and from shallow waters to deeper regions. This variability is influenced by factors such as the availability of sunlight and the accumulation of organic matter (Gomi et al. 2002). In shallower areas, greater light penetration supports more photosynthetic activity, while deeper regions tend to have higher concentrations of organic matter, creating distinct ecological niches (Khan et al. 2015). Biodiversity, combined with the geomorphological, hydrological and soil structures of humid zones, perform hydrological, biogeochemical and ecological ecosystem functions that provide a variety of local, regional, or even global benefits (Li 2011).

4.3.3 Threats to Forest Ecosystem

4.3.3.1 Deforestation and Illegal Logging

Different anthropogenic activities and other natural disasters severely threaten Pakistan's forest ecosystem, especially in KP. Due to these threats, biodiversity, local livelihood, and climate resilience are at risk. Among all, deforestation is one of the major hazards resulting from the irregular cutting of precious forest trees for different purposes, such as fuel, timber, medicines, and feed (Khalid et al. 2020). According to the PFI (Pakistan Forest Institute), there has been a substantial decline in the rate and amount of forest cover area due to illegal logging, especially in the Dir, Buner, Swat, and Chitral. Moreover, these forests are also a habitat for numerous organisms, and deforestation worsens their livelihood and unfortunately decreases and/or exterminates their populations. In the same way, deforestation causes soil erosion, which directly and indirectly increases the probability and frequency of land sliding and flooding. For example, an interesting scenario was taken into account in the 2010 floods in KP that caused extensive distractions (ul Haq 2022).

4.3.3.2 Population Growth and Urbanization

These wooded regions are further encroached on by population increase and the ensuing urbanization, which includes the construction of more roads and housing infrastructure. Unchecked urban growth in places such as Abbottabad has caused forest ecosystems to fragment, making it harder for animals to survive. Similarly, the conversion of forests into arable lands due to the expansion agricultural production in formerly wild regions—especially in the fertile valleys of KP—has reduced the area's capacity to absorb carbon dioxide naturally and changed local microclimates (Uddin et al. 2023). The expansion of roads, housing developments, gardens, and parks, combined with population growth and urbanization, is steadily encroaching upon these forested areas (Alam et al. 2024; Basit et al. 2018).

4.3.3.3 Overgrazing and Invasive Species

Another serious problem is overgrazing by cattle in Pakistan, and particularly in KP's rangelands. The KP areas include those of Kohistan, Dir, Swat, and Chitral, respectively. In the same way, due to the degrading effect of this activity, forests are more vulnerable to invading species and find it more difficult to regrow. For example, according to some studies, the area is being overtaken by alien species such as *Parthenium heterosporous* L., *Lantana camara* L., *Xanthium strumarium* L., *Eichhornia crassipes* (Mart.), *Imperata cylindrica* (L.), *Euphorbia helioscopia* L., and *Ageratum conyzoides* L., which displace local flora and modify the natural equilibrium of forest ecosystems (Junaid et al. 2010; Perveen and Khan 2015).

4.3.3.4 Community Conflict and Forest Fires

One of the main obstacles to sustainable forest management is disagreement between communities over ownership and tenure. These disputes have undermined conservation efforts in some regions, such as the Hazara division, Buner, Dir (upper and lower), Shangla, Swat, Chitral, Malakand, etc., by causing unchecked exploitation of forest resources (Yaseen et al. 2018). According to the KP Forest Department's yearly reports, both naturally occurring and man-made forest fires, which are frequently started on purpose to make way for agriculture or as a result of carelessness, annually burn enormous tracts of forest. Unfortunately, these fires persist for weeks and even months due to inadequate facilities and the government's inability to promptly control them, resulting in the loss of a significant number of valuable trees, shrubs, birds, and other wildlife (Jabeen et al. 2022).

4.3.3.5 Agriculture Expansion and Infrastructure Development in Forest Areas

The expansion of agricultural land and other building infrastructures like hotels, parks, houses, gardens, etc. in KP are the main causes of deforestation and serious deterioration of ecosystems (Woldearegay 2013). The rapid expansion of these infrastructures, driven by economic demands and population growth, is causing a staggering rate of forest destruction in KP, as land is cleared to establish various centers for agricultural activities such as animal farming and crop production. This destruction is seen in the lower mountainous and plane areas; however, there is an increasing tendency toward the removal of valuable trees in the higher areas too. For example, the removal of large tracts of forest for agriculture in the swat, Dir, Chitral, Buner, and Hazara divisions of KP has caused significant habitat loss and soil degradation (Khan et al. 2019a, b; Rauf et al. 2019).

Similarly, the problem of deforestation is also worsened by infrastructure development, which includes building homes, roads, and hydropower facilities (Munro 2009). It is now more difficult for wildlife to control and survive as a result of habitat fragmentation caused by the development of the Swat Expressway, which passes through deep forests and various mountainous regions. In addition to upsetting the surrounding environment, this road construction allows unlawful logging and land conversion to occur in previously inaccessible locations. The construction of roads through wooded regions can have a cascading impact, with initial clearance for infrastructure usually resulting in more deforestation and changes in land use, according to a World Wildlife Fund (WWF) analysis (Zada et al. 2021). The impact of road development on forest ecosystems is demonstrated through a case study from the Kaghan Valley. Due to the increased agricultural encroachment, illegal logging, human settlement, and hotels brought about by road construction, this region has lost important habitats for species like the western Tragopan, an endangered native bird. Along with increasing soil erosion and the likelihood of landslides during the monsoon season, this has also affected the hydrological balance in the area (Goheer et al. 2023).

4.3.3.6 Crude Methods for Collecting Medicinal Plants

The traditional and unregulated collection of medicinal plants in the KP seriously threatens the biodiversity and sustainability of some valuable plant species in the region (ABBAS et al. 2024). In KP, particularly in remote areas, people collect precious and unique medicinal plants, such as *Berberis lycium* (Royal), *Viola serpens* Wall., and *Valeriana wallichii* DC for their treatments (especially, healings) (Begum et al. 2022). Very often, these plants are removed following unsustainable methods such as uprooting the entire plant, rather than collecting a specific tissue such as root, leaves, fruit, stem, and flowers. These practices also cause a decrease in plant populations, threatening the survival of already endangered species (Khan et al. 2022). For example, the natural occurrence and population of a medicinal plant known as *Podophyllum hexandrum* Royal. Has been sharply reduced in the swat and Dir districts due to the inappropriate and over-harvesting of this plant. However, since this plant grows in specific regions, such as high-altitude areas, over-harvesting has caused this plant to be listed as endangered. Additionally, the local inhabitants of these areas, who rely on such precious plants both for medicinal purposes and livelihoods, are negatively impacted by over-harvesting (Khan et al. 2021a, b, c; Shaheen et al. 2017).

4.3.4 Threats to Aquatic Ecosystems

Fish are not only the most significant, diverse, and vital biotic component of an ecosystem, but they also directly and indirectly contribute to human well-being. They are a rich source of essential vitamins, proteins, and minerals, making them an important part of a nutritious diet. Moreover, fishing is vital as it provides job opportunities and income to people around the world (Prakash 2021). The global fisheries industry, which supports 10–12% of the world population, is projected to decline by 35% in 2050 due to climate change (Lam et al. 2016). Similarly, approximately 20.9% of people were associated with agriculture in 2014–2015 and 26.0% did so in 1997–1998 (Spielman et al. 2016). Agriculture is a key sector in Pakistan, employing over 43.5% of the workforce, particularly in rural areas, while the forestry and fisheries sectors employ 2.0% and 2.1% of workers, respectively (Chandio et al. 2021). Like other departments in Pakistan, the fisheries sector also plays a key role in the economy of emerging nations. Pakistan's coastline spans more than 300,270 km^2, and the nation receives a substantial amount of its food, employment, and income from the fishing sector and fisheries. The KP Department of Fisheries has also a key role in the sustainable management, preservation, and usage of province resources, accordingly. Several species of fish, both native and invasive, were found to be abundant across KP's diverse aquatic environments, including lakes, dams, rivers, and reservoirs. To maintain local livelihoods and food security, the department endeavors to protect these habitats, increase fish productivity, and promote sustainable fishing (Javed and Abbas 2018; Tagar et al. 2021).

Pakistan's freshwater habitats are home to more than 180 different fish species, like catfish, carp, loaches, etc. Among all, 28 species of fish exist only in fresh water in the northern regions of KP. Most trout were found to be confined in the Trans-Himalayan regions of the Indus system. Similarly, Pakistan's freshwater fish fauna consists of 179 species, classified into 26 families, 10 orders, 5 suborders, 82 genera, and 3 cohorts, according to recent updates (Rafique and Khan 2012). Preliminary investigation, from March 2013 to February 2016, of water bodies in the Hazara division, KP demonstrate that five rivers, including Indus, Harro, Dor, Siran, and Kunar, were considered for fish collection. A total of 33 fish species were identified by comparing them with published records/data in various databases. It was found that these fishes have 6 orders, 10 families, 22 genera, and 30 species. The research conducted in this study revealed that the Cyprinidae family was the most diverse, with 16 species represented. The Bagridae family included 3 species, while the Nemacheilidae, Siluridae, and Salmonidae families had 2 species each. Additionally, the Schilbeidae, Channidae, Mastacembelidae, and Cichlidae families collectively accounted for a total of 3 species (Usman et al. 2016). A total of 18 fish species were identified belonging to 3 orders and 3 families, respectively, from the river Swat. Moreover, one of the most famous game fish 'Mahseer and Schizothoracines' is increasingly rare due to the disappearance of the breading sites in the lake and are submerged by dams/dams like Ghazi Barotha, Tarbela, etc. (Akhtar et al. 2014). In a similar context, Mirza (2007) also reported 39 different types of fish in the lower Swat River. Currently, the Kabul River and its tributaries are home to more than 54 different species of fish. Because it shares its ichthyofauna with the Kabul River, the lower portion of the River has a greater concentration of fish fauna than the upper half of the river.

4.3.4.1 Illegal, Overfishing, and Unrecommended Ways of Fishing

Unsustainable and illicit fishing methods pose serious risks to aquatic ecosystems and wildlife. It has been shown that certain native fishes like Mahsheer (*Tor putitora*), carp (*Cyprinus carpio*), trout (*Oncorhynchus mykiss),* catfish (*Glyptosternum reticulatum*), etc. found in the KP's rivers and lakes are severely reduced due to overfishing, as a result of the growing demand for fish in local markets (Akhtar et al. 2015). Similarly, destructive fishing methods, like the use of dynamites, electric shocks, chemicals/poisons, explosives, and nets, are found to worsen the situation. In addition, such methods indiscriminately destroy the ecological balance of water bodies and other aquatic life, including endangered species, in addition to killing the target fish species. For example, dynamite fishing has been prohibited in the Swat River, which seriously harmed the fish population and more specifically affected certain species like *Schizothrax* and many more. On the other hand, by using certain chemicals and/or poisons, it remains in the water and speared along the water flow killing every type of fishlike eggs and small-size fish communities. Additionally, when fish habitats and breeding grounds are destroyed once, it might take decades for the environment to recover from the long-term harm caused by these practices (Ahmad 2020; Mahmood et al. 2021).

4.3.4.2 Water Pollution/Contamination

Due to rapid industrialization, infrastructure development, improper waste management, and lack of authority instructions serious hazards jeopardize aquatic ecosystems (Muhammad et al. 2023). The presence of contaminants such as heavy metals, industrial waste, raw sewage, and chemical effluents in lakes and rivers degrades water quality, leading to the death of many aquatic species, while the survivors are forced to migrate to other areas. In addition to the harm that water pollution does to fish, the health of communities relying on these sources for drinking and irrigation is also threatened (Gul et al. 2014; Sanchez-Triana et al. 2014).

4.3.4.3 Impact of Climate Change on Aquatic Fauna

Among others, climate change is also a key factor putting the aquatic habitat in the KP at greater risk. It was taken into account that the river and lake levels are drastically reduced as a result of low precipitation caused by climate change (Hayat et al. 2021). Due to this, their habitats are becoming even smaller and much more concentrated, resulting in a decline in water volume; hence, the fish population is becoming more vulnerable to overfishing and other pressures. Furthermore, the thermal regimes of water bodies are changing as a result of climate change, which may affect the survival and reproduction of cold-water fish species such as trout. For example, the decrease in water flows in the Chitral and Swat rivers has recently led to a decline in fish populations (Israr et al. 2016; Maryam et al. 2014).

4.3.4.4 Introduction of Exotic Fish Species

In Pakistan, fisheries are one of the main sectors with the quickest growth rate and represent a significant contribution to the country's economy. To improve its qualitative and quantitative production, fast-growing species, disease resistance, and less consumable food are the preferred species on the market (Shah 2011). Unfortunately, naturally occurring fish species do not meet these desired traits. To boost profits, the fisheries department and farmers often consider introducing exotic species that have a proven track record of higher productivity within a shorter time frame. Despite the introduction of several fish species, only a few of them persist in the waterbodies, occur naturally, and spread beyond their native range. Certain newly imported species have the potential to become invasive if they damage or compete with native fish, as well as other wildlife and plants, and they have the potential to spread far (Ullah and Khan 2022). The introduction of exotic fishes into the KP's water bodies seriously impacts the native fish populations. These indigenous species are in danger of declining and/or becoming extinct due to the introduction of non-native species that frequently outcompete them for resources like food, minerals, and habitat, respectively.

Similarly, some native fish species have already been affected by the introduction of common carp (*Cyprinus carpio*) into Khyber Pakhtunkhwa (KP) waterbodies, distressing the natural equilibrium (Jamshed et al. 2014). For example, two invasive fish species, namely brown trout (*Salmo trutta*) and rainbow trout (*Oncorhynchus mykiss),* were introduced to some districts of KPK. Similarly, the introduction of various exotic fish species in Pakistan during the 1960s, such as Mozambique tilapia (*Oreochromis mossambicus*), common carp (*Cyprinus carpio*), goldfish (*Carassius auratus),* and grass carp (*Ctenopharyngodon idella*), was intended to increase fish production and manage unwanted weeds (Moazzam & Osmany). However, these fish species quickly became invasive and negatively impacted the local fish fauna. Although grass carp was restored in the 1980s to improve aquaculture productivity, other species of fish were added, including silver carp (*Hypophthalmichthys molitrix*), bighead carp (H. nobilis), Nile tilapia (*O. niloticus*), and blue tilapia (*Oreochromis aureus*). Each of these species has also established itself in the natural ecology, influencing other animals (Bilal and Khan (n.d.); Said et al. 2022).

4.3.4.5 Threat to the Agrobiodiversity Ecosystem

The concept of agrobiodiversity in KP is important for the maintenance and flexibility of agriculture systems and guarantees food security for the local population in the region (POLICY 2016). However, the diverse agroecological zones support numerous traditional crop breeding, livestock, and farming approaches that are suitable for the area and inspire sustainable farming practices. Several key factors are taken under consideration, for example, abrupt usage of monoculture crops, climate change, and unsustainable agricultural practices. However, maintaining and improving agrobiodiversity is essential for the adaptability and resistance to pests and diseases and ensures good quality and diverse food for the upcoming generations. The adoption of some sustainable agricultural practices that ensure traditional crop varieties is mandatory for the preservation of this dynamic biodiversity (Amir Khatam et al. 2014).

4.3.4.6 Introduction of New Species (Hybrids), Hybrid Seeds, and GMOs

The introduction of genetically modified organisms (GMOs) and other hybrid seeds and species has greatly influenced traditional farming approaches. On the other hand, such developments have significantly improved agriculture production and also strengthened resistance to some diseases and pests, sparking reservations regarding the ecosystem complications and ultimately loss of biodiversity (Khan et al. 2022). For now, different maize varieties that are more suitable for the region and have specific genetic characteristics have been evacuated as consequences of hybrid farming in the swat valley. Moreover, discussions regarding the long-term impacts of GMOs on nontarget products, soil fertility, and health have been triggered by the introduction of Bt cotton. Similarly, such crops become less resistant to pests and environmental stresses when they depend too heavily on a small number of hybrid species and show lower genetic diversity, accordingly (Rana 2014; Yazdani and Ali 2017).

Over the past few decades, plant biotechnology has played a crucial role in addressing environmental challenges such as climate change, drought, and salt stress, helping to meet crop production and nutritional needs. Certain gene regulations (down/up) affect different pathways linked to tolerance to abiotic stress at the molecular level, such as cell signaling, the elimination of hazardous chemicals and free radicals, and the preservation of proteins and membranes (Altman and Hasegawa 2011). Research on these stress-responsive genes and pathways is an efficient tool because it allows the use of specific genetic modifications to confer stress tolerance on crops that are significant for agriculture and the environment. Examples of various stress-tolerant GMOs are typically used for horticultural and agronomic crops such as tobacco, maize, wheat, rice, mustard, sugarcane, soybean, cotton, banana, citrus, grapes, eggplant, and potato (Upadhyay et al. 2024).

4.3.4.7 Water Scarcity/Drought in Khyber Pakhtunkhwa

Among other environmental stresses, drought is a serious issue that affects the whole world, but the situation is getting worse, particularly in developing countries such as Pakistan (Xie et al. 2013). Currently, one in five people lacks access to clean drinking water. The world's current population will likely increase to 8.3 billion by 2025, bringing the issue of water shortage to a peak level. Similarly, developing countries such as Pakistan are experiencing increasing water shortages, due to huge losses in the economy of different forms of life, i.e., livestock crops, human beings, forests, etc. (Rajan and Verma 2017).

KP is located the northern region of Pakistan, with a peculiar topology, i.e., extending from the southern to the northern region with diverse climatic conditions. Similarly, the impact of water stress on biodiversity, i.e., agriculture, horticulture, forest, and food security will be more severe (Olivieri 2021). Generally, KP makes up 6.77% of Pakistan's total cultivated land area and rain-fed agriculture is the primary method used on more than half of the province's cultivatable territory, according to the Planning and Development Department (Zaidi 2020). However, lowering KP's water supply might have detrimental consequences for both KP and the surrounding regions. This may have detrimental effects on the home, industrial, livestock, forests, and agricultural sectors, which could lead to revenue losses and encourage migration. Furthermore, improving the management of water resources has the potential to boost agricultural land revenue and alleviate water scarcity (Awan and Abro 2021). Droughts occur more frequently, especially in the southern regions of KP than in the northern ones. For example, agricultural land consists of the Central Valley Plain, followed by Piedmont Plain and Suleiman Piedmont, respectively. It was found that KP has around 1.65 million acres of land available for cultivation (Khan et al. 2021a, b, c). This illustrates how much a drought may affect the province's ability to secure food. In the future, arid regions, especially those in the south, will probably experience more severe water stress, due to rising temperatures and declining rainfall. However, the plain valley has a warm and subhumid climate, whereas the southern regions have hot and dry conditions. It is interesting to note that D. I. Khan is the region most likely to experience droughts than other districts in KP (Awan and Abro 2021).

4.3.4.8 Climate Change and Variable Precipitation Rate

Climate change is considered to be one of the key environmental stresses that have caused extreme weather events, like variations in humidity, temperature, and precipitation, resulting in important economic losses. The environment and society are now faced with difficulty as a result of alterations in the hydrological cycles. One of the main hydroclimatic variables is precipitation, which has an immediate impact on ecosystems, floods, droughts, and water supplies, respectively (Babar et al. 2014). Due to climate change and diverse topology, the majority of the KP areas are affected by flooding due to their vulnerability to flash floods during the intense monsoon season because of these river systems. KP province has seen an increase in the frequency of floods in recent years. The area floods almost annually as a result of intense rains (Ali Shah et al. 2022).

Kohistan experienced severe flooding in 2007 as a result of glacier melting and heavy rain. The greatest floods in recent memory hit KP in 2010, forcing hundreds of thousands of people to flee their homes. The pattern was seen again in the next years (2011, 2012, 2013, 2014, and 2015); however, much less intensely. Years of flooding coupled with the ongoing repair of previous floods made the situation even more difficult and expensive (Israr et al. 2016). For example, on July 29, 2010, Pakistan experienced a catastrophic flood due to excessive rain in several parts, especially in the northern regions (Bajwar, Dir, Swat, Kalam, Chitral, and Buner) of KP (Fig. 4.5). It was estimated that approximately 2000 people died, and 700,000 houses were damaged. Similarly, in April 2009, the biggest rainfall of 187 mm was exceeded by the record-breaking 274 mm that fell in Peshawar in 24 h. On the contrary, from 1998 to 2001, there was a severe drought in Pakistan's central and southern regions of Pakistan (Khan 2013; Shah et al. 2023).

4.3.4.9 Unauthorized and Excessive Use of Agrochemicals

The abrupt usage of agrochemicals such as herbicides, pesticides, and insecticides have detrimental effects on soil health, biodiversity, and human health. Bees, beneficial pollinators, and predators that are vital in pollination have been reported to have decreased significantly due to unauthorized and excessive use of pesticides in KP, poisoning the water sources and soil (Ali et al. 2023). It was found that the bee population in the Hazara region is significantly declining due to the overuse of pesticides in apple orchards. Furthermore, certain pests showed greater resistance to pesticides, which also amplified due to the increased accumulation of these poison chemicals in the environment, generating a more vicious cycle of dependency on stronger and more effective chemical treatments (Jaspal and Khan 2017; Ullah and Nawab 2019).

4.4 Food Security: An Overview

Food security refers to 'the availability, accessibility of adequate and nutritionally sufficient food for each member of the home to live a healthy life in a socially acceptable way' (Capone et al. 2014). Food scarcity is a serious issue that affects everyone and everywhere; however, it is particularly worsened in underdeveloped countries (Malaysia 2010). Like other issues, food insecurity has also become a global problem due to the uncontrollable growing population, especially those undernourished (Hwalla et al. 2016). Currently, approximately 842 million people and/or nearly 12% of the entire world population are affected, especially in developing countries (Estrada et al. 2020). For example, Asia and Africa are home to more than 92% of the world's undernourished population, that is, 552 and 226.6 million, respectively. Among all, in South Asia alone, 294.7 million people face food insecurity, representing more than 35% of the global under-nutrition population (Eini-Zinab et al. 2020; Standl et al. 2019).

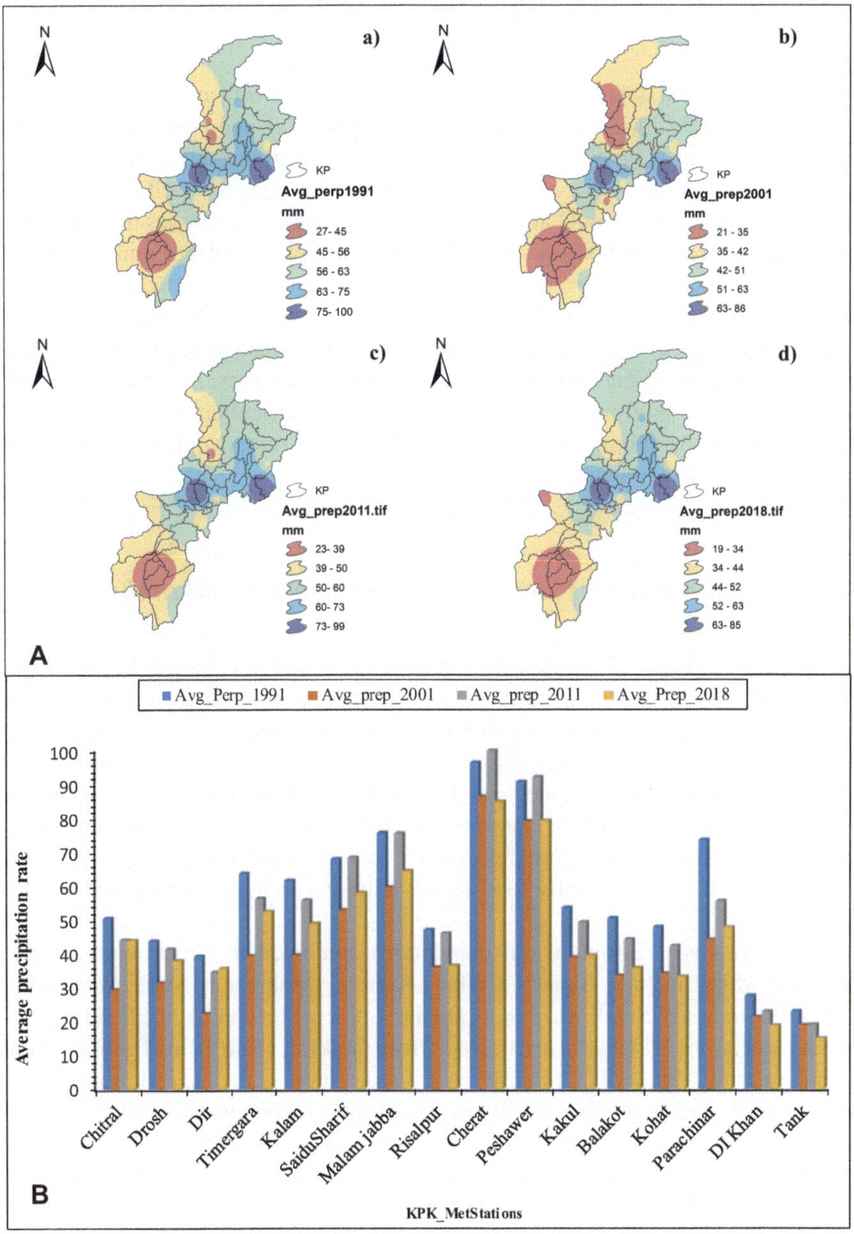

Fig. 4.5 Figure illustrates changes in precipitation rates across KP over time. Panel **A** shows the average precipitation for the years **a** 1991, **b** 2001, **c** 2011, and **d** 2018, with a comparison of precipitation levels at various meteorological stations across the province. The map was generated using ARC-GIS software (version 10.8)

4.4.1 *Food Security in Pakistan*

Among other South Asian countries, Pakistan is a country with an average per capita income of US$1512 in 2015. Pakistan's government executed a variety of contemporary and scientific procedures and strategies to support the agricultural and food industries. In Pakistan, agriculture accounts for about 20.88% of GDP and 43.5% of employment (Ahmad and Farooq 2010). The agriculture sector has mediated and unmediated links to other sectors and plays an important role in the country's socio-economic development. Currently, Pakistan is the sixth most populated country, with an estimated population of 191.71 million. In 2015, around 116.52 million people lived in rural regions and 75.19 million people lived in urban areas. The rural population was noted to decrease from 61.4% to 60.8% per year, however, the urban population increased from 38.5% to 39.2% by 2015. Agriculture is the only source that provides a living option directly and indirectly for rural people (Iqbal and Arif 2010; Shahzad and Amjad 2022).

4.4.2 *Food Security: Current State in Khyber Pakhtunkhwa*

KP is facing a serious food security challenge as it imports subsidized food from other provinces to meet demand because it cannot produce enough food to meet its own needs (Ahmad and Farooq 2010). To close the gap between the supply and demand for standard food, it is highlighted that increasing agricultural yields through the use of cutting-edge instruments and technology is mandatory. According to the integrated food security phase classification, KP is categorized as moderately food insecure, with four districts classified as very food insecure, ten as moderately food insecure, and eleven as usually food secure (Eini-Zinab et al. 2020; Hashmi et al. 2021).

Recently, seven districts (rural population) combined in KP province. These districts, which include Bajaur, Khyber, Kurram, Mohmand, North Waziristan, Orakzai agency, and South Waziristan, are the subject of this IPC Acute Food Insecurity investigation. All these districts, aside from Orakzai, share a direct geographical boundary with Afghanistan; almost every district surveyed is located in a rain-fed area. In total, 0.36 million people are in IPC Phase 4 (Emergency) and 1.08 million individuals are in IPC Phase 3 (Crisis). Furthermore, 1.92 million individuals are classified as stressed in IPC Phase 2. Each of the seven districts that were examined falls within IPC Phase 3 (Crisis). Between 25 and 30% of the people in the districts of Bajaur, Khyber, Mohmand, North Waziristan, and South Waziristan reside in Phase 3 and/or higher, whereas Kurram and Orakzai hold 40%, respectively (Ramay 2014; Rasheed et al. 2022).

4.4.3 Climate Change, Food Security, and Challenges Across Khyber Pakhtunkhwa

In KP, most agricultural activities occur in the central valley and the southern regions, and due to climate change, these areas are likely affected. As temperatures increase, precipitation becomes more erratic and weather patterns change, which result in changes in crop variety and quantity (Ramay 2014). Additionally, viral infections, weed growth, and insect attacks are predicted to increase due to the anticipated changes in climate. Additionally, farmers have to deal with diminishing water supplies and extremely high temperatures that are detrimental to crop growth. Furthermore, due to the warmer, more humid climate that attracts insects, they frequently deal with insect infestation using expensive chemical sprays. The majority of crops, including wheat, maize and rice, can be grown in the north due to rising temperatures, while the southern Piedmont region and the Central Valley Plain, which is currently the primary agricultural region, experience a shortage of water as a result of falling rainfall, which reduces crop yield and quality. These alterations, in addition to natural disasters such as droughts and floods that devastate agricultural fields and ready-to-harvest crops, worsen the problem of food security in the province (Akbar et al. 2020; Hussain et al. 2020).

4.4.3.1 Agriculture

In KP, agriculture is one of the main sources of income for rural households, especially rural households. However, food production has suffered in some flood-affected areas including Mohmand, Bajaur, Charsadda, Swat, Dir, Chitral, Khyber, and North Waziristan as a result of the localized flooding (Hussain et al. 2022). However, the majority of small landholding farmers produce small amounts of crops for subsistence, due to water shortage, quality seeds, proper guidelines, and chemical sprays. KP is experiencing a significant reduction in annual wheat production. It was taken into account that approximately 25% of the agricultural households reported a decrease in the agricultural land used for the planting of wheat, which is the main primary staple crop (Arshad and Shafqat 2012).

4.4.3.2 Livestock

In KP, livestock is one of the most important resources for rural households, which is raised both for household consumption and as a source of income. However, due to 2022 monsoon rains and flooding, livestock has also been impacted in some areas of Bajaur, Swat, Charsadda, Mohmand, Nowshehra, Khyber, Banu, D.I. Khan, and North Waziristan (Shahbaz et al. 2012). The four most common types of livestock owned by households, according to the household assessment, are cattle (69%), goats (23%), poultry (5%), and sheep (3%), respectively. It was found that approximately

21% of the livestock owners reported the death of their primary livestock, 50% of the households reported the death of their secondary livestock, and 59% of the households reported the death of other livestock (Hussain et al. 2022).

The three primary causes of deaths are serious diseases of the livestock, a lack of feed or fodder, and a restricted supply of drinking water for the animals (Nieto et al. 2012). It was observed that 14% of households sold livestock in distress to pay for food and other necessities; similarly, 8% reported selling their animals in distress due to poor health, 4% reported selling their animals in distress owing to a lack of food, 1% reported selling their animals in distress owing to a lack of drinking water, and 16% reported selling their animals normally to make a living (Shahbaz et al. 2012).

4.4.3.3 Wildlife and Biodiversity Losses

The diversity of livestock species is also threatened by climate change. Among 90% of Pakistan's 100 endemic species, the majority are located in the province of KP (Nawab et al. 2018). The Himalayan range and subhumid forests of KP are home to several mammal species, seven bird species, and 12 migratory and endemic species of birds that are internationally endangered. Consequently, mountainous areas are more susceptible to biodiversity loss caused by climate change (Braaksma and Tadema 2022). Furthermore, ecosystems that serve as habitats for a variety of forest birds, mammals, and insect species are impacted by variations in temperature and precipitation patterns. Numerous plant species are unable to the swift changes in climate. Additionally, alterations in ecosystems can result in modifications to animal feeding habits, which may weaken animals and ultimately lead to direct loss of biodiversity (Aslam et al. 2022; Junaid et al. 2010).

4.4.3.4 Fisheries

In KP, the majority of fish reside in lakes, rivers, and natural ponds. Additionally, the fisheries department runs hatcheries, which are essentially ponds used to raise trout fish. In addition, the fisheries department is in charge of offering extension services to commercial and public fish farming enterprises. Fishing is primarily a recreational and athletic activity, especially when it comes to trout (Khan and Baig 2020). There is competition between native and exotic fish species in the area. As a bioindicator of the aquatic ecosystem, trout fish, for example, are much more sensitive to variations in the water's temperature, geographical aspects of the area, turbidity, and quantity of water as well as dissolved oxygen. On the contrary, rising temperatures, fluctuating turbidity, and pollutants in the water body are harmful to it. Variations in their rivers and other bodies of water flow are the result of climate change, which directly and indirectly affects the quality and quantity (Aslam et al. 2022).

4.5 Conclusions

KP's topography is diverse and relates to the problem posed by climate change, which significantly affects food security and biodiversity. Climate change intensifies the loss of species, frequent and severe flood conditions, danger to wild animals and plants, deforestation and fluctuating weather or climate change, as a result of the ecosystem's instability. The interactions among climate change, biodiversity, and food security require interventions that may be immediate and synchronized. As agriculture continues to be the main source of livelihood for KP's population, these interrelated disasters must be combatted in a coordinated and holistic manner if future stability in the region is to be achieved. Similarly, to enhance the sustainable gains of the ecosystem and people living in KP, there is a need for cross-cutting climate change adaptation plans that consider both the positive synergies and negative impacts of mitigation measures related to climate change, biodiversity, and food security improvement. With the same motto, reducing emissions and managing climate change risks must be accompanied by encouraging and supporting sustainable agricultural practices and proper management of the environment to reduce the negative impacts of climate change that currently threaten populations. Moreover, education and awareness, the enhancement of innovation and capacity building for better resource management of local communities are critical.

To manage identified challenges, policymakers, researchers, and practitioners need to take the following actions to develop policies promoting sustainable development and climate resilience. It is therefore important to extend the collaboration to the local level as well as to the international one regarding the mobilization of resources and the exchange of knowledge. The governments should expand on integrating climate adaptation measures, rehabilitation of the environment and sustainable farming to safeguard the local people. Furthermore, there is a need for backing KP through funding, technical support, and technology transfer from the international community to strengthen the local adaptive capacities.

Therefore, in the future, it is recommended to expand upon recent studies included herein and incorporate research conducted elsewhere that has been published since 2020 to identify newer and more advanced methodologies. Other reports, including (Organization et al. 2022) established the need for climate resilience, especially in areas prone to climatic and socio-economic shocks including KP. Absorption of these findings will inform policies to help adjust to future climate situations in the region. To effectively assist KP, particular interventions and frameworks to address emerging challenges need to be created using available literature. Policymakers/researchers and community engagements are vital for developing resilience and preserving future resources such as food biodiversity.

Recommendations to Minimize Food Inflation

- The local inhabitants, particularly, the farming households need agricultural support from the government sector. This might be in the form of cash or food assistance, seeds, fertilizers, veterinary services, animal feed, veterinary inputs, agricultural loans, pesticides, irrigation water access, tools and animal restocking to increase crop and livestock production in the next three to six months.

- Initiate livelihood support interventions and recovery initiatives that help households affected by various shocks, including monsoon flooding, water scarcity, and/or prolonged droughts and price hikes.
- Give subsistence farmers in particular access to high-yielding crop and vegetable seeds as well as agricultural implements in a timely way.
- Establish educational institutions to educate farmers about financing options for new crops, technology, and equipment that will be needed in their areas due to climate change scenarios.
- Improve research facilities at specific agricultural research institutes to estimate agricultural productivity and evaluate the effects of climate change.
- Construct an efficient agriculture extension system to create a strong and timely communication channel among farmers and government agencies.
- To mitigate slope degradation, it is crucial to avoid cultivating water-intensive crops in these areas. It is important to raise awareness of gender-specific considerations in the use and storage of pesticides, herbicides, insecticides, seeds, and crops. The government needs to invest in infrastructure that will allow the use of genetic engineering and biotechnology to raise livestock and enhance breeds and types resistant to droughts.
- Encourage the use of urea, molasses, vitamins, and minerals to create multinutrient blocks (MNB) supplements that will improve the quality of feed for cattle.

References

Abbas A, Sajid M, Yar S, Khan A, Rahman AU, Hamayun M, Ali S, Moon YS, Alrefaei AF (2024) Phytochemical profiling and multifaceted biological activities of *Persicaria Hydropiper* L.: a promising medicinal plant. Pak J Bot 56(4):1547–1557

Abid S, Raza I, Khalil A, Khan MN, Anwar S, Masood MA (2014) Trend analysis and forecasting of maize area and production in Khyber Pakhtunkhwa, Pakistan. Eur Acad Res 2(4):4653–4664

Acreman M, Blake J, Booker D, Harding R, Reynard N, Mountford J, Stratford C (2009) A simple framework for evaluating regional wetland ecohydrological response to climate change with case studies from Great Britain. Ecohydrol Ecosyst Land Water Process Interact Ecohydrogeomorphology 2(1):1–17

Ahmad M (2020) Ichthyofaunal diversity and conservation status in rivers of Khyber Pakhtunkhwa, Pakistan. Proc Int Acad Ecol Environ Sci 10(4):131

Ahmad M, Farooq U (2010) The state of food security in Pakistan: future challenges and coping strategies. Pak Dev Rev 903–923

Ahmed W, Tan Q, Shaikh GM, Waqas H, Kanasro NA, Ali S, Solangi YA (2020) Assessing and prioritizing the climate change policy objectives for sustainable development in Pakistan. Symmetry 12(8):1203

Akbar M, Niaz R, Amjad M (2020) Determinants of households' food insecurity with severity dimensions in Pakistan: varying estimates using partial proportional odds model. Health Soc Care Community 28(5):1698–1709

Akbar G (2020). Problems and potential of agriculture for improving livelihood in Malakand division, Pakistan

Akhtar N, Khan S, Saeed K (2014) Exploring the Fish Fauna of River Swat, Khyber Pakhtunkhwa, Pakistan. World J Fish Mar Sci 6(2):190–194

Akhtar N, Saeed K, Khan S, Rafiq N (2015) Exploring the Icthyofaunal Diversity of River Chagharzi District Buner, Khyber Pakhtunkhwa Pakistan. World 7(4):228–236

Alam N, Ullah Z, Ahmad B, Ali A, Syed K (2024) Population growth poses a significant threat to forest ecosystems: a case study from the Hindukush-Himalayas of Pakistan. bioRxiv 2024.2004. 2001.587657

Ali A, Erenstein O (2017) Assessing farmer use of climate change adaptation practices and impacts on food security and poverty in Pakistan. Clim Risk Manag 16:183–194

Ali N, Kalsoom, Khan S, Li G, Ali M, Nazneen S, Ali L, Li Y, Samiullah, Ihsanullah (2023) Exposure to multiple pesticides in drinking water and potential health risks: a case study of selected districts from Khyber Pakhtunkhwa Province, Pakistan. Environ Earth Sci 82(20):460

Ali Shah I, Muhammad Z, Khan H (2022) Impact of climate change on spatiotemporal variations in the vegetation cover and hydrology of district Nowshera. J Water Clim Chang 13(11):3867–3882

Ali Z (2014) Spatio-temporal analysis of drought in northern area and KPK

Altman A, Hasegawa PM (2011) Plant biotechnology and agriculture: prospects for the 21st century. Academic press

Amir Khatam AK, Sher Muhammad SM, Chaudhry KM, Mumtaz Khan MK (2014) Impact of farmer field schools on skill development of farming community in Khyber Pakhtunkhwa Province, Pakistan

Arshad S, Shafqat A (2012) Food security indicators, distribution and techniques for agriculture sustainability in Pakistan. Int J Appl Sci Technol 2(5)

Ashraf A, Iqbal MB, Mustafa N, Naz R, Ahmad B (2021) Prevalent risk of glacial lake outburst flood hazard in the Hindu Kush–Karakoram–Himalaya region of Pakistan. Environ Earth Sci 80(12):451

Aslam S, Siddiqui S, Ullah U, Manzoor U, Lateef T, Samreen N, Nasir P, Khan S, Noor L, Ghalib SA (2022) Vertebrate wildlife of Pakistan: a review. Canadian J Pure Appl Sci 16(2):5483–5495

Awan NW, Abro AA (2021) Do environmental degradation and agricultural accessories impact on agricultural crops and land revenue? Evidence from Pakistan. Sarhad J Agric 37

Ayub S, Huang S, Raza SH (2021) Effect of climatic vulnerabilities on sugarcane production in Punjab, Sindh and KPK: a case study from Pakistan. J Agric Res (03681157):59(3)

Babar S, Amin A, Azam M, Khan AQ (2014) Empirical assessment of region-specific climate impact on crops production in Khyber Pukhtunkhawa, Pakistan. Int Postgrad Bus J (IPBJ) 6(2):45–63

Barman BK, Rao KS, Sonowal K, Prasad N, Sahoo UK (2020) Soil erosion assessment using revised universal soil loss equation model and geo-spatial technology: a case study of upper Tuirial river basin, Mizoram, India. AIMS Geosciences 6(4):525–545

Basit M, Khurshid SK, Ali A (2018) Trends and patterns of population growth in Pakistan

Begum HA, Hamayun M, Khan A, Yaseem T, Bussmann RW, Murad W (2022) Quantitative ethnobotanical appraisal of medicinal plants used by indigenous communities of District Malakand, Pakistan. Ethnobot Res Appl 24:1–14

Bertrand JC, Bonin P, Caumette P, Gattuso JP, Grégori G, Guyoneaud R, Le Roux X, Matheron R, Poly F (2015) Biogeochemical cycles. Environ Microbiol Fundam Appl Microb Ecol 511–617

Bilal W, Khan AM (n.d.) Pros and cons of alien fish introductions: a case scenario from Pakistan

Braaksma L, Tadema B (2022) Country report: The Netherlands: two developments: standing in environmental matters and environmental (impact) assessments for general rules. IUCN AEL J Environ Law 12:176–183

Cantonati M, Poikane S, Pringle CM, Stevens LE, Turak E, Heino J, Richardson JS, Bolpagni R, Borrini A, Cid N (2020) Characteristics, main impacts, and stewardship of natural and artificial freshwater environments: consequences for biodiversity conservation. Water 12(1):260

Capone R, El Bilali H, Debs P, Cardone G, Driouech N (2014) Food system sustainability and food security: connecting the dots

Carpenter SR, Benson BJ, Biggs R, Chipman JW, Foley JA, Golding SA, Hammer RB, Hanson PC, Johnson PT, Kamarainen AM (2007) Understanding regional change: a comparison of two lake districts. Bioscience 57(4):323–335

Chandio TA, Khan MN, Muhammad MT, Yalcinkaya O, Wasim AA, Kayis AF (2021) Fluoride and arsenic contamination in drinking water due to mining activities and its impact on local area population. Environ Sci Pollut Res 28:2355–2368

Chuenchum P, Xu M, Tang W (2023) Assessment of reservoir trapping efficiency and hydropower production under future projections of sedimentation in Lancang-Mekong River Basin. Renew Sustain Energy Rev 184:113510

Clark JS, Iverson L, Woodall CW, Allen CD, Bell DM, Bragg DC, D'Amato AW, Davis FW, Hersh MH, Ibanez I (2016) The impacts of increasing drought on forest dynamics, structure, and biodiversity in the United States. Glob Change Biol 22(7):2329–2352

Cruz MG, Alexander ME, Fernandes PM (2022) Evidence for lack of a fuel effect on forest and shrub land fire rates of spread under elevated fire danger conditions: implications for modelling and management. Int J Wildland Fire 31(5):471–479

Dobson A (2005) Monitoring global rates of biodiversity change: challenges that arise in meeting the Convention on Biological Diversity (CBD) 2010 goals. Philos Trans R Soc B Biol Sci 360(1454):229–241

Dore MH (2005) Climate change and changes in global precipitation patterns: what do we know? Environ Int 31(8):1167–1181

Eini-Zinab H, Edalati S, Sobhani S, Kezabi M, Hosseini S (2020) Undernourishment trends and determinants: an ecological study of 76 countries. Public Health 186:230–239

Estrada A, Garber PA, Chaudhary A (2020) Current and future trends in socio-economic, demographic and governance factors affecting global primate conservation. PeerJ 8:e9816

Fahim AM, Ali SM, Runping S, Zhang J (2016) Characteristics of drought variation in winter using drought Indices during the period 1971-2010: a case study of Khyber Pakhtunkhwa (Pakistan). Mausam 67(3):697–708

Farooq H, Naazer MA, Kafeel M (2021) The geostrategic significance of Hazara region in Khyber Pakhtunkhwa, Pakistan. SHAJAR 3(1):28–44

Field CD, Dise NB, Payne RJ, Britton AJ, Emmett BA, Helliwell RC, Hughes S, Jones L, Lees S, Leake JR (2014) The role of nitrogen deposition in widespread plant community change across semi-natural habitats. Ecosystems 17:864–877

Fisher J, Scholes R, Montanarella L (2018) Assessment report on land degradation and restoration of the Intergovernmental Science-Policy Platform on Biodiversity and Ecosystem Services

Forward W (2018) 18 forestry and range. In: Developing sustainable agriculture in Pakistan, p 359

Goheer MA, Fatima L, Farah H, Hassan SS, Abbas N (2023) Assessment of change in forests land, carbon stock and carbon emissions of KPK, Pakistan for past three decades using geospatial techniques. J Water Clim Chang 14(2):442–453

Gomi T, Sidle RC, Richardson JS (2002) Understanding processes and downstream linkages of headwater systems: headwaters differ from downstream reaches by their close coupling to hill-slope processes, more temporal and spatial variation, and their need for different means of protection from land use. Bioscience 52(10):905–916

Gul N, Shah MT, Khan S, Muhammad S (2014) Quantification of the heavy metals in the agricultural soils of Mardan District, Khyber Pakhtunkhwa, Pakistan. J Glob Innov Agric Soc Sci 2:158–162

Gul F, Jan D, Ashfaq M (2019) Assessing the socio-economic impact of climate change on wheat production in Khyber Pakhtunkhwa, Pakistan. Environ Sci Pollut Res 26:6576–6585

Hashim M (2020) Impact of climate change on major food crops productivity in Khyber Pakhtunkhwa: a regional analysis Quaid-i-Azam University Islamabad

Hashmi HA, Belgacem AO, Behnassi M, Javed K, Baig MB (2021) Impacts of climate change on livestock and related food security implications—overview of the situation in Pakistan and policy recommendations. In: Emerging challenges to food production and security in Asia, middle east, and Africa: climate risks and resource scarcity, pp 197–239

Hayat K, Khan A, Bibi F, Murad W, Fu Y, Batiha GES, Alqarni M, Khan A, Al-Harrasi A (2021) Effect of cadmium and copper exposure on growth, physio-chemicals and medicinal properties of *Cajanus cajan* L. (Pigeon pea). Metabolites 11(11):769

Hina S, Saleem F (2019) Historical analysis (1981–2017) of drought severity and magnitude over a predominantly arid region of Pakistan. Climate Res 78(3):189–204

Hoffmann P, Spekat A (2021) Identification of possible dynamical drivers for long-term changes in temperature and rainfall patterns over Europe. Theoret Appl Climatol 143(1):177–191

Hussain A, Bangash R (2017) Impact of climate change on crops' productivity across selected agro-ecological zones in Pakistan. Pak Dev Rev 56(2):163–187

Hussain H, Hussain S, Nisar M, Zubair S (2020) Economic analysis of food security in Peshawar, Pakistan. Lib Arts Soc Sci Int J (LASSIJ) 4(1):149–160

Hussain MA, Zhang S, Muneer M, Moawwez MA, Kamran M, Ahmed E (2022) Assessing and mapping spatial variation characteristics of natural hazards in Pakistan. Land 12(1):140

Hussain A, Mahapatra B, Rasul G (2019) Adaptation in mountain agriculture: Food security in the Hindu-Kush Himalayan (HKH) region. In: Status of climate change adaptation in Asia and the Pacific, pp 211–236

Hwalla N, El Labban S, Bahn RA (2016) Nutrition security is an integral component of food security. Front Life Sci 9(3):167–172

Ikram F, Afzaal M, Bukhari S., Ahmed B (2016) Past and future trends in frequency of heavy rainfall events over Pakistan. Pak J Meteorol 12(24)

Imran Z (2013) Climate change and its impact on the political dynamics of Pakistan. JSTOR

IPCC I (2006) IM. IPCC. In: IGES Japan

Iqbal MM, Arif M (2010) Climate-change aspersions on food security of Pakistan. J Sci Dev 15(1):15–23

Iqbal Z, Kayani AR, Akhter A, Qayyum M (2022) Prevalence and distribution of hard ticks and their associated risk factors in sheep and goats from four agro-climatic zones of Khyber Pakhtunkhwa (KPK), Pakistan. Int J Environ Res Public Health 19(18):11759

Iqbal M, Ahmad M (2005) Science and technology based agriculture vision of Pakistan and prospects of growth. In: Proceedings of the 20th Annual General Meeting Pakistan Society of Development Economics, Islamabad. Pakistan Institute of Development Economic (PIDE), Islamabad, Pakistan

Israr M, Faraz M, Jan D, Ahmad N, Ahmad S (2016) Farming community perceptions about climate change in Khyber Pakhtunkhwa Pakistan. World J Agric Res 4(3):70–76

Jabeen MR, Elahi N, Alam A, Khattak BN (2022) The practices and prospects of community participation in joint forest management in Khyber Pakhtunkhwa Pakistan. Indian J Econ Bus 21(1)

Jamshed S, Ullah A, Rashid A (2014) Tree species and their associated conservation issues at Hayatabad township, Khyber Pakthunkhwa Peshawar, Pakistan. Int J Biol Biotechnol 11(2):309–317

Jaspal ZN, Khan AU (2017) Plant biosecurity governance dilemma in Pakistan: the case study of Khyber Pakhtunkhwa. J Pol Stud 24:245

Javed S, Khan D, Khan A (2023) Impact of certified wheat seeds on food security: a case study of Karak, KP, Pakistan. Qlantic J Soc Sci 4(4):247–258

Javed M, Abbas K (2018) Inland fisheries and aquaculture in Pakistan. In: Developing sustainable agriculture in Pakistan, pp 543–559. CRC Press

Jiongxin X (2003) Sediment flux to the sea as influenced by changing human activities and precipitation: example of the Yellow River, China. Environ Manage 31:0328–0341

Jordano P (2016) Chasing ecological interactions. PLoS Biol 14(9):e1002559

Junaid A, Kashaf M, Naseem N, Arif M, Noor R, Saeed M, Rais M (2010) A review of research on Ayubia National Park, Khyber Pakhtunkhwa, Pakistan. Biharean Biologist 17(2):92–97

Kakakhel S (2018) Pakistan's water-related challenges. Pakistan Horizon 201–212

Khalid F, Taj MB, Jamil A, Kamal H, Afzal T, Iqbal MJ, Khan T, Ashiq M, Raheel A, Sharif M (2020) Deforestation dynamics in Pakistan: a critical review: deforestation dynamics. Proc Pak Acad Sci B Life Environ Sci 57(3):27–34

Khan AN (2013) Analysis of 2010-flood causes, nature and magnitude in the Khyber Pakhtunkhwa, Pakistan. Nat Hazards 66(2):887–904

Khan H, Baig SU (2020) Biodiversity conservation in the Hindu Kush-Karakoram-Himalayan mountain region of northern Pakistan: overview of big mammal protection. J Mt Sci 17(6):1360–1373

Khan MT, Hashim S, Ayub S, Jan A, Marwat KB (2015) A case study of ethnobotany and biodiversity conservation from tehsil barawal, Upper Dir, Khyber pakhtunkhwa, Pakistan. Pak J Bot 47:7–13

Khan A, Ali S, Shah SA, Khan A, Ullah R (2019a) Impact of climate change on maize productivity in Khyber Pakhtunkhwa, Pakistan. Sarhad J Agric 35(2):594–601

Khan N, Shah SJ, Rauf T, Zada M, Yukun C, Harbi J (2019b) Socioeconomic impacts of the billion trees afforestation program in Khyber Pakhtunkhwa Province (kpk), Pakistan. Forests 10(8):703

Khan A, Ali S, Murad W, Hayat K, Siraj S, Jawad M, Khan RA, Uddin J, Al-Harrasi A, Khan A (2021a) Phytochemical and pharmacological uses of medicinal plants to treat cancer: a case study from Khyber Pakhtunkhwa, North Pakistan. J Ethnopharmacol 281:114437

Khan T, Nouri H, Booij MJ, Hoekstra AY, Khan H, Ullah I (2021c) Water footprint, blue water scarcity, and economic water productivity of irrigated crops in Peshawar Basin, Pakistan. Water 13(9):1249

Khan A, Ali S, Khan M, Hamayun M, Moon Y-S (2022) *Parthenium hysterophorus's* endophytes: the second layer of defense against biotic and abiotic stresses. Microorganisms 10(11):2217

Khan MZ, Babar Khan BK, Ejaz Ahmed EA, Garee Khan GK, Anila Ajmal AA, Rehmat Ali RA, Saeed Abbas SA, Muhammad Ali MA, Ejaz Hussain EH (2014) Abundance, distribution and conservation of key ungulate species in Hindu Kush, Karakoram and western Himalayan (HKH) mountain ranges of Pakistan

Khan H, Ullah I, Ali I, Muhammad I, Ahmad T, Xing L (2021) Environmental and soil gradient effects on biodiversity of butterflies in Khyber Pakhtunkhwa, Pakistan. Arthropods

Khattack A (1992) Development of a model forest management plan for the Panjul Forest in Western Himalaya (Pakistan)

Koellner T, De Baan L, Beck T, Brandão M, Civit B, Margni M, I Canals LM, Saad R, De Souza DM, Müller-Wenk R (2013) UNEP-SETAC guideline on global land use impact assessment on biodiversity and ecosystem services in LCA. Int J Life Cycle Assess 18:1188–1202

Kuiper JJ (2023) A plea for more freshwater experts in IPBES. Aquat Conserv Mar Freshwat Ecosyst 33(6):620–628

Kundzewicz ZW (2008) Climate change impacts on the hydrological cycle. Ecohydrol Hydrobiol 8(2–4):195–203

Lam VW, Cheung WW, Reygondeau G, Sumaila UR (2016) Projected change in global fisheries revenues under climate change. Sci Rep 6(1):32607

Lee H, Markham SK (2016) PDMA comparative performance assessment study (CPAS): methods and future research directions. J Prod Innov Manag 33:3–19

Li XY (2011) Hydrology and biogeochemistry of semiarid and arid regions. In: Forest Hydrology and biogeochemistry: synthesis of past research and future directions, pp 285–299

Ma X, Zhao C, Zhu J (2021) Aggravated risk of soil erosion with global warming–a global meta-analysis. CATENA 200:105129

Maan YA, Jameel M (2021) The role of institutions in disaster risk management and mitigation in Punjab Pakistan. Indian J Econ Bus 20(1):269–279

Mahmood SA, Tahir Z, Batool S, Masood A, Haseeb M, Muhammad Ali M (2023) Appraisal of climate change disaster and food security in Khyber Pakhtunkhaw Pakistan using geospatial technologies. Int Arch Photogramm Remote Sens Spat Inf Sci 48:425–432

Mahmood T, Khan LU, Naeem M (2021) Diversity and abundance of Avifauna of Manglot Wildlife Park, Nowshera District, Khyber Pakhtunkhwa, Pakistan. Pakistan J Zool 53(5)

Malaysia P (2010) Food security: concepts and definitions. J Community Health 16(2):2

Mall RK, Gupta A, Sonkar G (2017) Effect of climate change on agricultural crops. In: Current developments in biotechnology and bioengineering. Elsevier, pp 23–46
Maryam A, Khan S, Khan K, Khan M, Rabbi F, Ali S (2014) The perception of local community about the effects of climate change in Upper Swat, Khyber Pakhtunkhwa, Pakistan. J Earth Sci Clim Change 5(183):2
Menzel L, Bürger G (2002) Climate change scenarios and runoff response in the Mulde catchment (Southern Elbe, Germany). J Hydrol 267(1–2):53–64
Michel Devadoss PS, Pariatamby A, Bhatti MS, Chenayah S, Shahul Hamid F (2021) Strategies for reducing greenhouse gas emissions from municipal solid waste management in Pakistan. Waste Manage Res 39(7):914–927
Miller V, Giles J, Khan M, Mumtaz H, Savelli A, Grosjean G (2021) Climate-smart agriculture in Khyber Pakhtunkhwa, Pakistan. CSA Country Profiles for Asia
Mirza M (2007) A note on the fishes of Swat river, NWFP Pakistan. Biologia Pakistan 53(2):109–112
Moazzam M, Osmany HB (n.d.) Exotic and invasive fish species: a major threat to aquatic biodiversity of Pakistan
Muhammad L, Khan A, Zhou Y, He M, Alrefaei AF, Khan M, Ali S (2023) Physiological and ultrastructural changes in *Dendranthema morifolium* cultivars exposed to different cadmium stress conditions. Agriculture 13(2):317
Mumtaz M, de Oliveira JAP, Ali SH (2019) Climate change impacts and adaptation in agricultural sector: the case of local responses in Punjab, Pakistan. In: Climate change and agriculture
Munawar S, Udelhoven T (2020) Land change syndromes identification in temperate forests of Hindukush Himalaya Karakorum (HHK) mountain ranges. Int J Remote Sens 41(20):7735–7756
Munir MA, Munir K (2023) Impacts of climate change on various sectors in Khyber Pakhtunkhwa Province of Pakistan; a review
Munro PG (2009) Deforestation: constructing problems and solutions on Sierra Leone's Freetown Peninsula. J Polit Ecol 16(1):104–122
Nasir J (2024) Development challenges of Pakistan. Springer Books
Navarro-Perez M, Tidball K (2012) Challenges of biodiversity education: a review of education strategies for biodiversity education. Int Electron J Environ Educ 2(1)
Nawab B, Alam J, Ali H, Hussain M, Shah M, Ahmad S, Shah AH, Mehmood A (2018) Conservation status assessment of native vascular flora of Kalam Valley, Swat District, Northern Pakistan. Int J Biodivers Conserv 10(11):453–470
Nazeer M (2012) Potential evapotranspiration trend analysis for different climatic zones in Khyber Pakhtunkhwa, Pakistan. The Nucleus 49(2):101–106
Nieto NC, Khan K, Uhllah G, Teglas MB (2012) The emergence and maintenance of vector-borne diseases in the Khyber Pakhtunkhwa province, and the federally administered tribal areas of Pakistan. Front Physiol 3:250
Olivieri LM (2021) J Asian Civil 44(1):1–286
Organization WH, Programme UUNE, Health WOFA (2022) One health joint plan of action (2022–2026): working together for the health of humans, animals, plants and the environment. World Health Organization
Panagos P, Imeson A, Meusburger K, Borrelli P, Poesen J, Alewell C (2016) Soil conservation in Europe: wish or reality? Land Degrad Dev 27(6):1547–1551
Pearman PB, Lavergne S, Roquet C, Wüest R, Zimmermann NE, Thuiller W (2014) Phylogenetic patterns of climatic, habitat and trophic niches in a European avian assemblage. Glob Ecol Biogeogr 23(4):414–424
Perveen F, Khan N (2015) Checklist of the first recorded spider fauna in Sheringal, Khyber Pakhtunkhwa, Pakistan. Adv Entomol 3(3):94–100
POLICY CC (2016) Khyber Pakhtunkhwa climate change policy
Prakash S (2021) Impact of climate change on aquatic ecosystem and its biodiversity: an overview. Int J Biol Innov 3(2)
Rafique M, Khan NUH (2012) Distribution and status of significant freshwater fishes of Pakistan. Rec Zool Surv Pakistan 21:90–95

Rahman G, Rahman A-U, Ullah S, Dawood M, Moazzam MFU, Lee BG (2021) Spatio-temporal characteristics of meteorological drought in Khyber Pakhtunkhwa, Pakistan. PLoS ONE 16(4):e0249718

Rahman G, Rahman A-U, Munawar S, Moazzam MFU, Dawood M, Miandad M, Panezai S (2022) Trend analysis of historical and future precipitation projections over a diverse topographic region of Khyber Pakhtunkhwa using SDSM. J Water Clim Chang 13(11):3792–3811

Rajan A, Verma S (2017) Evolving nature of India's irrigation economy: insights from the fifth minor irrigation census, vol 7. IWMI

Ramay SA (2014) State of food security in Pakistan and policy options. In: Sustainable development policy Institute of Pakistan, pp 1–12

Rana, M. A. (2014). *The seed industry in Pakistan: Regulation, politics and entrepreneurship* (Vol. 19). Intl Food Policy Res Inst

Rasheed R, Ishaq MN, Akbar M (2022) A correlation of socio-economic determinants and food security status in Pakistan. Pak J Hum Soc Sci 10(1):416–425

Rashid MH-O, Chung YR (2017) Induction of systemic resistance against insect herbivores in plants by beneficial soil microbes. Front Plant Sci 8:1816

Rauf T, Khan N, Shah SJ, Zada M, Malik SY, Yukun C, Sadique A (2019) Poverty and prosperity: impact on livelihood assets of billion trees afforestation program in Khyber Pakhtunkhwa (KPK), Pakistan. Forests 10(10):916

Rehman K, Shah SM, Badshah L, Abdullah (2022) Biodiversity and freshwater ecosystem services: a case study of the Hamzakot Area of Mardan, Pakistan. In: Biodiversity, conservation and sustainability in Asia: volume 2: prospects and challenges in South and Middle Asia. Springer, pp 163–179

Sabir MA, Shafiq-Ur-Rehman S, Umar M, Waseem A, Farooq M, Khan AR (2013) The impact of suspended sediment load on reservoir siltation and energy production: a case study of the Indus River and its tributaries. Pol J Environ Stud 22(1):219–225

Said A, Imran M, Waseem MT, Khan AM, Khalique N, Sarwar G, Ahmad RM (2022) Feeding niche overlap between native and alien fishes in Swat River, Khyber Pakhtunkhwa, Pakistan. Environ Biol Fishes 105(4):509–518

Saleem T, Shrestha RB (2019) Agricultural research and development: policy and program priorities in Pakistan. In: Agricultural policy and program framework: priority areas for research and development in South Asia, p 151

Salik KM, Jahangir S, ul Hasson S (2015) Climate change vulnerability and adaptation options for the coastal communities of Pakistan. Ocean Coast Manag 112:61–73

Sanchez-Triana E, Biller D, Nabi I, Ortolano L, Dezfuli G, Afzal J, Enriquez S (2014) Revitalizing industrial growth in Pakistan: trade, infrastructure, and environmental performance. World Bank Publications

Sawa JN (2023) New waves of flash floods and risks perception of climate change: an overview of the 2022 flood disaster in Pakistan. Pakistan J Hum Soc Sci Res 6(2):72–82

Schwitzer C, Glatt L, Nekaris KA-I, Ganzhorn JU (2011) Responses of animals to habitat alteration: an overview focussing on primates. Endanger Species Res 14(1):31–38

Shah S (2011) Conservation of endangered species in Khyber Pakhtunkhwa. J Anim Plant Sci 21(2):400–404

Shah S, Khalil AK (2017) Growth and variability of major food crops production in Khyber Pakhtunkhwa, Pakistan. J Appl Adv Res 2(1):6–12

Shah A, Khan N, Gong Z, Ahmad I, Naqvi S, Ullah W, Karmaoui A (2023) Farmers' perspective towards climate change vulnerability, risk perceptions, and adaptation measures in Khyber Pakhtunkhwa, Pakistan. Int J Environ Sci Technol 20(2):1421–1438

Shah SA, Ali S, Ali A, Baig A (2020) Economic analysis of maize production in central Khyber Pakhtunkhwa, Pakistan. Sarhad J Agric 36(3)

Shahbaz B, Shah QA, Suleri AQ, Commins S, Malik AA (2012) Livelihoods, basic services and social protection in north-western Pakistan

Shaheen H, Ullah Z, Khan SM, Harper DM (2012) Species composition and community structure of western Himalayan moist temperate forests in Kashmir. For Ecol Manage 278:138–145

Shaheen S, Abbas S, Hussain J, Mabood F, Umair M, Ali M, Ahmad M, Zafar M, Farooq U, Khan A (2017) Knowledge of medicinal plants for children diseases in the environs of district Bannu, Khyber Pakhtoonkhwa (KPK). Front Pharmacol 8:430

Shahzad N, Amjad M (2022) Climate change and food security in Pakistan. In: Sustainable agriculture and food security. Springer, pp 579–594

Shehzad K (2023) Extreme flood in Pakistan: is Pakistan paying the cost of climate change? A short communication. Sci Total Environ 880:162973

Shen C, Wang W-C, Hao Z, Gong W (2007) Exceptional drought events over eastern China during the last five centuries. Clim Change 85(3):453–471

Siddiqui MT, Nawaz MF, Khan RA, Khan ZH (2018) Forestry and range management in Pakistan present potential and way forward. In: Developing sustainable agriculture in Pakistan. CRC Press, pp 359–374

Sivakumar MV, Stefanski R (2011) Climate change in South Asia. In: Climate change and food security in South Asia, pp 13–30

Solomon S (2007) Climate change 2007-the physical science basis: working group I contribution to the fourth assessment report of the IPCC, vol 4. Cambridge University Press

Sonia K (2019) Sadozai, NP Khan, AU Jan and G. Hameed. 2019. Assessing the impact of climate change on wheat productivity in Khyber Pakhtunkhwa, Pakistan. Sarhad J Agric 35(1):284–292

Spielman DJ, Malik SJ, Dorosh P, Ahmad N (2016) Food, agriculture, and rural development in Pakistan. In: Agriculture and the rural economy in Pakistan: Issues, outlooks, and policy priorities, pp 1–40

Standl E, Khunti K, Hansen TB, Schnell O (2019) The global epidemics of diabetes in the 21st century: current situation and perspectives. Eur J Prev Cardiol 26(2_suppl):7–14

Stocker BD, Roth R, Joos F, Spahni R, Steinacher M, Zaehle S, Bouwman L, Prentice IC (2013) Multiple greenhouse-gas feedbacks from the land biosphere under future climate change scenarios. Nat Clim Chang 3(7):666–672

Tagar HK, Shaikh SM, Tagar AK, Bijarni GA (2021) Economic growth of Pakistan in 21st century and the use of 'blue economy' resources for sustainability. Eur J Soc Sci 62(2):38–49

Tahir M, Khaliq T (2018) Land use in Pakistan. In: Developing sustainable agriculture in Pakistan. CRC Press, pp 33–57

Tang B, Zhao X, Zhao W (2018) Local effects of forests on temperatures across Europe. Remote Sens 10(4):529

Uddin S, Kausar A, Afsar S, Afzal A, Lahori AH, Stepova O, Anwar MM, Bredun V (2023) Exploring the Complexities of Urbanization and Socio-Ecological Challenges in the High Mountainous Region of Chitral, Khyber Pakhtunkhwa (KPK), Pakistan. *Proceedings of the Pakistan Academy of Sciences: A. Physical and Computational Sciences*, *60*(1), 47–54.

ul Haq MI (2022) Deforestation in Khyber Pakhtunkhwa Province and government response–an auditorial analysis. Asian J 45

Ullah R, Nawab K (2019) Pesticides use in Khyber Pakhtunkhwa Province Pakistan: present scenario. Int J Biosci 14:197–208

Ullah L, Bangash AK, Khan M, Khan Y (2023) Challenges and potential opportunities of tourism in Kumrat Valley, Khyber Pakhtunkhwa. Qlantic J Soc Sci 4(4):108–120

Ullah R, Khan N (2022) *Xanthium strumarium* L. an alien invasive species in Khyber Pakhtunkhwa (Pakistan): a tool for biomonitoring and environmental risk assessment of heavy metal pollutants. Arab J Sci Eng 47(1):255–267

Ullah N, Chao L, Khan TU, Sai WL, Yazhuo Z, Khan IA, Hassan MA, Hu Y (2024) Insights into climate change dynamics: a tourism climate index-based evaluation of Gilgit-Baltistan, Pakistan. Heliyon

Upadhyay A, Singh R, Vivekanand V (2024) Gene editing for abiotic stress resistance in horticultural crops. In: CRISPRized horticulture crops. Elsevier, pp 337–348

Usman K, Pervaiz K, Hassan Z, Rehman HU, Ullah S, Haider I, Sadia H, Kiani BH, Khan MY, Nawaz A (2016) Exploring the fish fauna in water bodies of Hazara Division, Khyber Pakhtunkhwa, Pakistan. J Entomol Zool Stud 4:848–851

Van der Putten WH, Macel M, Visser ME (2010) Predicting species distribution and abundance responses to climate change: why it is essential to include biotic interactions across trophic levels. Philos Trans R Soc B Biol Sci 365(1549):2025–2034

Williams A, Pruski FF, Nearing MA (2002) Indirect impacts of climate change that affect agricultural production: soil erosion. In: Effects of climate change and variability on agricultural production systems. Springer, pp 249–264

Woldearegay K (2013) Review of the occurrences and influencing factors of landslides in the highlands of Ethiopia: with implications for infrastructural development. Momona Ethiop J Sci 5(1):3–31

Woodward A, Smith KR, Campbell-Lendrum D, Chadee DD, Honda Y, Liu Q, Olwoch J, Revich B, Sauerborn R, Chafe Z (2014) Climate change and health: on the latest IPCC report. The Lancet 383(9924):1185–1189

Xie H, Ringler C, Zhu T, Waqas A (2013) Droughts in Pakistan: a spatiotemporal variability analysis using the Standardized Precipitation Index. Water Int 38(5):620–631

Yaseen M, Saad M, Salim M, Saeed M, Suliman M, Khan S, Khan SM, Khan MS, Amir HH (2018) Impact of joint forest management on the insect population and socio-economics of local communities of village Miandam district swat, Pakistan

Yazdani AT, Ali N (2017) Seed policy in Pakistan: the impact of new laws on food sovereignty and sustainable development. Lahore J Policy Stud 7(1):77

Yousaf R, Ali S, Ullah I, Sha SA, Uçak H (2023) Challenges and prospects for tomato productivity in response to climatic variations: evidences from Khyber Pakhtunkhwa-Pakistan. Економика Пољопривреде 70(2):377–394

Zada M, Zada S, Ali M, Zhang Y, Begum A, Han H, Ariza-Montes A, Vega-Muñoz A (2021) Development of local economy through the strengthening of small-medium-sized forest enterprises in KPK, Pakistan. Sustainability 13(19):10502

Zada M, Zada S, Ali M, Zhang Y, Begum A, Han H, Ariza-Montes A, Araya-Castillo L (2022) Contribution of small-scale agroforestry to local economic development and livelihood resilience: evidence from Khyber Pakhtunkhwa Province (KPK), Pakistan. Land 11(1):71

Zaidi HA (2020) Disclosable Restructuring Paper-Khyber Pass Economic Corridor Project-P159577

Zhao T, Gong S, Huang P, Lavoué D (2012) Hemispheric transport and influence of meteorology on global aerosol climatology. Atmos Chem Phys 12(16):7609–7624

Asif Khan earned a Bachelor of Science in Botany from Abdul Wali Khan University Mardan and an M.Phil. in Botany from the University of São Paulo, Brazil. He is currently a Ph.D. scholar at the Universidade Estadual de Maringá, Brazil, focusing on identifying potential inhibitors of human caspases derived from Brazilian flora. His research is funded by CAPES. His doctoral research centers on investigating the inhibition of human caspasees enzymes, a critical target in inflammatory pathways, to discover novel anti-inflammatory compounds. Asif employs a multidisciplinary approach, combining advanced in vitro experiments with in silico computational techniques, such as molecular docking and molecular dynamics simulations, to identify and understand the mechanisms of bioactive compounds. His work not only contributes to the fields of pharmacology and biochemistry but also emphasizes the significance of biodiversity in the development of innovative therapies. In addition to his research, Asif is passionate about advancing scientific knowledge and fostering collaboration among researchers in pharmacology, botany, and bioinformatics. His work represents a commitment to bridging the gap between traditional knowledge of medicinal plants and modern drug discovery, with the ultimate goal of addressing global health challenges through sustainable and nature-inspired solutions. Asif has published several articles in reputable international journals and serves as a reviewer for multiple prestigious publications.

Muhammad Jawad holds a Bachelor of Science in Geomatics and Remote Sensing from the University of Peshawar and an M.Phil. in GIS and Remote Sensing from the University of the Punjab, Lahore. He is a seasoned geospatial professional with expertise in GIS and remote sensing technologies.

Sajid Ali earned his Ph.D. from the Center of Biotechnology and Microbiology (COBAM), University of Peshawar, Pakistan. He completed a postdoctoral fellowship at Kyungpook National University, South Korea, and is currently an assistant professor in the Department of Horticulture and Life Science at Yeungnam University, South Korea. His research focuses on Microbial Plant Biostimulants, and he aims to explore plant–microbe interactions to enhance agricultural practices under stressful conditions.

Asad Ali is a professor of Entomology at Abdul Wali Khan University Mardan. With an impact factor of around 80, he has published about 50 research articles. He has also participated in numerous international conferences and has supervised many graduate students, contributing significantly to the field of entomology.

Zaheer-ud-din Khan serves as a distinguished professor at the Department of Botany, GC University Lahore Pakistan. He did his master degree with distinction from the University of the Punjab in 1974 and started teaching and supervising graduate and postgraduate students of Plant Diversity in GC University Lahore. He did his Ph.D. in 1990 from Kings College, University of London in the field of Food Microscopy. Dr. Khan has a strong research background in Plant Systematics, Advance Plant Anatomy, and Ethnobotany. Some of his notable research works include Ethnobotanical Studies (He has conducted extensive research on the ethnobotanical aspects of various plant species in Pakistan, including studies on the useful shrubs of district Kotli, Azad Jammu & Kashmir.); DNA Barcoding (including the identification of the "Lahore District Flora" using rbcl and matK markers); Taxonomic Reaffirmation (He has contributed to the taxonomic reaffirmation of some members of family Cannabaceae, Moraceae, Rhamnaceae, Rosaceae, and Urticaceae of order rosales using DNA barcoding markers and Anatomical studies.). He worked out the microstructural/Pharmacognostic and Phytochemical evaluation of adulteration or admixturing in the processed food and herbal products using simple and electron microscopy and chemotaxonomic evaluation techniques including DNA barcoding. He successfully completed HEC research projects on the "Conservation Status of Flora of Punjab" and Ethnofloristic Study of Kotli, AJK". Dr. Khan has also helped WWF-Pakistan, World Bank, Pakistan WAPDA Tarbela Dam, EPA Punjab, and Punjab Small Industries Corporation in completing their projects. His research has been published in various journals, including the Pakistan Journal of Botany and Journal of Biodiversity and Environmental Sciences. He has published 134 research articles in National and International Journals, with an impact factor of 83.23 to his credit. He has a significant number of citations, with over 1767 and an h-index of 21. He is also the author of text books of BSc level. Dr Khan won HEC cash awards like: Best University Teacher Award, 2009; Productivity Allowance for Publications of Research articles, 2009-2011; and Productive Science Award, 2012 by PCST. Recently, he is bestowed with "Life Time Achievement Award" by his Alma Mater, GC University Lahore. He is the member, life member, and life fellow of various professional organizations and societies of Pakistan, such as Pakistan Botanical Society, Biological Society of Pakistan, and Pakistan Association for the Advancement of Science. Dr. Khan has represented Pakistan at many International fora and has attended several academic conferences and meetings as a resource person.

Waheed Murad is a professor of Botany at Abdul Wali Khan University, Mardan, Pakistan. He earned his Ph.D. in Palaeoclimatology and Biodiversity Conservation from Georg-August University, Göttingen, Germany. Dr. Murad's research focuses on plant and fungal ecology, climate change, and ethnomedicinal studies. He has published over 90 research articles and supervised numerous M.Phil. and Ph.D. students. His contributions to biodiversity conservation and phytodiversity have garnered national and international recognition.

Mirza Barjees Baig is a professor at the Prince Sultan Institute for Environmental, Water, and Desert Research, King Saud University, Saudi Arabia. He earned his MS degree in International Agricultural Extension in 1992 from Utah State University, Logan, Utah, USA, and was placed on the "Roll of Honor". He completed his Ph.D. in Extension for Natural Resource Management from the University of Idaho, USA, and was honored with the "1995 Outstanding Graduate Student Award". He has published extensively on the issues associated with natural resources in national and international journals. He has also made oral presentations about agriculture, natural resources, and the role of extension education at various international conferences. He serves on the editorial boards of many international journals and is a member of many international professional organizations. He has produced six books on climate change, food security, and climate-smart agriculture. Food waste, water management, degradation of natural resources, the deteriorating environment, and their relationship with society and the community are his areas of interest. He has attempted to develop strategies to conserve natural resources, promote the environment, and develop sustainable communities. He started his scientific career in 1983 as a researcher at the Pakistan Agricultural Research Council, Islamabad, Pakistan. He served at the University of Guelph, Ontario, Canada, as a Special Graduate Faculty member from 2000 to 2005. He served as a foreign professor at the Allama Iqbal Open University (AIOU), Pakistan, through the Higher Education Commission from 2005 to 2009.

Chapter 5
Agroforestry: Climate-Smart Agriculture for Increased Resilience and Food Security in a Changing Climate

Jagdish Chander Dagar and Sharda Rani Gupta

Abstract The atmospheric concentration of greenhouse gases due to anthropogenic activity has steadily increased throughout time, contributing to global warming. The most recent nine years have been among the warmest on record. This temperature increase is anticipated to affect agriculture in a number of ways, including increasing crop evapotranspiration, altering the salinity of the soil, reducing crop production and nutrient use efficiency, and changing insect-pest distribution. Climate change will put additional strain on already scarce water resources in some areas, especially in dry regions. Agroforestry has emerged as a field that addresses biodiversity conservation, carbon sequestration, climate change adaptation and mitigation, landscape restoration, as well as livelihood stability for communities. Agroforestry improves ecological processes and regulates ecosystem services such as erosion control, nutrient retention, water and air quality, pollination, and insect control. The contribution of agroforestry practices to improving food security, agroforestry-based solutions for coping with climate change, and the optimum use of natural resources through agroforestry interventions are discussed in view of the developing issues of climatic variability. Techniques for agroforestry have also been suggested as a strategy to boost agricultural production in degraded ecosystems. The restoration of sodic and saline waterlogged areas, severely degraded ravine lands, mine spoil areas, stabilizing sand dunes, and utilization of low-quality waters using agroforestry technologies are highlighted.

Keywords Agroforestry · Climate change · Sustainable agriculture · Climate-smart agriculture · Degraded landscapes · Carbon sequestration · Environmental services · Livelihood security

J. C. Dagar (✉)
Natural Resource Management Division, Krishi Anusandhan Bhavan—II, Indian Council of Agricultural Research, Pusa, New Delhi, India
e-mail: dagarjc@gmail.com

S. R. Gupta
Faculty of Life Sciences, Kurukshetra University, Kurukshetra, Haryana, India

M. Behnassi et al. (eds.), *Food Systems and Biodiversity in the Context of Environmental and Climate Risks*, https://doi.org/10.1007/978-3-031-89167-0_5

5.1 Introduction

The effects of global warming represent a serious danger to secure food supply and sustainable agricultural output. Unavoidable climate change is now a reality, especially for developing countries in Asia and Africa. The IPCC analysis from 2014 indicates that these regions are particularly vulnerable to the adverse effects of climate change, which has already affected agricultural productivity. The prevalence of land degradation is a challenge to achieving sustainable food security. Deforestation, soil erosion, salinity, land fragmentation, inappropriate farming practices, labor issues, and over-exploitation of natural resources are some of the important issues posing a threat to food and livelihood security (FAO and ITPS 2015; Dagar et al. 2020a, b; UNCCD 2022). Studies show that the soil organic carbon content in tropical agricultural systems has rapidly decreased as a result of the conversion of natural ecosystems into arable and pastoral land usage (Hillbrand et al. 2017). The impact of global warming poses a significant threat to sustainable agricultural production and food security. The consequences of climate change on land degradation, salinity, waterlogging, and flooding have been found to vary according to the hydrological basins. The study shows that if the yearly water availability per person is less than 1000 m^3 or the ratio of withdrawals to long-term average annual runoff exceeds 0.4, the basins become water-stressed (Mullen and Barett-Lennard 2010). Africa has experienced a plateau in agricultural production over the past few decades, and predictions indicate that due to the effects of climate change, agricultural yields would dramatically decline in the twenty-first century (IPCC 2014). These issues require immediate attention and focused efforts in order to implement sustainable practices and decrease the effects of climate change on agriculture and food security in sensitive regions.

Globally, sea levels are predicted to increase by 2–3 m over the next 2,000 years, even at 1.5 °C of warming, and up to 6 m with 2 °C of warming (Tollefson 2021; IPCC 2021). This would have an influence on the entire coastlines that are densely populated. Sea-level rise is expected to bring about significant consequences in various aspects, including salinity levels, waterlogging, inundation severity, future land use, and the need for resilient crops. As sensitive areas experience hotter and dryer conditions, there will be a decrease in the supply of good-quality water, which will lead to the adoption of saline or high-sodium subsurface water and wastewater for irrigation. Salinity will build up in the soil as a result of this use of saline water. Additionally, the increased frequency of heavy rainfall events may cause excessive runoff and more frequent floods in low-lying areas and valley floors. Extreme weather events will become more frequent as temperatures rise in the future. According to the IPCC assessment report of 2021, high-temperature events that occurred once every 50 years in previous eras will probably occur every three to four years. If global temperatures increase by 2 °C over the pre-industrial levels, heatwaves and prolonged droughts are expected to happen more regularly. Additionally, there will be more tropical cyclones than there are now. Reduced exposure to risks and shocks associated with climate change is important in order to adapt more effectively to longer-term stressors including shortened growing seasons and variable weather patterns.

Over the past two decades, the issue of global warming has given impetus to new research in agroforestry systems, and carbon sequestration for climate change mitigation (van Noordwijk et al. 2020). For adopting agroforestry, some of the major concerns include abiotic stresses, the spread of pests and diseases, the degradation of ecosystems, the loss of biodiversity, competing resource demands, and the potential impact on the availability of nutrient-rich food. According to Dagar et al. (2020a, b), agroforestry has been promoted as a multifunctional system that enhances agricultural sustainability, nutrient and livelihood security, the provision of environmental services, and climate change mitigation in degraded landscapes. The primary strategy for preventing future global food insecurity is adaptation to climate change (Porter et al 2014).

This chapter's objectives are to give a broad overview of climate variability, the role that agroforestry practices play in enhancing food security, agroforestry-based solutions for addressing climate change, and the most efficient use of natural resources through agroforestry interventions. Climate change-related agroforestry practices to boost agricultural production in degraded environments have also been discussed.

5.2 Emerging Challenges of Climate Variability

Throughout the Earth's 4.5 billion years of existence, its climate has consistently exhibited a dynamic nature, characterized by natural cyclic changes (Crowley 2002). Climate change has been a natural process occurring over extended periods of time, exerting significant impacts on sea levels, rainfall patterns, and temperature-related weathering phenomena. While evidence of past climate changes is mostly indirect, recent times have witnessed a remarkable increase in greenhouse gases within the troposphere, particularly since the pre-industrial era (pre-1750). Carbon dioxide levels have increased by 49% (from 280 in 1750 to 417 ppm in April 2020), methane by 164% (from 700 to 1845 ppb during the same period), and oxides of nitrogen by 21% (from 270 to 328 ppb) (IPCC 2007a, b; SCCRIPS 2020). The recorded greenhouse gas patterns of the last 2000 years (Fig. 5.1) reveal a substantial increase primarily occurring during the industrial era (post-1750), providing compelling evidence of the significant role of human activities in the emission of these gases.

The levels of methane, nitrous oxide, and CO_2 in the atmosphere in 2019 were the highest in 800,000 years. IPCC report shows that temperatures between 2011 and 2020 are higher than the most recent warm period, which lasted for around 6500 years (IPCC 2021). N_2O has a 298-fold greater global warming potential (GWP) than CO_2, while CH_4 has a 25-fold greater GWP than CO_2 over a 100-year time horizon, according to Forster et al. (2007). Anthropogenic activities have resulted in emissions of four primary greenhouse gases. When considering all sources combined, CO_2

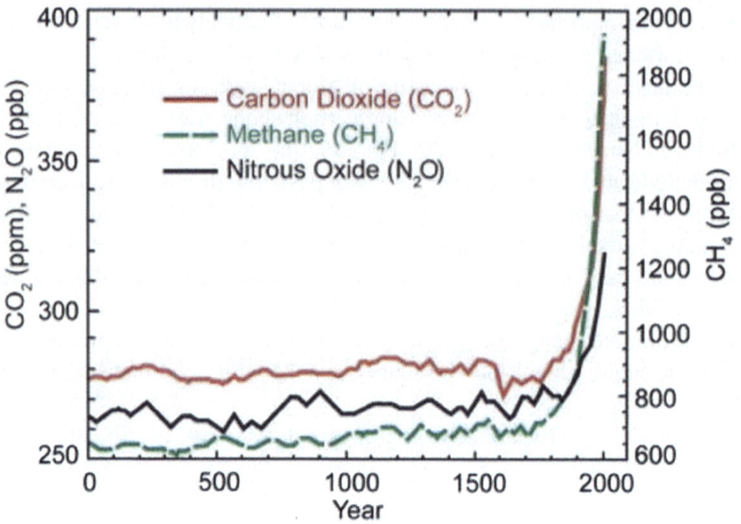

Fig. 5.1 The GHG emissions pattern during the past 2000 years, marked increase began after the year 1750. Parts per million (ppm) or parts per billion (ppb) are concentration units. *Source* Forster et al. (2007) and Blasing (2008)

accounts for 65% of GHG emissions, CH_4 for 17%, N_2O for 6%, and halocarbons (including CFCs, HFCs, PFCs, and SF6) for 12% (Stocker et al. 2013). It's worth noting that before 1750, the concentration of halocarbons was negligible, but they are now commonly used in various applications such as air conditioning, refrigeration, and insulation. Though the concentration of these compounds during the last decade is stabilized still these are present in sufficient quantity in the atmosphere. Sector-wise contribution of GHGs is estimated to be 25% from electricity and heat production, 24% from agriculture, forestry and other land uses, 21% from industry, 14% from transportation, 6% from building construction activities, and 10% from other energy sources (IPCC 2014).

Most of the warmest years on record come from this millennium, with 2020 and 2016 with a 1.02 °C rise in temperature from the 1951–80 baseline. Within the last 20 years, the years 2015, 2017, 2018, 2014, 2010, 2013, and 2005 are not far behind. NASA data indicates that 2016, continuing a long-term trend of rising global temperatures, was the warmest year since 1880 (Fig. 5.2).

According to the most recent IPCC assessment, the Earth's surface temperature increased by about 1.1 °C since 1850–1900 (IPCC 2021). The IPCC analysis predicts that global warming will probably approach or exceed 1.5 °C between 2030 and 2052 if it continues to increase at the current rate. Under a high-emission scenario, it is predicted that the world may warm by 4.4 °C by 2100, which might have catastrophic implications for how the planet's system function (IPCC 2021).

Since at least the nineteenth century, tide gauge records and satellite laser altimetry have both directly tracked changes in sea level. Ocean temperatures are increasing

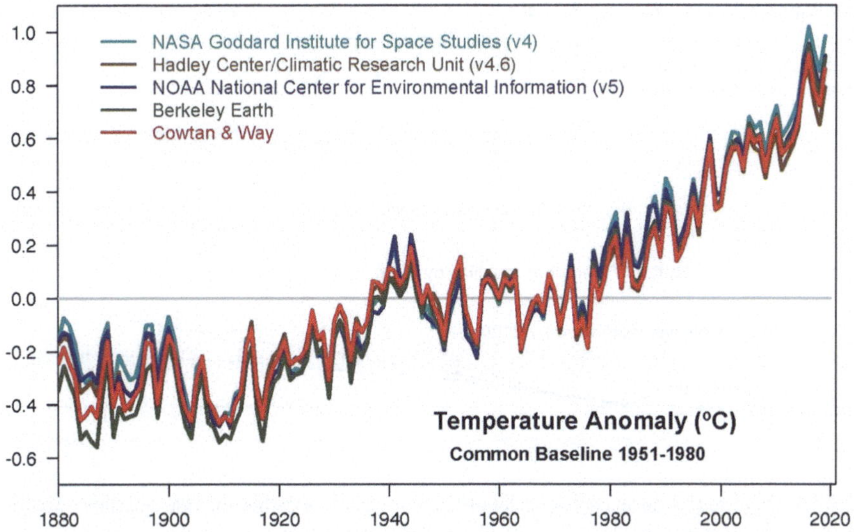

Fig. 5.2 Temperature anomaly showing that the last two decades have been the warmest ever since 1880. *Source* NASA

and causing the ocean to expand as ice sheets and glaciers melt and add more water to the ocean. Since 1900, the global mean sea level (GMSL) has been increasing irregularly at a rate of roughly 1.56 mm per year. However, the underlying reasons for this long-term trend and the variations observed over multiple decades are still not well understood, as highlighted in the study by Frederikse et al. in 2020 (Table 5.1). The historical changes in sea level have been monitored through tide gauges prior to 1992, and altimeters have been used for observations thereafter. To assess future changes in sea level, consistent methodologies have been employed, considering observational constraints and incorporating models related to CMIP (Coupled Model Inter-comparison Project), ice sheets, and glaciers.

Based on information from satellite altimeters and as published by Beckley et al. in 2017, the rate of global mean sea-level (GMSL) increase since 1993 has been calculated to be 3.34 mm per year. In its assessments, the Intergovernmental Panel

Table 5.1 Global mean sea-level change in meters relative to 1900

Period	Estimate (mm yr^{-1})	Uncertainty (± mm yr^{-1})	References	Measurement/ forecast model
1993–2020	3.34	0.40	Beckley et al. (2017)	Satellite altimetry
1993–2018	3.35	0.47	Frederikse et al. (2020)	Tide gauges
1900–2018	1.56	0.30	Frederikse et al. (2020)	Tide gauges

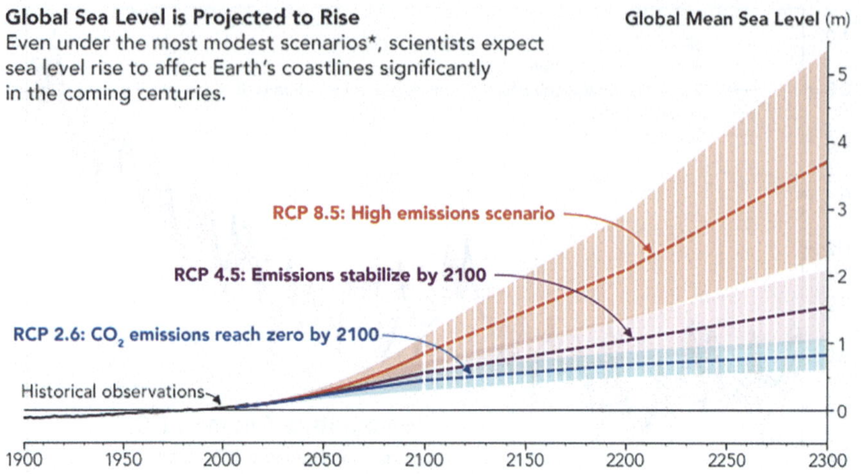

Fig. 5.3 Projected sea-level change under different scenarios of carbon dioxide emissions. *Source* https://earthobservatory.nasa.gov/images/148494/anticipating-future-sea-levels

on Climate Change (IPCC) projected that with high greenhouse gas (GHG) emissions (RCP8.5 scenario), the global sea level could rise by 0.6 to 1.1 m by 2100 (equivalent to about 15 mm per year). Under the worst-case scenario, by 2300, the seas could potentially rise as much as 5 m. However, if countries take significant actions to cut emissions (RCP2.6 scenario), the IPCC expects a lower sea-level rise of 0.3 to 0.6 m by 2100. The projections are based on different Representative Concentration Pathways (RCPs), which are used to calculate future scenarios depending on near-term emissions strategies and their expected radiative forcing levels (in W m^{-2}) in the year 2100 (Fig. 5.1). These pathways provide insights into how different emissions scenarios can influence the sea-level rise and other climate-related outcomes (Fig. 5.3).

There is no discernible trend in the total annual rainfall at the all-India level from 1813 to 2006, and the examination of rainfall data from the Indian Institute of Tropical Meteorology Pune (Dagar et al. 2012). The coefficient of variation is considerable, especially in arid and dry semi-arid environments. Fewer wet days but higher intensity rainfall indicate a skewed distribution of rainfall. It is anticipated that these changes in rainfall patterns will have a considerable impact on agriculture, increasing the likelihood of abiotic pressures like drought and salinity that will reduce crop productivity. These factors have an impact on ecosystems' capacity to produce enough biomass, including food for people and other animals.

According to Easterling et al. (2007), even modest temperature rises put developing countries' food security in danger. Additionally, the diverse effects of climate change on food production might have an impact on global food prices and commerce. Climate change is expected to result in a decrease in the global food supply, including potential C_4 crops. Berg et al. (2013) reported that the potential productivity of staple crops like millet is anticipated to decline by an average of 6% in tropical regions using an agro-DGVM (dynamical global vegetation model).

Rising sea levels have various consequences, such as more frequent and severe flooding in low-lying regions, beach erosion, and detrimental effects on infrastructure, the environment, and ecosystems like wetlands, inter-tidal zones, and mangroves (IPCC 2021). If sea levels continue to rise, experts anticipate that approximately 80% of wetlands globally could disappear by the century's end. Coastal wetlands play a crucial role in mitigating the effects of typhoons, tsunamis, and climate change, thereby enhancing resilience in the face of rising sea levels.

5.3 Relevance of Agroforestry for Climate Change

Across the globe, numerous traditional land-use practices have involved the simultaneous cultivation of trees and agricultural crops on the same land, as evidenced by studies conducted by Garrity et al. (2010) and Nair (1993). These farming systems recognized the importance of trees and intentionally preserved them within agricultural landscapes to support farming activities. While agroforestry systems have been traditionally implemented in various forms worldwide, they gained significant attention from researchers and practitioners during the 1970s and 1980s. During this period, there was a growing emphasis on exploring alternative approaches to enhance agricultural productivity, restore degraded lands, and improve the well-being of small-scale farmers, particularly in economically disadvantaged tropical regions.

The International Centre for Research in Agroforestry (ICRAF) conducted a survey that found that between 1.0 and 1.2 billion ha of agricultural landscapes now have trees on them and that about 1.2 billion people (20% of the world's population) are heavily dependent on agroforestry goods and services for their survival (ICRAF 2000; Zomer et al. 2016). In some areas, especially Southeast Asia and Central America, trees cover more than 30% of the croplands.

Agroforestry not only ensures food and nutritional security for marginalized farmers but also provides valuable environmental services such as erosion control, carbon sequestration, water drainage, soil improvement, biodiversity conservation, and climate change mitigation. It promotes soil health through processes like nitrogen fixation, nutrient recycling, organic matter accumulation, increased microbial and enzyme activity, and enhanced symbiotic relationships with mycorrhizal fungi (Dagar et al. 2016a, b; Dollinger and Jose 2018; Gupta et al 2023). According to Torralba et al. (2016), agroforestry increases biodiversity on degraded lands by enhancing ecological functions, regulating ecosystem services, and cultural services like erosion control, nutrient retention, water and air quality, pollination, and pest control. It also provides benefits like land regeneration and food production, supporting local economies and subsistence livelihoods (Dagar et al. 2020a, b; Hillbrand et al. 2017). In order to accomplish sustainable resource use, agroforestry responds to risks, collects and stores carbon, and restores multifunctionality in landscapes (van Noordwijk 2018).

Dagar et al. (2020a), while reporting traditional and improved agroforestry systems in Africa, have suggested policy guidelines and ways to use these systems for livelihood and nutritional security. There are many fruit tree species found in forests which can successfully be domesticated and blended as part of agroforestry systems. When commercializing a fruit product, care must be taken to grow (domesticate) the specific fruit tree in a sizable region to prevent exploitation of the tree's native environment. Akinnifesi et al. (2008) and Dagar et al. (2020a, b) have covered in-depth domestication processes, transformation, goods marketed and marketing systems, nutritional benefits of significant fruits, and research and policy challenges.

5.4 Agroforestry Systems for Sustained Food Production and Enhanced Resilience

Technologies based on agroforestry have been created to restore sodic and saline waterlogged areas, heavily degraded ravine lands, mine spoil areas, stabilize sand dunes, and utilize low-quality waters. This section discusses agroforestry systems, particularly those that deal with degraded environments.

5.4.1 Agroforestry Systems for Degraded Landscapes

The annual cost of lost ecosystem services as a result of land degradation is projected to be between USD 6.3 and 10.6 trillion, or 10–17% of global GDP, according to the Economics of Land Degradation (ELD) Initiative in 2015 (ELD and UNEP 2015). Various estimates indicate that a mix of physical, chemical, biological, and human factors may be responsible for the degradation of about 25% of the world's land (FAO and ITPS 2015). There is a serious problem of soil degradation because of rapid water and wind erosion, desertification, salinization, waterlogging, and loss of soil fertility. In the Himalayan region and other sloping plains, soil erosion by water is a severe issue, whereas wind erosion and desertification happen in dry areas. Due to salinity or water constraint, salt-affected regions remain desolate, and 20% of the irrigated area is at risk of secondary salinization (FAO 2022). In India, soil salinity and sodicity have an impact on around 7 million ha of land.

Throughout history, farmers, especially smallholders, have developed numerous agroforestry systems in diverse environments. These systems represent the collective wisdom and adaptive strategies of millions of farmers to meet their essential needs such as food, nutrition, fodder, fuelwood, medicinal plants, and income. Over time, many agroforestry practices have emerged as effective solutions, particularly in rehabilitating degraded habitats. Some of these practices are described briefly below.

5.4.1.1 Agroforestry Systems for Eroded Lands

There are socioeconomic, environmental, and technical aspects to soil erosion. Poor farmers and landless laborers suffer the most since they are least able to adopt traditional control methods. From an ecosystem perspective, developing a multifunctional land-use system is a more favorable response in degraded ecologies, including both high rainfall and semi-arid regions.

a. ***Dry ecologies***

Covering the space under trees, providing grasses for a surface layer, and protecting the sand dunes from biotic interference are the most crucial steps in stabilizing sand dunes. In addition to repairing the sand dunes, it is crucial to use windbreaks and mulch to monitor the movement of loose sand. The vegetation for sand dune stabilization is highly drought-tolerant with a deep root system capable of extracting moisture from lower soil depths. Among other trees include *Acacia tortilis, A. nilotica, A. leucophloea, A. senegal, Azadirachta indica, Ailanthus excelsa, Tecomella undulata* (3 forms), *Hardwickea binata, Cassia siamea, Cordia rothii, Balanites roxburghii, Prosopis cineraria, P. juliflora* (invaded in India during the nineteenth century). The most effective plants for stabilizing sand dunes have been identified as *Tamarix articulata/aphylla, Tamarix dioica, Tamarix traupii,* and *Holoptelia integrifolia,* as well as fruit trees like *Ziziphus mauritiana, Carissa carandas, Emblica officinalis, Aegle marmelos,* and *Phoenix dactylifera* (see Dagar 2014). These can withstand drought well and are used in both conventional and advanced agroforestry systems. Many of them serve as shelter belts as well.

For hot, arid climates, diverse production systems seem to be particularly sustainable. *Prosopis cineraria,* like *Faidherbia albida* in Africa, is well-known for its function in enhancing soil conservation and improvement, as well as crop yield (Table 5.1). The allied crops, such as pearl millet (*Pennisetum typhoides*), cluster bean (*Cyamopsis tetragonoloba*), green gram (*Vigna radiata),* black gram (*Vigna mungo*), and moth bean (*Vigna acontifolia*), are most usually produced in planted trees or in areas where *P. cineraria* trees are present (Table 5.2).

Table 5.2 Available macro- and micronutrients under *Prosopis cineraria* as compared to without trees

Treatment	Macro-nutrients (kg ha^{-1})			Micro-nutrients (ppm)			
	N	P	K	Zn	Mn	Cu	Fe
With tree	221	11	479	1.44	10.8	0.89	2.8
Without tree	199	6	3	0.19	7.0	0.38	3.5

*Source*Tewari et al. (2014)

In Pakistan, about 11 million ha of land resources are desert (Thal, Thar, Choliston and Chaki-Kharan). These areas can be brought under silvopastoral systems utilizing the local vegetation as well as saltbushes consisting of trees and using saline aquifers (Qureshi et al. 1993; Tewari et al. 2014).

In Kuwait, where the sand movement has seriously impacted the desert ecosystem, Al Muzaini (2003) proposed environmental remedies, like tree plantation techniques. Eucalyptus and date palms (*Phoenix dactylifera*) have reportedly done well in Kuwait. Jaradat (2003) suggested growing over 200 species of halophytes (used as fuel, pulp and fiber, fruits, pasture crops, grains and oil seeds, and as bioactive derivatives) in order to develop sustainable biosaline agricultural systems in the Middle East.

In Zambia, nitrogen-fixing plants like *Sesbania sesban* and *Tephrosea vogelii* may provide crops with yields comparable to those of properly fertilized fields, and in western Kenya, the same plants combined with *Crotalaria grahamiana* increased maize yields by twofold (Garrity et al. 2010). Furthermore, it has been demonstrated that using *Faidherbia albida* in a variety of combinations across Africa increases maize yields, particularly on soils with low fertility (Garrity et al. 2010). Faidherbia is the most prevalent tree species in parklands, which are abundant in many parts of Africa. It can also be found in a livestock system where the tree is used as shade and fodder. It is present along field boundaries or inside of fields without the presence of other tree species or livestock. The tree is also being pushed more and more in efforts to stop desertification, regreen the Sahel.

Kuyah et al. (2020) and Dagar et al. (2020a) have given details of various agroforestry systems (shifting cultivation, taungya and shamba systems, alley cropping, relay inter-cropping, sequential fallows, parklands, cut-and-carry systems, fodder protein banks, silvopasture, improved fallows, perennial crop orchards, homegardens, rotational woodlots, live-fences, smallholder timbers, etc.) and their role in enhancing livelihood and nutritional security of poor people.

According to Sileshi et al. (2020), *Faidherbia albida* is important for both crop and animal production systems on the African continent. They found that Faidherbia-based agroforestry systems are sustainable in terms of production, human well-being, and economic and environmental sustainability by using the sustainable intensification index. In areas with more open space, farmers have kept the trees (the number of trees varies by region). When compared to open regions, the production of corn and sorghum rose under the canopy by 150 and 73%, respectively, under the canopy compared to open regions; soil organic carbon was observed to rise in 96% of the cases. The tree greatly raises the nutritional value of the soil (Table 5.3), which promotes crop growth. In terms of production risks, the maize and sorghum yields in the open regions had greater CVs than those beneath the tree canopy, indicating higher production risks there (Table 5.3). Some of the methods for soil enrichment include deep nutrient collection and recycling, biological nitrogen fixation, nutrient subsidy, and atmospheric deposition. Through the breakdown of the litter, the nutrients so trapped are added to the soil (Sileshi et al. 2020).

b. ***Agroforestry for eroded soils in semi-arid and humid ecologies***

Table 5.3 Mean crop yield and soil nutrients under *Faidherbia* canopy as compared to the open area across Sub-Saharan Africa

Crop	Under canopy		Open area		Response ratio
	Yield (Mg ha^{-1})	CV (%)	Yield (Mg ha^{-1})	CV (%)	
Maize	3.4	66.2	1.7	85.5	2.5[a]
Sorghum	1.4	41.3	1.0	53.5	1.7[a]
Soil nutrients (Percent)					
SOC	1.8	44.6	1.3	48.2	1.5[a]
Total N	0.5	106.2	0.3	104.0	1.5[a]
Extractable P	28.4	105.9	27.6	108.7	1.2[a]
Extractable K	1.3	138.8	1.1	144.4	1.3[a]
Extractable Mg	3.1	117.9	3.3	130.5	1.0
Extractable Ca	10.6	127.7	10.5	156.9	1.1

[a] Indicates significant increases under canopy compared with open areas
Source Sileshi et al. (2020)

Water erosion is a big issue in humid and semi-arid regions with high rainfall. Globally, it is anticipated that soil erosion brought on by water is 20–30 Gt per year, whereas erosion brought on by tillage may be less than 5 Gt (FAO and ITPS 2015). With alternating dry and wet seasons, serious water erosion happens across South and East Asia. Water erosion and landslides are significant dangers in the sloping mountain terrain of Latin America and the Caribbean, particularly when vegetation is burned, as well as because of overgrazing and deforestation.

Native trees can be planted with shade-tolerant agricultural cash crops like coffee, cocoa, cardamom, ginger, or turmeric or medicinal plants in regions with heavy rainfall, particularly in South Asia. Plantation crops like coconut can be mixed well with fruits like pineapple and spices like clove, cardamom, and black pepper. Alley crops such *Gliricidia sepium, Leucaena leucocephala, Cassia siamea, Morus alba, Pithecellobium dulce,* and *Cajanus cajan*, as well as fodder grasses like inter-crops, may be grown on the sloping plains.

Agroforestry alternatives for Southeast Asian landscapes that have been degraded were discussed by van Noordwijk et al. (2020). While dealing with seven degradation syndromes (degraded hillslopes, fire-climax grasslands, over-intensified monocropping, forest classification conflicts, drained peatlands, converted mangroves, and disturbed soil profiles), including ash deposition and ex-mining sites, they emphasized that there are six appropriate actions for effective restoration, which include: (i) community involvement, (ii) rights, (iii) knowledge and know-how of sustainable land-use practices, (iv) and) markets for all inputs including soil amendments, tree germplasm, labor, crop protection means, (v) local environmental impacts, and (vi) global connectivity (van Noordwijk et al. 2020).

The use of water-harvesting technology led to an increase in irrigated land of 50–156% under various programs, which raised agricultural intensity by an average of 64% (NAAS 2009). In addition to boosting agricultural productivity, these projects assisted stakeholders in creating jobs and around 47% of degraded areas received rehabilitation treatment (Sharda et al. 2008). Trees may be planted as alleys, boundary plantations, or scattered trees in the field, depending on the slope, the severity of the issue, and the needs. However, all along the boundary facing gullies/ravines, hedges of the appropriate species, such as *Carissa carandas, Annona squamosa, Emblica officinalis, Aegle marmelos,* and *Cordia myxa* supported by perennial grasses, must be in place. Such systems have been demonstrated to provide greater total economic returns than solely growing crops in these settings, in addition to controlling soil erosion (Parandiyal et al. 2020).

Dagar (2018a, b) made the case in his study on the ecology and management of grazing regions in India that reducing grazing is the first step in recovering severely degraded ecosystems before adding native trees, perennial grasses, and legumes. In contrast to the 14.7% runoff and 3 Mg ha-1 from agricultural catchments, 5.8 and 2.6% of runoff and 1.26 and 0.62 Mg ha-1 of soil erosion, respectively, were created when *Acacia nilotica* + *Dichanthium annulatum* alone was planted in wild catchments. The production of 4.5 Mg ha^{-1} air-dried grass + firewood from such degraded lands revealed the possibility of grasses and trees as an alternative land use for conservation and productive use of degraded ravine lands (Sharda and Venkateswarlu 2007). For areas in ravines and extremely erodible soils, silvopastoral systems have been recommended as being ideal (Chaturvedi et al. 2014; Dagar 2018a).

5.4.1.2 Agroforestry Systems for Salt-Affected and Waterlogged Soils

Soil salinization is one of the main obstacles to achieving food security and halting environmental degradation and biodiversity loss. Due to low crop yields and low economic returns on salt-affected soils, conventional agriculture is typically not regarded as economically viable. Physical rehabilitation of salt-affected soils is also difficult, expensive, and time-consuming for farmers due to a lack of adequate advice and training. As a result, it has been reported that forestry and agroforestry systems are desirable options for these soils (Wicke et al. 2013; Dagar 2014; Gupta and Dagar 2016a, b).

Salt-tolerant tree plantations can be grown on salt-affected degraded areas to repair salt-induced soil deterioration and offer a source of energy and other products for rural people. Agroforestry interventions for reclaiming and rehabilitating salt-affected soils include trees for bio drainage, energy plantations, halophytic plants to remediate soil, and dryland salinity. For dryland salinity, more densely planted energy plantations, agricultural, silvopastoral, and fruit-based agroforestry systems have been successfully developed (Dagar et al. 2001a, b; Singh and Dagar 2005; Wicke et al. 2013; Dagar 2014; Gupta and Dagar 2016a, b, 2018a, b; Dagar and Minhas 2016; Dagar and Gupta 2020). Deep plowing, wide spacing, the use of salt-tolerant species, and careful management of soils and plantations are the key agrotechnical and ameliorative techniques used in agroforestry on salt-affected soils.

Dagar et al. (2016a) found timber and fuelwood biomass in clonal *Eucalyptus tereticornis* plantations at varied spacing in shallow water-table settings. These areas were flooded fields (due to seepage of canal water or rising of the water table due to improper irrigation practices). After 6 years, row plantations (1 m 1 m space) produced a total dry biomass production of 49.5 Mg ha^{-1}, while block plantations (2 m 4 m space) produced 193 Mg ha^{-1}, demonstrating the potential of clonal Eucalyptus on waterlogged farmlands. Table 5.4 displays significant plant species and systems.

The twin issues of salinity and waterlogging are present throughout Western Australia. In these locations, it is common to find and make use of a number of woody plant species (Mitchell and Wilcox 1994). The Australian salt land meadows where *Halosarcia pergranulata, H. lepidosperma, H. indica* subsp. *bideris,* and *Maireana brevifolia* are grown. The salt meadows are home to a variety of succulent perennial shrubs that are highly salt-tolerant. *H. pergranulata* is more suitable for sheep grazing since it has a crude protein content of 14% on an oven-dry basis.

5.4.1.3 Agroforestry for Acid Soils

In the tropics, the soils are often acidic and have low availability of nutrients. Some foods contain dangerous amounts of iron and aluminum. Woody perennials may recycle nutrients, conserve soil organic matter, and protect the soil from erosion and runoff, making them advantageous for the management of acidic soils. The usual agroforestry systems include backyard gardens, coffee and cocoa production systems, multi-tiered dense cropping systems based on plantations, and alley cropping on sloping terrain.

Over 90% of the land in the north-eastern Indian states is impacted by variable levels of soil acidity, which restricts crop options for more than 70% of the total cultivated area. As a result, cereal farming, especially rice farming and agroforestry systems based on Alder (*Alnus nepalensis*), is the main form of agriculture. These agroforestry systems, combining arable crops with high-value options like cardamom (*Elettaria cardamomum*), large cardamom (*Amomum subulatum*), pineapple (*Ananas sativum*), numerous fruit trees, and tuber crops like turmeric, ginger, colocasia, and taros, have proven to be effective and sustainable (Sharma et al 2000). These systems assist in soil improvement by fixing nitrogen and boosting the content of organic matter in addition to offering favorable economic yields.

5.4.1.4 Rehabilitation of Mined Areas

Careful planning, execution, and management are necessary because many aspects of the rehabilitation of mined areas must be handled concurrently, including soil conservation and improvement, planting and establishment of suitable species, mulching to prevent soil loss, soil and water conservation measures, and protection from cattle grazing (Raizada and Dhyani 2020). To speed up the growth of vegetation, improve

Table 5.4 Some prominent agroforestry practices on salt-affected lands in Indian sub-continent, and Central Asia

Saline environment and study location	Agroforestry Practices and effective role	Main components	References
Desert salinity in Central Asia and Transcaucasia	Rangeland management (silvopastoral) For wood and fodder production, gum, medicine	Woody species *Populus, Haloxylon, Salix, Elaeagnus angustifolia, Morus, Atriplex, Berberis, Kochia, Salsola, Alhagi, Hippophae rhamnoides*	Khamzina et al. (2008), Toderich et al. (2009)
	Agri-silvicultural practices	Multipurpose salt-tolerant trees, Atriplex and crops such as sorghum, millets, triticale, sesame, quinoa, amaranthus, Jerusalem artichoke, green gram, and peas	Toderich et al. (2018)
Dryland salinity in Indian sub-continent	Silvopastoral and energy plantations for fuelwood and small timber	Tree and shrubs *Eucalyptus, Acacia, Prosopis, Tamarix Atriplex (amnicola, lentiformis, undulata*) and grasses *Leptochloa fusca, Sporobolus helvolus, Brachiaria ramosa, B. mutica, Aeluropus lagopoides, Dichanthium annulatum, Panicum maximum*, etc.	Dagar (2014), Qureshi (2016)
	Fruit-based AF system with saline irrigation	Fruit trees such as Carissa carandas, Aegle marmelos, Emblica officinalis, Ziziphus mauritiana, Cordia rothii Arable crops barley, mustard, pearl millet, cluster bean	Dagar et al. (2016b)
	Silvopastoral established with saline water	Trees species plus grasses such as *Cenchrus ciliaris, C. setigerus, Panicum antidotale, P. coloratum, P. laevifolium, P. virgatum, P. maximum, Cynodon dactylon*	Tomar et al. (2003a, b), Dagar and Minhas (2016)
Salinity Management in Pakistan	Bioremediation and biodrainage plantations	Trees such as *Eucalyptus camaldulensis, Prosopis juliflora, Acacia nilotica, A. ampleceps, Populus deltoides, Tamarix articulata, Atriplex amnicola, A. lentiformis*, grasses *Leptochloa fusca, Chloris gayana, Elytrigia elongata*	Qureshi (2016)
Highly alkali soils (pH > 10) in India	Energy plantations	*Prosopis juliflora, Tamarix articulata, Acacia nilotica, Terminalia arjuna, Eucalyptus tereticornis, Casuarina equisetifolia (frost-sensitive), Pongamia pinnata, Pithecellobium dulce, Cassia siamea, Parkinsonia aculeata*	Dagar et al. (2001b), Singh et al. (2008)

(continued)

Table 5.4 (continued)

Saline environment and study location	Agroforestry Practices and effective role	Main components	References
	Silvopastoral systems	Above tree species plus grasses such as *Leptochloa fusca, Brachiaria mutica, Chloris gayana, Panicum maximum, P. laevifolium, P. antidotale*	Dagar et al. (2001b), Singh and Dagar (2005), Dagar (2014)
Moderate alkali soils (pH 9.0–9.4) in Indo-Gangetic Plains	Agri-silvicultural system	Trees like *Populus deltoides and Eucalyptus tereticornis* with rice–wheat cropping system	Singh et al. (1995)
	Fruit-based AF systems	Fruit trees *Aegle marmelos, Emblica officinalis, Psidium guajava, Syzygium cuminii, Carissa carandas, Achrus zapota* (frost sensitive), *Punica granatum, Ziziphus maurtiana* plus arable crops like wheat, mustard, Egyptian clover (fodder), rice, onion, medicinal like *Ocimum, Matricaria, Vetiver*	Singh et al. (1997), Dagar et al. (2001a, b), Tomar et al. (2004), Singh and Dagar (2005), Dagar (2014)
Waterlogged saline soils in Indo-Gangetic plains	Biodrainage plantations	Cloned *Eucalyptus tereticornis* in wider spaces (plant to plant 1 m; row to row 4–5 m or on acre-line in wheat-rice cropping system	Ram et al. (2011), Dagar et al. (2016b)

soil fertility and moisture retention, stabilize slopes, and encourage the natural establishment of native tree and shrub saplings, fast-growing grasses can be artificially seeded.

Mixed tree species planting may be done two to three years after the grass has grown. According to Juyal et al. (2007), it is advantageous to directly plant tree species along with grasses and leguminous forbs for a period of three years. Extensive work has been done on the reclamation of rock phosphate mines, limestone mines, gypsum mines, dolomite mines, iron ore mines, lignite mines, and coal mine spoils. (Prasad 1989; Soni et al. 1989; Raizada et al. 2002; Raizada and Julal 2012; Manivanan et al. 2016).

In Amarkantak, *Eucalyptus camaldulensis, Grevillea pteridifolia, Pinus caribaea,* and *Acacia auriculiformis* were planted in an effort to successfully recover a bauxite-mined region. Manivanan et al. (2016) revealed that in Goa, out of eight species examined for the rehabilitation of iron ore overburdens, cashew, gooseberry, mango, and Jamun (*Syzygium cuminii*) outperformed the other species. Fodder legume Stylo (*Stylosanthese hamata*) was determined to be the best cover crop. Dhyani et al. (2007) reviewed some case studies of the afforestation of mines in India while

Chaturvedi et al. (2014), Dagar and Gupta (2016), and Raizada and Dhyani (2020) reported different plant species suitable for the revegetation of different mine spoils (Table 5.5).

5.5 Carbon Sequestration Potential of Agroforestry Systems for Climate Change Mitigation

According to estimates of C accumulation in different AF systems, soils up to 1 m deep accumulated between 30 and 300 Mg C ha^{-1} yr^{-1} while aboveground plant components accumulated between 0.29 and 15.2 Mg C ha^{-1} yr^{-1} (Nair et al. 2009a, b). According to estimates of C accumulation in various AF systems, aboveground plant components accumulated between 0.29 and 15.2 Mg C ha^{-1} yr^{-1} while soils up to 1 m deep accumulated between 30 and 300 Mg C ha^{-1} yr^{-1} (Nair et al. 2009a, b). The efficiency of agroforestry systems as carbon sinks depends on a variety of factors, including species composition, tree age, location, regional weather patterns, and management techniques. For instance, at the age of 11 years, *Dalbergia sissoo* was able to amass 48–52 Mg ha^{-1} of biomass in an agri-silvicultural system (Newaj and Dhyani 2008).

In cloned *Eucalyptus tereticornis*, Dagar et al. (2016a) showed 33.2 Mg wood biomass per ha sequestering 15.2 Mg C ha^{-1} after 6 years of growth. According to national inventory, Newaj et al. (2020) showed that the total carbon stock under baseline in various Indian states ranged from 14.5 to 33.48 Mg C ha^{-1} and that the net carbon sequestered over a simulated period of 30 years was 11.35 Mg C ha^{-1}. The carbon sequestration potential of agroforestry systems on farmers' fields ranged from 0.11 to 0.82 Mg C ha^{-1} yr^{-1}, with a mean value of 0.35 Mg C ha^{-1} yr^{-1} across different states in India. The ability of agroforestry systems to store carbon is primarily controlled by the climate of a given region or state, the density of trees, the types of trees selected, and the management techniques used to grow both trees and crops. Higher SOC stocks are seen in India's silvopastoral systems (33.5 Mg C ha^{-1}), according to a recent investigation by Nath et al. (2021).

According to a well-conducted nationwide study, Newaj et al. (2020) found that the net carbon sequestered over a simulated period of 30 years was 11.35 Mg C ha^{-1}, and that the total carbon stock under baseline in different states of India ranged from 14.5 to 33.48 Mg C ha^{-1}. With a mean value of 0.35 Mg C ha^{-1} yr^{-1} across different states, the carbon sequestration potential of agroforestry systems on farmers' fields ranged from 0.11 to 0.82 Mg C ha^{-1} yr^{-1}. The carbon sequestration potential of agroforestry systems is primarily influenced by specific district or state climatic conditions, tree density, the nature of tree species chosen, and the management practices employed in cultivating both trees and crops. Agrosilvopastoral systems in India have higher SOC stocks than agri-silvicultural systems (44.1 Mg C ha^{-1}) or silvopastoral systems (33.5 Mg C ha^{-1}), according to a recent investigation by Nath et al. (2021).

Table 5.5 Plant species suitable for revegetation of some mine spoils in India

Category of mine spoils	Suitable plant species
Bauxite mined area of Madhya Pradesh (MP), Gujarat, Maharashtra, Karnataka	*Acacia auriculiformis, Albizia procera, Eucalyptus camaldulensis, Grevillea pteridifolia, G. robusta, Emblica officinalis, Pongamia pinnata, Melia azedarach, Pinus* spp, *Shorea robusta, Dendrocalamus strictus*
Coal mine spoils of Singrauli and in MP and West Bengal	*Ziziphus mauritiana, Dodonaea viscosa, Shorea robusta, E. camaldulensis, E. hybrid, Pongamia pinnata, Leucaena leucocephala, Azadirechta indica, Terminalia arjuna, Pongamia pinnata, Butea monosperma, Dendrocalamus strictus* bamboo
Limestone mine spoils of outer Himalayas, Uttarakhand	*Acacia catechu, Agave americana, Arundo donax, Bauhinia retusa, Buddleja asiatica, Boehmeria rugulosa, Woodfordia exserta D. sissoo, Emblica officinalis, Pongamia pinnata, Acacia auriculiformis, Erythrina suberosa, Ipomoea carnea, Leucaena leucocephala, Salix tetrasperma, Vitex negundo, Mimosa himalayana, Rumex hastatus* and grasses *Chrysopogon fulvus, Eulaliopsis enega Pennisetum purpureum*
Mica, copper, tungsten, marble, dolomite, limestone, etc. mine spoils of Rajasthan and MP	Some tree species like *Eucalyptus hybrid, Grewia tenax, P. juliflora, Salvadora oleoides, Tamarix articulata, Ziziphus nummularia*, and grasses *Cenchrus setigerus, Cymbopogon* spp, *Cynodon dactylon, Dichanthium annulatum, Sporobolus marginatus*, etc.
Iron ore wastes in Orissa, Chattisgarh, Durg (MP), and Goa	Some tree species like *Albizia lebbeck*, bamboos, *Dalbergia sissoo, Emblica officinalis, Eucalyptus* spp, *Casuarina equisetifolia, Tamarindus indica, Azadirechta indica, Mangifera indica, Garcinia indica, Syzygium cuminii, Artocarpus heterophyllus, Anacardium occidentale, Leucaena leucocephala* and local plants
Hematite/magnetite, manganese spoils from Karnataka	*Albizia lebbeck, Tectona grandis, Gmelina arborea, Azadirechta indica, Dalbergia sissoo*, and local plant species
Zinc mines in AP, Rajasthan, Bihar, Maharashtra, Karnataka	*Anogeissus pendula, Acacia catechu, A. pendula, Holoptelia integrifolia, Butea monosperma, Ziziphus mauritiana, and grasses Dichanthium annulatum, Cenchris ciliaris, Paspalidium geminatum*

Source Compiled from different sources

5.6 Adaptation Strategies to Climate Change

Rockstorm et al. (2009) proposed a set of nine planetary sustainable boundaries within which humanity can continue to develop and thrive for generations to come. Recently, scientists quantified and updated all nine Planetary Boundaries that regulate Earth's stability and resilience in September 2023 (Richardson et al. 2023;Fig. 5.4). This update revealed that six of the nine boundaries have been crossed, increasing the risk of significant environmental changes (Richardson et al. 2023; Fig. 5.4). Although drastic shifts may not occur immediately, these boundaries signify a critical threshold for risks to be faced by both humans and ecosystems. Therefore, comprehensive policies must be put into place on a worldwide scale in all sectors to reduce emissions (mitigation) and create plans for surviving in the face of climate change (adaptation). Agriculture, forestry, and other related land-use systems play a significant role in uplifting the livelihood and ensuring food security; hence, it is essential that they are developed properly while employing cutting-edge, environmentally friendly technologies. For sustainable food production in the face of a changing environment, agroforestry may be essential.

The creation of new land-use systems, the evolution of climate-smart agronomic management strategies for climate change scenarios, the investigation of opportunities for soil property maintenance, restoration, and enhancement, and the creation of new resource use such as agroforestry-based cropping systems should be the focus of adaptation strategies to address climate change. There is a need to investigate ways to increase the value of agricultural products, as well as strategies to introduce nitrogen-fixing trees and crops and preserve, improve, and enhance soil characteristics, popularizing technologies for resource conservation.

5.6.1 Developing Trees and Crops for Stress Tolerance

The creation of diverse stress-tolerant cultivars, including those that can withstand heat, drought, salinity, waterlogging, inundation, cold, frost, elemental toxicities, and cultivars resistant to pests and diseases, is necessary to meet the demands imposed by a changing climate. This would call for intensive breeding activities involving the gathering, preservation, and exchange of suitable agricultural genetic material among plant breeders and other researchers. The genetic resources, particularly landraces from regions with past climates similar to those predicted for the future, could be used as the seed stock for creating genes that can withstand stress.

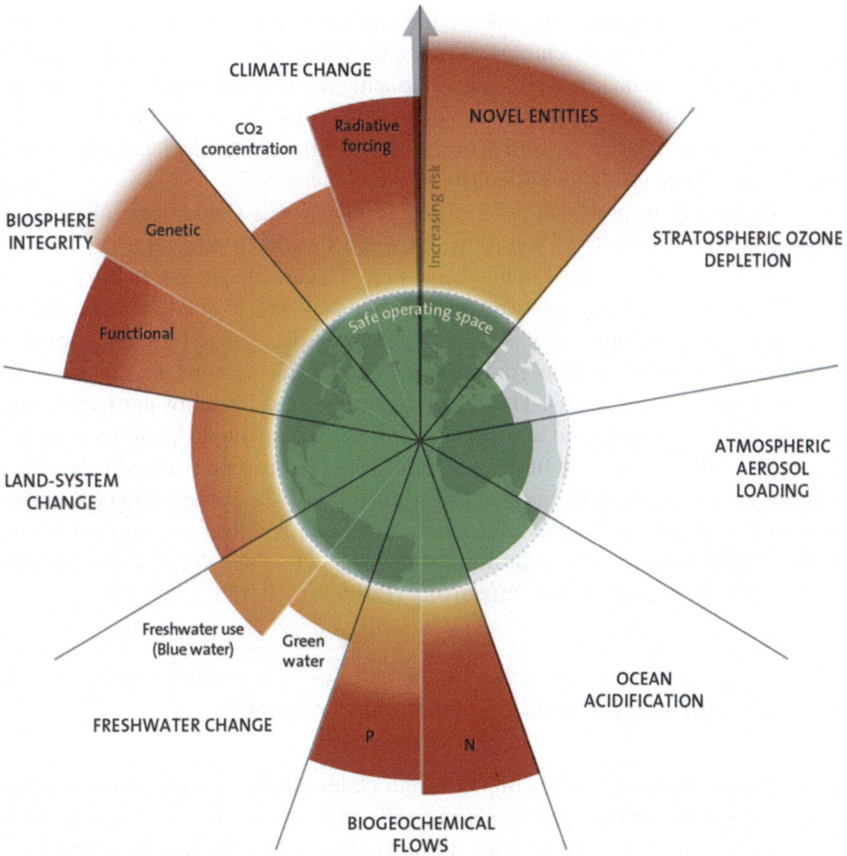

Fig. 5.4 The 2023 update to the Planetary Boundaries, as originally proposed by Rockström et al. (2009). Licensed under CC BY-NC-ND 3.0. Azote for Stockholm Resilience Centre, based on analysis in Richardson et al. (2023)

5.6.2 Development of New Land-Use Systems

Salinity and flooding provide substantial difficulties for farmers in West Bengal and Bangladesh's coastal regions, making using these lands for crop production impossible. However, there have been new initiatives to reconfigure these lands and boost agricultural output. To test several land-shaping strategies, Burman et al. (2013, 2015) undertook a study in the Sundarbans region of the Ganges Delta (West Bengal) and tsunami-affected areas in the Andaman and Nicobar Islands. The land-shaping techniques based on Burman et al. (2013) are briefly discussed.

The land-shaping technologies developed and implemented by Burman et al. (2013) focused on improving drainage facilities to effectively remove excess water from the fields, mitigating the risk of inundation and waterlogging. Rainwater-harvesting techniques were employed to capture and store rainwater, providing a valuable resource during dry periods and reducing reliance on saline water sources. The outcomes of this research could provide valuable insights for addressing similar challenges in other coastal areas of West Bengal and Bangladesh, where salinity and inundation continue to hinder agricultural development.

It is interesting to note that 50% of the farmland was slanted into alternating furrows (3 m top width, 1.5 m bottom width, and 1.0 m depth) and ridges (1.5 m top width, 3 m bottom width, and 1 m height) as part of a land-shaping strategy (Burman et al. 2013). The ridges' layout made for effective drainage, avoiding water ponding and reducing the development of soil salinity. These ridges proved suitable for cultivating plantation crops such as fruits and vegetables during both the rabi (winter) and kharif (monsoon) seasons. The furrows were utilized for rainwater harvesting, allowing water to be collected and stored for use as life-saving irrigation during the rabi season. Additionally, the furrows were used for cultivating rice and fish, effectively utilizing the remaining parts of the original field. The land-shaping technique provided a well-balanced approach to utilize the land for diverse crop cultivation, rainwater harvesting, and fish farming, addressing the challenges of salinity and inundation in the coastal regions effectively (Burman et al. (2013, 2015).

5.6.3 Agroforestry-Based Agricultural Systems

Recent research has significantly improved our understanding of (agro)forestry plantations, particularly on degraded lands and with saline waters. It is evident that, with appropriate reclamation technologies, these lands can be successfully utilized for agroforestry programs. However, the implementation of reclamation solutions for degraded lands is hindered by their high cost, time-consuming nature, and socio-political constraints. Consequently, the rehabilitation of degraded and waterlogged lands is progressing slowly, despite the available technical knowledge.

With the addition of trees to farms and agricultural landscapes through the practice of agroforestry, there are more social, economic, and environmental advantages. Twenty percent of the world's population, or 1.2 billion people, severely depends on agroforestry goods and services for survival (ICRAF 2000). Marketable products, additional income, and advantages like enhanced soil fertility due to nitrogen fixation and nutrient recycling, which leads to higher crop yields and stable production over the long run, are all provided to farmers by including trees in farming operations. Agroforestry is a beneficial adaptation for land-use systems that offers a solution to reduce hazards associated with the climate.

Agroforestry improves the functions and services provided by ecosystems, increasing the resilience of agricultural systems (Gupta et al 2023). Agroecosystems are strengthened through restoration or creation, which also enhances nutrient cycling, pest control, and ecosystem health. Agricultural systems can better tolerate climate-related shocks. There are numerous advantages of trees on farms including moisture retention, a reduction in soil erosion, and topsoil protection. By storing carbon in their biomass and soil, trees contribute to the biodiversity of ecosystems and are a key factor in combating global warming. Within agroforestry, agricultural/silvopastoral systems serve as carbon sinks by temporarily storing carbon (see Gupta et al 2023). Agroforestry systems with trees reduce the effects of temperature extremes by providing shade and altering microclimates. They may protect crops and cattle from heat stress while also giving shelter to pollinators and helpful insects. In regions where temperatures are rising as a result of climate change, this control may be especially crucial.

5.6.4 Conservation Agriculture vis-à-vis SMART Agriculture

The goal of SMART agriculture, or the sustainable management of agricultural resources and practices, is to maximize the use of the agricultural resources presently available while considering specific soil and environmental conditions. SMART agriculture, when combined with conservation agriculture (CA), uses adaptive and mitigating techniques for sustainable agricultural output, assisting in ensuring food security (FAO 2014).

Sustainable Management of Agrarian Resources and Technologies, or SMART agriculture is an approach to crop production that deals with the management of existing agricultural resources using the most recent management techniques and farm equipment under a specific set of edaphic and environmental circumstances. Therefore, by employing adaptive and mitigating techniques/strategies for sustainable agricultural production, SMART will contribute to food security if CA is implemented at the appropriate moment with the necessary resources in a certain typological area.

By conserving crop leftovers and avoiding tilling the soil before each planting, CA holds up well as an adaptation method in the expected scenario of climate change for increasing organic matter in soils and creating a healthy soil ecosystem. CA fits into the sustainable intensification paradigm of producing more with fewer purchased inputs and progressively improving the resource base's productivity and capacity to provide ecosystem services. Numerous research in India's irrigated production systems shows that CA-based management practices aid in improving agriculture's resilience to climate variability through improved climate change adaptation (Jat et al. 2016). Compared to the traditional tillage-based wheat production system (CT; Aryal et al. 2016), CA is more resilient to climate extremes.

In the projected scenario of climate change, CA holds good as adapting strategy to build up organic matter in soils and create a healthy soil ecosystem by retaining the crop residues and not tilling the soil before each planting. CA fits within the sustainable intensification paradigm of producing more from less purchased inputs and enhancing the resource base, productivity, and ecosystem service provision capacity over time. In CA systems, the three interconnected principles of little or minimal mechanical soil disturbance, preservation of soil mulch cover, and diverse cropping system provide the foundation for the outputs of desired products and ecosystem services. This type of farming suits these methods. Any rainfed or irrigated production system, including arable, horticulture, agroforestry, plantations, pasture, crop-livestock, and mixed systems, will be better able to respond to any applied production inputs to achieve intensification when practices based on these principles and supported by other "good agricultural practices" are used.

Hobbs and Govaerts (2010) assert that CA reduces runoff and soil erosion by increasing the quantity of water-stable aggregates, enhancing water-holding capacity, and speeding up infiltration. Additionally, it enhances biological porosity, increases the number of microbes in the soil, especially earthworms, and makes the soil more permeable to roots. Studies have indicated that CA can increase soil carbon sequestration at a rate ranging from around 0.2 to 1.0 Mg ha-1 year-1, depending on the agroecological location and management practices (Corsi et al. 2012). Increased soil productivity, protection from future environmental damage due to unsustainable heavy tillage practices, and (i) aid in lowering greenhouse gas emissions that result in global warming are all benefits of sequestering soil organic carbon (SOC), as discussed by Jat et al. (2022).

5.6.5 Resource Conservation Technologies (RCTs)

The term "resource-conserving technologies" (RCTs) is used in South Asia to describe a group of procedures that act as first steps in the adoption of Conservation Agriculture (CA) concepts. There are a number of creative RCTs in South Asian agriculture. These include site-specific nutrient management, diversification/intensification, alternate land uses/agroforestry, bed planting, zero tillage, direct seeding rice (DSR), residue management utilizing turbo-seeder, alternating wetting and drying in rice, and alternate land uses. These technologies are intended to respond rapidly to urgent requirements and address problems with the region's climate change and agriculture economics.

5.6.6 Efficient Water Management

Climate change poses a significant challenge for irrigated areas, potentially surpassing current irrigation capacity due to water shortages. However, farmers without access to irrigation are even more vulnerable to these changing scenarios. Promoting micro-irrigation is vital to maximizing water productivity, making each drop count. Investment in integrated water management, including addressing inefficiencies in water delivery systems, requires farmer participation to be successful.

It is inevitable that low-quality water will be used in agriculture. Several salt-tolerant fruit and forest trees that can be grown in block plantations or as agroforestry trees have been identified based on a number of investigations. These trees can be established using saline water up to EC 10 dS m^{-1}. In addition, a number of agroforestry for salt-affected soils listed above arable and medicinal crops have been found and assessed for their viability to cultivate successfully utilizing saline water for irrigation. These plants and trees are used for making building materials, firewood, food, fruits, fodder, and medicine. Numerous halophytes can also be grown in saline water and used as energy crops, forages and animal feeds, oil seeds, and traditional foods. See Dagar and Minhas (2016) for more information.

5.7 Conclusion and the Way Forward

With regard to addressing challenges with food security, and adapting to climate change, and conserving the natural resources of tropical environments, agroforestry offers a lot of promise. For smallholder farmers in poor countries that face pressures from climate change and land degradation, agroforestry offers a realistic path to resilient livelihoods and food security. The effects of rising temperatures and atmospheric CO_2 concentrations on soil properties, including biological traits, as well as the physiology of trees and crops, require in-depth investigation. It is essential to consider the local climatic conditions, ecological context, and socio-economic considerations when implementing agroforestry-based climate change adaptation techniques. For agroforestry efforts to be scaled up and to be successful over the long term, it is crucial to have local knowledge, build local capacity, and support local policies.

Making decisions on policy at the local and regional levels is vital to advancing smart agriculture and promoting conservation practices. Measures including accurate land leveling, water harvesting, conservation and management techniques, and the effective use of low-quality water resources should be given priority in these strategies. Degraded lands should also be restored, site-specific nutrient management techniques should be used, integrated weed and pest control techniques should be adopted, and crop varieties with multiple stress tolerances should be developed. These policies must include capacity-building programs for risk and weather forecasting systems as well as the deployment of climate-resilient technologies. Field crop, vegetable, and horticultural crop genetic improvement can be crucial in developing successful adaption tactics. Investigating economically significant natural species that can survive in challenging environments.

Agroforestry is an effective method for raising the food security and climate change resilience of agricultural systems. Combining the farming of trees with the production of crops or livestock produces a more diverse and sustainable farming system. To encourage the adoption of agroforestry systems, appropriate legislation and regulations must be developed. Farmers who employ agroforestry practices require incentives from the government, such as subsidies, tax exemptions, and technical assistance. According to policy frameworks, agroforestry should be a part of national initiatives for forestry, agriculture, and climate change.

References

Akinnifesi FK, Leakey RRB, Ajayi OC, Silishi GW, Tchoundjeu Z, Matakala P, Kwesegi FR (eds) (2008) Indigenous fruit trees in the tropics: domestication, utilization and commercialization. CAB International, Oxfordshire. ISBN 978-1-8-84593-110-0

Al Muzaini S (2003) Environmental measures to control sand movement in Kuwait. In: Alsharhan AS, Wood WW, Goudie AS, Fowler A, Abdellatif EM (eds) Desertification in the third millennium. Swets and Zeitlinger Publishers, Lisse, The Netherlands, pp 309–313

Aryal JP, Sapkota TB, Stirling CM, Jat ML, Jat HS, Rai M, Mittal S, Sutaliya JM (2016) Conservation agriculture-based wheat production better copes with extreme climate events than conventional tillage-based systems a case of untimely excess rainfall in Haryana. India Agric Ecosyst Environ 233:325–335

Beckley BD, Callahan PS, Hancock DW, Mitchum GT, Ray RD (2017) On the "cal-mode" correction to TOPEX satellite altimetry and its effect on the global mean sea level time series. J Geophys Res Ocean 122:8371–8384. https://doi.org/10.1002/2017JC013090

Berg A, de Noblet-Ducoudre N, Sultan B, Lengaigne M, Guimberteau M (2013) Projections of climate change impacts on potential C4 crop productivity over tropical regions. Agric Meteorol 170:89–102

Blasing TJ (2008) Recent greenhouse gas concentrations. Dataset

Burman D, Mandal S, Bandopadhyay BK, Maji B, Sharma DK, Mahanta KK, Sarangi SK, Mandal UK, Patra S, De S, Patra S, Mandal B, Maitra NJ, Ghoshal TK, Velmurugan A (2015) Unlocking production potential of degraded coastal land through innovative land management practices: a synthesis. J Soil Salinity Water Qual 7(1):12–18

Burman D, Bandyopadhyay BK, Mandal, Subhasis, Mandal UK, Mahanta KK, Sarangi SK, Maji B, Rout S, Bal AR, Gupta SK, Sharma DK (2013) Land shaping—a unique technology for improving productivity of coastal land. Bulletin No. CSSRI/Canning Town/ Bulletin/2013/02. Central Soil Salinity Research Institute, Regional Research Station, Canning Town, 38 p

Chaturvedi OP, Kaushal R, Tomar JMS, Prandiyal AK, Panwar P (2014) Agroforestry for wasteland rehabilitation: Mined, ravine, and degraded watershed areas. In: Dagar JC, Singh AK, Arunachalam A (eds) Agroforestry systems in India: livelihood security and ecosystem services, advances in agroforestry, vol 10. Springer, New Delhi, pp 233–272

Corsi S, Friedrich T, Kassam A, Pisante M, de Moraes Sá J (2012) Soil organic carbon accumulation and greenhouse gas emission reductions from conservation agriculture: a literature review. Integr Crop Manage 16:89

Crowley TJ (2002) Cycles, cycles everywhere. Science 295:1473–1474

Dagar JC (2014) Greening salty and waterlogged lands through agroforestry systems for livelihood security and better environment. In: Dagar JC, Singh A, Arunachalam A (eds) Agroforestry systems in India: livelihood security & environmental servicesadvances in agroforestry, vol 10. Springer India, New Delhi/Dordrecht/New York, pp 273–332

Dagar JC (2018a). Ravines: formation, extent, classification, evolution and measures of prevention and control. In: Dagar JC and Singh AK (eds) Ravine lands: greening for livelihood and environmental security. Springer Nature Singapore Pte Ltd, pp 19–67

Dagar JC (2018b) Perspectives of vegetation ecology and biodiversity for management of ravine lands. In: Dagar JC and Singh AK (eds) Ravine lands: greening for livelihood and environmental security. Springer Nature Singapore Pte Ltd, pp 69–118

Dagar JC, Gupta SR (2016) Agroforestry: potentials for rehabilitation of degraded lands, constraints and the way forward. In: Dagar JC, Tewari JC (eds) Agroforestry research developments. Nova Publishers, New York, pp 47–97

Dagar JC, Sharma HB, Shukla YK (2001a) Raised and sunken bed technique for agroforestry on alkali soils of northwest India. Land Degrad Dev 12:107–118

Dagar JC, Singh G, Singh NT (2001b) Evaluation of forest and fruit trees used for rehabilitation of semi-arid alkali/ sodic soils in India. Arid Land Res Manage 15:115–133

Dagar JC, Singh AK, Singh R, Arunachalum A (2012) Climate change vis-a-vis Indian agriculture. Ann Agric Res New Ser 33(4):189–203

Dagar JC, Lal K, Jeet-Ram M-K, Chaudhari SK, Yadav RK, Sharif-Ahamad SG, Kaur A (2016a) Impact of eucalyptus geometry on biomass production, watertable drawdown, carbon sequestration and inter-crop yield on waterlogged saline soils of northwest India. Agric Ecosys Environ 233:32–45

Dagar JC, Yadav RK, Tomar OS, Minhas PS, Yadav G, Lal K (2016b) Fruit-based agroforestry systems for saline water irrigated semi-arid hyperthermic camborthids regions of north-west India. Agrofor Syst 90(6):1123–1132

Dagar, JC Gupta SR (2020) Agroforestry interventions for rehabilitating salt-affected and waterlogged marginal landscapes. In: Dagar JC, Gupta SR, Teketay D (eds) Agroforestry for degraded landscapes: recent advances and emerging challenges vol 2, Springer Nature Singapore Pte Ltd, pp 111–162

Dagar JC, Minhas PS (2016) Agroforestry for management of waterlogged saline soils and poor-quality waters. Advances in agroforestry series, vol 13. Springer India, New Delhi/Dordrecht/New York, p 210

Dagar JC, Singh AK, Arunachalam, A (eds) (2014) Agroforestry systems in India: livelihood security & ecosystem services. Advances in Agroforestry. https://doi.org/10.1007/978-81-322-1662-9

Dagar JC, Gupta SR and Teketay D (2020a) Agroforestry for degraded landscapes: recent advances and challenges, vol 1 Springer Nature Singapore Pte Ltd, p 554

Dagar JC, Gupta SR and Teketay D (2020b) Agroforestry for degraded landscapes: recent advances and challenges, vol 2. Springer Nature Singapore Pte Ltd, p 475

Dhyani SK, Samra JS, Ajit, Handa AK, Uma (2007) Forestry to support increased agricultural production: focus on employment generation and rural development. Agric Econ Res Rev 26(2):179–202

Dollinger J, Jose S (2018) Agroforestry for soil health. Agrofor Syst 92:213–219

Easterling WE, Aggarwal PK, Batima P, Brander KM, Erda L, Howden SM, Kirilenko A, Morton J, Soussana JF, Schmidhuber J, Tubiello FN (2007) Food, fibre and forest products. Climate change 2007: impacts, adaptation and vulnerability. Contribution of working group II to the 4th assessment report on the Inter-Governmental Panel on Climate Change, pp 273–313

ELD Initiative and UNEP (2015) The economics of land degradation in Africa: benefit of action out-weigh the costs. www.eld-initiative.org

FAO (2014) What is conservation agriculture? http://www.fao.org/ag/ca/1a.html. Accessed 05 July 2014

FAO (2022) Global symposium on salt-affected soils: outcome document. Rome. https://doi.org/10.4060/cb9929en

FAO and ITPS (2015) Status of the world's soil resources (SWSR): main report. Food and Agriculture Organization of the United Nations and Intergovernmental Technical Panel on Soils, Rome

Forster P, Ramaswamy V, Artaxo P, Berntsen T, Betts R, Fahey DW, Haywood J, Lean J, Lowe DC, Mhre G, Nganga J, Prinn R, Raga G, Schulz M, van Dorland R (2007) Changes in atmospheric constituents and in radiative forcing. In: Solomon S, Qin D, Manning M, Chen Z, Marquis M, Averyt KB, Tignor M, Miller HL (eds) Climate change 2007: the physical science basis. Contribution of working group 1 to the fourth assesment report of the intergovernmental panel on climate change. Cambridge University Press, Cambridge, pp 129–234

Frederikse T, Landerer F, Caron L, Adhikari S, Parkes D, Humphrey VW, Dangendorf S, Hogarth P, Zanna L, Cheng L, Wu YH (2020) The causes of sea-level rise since 1900. Nature 584(7821):393–397. https://doi.org/10.1038/s41586-020-2591-3

Garrity DP, Akinnifesi FK, Ajoyi OC, Weldesemayat SG, Mowo JG, Kalinganire A, Larwanou M, Bayalo J (2010) Evergreen agriculture: a robust approach to sustainable food security in Africa. Food Security 2(3):197–214

Gupta SR, Dagar JC (2016a) Agroforestry for ecological restoration of salt-affected lands. In: Dagar JC, Sharma PC, Sharma DK, Singh AK (eds) Innovative saline agriculture. Springer India, Dordrecht, pp 161–182

Gupta SR, Dagar JC (2016b) Enhancing environmental services of salt-affected lands through agroforestry. In: Dagar JC, Tewari JC (eds) Agroforestry research developments. Nova Publishers, New York, pp 209–244

Gupta SR, Dagar JC, Sileshi GW, Chaturvedi RK (2023) Agroforestry for climate change resilience in degraded landscapes. In: Dagar JC, Gupta SR, Sileshi GW (eds) Agroforestry for sustainable intensification of agriculture in Asia and Africa. Sustainability sciences in Asia and Africa. Springer, Singapore. https://doi.org/10.1007/978-981-19-4602-8_5

Hillbrand A, Borelli S, Conigliaro M, Olivier A (2017) Agroforestry for landscape restoration: exploring the potential of agroforestry to enhance the sustainability and resilience of degraded landscapes. Food and Agriculture Organization of the United Nations, Rome, Italy

Hobbs PR, Govaerts B (2010) How conservation agriculture can contribute to buffering climate change. In: Reynolds MP (ed) Climate change & crop production, CABI climate change series. CAB International, Oxfordshire, pp 177–199

ICRAF (2000) World centre of agroforestry. www.icraf.org

IPCC (2007a) Summary for policymakers. Climate change: the physical science basis. Contribution of working group I to the fourth assessment report of the Intergovernmental Panel on Climate Change, Cambridge University Press, Cambridge

IPCC (2007b) Cliate change: climate change impacts, adaptation and vulnerability. Intergovernmental panel on climate change report.

IPCC (2014) Climate change 2014: impacts, adaptation and vulnerability—summary for policymakers. IPCC WGII AR5, Phase I Report Launch, p 44

IPCC (2021) Summary for policymakers. In: Masson-Delmotte et al (eds) Climate change 2021: the physical science basis. Contribution of working group I to the sixth assessment report of the intergovernmental panel on climate change. Cambridge University Press

Jaradat AA (2003) Halophytes for sustainable biosaline farming systems in the Middle East. In: Alsharhan AS, Wood WW, Goudie WS, Fowler A, Abdellatif AM (eds) Desertification in the third millennium. Swets & Zeitlinger Publishers, The Netherlands, pp 187–204

Jat ML, Dagar JC, Sapkota TB, Singh Y, Govaerts B, Ridaura SL, Saharawat YS, Sharma RK, Tetarwal JP, Jat RK, Hobbs H, Stirling C (2016) Climate change and agriculture: adaptation strategies and mitigation opportunities for food security in South Asia and Latin America. Adv Agron 137:127–235

Jat ML, Chakraborty D, Ladha JK, Parihar CM, Datta A, Mandal B, Nayak HS, Maity P, Rana DS, Chaudhari SK, Gerard B (2022) Carbon sequestration potential, challenges, and strategies towards climate action in smallholder agricultural systems of South Asia. Crop Environ 1:86–101. https://doi.org/10.1016/j.crope.2022.03.005

Juyal GP, Katiyar VS, Dhadwal KS, Joshie P, Arya RK (2007) Mined area rehabilitation in Himalaya-Shastradhara experience. Central Soil and Water Conservation Research and Training Institute, Dehradun, India

Khamzina A, Lamers JPA, Vlek PLG (2008) Tree establishment under deficit irrigation on degraded agricultural land in the lower Amu Darya River region, Aral Sea Basin. For Ecol Manage 255:168–178

Kuyah S, Sileshi GW, Luedeling E, Akinnifesi FK, whitney CW, Bayala J, Kuntashula E Dimobe K, Mafongoya PL (2020) Potential of agroforestry to enhance livelihood security in Africa. In: Dagar JC, Gupta SR, Teketay D (eds) Agroforestry for degraded landscapes-recent advances and emerging challenges, vol 1. Springer Nature, Singapore Pte Ltd, pp 135–167

Manivanan S, Sikka AK, Priya Devi S, Ashok Kumar J, Singh NP, Manjunath BL (2016) Biological reclamation of iron ore overburden. Technical Bulletin No.57. ICAR-Central Coastal Agricultural Research Institute, Goa

Mitchell AA, Wilcox DG (1994) Arid Shrubland plants of Western Australia. University of Western Australia Press and Department of Agriculture, Western Australia, Perth, WA, p 478

Mullen DJ, Barett-Lennard EG (2010) Breeding crops for tolerance to salinity, waterlogging and inundation. In: Reynold MP (ed) Climate change and crop production. CABI, Wallingford, pp 92–114

NAAS (2009) State of Indian agriculture. National Academy of Agricultural Sciences, New Delhi, pp 256

Nair PKR (1993) An introduction to agroforestry. Kluwer Academic Publishers, Dordrecht, The Netherlands

Nair PKR, Kumar BM, Nair VD (2009a) Agroforestry as a strategy for carbon sequestration. J Plant Nutr Soil Sci 172:10–23

Nair PKR, Nair VD, Kumar BM, Haile SG (2009b) Soil carbon sequestration in tropical agroforestry systems: a feasibility appraisal. Environ Sci Pol 12:1099–1111

Nath AJ, Sileshi GW, Laskar SY, Pathak K, Reang D, Nath A, Das AK (2021) Quantifying carbon stocks and sequestration potential in agroforestry systems under divergent management scenarios relevant to India's nationally determined contribution. J Clean Prod 281:124831. https://doi.org/10.1016/j.jclepro.2020.124831

Newaj R, Dhyani SK (2008) Agroforestry for carbon sequestration: scope and present status. Indian J Agroforestry 10:1–9

Newaj R, Chaturvedi Op, Kumar D, Chavan SB, Rajawat BS, Yadav DK (2020) Carbon sequestration potential of agroforestry systems in rehabilitating degraded lands in India. In: Dagar JC, Gupta SR, Teketay D (eds) Agroforestry for degraded landscapes-recent advances and emerging challenges, vol 2. Springer Nature, Singapore Pte Ltd, pp 319–348

Parandiyal AK, Sethy BK, Somasunndaram J, Ali S, Meena HR (2020) Potential of agroforestry for the rehabilitation of degraded ravine lands. In: Dagar JC, Gupta SR, Teketay D (eds) Agroforestry for degraded landscapes-recent advances and emerging challenges, vol 2. Springer Nature, Singapore Pte Ltd, pp 229–251

Porter JJ, Dessai S, Tompkins EL (2014) What do we know about UK household adaptation to climate change? A systematic review. Clim Change 127:371–379

Prasad R (1989) Removal of forest covers through mining and technology for retrieval. J Trop for 5:109–116

Qureshi RH, Nawaz S, Mahmood T (1993) Performance of selected tree species under saline sodic field conditions in Pakistan. In: Leith H, Al Masoom A (eds) Towards the rational use of high salinity tolerant plants, vol 2. Kluwer Academic Publisher, Dordrecht, The Netherlands, pp 259–269

Qureshi AS (2016) Perspectives for bio-management of salt-affected soils in Pakistan. In: Dagar JC, Minhas PS (eds) Agroforestry for management of waterlogged saline soils and poor-quality waters. Advances in agroforestry series, vol 13. Springer India, New Delhi/Dordrecht/New York, pp 97–108

Raizada A, Juyal GP (2012) Tree species diversity, species regeneration and biological productivity of seeded *Acacia catechu* Willd in rehabilitated limestone mines in the North West Indian Himalayas. Land Degrad Dev 23:167–174

Raizada A, Singh C, Singh G (2002) Production, nutrient dynamics and breakdown of leaf litter in six forest plantations raised on gravelly flood plains in the lower western Himalayas. J Trop for Sci 14(4):499–512

Raizada A, Dhyani SK (2020) Agroforestry approach for the rehabilitation of mine spoils. In: Dagar JC, Gupta SR, Teketay D (eds) Agroforestry for degraded landscapes-Recent advances and emerging challenges, vol 2. Springer Nature Singapore Pte Ltd, pp 271–295

Ram J, Dagar JC, Lal K, Singh G, Toky OP, Tanwar RS, Dar SR, Mukesh K (2011) Biodrainage to combat water logging, increase farm productivity and sequester carbon in canal command area of north-west India. Curr Sci 100(11):1673–1680

Rockström J, Steffen W, Noone K, Persson Å, Chapin FS III, Lambin E, Lenton TM, Scheffer M, Folke C, Schellnhuber H, Nykvist B, De Wit CA, Hughes T, van der Leeuw S, Rodhe H, Sörlin S, Snyder PK, Costanza R, Svedin U, Falkenmark M, Karlberg L, Corell RW, Fabry VJ, Hansen J, Walker B, Liverman D, Richardson K, Crutzen P, Foley J (2009b) Planetary boundaries:

exploring the safe operating space for humanity. *Ecology and Society* 14:32. http://www.ecologyandsociety.org/vol14/iss2/art32/

SCRIPPS (2020) Credit to Scripps Institution of Oceanography at the University of California San Diego. https://scripps.ucsd.edu/programs/keelingcurve/wp-content/plugins/sio-bluemoon/graphs/mlo_full_record.png

Sharda VN, Juyal GP and Naik BS (2008) Watershed development in India: status and perspective. Allied Printers, Dehradun, India, p 219

Sharda VN, Venkateswarlu B (2007) Crop diversification and alternate land use systems in watershed management. In: Best-bet options for integrated watershed management. Proceedings of the comprehensive assessment of watershed programs in India, 25–27 July 2007. ICRISAT, Patancheru, pp 111–128

Sharma E, Sharma R, Singh KK and Sharma G (2000) A boon for mountain populations. Large cardamom farming in the Sikkim Himalayas. Mt Research Devel 20(2):108–111

Sileshi GW, Teketay D, Gebrekirstos A, Hadgu K (2020) Sustainability of *Faidherbia albida*-based agroforestry in crop production and maintaining soil health. In: Dagar JC, Gupta SR, Teketay D (eds) Agroforestry for degraded landscapes-recent advances and emerging challenges, vol 2. Springer Nature, Singapore Pte Ltd, pp 349–369

Singh G, Singh NT, Dagar JC, Singh H, Sharma VP (1995) An evaluation of agriculture, forestry and agroforestry practices in moderately alkali soil in northwestern India. Agrofor Syst 37:279–295

Singh G, Dagar JC, Singh NT (1997) Growing fruit trees in a highly alkali soils—a case study. Land Degrad Dev 8:257–268

Singh G, Dagar JC (2005) Greening sodic soils: Bichhian model. Technical Bulletin No.2 /2005. Cental Soil Salinity Research Institute, Karnal, 51 p

Singh YP, Sharma DK, Singh G, Nayak AK, Mishra VK, Singh R (2008) Alternate land use management for sodic soils. CSSRI Technical Bull No. 2/2008. CSSRI, Karnal, 16 p

Soni P, Vasistha HB, Kumar O (1989) Biological diversity in surface mined areas after reclamation. Ind for 115(7):475–482

Stocker TF, Qin D, Plattner G-K, Alexander LV, Allen SK, Bindoff NL, Bréon, F-M, Church JA, Cubasch U, Emori S, Forster P, Friedlingstein P, Gillett N, Gregory JM, Hartman DL, Jansen E, Kirtman B, Knutti R, Krishna Kumar K, Lemke P, Marotzke J, Masson-Delmott V, Meehl GA, Mokhov II, Piao S, Ramaswami V, Randall D, Rhein M, Rojas M, Sabine C, Shindell D, Talley LD, Vaughan DG, Xie S-P (eds) (2013) Technical summary. In: Climate change 2013: the physical science basis. Contribution of working group I to the fifth assessment report of the intergovernmental panel on climate change. Cambridge University Press, pp. 33–115. https://doi.org/10.1017/CBO9781107415324.005.tocker

Tewari JC, Ram M, Roy MM, Dagar JC (2014) Livelihood improvement and climate change adaptations through agroforestry in hot arid environments. In: Dagar JC, Singh AK, Arunachalam A (eds) Agroforestry systems in India: livelihood security and ecosystem services, advances in agroforestry, vol 10. Springer, New Delhi, Dordrecht, pp 155–184

Toderich KN, Shuyskaya EV, Ismail S, Gismatullina LG, Radjabov T, Bekchanov BB, Aralova DB (2009) Phytogenic resources of halophytes of Central Asia and their role for rehabilitation of sandy desert degraded rangelands. Land Degrad Dev 20:386–396

Toderich KN, Shoaib I, Khunshazarov TM, Khasankhanova GM (2018) Biosaline technologies and approaches to the management of salt-affected agricultural lands under arid climate conditions (by examples in Central Asia and Transcaucasia). In: Vargas R, Pankova EI, Balyuk SA, Krasilnikov PV, Khasankhanova GM (eds) Handbook for saline soil management. Food and Agriculture Organization of the United Nations and Lomonosov Moscow State University, Rome and Moscow, pp 60–72

Tollefson J (2021) IPCC climate report: earth is warmer than it's been in 125,000 years. Nature 596:171–172

Tomar OS, Minhas PS, Sharma VK, Gupta RK (2003a) Response of nine forage grasses to saline irrigation and its schedules in a semi-arid climate of north-west India. J Arid Environ 55:533–544

Tomar OS, Minhas PS, Sharma VK, Singh YP, Gupta RK (2003b) Performance of 31 tree species and soil condition in a plantation established with saline irrigation. For Ecol Manag 177:333–346
Tomar OS, Dagar JC, Singh YP (2004) Forest and fruit trees for alkali soils. Indian Farm 53(11):44–47
Torralba M, Fagerholm N, Burgess PJ, Moreno G, Plieninger T (2016) Do European agroforestry systems enhance biodiversity and ecosystem services? A meta-analysis. Agric Ecosyst Environ 230:150–161
UNCCD (United Nations Convention to Combat Desertification) (2022) The global land outlook, 2nd edn. UNCCD, Bonn
van Noordwijk M (2018) Integrated natural resource management as pathway to powerty reduction: innovating practices, institutions and policies. Agric Syst 172:60–71
van Noordwijk M, Ekadinata A, Leimona B, Catacutan D, Martini E, Tata HL, Oborn I, Hairiah K, Wangpakapattanawong P, Mulia R, Dewi S, Rahayu S, Zulkarnain T (2020) Agroforestry options for degraded landscapes in Southeast Asia. In: Dagar JC, Gupta SR, Teketay D (eds) Agroforestry for degraded landscapes-recent advances and emerging challenges, vol 1. Springer Nature Singapore Pte Ltd, pp 307–347
Wicke B, Edward SWM, Akanda R, Stille L, Singh RK, Awan AR, Mahmood K, Faaij AP (2013) Biomass production in agroforestry and forestry systems on salt affected soils in South Asia: exploration of the GHG balance and economic performance on three case studies. J Environ Manage 127:324–334
Zomer RJ, Neufeldt H, Xu J, Ahrends A, Bossio D, Trabucco A, van Noordwijk M, Wang M (2016) Global tree cover and biomass carbon on agricultural land: the contribution of agroforestry to global and national carbon budgets. Sci Rep 6:29987. https://doi.org/10.1038/srep29987. (Illustrations shared under the licence CCBY4.0)

Dr. Jagdish Chander Dagar is a former Assistant Director General and Emeritus Scientist at ICAR, internationally recognized for his research in agricultural and ecological sciences. He has authored over 300 research papers, 13 books, and numerous chapters on agroforestry, biosaline agriculture, and sustainable land management. Dr. Dagar has received numerous awards, including the Sajjad Memorial Gold Medal and the Hari Om Ashram Trust Award. He is a Fellow of several prestigious societies and has served as a consultant to FAO, CIMMYT, and other international organizations. His work on halophytes, climate change, and biodiversity conservation has earned him national and international acclaim.

Dr. Sharda Rani Gupta is a renowned ecologist and former Dean of Life Sciences at Kurukshetra University, India. She has made significant contributions to biodiversity and ecosystem functioning in grasslands and forests, soil biodiversity, and carbon sequestration in agroforestry systems. With 98 research papers and numerous projects completed, she has also guided 20 Ph.D. students. Her work has received several awards and she has collaborated with international institutions such as the Soil Science Department at Rothamsted Research, UK.

Chapter 6
Effects of Incentive Mechanisms for Promoting Sustainable Intensification Pathways Within Cocoa Agroforestry Landscapes in Efoulan, Cameroon

Dieudonne Alemagi, Lalisa Duguma, Peter A. Minang, and Kalame Fobissie

Abstract Today, incentive mechanisms are used to encourage practices that promote intensification of cocoa agroforestry landscapes (CAL) through the use of farm inputs, renewal efforts, as well the integration of timber and fruit-producing trees into these landscapes. In this book chapter, we profile the effects of incentives mechanisms (training schemes and the provision of seedlings) that have been formulated and implemented by the International Centre for Research in Agroforestry (ICRAF) to advance sustainable intensification pathways within CAL in the Efoulan municipality located in the South Region of Cameroon in Central Africa. Results indicate that these incentive mechanisms have had a great effect in the municipality as all respondents who had received seedlings, training on seedlings production techniques, and training on tree domestication and improvement techniques were putting into practice these techniques. However, this research identifies a considerable number of challenges that impede the implementation of these incentive mechanisms and proffer possible options for overcoming these challenges. It is hoped that an in-depth analysis of this nature would contribute in enhancing the role of these incentive mechanisms in advancing sustainable intensification of CAL, thereby addressing overarching issues like food insecurity, deforestation, forest degradation, and other developmental challenges.

Keywords Incentives mechanism · Cocoa agroforestry landscapes · Intensification · Efoulan

---

D. Alemagi (✉) · K. Fobissie
FOKABS INC., Ottawa, ON, Canada
e-mail: dalemagi@yahoo.co.uk

K. Fobissie
e-mail: kfobissie@fokabs.com

L. Duguma · P. A. Minang
World Agroforestry Centre, Nairobi, Kenya
e-mail: A.Minang@CGIAR.ORG

M. Behnassi et al. (eds.), *Food Systems and Biodiversity in the Context of Environmental and Climate Risks*, https://doi.org/10.1007/978-3-031-89167-0_6

6.1 Introduction

From a broader perspective, an incentive mechanism generally refers to anything that serves as motivation or a stimulus for people to act (FAO 1999; Enters 2004). Incentives can be both financial and non-financial and can be used by the government and other non-governmental organizations (NGOs) to lure, spur up, and encourage ecological stewardship (Alemagi et al. 2006). For example, in Southeast Asia, incentive mechanisms have been employed to promote agroforestry practices. For instance, in the Philippines, the total forested area is expanded by an additional 0.65 million hectares in 2003 due to incentive policies that facilitated tree planting practices in smallholder agroforestry systems (Pulhin et al. 2006; Catacutan et al. 2012).

In Central and West Africa, many governments, NGOs, and donor bodies have implemented a series of incentive mechanism (like training schemes, distribution of inputs,[1] and provision of credit schemes) in an ambitious attempt at promoting sustainable intensification pathways (tree improvement, tree planting, the use of inputs as well as farm renewal through the replacement of aging tree stocks with improved planting materials or quality germplasm) within cocoa agroforestry landscapes (CAL). For instance, as part of its policy to increase cocoa production, the government of Ghana in 2010 initiated an incentive program to rehabilitate old, destroyed, and abandoned CAL in the country. Implemented by the Cocoa Swollen Shoot Virus Disease Program (CSSVDP), it encouraged intensification of CAL by supplying tree seedlings—18–20 native timber tree species per hectare—which were planted with cocoa seedlings within CAL (DOREO PARTNERS 2012) by cocoa farmers. In the same vein, Cameroon's Ministry of Agricultural and Rural Development (MINADER) provides extension workers who disseminate information about proper agricultural practices and technologies to rural farmers.

While many authors (see, for example Feudjio et al. 2012; Alemagi et al. 2014a, b) have identified incentive mechanisms that are instrumental for the promotion of intensification and diversification schemes within cocoa agroforestry systems in the Efoulan municipality located in the South Region of Cameroon, little is known about the extent to which these incentive mechanisms are encouraging farmers to engage in tree domestication and improvement, the use of inputs, and farm renewal within CAL, if at all. Focusing upon the Efoulan as a case study, the aim of this chapter is to profile the effects of incentive mechanisms (like provision of seedlings, training on seedling production techniques, and training on tree domestication and improvement techniques) implemented in this municipality by the International Centre for Research in Agroforestry (ICRAF) to promote sustainable intensification pathways within CAL.

Structurally, the chapter is divided into six sections: The first section is the introduction; the second provides a rationale for sustainable intensification of CAL; research methods are presented in the third section, while the results are profiled in the fourth section; in the penultimate section, the results are discussed, while the last section provides concluding remarks.

[1] Inputs here include fertilizers and pesticides like fungicide and insecticides.

6.2 Background

In general, in most countries of the Congo Basin where primary and secondary forest is converted into CAL, the rate of deforestation remains high. A case in point is Cameroon where it is estimated that between 2000 and 2010, the annual rate of deforestation was 1.04% (FAO 2011) with agriculture being the main driver (Robiglio et al. 2010). However, sustainable intensification pathways can contribute to enhancing multi-functionality of CAL by providing the following climate change mitigation, adaptation, and conservation benefits:

- *First*, Gockowski and Sonwa (2011) showed that if intensification of CAL in Central and West Africa (Cote D'Ivoire, Ghana, Nigeria, Cameroon) through the use of farm inputs and the incorporation of timber producing trees was done in the late 1960s, about 21,000 km^2 of forests would have been preserved, thus contributing to a reduction in emissions of about 1.4 billion tons of CO_2. This is in contrast to the widespread extension of cocoa, cassava, and oil palm into forest landscapes by about 68,000 km^2 over the same period.
- *Second*, according to the 'land sparing' concept posited by Borlaug (2007): First, investments are made in agriculture that lead to an increase in productivity on a given piece of land through the use of inputs and better technology; once these interventions permit an adequate supply of food, fuel, and fiber, this leads to a decrease in deforestation rates resulting from agriculture thereby sparing more forest lands; third, the incorporation of timber and fruit-producing trees into the CAL, as well as the use of farm inputs within this landscape can increase productivity, the farmer thus has several income streams and can remain on the same land and in so doing reducing deforestation (Minang et al. 2014).
- *Fourth*, the intensification of CAL with shade producing trees can greatly improve the fertility of the soil and provide the soil with organic matter via litter and wastes emanating from wild animals that use these trees as their natural habitat (Hartemink 2005). These functions can enhance soil quality, thereby improving cocoa production.
- *Fifth*, it is posited that the intensification of CAL with different tree species could also be a means by which farmers can adapt or resist climatic shocks (Gockowski and Sonwa 2011). This is deeply rooted in the fact that if an epidemic eradicates the cocoa plants, which are the principal sources of income, a farmer with an intensified CAL has other alternative outputs which can be economically beneficial for the farmer—e.g., timber and fruits for sale.
- *Sixth*, it has been demonstrated that, comparatively, intensified CAL in Cameroon's forest-dependent communities are more profitable than less intensified ones (Eboutou 2009). As profitability increases, the demand for land reduces thus sparing the forest (Eboutou 2009; Alemagi et al. 2014a, b).
- *Seventh*, Minang et al. (2014) posit that the sustainable intensification of CAL delivers prospects for profitable labor absorption that would otherwise engage in forest clearing.

- *Finally*, the analysis of Robiglio et al. (2011) demonstrated that on-farm timber within CAL is increasing becoming one of the main sources of timber in some tropical countries like Cameroon. Therefore, with increasing demand for fuelwood and timber, if on-farm timber and fuelwood production is increased, this might result in a decrease in deforestation and forest degradation.

6.3 Methods

6.3.1 Case Study Municipality and the Selected Communities

The study was conducted in five communities in the Efoulan municipality (Fig. 6.1) in the South Region of Cameroon: Bongolo, Mengale, Minto, Ngonebok, and Binyina. The selection of the Efoulan municipality was based on the following considerations: an estimated 95% of the municipality's 250,000 people are currently engaged in shifting cultivation for small-scale traditional agriculture with cocoa being the main agricultural product (Feudjio et al. 2012; Yemefack et al. 2013); and key stakeholders such as the ICRAF have implemented a series of incentive mechanism (like training schemes) to encourage intensification of CAL in the municipality.

The sampling of the case study communities was nonrandom and purposive. We selected the case study communities with the help of ICRAF technicians (that had a mastery of the communities) as well as local collaborators with whom we had previous professional collaborations. The selection was done purposively in order to have a representation of a variety of several forest-dependent communities that had received training as an incentive mechanism for intensification of CAL. While the communities sampled may not be representative of all communities in the Efoulan, however, they reflect a diversity of experiences that offer interesting insights into the impacts of incentive mechanism for intensification of CAL.

6.3.2 Data Collection and Analysis

Field visits took place in December 2018, during which semi-structured interviews were conducted with community members. Key informants who had benefited from incentive mechanisms for intensification of CAL were identified with the help of local collaborators. Interviews with community informants were designed to elicit information on: training received on seed production techniques and other related information; training received on tree domestication and improvement techniques; commercialization of seedlings and plants produced as a result of training, as well as other related information; and training, seedlings, and material needs from ICRAF. A total of 36 interviews were conducted, alongside a review of relevant forestry and agriculture documents pertaining to incentive mechanisms. Again, the rationale

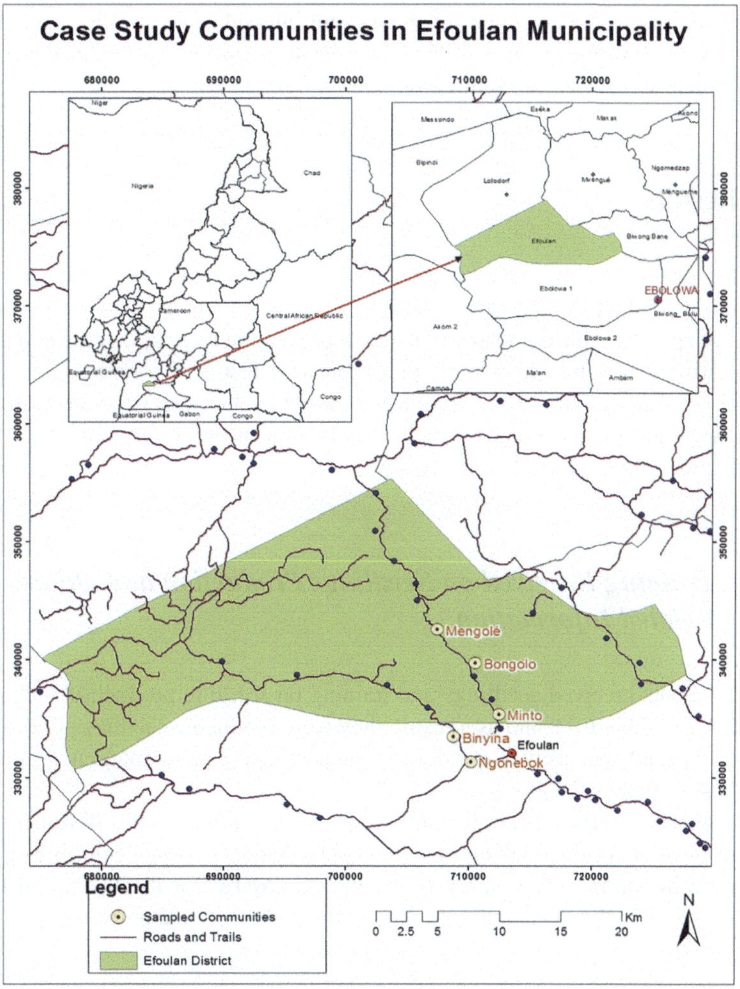

Fig. 6.1 Case study communities located in the Efoulan municipality in the South Region of Cameroon

for choosing these farmers was borne out of the consideration that there they had benefited from incentive mechanisms for intensification of CAL.

We recruited and trained two competent research assistants to administer the questionnaires and help conduct the interviews in all the case study communities. Interviews were conducted mostly at homes in French or the local dialect (*Bulu*), and each interview lasted about one hour, on average. To avoid the biased responses, interviewers avoided leading the interviewees during the interview process (Kvale 1996; Ammenberg 2003) and behaved in a neutral manner to avoid signaling preferred answers (Ammenberg 2003). Descriptive statistics (means and frequencies) were used for data analysis and were computed using Microsoft Excel 2010.

Mean responses were calculated for the 4-point interval scale question, and a one-way analysis of variance ($\alpha = 0.05$) was used to measure significant differences between means.

6.4 Results

The main results of this study are presented in five sections: (i) training received on seedling production and other related information; (ii) level of confidence in putting into practice tree domestication and improvement techniques received during training; (iii) extent to which trees produced as a result of training received on marcotting, grafting, and cutting are integrated into the farm; (iv) commercialization of seedlings and plants produced and other related information; and (v) training, seedlings, and material needs from ICRAF.

6.4.1 Training Received on Seedlings Production and Other Related Information

All respondents received seedlings and training on seedling production techniques from ICRAF (Table 6.1), and as a result, they were producing seedlings (Table 6.2). A variety of seed species were produced, and some were being integrated into their cocoa farms (Table 6.2).

It is important to underscore that all the species mentioned in the aforementioned table are used for commercialization and for subsistence purposes. Simply put, these species are important agroforestry tree products (AFTs) for the livelihood of the farmers, especially for consumption, marketing, and medicinal purposes.

6.4.2 Level of Confidence in Putting into Practice Tree Domestication and Improvement Techniques Received During Training

Respondents were asked to rate their level of knowledge of various opinions regarding training on tree domestication and improvement techniques received from ICRAF. A four-point interval scale varying from '1 = strongly agree' to '4 = completely disagree' was used (Fig. 6.2). A one-way ANOVA test ($\alpha = 0.05$) supports the claim that there is no significant difference between the means of the responses. The results indicate that farmers had a good knowledge regarding most of the tree domestication and improvement techniques received from ICRAF.

Table 6.1 Number of trainings received on seedling production by respondents and area of cocoa farm

Number of respondents	Number of trainings received	Estimated area of cocoa farm (ha)
1	1	0.5
2	2	0.75
3	3	1
4	2	0.7
5	1	4
6	2	1
7	4	2
8	1	0.7
9	5	2
10	4	2
11	5	1.5
12	7	3
13	3	4
14	2	1.5
15	2	1
16	4	1.5
17	5	1.5
18	1	2
19	1	1
20	1	1
21	1	1
22	1	2.5
23	1	0.5
24	2	6
25	1	2
26	2	2.5
27	2	3.5
28	2	2
29	3	3
30	3	3
31	3	6
32	3	1
33	2	1.5
34	3	2
35	3	3
36	1	1.5

Table 6.2 Number of seedlings produced by respondents, species produced, as well as the number and species integrated into cocoa farms

Number of respondents	Number of seedlings produced	Species produced	Number of seedlings integrated into farm	Species integrated into the farm
1	90	*Theobroma cacao*	13	*Citrus sinensis*
		Persea americana		*Persea americana*
		Citrus sinensis		
2	46	*Dacryodes edulis*	5	*Persea americana*
		Mangifera indica		*Citrus reticula*
		Persea americana		
		Citrus reticula		
3	100	*Dacryodes edulis*	30	*Persea americana*
		Mangifera indica		*Mangifera indica*
		Persea americana		*Dacryodes edulis*
		Citrus sinensis		
		Irvingia gabonensis		
4	70	*Theobroma cacao*	0	*0*
		Persea americana		
		Irvingia gabonensis		
5	835	*Theobroma cacao*	75	*Theobroma cacao*
		Mangifera indica		*Mangifera indica*
		Persea americana		*Citrus reticula*
		Citrus reticula		*Citrus sinensis*
		Citrus sinensis		
		Irvingia gabonensis		
6	345	*Dacryodes edulis*	245	*Dacryodes edulis*
		Theobroma cacao		*Theobroma cacao*
		Irvingia gabonensis		*Irvingia gabonensis*

(continued)

Table 6.2 (continued)

Number of respondents	Number of seedlings produced	Species produced	Number of seedlings integrated into farm	Species integrated into the farm
7	385	*Dacryodes edulis*	275	*Dacryodes edulis*
		Theobroma cacao		*Theobroma cacao*
		Persea americana		*Persea americana*
		Citrus sinensis		*Citrus sinensis*
8	340	*Dacryodes edulis*	330	*Dacryodes edulis*
		Theobroma cacao		*Theobroma cacao*
		Persea americana		*Persea americana*
9	375	*Dacryodes edulis*	156	*Dacryodes edulis*
		Mangifera indica		*Mangifera indica*
		Theobroma cacao		*Theobroma cacao*
		Persea americana		*Persea americana*
		Moringa oleifera		*Citrus reticula*
		Citrus reticula		*Irvingia gabonensis*
		Irvingia gabonensis		*Annona muritica*
		Annona muritica		
10	500	*Dacryodes edulis*	250	*Dacryodes edulis*
		Mangifera indica		*Mangifera indica*
		Theobroma cacao		*Theobroma cacao*
		Citrus sinensis		*Citrus sinensis*
		Cola nitida		*Cola nitida*
		Irvingia gabonensis		*Irvingia gabonensis*
11	1291	*Dacryodes edulis*	1290	*Dacryodes edulis*
		Mangifera indica		*Mangifera indica*
		Theobroma cacao		*Theobroma cacao*
		Persea americana		*Persea americana*

(continued)

Table 6.2 (continued)

Number of respondents	Number of seedlings produced	Species produced	Number of seedlings integrated into farm	Species integrated into the farm
		Cola nitida		*Cola nitida*
		Citrus reticula		*Citrus reticula*
		Irvingia gabonensis		*Irvingia gabonensis*
		Garcinia kola		*Garcinia kola*
12	403	*Dacryodes edulis*	323	*Dacryodes edulis*
		Theobroma cacao		*Theobroma cacao*
		Persea americana		*Persea americana*
		Cola nitida		*Cola nitida*
		Citrus reticula		*Citrus reticula*
13	925	*Dacryodes edulis*	35	*Dacryodes edulis*
		Theobroma cacao		*Persea americana*
		Persea americana		*Moringa oleifera*
		Moringa oleifera		*Cola nitida*
		Cola nitida		*Citrus sinensis*
		Citrus sinensis		*Garcinia kola*
		Garcinia kola		
14	53	*Dacryodes edulis*	32	*Dacryodes edulis*
		Theobroma cacao		*Persea americana*
		Persea americana		*Citrus reticula*
		Citrus reticula		*Irvingia gabonensis*
		Irvingia gabonensis		
15	260	*Dacryodes edulis*	90	*Dacryodes edulis*
		Persea americana		*Persea americana*
		Citrus sinensis		*Citrus sinensis*
16	351	*Theobroma cacao*	293	*Persea americana*

(continued)

Table 6.2 (continued)

Number of respondents	Number of seedlings produced	Species produced	Number of seedlings integrated into farm	Species integrated into the farm
		Persea americana		*Cola nitida*
		Cola nitida		*Citus aurantifolia*
		Citus aurantifolia		
17	1708	*Dacryodes edulis*	1473	*Dacryodes edulis*
		Theobroma cacao		*Theobroma cacao*
		Persea americana		*Persea americana*
		Citrus sinensis		*Citrus sinensis*
		Irvingia gabonensis		*Irvingia gabonensis*
		Vitis vinifera		*Shea butter*
		Shea butter		
18	2074	*Dacryodes edulis*	424	*Dacryodes edulis*
		Mangifera indica		*Mangifera indica*
		Theobroma cacao		*Theobroma cacao*
		Persea americana		*Persea americana*
		Annona muritica		*Annona muritica*
19	250	*Persea americana*	250	*Persea americana*
		Dacryodes edulis		*Dacryodes edulis*
		Citrus sinensis		*Citrus sinensis*
		Theobroma cacao		*Theobroma cacao*
20	1352	*Theobroma cacao*	1352	*Theobroma cacao*
		Persea americana		*Persea americana*
		Irvingia gabonensis		*Irvingia gabonensis*
		Dacryodes edulis		*Dacryodes edulis*
21	415	*Dacryodes edulis*	80	*Dacryodes edulis*

(continued)

Table 6.2 (continued)

Number of respondents	Number of seedlings produced	Species produced	Number of seedlings integrated into farm	Species integrated into the farm
		Theobroma cacao		*Theobroma cacao*
		Persea americana		*Persea americana*
22	1183	*Theobroma cacao*	652	*Theobroma cacao*
		Irvingia gabonensis		*Irvingia gabonensis*
		Dacryodes edulis		*Dacryodes edulis*
		Dacryodes macrophyla		
		Baillonella toxisperma		
		Cola nitida		
		Psidium gojava		
		Mangifera indica		
		Garcinia kola		
23	600	*Theobroma cacao*	0	*0*
24	618	*Dacryodes edulis*	378	*Dacryodes edulis*
		Irvingia gabonensis		*Baillonella toxisperma*
		Baillonella toxisperma		*Theobroma cacao*
		Theobroma cacao		*Persea americana*
		Persea americana		*Citrus sinensis*
		Citrus sinensis		
25	525	*Theobroma cacao*	525	*Theobroma cacao*
		Dacryodes edulis		*Dacryodes edulis*
		Irvingia gabonensis		*Irvingia gabonensis*
26	275	*Theobroma cacao*	247	*Theobroma cacao*
		Persea americana		*Persea americana*

(continued)

Table 6.2 (continued)

Number of respondents	Number of seedlings produced	Species produced	Number of seedlings integrated into farm	Species integrated into the farm
		Mangifera indica		*Mangifera indica*
		Dacryodes edulis		*Dacryodes edulis*
		Annona muritica		*Annona muritica*
27	790	*Theobroma cacao*	765	*Theobroma cacao*
		Dacryodes edulis		*Dacryodes edulis*
		Persea americana		*Persea americana*
		Citrus sinensis		*Citrus sinensis*
		Mangifera indica		*Mangifera indica*
		Psidium gojava		*Psidium gojava*
		Irvingia gabonensis		*Irvingia gabonensis*
		Moringa oleifera		
28	147	*Dacryodes edulis*	122	*Dacryodes edulis*
		Citrus sinensis		*Citrus sinensis*
		Mangifera indica		*Mangifera indica*
		Cola nitida		*Cola nitida*
		Theobroma cacao		*Theobroma cacao*
		Persea americana		*Persea americana*
29	480	*Theobroma cacao*	347	*Theobroma cacao*
		Persea americana		*Persea americana*
		Dacryodes edulis		*Dacryodes edulis*
		Citrus reticula		*Citrus reticula*
		Cocos nucifera		*Cocos nucifera*
		Irvingia gabonensis		*Irvingia gabonensis*
30	1141	*Theobroma cacao*	111	
		Dacryodes edulis		*Dacryodes edulis*
		Citrus sinensis		*Citrus sinensis*
		Cola nitida		*Cola nitida*

(continued)

Table 6.2 (continued)

Number of respondents	Number of seedlings produced	Species produced	Number of seedlings integrated into farm	Species integrated into the farm
		Mangifera indica		*Mangifera indica*
		Citrus reticula		*Citrus reticula*
31	588	*Theobroma cacao*	588	*Theobroma cacao*
		Persea americana		*Persea americana*
		Baillonella toxisperma		*Baillonella toxisperma*
		Cola nitida		*Cola nitida*
		Dacryodes edulis		*Dacryodes edulis*
		Citrus sinensis		*Citrus sinensis*
		Mangifera indica		*Mangifera indica*
		Dacryodes Macrophylla		*Dacryodes Macrophylla*
32	445	*Theobroma cacao*	379	*Theobroma cacao*
		Citrus sinensis		*Citrus sinensis*
		Persea americana		*Persea americana*
		Dacryodes edulis		*Dacryodes edulis*
		Cola nitida		*Cola nitida*
		Baillonella toxisperma		*Baillonella toxisperma*
		Moringa oleifera		*Moringa oleifera*
		Irvingia gabonensis		*Irvingia gabonensis*
33	625	*Theobroma cacao*	625	*Theobroma cacao*
		Persea americana		*Persea americana*
		Citrus reticula		*Citrus reticula*
		Dacryodes edulis		*Dacryodes edulis*
		Moringa oleifera		
		Irvingia gabonensis		
34	80	*Theobroma cacao*	50	*Theobroma cacao*

(continued)

Table 6.2 (continued)

Number of respondents	Number of seedlings produced	Species produced	Number of seedlings integrated into farm	Species integrated into the farm
		Dacryodes edulis		*Dacryodes edulis*
		Persea americana		
35	1067	*Theobroma cacao*	1037	*Theobroma cacao*
		Dacryodes edulis		*Dacryodes edulis*
		Persea americana		*Persea americana*
		Citrus sinensis		*Citrus sinensis*
36	1648	*Theobroma cacao*	1620	*Theobroma cacao*
		Persea americana		*Persea americana*
		Dacryodes edulis		*Dacryodes edulis*
		Cola nitida		*Cola nitida*
		Irvingia gabonensis		*Irvingia gabonensis*
		Mangifera indica		*Mangifera indica*

6.4.3 Extent to Which Trees Produced as a Result of Training Received on Marcotting, Grafting, and Cutting Are Integrated into the Farm

Apart from one respondent, all respondents acknowledged that they had received training on marcotting, grafting, and cutting from ICRAF, and as a result were producing young trees from these techniques, which were being incorporated into their CAL (Table 6.3).

6.4.4 Commercialization of Seedlings and Plants Produced and Other Related Information

Young trees produced were not commercialized yet. However, when agroforestry tree products (AFTs) are eventually produced, they are mostly consumed or used for subsistence purposes, or commercialized when there is excess (Table 6.3). Respondents identified other reasons for not commercializing the seedlings and young plants

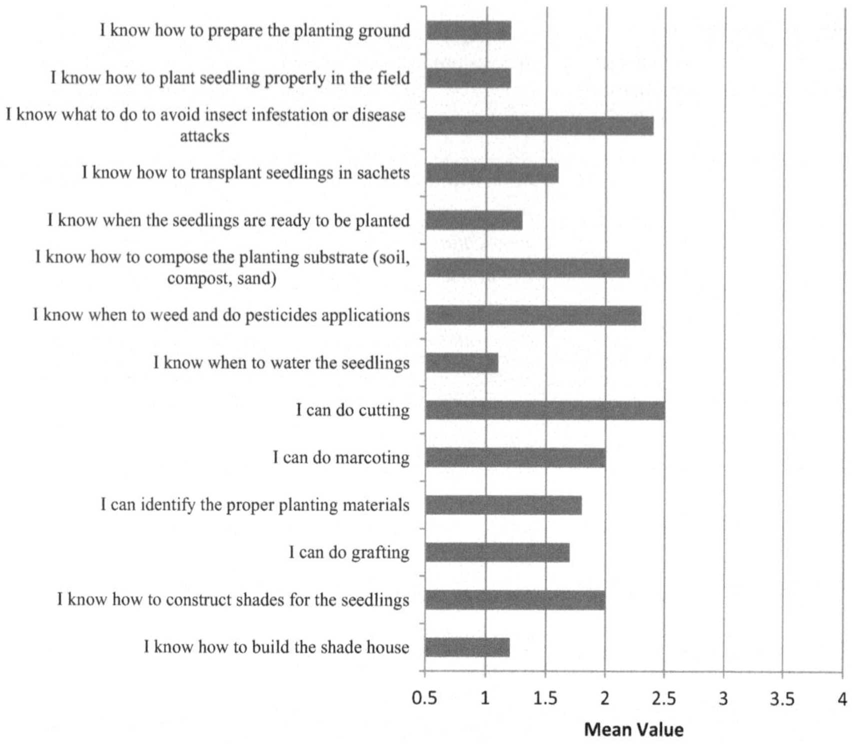

Fig. 6.2 Mean levels of knowledge regarding the mastery of tree domestication and improvement techniques received from ICRAF by respondents (*1 = Strongly Agree*; *4 = completely disagree*)

including limited market, poor quality of the seedlings, and low production. Respondents also acknowledged that during the production process, some of the seedlings and young plants are dead and provided several reasons to underpin seedlings and plant mortality (Table 6.4). Some of them had taken actions to combat seedlings and plant mortality including watering, changing planting materials, application of pesticides, and weeding.

6.4.5 Training, Seedlings, and Material Needs from ICRAF

Most respondent (64%) stated that while they were able to put into practice the training received thus far from ICRAF, they still needed some training from ICRAF. Some examples of areas where training was needed included grafting, cutting, marcotting, cocoa cultivation and production, construction of a shade house, input application, weed management, and disease management. That said, fewer respondents (41%) noted that they needed more seedlings from ICRAF perhaps as a result

Table 6.3 Plant species produced, number planted in CAL, and their use

Species produced	Number planted	Primary use	Secondary use
Citrus sinensis	156	Fruits for consumption, shade for cocoa plants	Medicinal value, fruits for commercialization
Persea americana	542	Fruits for consumption, shade for cocoa plants	Fruits for commercialization
Mangifera indica	115	Fruits for consumption, shade for cocoa plants	Medicinal value, fruits for commercialization
Dacryodes edulis	586	Fruits for consumption, shade for cocoa plants, fruits for commercialization	Fruits for commercialization, fruits for consumption,
Theobroma cacao	13,198	Cocoa bean for commercialization	–
Citrus reticulata	89	Fruits for consumption, shade for cocoa plants	Fruits for commercialization
Irvingia gabonensis	451	Shade for cocoa plants, fruits for consumption, fruits for commercialization	Fruits for consumption, fruits for commercialization
Cola nitida	72	Shade for cocoa plants, fruits for consumption	Fruits for commercialization
Baillonella toxisperma	14	Shade for cocoa plants	Fruit and seeds for commercialization and consumption, medicinal value
Garcinia kola	6	Shade for cocoa plants, fruits for consumption	Fruits for commercialization
Dacryodes macrophylla	20	Shade for cocoa plants, fruits for consumption	Fruit for commercialization
Moringa oleifera	4	Medicinal value	Fruits for commercialization
Psidium guajava	8	Fruits for consumption	Fruits for commercialization
Annona muritica	5	Shade for cocoa plants, fruits for consumption	Fruits for commercialization
Shea butter	20	Shade for cocoa plants	–
Citus aurantifolia	14	Fruits for commercialization	–
Annona muritica	4	Shade for cocoa plants	Fruits for consumption and commercialization
Cocos nucifera	1	Fruits for consumption	Shade for cocoa plants

Table 6.4 Dead tree species, their numbers, and reasons underpinning mortality

Species	Number dead	Reason for dying
Theobroma cacao	1225	No pesticides application, lack of water, delay in treatment with pesticides, application of wrong pesticides, grazing by animals, excess light
Persea americana	254	Poor water absorption, drought or excessive heat, poor transportation of marcots into sachets, lack of pesticides, lack of proper shade, lack of proper care because of traveling imperatives, grazing by animals
Dacryodes edulis	86	Did not weed on time, excessive drought conditions, lack of proper shade, poor seed quality, lack of proper care because of traveling imperatives, grazing by animals
Irvingia gabonensis	93	Drought conditions
Dacryodes macrophyla	90	Excessive drought conditions, lack of proper care because of traveling imperatives
Citrus reticulata	13	Poor management, drought, poor seed quality, and grazing by animals
Baillonella toxisperma	18	Excessive drought conditions, lack of proper care because of traveling imperatives
Mangifera indica	36	Did not weed on time, excessive drought conditions
Citrus sinensis	11	Excessive drought conditions, attack by pest and diseases
Annona muritica	25	Poor seed quality
Psidium gojava	18	Excessive drought conditions, lack of proper care because of traveling imperatives
Garcinia kola	5	Excessive drought conditions, lack of proper care because of traveling imperatives
Cola nitida	1	Excessive drought conditions

of the fact that most of them had received enough supply from ICRAF. In the same vein, only 44% of respondents said they needed more material support from ICRAF. Support in the form of sachets, watering cans, sprayers, pesticides, pruning shears, grafting material, and gloves were identified by respondents as examples of material they needed.

6.5 Discussion

This study attempted to profile the effects of incentives mechanisms implemented by ICRAF in the Efoulan municipality of the South Region of Cameroon to promote sustainable intensification of CAL. All respondents had received training on seedling production techniques and were producing and planting seedlings in their cocoa farms. In the same vein, these respondents had received training on tree domestication and improvement techniques (like marcotting, grafting, and cutting) and were producing young plants from these techniques.

That said, some respondents did not have a complete mastery of the training techniques. This is reflected in the extremely low plant productivity, the frequent plant mortality, and the fact that many respondents noted that they still need training from ICRAF. Indeed, this underscores the importance of continuous capacity building on tree domestication and improvement, a point that has been echoed by several scholars (see, for example, Tchoundjeu et al. 2010; Asaah et al. 2011). Thus, continuous training in this municipality in the domain of seedling production techniques and tree domestication and improvement techniques is considered vital.

Limited market was identified as a fundamental reason for the non-commercialization of seedlings and plants produced as a result of training received from ICRAF. This observation is not particularly surprising and further confirms previous studies that underscore the fundamental importance of rural infrastructure like good farm-to-market roads for promoting sustainable intensification pathways within agroforestry landscapes (Santos-martin et al. 2011). Additionally, as Alemagi et al. (2014a, b) explain, farm-to-market roads are limited in the case study municipality; all the roads in the municipality being earth roads are seasonal due to poor maintenance. In this poor road conditions, it is very difficult to transport cocoa which is produced in remote locations to the market. Thus, dis-enclaving agricultural areas like Efoulan and making them accessible so that agricultural products are easily transported from the farm to the market is central to economic development. Under such scenarios, buyers with adequate transport means go to the community and buy the produce from the farmers at low prices, putting the farmers at a disadvantage end.

Poor quality of the seedlings was also cited as another important rationale for the non-commercialization of seedlings and plants produced as a result of training received. In fact, several other studies have corroborated these results. For instance, it has been reported that one of the main challenges that affects agroforestry as a strategy or approach for promoting sustainable intensification pathways which, in turn, reduces pressure on the forest is the need for good quality planting material for desired species (Pretty et al. 2011; Jama and Zeila 2005). Therefore, one immediate option for overcoming this major hurdle is the provision of improved germplasm to farmers engaged in sustainable intensification of CAL in this municipality.

Several reasons like drought, lack of proper shade, and lack of pesticides were identified, among others, as reasons underpinning plant mortality. Additionally, the need for more material support from ICRAF to help address this problem was cited by respondents. One possible reason why community members depend on organizations like ICRAF for this support is that agriculture extension workers in Cameroon who provide extension services to farmers in rural areas have greatly decrease. Indeed, as Fonjong (2004) posits, most rural development agencies created by the government of Cameroon to provide planting materials to farmers in rural areas have all collapsed. Therefore, apart from ICRAF, there is a need for the State and all other relevant actors to provide cocoa farmers with the required technological and material assistance as an incentive scheme for sustainable intensification of CAL.

The outcomes of the study could support the national REDD+ initiative in Cameroon in that the results could be used by policymakers to promote sustainable intensification of CAL as a REDD+ strategy (which is both a climate change mitigation and adaptation strategy) in the country. Also, other national initiatives have proliferated in the country such as the Roadmap for Zero deforestation free Cocoa in Cameroon championed by the Sustainable Trade Initiative (IDH). The findings of their study align with or could support this initiative, which is geared at reducing deforestation from cocoa production, which is one of the leading drivers of deforestation and forest degradation in the country.

6.6 Conclusions

This chapter set out to explore the impacts of incentive mechanisms for intensification of CAL in the Efoulan municipality located in the South Region of Cameroon. We found that all respondents had received some forms of incentive mechanisms (like provision of seedlings, training on seedling production techniques, and training on tree domestication and improvement techniques) and, as a result, were producing and planting seedlings in their CAL. Moreover, these respondents were also producing young trees and incorporating them into their farms.

That said, a considerable number of challenges that hinder successful implementation of these incentive mechanisms still persist including continuous training in the domain of seedling production techniques and tree domestication and improvement techniques, dis-enclaving the municipality and making it fully accessible so that agricultural products are easily transported from the farm to the market, the provision of improved germplasm to cocoa farmers, and the need for all relevant actors to provide cocoa farmers with the required technological and material assistance. Advances on these key recommendations will help promote intensification of CAL as a climate mitigation, adaptation, and development strategy.

Acknowledgements This chapter is a research output of the SECURED LANDSCAPE PROJECT supported financially by the Norwegian Agency for Development Cooperation (NORAD). Additionally, the research was also made possible by the financial assistance from the CGIAR Research Program (CRP) 6.3, and 6.4. The authors also acknowledge helpful comments and suggestions from anonymous reviewers.

References

Alemagi D, Oben PM, Ertel J (2006) Implementing environmental management systems in industries along the Atlantic coast of Cameroon: drivers, benefits and barriers. Corp Soc Responsib Environ Manag 13(4):221–232

Alemagi D, Duguma LA, Minang PA, Nkeumoe F, Feudjio M, Tchoundjeu Z (2014a) Intensification of cocoa agroforestry systems as a REDD+ strategy in Cameroon: hurdles, motivations, and challenges. Int J Agric Sustain (online First). https://doi.org/10.1080/14735903.2014.940705

Alemagi D, Minang PA, Duguma AL, Kehbila A, Ngum F (2014b) Pathways for sustainable intensification and diversification of cocoa agroforestry landscapes in Cameroon. In: Minang et al (eds) Climate smart landscapes: multifunctionality in practice. The World Agroforestry Centre, Nairobi, Kenya, pp 347–359

Ammenberg J (2003) Do standardized environmental management systems lead to reduce environmental impacts? Doctoral thesis. Institute of Technology, Linkoping University, Sweden

Asaah EK, Tchoundjeu Z, Leakey RB, Takousting B, Njong J, Edang I (2011) Trees, agroforestry and multifunctional agriculture in Cameroon. Int J Agric Sustain 9(1):110–119

Borlaug N (2007) Sixty-two years of fighting hunger: personal recollections. Euphytica 157:287–297

Catacutan D, Lasco RD, Duque-Piñon C (2012) Incentive mechanisms for smallholder agroforestry: opportunities and challenges in the Philippines. In Nair PR, Garrity DP (eds) Agroforestry—the future of global land use. Springer, Dordrecht, The Netherlands, pp 497–514

DOREO PARTNERS (2012) Reducing cocoa and palm oil expansion into Ghanaian forests: analysis of implementation and financing gaps. Consultancy report. http://pcfisu.org/wpcontent/uploads/pdfs/Reducing_Cocoa_and_Palm_Oil_Expansion_into_Ghanaian_Forests_Analysis_Implementation_and_Financing_Gaps.pdf

Eboutou LY (2009) Rentabilité financière des agroforêts à base de cacao enrichies par des arbres domestiqués dans le bassin de production du centre Cameroun. Postgraduate Diploma Thesis. Faculty of Agronomy and Agricultural Science, University of Dschang, Cameroon

Enters T (2004) Incentives: key concepts, typology, and rationale. In Enters T, Durst P (eds) What does it take? The role of incentives in forest plantation development in Asia and the Pacific. Food and Agriculture Organization (FAO) Regional Office for Asia and the Pacific, Bangkok, Thailand

FAO (1999) Incentive systems for natural resource management. Environmental Reports Series 2. FAO Investment Centre. FAO, Rome

FAO (2011) The state of forests in the Amazon Basin, Congo Basin and Southeast Asia. Rome, Italy

Feudjio M, Magne AN, Alemagi D, Robiglio V, Ewane N, Yemefack M, Asaah E, Tchienkoua M (2012) Incentives measures for the REED+ mechanism and sustainable development in the Efoulan Municipality of the South Region of Cameroon. Alternative to Slash and Burn (ASB) Technical Report. ASB Partnership for the Tropical Forest Margins, Nairobi, Kenya

Fonjong L (2004) Changing fortunes of government policies and its implications on the applications of agricultural innovations in Cameroon. Nord J Afr Stud 13(1):13–29

Gockowski J, Sonwa D (2011) Cocoa intensification scenarios and their predicted impacts on CO_2 emissions, biodiversity conservation and rural livelihoods in the Guinea rainforest of West Africa. Environ Manage 48:307–321

Hartemink AE (2005) Nutrient stocks, nutrient cycling, and soil changes in cocoa ecosystems: a review. Adv Agron 86:227–253

Jama B, Zeila A (2005) Agroforestry in the drylands of eastern Africa: a call to action. ICRAF Working Paper Number 1. World Agroforestry Centre, Nairobi, Kenya

Kvale S (1996) Interviews. Sage Publications, Thousand Oaks, CA

Minang PA, Duguma LA, Bernard F, Mertz O, van Noordwijk M (2014) Prospect for agroforestry in REDD+ landscapes in Africa. Curr Opin Environ Sustain 6:78–82

Pretty J, Toulmin C, Williams S (2011) Sustainable intensification in African agriculture. Int J Agric Sustain 9:5–24

Pulhin J, Chokkalingam U, Peras RJ, Acosta R, Carandang A, Lasco R, Natividad M (2006) Historical overview. In: Chokkalingam U, Carandang A, Pulhin J, Lasco R, Peras RJ, Toma T (eds) One century of forest rehabilitation in the Philippines: approaches, outcomes, and lessons, pp 6–41. Jarkata, Indonesia

Robiglio V, Ngendakumana S, Gockowski J, Yemefack M, Tchienkoua M, Tchawa P, Tchoundjeu Z, Bolognesi M (2010) Reducing emissions from all land uses in Cameroon. Final national report. ASB partnership for the tropical forest margins, Nairobi

Robiglio V, Minang PA, Asare R (2011) On-farm timber production for emission reduction with sustainable benefits at the tropical forest margins. ASB Policy Brief No. 23. ASB Partnership for the Tropical Forest Margins, Nairobi, Kenya

Santos-Martin F, Bertomeu M, van Noordwijk M, Navarro R (2011) Why smallholders plant native timber trees: lessons from the Philippines. ASB Partnership for the Tropical Forest Margins, Nairobi; World Agroforestry Centre (ICRAF) Southeast Asia Program, Bogor, Indonesia

Tchoundjeu Z, Degrande A, Leakey RB, Nimino G, Kemajou E, Asaah E, Facheux C, Mbile P, Mbosso C, Sado T, Tsobeng A (2010) Impacts of participatory tree domestication on farmer livelihoods in West and Central Africa. For Trees Livelihoods 19(3):217–234

Yemefack M, Alemagi D, Duguma LA, Minang PA, Tchoundjeu Z (2013) Linking development pathways and emission reduction at local levels: An analysis of feasibility in the Efoulan Municipality, Cameroon. ASB Policy Brief No. 39, ASB Partnership for the Tropical Forest Margins, Nairobi, Kenya

Dr. Dieudonne Alemagi holds a Ph.D. in Environmental and Resource Management and is currently an Associate Partner at FOKABS in Ottawa, Canada. His research focuses on climate change mitigation and sustainability in forest-dependent communities across Central and West Africa. Previously, he was a Scientist with ICRAF, working on related environmental issues.

Dr. Lalisa Duguma holds a Ph.D. in Agricultural Sciences from the University of Natural Resources and Life Sciences, Vienna, Austria. He is an expert at the Global Evergreening Alliance, previously serving as a Senior Scientist at the World Agroforestry Centre. His research focuses on climate actions, multifunctional landscapes, and Low Emission Development Strategies, especially in the African context.

Mr. Peter A. Minang is the Director for Africa at CIFOR-ICRAF and the Global Coordinator of the ASB Partnership for the Tropical Forest Margins at the World Agroforestry Centre, Nairobi, Kenya. His research focuses on REDD+, Low Emission Development Strategies, and the synergies between mitigation, adaptation, and green growth in Africa.

Mr. Kalame Fobissie is the CEO of FOKABS and former Coordinator of the Climate Change Program at the University of Helsinki, Finland. He has represented Cameroon as a REDD+ negotiator in the UNFCCC and served as the Congo Basin Regional Forest and Climate Change Coordinator for WWF-CARPO. His expertise includes forest conservation and climate change mitigation.

Chapter 7
Exploring Biodiversity in Wild Edibles for Enhancing Food Security in Saudi Arabia Amidst Climate Change

Mumna Munir, Mushtaq Ahmad, Muhammad Manzoor, Syed Waseem Gillani, Shabir Ahmad, Amir Sohail, and Mirza Barjees Baig

Abstract Environmental degradation is recognized as a serious problem worldwide, and Saudi Arabia is not an exception. The study explores potential uses of Saudi Arabia's wild edible fruits to improve food security in the context of climate change. The nutritional worth, cultural significance, and climate adaptation of typical wild foods found in Saudi Arabia and their availability are the main focus of the current study. Many wild plants are consumed by local communities in Saudi Arabia, and these types of plants are sometimes undervalued in the broader context of modern-day agriculture. These are wealthy in nutrients in and demand less use of natural assets to thrive. The significance of native plant species that supply a substantial amount of vitamins and nutrients is also emphasized in the current scenario. By using indigenous foods in local dietary habits, one may reduce dependency on foreign food supplies. Significant findings indicate that a range of wild food products, with a substantial amount of essential nutrients, vitamins, and minerals, are viable alternatives to standard traditional crops. Additionally, some crops have the potential to resist extreme weather conditions enabling these crops to strategically negate the effects of

---

M. Munir · M. Ahmad · M. Manzoor · S. W. Gillani · S. Ahmad · A. Sohail
Department of Plant Sciences, Quaid-i-Azam University, Islamabad, Pakistan
e-mail: mumnamunir@bs.qau.edu.pk

M. Manzoor
e-mail: mmanzoor@bs.qau.edu.pk

S. W. Gillani
e-mail: sgillani@bs.qau.edu.pk

S. Ahmad
e-mail: shabir@bs.qau.edu.pk

A. Sohail
e-mail: amirsohail@bs.qau.edu.pk

M. B. Baig (✉)
Prince Sultan Institute for Environmental, Water and Desert Research, King Saud University, Riyadh, Saudi Arabia
e-mail: mbbaig@ksu.edu.sa

M. Behnassi et al. (eds.), *Food Systems and Biodiversity in the Context of Environmental and Climate Risks*, https://doi.org/10.1007/978-3-031-89167-0_7

extreme weather on the food chain. Through increasing knowledge and comprehension of wildlife foraging, Saudi Arabia may encourage nutrient diversity, help boost regional economies, and increase adaptability to food shortages brought about by climate change. By achieving greater ecological objectives and bettering nutritional outcomes, this study promotes an equitable relationship between humankind and its surroundings.

Keywords Climate change · Food security · Nutrients · Saudi Arabia · Wild edibles

7.1 Introduction

The intricate relationship between climate change and global food security is a complex issue that involves challenges such as disrupted agricultural productivity, food shortages, and malnutrition. Climate change disproportionately impacts marginalized communities, smallholder farmers, and low-income populations (Toromade et al. 2024; Gillani et al. 2024a). The worldwide food supply chain is under threat from climate change, especially in Saudi Arabia. Due to the nation's high per capita water demand, population expansion, scarcity of water resources, and restricted groundwater supplies, the food system is already under strain. However, the native wild edible plants can withstand temperature fluctuations and have adaptability properties. These wild edibles are an important source of food and medicines for natives and have several economic benefits highlighting their importance in the stock market (Rahman et al. 2022; Manzoor et al. 2023a).

Saudi Arabia's arid land area, characterized by its geologic structure, is home to diverse ecosystems with a wide range of plant species, including crop, medicinal, and xerophytic vegetation. The region's flora includes 2250 species, with 24.57% used for medicinal purposes. The western region, with its harsh climate and scarce water, is unique in its flora composition. The Aljumum area, one of the largest governorates, has a typical arid and desert floral composition, with varying species depending on rainfall (Al-Eisawi and Al-Ruzayza 2015; Manzoor et al. 2024a). Within four endemic families, researchers identified Fabaceae (32.35%), Poaceae (20.58%), Asteraceae and Brassicaceae (17.64%), and *Calotropis procera*, *Panicum turgidum*, and *Aerva javanica* (5.31%) (Qari et al. 2021; Manzoor et al. 2024b). Another study evaluated medicinal plants in the Al Riyad region compared to Saudi Arabia's total, revealing 108 species from 36 families and 94 genera. Ninety-seven percent have potential economic uses, with 165 having medicinal value (Shawky and Alzamel 2020).

Wild foods (WFPs) are essential to the survival of human communities, particularly in times of scarcity, low household incomes, or difficult market access. Since they make up 5% of the world's population, indigenous peoples are well versed in the uses of plants, traditional food systems, and the biodiversity of local foods. This information facilitates resource monitoring (Aryal et al. 2018; Gillani et al. 2024b).

International agreements such as the 2030 Agenda for Sustainable Development, the International Treaty on Plant Genetic Resources, the Second Global Plan of Action, and the Global Strategy for Plant Conservation of the Convention on Biological Diversity recognize the benefits of wild species in enhancing food and nutrition security. In addition to benefiting users and custodians, sustainable use promotes habitat and biological resource protection. However, the consequences of climate change, economic expansion, and population growth make sustainability difficult to maintain (Borelli et al. 2020; UN 1992).

The varied topography of Saudi Arabia consists of mountains, sand dunes, salt-marshes, and desert plains. Polar and tropical continental air masses affect the climate, with tropical maritime air affecting the southwest. There are hot, humid coastal regions, hot, dry deserts, and temperate mountainous regions, depending on the location and height. Climate, plant characteristics, physiology, and chorology are reflected in plant communities and species distribution. Consideration is given to nature reserves and hotspots for phytodiversity, with an emphasis on invasive, uncommon, and endemic plants (Sayed and Masrahi 2023; Gillani et al. 2024c).

Since ancient times, people living in arid and semi-arid regions have relied on plants as a significant renewable natural resource. It satisfies the majority of all living things' nutritional needs in their surroundings (Al-Munqedhi et al. 2022). Wild plants are rich in antioxidant and anti-inflammatory compounds, including phytochemicals like phenols, flavonoids, tannins, alkaloids, terpenoids, and saponins. These compounds can inhibit oxidative insults and inflammatory reactions, aiding in disease prevention and treatment. Plants also contain vitamins like vitamin B complex and vitamin C, which play a crucial role in body function and disease prevention. Trace elements like zinc and selenium are also important in medicine and food (Scheiermann et al. 2022; Shahid et al. 2023; Mirzaman et al. 2023).The utilization of wholesome edible wild flora can be researched to meet the nutritional and therapeutic needs around the world, as wild plant species are crucial to global nutrition and food security. The systematic creation of a food value inventory for wild edible plants is necessary to improve dietary diversity and manage hunger (Wooten et al. 2022; Manzoor et al. 2023b).

The biological examination of plant forms utilized in Saudi natural treatments revealed that the two most common forms, shrubs and herbs, accounted for 43% and 30% of the total. In Saudi alternative medicine, decoction and infusion are the most common methods of preparing plant-based medications. The most beneficial plant parts for natural preparation in Saudi traditional medicine were the whole plant, leaves, seeds, and aerial portions, which accounted for 29, 28, 7, and 5% of the total plant (Ullah et al. 2020; Alam et al. 2023).

Saudi Arabia is a superb illustration of both adaptation and mitigation strategies as well as how reckless consumption and the climate work together to put strain on and erode natural resources (DeNicola et al. 2015). Solutions to enhance food security resilience include climate-smart agricultural practices, access to climate information, sustainable infrastructure, and gender equity in agriculture. Policy interventions, international collaboration, and community engagement are crucial for addressing climate-related food security challenges and advancing sustainable development goals. By integrating climate adaptation measures, promoting research, and fostering multi-stakeholder partnerships, we can build more resilient and equitable food systems that can withstand climate change's impacts.

7.2 Overview of Environmental Degradation in Saudi Arabia

Global environmental deterioration is a problem that affects both wealthy and developing nations, and the biggest concerns come from rising CO_2 emissions and harmful components. Eighty-five percent of the world's CO_2 emissions, which hurt the water cycle, raise sea levels and spread deadly diseases, are caused by China, India, and the United States. Seven million deaths and seven million cases of diabetes are caused by air pollution, which is a result of industrialization and energy use. The degradation of the environment is bad for human welfare and can be fatal. The degradation of the environment is a result of human activity, such as burning fossil fuels for energy, transportation, and medical purposes. The usage of fossil fuels has contributed to respiratory issues, sea level rise, and global warming, resulting in a notable increase in CO_2 emissions in Saudi Arabia. However, research on the determinants of CO_2 emissions is under-researched, highlighting the need for a more comprehensive understanding of this global issue (Tahir et al. 2022). The Intergovernmental Panel on Climate Change (IPCC) identifies human factors such as population growth, deforestation, industrialization, agriculture, and trade as contributing to environmental degradation (Alshehry 2015).

7.2.1 Impact of Climate Change on Agriculture

Climate change threatens global agricultural production systems and food security, affecting crop productivity and causing undernutrition. Around 540–590 million people will be undernourished at 2 °C warming, and some parts of the world may lose up to 6% of their GDP due to climate-induced water scarcity. The most severe impacts will be in vulnerable regions (Shahzad et al. 2021).

Saudi Arabia's arid climate, with temperatures exceeding 50 degrees Celsius, and limited rainfall, makes it highly vulnerable to climate change. Over the last 50 years, a 1.9 °C increase in average temperature has been observed, with a faster rate in the dry season. Studies predict a 2–4 °C increase by the end of 2100 due to ongoing climatic changes (Sowers et al. 2011). The Kingdom lacks recurrent rivers and permanent water bodies, making it a water-scarce nation. Climate change impacts food production, agriculture, and other economic sectors, with a 3–5 °C rise posing challenges. Crop yields may range between 5–25%, and irrigation water requirements could increase by 602 to 3122 million cubic meters. Water scarcity could also lead to a change in the abundance and distribution of diurnal desert animals (Azeem and Alhafi Alotaibi 2023).

7.2.2 Current Food Security Challenges

Food security is a crucial global issue, ensuring access to sufficient, safe, and nutritious food for all people. It is a top priority for international policy. Economic expansion, rising incomes, urbanization, and a growing population are driving up demand for food, particularly in developing countries. The need to double food production within the coming decades is a major obstacle, as natural resources like land, water, and energy are subject to increasing competition (Lang and Barling 2012). Climate change poses challenges for agricultural production, particularly in developing nations, and some farming practices contribute to greenhouse gas emissions. The number of malnourished people has risen from 785 million in 2015 to 822 million in 2018, and children in developing countries face vitamin and nutritional deficiencies and underweight issues, making them at risk for various illnesses due to food insecurity (Pachapur et al. 2020). The Kingdom of Saudi Arabia faces challenges such as desertification, a hot dry climate, water scarcity, poor soil fertility, and labor shortages, putting food security at risk. Currently, one-fifth of food requirements are met through domestic production, while the rest is imported. To address these issues, the Kingdom is focusing on optimizing food imports and developing infrastructure to promote domestic agricultural production. Despite environmental challenges, the Kingdom is classified as a GOOD PROGRESS COUNTRY (Al-Khateeb et al. 2021).

7.3 Enhancing Food Security in Saudi Arabia

Considering an emphasis on environmentally conscious cultivation and efforts to preserve them, wild foods are seen as a viable remedy for communities experiencing food insecurity in these areas. The widening disparity between food availability and population growth poses a serious threat to human life in developing nations. People in these areas who experience food insecurity tend to consume fewer fruits and vegetables and become deficient in certain minerals and vitamins. Edible Wild Plants: Growing and utilizing wild plants is a different strategy for ensuring food security that aims to lower food insecurity and malnutrition. Inexpensive and abundant in antioxidants, vitamins, fiber, and minerals, wild edible plants are (Shaheen et al. 2017). The use of herbal medicine obtained from wild plants among Saudi patients for treating diseases like cancer, asthma, neurological, and hepatic conditions is reported at 55%, 80%, 42.3%, and 90%, respectively (Mohammad et al. 2015). 81.2% of Riyadh citizens believe herbal medicine is safe and harmless. In Jeddah, 64% of diabetic patients prefer herbs for control, with 55.1% preferring

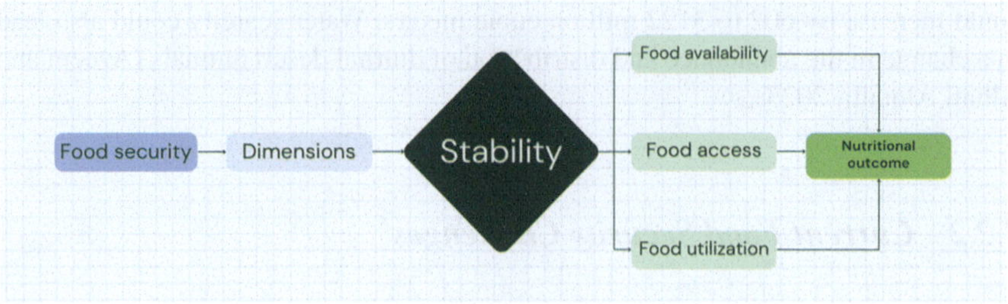

Fig. 7.1 Food security implications leading to nutritional outcomes

plants over synthesized drugs. Many others are considered promising sources of food for local inhabitants. These are definitely promoting food security issues among the inhabitants (Suleiman 2014) (Fig. 7.1)

7.3.1 The Transformative Role of Agriculture Extension and Innovation

Agri-food production has undergone significant technological changes over the past two centuries, including the introduction of mechanical modernizations like lawnmowers, mechanical harvesters, and threshers, as well as the introduction of canned food, industrial agriculture, and the transition to fuel-based energy sources (Movilla-Pateiro et al. 2021). The Green Revolution, a significant technological change in the past century, involved the use of high-yielding wheat varieties and chemical fertilizers, leading to significant increases in agricultural productivity. This revolution, characterized by capital-intensive production processes, has since evolved into biotechnology, nanotechnology, and digital technology, transforming the agricultural food sector (Sodano 2019).

Technological advances have significantly impacted human well-being since Neolithic times, including digital agriculture, cellular agriculture, food processing, gene technology, health, and resource use efficiency. However, environmental and health indices are declining, particularly in the twenty-first century due to increased air and water pollution, greenhouse gas emissions, and excessive nitrogen and phosphorus usage in agriculture. The development of inexpensive, fast, or discretionary foods has also contributed to significant malnutrition in many parts of the world (Swinburn et al. 2019). Innovation in food systems requires significant changes in infrastructure, technology, expertise, and capabilities, along with structural reforms in principles, legislation, policy, economies, and governance. Modern technology alone isn't enough; societal and structural forces are needed (Altshul et al. 2019).The wild edibles can be more sustainably used by their agricultural extension through innovative technologies.

7.4 Overview of Native Wild Edible Plants

The Kingdom of Saudi Arabia (KSA) has a diverse flora, including edible and medicinal plants, despite its arid climate. Traditional ethnomedical practices in the KSA are interconnected with familiar remedies, health, diet, and healing practices. A paper reviews 309 genera of medicinal plants used in ethnomedicine, covering 471 species from 2253 known species in 89 families. Strategies for sustainable use of these plants are crucial for the KSA's folk medicine. In Saudi Arabia, the Asteraceae and Fabaceae families include the greatest number of ethnomedicinal species—54 and 49, respectively (Aati et al. 2019). The Tihama coastal area in Yemen and East African countries like Ethiopia and Eritrea has sparse vegetation cover with eight major community types dominated by nine perennials. The region is known for producing high-quality tropical fruits and is a natural reservoir for wild medicinal plants. The west-facing slopes of mountains with frequent moisture-laden winds support a plant cover with endemic and endangered species. Terrace cultivation has been practiced for centuries, with natural vegetation dominated by *Acacia asak*, *Otostegia fruticose*, *Olea europaea*, *Dodonaea viscosa*, *Rhus retinorrhaea*, and *Pennisetum setaceum* (Tounekti et al. 2019) (Fig. 7.2).

Another study examined the floristic composition, vegetation structure, and species distribution in 16 sample plots in Wadi Fatimah, Mecca, Saudi Arabia. It found 100 plant species, with Fabaceae and Poaceae being the dominant families. Altitude significantly influenced species frequency distribution. The coastal plain had the highest species richness, while the littoral salt marsh had the lowest. The study classified four major vegetation groups into three main ecosystems: littoral salt marsh, coastal plain, and inland mountainous country. The study also found correlations between vegetation concentrations and heavy metal concentrations (El-Ghanim et al. 2010).

7.4.1 Cultural and Historical Significance

Saudi Arabians still follow many of their ancestors' customs since they are derived from tribes of migratory goats and sheep herders. Deserts and low biodiversity are acknowledged as the Arabian Peninsula's main arid features. The Kingdom of Saudi Arabia, sometimes known as the KSA, has a diverse range of flora, including many kinds of trees, shrubs, and herbs as well as a large number of edible and medicinal plants. The KSA is distinguished by a wide range of climates and geographical features. The distribution of flora throughout the Kingdom varies greatly as a result. There is a close relationship between common remedies, health, diet, and customary healing methods that are unique to individual cultures, as evidenced by the customary use of wild plants in the Kingdom of Saudi Arabia (Aati et al. 2019). Even though the majority of Saudis have settled in towns and cities and no longer lead a nomadic lifestyle, traditional foods like dates, fatir (flat bread), arikah (bread

Fig. 7.2 **a** *Vitis vinifera*, **b** *Solanum nigrum*, **c** *Rubus ellipticus*, **d** *Ziziphus spina-cristi*, **e** *Diospyrus lotus*, **f** *Prunus domestica*

from the southwest of the nation), and hawayij (a spice blend) are still consumed by Saudis. Mecca, the birthplace and holiest site of Islam, is also located in Saudi Arabia. Saudi Arabia's laws and culture are based on Islamic precepts, which include abstaining from alcohol and pork in the diet (El Sheikha 2015). Muslims hold *Ziziphus spina-christi* in high regard since it is twice mentioned in the Sunnah and the Holy Quran. Research reveals that locals value wild fruits more than vegetables, and some wild edibles are culturally stigmatized. However, some species are increasingly harvested for sale and become local delicacies. Gathering wild edible plants is also seen as a recreational activity. The divergent trends in the use of these plants in KSA are attributed to their cultural services, which include spiritual enrichment, cognitive development, reflection, recreation, and aesthetic experience.

7.4.2 Resilience of Wild Edibles to Extreme Weather

The exploration of new wild plant species tolerant to severe conditions is a global priority to meet food and feed demands in developing countries like Saudi Arabia. The country's arid climate offers advantages for incorporating native or locally suited species due to their ability to thrive under local conditions and their superior tolerance to extremes like heat, desert winds, and drought. Saudi Arabia's flora, which includes 2250 species in 142 families, is rich in genetic resources and xerophytic vegetation, encompassing elements from Asia, Africa, and the Mediterranean region (AlNafie 2008).The majority of plant species in Saudi Arabia are known to belong to the Poaceae, Leguminosae, and Asteraceae families. Their enormous capacity for seed distribution and their broad tolerance throughout an ecological range are the reasons for this. As in other tropical and subtropical deserts, the majority of plant species belong to a small number of plant families. This makes the floristic makeup of the Saudi Arabian flora particularly notable (Alsherif et al. 2012).

The Kingdom of Saudi Arabia (KSA) is divided into several regions, with the Asir highlands being a continuous chain of cliffs, while the eastern part is characterized by sand and lower mountains and plains. The central region is characterized by stone deserts, hillocks, and valleys, while the northern and northern central regions have metamorphic rocks. The KSA is located in a climatic transitional zone of tropical and temperate climates, with irregular rainfall in arid areas. Forests cover 2.7 million hectares, with most in the southwestern part of the Kingdom (Alfarhan et al. 1998).

7.5 Comparative Analysis of Wild Edibles Versus Conventional Crops

The proximate and mineral value of wild edibles is usually considered high and more effective due to their natural development under natural growth conditions. The fruit of the date palm from KSA contained phenols and flavonoids at higher rates, and the fruit contained a high percentage of carbohydrates, with sucrose being the highest sugar. The average content of fatty acids was 1.07mg/g dry matter content, with Vitamin E being the major content and Vitamin B being the minor vitamin (Ismail and Altuwairki 2016). The protein, fiber, crude fat, ash, and moisture content of the various cultivars of olives were found to differ significantly across the eight olives analyzed in Saudi Arabia. The study reveals that Picual has a higher protein content than other cultivars, while Kaissy has the lowest protein content. Kaissy has the highest fiber content, while Picual has the lowest. There is a negative correlation between protein and fiber content, but not in all other cases. K-18 has the lowest crude fat content, while Sorani has the highest fat content. K-18 has the highest ash content, while Koroneiki has the lowest. Moisture content is highest in Picual-H78, with the lowest in K-18 and highest in Picual-H78. The study also found that oleic acid content in olive oil varies based on variety, with Arbequina having the lowest oleic acid values. Moisture content affects the quality and shelf life of fruits (Al-Ruqaie et al. 2016). In contrast, a study conducted in Spain compared organic farming to mineral fertilization over nine years and found that organic crops had lower nitrate concentrations and lower nitrogen and phosphorus content in their edible parts. However, the study also revealed variability in nutritional parameters across different crop cycles. The researchers concluded that the nutritional quality of organic crops cannot be solely attributed to fertilizer type, but also to factors like fertilizer characteristics and crop cycle management (Herencia et al. 2011).

7.6 Potential for Climate Change Mitigation

A study highlights the importance of understanding biodiversity and ecosystem services in carbon sequestration to mitigate climate change. It highlights the role of biodiversity in climate regulation, carbon sequestration, nutrient cycling, food provision, clean air and water, water cycling, crop pollination, disease regulation, and economic value. The trade-offs and synergies between biodiversity and ecosystem services, their role in regulating greenhouse gas emissions, and the implications of biodiversity and ecosystem service provision in carbon sequestration are crucial. Managing biodiversity and natural ecosystems is essential for reducing greenhouse gas emissions and supporting sustainable societal adaptation (Daba and Dejene 2018).

7.7 Policy Recommendation and Strategic Implementation

Guidelines are available to help countries prepare a National Plan for the Conservation and Sustainable Use of Wild Food Plants and crop wild relatives, including a monitoring plan for WFP diversity. The plan's scope, application, and effectiveness depend on the national context, policy framework, institutional arrangements, stakeholders, and available resources (Beltrame et al. 2016). One potential solution is linking producers and collectors to institutional or private sector markets, enabling them to benefit from the authorized trade of WFPs. Brazil has used its Food Procurement Program and National School Feeding Program to engage family farmers and wild plant collectors in sustainable agriculture and food diversity management (Hunter et al. 2019). Food security has been a key concept in policy-making throughout the twentieth century, with two main schools of thought: one focused on increased production as a primary solution to under-consumption and hunger, and the other on social and environmentally conscious approaches. The central mission of the twenty-first century is to establish a sustainable food system, which requires a more concrete policy framework. Climate change is pushing millions into a cycle of food insecurity and poverty, but implementing climate-friendly agricultural production methods can combat both (Lang and Barling 2012). A comprehensive food security policy should consider resilience, trade, self-sufficiency, food waste, and sustainability. The four pillars of food security—availability, access, utilization, and stability—should be reconsidered to include additional factors such as climate change. The International Food Policy Research Institute has launched a 2020 Vision of Food Security to achieve food security (Fanzo 2015), stating that a world where every person has economic and physical access to sufficient food to sustain a healthy and productive life, where malnutrition is absent, and where food originates from efficient, effective, and low-cost food and agricultural systems that are compatible with sustainable use and management of natural resources (Clapp et al. 2022). Moreover, Saudi Arabia (KSA) is implementing a plan to regulate and organize the use of traditional medicines from wild edibles, known as Complementary and Alternative Medicine (TM/CAM), through organizing acts and laws. The National Office was established in 1995, and the TM/CAM Committee collaborated with the National Research Institute on herbal medicine studies (Mossa et al. 1987). Traditional medicine practices in KSA date back to ancient times but are beginning to disappear. Conducting folk and ethnobotanical surveys is crucial to preserving these practices and discovering new drug resources.

7.7.1 Successful Applications of Wild Edibles

Wild plants enhance global food diversity, promote eco-friendly practices, and offer essential micronutrients. They are traditional dietary staples in cultures, addressing malnutrition and demonstrating resilience against climatic fluctuations. Promoting their nutritional advantages is crucial for global food security (Belhaj et al. 2021). Worldwide, a large portion of the human diet has consisted of foraging for food and water from wild edible plants. They are still consumed around the world, including in affluent countries' rural and urban areas and in economies focused on subsistence, despite a fall in gathering and consumption. According to major categories of ecosystem services, these plants are regarded as provisioning services (Fig. 7.3).

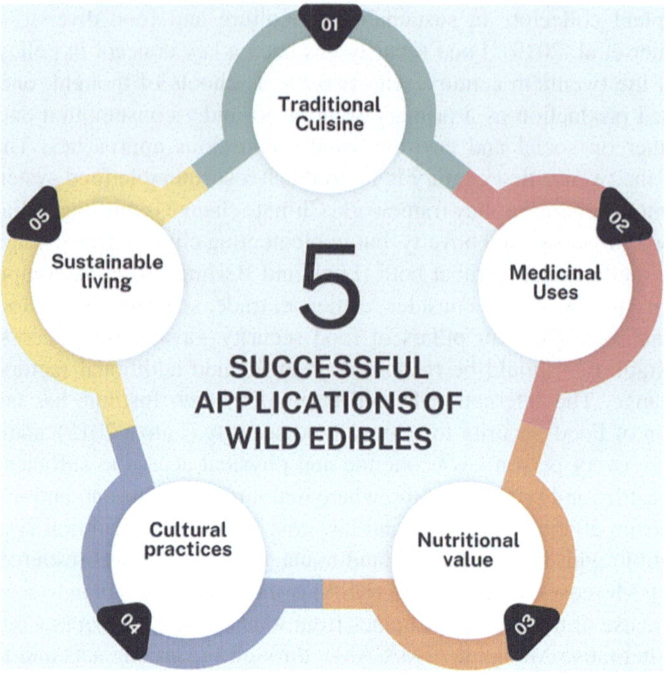

Fig. 7.3 Applications of wild edibles

7.7.2 *Economic Benefits of Wild Edibles*

The economic importance of biodiversity is crucial for sustainable development and use. Its services, such as edible wild species, contribute to human nutrition and are easily accessible and low-cost. However, determining sustainable use strategies for these products can be challenging due to their unknown supply and demand quantities and economic value. To ensure maximum benefit from limited resources, it is essential to diversify the benefits offered and conduct a financial evaluation for each. Gender-sensitive value chain analysis is also needed to enhance economic and socio-cultural research (Karabak 2017). A study collected 236 species from various habitats in Saudi Arabia, revealing the Asteraceae family with the highest number (55), followed by Fabaceae, Chenopodiaceae, Poaceae, Lamiaceae, Zygophyllaceae, Brassicaceae, Asclepiadaceae, Solanaceae, and Euphorbiaceae. The recorded taxa are categorized into nine economic uses: fodder, edible, medicinal, fuel, ornamental, sand accumulator, dye, oil soup, honey, fiber production, and carpentry. Fodder and multipurpose use have the most species, while ornamental enriches soil fertility (Al Masoudi 2024). The economy and sustainability are crucial aspects of society, and the economic value of wild edible plants (WEPs) is essential. Edible flowers, available in various formats like fresh, dried, or candied, are highly attractive and a good business proposition in the flower market (CENTER 2014). Globalization and online marketing have created an emerging market for these products worldwide. In recent years, haute cuisine has valued these flowers, with restaurants like "Mugaritz" using edible flowers in their dishes. Other ways to contribute to the economic value of WEPs include selling them directly, through books, and the distribution of information. Studies have highlighted the need for strategic plans at the national level in Ethiopia for the sustainable use and domestication of these plants, aiming to improve socio-economic conditions and achieve the Sustainable Development Goals (SDGs). In northeast India, WEPs are critical for the survival of ethnic communities and the Japanese population, as harvesting these plants is not possible in rural areas near Fukushima due to safety reasons (Chaudhury et al. 2021).

7.7.3 Environmental Implications and Sustainability

Improving environmental sustainability and reducing diet-related diseases are two of the biggest concerns facing food policy in the twenty-first century. Wild edible plants can significantly contribute to environmental sustainability, particularly in the face of climate change and food insecurity. They offer diverse, easily accessible, and reliable resources, making them a sustainable strategy. Wild food also addresses the issue of micronutrient deficiency, which cannot be eliminated by staple crops (Pingali 2015). Fortification of commercial food and bio-fortified food crops with increased micronutrient content have been proposed to combat hidden hunger. Wild food has untapped nutritional potential, as it is a reservoir of vitamins and trace elements. Health benefits from eating wild edibles have been widely reported, but the practice is currently at stake due to various factors.

Agricultural intensification has led to increased staple crop production but reduced farmland biodiversity and dietary diversity. This is due to the intensive use of agro-chemicals, land-use change, forest fragmentation, and governance changes. Cultural factors, such as changing food habits and loss of traditional knowledge systems, have also eroded the culture of eating wild food. This results in a loss of dietary diversity and disruption of sustainable food systems, particularly for lower economic groups in developing countries like India, where access to essential nutrients is limited due to price (Von Grebmer et al. 2014). The food system, a vital component of human activities, faces significant environmental sustainability challenges due to its dependence on healthy ecosystems. Poor diets are a leading risk factor for disease, and these challenges are exacerbated by global environmental change. Traditional and indigenous food systems, which involve communities procuring wild and cultivated foods from their surroundings, can provide sustainability solutions for reconciling food production with human and planetary health. These systems, which include natural food environments like forests, fields, and gardens, influence the availability, affordability, convenience, desirability, and sustainability of food (Ahmed et al. 2022).

7.7.4 Implications of Food Security and Climate Adaptation

Climate change impacts food systems through direct impacts on crop production, markets, food prices, and supply chain infrastructure. Direct impacts include changes in rainfall, temperature, and CO_2 concentrations, which can cause sudden changes in environmental conditions (Rosenzweig et al. 2002). Increased CO_2 concentrations can benefit crops by increasing biomass buildup and yield, but not all respond favorably. Climate change may also indirectly impact food production by altering species behavior, resulting in diseases, new crop and livestock species, changes in pesticide usage, and changes in pollution transfer routes (Haque and Khan 2022; Lake et al. 2012).

7.8 Conclusion

In a nutshell, Saudi Arabia's research on wild edibles offers a workable plan for enhancing food security despite climate change. Certain native plant species are resilient to changes in the weather. These plants consume the nutrients to obtain their nutrient value, which plays a major economic role. The inhabitants' economic empowerment can be ensured by these naturally occurring edible resources. The long-term sustainability of wild edibles is enhanced by the use of sustainable foraging techniques. To ensure the diversity and complexity of these natural treats, we also consider the inventive ways in which they are produced. Examining and advocating for these foraged foods can assist us in utilizing a sustainable future. Saudi Arabia exemplifies adaptation and mitigation strategies, highlighting the impact of reckless consumption and climate on natural resources. Solutions to enhance food security resilience include climate-smart practices, climate information access, sustainable infrastructure, and gender equity, requiring policy interventions and community engagement.

References

Aati H, El-Gamal A, Shaheen H, Kayser O (2019) Traditional use of ethnomedicinal native plants in the Kingdom of Saudi Arabia. J Ethnobiol Ethnomed 15:1–9

Ahmed S, Warne T, Stewart A, Byker Shanks C, Dupuis V (2022) Role of wild food environments for cultural identity, food security, and dietary quality in a rural American state. Front Sustain Food Syst 6:774701

Al-Eisawi DM, Al-Ruzayza S (2015) The flora of holy Mecca district, Saudi Arabia. Int J Biodivers Conserv 7:173–189. https://doi.org/10.5897/IJBC2014.0773

Alam NM, Shaheen H, Manzoor M, Tinghong T, Arfan M, Idrees M (2023) Spatial distribution and population structure of Himalayan Fir (Abies pindrow (Royle ex D. Don) Royle) in moist temperate forests of the Kashmir region. Forests 14(3):482

Al-Khateeb SA, Hussain A, Lange S, Almutari MM, Schneider F (2021) Battling food losses and waste in Saudi Arabia: mobilizing regional efforts and blending indigenous knowledge to address global food security challenges. Sustainability 13:8402. https://doi.org/10.3390/su13158402

Al-Munqedhi BM, El-Sheikh MA, Alfarhan AH, Alkahtani AM, Arif IA, Rajagopal R, Alharthi ST (2022) Climate change and hydrological regime in arid lands: impacts of dams on the plant diversity, vegetation structure and soil in Saudi Arabia. Saudi J Biol Sci 29:3194–3206

Al-Ruqaie I, Al-Khalifah N, Shanavaskhan A (2016) Morphological cladistic analysis of eight popular Olive (Olea europaea L.) cultivars grown in Saudi Arabia using Numerical Taxonomic System for personal computer to detect phyletic relationship and their proximate fruit composition. Saudi J Biol Sci 23:115–121

Al Masoudi LM (2024) Plant diversity in some habitats in Saudi Arabia with its economic uses. Sohag J Sci 9:439–442

Alfarhan AH, Chaudhary SA, Thomas J (1998) Notes on the flora of Saudi Arabia. J King Saud Univ Sci 10:31–40

AlNafie AH (2008) Phytogeography of Saudi Arabia. Saudi J Biol Sci 15

Alshehry AS (2015) Economic growth and environmental degradation in Saudi Arabia. Arabia 6

Alsherif EA, Ayesh AM, Allogmani AS, Rawi SM (2012) Exploration of wild plants wealth with economic importance tolerant to difficult conditions in Khulais Governorate. Saudi Arabia Sci Res Essays 7:3903–3913

Altshul HJ, McMillan L, Hall A (2019) The role of public research agencies in building agri-food bioscience impact and innovation capacity in sub-Saharan Africa: the challenge beyond science capability. Int J Technol Manag Sustain Dev 18:105–125

Aryal KP, Poudel S, Chaudhary RP, Chettri N, Chaudhary P, Ning W, Kotru R (2018) Diversity and use of wild and non-cultivated edible plants in the Western Himalaya. J Ethnobiol Ethnomed 14:1–18

Azeem MI, Alhafi Alotaibi B (2023) Farmers' beliefs and concerns about climate change, and their adaptation behavior to combat climate change in Saudi Arabia. PLoS One 18:e0280838

Belhaj S, Chaachouay N, Zidane L (2021) Ethnobotanical and toxicology study of medicinal plants used for the treatment of diabetes in the High Atlas Central of Morocco. J Pharm Pharmacogn Res 9:619–662. http://jppres.com/jppres

Beltrame D, Oliveira C, Borelli T, de Andrade Cardoso Santiago M, Monego ET, de Rosso VV, Coradin L, Hunter D (2016) Diversifying institutional food procurement–opportunities and barriers for integrating biodiversity for food and nutrition in Brazil. Revista Raízes 36:55–69

Borelli T, Hunter D, Powell B, Ulian T, Mattana E, Termote C, Pawera L, Beltrame D, Penafiel D, Tan A (2020) Born to eat wild: an integrated conservation approach to secure wild food plants for food security and nutrition. Plants 9:1299

CENTER B (2014) Capacitary Orlicz spaces, Calderón products and interpolation Pilar Silvestre. Banach Center Publ 102:215–228

Chaudhury G, Basumatari M, Darji CB, Ahmed AF, Borah D, Sah RK, Devi A, Hazarika N, Dutta G (2021) Economic significance of wild bioresources to rural communities in the Eastern Himalayan state of Assam, Northeast India. Trees For People 5:100102

Clapp J, Moseley WG, Burlingame B, Termine P (2022) The case for a six-dimensional food security framework. Food Policy 106:102164

Daba MH, Dejene SW (2018) The role of biodiversity and ecosystem services in carbon sequestration and its implication for climate change mitigation. Environ Sci Nat Resour 11:1–10

DeNicola E, Aburizaiza OS, Siddique A, Khwaja H, Carpenter DO (2015) Climate change and water scarcity: the case of Saudi Arabia. Ann Glob Health 81:342–353

El-Ghanim WM, Hassan LM, Galal TM, Badr A (2010) Floristic composition and vegetation analysis in Hail region north of central Saudi Arabia. Saudi J Biol Sci 17:119–128

El Sheikha AF (2015) Traditional bread in Arab countries... key of nutrition. In: Bread and its fortification: nutrition and health benefits, p 306

Fanzo J (2015) Ethical issues for human nutrition in the context of global food security and sustainable development. Glob Food Sec 7:15–23

Gillani SW, Ahmad M, Manzoor M, Waheed M, Iqbal Z, Ullah R, Alrhmoun M et al (2024a) The nexus between ecology of foraging and food security: cross-cultural perceptions of wild food plants in Kashmir Himalaya. J Ethnobiol Ethnomed 20(1):77

Gillani SW, Ahmad M, Zafar M, Haq SM, Waheed M, Manzoor M, Makhkamov T et al (2024b) An insight into indigenous ethnobotanical knowledge of medicinal and aromatic plants from Kashmir Himalayan region. Ethnobot Res Appl 28:1–21

Gillani SW, Ahmad M, Zafar M, Manzoor M, Shah GM, Shaheen H, Khishlatovna KK (2024c) Ethnobotanical exploration of traditional medicinal plants among the rural inhabitants of District Muzaffarabad, Kashmir Himalayan Region. Plant Science Today 11(sp1)

Haque MI, Khan MR (2022) Impact of climate change on food security in Saudi Arabia: a roadmap to agriculture-water sustainability. J Agribus Dev Emerg Econ 12:1–18

Herencia JF, García-Galavís PA, Dorado JAR, Maqueda C (2011) Comparison of nutritional quality of the crops grown in an organic and conventional fertilized soil. Sci Hortic 129:882–888. https://doi.org/10.1016/j.scienta.2011.04.008

Hunter D, Borelli T, Beltrame DM, Oliveira CN, Coradin L, Wasike VW, Wasilwa L, Mwai J, Manjella A, Samarasinghe GW (2019) The potential of neglected and underutilized species for improving diets and nutrition. Planta 250:709–729

Ismail I, Altuwairki D (2016) Chemical composition and antimicrobial efficacy of date palm fruit of Saudi Arabia. World Appl Sci J 34:140–146

Karabak S (2017) Economic and socio-cultural importance of edible wild species. Anadolu Ege Tarımsal Araştırma Enstitüsü Dergisi 27:26–38

Lake IR, Hooper L, Abdelhamid A, Bentham G, Boxall AB, Draper A, Fairweather-Tait S, Hulme M, Hunter PR, Nichols G (2012) Climate change and food security: health impacts in developed countries. Environ Health Perspect 120:1520–1526

Lang T, Barling D (2012) Food security and food sustainability: reformulating the debate. Geogr J 178:313–326

Manzoor M, Ahmad M, Zafar M, Gillani SW, Shaheen H, Pieroni A, Khaydarov K et al (2023a) The local medicinal plant knowledge in Kashmir Western Himalaya: a way to foster ecological transition via community-centred health seeking strategies. J Ethnobiol Ethnomed 19(1):56

Manzoor M, Ahmad M, Zafar M, Haq SM, Shaheen H, Waheed M, Makhkamov T et al (2023b) Unveiling the indigenous ethnomedicinal knowledge of genus nepeta from division Muzaffarabad, Azad Jammu & Kashmir, Pakistan. Ethnobot Res Appl 26:1–15

Manzoor M, Ahmad M, Zafar M, Gillani SW, Shah GM, Shaheen H, Khishlatovna KK (2024a) Exploration of traditional ethno-gynaecological knowledge: advances to ethnobotanical studies from indigenous communities of Neelum Valley in the Himalayan Region. Plant Sci Today 11(sp1)

Manzoor M, Shaheen H, Attique A, Khan RWA, Gillani SW, Khan MS (2024b) Distribution pattern, population structure and habitat geography of Betula utilis subalpine forests of Kashmir region in the western Himalayas: implications for conservation and sustainable management. Environ Dev Sustain, 1–26

Mirzaman Z, Kayani S, Manzoor M, Jameel MA, Waheed M, Gillani SW, Bussmann RW et al (2023) Ethnobotanical study of Makra Hills district Muzaffarabad, Azad Jammu and Kashmir, Pakistan. Ethnobot Res Appl 26:1–17

Mohammad Y, Al-Ahmari A, Al-Dashash F, Al-Hussain F, Al-Masnour F, Masoud A, Jradi H (2015) Pattern of traditional medicine use by adult Saudi patients with neurological disorders. BMC Complement Altern Med 15:1–5

Mossa JS, Al-Yahya MA, Al-Meshal IA (1987) Medicinal plants of Saudi Arabia

Movilla-Pateiro L, Mahou-Lago X, Doval M, Simal-Gandara J (2021) Toward a sustainable metric and indicators for the goal of sustainability in agricultural and food production. Crit Rev Food Sci Nutr 61:1108–1129

Pachapur PK, Pachapur VL, Brar SK, Galvez R, Le Bihan Y, Surampalli RY (2020) Food security and sustainability. In: Sustainability: fundamentals and applications, pp 357–374

Pingali P (2015) Agricultural policy and nutrition outcomes–getting beyond the preoccupation with staple grains. Food Secur 7:583–591

Qari SH, Alrefaei AF, Filfilan W, Qumsani A (2021) Exploration of the medicinal flora of the aljumum region in saudi arabia. Appl Sci 11:7620

Rahman MM, Akter R, Abdul Bari JB, Hasan MA, Rahman MS, Abu Shoaib S, Shatnawi ZN, Alshayeb AF, Shalabi FI, Rahman A (2022) Analysis of climate change impacts on the food system security of Saudi Arabia. Sustainability 14:14482

Rosenzweig C, Tubiello FN, Goldberg R, Mills E, Bloomfield J (2002) Increased crop damage in the US from excess precipitation under climate change. Glob Environ Chang 12:197–202

Sayed OH, Masrahi YS (2023) Climatology and phytogeography of Saudi Arabia. A review. Arid Land Res Manag 37:311–368

Scheiermann E, Puppa M-A, Rink L, Wessels I (2022) Zinc status impacts the epidermal growth factor receptor and downstream protein expression in A549 cells. Int J Mol Sci 23:2270

Shaheen S, Ahmad M, Haroon N (2017) Edible wild plants: an alternative approach to food security. Springer

Shahid M, Singh RK, Thushar S (2023) Proximate composition and nutritional values of selected wild plants of the United Arab Emirates. Molecules 28:1504

Shahzad A, Ullah S, Dar AA, Sardar MF, Mehmood T, Tufail MA, Shakoor A, Haris M (2021) Nexus on climate change: agriculture and possible solution to cope future climate change stresses. Environ Sci Pollut Res 28:14211–14232

Shawky RA, Alzamel NM (2020) Survey on medicinal plants in the flora of Al Riyadh Region, Saudi Arabia. EurAsian J BioSciences 14:3795–3800

Sodano V (2019) Innovation trajectories and sustainability in the food system. Sustainability 11:1271

Sowers J, Vengosh A, Weinthal E (2011) Climate change, water resources, and the politics of adaptation in the Middle East and North Africa. Clim Change 104:599–627

Suleiman A (2014) Attitudes and beliefs of consumers of herbal medicines in Riyadh, Saudi Arabia. J Community Med Health Educ 4:269

Swinburn BA, Kraak VI, Allender S, Atkins VJ, Baker PI, Bogard JR, Brinsden H, Calvillo A, De Schutter O, Devarajan R (2019) The global syndemic of obesity, undernutrition, and climate change: the Lancet commission report. The Lancet 393:791–846

Tahir M, Hayat A, Burki U (2022) Determinants of environmental degradation in Saudi Arabia: exploring the unexplored. Int J Energy Sect Manage 16:129–148

Toromade AS, Soyombo DA, Kupa E, Ijomah TI (2024) Reviewing the impact of climate change on global food security: challenges and solutions. Int J Appl Res Soc Sci 6:1403–1416

Tounekti T, Mahdhi M, Khemira H (2019) Ethnobotanical study of indigenous medicinal plants of Jazan region, Saudi Arabia. Evid Based Complement Altern Med 2019:3190670

Ullah R, Alqahtani AS, Noman OM, Alqahtani AM, Ibenmoussa S, Bourhia M (2020) A review on ethno-medicinal plants used in traditional medicine in the Kingdom of Saudi Arabia. Saudi J Biol Sci 27:2706–2718

UN I (1992) Convention on biological diversity. Treaty Collection

Von Grebmer K, Saltzman A, Birol E, Wiesman D, Prasai N, Yin S, Yohannes Y, Menon P, Thompson J, Sonntag A (2014) Synopsis: 2014 global hunger index: the challenge of hidden hunger. Intl Food Policy Res Inst

Wooten DJ, Sinha I, Sinha R (2022) Selenium induces pancreatic cancer cell death alone and in combination with gemcitabine. Biomedicines 10:149

Mr. Mumna Munir is a Ph.D. scholar at Quaid-i-Azam University, Islamabad, with expertise in biodiesel synthesis, nanocatalysts, and the nutritional analysis of wild edibles. Her research focuses on developing sustainable energy solutions through efficient biodiesel production, utilizing nanocatalysts to enhance the process. Additionally, Mumna investigates the nutritional value of wild edible plants, aiming to highlight their potential for improving food security and health. Through her interdisciplinary work, she strives to contribute to both environmental sustainability and the promotion of indigenous plant resources for human well-being.

Prof. Dr. Mushtaq Ahmad is a distinguished academic and researcher, currently serving as the Director of the QAU Botanical Garden and Herbarium, Director of the Technology Science Park, and Chairman of the Department of Plant Sciences at Quaid-i-Azam University, Islamabad. He has pioneered the establishment of several cutting-edge research facilities, including the modern digital Herbarium, Botanical Garden, and labs focused on Melissopalynology, Aerobiology, and Green Biofuels. Prof. Ahmad's expertise spans botany, bioengineering, and environmental engineering, with significant contributions to medicinal plants, nanomaterials, natural product chemistry, and biodiesel production. His interdisciplinary research integrates phytochemical analysis, herbal medicine, and biofuel development to advance sustainable solutions in health and energy.

Mr. Muhammad Manzoor is a dedicated Ph.D. research scholar at Quaid-i-Azam University, Islamabad, specializing in Ecology, Systematics, and Ethnobotany. With a deep passion for understanding the intricate relationships between plants, ecosystems, and human cultures, his research spans a wide range of fields, including biodiversity conservation, plant systematics, and traditional ecological knowledge. His academic interests also encompass plant morphology, anatomy, and palynology, with a strong focus on biological conservation and the sustainable use of plant resources.

Mr. Syed Waseem Gillani is a research scholar at Quaid-i-Azam University, Islamabad, with a strong focus on Ecology, Systematics, and Ethnobotany. His research interests include biodiversity conservation, plant systematics, and integrating traditional ecological knowledge with modern scientific approaches. Waseem is committed to advancing scientific understanding of plant diversity and its crucial role in environmental balance and cultural heritage. Through his work, he aims to contribute to the preservation of both natural ecosystems and indigenous knowledge systems.

Mr. Shabir Ahmad is a Ph.D. scholar at Quaid-i-Azam University, Islamabad, specializing in Ecology, Systematics, Ethnobotany, and Paleobotany. His research integrates plant biodiversity, systematics, and the study of ancient plant life to understand the evolution of plant species and their ecological roles over time. Shabir's work also focuses on the conservation of plant diversity and the application of traditional ecological knowledge. Through his interdisciplinary approach, he aims to contribute to both modern conservation efforts and the understanding of historical plant environments.

Mr. Amir Sohail is an M.Phil. Scholar at Quaid-i-Azam University, Islamabad, specializing in Plant Systematics and Biodiversity. His research focuses on the classification, identification, and conservation of plant species, with an emphasis on understanding plant diversity and its ecological roles. Amir is passionate about exploring the connections between plant morphology and biodiversity, and he aims to contribute to the conservation of plant resources through his studies. His work seeks to advance knowledge in plant taxonomy and to support sustainable biodiversity management in changing environments.

Dr. Mirza Barjees Baig is a professor at the Prince Sultan Institute for Environmental, Water, and Desert Research, King Saud University, Saudi Arabia. He earned his MS degree in International Agricultural Extension in 1992 from Utah State University, Logan, Utah, USA, and was placed on the "Roll of Honor". He completed his Ph.D. in Extension for Natural Resource Management from the University of Idaho, USA, and was honored with the "1995 Outstanding Graduate Student Award". He has published extensively on the issues associated with natural resources in national and international journals. He has also made oral presentations about agriculture, natural resources, and the role of extension education at various international conferences. He serves on the editorial boards of many international journals and is a member of many international professional organizations. He has produced six books on climate change, food security, and climate-smart agriculture. Food waste, water management, degradation of natural resources, the deteriorating environment, and their relationship with society and the community are his areas of interest. He has attempted to develop strategies to conserve natural resources, promote the environment, and develop sustainable communities. He started his scientific career in 1983 as a researcher at the Pakistan Agricultural Research Council, Islamabad, Pakistan. He served at the University of Guelph, Ontario, Canada, as a Special Graduate Faculty member from 2000 to 2005. He served as a foreign professor at the Allama Iqbal Open University (AIOU), Pakistan, through the Higher Education Commission from 2005 to 2009.

Chapter 8
Implications of Deforestation on Carbon Sequestration Potential of Tropical Forests

Gyanaranjan Sahoo, Asim Chandra Dash, Meenakhi Prusty, and Amita Sharma

Abstract Deforestation is one of the major causes of emission of large amounts of CO_2 and other greenhouse gases into the atmosphere. Global warming, brought on by increased anthropogenic greenhouse gas (GHG) emissions (421 parts per million compared to 280 parts per million in the pre-industrial era), has altered the climate, reducing ecosystem productivity and water availability, and harming people's livelihoods. Burning fossil fuels, electricity generation, agriculture, industry, polluting water bodies, and urban activities all contribute to a rise in GHG footprint, with CO_2 accounting for 72% of the total. To maintain ecological services, GHG emissions must be balanced with carbon sequestration. Forests are the most important carbon sinks (about 45%) for reducing global warming. Forestlands have been substantially destroyed for agriculture around the world, and deforestation is still occurring today. Afforestation initiatives are emerging as key policy interventions around the world to boost carbon sequestration, but there has been little systematic research on their effects on the livelihoods of forest-dependent communities. Developing strategies to stimulate the use of carbon-sequestering practices face a number of obstacles, including demonstrating additionality, addressing the risk of stored carbon losses, and involving smallholders and pastoralists with shaky land tenure. Carbon sequestration, if appropriately designed, can conserve and/or increase carbon stocks while also improving rural livelihoods.

G. Sahoo (✉)
Krishi Vigyan Kendra, Odisha University of Agriculture and Technology, Nayagarh, Odisha, India
e-mail: gsahoo.forestry@ouat.ac.in

A. C. Dash
Department of Agronomy, Odisha University of Agriculture and Technology, Bhubaneswar, Odisha, India

M. Prusty
AICRP on LTFE, College of Agriculture, Odisha University of Agriculture and Technology, Bhubaneswar, Odisha, India

A. Sharma
Krishi Vigyan Kendra, Rajmata Vijayaraje Scindia Krishi Vishwa Vidyalaya, Gwalior, Madhya Pradesh, India

M. Behnassi et al. (eds.), *Food Systems and Biodiversity in the Context of Environmental and Climate Risks*, https://doi.org/10.1007/978-3-031-89167-0_8

Keywords Afforestation · Carbon sequestration · Deforestation · Environmental implications · Forest livelihood security

8.1 Introduction

The adverse impacts of deforestation make it primarily a problem for developing countries, which include biodiversity loss and increased greenhouse gas emissions (Sahoo et al. 2021a, b). Trees are the most ancient, dependable, incredibly valuable, and widely utilized raw resources, and they are essential for the provision of oxygen and the uptake of CO_2. Forests cover 30% of the earth's geographical surface, or around 3.9 billion hectares. The initial forest cover was believed to be over six billion hectares (Buotte et al. 2017). Only 6% of the Earth's land surface is covered by tropical forests. Trees are said to house between 70 and 90% of all species on the planet. Each day, deforestation results in the extinction of 50–100 animal and plant species. Despite their usefulness to humanity, several of these genera are currently critically endangered, particularly plants having diverse values.

The fundamental source of the issues with rising temperatures is the accumulation of human emissions. When it comes to global warming, carbon is a significant greenhouse gas. Forests, like other biological systems, may store, collect, and release carbon (Food and Agriculture Organization 2020). Carbon is drawn from the atmosphere via biological growth. Because of their long lifespan and potential bulk, forests can and do store massive quantities of carbon in their bodies. Forests are carbon silos in principle, with cumulative biological growth increasing the amount of carbon stored (Buotte et al. 2019a). Biological decomposition or loss, such as decomposition or fire, releases carbon, which enters into the atmosphere (Dass et al. 2018). Soils may also act as a carbon sink, capturing and storing carbon. Indeed, carbon-rich soils are typically well-suited to biological development, especially in agriculture.

The bulleted style of presentation in Sects. 8.1 and 8.2 is unacceptable for a book chapter. Moreover, the parenthetical acronyms/references are not clear to me (although I am familiar with words like WWF). If these are indeed references, please include proper citations and rewrite the text as full sentences (instead of bullet points) so that the reader gets a clear understanding of what the authors want to state. The FAO series "The State of the World's Forests" is an appropriate reference for this purpose. In any case, it is customary to write the expanded form of any abbreviation/acronym at first mention in the text, and the abbreviated forms can be subsequently used.

The parenthetical References should be specific so that the reader is able to access them. Citing Newspapers or other similar publications as in the preceding section will be of no use to the reader.

8.2 Deforestation's Consequences

From the equator to the poles, forests are found all across the planet. According to the World Bank, a global financial institution, they make up 80% of the world's species diversity, which includes a vast range of trees, plants, animals, and bacteria. Some places are incredibly varied; for example, the tropical forests of New Guinea are inhabited by further than 6% of the world's biodiversity (Sahoo and Wani 2019).

Trees are more than simply a habitat for a diverse range of organisms; they are also an essential commodity for several people across the world. Trees provide fuel, lumber, and charcoal in nations such as Uganda (Barnosky et al. 2012). According to a news organization, Uganda has lost 63% of its woodland cover during the past 25 years. Families send their kids, generally females, to gather kindling, and the youths should venture out longer distances to arrive at the trees. Children miss school because social events might take up a significant amount of time (Berner and Law 2016).

Forestry generates three-quarters of the globe's surface water, as per the 2018 Food and Agriculture Organization report, therefore tree loss can have an impact on water quality. According to the United Nations' 2018 State of the World's Forests statement, over half of the inhabitants of the globe rely on aquifers for potable water in wooded areas, agriculture, and industry (Food and Agriculture Organization 2020). In tropical areas, deforestation can alter the way water vapor develops above the canopy, resulting in less rainfall. The land and atmospheric conditions were greater in parts of the Amazon basin that were transformed into arable land, according to the 2019 research published in the journal Eco-hydrology, which can worsen dry conditions. Evapotranspiration rates were roughly three times higher on forested terrain, resulting in more water vapor in the atmosphere (Food and Agriculture Organization 2020).

Trees also sequester, reducing the amount of CO_2 emitted by people. As the global climate worsens, trees will play an extremely vital function in reducing emissions, or the collection and management of carbon dioxide emissions. Tropical forests alone are expected to offer about 23% of the emissions reductions required to realize environmental issues, (Creed et al. 2016).

Deforestation not only damages flora that helps absorb CO_2, However, it also emits carbon dioxide into the atmosphere (Dass et al. 2018). Deforestation is the second most crucial source of climate change, as per the UN's Food and Agriculture Organization. (The combustion of fossil fuels is the first.) Deforestation is responsible for around a quarter of all CO_2 emissions (Buotte et al. 2019b).

8.3 Approaches to Reforestation

To help lessen the necessity of tree felling, remedies to deforestation are being explored. For instance, a strong motivation to deforest a region is to improve the quantity of land available for cultivation. Using contemporary farming methods and crops or adopting sustainable agricultural practices could help communities reduce their need for additional land, according to the UN's Sustainable Forest Management Framework.

Forests may be restored by replacing species in deteriorated places or just allowing the forest ecosystem to regenerate over time. The goal of rehabilitation, according to the US Forest Service (Buotte et al. 2019b), is to return the forest to its previous state before it was degraded. The quicker a devastated region is restored, the faster the ecosystem recovers (Ferreira et al. 2018). As a consequence of this, biodiversity will return, aquifers will be regenerated, carbon will be retained, and soils will be restored. (Brandt et al. 2014).

We can stop deforestation if we all cooperate. We may limit our use of products made with palm oil, purchase items made of known wood, avoid using paper as much as possible, and plant trees whenever we can (Brandt et al. 2014).

8.4 Issues

8.4.1 Poverty

Poverty is also a major contributor to deforestation. Poverty causes trees to be felled and burned for fuel. To make ends meet, the charcoal created from the burned trees is sold for money. On the other side, felled trees are sold as lumber, while cleared land is utilized for cattle grazing, community plantings, and towns (Baker and Williams 2015). Agriculture, mining and logging, together with industrialization processes, are the causes of deforestation (Houghton and Naasikas 2017). During the conflict, (cutting and felling of trees) deforestation was used as a defensive tactic and a purposeful action to open up tight regions with trees that may have acted as a hiding spot for the enemy (Baker and Williams 2015). Deforestation is frequent in populated countries like Nigeria, according to FAO. Nigeria's deforestation processes are exacerbated by increasing population expansion and demographic pressure. Nigeria is Africa's most populous country, with a population of 162.5 million people (Keenan et al. 2013). When a growing population is combined with a high degree of poverty, this creates a key issue. According to Mackey et al. (2015) and (Sabatini et al. 2019), around 70% of Nigerians (105 million) are poor. Overcrowding leads to a rise in the building of residential and public spaces. This loosens the soil, making it more prone to runoff and flooding. The consequences might be disastrous. According to the United Nations Environment Programme (UNEP), Africans are experiencing deforestation at double the global rate (Sabatini et al. 2019). Insert proper citations!

8.4.2 Agricultural Practices

Deforestation is also aided by poor agricultural practices such as slashing and burning (Riitters and Wickham 2012). According to a study, nearly 60% of Nigerians use firewood for cooking due to the high expense of kerosene (UNFCCC 2002). Inadvertent forest fires started by some people result in deterioration. According to the FAO, tropical emerging nations suffered the most deforestation between 2000 and 2005. This shows a link between deforestation and poverty. The main causes of deforestation in Nigeria are poverty-induced human activities (Riitters and Wickham 2012). Is this not applicable to other tropical countries as well? Poverty is often considered a driver of tropical deforestation for a long time pan-tropically.

Lawrence and Vandecar (2015) project that the complete loss of tropical forest cover would, on its own, increase mean global temperatures by 0.1–0.7 °C. Because of habitat destruction, fragmentation, and degradation, tropical deforestation significantly reduces biodiversity, which in turn lowers species richness and raises the danger of extinction (Bradshaw et al. 2009). Moreover, recent studies have documented tropical deforestation's impact on a wide range of human health outcomes, including infant mortality due to smoke from forest fires in Indonesia and increased malaria prevalence in Nigeria (Berazneva and Byker 2017) and the Amazon (MacDonald and Mordecai 2019). Poverty-reduction strategies include the intensification of agriculture, the development of agricultural exports, and the provision of education and medical support. However, poverty-reduction options have been rarely used as strategies for reducing deforestation, although many developing countries consider both poverty alleviation and environmental protection. Poverty and deforestation reductions have been considered to be trade-offs (Fischer et al. 2021).

8.4.3 Urbanization

Another important component in the issue of deforestation is the urbanization. Due to a lack of understanding about the negative impacts of deforestation, every year, almost 8.5 million hectares of tropical forest are irreparably damaged for the building of new cities and housing (Ferreira et al. 2018). As a result, forest resources are being depleted in an uncontrolled and unending manner. In Nigeria, 81% of the natural forest cover has been destroyed. Forest clearing for agriculture operations and deforestation, rural people felling trees for sale as a strategy for survival because of scarcity, and the use of trees as wood for fuel are all causes of deforestation in Nigeria, all of which cause significant harm and lead to desertification (Riitters and Wickham 2012).

8.5 The Environmental Effects of Deforestation and Forest Degradation

8.5.1 Deforestation

Deforestation causes a host of societal and affordable problems that could make living miserable for people. It is connected to issues such as biodiversity loss, destruction of forest-based civilizations, and climate change. Deforestation is a worldwide issue that endangers the environment's viability, with Nigeria carrying a disproportionately significant share of the cost due to its high pace. Deforestation has negative consequences for the environment, the economy, and the population. When natural vegetation is lost, livestock suffers shelter and becomes more prone to trafficking. Malaysia's biodiversity is threatened by deforestation (UNFCCC 2015). Malaysia is home to 15,500 higher plant species, 746 bird species, 300 animal species, 379 reptiles, 198 amphibian species, and 368 fish species (Food and Agriculture Organization 2020).

8.5.2 Release of Greenhouse Gas Emissions

Despite the fact that forests store the most carbon on the earth, deforestation, which represents 15% of global greenhouse gas emissions after fossil fuel burning, is the second-largest contributor to climate change. Planting palm oil in upland moist deciduous forests, on the other hand, does not result in as much carbon sequestration over time (UNFCCC 2007).

Thousands of hectares of forest are burned across Malaysia, notably on Borneo's island, as a result of forest fires that commonly coincide with *el nino* occurrences (Schoene 2002). Smoke from these fires, as well as fires in Kalimantan (Indonesia), pollutes Malaysian air and causes health issues. The Malaysian government-imposed media prohibitions in the 1990s to warn tourists about the health risks associated with fires (Hoogeveen 2007).

This has changed recently, with the administration increasingly blaming Indonesia for wildfire spread. Although ASEAN is striving to bring the parties concerned together, such transboundary pollution is particularly challenging to regulate (Houghton and Naasikas 2017).

Recent apparent environmental weather shifts and monsoon patterns in specific locations, sea level rise, photochemical haze, carbon emissions, and habitat degradation are all examples of anthropogenic-caused climatic and ecological issues (Griscom et al. 2017). Tropical forests are being cut and cleared at a pace of around 5% every decade to service local, regional, national, and emerging economies for timber production, animals, food production, and bioenergy (IPCC 2018). The impact of deforestation on the global atmosphere is one of the most significant consequences. Because forests are the principal carbon sink on the ground, deforestation contributes to comprehensive warming, it is produced by increasing greenhouse gas (GHG) emissions in the environment, causing a net rise in global temperature rise (Sahoo and Wani 2020). As a consequence, deforestation disturbs the global carbon cycle, resulting in higher quantities of carbon dioxide in the atmosphere. Forest conversion and burning for agricultural purposes, as well as forest harvesting for lumber and fuelwood, result in a net CO_2 emission from the biosphere into the atmosphere.

Two billion tons of carbon dioxide (CO_2) are emitted annually by tropical deforestation, albeit unevenly (Houghton and Naasikas 2017).

8.5.3 Water Cycles that Have Been Interrupted

Because trees no longer drain groundwater and precipitation rates are lowered as a result of deforestation, the surrounding ecosystem may become much drier. In 1998, excessive deforestation in Kuala Lumpur, Malaysia's capital, combined with an El Nino-stimulated dryness impelled the city to import water for the first time. The soil can no longer contain as much water as it once is due to the destruction of a chunk of forest, as a result of which the environment will be drier. Deforestation has an impact on potable water, seafood, and marine environments, as well as flood and drought abatement, river and reservoir coastal erosion, less desirable water-related activities, and erosion and turbidity impact farming and land management. (Bruijnzeel et al. 2005). The preservation of urban water supply is one of the most important roles that woods provide (Segan et al. 2016). Water filtration and purification is costly. Forests can either lower costs directly by absorbing runoff or indirectly by relieving runoff-producing dwellings or crops (Kremen and Merenlender 2018). Deforested ecosystems may lose their ability to control and manage river and stream flow. Too much water can produce downstream floods, resulting in innumerable catastrophes across the world. Silting of streams, lakes, and reservoirs, as well as land degradation, are all consequences of this downstream flow. Overflowing is aggravated by deforestation for two reasons (Nolan et al. 2018a). Initially, with less of a 'tree fountain' outcome, groundwater table is more prevalent in soils. During the rainy season, the' sponge' goes up quickly, allowing more precipitation to run off and increasing flood risk. Second, deforestation often fallouts in soil texture, which makes it difficult for rain to soak. Deforestation has the potential to have a significant protracted influence on soil properties. By eliminating the forest cover, slash-and-burn farming exposes the soil to the extremes of tropical heat and rain (Nolan et al. 2018b). Forests deck can endure severe rains because of their plant debris and porous soils. Which happens to these opposing forces of infiltration aim to assess the effect of deforestation on rainfall patterns, river flooding and parched fluxes, and evapotranspiration (the sponge vs. the fountain) (Bruijnzeel et al. 2005). Deforestation and other agricultural expansion have increased the subsidence area of the reservoir, causing protracted erosion and sedimentation. River bottoms of China's Yangtze River Basin, East Asia's major humid tropics lakes and rivers, and the Amazonian Basin have all risen due to heavy siltation, increasing the probability of flooding (Rogelj et al. 2016; Sahoo et al. 2020).

8.5.4 Increased Soil Erosion

Soil erosion is exacerbated by deforestation, which increases precipitation and reduces the shielding of dirt from tree litter. Only a small percentage of the areas have good soils, which are quickly swallowed up after clearing owing to heavy rain (Sahoo et al. 2021a, b). Crop yields suffer as a result, forcing farmers to spend more money on imported fertilizers or clearing additional land. The problem is made worse by the failure to use erosion control technologies. Cleared land is often kept exposed for long periods of time until erosion control work is completed (Sabatini et al. 2019; Schimel et al. 2015). In highland sites where high-value sub-tropical crops are cultivated, substantial amounts of manure and fertilizer can help offset the effects of erosion. Sediment, mineral, and chemical losses due to unregulated runoff result in subsequent erosion and compaction, despite the fact that applying high rates of fertilizer and manure helps to maintain adequate levels of agricultural production (Trexler and Haugen 1994).

8.6 Native Communities' Impacts on Deforestation

8.6.1 Homelands Are Being Destroyed

When large portions of forest are chopped down, vulnerable land squashes and perish, and many species' ecosystems disappear, indigenous communities that live there and are reliant on the forest to maintain their style of living are also endangered (Sabatini et al. 2019; Schimel et al. 2015).

People in developed countries of the globe will never realize how the destruction of forests impacts their style of living, regardless of our own dependency on what the rainforest provides. Indigenous individuals live in much closer proximity to one another than non-indigenous peoples (Sahoo et al. 2021a, b). Governments of countries with rainforests on their limitations routinely deport indigenous tribes previously clear-cutting. This is one of the prophylactic effects of deforestation.

8.6.2 Local and Regional Consequences of Deforestation

Deforestation has the greatest direct impact at the local level. The local community loses a system that provides crucial but generally disregarded tasks such as ensuring a regular supply of clean water and shielding the village from floods and drought when the forest is destroyed (Schuur et al. 2015). Much like a sponge, the forest absorbs tropical storm rainfall while anchoring soils and releasing water at regular intervals. Tropical rainforests play a variable function in the environment, which can support mitigating the disastrous flood and drought cycles that can happen when forests are chopped down (Sahoo et al. 2020b).

Runoff reaches streams fast when forest cover is removed, raising flood waters and creating floods in nearby communities, cities, and agricultural areas, particularly during the rainy season. Communities downstream of deforestation are susceptible to months-long droughts that disrupt river traffic, food production, and economic operations throughout the dry season (Nilsson and Schopfhauser 1995).

Montane and watershed forests, which are located on steep hills, are particularly critical for maintaining surface runoff and flood protection; nonetheless, montane forests saw the highest rate of loss of all tropical forests during the 1980s. (In the late 1990s and early 2000s, this pattern reversed, with highland forests regenerating but lowland regions bearing the brunt of deforestation, owing mostly to agricultural expansion (Sahoo et al. 2021a, b).

The forest also aids in the growth of local humidity, which in turn aids in the increase of rainfall (the process by which plants release water through their leaves). The ecosystem water cycle in the central and western Amazon, for example, maintains 50–80% of the moisture (Law et al. 2018). As part of the water cycle, moisture is traversed and dissipated into the sky, forming rain clouds, and then precipitated as rain back onto the forest. Less moisture is evapo-transpired into the sky when trees are chopped down, resulting in fewer rain clouds forming. As a result of the decrease in rainfall, the area is at risk of drought. The area will dry out in a few years if the rains stop falling because the intense, tropical sun will sear the scrubland. Because of millennia of forest clearance by fire, Madagascar is now primarily a red, treeless desert (MEA 2005; Brown et al. 1996). River flows are decreasing, and less high-quality water reaches cities and agricultural fields. Excessive destruction of coastal rainforests has been blamed in part for diminishing rainfall in West African interior countries (Bass et al. 2000; Brown 1996). Similarly, a new study in Australia (Canadell and Raupach 2008) reveals that the barren desert would be a wetter, more pleasant location if it weren't for human activities, particularly frequent agricultural fires. Transitions in vegetation, such as the transition from rain-loving forests to grassland and shrubs, can affect precipitation patterns. Colombia, which was once ranked second in the world for freshwater reserves, is currently ranked 24th as a result of massive deforestation over the last 30 years. In 1998, the Malaysian capital of Kuala Lumpur was compelled to import water for the first time due to excessive deforestation and an el Nino-induced drought (Canadell and Raupach 2008).

There is great fear that forest degradation would result in a major decrease in rainfall, generating a positive-feedback cycle of rising decomposition for nearby forest cover; lowering the region's moisture supply and vegetation will exacerbate the region's desiccation impact (Berner et al. 2017). The effect might eventually spread beyond the region, wreaking havoc on major agricultural zones and surrounding watersheds.

Britain (Law et al. 2018) issued a warning that due to changes in climate and forest degradation, the Amazon rainforest may be gone in 50 years. Using a worrisome analysis from the Institute of Ecology in Edinburgh, it was said during the 1998 global climate treaty conference in Buenos Aires (FAO 2004).

Desiccated forest fires are more likely to be disastrous. Such flames erupted in 1997 and 1998 as a result of El Nino-induced dry weather. Across Indonesia, Brazil, Colombia, Central America, Florida, and other countries, hundreds of hectares were destroyed (Food and Agriculture Organization 2003). In 1998, the Woods Hole Research Center cautioned that the Brazilian Amazon, which covered over 400,000 km^2, was extremely prone to fire. When the Amazon was inundated by even harsher droughts in 2005 and 2010, the scope of the destruction escalated much more (Dass et al. 2018).

8.7 Deforestation's Effects on Sustainable Growth

Although it is indisputable that millions of people's reliance on trees highpoints to deterioration and deforestation, there are a number of assistance programs for indigenous people's sustenance that result from such losses (in terms of livelihoods, money, and employment). According to the FAO (2005), around 60 million indigenous peoples rely almost totally on forests, 350 million people rely substantially on forests for life and income, and 1.2 billion people depend on agricultural practices based on rainfed agriculture (Rahman et al. 2015). According to the United Nations Federation on Climate Change (UNFCC) 2007, sustainable livelihoods provide access and rights to a diverse combination of commodities and prospects critical to human well-being. This is extremely crucial for most indigenous peoples, particularly those living in forest edges, who often lack basic requirements including appropriate and healthy environment, proper nutrition, safe housing, health care, renewable energies, proper sanitation, and learning (Le Quere et al. 2018). The trade of wood products appears to be a substantial source of money for regional and community agencies, as well as traditional rulers and individuals. Examples include export income, tariffs, copyrights, and financial enrichment for people in many ways involved in the utilization of forest products. Tropical wood exports generate around US $ 100 billion per year, or about 0.5% of world GDP, according to (Kaimowitz and Angelsen 1998). Inside national economies, it is alleged that the sale and distribution of chainsaw lumber provide unaccounted-for money in the form of market tolls, income taxes, waybill taxes, and customs charges (Keenan et al. 2013). According to CIFOR's Poverty Environment Network (PEN), forest revenue represents approximately a quarter of total family revenue for rural societies existing in or near forests (Rahman et al. 2015).

Forest exploitation may also aid in the strengthening of dependent communities' social asset base. Because woods are often located in rural areas where national government investments are modest, people frequently lack basic utilities and are characterized by poor development (UNFCCC 2007). However, in forest edge settlements, the social rapport developed via collaboration has served as the foundation for the notion of social facilities such as good roads, schools, electricity, and a sufficient source of water (Lecina-Diaz et al. 2018). Illegal logging sabotages the government's finances and exposes the country to unavoidable shortages and low-quality planks (Griscom et al. 2017). Because of the possible influence on plant life and the necessity of global biota, sustainable forest resource management has been a key issue for decades (Brown 1996). Despite its significance, Africa's natural tropical high forest has continued to decrease, posing a danger to long-term forest management. Due to expanding population, existing wood supplies are being overexploited in many areas. As a result of substantial deforestation, villagers are forced to journey considerable distances to obtain firewood and are eventually encouraged to substitute dried animal dung and agricultural residues for fuel wood. Because rural people rely on this replaced resource? (dried animal dung and agricultural residues) for increasing soil fertility, this has major ramifications for local agricultural production and productivity. Apart from the evident reduction in quality of life brought on by forest degradation, there are other, less visible consequences that threaten humanity's existence on this planet. Forests provide a living for almost 500 million people throughout the world, with a big number of forest and woodworkers among them (Berner et al. 2017).

In general, a lot of factors contribute to deforestation (Canadell and Raupach 2008). Deforestation, on the other hand, is described as the removal of any region's natural vegetation cover, which often leads to a decrease in plant population and loss of plant biodiversity (Brown et al. 1996). Long-term deforestation generates a host of environmental difficulties in today's society. Hunting, brush incinerating, and land clearance for economic growth are all examples of deforestation that are almost guaranteed to imperil human life on our planet (Sahoo et al. 2020b). As a result of increased aridity produced by deforestation, the rate of deforestation is increasing at an alarming rate, resulting in a decline in land productivity. As a result of growing aridity caused by deforestation, fertile agricultural land is rapidly being replaced by unfertile land (Morin et al. 2018). The strong harmful wind that destroys and damages constructing roofs and cultivating plants is slowly becoming a common occurrence after being an unusual occurrence for a long time. Furthermore, as a result of urbanization and the rapid development of agricultural activities, the problem of flooding, which is not new, has taken on a more prominent role. Deforestation is the systematic removal of plants without the replacement of new ones for pecuniary or societal causes (Maron et al. 2018). As a result, erosion affects the soil, mountains, hills, and even valleys. Floods, landslides, and mudslides occur as a result, as does the extinction of wild animals and the shrinkage of deserts. In the long term, human health and life are significantly harmed.

8.8 Suggestions

Illegal loggers should be prosecuted alongside corrupt government officials in control of forestry laws and regulations. The entire public should be educated about the devastating effects of deforestation on people and society as a whole. The government should start a tree-planting campaign by informing the people that we only have one planet. A skills development programme for rural women and uninformed youth should be devised to slow the rate of deforestation (Maron et al. 2018). Reduced greenhouse gas emissions from deforestation and forest degradation are necessary for ensuring that tropical forests keep up their function in sequestering carbon. A significant part of achieving this can be played by global programs like Reducing Emissions from Deforestation and Forest Degradation (REDD+). Though this new study emphasizes the need to develop strategies to protect areas recovering from human activity, the researchers still believe that protecting old tropical forests should be the main priority for policymakers (Goetz et al. 2015).

8.9 Carbon Sequestration Potential of Forestry Sector: Global Context

By both preserving and sequestering carbon from the atmosphere, forests can play a substantial role in the global carbon budget. Carbon sequestration is a technique for managing forest carbon to improve the amount or pace of carbon accumulation by constructing or improving carbon sinks (Bass et al. 2000). Agroforestry activities on agricultural fields, enhanced silvicultural techniques to promote growth rates and afforestation, reforestation, and restoration of degraded areas are all part of this strategy. In both natural forest and plantation environments, carbon sequestration has been associated with tree planting. Despite the fact that there are many more choices beyond afforestation and reforestation, estimations of the worldwide potential for carbon sequestration have used the area of land accessible for afforestation as a starting point (Bass et al. 2000). The global forest includes 340 Pg C in both living and dead along each ground plants, as well as 618 Pg C in mineral soils and the O horizon (Brown 1996). Globally, 700 million hectares of land could be available for a carbon sustainability and absorption programme, including 345 million hectares of currently unforested or under-stocked land for plantation and agroforestry, 138 million hectares through tropical deforestation slowed, and 217 million hectares through natural and assisted tropical forest regeneration (Trexler and Haugen 1994; Nilsson and Schopfhauser 1995). By 2050, this amount of land might save 60–87 Pg C (Table 8.1).

Between 1995 and 2050, worldwide carbon-equivalent assumptions that might be stored and saved by forest management methods are shown in Table 8.1. The tropics have the greatest capacity to store and sequester carbon (80%), trailed by the Mediterranean zone (17%) and the coniferous zone (3%). By 2045, the yearly rate of carbon sequestration from all activities would have increased to around 202 Pg/year (Brown 1996). Brown et al. (1996) predict that during the next 50 years, roughly 38 Gt of carbon might be stored, with 30.6 Gt coming through afforestation/reforestation and 7 Gt coming from the increasing use of agroforestry methods. From 1995 to 2050, the UNFCCC's goal was to keep CO_2 levels in the atmosphere

Table 8.1 Global estimates of the potential amount of C that could be sequestered and conserved by forest management practices between 1995 and 2050

Latitudinal belt	Practice	Area (Million hectares)	Carbon sequestered and conserved (Pg[c])
High	Forestation	95.21	2.4
	Forestation	113	11.8
Mid	Agroforestry	6.5	0.7
	Forestation	66.9	16.4
	Agroforestry	63.2	6.3
Low	Regeneration[b]	217	11.5–28.7
	Slow reforestation[a]	138	10.8–20.8
	Total	**700**	**60–87**

Source Brown et al. (1996)
Bold represents global estimates of the potential amount of Carbon that could be sequestered and conserved by forest management practices between 1995 to 2050
[a]Includes not satisfactorily restocked forest lands in Canada
[b]Includes an additional 25% of aboveground Carbon to account for Carbon belowground in roots, litter, and soil [based on data in Nilsson and Schopfhauser (1995)]; the range in values is based on the use of low and high estimates of biomass C density resulting from the uncertainty in these estimates
[c]1 Pg = 1015 g or billion metric tons

"at a level that prevents hazardous influence on the climate system." As a result of the Kyoto Protocol (of the UNFCCC), which set GHG emission limits for amending provinces and methods for decreasing global GHG at about 5% below 1990 echelons by the end of the century, a global carbon market has arisen (UNFCCC 2002). As a consequence, the notion of 'carbon trading' emerges. The notion of carbon trading offers developing nations a chance. The Kyoto Protocol has devised a plan in which wealthier countries subsidize the creation of new forests in underdeveloped countries. Several funds have been formed as a result of this, through which project implementers will get forest credits (Schoene 2002). A system known as CDM is developed to attain this purpose. The CDM's goal is to assist poor nations in attaining sustainable development while also assisting wealthier countries in meeting their Kyoto Protocol commitments (KP). However, there is a significant constraint on the participation of Carbon Sequestration (CS) in the CDM during the first commitment term of the KP (Food and Agriculture Organization 2003, 2004). The CDM was created to oversee the process of establishing a new forest (Schoene 2002). Climate change may be mitigated by using the CDM for carbon trading. The cost of mitigation through forestry can be relatively low in some tropical poor nations, ranging from US$ 0.1 to US$ 20/t C, and slightly more in industrialized countries, ranging from US$ 20 to US$ 100/t C (Hoogeveen 2007).

The Paris Agreement asks nations to create and submit long-term development plans with low greenhouse gas emissions in order to more effectively define activities toward the long-term objective.

Even if the Paris Agreement's objectives require a massive increase in climate change activity, low-carbon solutions, and new markets have emerged in the years since it went into effect. An increasing number of nations, areas, towns, and businesses are setting goals to become carbon neutral. Across all economic sectors, which account for 25% of emissions, zero-carbon solutions are starting to compete. For early movers, this tendency has offered numerous new business opportunities. It is especially apparent in the power and transportation sectors.

8.10 Carbon Sequestration Practices in Soils and Forests

Changing climate conditions have an impact on forests and their function in the carbon cycle. Variations in rainfall and temperature can have both positive and negative implications on forest production and health, that are difficult to predict (Hoogeveen 2007). Climate change will either diminish or boost carbon absorption in forests, depending on the conditions, raising questions about how much in the long term, the world's forests will be willing to impart to climate change stabilization. Restoring ecosystems can impact carbon sequestration by promoting certain processes while reducing the negative consequences of others (Food and Agriculture Organization 2003, 2004).

In the European Union, forest ecosystems play a variety of important services, including carbon sequestration. The EU27 countries' forest biomass is expected to hold 9.8 billion tonnes of carbon (tC). In 2004, the EU27 nations' total CO_2 emanations were 1.4 billion tonnes of carbon.

Soils and forests are both capable of storing enormous amounts of carbon. Soil management practices have a considerable influence on the quantity of carbon retained if the land is utilized for farming. Tilling is used in agriculture to remove weeds, combine fertilizers, and prepare the ground for sowing (Rahman et al. 2015). On the other hand, tilling the soil and adding fertilizer both lower the amount of carbon contained in the soil. Bacteria come into touch with previously undisturbed humus when the soil is tilled, which quickly decomposes and releases carbon (Segan et al. 2016). In addition, the soil is frequently amended with limestone and dolomite to avoid acidification, which causes the soil to dissolve and produce CO_2. Eutrophication of soils releases roughly 4 million tonnes of CO_2 equivalents per year in the United States (FAO 2020). Carbon may be harvested in a variety of ways in forestry.

Agricultural practices, on the contrary, can aid in carbon sequestration in the soil (UNFCCC 2002). Intercropping, shelter gardening, and crop rotation are examples of these techniques. Conservation tillage aims to reduce soil change to the bare minimum. This method, known as no-till agriculture, avoids the traditional ploughing portion of the planting and harvesting process, which disturbs the soil and increases carbon discharge (Nolan et al. 2018a). Ploughing is done primarily to decrease weeding, and herbicides generally replace much of the ploughing in a no-till system. Growing conservation tillage like pasture and foxtail in between traditional agricultural years to maintain the soil is known as protection farming (Lecina-Diaz et al. 2018).

The expense of no-till procedures is frequently equivalent in relation to existing approaches (Creed et al 2016; Ferreira et al. 2018). Because keeping the right amount of carbon in the soil enhances crop yield, no-till farming may eventually pay off for the farmer. Carbon in the soil enhances the state of the water, saves the environment, and depletion of the soil. Finally, if the farmer receives reimbursement for carbon trapped in soils, he or she may be entitled to further financial advantages. While no-till may be less expensive, Härtl et al. (2017) discovered that it appears to have negative yield consequences in maize. New information on crop management nutrient losses also shows that no-till lowers connected phosphorus, however, it raises phosphorus outputs from dissolved sources, which is a completely separate and potentially more severe issue.

The "4 per mille Soils for Food Security and Climate" of the Lima-Paris Action Agenda is an ambitious multinational research initiative that was set at COP21 by French Minister of Agriculture Stéphane Le Foll. The 4 per mille or 4 per 1000 initiative aims to offset global greenhouse gas emissions from anthropogenic sources by increasing soil organic matter stocks by 0.4% annually. It was introduced in December 2015 during COP21, with the backing of nearly 150 signatories, including nations, regions, businesses, NGOs, and international organizations. In a voluntary action plan, stakeholders pledge to preserve carbon-rich soils and apply farming techniques that maintain or increase soil carbon stores in agricultural soils (Chambers et al. 2016; Lal 2016).

One potential way to slow down climate change is through soil organic carbon (SOC) sequestration, which transforms atmospheric CO_2 into long-lasting soil carbon. Given that soil holds two to three times the amount of carbon as the atmosphere, even a slight increase in the stocks might have a big impact on reducing greenhouse gas emissions. According to Paustian et al. (2016), the worldwide estimate of soil carbon stored at a depth of 2 m is 2400 giga tonnes (2400×10^{15} g), while the annual greenhouse gas emissions from fossil carbon are projected to be 8.9 giga tonnes C (8.9×10^{15} g). Taking the ratio of global anthropogenic C emissions and the total SOC stock (8.9/2400), results in the value of 0.4% or 4‰ (4 per mille). Increasing SOC has been proposed to mitigate climate change with an additional benefit of improving soil structure and conditions (Lal 2016).

8.11 Carbon Sequestration Presents Unique Challenges and Policy Constraints

Additionality, permanence, and leakage issues arise when carbon is stored in forests or soils. There's also the issue of measuring, monitoring, and verification costs and feasibility. The great majority of proposals for assigning incentives to regulate forest carbon seek to encourage activities that would not have happened alternatively, behaviors that are not part of regular operational processes (Morin et al. 2018).

The word "additionality" refers to whether the carbon sequestered as a result of the programme is truly more than what would have been sequestered without it, or if the carbon would have been sequestered even if the programme had not been implemented. For example, even when a forest was established for commercial timber production, the carbon stored as the tree developed would be absorbed without the need for any carbon sequestration schemes or inducements (Barnosky et al. 2012). However, unless a forest is planted solely to trap carbon, the carbon will be considered supplemental and, in some cases, may be eligible for carbon credits and rewards. In order to detect changes, additionality necessitates a baseline inventory evaluation as well as ongoing inventory monitoring (Mackey et al. 2015).

The term "permanence" refers to whether the carbon is sequestered permanently or merely temporarily. The majority of biological sequestration mechanisms are considered transient. Forests, for example, are prone to both natural disturbances (such as parasites and burning) and timber extraction (Mackey et al. 2015). Compensation for momentary confiscation through carbon offset programmes is frequently lower than payments for permanent sequestration or emissions reductions. In addition, payments for carbon offsets require restitution if the trees are taken down or destroyed (Morin et al. 2018).

Leakage is the process of compensating for carbon leaks caused by a sequestration project. There is no net sequestration and permeability if a forest is retained to avoid degradation in order to keep the stored carbon presently held hostage in the jungle, but the degeneration is just relocated (Sahoo et al. 2021a, b). The term "outflow" does, in fact, have a broad definition. Another example is market-generated leakage, which happens when carbon is stored in one forest but the supply of lumber is limited, raising the market price and leading to increased extraction from other forests, resulting in no net sequestration. Leakage is difficult to identify and contain since it has no geographical borders. For diverse operations within the United States, Morin et al. 2018) evaluate outflow at 10–90%, and Sohngen and Brown (2004) investigate leakage in an international setting. To account for the risk of their impermanence, several forest carbon management ideas enable overlooking or leasing of forest assets, as well as the assets' interoperability (Bass et al. 2000).

Future CO_2 emissions from harvesting are prevented by protecting high-carbon-priority forests while existing CO_2 emissions are offset by carbon sequestration. According to the IPCC (2018), regional fossil fuel emissions averaged 260 Tg C/year from 2003 to 2012. Degraded forest soils are responsible for around 11% of international disposable releases of carbon dioxide (Houghton and Nassikas 2017). Because the soil carbon in high-carbon-priority forests is greatest, because surface litter and root debris disintegrate after harvest, maintaining these forests reduces the environmental impact of the soil.

Measurements, validation, and surveillance are all required components of any carbon sequestration plan. Researchers typically employ sampling procedures and regression models to quantify volume in big, varied forests. For decades, if not centuries, on-the-ground inventorying methods have been created and used to assess the number of forest biomass. These techniques, on the other hand, were largely employed to identify whether or not timber was available for harvesting (Law et al. 2018). Furthermore, despite the fact that the US Forest Service used this strategy to assess US forest stocks during the duration of the twentieth century, these methods were often administered to individual forest stands on a local level. Following the 1980s, the United Nations Food and Agricultural Organization performed yearly evaluations of global forestlands, relying mostly on data provided by individual countries. Aerial or satellite techniques are used in a variety of remote sensing technologies that have lately been developed, habitually in combination with facts on the field (Nolan et al. 2018b).

The most common method for assessing carbon levels and/or changes is to use an on-the-ground surveying process to evaluate forest biomass. The capacity is then transformed into bioenergy, which is the forest's total mass (Buotte et al. 2019a). The link between forest volume and biomass exists, albeit it varies widely depending on tree species, forest height, and other variables. Furthermore, several forest volume-to-biomass ratios solely include growing stock volume's aboveground biomass (Candell and Raupach 2008). When it comes to carbon measurements, however, more biomass is important. Techniques for assessing forest biomass that is being developed, which consider the biomass of standing dead trees, root system biomass, and species variations (Food and Agriculture Organization 2003). Such strategies, on the other hand, solely use conversion factors to estimate biomass, leaving the carbon component to be estimated.

Calculating carbon in soil is far more complicated than calculating carbon in aboveground forest plants. Large-scale methodologies for detecting soil carbon and changes in soil carbon over time have yet to be established in the case of soils. As a result, determining the amount of carbon in soil requires the use of a sampling system (Sahoo and Wani 2019). Soil carbon levels can alter as a result of agricultural strategies like no-till. Regular on-the-ground sampling is required for such approaches, which may be an expensive procedure (Millenium Ecosystem Assessment 2005). A comparable on-the-ground approach is required to test for carbon changes in forest soils.

Finally, sampling, surveillance, and validation were neither easy nor inexpensive at the time. As distant sensing technologies evolve, emerging innovation has the capacity to substantially enhance the process. (Creed et al. 2016). Radar (which sends and receives energy pulses) is a sort of remote sensing that may produce quite accurate volume estimates in woodlands with a low population density (When the forest is too lush, the radar inundates and is unable to penetrate it). LIDAR (light detection and ranging) is a current method of determining an object's distance using scattered light. LIDAR can pierce the forest canopy with good precision and offer data on the underlying terrain's topography (Berner and Law 2016). Despite this, although technological advancements may enhance estimated assessment and validation, it would be difficult and costly to transform forest volume predictions into precise biomass assessments.

8.12 The Journey Ahead

National policy responses: national policies appear to be inclusive in all participating countries, but there is much more work to be done at lower levels (provincial, district). Many policy initiatives, to a large degree, need "surgery" to establish harmony and synergy. The capability at the local level is limited, and solving this problem will need a dedicated effort. As a result, participation in LULUCF operations in Asia and other continents are restricted due to: a lack of capacity at all levels of government; and a lack of capacity to execute, awareness of the international agreement/national laws, or local ramifications.

Climate Change (COP 26), India declared its intention to attain net-zero emissions by 2070. Reiterating the objective of achieving net-zero emissions by 2070, India's long-term low-carbon development strategy has been presented to the United Nations Framework Convention on Climate Change in respect of Paragraph 19 of Article 4 of the Paris Agreement. The tenets of equity, climate justice, and common but differentiated responsibilities and respectful capacities form the foundation of India's long-term low-carbon development plan. With an investment of $25 million over a five-year period, the MoEFCC, the Indian government, and USAID have launched the Trees Outside Forests in India (TOFI) program, which will be implemented in seven states: Andhra Pradesh, Assam, Haryana, Odisha, Rajasthan, Tamil Nadu, and Uttar Pradesh. The program's goal is to increase the number of trees planted outside of forests (TOF) in order to improve the provision of ecosystem services, particularly carbon sequestration, and to provide more inclusive lives and economic possibilities for the rural populace.

8.12.1 Integrated Actions

Mechanisms, rules, and procedures should all take class of people into account. These aims must be met by guidelines within the climate change discussions. Species diversity and drainage management are two more environmental services, that might be integrated to lower transaction costs and increase revenue for local residents. Human cultures' response to climate change will likewise put more pressure on trees (Sahoo et al. 2021a, b). In terms of institutions, no comparable precedent in the poor world that can illustrate a quick reaction to climate change. Starting from the outset and learning from other nations or natural catastrophes experiences is essential.

8.12.2 Adaptation

Although the net loss of carbon sinks may impact climate, the changing climate has jeopardized certain sensitive forest ecosystems. Adaptation strategies are required.

Human cultures' response to climate change will likewise put more pressure on trees. Historically, the societal repercussions of food shortages and water stress have frequently been mitigated at the expense of forests (Sahoo et al. 2020b). Although climate change may hasten these processes, the government, civil society, and resource management should consider adopting suitable and cost-effective adaptation strategies. Integration of adaptation and mitigation methods is critical, particularly in the most sensitive forest ecosystems.

Furthermore, while preserving energy core (Mackey et al. 2015) and intact forest (Ferreira et al. 2018) ecosystems is necessary, it is not sufficient for conservation efforts. Secondary forests, as opposed to primary forests, can support a larger diversity of species and species assemblages (Barnosky et al. 2012; Brandt et al. 2014). Certain locations were identified as having a medium to high potential fire vulnerability (Schuur et al. 2015; for example, the Klamath-Siskiyou region in southwest Oregon). As a result, distressed forests must be included in regional goals where biodiversity is a sustainably managed concern (Maron et al. 2018). To evaluate prospective species diversity and ascertain preservation key focus standings of encroachment regions, provincial analyses (Dass et al. 2018) that replicate plant community conversion on multi-decadal timelines are needed to determine the effect of future climate variability on plant community composition.

Controlled landscape conservation will be an essential aspect of strategies aimed at maintaining biodiversity and boosting climate mitigation because secondary forests are also a product of human activity (Kremen and Merenlender 2018). In regional studies, longer harvest cycles have been found to boost carbon sequestration (Law et al. 2018) and biodiversity (Segan et al. 2016), and hence present chances for greater climate extenuation (Griscom et al. 2017). Young trees may have been crucial in shielding against a variety of shocks, according to historical stand structure analyses (Schimel et al 2015). Even though young trees are more susceptible to stress conditions than older trees (UNFCCC 2002), future climate sensitivity studies of young forests will play an important role in determining harvest methods (Nolan et al. 2018b). To analyze the effects of land management possibilities and develop plans for regulated regions, local thresholding ecosystem models with thorough handling of natural vegetation are required.

Nonentities regarding future temperature, forest growth, and forest death causes make estimating future forest carbon sequestration capability difficult. People confront future climate ambiguity using two forecast models (Buotte et al. 2019b) that deal with various temperature and precipitation variations, though researchers recognize that future climate remains uncertain due to carbon emissions trajectory, climate sensitivity to these emissions, and climate feedback (Schuur et al. 2015). Furthermore, the model's parameterization affects the emulated forest growth (Bass et al. 2000). Especially relative to more broad forest-type scenarios, researchers adopted simulations built particularly for forest types in the western United States (Berner et al. 2017; Law et al. 2018; Buotte et al. 2019b), which enhanced model agreement with statistical information (Buotte et al. 2019b). Trees may enhance their water usage efficiency in response to higher CO_2 concentrations (Keenan et al. 2013; Schimel et al. 2015), albeit this reaction may be influenced by fertilizer availability

(Morin et al. 2018). Because the CLM includes nitrogen limitation (Keenan et al. 2013), it can properly mimic present variations in NPP as CO_2 levels rise (Segan et al. 2016).

The world is paying more and more attention to climate change adaptation as confidence in climate change projections grows. Despite the fact that industrialized countries have the majority of the blame for growing atmospheric greenhouse gas concentrations, developing countries will bear a disproportionate share of the global costs associated with climate change due to their unique adaptation demands resulting from their high levels of vulnerability. Climate change adaptation is a must for all nations, wealthy or not.

The United Nations Intergovernmental Panel on Climate Change has released a study (Georgieva et al. 2022) outlining the dire repercussions of not slowing down global temperature rise and adjusting to a hotter planet. Risks from extreme weather and climate change should be addressed through adaptation, such as by preserving agriculture, controlling the effects of rising sea levels, and strengthening infrastructure.

Because they depend on certain characteristics, such as how well-adapted a nation is to its current environment, the benefits of adaptation can be hard to quantify at times. Through institutional capacity building and financial support, the international community may assist poor and vulnerable nations in adapting. Despite not being its cause, these nations will bear the brunt of climate change's catastrophic effects. Ensuring that climate change does not imperil growth and stability in less developed nations is also a global priority. Since upfront investments in protection might be less expensive than humanitarian relief and post-disaster reconstruction, development partners may find that investing in climate resilience can also be financially advantageous. For adaptation support to be effective, it must be added to current aid and come with simplified conditions that match the institutional capability of the recipient nation.

8.13 Conclusions

Land management must become a greater social priority if we are to avoid the current path of catastrophic global change. Temperate forests in the western United States with moderate to high carbon sequestration capacity and minimal projected climate sensitivity might cover around 8 years of regional fossil fuel usage, or 27–32% of the global adaptation strategies positively found for subalpine forests, while also boosting ecological integrity and species diversity maintenance. When establishing safeguards, biodiversity metrics must be considered to ensure that species-rich habitats generated by regular disturbance regimes are not neglected. Climate change's long-term consequences, as well as related pressures as the global race is increasing at an exponential rate, make it critical to assess mitigation and restoration viable alternatives over timeframes that span several decades, with the common aim of reducing future emissions, significantly reducing committed CO_2 emissions, and limiting ecological transition and deterioration through conserving plant and animal variability.

Forests are extremely important in the global carbon cycle. Climate change both affects and is impacted by them and how they are managed or destroyed in the twenty-first century will have a huge impact on the direction of global warming. Forest management can help with carbon sequestration and emissions reductions. We can reduce the impact of climate change by properly managing forests. Climate change mitigation through forest carbon sequestration is a low-cost strategy that will open the door for development activities since securing funds for carbon sequestration is extremely quick and simple. To improve forest capacity for carbon sequestration, the following steps should be considered: increasing funding for the survival by focusing on innovative sustainable forest attributes in contexts that prevent the transformation of forests to other land uses, amplifying restoration to enhance species diversity and assets, increasing people participation in forest management, slowing carbon sequestration hotspots and protected areas, and a training programme for government officials.

References

Baker WL, Williams MA (2015) Bet-hedging dry-forest resilience to climate-change threats in the western USA based on historical forest structure. Front Ecol Evol 2:1–7

Barnosky AD et al (2012) Approaching a state shift in Earth's biosphere. Nature 486:52–58

Bass S, Dubois O, Moura CP, Pinard M, Tipper R, Wilson C (2000) Rural livelihood and carbon management. IIED natural resources issues paper no. 1. London, IIED, 120pp

Berazneva J, Byker TS (2017) Does Forest loss increase human disease? Evidence from Nigeria. Am Econ Rev 107(5):516–521

Berner LT, Law BE (2016) Plant traits, productivity, biomass and soil properties from forest sites in the Pacific Northwest, 1999–2014. Nat Sci Data 3:1–14

Berner LT, Law BE, Hudiburg TW (2017) Water availability limits tree productivity, carbon stocks, and carbon residence time in mature forests across the western US. Biogeosciences 14:365–378

Bradshaw CJA, Sodhi NS, Brook BW (2009) Tropical turmoil: a biodiversity tragedy in progress. Front Ecol Environ 7(2):79–87

Brandt P, Abson DJ, DellaSala DA, Feller R, von Wehrden H (2014) Multifunctionality and biodiversity: ecosystem services in temperate rainforests of the Pacific Northwest, USA. Biol Conserv 169:362–371

Brown S (1996) Present and potential roles of forests in global change debate. Unasylva 47(2):3–10

Brown S, Sathaye J, Cannel M, Kauppi P (1996) Management of forests for mitigation of greenhouse gas emissions. In: Watson RT, Zinyowera MC, Moss RH (eds) Climate change 1995, impacts, adaptations and mitigation of climate change: scientific-technical analyses. Report of working group II, assessment report, IPCC, Cambridge, UK, Cambridge University Press, pp 773–797

Bruijnzeel LA, Bonell M, Gilmour DA, Lamb D (2005) Forest, water and people in the humid tropics: an emerging view Forest. In: Bonell M, Bruijnzeel LA (eds) Water and people in the humid tropics. Cambridge University Press, Cambridge

Buotte PC, Hicke JA, Preisler HK, Abatzoglou JT, Raffa KF, Logan JA (2017) Recent and future climate suitability for whitebark pine mortality from mountain pine beetles varies across the western US. For Ecol Manag 399:132–142

Buotte PC, Levis S, Law BE (2019a) NACP: forest carbon stocks, fluxes and productivity estimates, western USA, 1979–2099. ORNL DAAC, Oak Ridge, Tennessee, USA. https://doi.org/10.3334/ornldaac/1662

Buotte PC, Levis S, Law BE, Hudiburg TW, Rupp DE, Kent JJ (2019b) Near-future forest vulnerability to drought and fire varies across the western United States. Glob Change Biol 25:290–303

Canadell JG, Raupach MR (2008) Managing forests for climate change mitigation. Science 320:1456–1457

Chambers A, Lal R, Paustian K (2016) Soil carbon sequestration potential of US croplands and grasslands: implementing the 4 per thousand initiative. J Soil Water Conserv 71:68A-74A

Creed IF, Weber M, Accatino F, Kreutzweiser DP (2016) Managing forests for water in the anthropocene—the best kept secret services of forest ecosystems. Forests 7:60

Dass P, Houlton BZ, Wang YP, Warlind D (2018) Grasslands may be more reliable carbon sinks than forests in California. Environ Res Lett 13:074027

FAO (2003) Carbon Sequestration in dry land soil. FAO Corporate Document Repository, Rome, Italy, pp 75–80

FAO (2004) Assessing carbon stocks and modeling win-win scenarios of carbon sequestration. FAO Corporate Document Repository, Rome, Italy, 55pp

FAO (2005) Annual rate of deforestation. Retrieved from http://green.wikia.com/wiki/Deforestation_in_Nigeria

FAO (2020) The state of food and agriculture. Overcoming water challenges in agriculture, Rome, pp 111–120

Ferreira J et al (2018) Carbon-focused conservation may fail to protect the most biodiverse tropical forests. Nat Clim Chang 8:744

Fischer R, Taubert F, Müller MS, Groeneveld J, Lehmann S, Wiegand T, Huth A (2021) Accelerated forest fragmentation leads to critical increase in tropical forest edge area. Sci Adv 7(37):eabg7012

Georgieva K, Gaspar V, Pazarbasioglu C (2022) Poor and vulnerable countries need support to adapt to climate change. IMF Blog

Goetz SJ, Hansen M, Houghton RA, Walker W, Laporte N, Busch J (2015) Measurement and monitoring needs, capabilities and potential for addressing reduced emissions from deforestation and forest degradation under REDD+. Environ Res Lett 10(12):123001

Griscom BW et al (2017) Natural climate solutions. Proc Natl Acad Sci USA 114:11645–11650

Härtl FH, Höllerl S, Knoke T (2017) A new way of carbon accounting emphasises the crucial role of sustainabletimber use for successful carbon mitigation strategies. Mitig Adapt Strateg Glob Chang 22:1163–1192

Hoogeveen H (2007) Forest and climate change. In: Complex problem to integrated solution. UN Chronicle Home Page.

Houghton RA, Nassikas AA (2017) Global and regional fluxes of carbon from land use and land cover change 1850–2015. Glob Biogeochem Cycl 31:456–472

IPCC (2018) Summary for policymakers. In: Masson-Delmotte V, Zhai P, Pörtner H-O, Roberts D, Skea J, Shukla PR, Pirani A, Moufouma-Okia W, Péan C, Pidcock R, Connors S, Matthews JBR, Chen Y, Zhou X, Gomis MI, Lonnoy E, Maycock T, Tignor M, Waterfield T (eds) Global Warming of 1.5 °C. An IPCC special report on the impacts of global warming of 1.5 °C above preindustrial levels and related global greenhouse gas emission pathways, in the context of strengthening the global response to the threat of climate change, sustainable development, and efforts to eradicate poverty. World Meteorological Organization, Geneva, Switzerland, 32pp

Kaimowitz D, Angelsen A (1998) Economic models of tropical deforestation: a review. CIFOR

Keenan TF, Hollinger DY, Bohrer G, Dragoni D, Munger JW, Schmid HP, Richardson AD (2013) Increase in forest water-use efficiency as atmospheric carbon dioxide concentrations rise. Nature 499:324

Kremen C, Merenlender AM (2018) Landscapes that work for biodiversity and people. Science 362:eaau 6020

Lal R (2016) Beyond COP 21: potential and challenges of the "4 per thousand" initiative. J Soil Water Conserv 71:20A-25A

Law BE, Hudiburg TW, Berner LT, Kent J, Buotte PC, Harmon ME (2018) Land use strategies to mitigate climate change in carbon dense temperate forests. Proc Natl Acad Sci USA 115:3663–3668

Lawrence D, Vandecar K (2015) Effects of tropical deforestation on climate and agriculture. Nat Clim Change 5(10):27–36

Le Quere C et al (2018) Global carbon budget 2018. Earth Syst Sci Data 10:2141–2194

Lecina-Diaz J, Alvarez A, Regos A, Drapeau P, Paquette A, Messier C, Retana J (2018) The positive carbon stocks biodiversity relationship in forests: co-occurrence and drivers across five sub-climates. Ecol Appl 28:1481–1493

MacDonald AJ, Mordecai EA (2019) Amazon deforestation drives malaria transmission, and malaria burden reduces forest clearing. Proc Natl Acad Sci 116(44):22212–22218

Mackey B et al (2015) Policy options for the world's primary forests in multilateral environmental agreements. Conserv Lett 8:139–147

Maron M, Simmonds JS, Watson JEM (2018) Bold nature retention targets are essential for the global environment agenda. Nat Ecol Evol 2:1194–1195

Millennium Ecosystem Assessment (MEA) (2005) Ecosystems and human well-being: biodiversity synthesis. World Resources Institute, Washington, D.C., USA. Moomaw WR, Masino S, Falson EK (2019) Intact forests in the United States: proforestation mitigates climate change and serves the greatest good. Front For Glob Change 2:1–10

Morin X, Fahse L, Jactel H, Scherer-Lorenzen M, Garcia-Valdes R, Bugmann H (2018) Long-term response of forest productivity to climate change is mostly driven by change in tree species composition. Sci Rep 8:1–12

Nilsson S, Schopfhauser W (1995) The carbon sequestration potential of a global afforestation program. Clim Change 30:267–293

Nolan C et al (2018a) Past and future global transformation of terrestrial ecosystems under climate change. Science 361:920–923

Nolan RH, Drew DM, O'Grady AP, Pinkard EA, Paul K, Roxburgh SH, Mitchell PJ, Bruce J, Battaglia M, Ramp D (2018b) Safeguarding reforestation efforts against changes in climate and disturbance regimes. For Ecol Manag 424:458–467

Paustian K, Lehmann J, Ogle S, Reay D, Robertson GP, Smith P (2016) Climate-smart soils. Nature 532:49–57

Rahman MS, Akter S, Al-Amin M (2015) Forest and agro-ecosystem productivity in Bangladesh: a climate vegetation productivity approach. For Sci Technol 11:126–132

Riitters KH, Wickham JD (2012) Decline of forest interior conditions in the conterminous United States. Scien Rep 2:1–4

Rogelj J, den Elzen M, Hohne N, Fransen T, Fekete H, Winkler H, Chaeffer RS, Ha F, Riahi K, Meinshausen M (2016) Paris Agreement climate proposals need a boost to keep warming well below 2 degrees C. Nature 534:631–639

Sabatini FM et al (2019) Trade-offs between carbon stocks and biodiversity in European temperate forests. Glob Change Biol 25:536–548

Sahoo GR, Wani AM (2019) Multifunctional agroforestry systems in India for livelihoods. Ann Hortic 12(2):139–149

Sahoo GR, Wani AM (2020) Effect of climate change on land degradation. Int J Innov Engin Manage Res SSRN Elsevier 09(12):483–494

Sahoo GR, Wani AM, Satpathy B (2020a) Greening wastelands for environmental security through agroforestry. Int J Adv Res Sci Technol (IJARST) 7:2581–9429

Sahoo GR, Wani, AM Sharma A (2020b) Enhancing food security through agroforestry for sustainability—a review. Int J Curr Microbiol Appl Sci Spec 11:2001–2020

Sahoo GR, Wani AM, Gupta S, Prusty M (2021a) Soil and water conservation measures through agroforestry—a review. PLANTA Res Book Ser 2:251–262. ISBN 978-81-953419-4-8

Sahoo GR, Wani AM, Prusty M, Ray M (2021b) Effect of globalization and climate change on forest—a review. Mater Today Proc. Article in Press. https://doi.org/10.1016/j.matpr.2021.06.113
Schimel D, Stephens BB, Fisher JB (2015) Effect of increasing CO_2 on the terrestrial carbon cycle. Proc Natl Acad Sci USA 112:436–441
Schoene D (2002) Assessing and reporting forest carbon stock changes: a concerted effort. Unasylva 210(53):76–81
Schuur EAG et al (2015) Climate change and the permafrost carbon feedback. Nature 520:171–179
Segan DB, Murray KA, Watson JEM (2016) A global assessment of current and future biodiversity vulnerability to habitat loss-climate change interactions. Glob Ecol Conserv 5:12–21
Trexler MC, Haugen C (1994) Keeping it green: evaluating tropical strategies to mitigate global warming. World Resource Institute, Washington, DC, p 77
United Nations Framework Convention on Climate Change (UNFCC) (2015) The Paris agreement. UNFCC, Paris, France. United Nations General Assembly. 2015. Transforming our world: the 2030 agenda for sustainable development. United Nations General Assembly, New York, New York, USA. U.S. Energy Information Administration (2015) Energy-related carbon dioxide emissions at the state level, 2000–2013. U.S. Department of Energy, Washington, D.C., USA. U.S. Fish and Wildlife Service (2018) U.S. FWS threatened & endangered species active critical habitat report. U.S. Fish and Wildlife Service, Washington, DC, USA
UNFCCC (2002) Report of the conference of parties on its Seventh session held at Marrakesh from 29th October to 10th November 2001: part II—action taken by the conference of parties, p 54
UNFCCC (2007) Status of ratification of the UNFCCC. Available online at http://unfccc.int/essential_background/convention/items/2627.php. Last accessed 3 Jan 2009

Dr. Gyanaranjan Sahoo is a scientist in Forestry at Krishi Vigyan Kendra (KVK), Odisha University of Agriculture & Technology (OUAT), Nayagarh, India. He provides training and demonstrations to farming communities and has received the Best KVK Scientist Award at an international conference in Kathmandu, Nepal. Dr. Sahoo has published several national and international research papers and has actively participated in numerous conferences. His research focuses on forestry and agriculture, with a commitment to supporting rural and agricultural development.

Chapter 9
Sustainable Use of Marine Resources and Perspectives of Blue Economy in Sustainable Development in Developing Countries

Alavudeen Asmiya, Muneeb M. Musthafa, Rosa María Martínez-Vázquez, Mohamed Mujithaba Mohamed Najim, and Faiz Mohideen Mohamed Thassim Marikar

Abstract For centuries, oceans have supported human populations by providing food, energy, biodiversity, and recreation, thanks to their natural ability to regenerate. Harnessing the oceans without depleting their potential for future generation, and utilizing resources economically as part of a strategic approach to the global economy, defines what is now referred to as the 'Blue Economy' (BE). BE models enhance the potential of sectors such as fisheries, aquaculture, biotechnology, tourism, transport, and energy production. An appropriate regulatory framework for managing the sustainability of coastal zones through these models will be crucial for states that rely on the oceans. Small Island Developing States (SIDS) pioneered the global drive toward BE. Sri Lanka, strategically located at the heart of the Indian Ocean's maritime trade, has emerged as a leading nation in capitalizing on this potential. However, despite its advantageous geographic and territorial position, Sri Lanka

A. Asmiya · M. M. Musthafa (✉)
Department of Biosystems Technology, Faculty of Technology, South Eastern University of Sri Lanka, University Park, Oluvil #32360, Sri Lanka
e-mail: muneeb@seu.ac.lk

A. Asmiya
e-mail: asmiya93@seu.ac.lk

R. M. Martínez-Vázquez
Faculty of Economics and Business, University of Almería, Ctra. De Sacramento, s/n, 04120 Almería, Spain
e-mail: rosamaria@ual.es

M. M. M. Najim
Faculty of Agriculture, Sultan Sharif Ali Islamic University, Sinaut Campus, Km 33, TB1741, Jln Tutong, Kampong Sinaut, Brunei
e-mail: najim.mujithaba@unissa.edu.bn

F. M. M. T. Marikar
General Sir John Kotelawala Defence University, Ratmalana 10350, Sri Lanka
e-mail: faiz@kdu.ac.lk

M. Behnassi et al. (eds.), *Food Systems and Biodiversity in the Context of Environmental and Climate Risks*, https://doi.org/10.1007/978-3-031-89167-0_9

faces substantial challenges in sustainably harnessing its marine resources. These include forging robust regional agreements for marine conservation, establishing consistent revenue streams from ocean-based activities, and securing comprehensive government and public support. Consequently, this chapter aims to examine the existing literature on the potential of BE models in Less Developed Countries (LDCs), Small Island Developing States (SIDS), and developing nations such as Sri Lanka. The goal is to identify opportunities for further expanding the benefits of the Blue Economy in coastal states through the development of innovative approaches. This review will provide a global perspective on the demands, opportunities, and challenges associated with maximizing the potential of BE approaches. By identifying limitations, this context will contribute to a deeper understanding of the factors hindering the successful implementation of BE initiatives.

Keywords Development · Economic growth · Ecosystem · SIDS · United Nations

9.1 Understanding the Concept and History of the Emergence of the Blue Economy

The 'Blue Economy' concept emerged in response to the need for sustainable management of the world's oceans, seas, and coastal areas. Traditionally, economic activities related to the ocean, such as fishing, shipping, and tourism, were not considered through a sustainability lens. By the late twentieth century, overexploitation of marine resources and degradation of aquatic ecosystems had led to global recognition of the need for a more balanced approach. The term 'Blue Economy' gained popularity in the early 2010s, particularly after the 2012 United Nations Conference on Sustainable Development (Rio + 20), where it was emphasized as a pathway to sustainable development by integrating economic growth, social inclusion, and environmental sustainability within the marine context (Leggett and Carter 2012). This conference, which aimed to build upon the Millennium Development Goals (MDGs) by establishing the UN High-level Political Forum on Sustainable Development, officially prioritized the sustainability of marine resources as essential to global food security and energy provision (UNCTAD 2014; Silver et al. 2015). The Rio + 20 conference not only underscored the need to shift focus from resource scarcity to abundance but also laid the groundwork for the Blue Economy by addressing environmental challenges and fostering a more sustainable relationship with the world's oceans (Kathijotes 2013).

The Blue Economy concept is heavily inspired by the 'green economy', but with a specific focus on marine and coastal ecosystems. It promotes the sustainable use of ocean resources for economic growth, improved livelihoods, and job creation, while preserving the health of ocean ecosystems. This approach covers a wide range of industries, including fisheries, aquaculture, maritime transport, tourism, offshore renewable energy, and marine biotechnology. The Blue Economy emphasizes the importance of maintaining healthy marine ecosystems, which are vital for the resilience and productivity of ocean and coastal regions. Concept of Blue Economy is promoted by different organizations such as the United Nations (UN) (including the United Nations Development Programme (UNDP), United Nations Environment Programme (UNEP), United Nations Conference on Trade and Development (UNCTAD), United Nations Educational, Scientific and Cultural Organization (UNESCO), Intergovernmental Oceanographic Commission (IOC), World Bank, European Union (EU), African Union (AU), Commonwealth Secretariat, The World Wildlife Fund (WWF), International Union for Conservation of Nature (IUCN), The Ocean Foundation, World Ocean Council (WOC), Global Environment Facility (GEF), Food and Agriculture Organization (FAO) of the United Nations (FAO), and Indian Ocean Rim Association (IORA) (Blue Economy Concept Paper 2014).

The development of the Blue Economy has been driven by international organizations, governments, and NGOs to address the growing challenges facing the world's oceans. The World Bank and the UN have played significant roles in promoting the Blue Economy through various initiatives and frameworks. For instance, the UN's Sustainable Development Goal 14 (SDG 14) specifically focuses on the conservation and sustainable use of the oceans, seas, and marine resources. The Blue Economy is increasingly recognized as a critical component of global efforts to achieve sustainable development, especially for coastal and island nations that rely heavily on marine resources.

Heading toward nine billion of the human population, oceans play an important role in maintaining biodiversity, ecosystems, food chains, and climate regulations. Economy-oriented biological concepts related to oceans include Blue Economy, blue growth, marine economy, and ocean economy, which are often used interchangeably (Kaczynski 2011). Phelan et al. (2020) suggested that the Blue Economy is a subset of the ocean economy focused on generating wealth from marine activities while safeguarding and enhancing the health of marine ecosystems. Meanwhile, the efficient utilization of ocean space to accommodate multiple economic activities is termed 'blue growth' (Dalton et al. 2019). The concepts of 'Blue Economy' and 'Blue Growth' have been used as a political discourse around the world and as a tool to represent the potential contribution to human well-being made by aquatic and marine spaces. In this sense, it presents important challenges such as balancing economic development in a blue context with the preservation of ocean and marine resources for future sustainability (Martínez-Vázquez et al. 2021), where the marine economy plays an important role.

The concept of the 'marine economy' is defined as a new economic model that seeks to introduce innovative development concepts, operating mechanisms, and management models (Xiao et al. 2016). In contrast, the 'ocean economy' encompasses activities related to marine construction, resources, shipping, recreation industries, and tourism (Colgan 2013). The term 'Blue Economy' first gained attention during a 2009 session of the U.S. Congress Senate Committee on Commerce, Science, and Transportation, which highlighted its significance to the US economy by providing substantial business opportunities, addressing climate change, and creating jobs in the renewable energy sector (Cantwell 2009). In the same year, Korea participated in an International Symposium on the 'Blue Economy Initiative for Green Growth', focusing on the sustainable use of ocean resources and strategies for future sustainability (Joroff 2009).

Following the introduction of the Blue Economy's economic models by Pauli (2010), countries with oceanic ties began prioritizing economic growth and development within their respective regions. The Blue Economy encouraged a focus on low-cost supply chains combined with technical innovations, aiming to promote local job creation in competitive markets (Balboa and Somonte 2014). The 1992 Earth Summit had previously examined the complex interconnections between poverty, population, food security, social development, human settlements, and human rights (Leggett and Carter 2012). However, it wasn't until the concept of the green economy was revitalized in 2012—emphasizing a system of economic activities that balances improved living standards with environmental sustainability (D'amato and Korhonen 2021)—that the Blue Economy began to take shape as a distinct focus.

Blue Economy implies efficient planning, management, and governance of marine resources, using inclusive methods to preserve ocean ecosystem health, while fostering economic growth, improved livelihoods, and job creation (World Bank 2017). The Sustainable Development Goals (mainly underwater life, land ecosystems, justice, and peace) interconnected with the Blue Economy (Lee et al. 2020), conceptualized, and targeted to increase the economic benefits of sustainable marine resources to SIDS and LDC, especially in fisheries, aquaculture, and tourism (Spalding 2016). Oceans are a defining feature of the Earth and the foundation of life; thus, the Blue Economy, following the green economy, has become a milestone in the economies of states, beyond just ocean health. But Rio + 20 emphasized the conservation of oceans and seas with the healthy functioning of earth's ecosystems to eradicate poverty, sustain ocean-based economies, enhance social and sustainable management, improve societal welfare, create job opportunities, ensure food security, and provide decent work for all (UNCSD 2012).

The Blue Economy is an economic framework that emphasizes the sustainable use and management of ocean and coastal resources to drive economic growth, improve livelihoods, and protect the health of marine ecosystems. This concept extends the principles of the green economy to the marine environment, recognizing the ocean as a critical source of economic opportunity and ecological wealth. It aims to balance the benefits derived from ocean resources with the need to maintain the health of marine and coastal ecosystems, ensuring that these resources can be sustainably utilized for future generations. The Blue Economy is a sustainable, ocean-based economic system that drives national growth by creating jobs and economic opportunities while preserving marine and coastal ecosystems. As a cornerstone of the green economy, it encourages addressing environmental challenges, such as climate change, through low-carbon and resource-efficient practices in shipping, fisheries, marine tourism, and renewable energy (UNEP 2012).

9.2 Key Components of the Blue Economy

These components include:

- Fisheries and aquaculture: Sustainable fishing practices and responsible aquaculture are central to the Blue Economy, ensuring that seafood production meets global demand without depleting fish stocks or harming marine environments. This involves managing fisheries to prevent overfishing and adopting aquaculture techniques that minimize environmental impacts.
- Maritime transport and ports: Shipping is a vital component of global trade, and the Blue Economy promotes the development of greener shipping technologies and practices to reduce pollution, greenhouse gas emissions, and ocean noise. Sustainable port development is also crucial, with a focus on reducing the environmental footprint of port operations while supporting economic growth.
- Tourism and coastal development: Coastal and marine tourism is a significant economic driver, particularly for many small island and coastal nations. The Blue Economy advocates for sustainable tourism practices that protect natural habitats, promote local cultures, and ensure that tourism benefits are equitably shared among communities.
- Marine renewable energy: Harnessing renewable energy from the ocean, such as offshore wind, wave, and tidal energy, is a growing sector within the Blue Economy. These technologies offer the potential to provide clean energy while reducing reliance on fossil fuels and minimizing environmental impacts.
- Marine biotechnology: The Blue Economy also includes the development of marine biotechnology, which involves using marine organisms for products and services in areas like medicine, food production, and environmental management. This sector emphasizes the sustainable use of marine biodiversity to create innovative solutions while preserving ecosystems.
- Conservation and ecosystem services: Protecting and restoring marine and coastal ecosystems is a fundamental component of the Blue Economy. This includes initiatives such as establishing marine protected areas, promoting sustainable practices in coastal development, and recognizing the value of ecosystem services provided by healthy oceans, such as carbon sequestration, storm protection, and biodiversity conservation.

The sectors of the Blue Economy (Table 9.1), as identified by the World Bank and UNDESA (2017), are fisheries, maritime transport and tourism, and traditional sectors while aquaculture, bioprospecting of biological resources and biotechnology, seabed extractive activities, offshore renewable energy are emerging sectors (Bhattacharya and Dash 2020). The Blue Economy seeks to integrate these components into a cohesive strategy that maximizes economic and social benefits while ensuring the long-term health and resilience of the world's oceans.

Table 9.1 Sectors of the Blue Economy

Sectors	Activities
Traditional sectors	
Fisheries	Capture (primary) and secondary fish production (net and gear making, boat construction and maintenance, manufacturing of fish processing equipment, packaging, marketing, distribution, etc.), seafood harvesting and processing, trade of seafood and non-edible seafood products
Maritime transport	Seafaring, shipping, shipbuilding, ship repairing, ship scrapping and supplying, and shipping services (custom clearance, actions of port companies, ship management, activities of shipping agents and brokers, container shipping services, roll-on roll-off operations, safety and training)
Tourism	Boating at sea, sea angling from boats and shore, dive tourism, sailing, jet and water skiing, bird watching in coastal areas, swimming, visiting coastal nature reserves, beach, seaside, and islands
Emerging sectors	
Aquaculture	Site selection, culture of fish species, hatchery, live and pellet feeding, disease management, trading of culture fish, seaweed aquaculture, shellfish aquaculture, and well-planned fish rearing based
Bio prospecting of biological resources and biotechnology	Pharmaceuticals, cosmeceuticals, chemicals, seaweed harvesting, seafood products, marine-derived bio-products, biofuels, and human foods
Seabed extractive activities	Oil and gas extraction, deep-sea mining, exploration of the rare earth metals and hydrocarbons, extraction of minerals, energy sources, and freshwater generation,
Offshore renewable energy	Generation of (offshore) renewable energy, carbon sequestration, offshore wind energy production, wave energy, production, and tidal energy production

Source Mohanty et al. (2015); World Bank and UNDESA (2017)

9.3 Marine Resources and the Need for Sustainability and Blue Economy Models

Oceans are the Earth's life support system, providing essential services such as oxygen production, carbon dioxide absorption, climate regulation, nutrient cycling, and food and energy resources. They also facilitate global trade and transportation. Effective governance of the oceans is essential for poverty eradication, sustainable livelihoods, food security, and weather pattern regulation (UNCTAD 2016). In addition, marine energy resources, such as tidal and thermal, are crucial for social and economic development. The demand for ocean resources and services is rapidly increasing, leading to heightened concerns about resource depletion. Threats to marine environmental health extend beyond well-known issues such as plastic and chemical pollution, overfishing, nutrient depletion, and invasive species (United Nations 2016). Factors such as rapid population growth, technological advancements (Bax et al. 2021), and development strategies in Global South countries pose significant risks to the sustainability of marine and coastal ecosystems. While these factors are crucial, other determinants also contribute to resource scarcity and the degradation of marine ecosystems. Preserving marine ecosystems requires a delicate balance between economic, social, and environmental interests. It is imperative that economic growth and development are both inclusive and sustainable, avoiding the depletion of natural resources (Griggs et al. 2013).

The mid-twentieth century saw an accelerated exploitation of marine resources, largely driven by the Industrial Revolution. Coastal regions expanded their territorial reach to facilitate maritime trade, resource extraction, and migration (Zacharias 2014; Alexander 2019; Jouffray et al. 2020). As a result, marine resources have been significantly depleted (World Bank 2017; Smith and Zeder 2013; Zacharias 2014; Novaglio et al. 2018). Wilkinson and Souter (2008) estimated that approximately 20% of the world's coral reefs have been lost and another 20% have been degraded. Similarly, 20% of mangroves have been reduced to 30–50% of their historical cover, and it is estimated that 29% of seagrass habitats have disappeared since the late 1800s (Nellemann and Corcoran 2009). In 2012, a global effort was made to protect 10% of ocean areas; however, currently, only 2% of the world's oceans are safeguarded from overexploitation (Benzaken et al. 2007; UNEP 2016). The Sustainable Development Goals (SDGs) offer a potential solution to this crisis. By implementing appropriate management strategies, these goals can help ensure the long-term sustainability of marine resources. Full restoration of depleted resources is challenging. Implementing effective measures to maintain and restore marine and coastal ecosystems, along with stringent resource extraction regulations, can help mitigate further oceanic losses. For instance, some reputed species of fish could be restored by adding an efficient, productive juvenile that can make sure the sustainability of fisheries and food security (FAO/IBRD 2009; UNEP 2016). At contrast, some well-intentioned strategies for natural resource protection have inadvertently contributed to the current challenges posed by the brown economy. This economic model, heavily reliant on the environmentally destructive exploitation of fossil fuels such as coal, oil, and gas (Shaikh 2020), has yielded both benefits and significant negative impacts over centuries. Excessive greenhouse gas emissions, including carbon dioxide and methane and widespread pollution of water and air resources, have harmed biodiversity and strained finite resources (Matthews 2013).

To build a sustainable marine resource base and a thriving marine economy, it is essential to adopt the Blue Economy model. This model serves as an effective platform that balances the health and sustainability of marine resources with the promotion of economic growth and development. By focusing on both living marine resources, such as traditional fisheries and aquaculture, and non-living resources like seabed mining, offshore oil, and gas, the Blue Economy aims to enhance the standard of living, particularly for poorer populations. This approach acts as a catalyst for sustainable economic growth in the future, ensuring that marine resources are managed responsibly while contributing to overall economic prosperity (World Bank and UNDESA 2017). To foster a sustainable marine economy, a complex system encompassing diverse industries and regions (Qi and Xiao 2019), the Blue Economy model is essential. This approach prioritizes marine ecosystem health and sustainability while driving economic growth and development. By maximizing the potential of both living marine resources and non-living resources, the Blue Economy can improve the livelihoods of marginalized populations (Ranasignhe 2017) and catalyze sustainable economic growth (World Bank and UNDESA 2017).

9.4 Challenges Faced in Emerging and Implementing the Concept of the Blue Economy

The development of long-term strategic marine economic frameworks has been a challenging process, requiring concerted efforts to address the complex interplay between maximizing ocean benefits, mitigating negative impacts, and driving national economic growth (UNEP 2013; UNCTAD 2016). Legal rules and regulations regarding the exploitation, exploration, and conservation of marine resources, initiated by the 1982 United Nations Convention on the Law of the Sea (UNCLOS), prioritized the obligation of countries to protect and preserve marine resources. UNCLOS has been prepared based on the spatial distribution of oceans, sovereign rights, sovereignty, and jurisdiction. Consequently, two implementations were made, one was in 1994, associated with the exploration and exploitation of resources in international seabed areas, and the rest was in 1995, UN Fish Stock Agreement, related to migratory fish stocks (FICCI 2017). The Commission on the Limits of the Continental Shelf (CLCS), The International Maritime Organization (IMO), the Ramsar Convention on Wetlands of International Importance, and the Convention on Biological Diversity (CBD) for protecting ocean biotechnology are some of the amendments associated with the Blue Economy model (FICCI 2017).

While various Blue Economy (BE) models have been proposed, their successful implementation is hindered by numerous country-specific challenges. Furthermore, the uneven distribution of benefits poses a significant issue, as Small Island Developing States (SIDS) and Least Developed Countries (LDCs) are often better positioned to capitalize on low-carbon, resource-efficient growth paths (World Bank and UNDESA 2017). This disparity can create tensions among nations.

Challenges:

- The rapid depletion of ocean resources in developing nations like Sri Lanka and Bangladesh is driven by unsustainable economic and political pressures.

Human activities including overexploitation, habitat destruction, physical alterations, pollution, and climate change are accelerating this decline (World Bank 2017). It is due to poor governance and public awareness, ineffective government institutions, technological deficiencies, and mismanagement. These factors, coupled with increasing demands, have led to irreversible resource exploitation. For instance, the maritime boundary dispute between Sri Lanka and India in the Palk Bay and Gulf of Mannar hinders the realization of Blue Economy benefits for both nations due to the destructive impacts of bottom trawling on marine ecosystems (Goonetilleke and Colombage 2017).

Overfishing, exacerbated by technological advancements and inadequate fisheries management, poses a critical threat to marine ecosystems. With 57% of fish stocks fully exploited and 30% overexploited (FAO 2016), the sector is facing a sustainability crisis. Illegal, unregulated, and unreported fishing, amounting to 11–26 million annual catches, further undermines the potential of the Blue Economy to alleviate poverty and improve livelihoods. Coastal development, mining, and deforestation are destroying critical marine and coastal habitats, impairing ocean health and economic productivity. Unplanned and unregulated development near coastal areas is causing environmental degradation and harming marine ecosystems (World Bank and UNDESA 2017).

Climate change, driven by human activities, poses a significant challenge to the success of Blue Economy models by diminishing the economic value of marine resources. Long-term shifts in sea temperature, sea level, acidity, and ocean currents, resulting from climate change, disrupt marine ecosystems, affecting the behavior and reproductive patterns of marine organisms and the livelihoods of dependent communities (World Bank and UNDESA 2017; Ranasignhe 2017).

- Limited investment in human capital and difficulties in attracting private and public sector involvement.

Many potential investors lack awareness of the Blue Economy and its potential benefits. Additionally, strengthening Blue Economy models across different states and addressing the undervaluation of marine resources pose significant obstacles (World Bank and UNDESA 2017).

- Failure of agreements or amendments signed between authorities

International agreements governing marine resource utilization can be undermined by non-compliance. For example, India's contracts with the International

Seabed Authority for polymetallic nodules (2002) and exploration of sulfides (2016) highlight the potential for resource exploitation. Competition, lack of cooperation, and economic instability among nations can hinder adherence to these agreements, jeopardizing the sustainable management of shared marine resources (World Bank and UNDESA 2017).

Anyhow, trades within Exclusive Economic Zones (EEZs) based on the principle of neutral resource utilization are insufficient for ensuring the sustainable and economically efficient use of marine resources. Promoting neutral trade practices can enhance a nation's ability to exploit these resources.

- Unsustainable maritime and coastal tourism practices

Prasetyo et al. (2018) highlight the significant challenges in balancing the economic benefits of maritime and coastal tourism with its environmental sustainability. Insufficient investment, the COVID-19 pandemic, limited financial resources, and inequitable benefit sharing have hindered the sustainable growth of the tourism sector and, consequently, the Blue Economy. Moreover, marine tourism has adverse environmental impacts, including increased greenhouse gas emissions, sewage and waste pollution, habitat destruction, and water scarcity. The challenges and issues in managing tourism can be categorized under three concepts: planning and management, governance, and effective implementation as shown in Table 9.2.

At present, coastal tourism in regions reliant on coastal areas faces numerous challenges, including a lack of regional cooperation, rapid and uncontrolled urbanization, overdependence on tourism, transportation issues, negative impacts on other industries, revenue leakage, offshore outsourcing, ineffective multi-stakeholder engagement, limited innovation, and inadequate community involvement and benefits (Dwyer 2017).

However, strengthening the conceptual framework of Blue Economy models, particularly those focused on resource sustainability and inter-state cooperation, is imperative. Addressing the undervaluation of marine ecosystems and their resources must be a priority (World Bank and UNDESA 2017).

9.5 Marine Pollution and the Blue Economy

The Blue Economy emphasizes two key goals: economic growth and development, and the protection and sustainability of resources for future generations. However, marine pollution significantly undermines the latter by accumulating waste from both living and non-living sources, drastically reducing the value of marine and coastal ecosystems (Beaumont et al. 2019). According to the World Bank and UNDESA (2017), coastal areas are often treated as cost-free waste repositories, where millions of liters of waste are disposed of annually, disrupting land and marine ecosystems, impairing their ability to regenerate, and leading to numerous health issues for the growing population (European Commission 2021).

Table 9.2 Issues and challenges in managing tourism in coastal zones

Major concepts	Issues and challenges
Planning and management	Development planning does not consider environmental concerns
	Less effective national- and local-level administrative structures
	Management focus: providing priorities to traditional land-use patterns than focusing on sustainable use of scarce land resources
	Resource management issues: habitat conservation and restoration, coastal hazards, promoting sustainable development
	Support management decision-making: scientific research and information/intelligence, providing tools and supporting technologies
	Predominantly sectorial approach not contributing to an integrated management
Governance	Insufficient stakeholder involvement support
	Information and capacity building through education
	Lack of public participation and awareness: difficulties in communicating local stakeholders
	In-efficient institutions
Effective implementation	The implementation of plans in coastal areas is not effective due to: 1. Limited economic analysis of costs, benefits, and their impact on stakeholders 2. Monitoring of progress and performance evaluation with specific key performance areas (KPAs and indicators) 3. Partial and incomplete implementation 4. Insufficient national financial support for coastal programs

Source IOC/UNESCO (2001); Vandeweerd et al. (2002); Sorensen (2002)

Marine pollution originates from both land-based and sea-based activities. Land-based sources include population growth, industrial emissions, landfills, intensified agriculture, untreated sewage, rapid coastal urbanization, and chemical discharges from rivers and floodwaters. Sea-based pollution arises from offshore mining, plastic and nutrient discharge, aquaculture, oil spills, chemical accumulation, airborne contaminants like pesticides and microplastics, accidental dumping, and abandoned fishing gear. The problem is exacerbated by increasing maritime shipping and submarine hydrocarbon and mineral exploration (Tornero and Hanke 2016).

Dead zones in the oceans, where life is unsustainable due to low oxygen levels, have increased to 400 areas covering 240,000 km^2, including productive regions like estuaries and shelves (UNEP 2016). The enrichment of nitrogen and phosphorus in the oceans has tripled since pre-industrial times (UNESCO, IMO, FAO and UNDP 2011). Sherman and Adams (2010) predicted that if current practices continue, the amount of inorganic nitrogen entering the ocean could increase by 50% by 2050.

Ocean acidification, also known as the 'CO_2 Problem', is another significant source of pollution. Over the past 100 years, increasing CO_2 levels has caused a decline in ocean pH (World Tourism Organization 2012). This has a negative impact on both fisheries and tourism, as ocean-absorbed CO_2 harms marine life and habitats (Cheung and Law 2001; Dong et al. 2019). Plastic pollution is particularly harmful to marine ecosystems. The durability, versatility, and lightweight nature of plastics have contributed to a crisis, with an estimated 19–23 million metric tons of plastic waste entering aquatic ecosystems in 2016 alone—a figure projected to double by 2050 (Geyer et al. 2017; Lau et al. 2020; Lebreton and Andrady 2019).

Microplastics, particles smaller than 5 mm, are especially damaging because they infiltrate marine environments through rivers, the atmosphere, and coastlines, including via sewage. These tiny particles are ingested by aquatic organisms, disrupting ecosystems and harming species such as fish, corals, seabirds, and turtles (ICC 2016). Without intervention, 99% of seabird species are expected to ingest microplastics by 2050 (ICC 2016). Koongolla et al. (2018) found significant microplastic contamination along a 91-km stretch of Sri Lanka's southern coastline, especially near fishing harbors. The predominant microplastics were polyethylene (PE), polypropylene (PP), and polystyrene (PS), which are known to absorb pollutants (Karapanagioti and Klontza 2008). However, the levels of microplastics in Sri Lanka were lower compared to other countries with similar per capita GDP. Another study found 13, 25, and 36 plastic items per cubic meter of sand on three western Sri Lankan beaches (Ranatunga 2018). This aligns with regional data, placing Sri Lanka fourth among nine countries studied, with an average of 10 microplastics per 25 g of beach sand (Balasubramanim and Phillott 2016). As an Indian Ocean island, Sri Lanka is categorized as a major plastic polluter, alongside China, Indonesia, the Philippines, and Vietnam, contributing 56% of global plastic waste (Rhodes 2018).

Addressing the different sources of pollution requires tailored mitigation strategies to protect marine resources and sustain the Blue Economy. The relationship between economic growth and marine pollution is complex, as activities that drive economic expansion often contribute to pollution, which in turn diminishes the economic value of marine resources (European Commission 2021).

9.6 Food Security and Blue Economy

The concept and framework of the BE are intrinsically linked to ensuring the sustainability of its components, including fisheries and aquaculture (Boteler 2014), while providing stable income across generations. According to the World Bank (2012), marine fisheries contribute more than 270 billion US$ to world GDP per year. Being a vital source of animal protein, nearly 1 billion people depend on seafood as their primary source of protein in developing countries, especially in low-income food-deficit economies (GPO 2013). Especially the people in Asia–Pacific region consume 70% (FAO 2009) of the world's fish products, whereas 300 million people are engaged in fishing for their livelihoods (FAO 2016). However, securing food for the whole human population is a critical challenge and alleviating poverty has been a matter of concern for years. UNCSD (2012) estimated that over 1 billion people on this planet are under extreme poverty and 14% of them are undernourished.

Adopting Blue Economy approaches is crucial for reducing poverty, ensuring long-term food security, and boosting economic growth, as these approaches play a key role in providing food and nutrition. As billions rely on the ocean for livelihoods, fisheries and aquaculture management are crucial, if not it will exacerbate poverty and economic instability. Fishing for food within Exclusive Economic Zones (EEZs), governed by UNCLOS regulations, is commonplace. However, maintaining sustainable fishing practices and preventing overexploitation remains a challenge in both oceanic and coastal environments, which have already been adversely impacted by climate change (UNGA 2014).

Global climate change, characterized by rising temperatures, extreme weather events, ocean acidification, and sea-level rise, has severely disrupted fisheries worldwide. These impacts, including altered reproductive rates, changes in community structure, and fluctuations in species populations, have been well documented (Cheung et al. 2016). The Indian Ocean region, renowned as a global seafood hub, has not been immune to these challenges. Despite its rich marine resources, the region faces significant pressure from overexploitation. Meantime, UNEP (2016) reported that between the periods of 1970 to 2008, under-exploitation or moderately exploited fish stocks declined from 40 to 15%, while overexploited recovering fish stocks increased from 10 to 32% between the same periods (FAO 2010). In addition, subsidies for fishing fleets have further enhanced overfishing activities, resulting in an estimated loss of nearly US$50 billion per year (FAO/IBRD 2009). However, the majority of people depend on it, if fishing in coastal areas fails, moving toward other employment opportunities would be critical especially for small-scale fishers as alternative livelihoods of people dwelling near sea areas are limited (Daw et al. 2012). Aquaculture, a rapidly expanding BE component, now contributes 47% of global fish production for human consumption (FAO 2010), though feed supply remains a significant challenge for the industry. From an economic perspective, aquaculture integrated into the BE can enhance natural resource value, maintain ecological balance, provide stable employment, and generate export revenue through high-quality products (UNEP 2016).

9.7 Economic Growth Comparisons in Asian Pacific Countries Reference the Blue Economy

The Asia–Pacific region encompasses countries situated in or near the Western Pacific Ocean. With an ocean area 80% larger than its landmass (Stone and Degnarian 2017), the region comprises East and Southeast Asia, as well as Oceania. Among these, India, Indonesia, Mongolia, and China—with their vast populations and extensive coastlines—are the most populous nations and hold significant weight in the global economy today. World Bank (2018) stated that 28.4% of total world GDP increased to

33.5% from 1995 to 2016 shared by Asian Pacific Region and world trade increases gradually year by year (ADB 2017). However, these countries have lowered the average real GDP growth rate compared to other global economies as majority of Asian Pacific countries are classified as Least Developed Countries (LDCs) (IMF 2018). In addition, a study of 19 Asian Pacific Island nations concluded that the size of the Blue Economy is positively correlated with gross fixed capital formation and technological advancement. Consequently, increased investment in physical capital and ICT promotion are crucial for its growth (Bhattacharya and Dash 2020).

In this sense, fisheries and tourism have significantly contributed to economic output. Effective governance and policies in these sectors have been instrumental in determining their overall impact. While fisheries output is influenced by factors such as investment, electricity access, and export opportunities, tourism income is primarily determined by relative prices, exchange rates, and global economic conditions. New Delhi, the capital of India, is exploring significant aspects of the BE due to its strategic location in the Indian Ocean, available infrastructure, and connectivity initiatives. India has found a place in the blue economic approach by creating relationships among other countries bilaterally as well as multilaterally. It is projected that ocean-based industries in the Asian Pacific region will outperform the global growth by 2030 in terms of value, creating 40 million full-time jobs and doubling the contribution of the Blue Economy. Predictions suggest that offshore wind will experience the strongest growth (24.5%), followed by fish processing (6.26%), industrial marine aquaculture (5.69%), seaport activities (4.58%), industrial capture fisheries (4.1%), and maritime and coastal tourism (3.51%), considered to be potential sectors for creating jobs (FICCI 2017). Likewise, it has been concluded that ocean-based trade also quadruples by 2050, and two-thirds of food production will be farmed from the sea by 2030 (OCED 2016).

The EU's 2018 annual report highlighted that the Blue Economy created 3.5 million jobs, generating 174 billion euros in value-added and 566 billion euros in annual turnover. Spain contributed to one-fifth of total employment, while the UK and Greece demonstrated notable economic growth and development in their marine-based economies. Following the 2014 proposal of the 'Bay of Bengal Partnership for Blue Economy' by Dhaka, Bangladesh expressed keen interest in the Blue Economy's potential. While remote South Pacific islands generally exhibit strong ocean health, countries in Central Africa and the Caribbean lag behind in ocean ecosystem health (Bhattacharya and Dash 2020). Sri Lanka, an Asian Pacific nation strategically positioned in the Indian Ocean, is not classified as a Small Island Developing State (SIDS). With an ocean area seven times its landmass, the country possesses immense potential for Blue Economy activities and trade within South Asia. To fully realize this potential, strict adherence to sustainable ocean economic practices is imperative.

9.8 Experienced Financial, Economic Benefits, and Costs in the Blue Economy

More than three-quarters of the planet's earth is covered by oceans and the economic richness of the marine environment is represented by millions of living organisms (animals and vegetation, material goods), limitless services (fishing, shipbuilding, tourism, ports), and renewable energy (tidal and thermal energy, wind, wave, biomass) provided to the human population. United Nations estimated that the total commercial value of the world's ocean activities is three to six trillion US$, approximately 20 to 40 lakh cores, and the annual value of goods and services approximated as 20,000 core US$ (UN Working Paper 2014). The estimated commercial value of oceans accrued from marine transport (90% of global trade), providing food (both aquaculture and fisheries provide food to 4.3 billion people and annual protein consumption is 15%), worldwide telecommunication, marine tourism (5% of world GDP), and coastal shore-based commercial activity. Anyhow oceans are included among the world's top ten economies, unfortunately, they are undervalued of all ecosystems and less understandable (UN Working Paper 2014).

9.9 Job Opportunities Available from the Implementation of the Blue Economy

The consistent job creation potential of the BE concept offers a significant opportunity for countries struggling to alleviate poverty. According to Pauli (2010), the BE has the potential to generate 100 million jobs worldwide by 2030, while booming economic growth. All components of the Blue Economy have the potential to generate employment, necessitating labor and capital inputs. Fishing is integral to all marine-dependent societies and is associated with over 350 million jobs globally. However, the main restriction in providing jobs in the fishing sector is the deep-sea fishing of commercial fishers, aiming for high profit (Mohanti et al. 2015).

Tourism, one of the major sectors in the Blue Economy, also holds significant potential for job creation, contributing 9.9% to the global GDP (Chattoraj 2020). Meanwhile, as a backbone of the worldwide economy, maritime transport ensures job availability for both inland and outland people annually. For instance, nearly 100,000 ships transfer with one-third of the world's cargo, two-third of oil containers between countries annually and millions of workers are in operation (Valles 2014). In addition, scrapping of old vessels is an imperative process known as 'ship breaking' providing jobs including steel and iron for further processing of ship construction (Bari 2017).

9.10 Potentials Achievable from the Blue Economy in the Economic Growth of a Country like Sri Lanka

Sri Lanka, known as the 'Pearl of the Indian Ocean', is strategically positioned in the center of the Indian Ocean, strengthening its deep connection with marine resources. Due to its location, Sri Lanka has enormous opportunities to enhance economic growth by capitalizing on the potential of the Blue Economy. Not only in the case of Sri Lanka, most of the countries in South Asia (Bangladesh, India, Maldives, and Pakistan), occupying less than 2% of the world's total coastline, dealing with 40% of its economic activities with the ocean as the countries' coastal line is rich in ecological diversity (Bari 2017). For instance, coastal tourism shows an economic boom in these regions as South Asia consists of more than 8% of the world's mangrove area (Bari 2017). Combining all the economic approaches related to the sea, the ocean economy is positioned to be the world's seventh largest economy, providing an approximate estimated benefit of $1.5 trillion per year and expected to be doubled by 2030 (Commonwealth Blue Charter 2018).

Anyhow, the countries allowed to utilize and benefit from the marine resources, the extent to which countries can be able to access is an imperative thing, called as 'Exclusive Economic Zone (EEZ)', determines the degree of potential drawn from blue and ocean economy. In the case of Sri Lanka, its Exclusive Economic Zone (EEZ) extends 200 nautical miles from the baseline, covering approximately 517,000 km^2—about eight times larger than the country's land areas (Azmy 2011). With 25% of Sri Lanka's population relying on coastal livelihoods, particularly fishing, there is a critical need to preserve coastal ecosystems and sustainably harness marine resources. Implementing Blue Economy principles is essential to uplift the country's economic standing (Ranasignhe 2017). However, Small Island Developing States like Sri Lanka face significant challenges in achieving Blue Economy goals due to limited resources, a small population, heavy reliance on international trade, and vulnerability to natural disasters (FAO 2016).

A. **Fisheries and Aquaculture**

The fishing sector supports 5–8% of livelihoods in South Asia, with India and Myanmar being the region's top fish producers, capturing approximately 1.2 and 1.1 million tons annually, respectively (Funge-Smith et al. 2012). In contrast, Sri Lanka, Bangladesh, and the Maldives have significantly lower fish production, with annual catches of 0.12, 0.16, and 0.6 million tons, respectively (Krishnamurthy et al. 2018). Global export of fish and other fishing products to other countries estimated in 2014 was 132 billion dollars, while exports from SIDS were just 1.7 billion dollars as it contributes just over 1% to the global exports (Day et al. 2016). Reports from UNCTAD (2016) indicate that in SIDS, 3% of GDP is derived solely from fisheries.

Reports from Commonwealth Blue Charter Action Group on Coral Reef Protection and Restoration (2021) rely that, more than 15 tons of fish per year are provided by at least one square km of well-managed, healthy coral reef. Since 250 million people in developing countries, including Sri Lanka, depend on coral reefs for income and food, these ecosystems must be carefully managed to prevent damage. Coral reefs, existing within the regions of Sri Lanka, India, and Maldives, have higher economic values and protect thousands of faunal and floral species (South Asia Region, World Bank). Eliminating harmful, illegal, and unreported fishing activities, and improving conservation approaches, will help maximize the economic value derived from marine resources (Bari 2017).

B. **Marine Biotechnology**

Marine biotechnology, also known as 'blue biotechnology', involves the application of science and technology to modify living and non-living marine materials for the extraction and utilization of goods and services while gaining knowledge (OECD 2005). Blue biotechnology is advancing rapidly, becoming an anchor for addressing global challenges related to food supply, clean energy production, and environmental protection (EBI 2010). As oceans represent diverse ranges in temperatures, pressure, and light, organisms living underwater are rich in bio and genetic diversity, and can be applicable to produce pharmaceuticals, cosmeceuticals, animal and fish feed, functional processed foods, biofuels, and human foods.

Bioprospecting is the major driving factor in biotechnology in the future. It is the systemic, continuous search for marine organisms with the potential to produce or extract marketable products from beneath and along the seabed. Despite developing countries, developed countries (North America) and the EU are also interested and focused on blue biotechnology as a stable technical and economic resource for the growth and development of a country. Further, the European Science Foundation estimated and revealed that one hectare of microalgae culture can provide 20–80 L of biofuel (EBI 2010; UNEP 2016). Querellou et al. (2010) estimated that marine biotechnology outputs approximately valued at €2.8 billion (US$3.16 billion) have been commercialized in the global market by the year 2010 with a growth rate of 5–10% per year. However, another estimation done in 2013 proves that it could be more than € 2.8 billion (US$6.7–29 billion) if antioxidants and other functional ingredients are accounted for (Day et al. 2016). However, biological materials extracted from aquatic and freshwater environments contributed US$ 4.6 billion to the global economy in 2017 (Chattoraj 2020).

Pharmaceuticals obtainable from marine living and non-living materials evidenced that 39 marine-derived drugs are used for clinical treatment while 15 of them are used to treat cancer (Commonwealth Secretariat 2021). Because the steps in drug discovery are complex, long term (10–15 years), and resource-intensive process (hundreds of millions of dollars) (Simpkin et al. 2017), only 15 drugs were clinically produced over 70 years of bioprospecting. It is to treat diseases such as heart problems, chronic neuropathic pain, viral infections and especially cancer. An anti-cancer agent known as 'Yondelis' is a product of blue biotechnology approved to treat soft tissue sarcoma, and was the first approved drug over 25 years (Querellou et al. 2010). Anyhow, many pharmaceuticals have been obtained from oily fish, by the research conducted on coral reef organisms such as cyanobacteria and sponges (Commonwealth Blue Charter 2021).

Marine genetic resources (MGR) available from the sea also provide benefits in both monetary and non-monetary terms to the state. Royalty and milestone payments are financial and non-financial incentives associated with capacity building initiatives, including scientist training and infrastructure development. Some countries set their royalty and milestone payments based on their annual net revenue (Brazil), and some are negotiable and non-monetary benefits depend on significant resources based on where and when the bioprospecting and scientific research is processed (Commonwealth Secretariat 2021). Records state that coastal states gained nearly £250 million (215.5 million pounds) from the sales of the anti-cancer agent Halaven produced from marine genetic resources per annum (Tiller et al. 2020).

C. **Tourism**

Tourism, especially marine-related tourism, plays a crucial role in strengthening the economies of developing islands, especially in the South Asian region. Despite its naturally abundant resources, the South Asia region ranked as the poorest continent in the world due to the underutilization of resources, poor living standards, and high infant mortality rates (Khan 2005; Jayawardena et al. 2013). Therefore, small states in these regions focus and invest in marine tourism for better growth in GDP and economic development. According to Eikeset et al. (2018), economic growth and the Blue Economy are positively correlated if continuous modifications and careful considerations are taken as a series. As a key component of the Blue Economy, tourism is growing into one of the largest global industries, employing one in every 11 people. Based on the estimated results in 2014, when global tourist arrival increased by 5%, small states showed a 9% increase in marine tourism, accounts to 24 to 34% of total exports (World Bank 2016). Ocean-related and coastal tourism, encompassing activities such as surfing, diving, recreational fishing, and maritime archeology, is a rapidly expanding sector with an average annual growth rate of 3.3%. Notably, this sector holds a 50% share of the global market in SIDS and LDCs.

World Travel and Tourism Council say that 9.8% of GDP contribution was enhanced by tourism from the periods of 2000 to 2017, and the approximated monetary value was US$7.2 trillion (WTTC 2016). Further justifies that international tourist arrivals to SIDS also increased by 13 million within the first 3 years (UNWTO 2014) between 2000 and 2017, South Asian countries experienced a significant increase in tourist arrivals, with India leading in 2014 and Bangladesh ranked at last (Alam 2016). Now, it is observable that Bangladesh is putting efforts in prioritizing the tourism sector by including tourism in its Global Development Strategies (GDS) (Bari 2017).

Since the role of tourism in a country's economy is undeniable and a key source of foreign exchange, small states and least developed countries (LDCs) are actively seeking ways to improve their tourism sectors. In the way forward, challenges and barriers should be overcome. Climate change, sea-level rise, and marine pollution are the foremost long-term threats to marine tourism. Monitoring resource sustainability is crucial for ensuring the long-term viability of tourism. Conversely, unrestricted bulk agreements can devastate coastal ecosystems (Roberts and Ali 2016). It is interesting to assess the opportunities offered by ecotourism in the Blue Economy, as this model has demonstrated important interactions with the local community in coastal environments (Phelan et al. 2020). Proper management of local coastal tourism development is paramount to preserving coastal green infrastructure, which is vital for the health and well-being of local communities. In the Sri Lankan context, it has the greatest opportunities and potentials from the sea as the abundant availability of flora and fauna from the Indian Ocean acts as a key for tourist investment. Without breaking the marine cycle and natural habitats coastal tourism in Sri Lanka can go far and contribute more to GDP.

D. **Marine Transport**

Marine transport, which connects multiple states via sea routes, is the foundation for global trade and a key driver of economic growth. Being a common way to connect and communicate, marine transport is linked with 90% of seaborne global trade and foreign exchanges (Day et al. 2016). In 2014, worldwide shipments increased by 3.2%, and it was a huge raise (9.84 billion tons) than 2013 (UNCTAD 2016) while 2015 estimated the amount to be 10 billion tons. Large buildings, hotels, and industries often concentrated near coastal areas, where over 50% of the global population resides (Day et al. 2016). This proximity facilitates the global trade of goods and services. It is a great opportunity for small states and developing countries in the South Asian region as the region is rich with waterfront districts, vessels, and other technical facilities with sufficient labors.

Sri Lanka's location in the Indian Ocean positions it strategically as a potential maritime hub. With approximately two-thirds of global marine trade passing through the Indian Ocean, the country can capitalize on this advantage by developing its port and shipping services (Ranasignhe 2017). As the capital and main business port, Colombo can utilize sophisticated Triple E-class vessels (Energy-efficient, Economy of scale, and Environmentally friendly), offering a competitive edge over other countries (Ranasignhe 2017). Integrating Blue Economy policies and strategies is crucial to fostering robust business and international agreements. This necessitates a review and implementation of existing regulations to optimize transport services. For instance, retiring outdated vessels can enhance efficiency and productivity. Moreover, vessel registration and sales can boost national revenue and reduce costs for shipowners. Notably, Commonwealth small states have achieved significant merchant fleet registration, highlighting the sector's potential for economic growth. To maximize this potential while safeguarding marine resources, strategic investments in transport are imperative.

E. **Energy Production**

Economic growth is a key indicator of the well-being of a country's population. Historically, a country's economic growth was measured by factors such as coral and steel consumption, but today, it is primarily measured by per capita energy consumption. Bari (2017) says, at present worldwide consumption of average per capita energy is lower, especially in developing and least developed countries. But exactly there are chances, to demand increase by 40% by 2030 (Commonwealth Blue Charter 2021) with unaffordable energy prices. Since global energy production is heavily influenced by geopolitical factors, energy prices are prone to volatility, which disproportionately affects vulnerable populations. Consequently, harnessing renewable inland and coastal resources, as well as investing in ocean-based energy generation, becomes increasingly crucial, particularly for SIDS and LDCs. Ocean energy extraction can be categorized into two primary methods: harnessing continuous power sources from the ocean, such as salinity gradients, waves, thermal energy, and tides; or utilizing marine space for energy generation, primarily through offshore wind power.

Atherley-Ikechi (2016) says that offshore wind plants are evidenced that it has the potential to create a high amount of energy sufficient to distribute a dense population. Offshore regions are rich in methane hydrate and crude oil, with reports indicating that nearly 32.6% of the world's crude oil has been extracted from offshore mines (Chattoraj 2020). Whatever it is, exploitation of oil and gas from offshore should be stopped if the negative externalities on marine resources exceed the economic benefits (Roberts and Ali 2016). Small states in the South Asian region are dependent on fuel imports for their own generation of energy for the population, directly to response to price fluctuations. Therefore, it is much needful for LDC and SIDS to generate their own, utilizing available technology and marine resources even though the process can be restricted by lack of investment and physical limitations.

Sri Lanka has the potential and the necessary technology to generate energy from hydroelectric, solar, wind, and wave sources. Further, with the help of two defined monsoons in Sri Lanka, there are possibilities for generating tidal and wave energy along the North-West and South-West coastal areas (Commonwealth blue charter 2021). Overall, potentials from the Blue Economy can be drawn in ways of fishing and aquaculture, blue biotechnology, tourism, maritime transport, and clean energy production, etc. However, small states like Sri Lanka face challenges in involving all segments of the Blue Economy approach due to its small size, climate change, narrowed funding, limited economies of scale, and remoteness. Especially, a small population in small states lowers the potential number of qualified servants, leads to the declined workforce in the coastal region (Robert and Ali 2016).

9.11 Conclusion

In conclusion, the Blue Economy concept emphasizes the sustainable exploitation of ocean resources for economic progress, with a particular focus on assisting SIDS and LDCs. It seeks to strike a balance between the need to preserve marine resources for future generations and economic rewards. The implementation of successful Blue Economy models faces several hurdles, including political obstacles and a deficiency in human capital investment. The Blue Economy has significantly increased economic growth in the Asia–Pacific area, especially in industries like transportation, fishing, and marine tourism, which includes Sri Lanka. The long-term advantages are, however, limited by issues like marine pollution, overfishing, and climate change. To fully achieve economic growth while protecting marine resources and realizing the potential of the Blue Economy, Sri Lanka must invest in technology, adopt sustainable practices, and collaborate with other countries.

References

ADB (2017) Asian Development Bank Annual Report-2017. https://www.adb.org/sites/default/files/institutional-document/411996/adb-annual-report-2017.pdf

Alam MK (2016) Ocean/blue economy for Bangladesh, Ministry of Foreign Affairs, Government of the peoples' republic of Bangladesh

Alexander KA (2019) Conflicts over marine and coastal common resources: causes, governance and prevention. Routledge

Atherley-Ikechi M, Rogers T (2016) Energy: the key to a cleaner, more prosperous Caribbean. In: Achieving a resilient future for small states: Caribbean 2050, pp 73

Azmy SAM (2011) Sri Lanka report on coastal pollution loading and water quality criteria, s.l.: Bay of Begal Large Marine Ecosystem Project

Balasubramanim M, Phillott A (2016) Preliminary observations of microplastics from beaches in the Indian Ocean. Indian Ocean Turtle Newsletter 23:13–16

Balboa CH, Somonte MD (2014) Economia circular como marco para elecodiseno: el modelo ECO-3. Informador Tecnico 78(1):82–90

Bari A (2017) Our oceans and the blue economy: opportunities and challenges. Proc Eng 194:5–11

Bax N, Novaglio C, Maxwell KH, Meyers K, McCann J, Jennings S, Frusher S, Fulton EA, Nursey-Bray M, Fischer M, Anderson K (2021) Ocean resource use: building the coastal blue economy. Rev Fish Biol Fisheries 1–19

Beaumont NJ, Aanesen M, Austen MC, Borger T, Clark JR, Cole M, Hooper T, Lindeque PK, Pascoe C, Wylesd KJ (2019) Global ecological, social and economic impacts of marine plastic. Mar Pollut Bull 142:189–195

Benzaken D, Miller-Taei S, Wood L (2007) Status of policy and target development and implementation for marine protected areas/marine managed areas in the Pacific Island Region—a preliminary assessment and future directions. SEREP

Bhattacharya P, Dash AK (2020) Drivers of blue economy in Asia and pacific island countries: an empirical investigation of tourism and fisheries sectors

Bimali Koongolla J, Andrady AL, Terney Pradeep Kumara PB, Gangabadage CS (2018) Evidence of microplastics pollution in coastal beaches and waters in southern Sri Lanka. Marine Pollut Bullet 137:277–284.https://doi.org/10.1016/j.marpolbul.2018.10.0

Blue Economy Concept Paper (2014) Sustainable Development Knowledge Platform, United Nations (UN), January 2014

Boteler B (2014) Coastal zones: achieving sustainable management—Guest editorial. Sci Environ Policy 463:4–5

Cantwell M (2009) The blue economy: the role of the Oceans in our Nation's Economic Future. U.S. Committee on Commerce, Science, and Transportation United States Senate One Hundred Eleventh Congress. Government Printing Office, Washington, 9 June 2009

Chattoraj KK (2020) The blue economy: sustainability, opportunity and future. Econ Polity Environ: an Int Peer-Rev J Soc Stud 2(1):30–34

Cheung C, Law R (2001) The impact of air quality on tourism: the case of Hong Kong. Pacific Tourism Rev 5(1–2):69–74

Cheung WDL, Jones MC, Reygondeau G, Stock CA, Lam VWY, Frolicher TL (2016) Structural uncertainty in projecting global fisheries catches under climate change. Ecol Model 325:57–66

Colgan CS (2013) The ocean economy of the United States: measurement, distribution, and trends. Ocean Coast Manag 71:334–343. https://doi.org/10.1016/j.ocecoaman.2012.08.018

Commonwealth Blue Charter (2021). https://bluecharter.thecommonwealth.org/action-groups/coral-reef-restoration/
Commonwealth Blue Charter (2018). https://bluecharter.thecommonwealth.org/
Commonwealth Secretariat (2021) Tracing options for marine genetic resources from within national jurisdictions. https://bluecharter.thecommonwealth.org/documents/
D'amato D, Korhonen J (2021) Integrating the green economy, circular economy and bioeconomy in a strategic sustainability framework. Ecol Econ 188:107143
Dalton G, Bardocz T, Blanch M, Campbell D, Johnson K, Lawrence G, Lilas T, Friis-Madsen E, Neumann F, Nikitas N, Ortega ST (2019) Feasibility of investment in blue growth multiple-use of space and multi-use platform projects; results of a novel assessment approach and case studies. Renew Sustain Energy Rev 107:338–359
Daw TM, Cinner JE, McClanahan TR, Brown K, Stead SM, Graham NAJ, Clifton J (2012) To fish or not to fish: factors at multiple scales affecting artisanal fishers' readiness to exit a declining fishery. PLoS ONE 7(2):e31460. https://doi.org/10.1371/journal.pone.0031460
Day J, Hughes A, Greenhill L, Stanley MS (2016) Blue biotechnology. Commonwealth Blue economy series, No. 5. Commonwealth Secretariat, London
Dong D, Xu X, Wong YF (2019) Estimating the impact of air pollution on inbound tourism in China: an analysis based on regression discontinuity design. Sustainability 11(6):1682
Dwyer L (2017) Coastal and marine tourism in the Indian Ocean rim: challenges and opportunities. J Indian Ocean Rim Stud 1(1):38–55
EBI (2010) European bioinformatics institute. https://www.ebi.ac.uk/GOA/publications
Plastic Europe (2017) Plastics—the facts 2017: an analysis of European plastics production, demand and waste data. www.plasticseurope.org
European Commission (2021) The EU blue economy report. 2021. Publications Office of the European Union. Luxembourg
FAO (2010) The state of world fisheries and aquaculture. FAO, Rome
FAO (2009) The state of food insecurity in the World 2009 in the Niigata declaration on APEC Food Security-2010
FAO (2016) Seychelles and FAO: partnering for sustainable agricultural development and food security. Retrieved from http://www.fao.org/3/a-au744e.pdf
FAO/IBRD (2009) The Sunken billions—the economic justification for fisheries reform ISBN: 978-0-8213-7790-1
FICCI (2017) Blue economy vision 2025: harnessing business potential for India Inc. and international partners. https://ficci.in/spdocument/20896/Blue-Economy-Vision-2025.pdf
Funge-Smith S, Briggs M, Miao W (2012) Regional overview of fisheries and aquaculture in Asia and the Pacific 2012. RAP Publication (FAO)
Geyer R, Jambeck JR, Law KL (2017) Production, use, and fate of all plastics ever made. Sci Adv 3:e1700782. http://advances.sciencemag.org/content/3/7/e1700782, abstract, https://doi.org/10.1126/sciadv.1700782
Goonetilleke B, Colombage ADJ (2017) Indo-Sri Lanka fishery conflict: an impediment to development and human security (Online)
GPO (2013) Global ocean partnership. http://www.globalpartnershipforoceans.org/
Griggs D, Stafford-Smith M, Gaffney O, Rockström J, Öhman MC, Shyamsundar P, Noble I (2013) Sustainable development goals for people and planet. Nature 495(7441):305–307
ICC (2016) International Coastal Cleanup, Annual Report. Technical Report
IMF (International Monetary Fund) (2018) Regional economic outlook. Asia Pacific: good times, uncertain times, a time to prepare, Washington, DC, US. Available at https://www.imf.org/en/Publications/REO/APAC/Issues/2018/04/16/areo0509
IOC/UNESCO (2001) A methodological guide: steps and tools towards integrated coastal area management. Paris, United Nations Educational, Scientific and Cultural Organization
Jayawardena R, Byrne NM, Soares MJ, Katulanda P, Hills AP (2013) Prevalence, trends and associated socio-economic factors of obesity in South Asia. Obes Facts 6(5):405–414

Joroff M (2009) The blue economy: sustainable industrialization of the oceans (at) proceedings. In: International symposium on blue economy initiative for green growth, Massachussets Institute of Technology and Korean Maritime Institute, Seoul, Korea, May 7, pp 173–181
Jouffray JB, Blasiak R, Norström AV, Osterblom H, Nyström M (2020) The blue acceleration: the trajectory of human expansion into the ocean. One Earth 2(1):43–54
Kaczynski W (2011) The future of blue economy: lessons for European Union. Found Managem 3(1):21–32
Karapanagioti HK, Klontza I (2008) Testing phenanthrene distribution properties of virgin plastic pellets and plastic eroded pellets found on Lesvos island beaches (Greece). Mar Environ Res 65:283–290
Kathijotes N (2013) Keynote: blue economy-environmental and behavioural aspects towards sustainable coastal development. Proc Soc Behav Sci 101:7–13
Khan HA (2005) Assessing poverty impact of trade liberalization policies: a generic macroeconomic computable general equilibrium model for South Asia (No. 22), ADB Institute Discussion Papers
Krishnamurthy A, Nagel S, Vissing-Jorgensen A (2018) ECB policies involving government bond purchases: impact and channels. Rev Finan 22(1):1–44
Lau WWY, Shiran Y, Bailey RM, Cook E, Stuchtey MR, Koskella J, Velis CA, Godfrey L, Boucher J, Murphy MB, Thompson RC, Jankowska E, Castillo A, Pilditch TD, Dixon B, Koerselman L, Kosior E, Favoino E, Gutberlet J, Baulch S, Atreya ME, Fischer D, He KK, Petit MM, Sumaila UR, Neil E, Bernhofen MV, Lawrence K, Palardy JE (2020) Evaluating scenarios toward zero plastic pollution. Science 36:1455 LP–1461. http://science.sciencemag.org/content/369/6510/1455.abstract, https://doi.org/10.1126/science.aba9475
Lebreton L, Andrady A (2019) Future scenarios of global plastic waste generation and disposal. Palgrave Commun 5:6. https://doi.org/10.1057/s41599-018-0212-7, https://doi.org/10.1057/s41599-018-0212-7
Lee K, Noh J, Khim JS (2020) The blue economy and the United Nations' sustainable development goals: challenges and opportunities. Environ Int. https://doi.org/10.1016/j.envint.2020.105528
Leggett JA, Carter NT (2012) Rio+ 20: the united nations conference on sustainable development, library of congress, congressional research service, June
Martínez-Vázquez RM, Milán-García J, de Pablo Valenciano J (2021) Challenges of the blue economy: evidence and research trends. Environ Sci Eur 33(1):1–17
Matthews R (2013) Environmental implications of three types of economies: brown, blue and green. https://thegreenmarketoracle.com/2013/10/04/environmental-implications-of-three/
Mohanty SK, Dash P, Gupta A, Gaur P (2015) Prospects of blue economy in Indian ocean, New Delhi, Research and Information System for Developing Countries
Nellemann C, Corcoran E (Eds) (2009) Blue carbon: the role of healthy oceans in binding carbon: a rapid response assessment. UNEP/Earthprint. ISBN: 978-82-7701-060-1
Novaglio C, Smith ADM, Frusher S, Ferretti F, Klaer N, Fulton EA (2018) Fishery development and exploitation in South East Australia. Front Mar Sci 5:145. https://doi.org/10.3389/fmars.2018.00145
OECD (2005) Statistical definition of biotechnology. Available at: www.oecd.org/sti/biotech/statisticaldefinitionofbiotechnology.htm. Accessed 10 Nov 2021
OECD (2016) The ocean economy in 2030. OECD Publishing, Paris.https://doi.org/10.1787/9789264251724-en
Pauli G (2010) The blue economy: 10 years, 100 innovations, 1000 million jobs. Paradigm Publication
Phelan A, Ruhanen L, Mair J (2020) Ecosystem services approach for community-based ecotourism: towards an equitable and sustainable blue economy. J Sustain Tour 28:1–21
Prasetyo JB, Muhammad F, Sugianto DN (2018) An overview of encouraging sustainable tourism in the coastal tourism of Karang Jahe beach: issues and challenges. E3S web of conferences 73 ICENIS 0205. https://doi.org/10.1051/e3sconf/20187300215 (Accessed 10 Mar 2020)
Qi X, Xiao W (2019) Identifying the evolutionary path of maritime industries from the perspective of super network. J Coast Res 97(sp1):131–135

Querellou J, Børresen T, Boyen C, Dobson A, Höfle M, Ianora A, Jaspars M, Kijjoa A, Olafsen J, Rigos G, Wijffels RH (2010) Marine biotechnology: a new vision and strategy for Europe. Drukkerij De Windroos NV
Ranasignhe D (2017) Strategic importance of blue economy to Sri Lanka and challenges
Ranatunga RRMKP (2018) First evidence of micro plastics in beach sand from Negombo. In: Sri Lanka and the potential accumulation in marine fish
Rhodes CJ (2018) Plastic pollution and potential solutions. Sci Prog 101(3):207–260
Roberts J, Ali A (2016) The blue economy in small states. Commonwealth Blue Econ Series 1:1–51
Shaikh V (2020) The irreplaceable technologies of brown economy. Proceedings of innovation summit Sindh. http://www.innovationsummit.net/
Sherman K, Adams S (Eds.) (2010) Sustainable development of the World's large marine ecosystems during climate change: a commemorative volume to advance sustainable development on the occasion of the presentation of the 2010 Göteborg Award. International Union for Conservation of Nature and Natural Resources
Silver JJ, Gray NJ, Campbell LM, Fairbanks LW, Gruby RL (2015) Blue economy and competing discourses in international oceans governance. The J Environ Developm 24(2):135–160
Simpkin VL, Renwick MJ, Kelly R, Mossialos E (2017) Incentivising innovation in antibiotic drug discovery and development: progress, challenges and next steps. J Antibiot 70(12):1087–1096
Smith BD, Zeder MA (2013) The onset of the Anthropocene. Anthropocene 4:8–13
Sorensen J (2002) Baseline 2002 background report: the status of integrated coastal management as an international practice. Boston, Massachusetts: Harbour and Coastal Centre, University of Massachusetts. http://www.eucc-d.de/ (Accessed 25 Oct 2019)
Spalding MJ (2016) The new blue economy: the future of sustainability. J Ocean and Coastal Econ 2(2):8–21
Stone GS, Degnarian N (2017) The soul of the sea in the age of the algorithm. Leete's Island Books/ World Ocean Observatory, Sedgwick, Maine, p 189
Tiller R, De Santo E, Mendenhall E, Nyman E, Ralby I (2020) Wealth blindness beyond national jurisdiction. Mar Pollut Bull 151:110809
Tornero V, Hanke G (2016) Identification of marine chemical contaminants released from sea-based sources: a review focusing on regulatory aspects. Publications Office of the European Union, Luxembourg
UN Working Paper (2014) Goal 14—Conserve and sustainably use Oceans, Seas and marine resources for sustainable development. https://unchronicle.un.org/article/goal-14-conserve-and-sustainably-use-oceans-seas-and-marine-resources-sustainable (Accessed 07 Nov 2021)
UNCSD (2012) United Nations conference on sustainable development, Rio 20+. https://sustainabledevelopment.un.org/rio20
UNCTAD (2014) The state of commodity dependence 2014. United Nations Conference on Trade and Development, Geneva
UNCTAD (2016) Trade and development report -2016. https://unctad.org/en/PublicationsLibrary/tdr2016_en.pdf
UNEP (2012) United Nations environmental program 2012 annual report. https://www.unep.org/resources/annual-report/unep-2012-annual-report
UNEP (2013) Green economy definition. https://www.unep.org/regions/asia-and-pacific/regional-initiatives/supporting-resource-efficiency/green-economy
UNEP (2016) Blue economy concept paper. Information document of the scoping meeting on collaboration between regional Seas Programmes and regional fisheries bodies in the Southwest Indian Ocean. https://wedocs.unep.org/20.500.11822/11129
UNESCO, IMO, FAO, & UNDP (2011) A blueprint for ocean sustainability, Paris
UNGA (2014) Transforming our world: the 2030 agenda for sustainable development A/RES/70/1
United Nations (2016) The first global integrated marine assessment: World Ocean assessment I by the group of experts of the regular process. New York. https://sdgs.un.org/publications/unwto-2014-report-17916 (Accessed on 11 Nov 2021)
UNWTO (2014) United Nations World Trade Organization annual report-2014

Valles G (2014) The oceans economy: opportunities and challenges for small island developing states. United Nations Conference on Trade and Development, Geneva

Vandeweerd V, Bernal P, Belfiore S, Goldstein K, Cicin-Sain B (2002) A guide to oceans, coasts, and Islands at the World summit on sustainable development. Delaware, US: Centre for the Study of Marine Policy

Wilkinson CR, Souter D (2008) Status of Caribbean coral reefs after bleaching and hurricanes in 2005

World Bank & UNDESA (United Nations Department of Economic and Social Affairs) (2017) The potential of the blue economy: increasing long-term benefits of the sustainable use of marine resources for small island developing states and coastal least developed countries

World Bank (2012) The World Bank Annual Report 2012. Volume 1. World Bank Annual Report, Washington. https://openknowledge.worldbank.org/handle/10986/11844 License: CC BY 3.0 IGO

World Bank (2016) World Development Indicators. Available at: http://databank.worldbank.org/data/reports.aspx?source=2&country=&series=ST.INT.ARVL&period=# (Accessed 10 Nov 2021)

World Bank (2017) The World Bank Annual Report-2017. Washington, DC, World Bank. © World Bank. https://openknowledge.worldbank.org/handle/10986/27986 License: CC BY-NC-ND 3.0 IGO

World Bank (2018) World Development Indicators. Retrieved from https://data.worldbank.org/indicator/ny.gdp.mktp.kd (Assessed in Nov 2021)

World Tourism Organization (2012) Challenges and opportunities for tourism development in small island developing states. UNWTO, Madrid

World Travel and Tourism council (WTTC) (2016) Travel and tourism economic impact 2016 World. http://www.arab-tourismorg.org/images/pdf/World2016.pdf (Accessed on 11Nov 2021)

Xiao W, Zhao B, Wang L (2016) Marine industrial cluster structure and its coupling relationship with urban development: a case of Shandong Province. Polish Maritime Res

Zacharias M (2014) Marine policy: an introduction to governance and international law of the oceans. Routledge

Alavudeen Asmiya is an agricultural economist with a BSc Honors in Agriculture and a Master's degree in Agricultural Economics from Sri Lankan institutions. With over three years of teaching experience at state universities, her research has focused on rice processing efficiency, the impact of the chemical fertilizer import ban, and sustainability challenges in agriculture. Ms. Asmiya is pursuing PhD opportunities to further explore global issues such as climate change and sustainable agriculture.

M. Musthafa Muneeb is a Senior Lecturer at the Biosystems Department of South Eastern University of Sri Lanka. He earned his PhD in Ecology and Biodiversity from the University of Malaya, Malaysia, and has published over 70 peer-reviewed papers and seven book chapters. Dr. Muneeb has extensive research experience in ecology, biodiversity, genetic diversity, and environmental conservation. He has contributed to the establishment of postgraduate research programs at South Eastern University and is a member of several international academic societies.

Rosa Maria Martinez Vazquez is an Assistant Professor of Applied Economics at the University of Almería, Spain. She holds multiple degrees, including a PhD in Applied Economics and two Master's degrees in business administration. Her research focuses on the Blue Economy, nautical tourism, and local development, with numerous publications in prestigious journals such as Marine Pollution Bulletin and Frontiers in Marine Science.

MM Mohamed Najim is a Professor of Agritechnology at Sultan Sharif Ali Islamic University, Brunei. He holds a B.Sc. in Agriculture from the University of Peradeniya, Sri Lanka, an M.Eng. in Irrigation Engineering and Management from the Asian Institute of Technology, Thailand, and a PhD in Water Resources Engineering from University Putra Malaysia. With 30 years of academic experience, Dr. Najim has held key positions, including Vice-Chancellor of South Eastern University of Sri Lanka. He has published extensively, including 37 peer-reviewed journal articles, 11 book chapters, and numerous short scientific communications. He is also a trained reviewer for Sri Lanka's Quality Assurance Council and the National Science and Technology Commission.

Faiz MMT Marikar is the Director of the Staff Development Centre at General Sir John Kotelawala Defence University, Sri Lanka. He holds a B.Sc. from the University of Ruhuna, an MSc and PhD from the University of Nanjing, China, and a DSc. Dr. Marikar has conducted postdoctoral research at Johns Hopkins University and Harvard University, and his work has earned him numerous awards, including the Presidential Award.

Chapter 10
Insect Pollinators and Food Security Under the Changing Climate

Mudssar Ali, Fawad Zafar Ahmad Khan, and Muhammad Awais Ahmad

Abstract The majority of angiosperms require pollinators for successful reproduction. Crops pollinated by insects yield a higher number of seeds and larger fruits. However, pollinators are facing decline all over the world, primarily due to climate change. Factors such as rising temperatures, changing rainfall patterns, and soil erosion caused by floods contribute to this decline. Global warming significantly impacts plant–pollinator interactions, altering the biology of both plants and insects. Rising temperatures lead to early insect maturation, resulting in reduced plant foraging and fewer flower visits. Additionally, higher temperatures affect nectar sugar concentrations, reducing flower attraction for pollinators. Changes in water availability also affect pollinator interactions with flowers. Drought stress reduces floral units, increasing insect mortality due to decreased food resources. Conversely, heavy rains restrict bee activities and wash away pollen from flowers. To address the severe effects of climate change on pollinators, several remedies can be implemented. These include developing drought- and heat-resistant plant varieties, installing bee hotels, and planting crops that offer year-round nectar resources for both pollinators and humans. The current chapter gives a detailed overview of the impact of climate change on pollinators and outlines different policies and strategies to safeguard them in the face of these environmental challenges.

Keywords Bee hotels · Conservation · Global warming · Honey bees · Pollination deficit · Solitary bees

M. Ali
Institute of Plant Protection, MNS University of Agriculture, Multan 60000, Pakistan
e-mail: mudssar.ali@mnsuam.edu.pk

F. Z. A. Khan (✉) · M. A. Ahmad
Institute of Plant Protection and Department of Outreach and Continuing Education, Muhammad Nawaz Shareef University of Agriculture, Multan, Pakistan
e-mail: fawad.zafar@mnsuam.edu.pk

M. Behnassi et al. (eds.), *Food Systems and Biodiversity in the Context of Environmental and Climate Risks*, https://doi.org/10.1007/978-3-031-89167-0_10

10.1 Introduction

Human malnutrition will be a major concern in the coming decades (Poole et al. 2018), mainly due to the rapid increase in the global population (Cassils 2004). Pollination plays a critical role in food production systems (Roberts and Mattoo 2019). It occurs through both abiotic agents (such as wind and water) and biotic agents (mainly insects) (Lei et al. 2019). However, insect pollinators are facing a constant decline, largely driven by climate change, which is expected to negatively impact crop yields and, consequently, human nutrition (Lebuhn et al. 2013).

Pollination deficits, resulting from insufficient pollen reaching the stigmas, reduce sexual reproduction in plants (Allsopp et al. 2008). These deficits occur due to several factors, such as low abundance or activity of pollinators, and inefficient pollen transfer. Agricultural intensification, pesticide use, eutrophication, and biological invasions further threaten pollination services that underpin food production (Bailes et al. 2015). Furthermore, the decline in pollinator populations due to environmental degradation has far-reaching consequences, affecting global food supply and local agricultural productivity (van der Sluijs and Vaage 2016). Notably, crops such as apples, apricots, mangoes, soybeans, and sunflowers, face yield reduction due to pollinator deficits (Garratt et al. 2014; Oprita and Gavat 2016; Rafique et al. 2016; Chabert et al. 2022).

Climate change has led to rising temperatures, increased dioxide carbon (CO_2) concentrations, seawater changes, and ocean acidification (Naz et al. 2022). Consequently, the distribution of pollinators and plants may be affected (Renner and Zohner 2018). Pollinators face habitat degradation, loss of food resources, harsh environmental conditions, and health risks due to pollutants.

Sudden climatic shifts disrupt plant–pollinator interactions, leaving many crops pollinator-deficient and resulting in low yields (Hegland 2009). To address this, safeguarding nectar corridors and recognizing migratory pollinators (such as bats, hummingbirds, and monarch butterflies) become essential. Butterflies provide compelling evidence of pollinator extirpations linked to climate change. For instance, three of the four alpine-specific butterfly species tracked in California's Sierra Nevada mountains have decreased in frequency since the 1970s, possibly due to environmental changes (Forister et al. 2010). Similarly, local extinctions of the Bay checkerspot butterfly (*Euphydryas editha bayensis*) along the California coast were linked to increased precipitation variability, a consequence of global climate change compounded by regional habitat loss (McLaughlin et al. 2002). Additionally, climate change has been linked to declining fly populations, with a > 60% drop in frequent flower visitors (Muscid and Chironomid, Diptera) in Greenland, where summer temperatures have increased by about 2.5 °C (Høye et al. 2013).

This chapter emphasizes the vital role of insect pollinators in our food supply chain. It assesses the impact of climate change on pollinator decline, explores potential reasons behind this decline, and discusses its implications for human nutrition, food security, and the overall food supply chain.

10.2 Pollination and Insect Pollinators

Pollination is a vital ecological service provided by angiosperms (flowering plants) that contributes to ecosystem sustainability (Gill et al. 2016). Angiosperms rely on either biotic or abiotic agents for pollination, especially due to self-incompatibility mechanisms that prevent self-pollination. Biotic pollination involves a mutualistic relationship: plants ensure their reproductive success, while pollinators receive nectar or pollen (or both) as a nutritive reward. These pollinators can be categorized based on their taxonomic classification and functional traits. Their role is crucial for cultivating a wide range of crops, including both agricultural and non-agricultural plants, such as commercial horticultural flowers (Ollerton et al. 2011).

The most efficient insect pollinators primarily belong to the orders Hymenoptera, Diptera, Lepidoptera, and Coleoptera (Cook et al. 2020). Bees, specifically honey bees (genus *Apis*) and solitary bees, are abundant and effective pollinators. Honey bees are not only essential but also widely managed for their efficiency in flower pollination (Willmer et al. 2017). They are generalists, visiting a wide variety of flowers. Solitary bees, on the other hand, exhibit high foraging efficiency and often specialize in specific pollen choices and flower types (Kocher and Paxton 2014). Bumblebee (genus *Bombus*) colonies are also maintained in temperate regions to increase flower pollination (An et al. 2014).

Dipteran flies constitute the second largest taxa of pollinators, with 71 families containing species that visit flowers (Orford et al. 2015). Flies from the family Syrphidae are the most frequent and efficient pollen vectors after bees. Non-Syrphid dipterans also contribute to pollination. These flies are attracted to yellow or white flowers (Upadhyay 2014), and can carry more pollen than bees. In certain ecological zones, such as alpine and subarctic regions, flies may serve as better alternatives for pollination.

Lepidopterans, including butterflies and moths, are common floral visitors. However, not all of them are equally effective as pollinators. Butterflies, which are diurnal in nature, prefer red flowers, while moths, which are nocturnal, also visit flowers (Van Zandt et al. 2020).

The contribution of coleopteran (beetle) pollination is often underestimated compared to other pollinator groups such as Hymenoptera, Lepidoptera, and Diptera (Wardhaugh 2015; Hahn and Brühl 2016). Beetles were historically diverse pollinators, but their role has shifted (Ollerton 2017). They now pollinate a larger variety of angiosperms, including recently split eudicot lineages. Unlike the 'typical' beetle pollination syndrome, characterized by plate- or bowl-shaped flowers, their interactions with flowers are more varied (Bernhardt 2000; Ollerton et al. 2003).

The mutualistic relationship between plants and pollinators involves an exchange of resources: food for the pollinators and effective sexual reproduction for the plants. Factors such as flower shape, color, fragrance, and pollen/nectar production have evolved due to competition among pollinators and flowering plants (Nicolson and Wright 2017). Natural habitat loss diminishes plant diversity and abundance, limiting pollinators' foraging options (Senapathi et al. 2015). Ensuring a rich environment with diverse flowering plants is critical for pollinator survival (Wray and Elle 2015).

10.3 Role of Insect Pollinators in Human Nutritional Security

In addition to affecting crop production, pollinator loss has an impact on both macro- and micronutrients in the human diet. Many micronutrients, including vitamins A and C, folic acid, calcium, and fluoride, are primarily obtained from pollinator-dependent crops (Smith et al. 2015). Vitamin A production is highly dependent on pollinators. Many studies conducted in Argentina, Mexico, Southeastern and North-Central India, Romania, Thailand, and Western Iran demonstrate a 50% dependence of vitamin A production on pollinators in different crops (Chaplin-Kramer et al. 2014a, b). Folic acid, which is necessary for bodily processes but cannot be manufactured by humans, is critical for the growth and development of our bodies (Mehmood 2014). A lack of pollinators may further threaten human health.

The risk of three categories of diseases—non-infectious chronic diseases (such as heart diseases, cancer, and diabetes), transmissible diseases (such as tuberculosis and influenza), and malnutrition-related diseases (such as rickets)—has been linked to changes in food and nutrient intake due to the decline in animal pollinators and pollination (Lim et al. 2012). If all pollinators disappeared, 2.2 billion people whose intake is currently below average would see additional drops in supply, and 71 million people in low-income nations could experience vitamin A insufficiency (Ellis et al. 2015). Approximately 70% of vitamin A is found in plants that are pollinated by insects, and pollination boosts crop yield by 43% (Eilers et al. 2011). Regarding folic acid, 1.23 billion people who are currently deficient in the vitamin could become even more deficient, while 173 million people could become newly deficient (Nicole et al. 2015). According to published research, oils from plants dependent on animal pollination contain approximately 74% of all lipids generated globally and are also major suppliers of fat-soluble vitamins. In regard to water-soluble vitamins, citrus trees and other fruits and vegetables provide 98% of the accessible vitamin C (Pinke 2013). Plants, specifically those rich in folic acid, are a beneficial source of B vitamins, which are especially important during pregnancy to reduce the risk of neural tube abnormalities in the offspring (Poletti et al. 2004).

Pollinator loss and associated services may also result in issues with human nutrition, although the severity of these issues may frequently vary by region and level of societal development. There is a negative impact on human health, particularly in developing nations where the population is typically poorer and more reliant on locally pollinated crops from insects. For example, a study conducted in four developing countries found a 0–56% risk of nutrient deficiency in the population (reduction in Vitamin A and iron) when pollinators were removed (Ellis et al. 2015). The impact of pollinator losses on human health may be less obvious, but it still has the potential to negatively impact dietary quality or increase reliance on artificial micronutrients, such as vitamin supplements.

A bioeconomic method was used in a study to estimate the global value of insect pollinators' contributions to the growth of crops that are directly utilized as human food. Approximately 9.5% of the overall world's human food production, valued at US$166 billion, depends on pollinators (Gallai et al. 2009).

10.4 Insect Pollinator Deficit and Food Security

There is a risk of nutrient deficiencies worldwide due to declines in pollinators. Areas with nutrient-rich and pollinator-dependent crops face more nutrient deficiencies in humans when pollinators decline (Smith et al. 2015). The deficiency of vitamin A is around three times higher in areas where crops depend on pollinators (Chaplin-Kramer et al. 2014a, b). Serious concerns arise about human health due to pollinator decline in areas with a deficit of pollinators. Mozambique and Uganda have increased the risk of vitamin A deficiency by approximately 56% and 15%, respectively, owing to the loss of pollinators (Ellis et al. 2015). One report stated that annually, 1.4 million deaths could occur due to malnutrition as a result of pollinator decline (Smith et al. 2015). Biotic pollination affects the commercial viability and nutritive value of some foods. For instance, cross-pollinated almond trees, compared to those that are self-pollinated, produce nuts with a healthier profile of fatty acids (Brittain et al. 2014).

There are multiple threats to both wild and managed pollinators, which affect different species in various ways. Pesticide exposure is a common threat that impairs the immune responses of pollinators such as wild bees, making them susceptible to pathogens (Pervez and Manzoor 2021). Additionally, pollinators respond differently to management strategies, and some species may benefit from anthropogenic-induced environmental changes (Vasiliev and Greenwood 2020). To safeguard pollinators, various strategies can be implemented to improve pollinator management and conservation. These include promoting extension services, conserving habitats, and providing more research opportunities for pollinators, which can help attain sustainable ecosystems and livelihoods (Mpondo et al. 2021).

10.5 Role of Pollinators in the SDG's Under Climate Change

Insect pollinators, particularly bees, make a significant contribution to the attainment of the 17 Sustainable Development Goals (SDGs). Pollinators also play a crucial role in achieving the thirteenth SDG, which pertains to climate change. Flowering trees and insect pollinators, like bees, are both important components of the ecosystem.

Flowering trees act as a habitat for wild pollinators and provide them with food resources such as nectar and pollen. Pollinators are an essential part of the ecosystem and help to mitigate the effects of climate change. Given that climate change is widely viewed as the most pressing challenge of our time, it is not unsurprising that SDG 13 calls for rapid action to address climate change and its repercussions (Stern 2016).

There are different indicators of SDG 13, which include:

- Boosting global adaptability and resistance to climate-related dangers and natural disasters.
- Incorporating climate change mitigation measures into national planning, strategy, and policies.
- Improving knowledge, awareness, and institutional and human capacity for early warning, damage reduction, and climate change adaptation.
- Implementing the pledge made by developed countries in the framework of effective mitigation measures and transparency regarding implementation and capitalizing on the Green Climate Fund to fully operationalize it.
- Promoting methods for increasing the ability of least developed nations and small island developing states to effectively plan for and manage climate change, with an emphasis on women, youth, and local and marginalized populations (Doni et al. 2020).

Furthermore, pollinators also play a crucial role in fulfilling all the SDGs (Fig. 10.1). For example, beekeeping can contribute to achieving the first SDG, "No Poverty" (Mugisha 2019). The second Sustainable Development Goal, "Zero Hunger," can be achieved with the assistance of pollinators. These organisms play a crucial role in producing healthy and nutritious food in sufficient quantities, as most food crops depend on insect pollinators (FAO 2018). Additionally, pollinators contribute to the attainment of the third SDG, "Good Health and Well-being," by increasing the nutritional value of food (Panța 2021). The fifteenth SDG, "Life on Land," can also be achieved through the promotion of pollinators, as they encourage individuals to plant more trees, thereby preserving these essential species. There are approximately 200,000 different pollinator species on Earth, and nearly 90% of wild plants rely on them for pollination, thereby promoting biodiversity. Ensuring the financial viability of beekeeping operations is crucial for food security and the ability of nations to produce their own food.

10.6 Economic Importance of Insect Pollinators

Insect pollination constitutes 80% of all pollination services provided. Previous studies have shown that the absence of pollinators can result in a reduction of yields for various nuts, fruits, and seed crops by up to 90% (Abrol et al. 2019). While other pollinators are also necessary, bees are considered the most important (Decourtye et al. 2019; Nayak et al. 2020). Several studies have found that insect pollinators are essential for fruit set and seed formation in 75% of important crops (Sgolastra

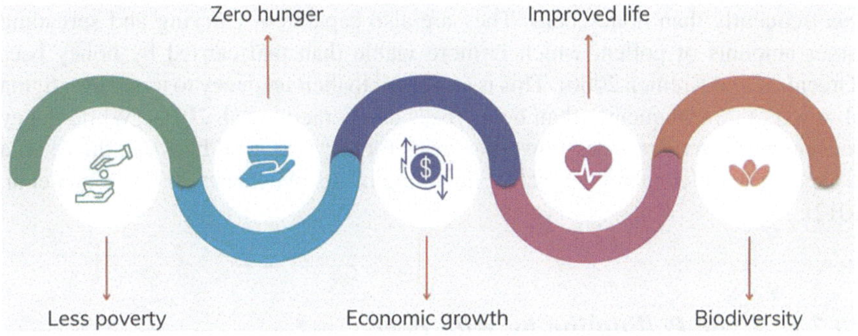

Fig. 10.1 Role of pollinators in sustainable development

et al. 2018). Bees can provide optimal pollination services, offering greater economic value (Quigley et al. 2019).

The global value of pollinators has been estimated at US$794 billion, accounting for approximately 9.5% of the global agricultural products used for human consumption. This includes €50 billion for vegetables and fruits, €39 billion for oil crops, €7.0 billion for stimulants, €4 billion for nuts, €0.2 billion for spices, and €1.0 billion for pulses. Pollination is valued at US$4.5 billion in the United States, €7 billion in Brazilian agricultural exports, €900 million in East Africa, €370 million in Uganda, and €1 billion in the Netherlands (Munyuli 2011). In the absence of insects, pollination must be accomplished through modern technology, such as Edete Precision Technologies and pollinating robots, or by hand. This can result in market losses and increased costs for customers, with an estimated loss of US$2 billion per year (Settele et al. 2016).

Several factors contribute to the decline of pollinators, including habitat loss, landscape degradation, excessive use of agrochemicals, climate change, air pollution, insect predation, invasive species, and exposure to pesticides and pathogens (Vasiliev and Greenwood 2021). Among plants, the production of fruit and vegetables is particularly affected, with losses estimated to be EUR 50 billion per crop and EUR 39 billion in losses from edible seed crops. Reduced pollination services can result in food shortages or limited food diversity (Shivanna et al. 2020).

10.7 Importance of Wild Pollinators for Food Security and Nutrition

In contrast to honey bees, which live and work in colonies, wild bees are solitary. These bees have superior nesting abilities, as they can locate their nests in holes, burrow tunnels in wood, or even under the ground (Woodcock et al. 2013). Some wild bees have also been known to use old beetle tunnels in wood as their nests (Everaars et al. 2018). Additionally, wild bees are less aggressive and tend to sting less frequently than honey bees. They are also capable of carrying and spreading larger amounts of pollen, which is more viable than that carried by honey bees (Greenleaf and Kremen 2006). This is attributed to their tendency to touch the stigma of flowers more frequently than honey bees do (Eeraerts et al. 2020). While honey bees primarily require mass-flowering crops for survival, wild bees can thrive on a diverse array of less abundant native flowering plants (Michener 2007; Odoux et al. 2012).

10.7.1 Crop Pollination by Wild Bees

Wild bees are an important group of pollinators (Osterman et al. 2021) and make significant contributions to food production (Kleijn et al. 2015; Stout et al. 2018). Previous studies have shown that wild bees can be more efficient pollinators than honey bees in certain crops, such as strawberries (MacInnis and Forrest 2019), tomatoes (Kgaphola 2021), soybeans (Roubik 2018), and apples (Mallinger and Gratton 2015). Additionally, research has demonstrated that wild bees can enhance the efficiency of honey bees and that wild bees and honey bees working together can produce a higher yield in sunflowers (Estravis et al. 2019).

In comparison with self-pollinated or wind-pollinated flowers, strawberries pollinated by wild bees are typically heavier, have fewer deformities, have a longer shelf life, have a firmer consistency, and last longer (Klatt et al. 2014). Furthermore, wild pollinators are responsible for the increase in quality and yield of many crops. For example, a study conducted in Brazil found that plants pollinated by wild bees produced tomatoes that were significantly heavier and had more seeds than plants that were not visited by pollinators (Bergamini et al. 2013). Similarly, research in Argentina found that soybean production was higher near forest areas, which provided nesting or habitat sites for wild pollinators (Monasterolo et al. 2015).

Pollinator-dependent crops are up to five times more valuable than those that do not depend on pollinators (FAO 2018). Millions of people rely on jobs and income from high-value commodity crops that depend on pollinators, such as cacao, oilseed rape, and coffee, which make major economic contributions to developing countries (USDA 2018). Wild pollinators are sometimes more effective in providing pollination services to some crops than managed bees (Mallinger and Gratton 2015; MacInnis and Forrest 2019; Kgaphola 2021). They also provide complementary services, as different wild bees are present throughout the year and can visit a variety of plants at different times of the year, even when weather conditions are not favorable for other pollinators (Brittain et al. 2014; Rader et al. 2016). Complementary pollination by wild bees ensures a more stable crop yield and improves crop quality (Rader et al. 2016).

10.7.2 Understanding the Nutritional Ecology of Bees

Approximately 80% of flowering crops rely on bees for their pollination, while bees depend on these flowering crops to meet their nutritional needs (Jordan et al. 2021; Rodger et al. 2021). There are several approaches to understanding the nutritional ecology of bees. It is crucial to understand the preferences and nutritional requirements of bees for different species of plants. To attract bees to agricultural crops, it is necessary to maintain, create, and improve bee habitats (Vaudo and Leonard 2021). However, applying current methods to study the nutritional ecology of wild bee species is challenging. This is especially true for solitary, ground-nesting species, which comprise approximately 70% of the 20,000 or more wild bee species worldwide (Harmon-Threatt 2020).

10.7.3 Evaluating the Primary Diet of Bees

It is important to study the dietary preferences and nutritional content provided by pollen to bees, as pollen is the primary diet of bees (Trinkl et al. 2020; Ruedenauer et al. 2020; Austin and Gilbert 2021; Lawson et al. 2021). The health outcomes and dietary preferences of bees are influenced by the composition of pollen (Vaudo and Leonard 2021). Sugars are the primary constituents of both nectar and pollination drops, but their total content and comparative distribution vary. They also include amino acids, with proline being one of those found most commonly. Proteomic research has revealed the presence of invertases and defense-related proteins, as well as other common functional classes of proteins, in nectar (floral and extrafloral) and pollination drops (Nepi et al. 2009). Furthermore, nectar is also an important component of bee diets, influencing pollinator preferences. Different approaches have been used to study the nutritional ecology of pollinators, particularly bees (Leach and Drummond 2018).

Not all plants produce pollen that satisfies the nutritional needs of bees in terms of quality and quantity. However, when selecting bee-friendly plants for planting, we often overlook the fact that they provide food for bees (Filipiak and Weiner 2017; Filipak et al. 2017). For example, we often overlook the fact that the nutritional needs of larvae are different from those of adults. Bee-friendly plants are often identified based on the amounts of nectar and pollen they produce and the frequency with which adult pollinating insects visit them. However, bees seek out nectar and pollen from a variety of plant types (Kraemer et al. 2014). Macronutrients (e.g., calcium, potassium, magnesium, iron, and phosphorus), lipids, and proteins are provided by pollen. The amount of nutrients can vary among plant species, and the requirements of bees also differ. Therefore, it is necessary to study the preferences and nutritional requirements of bees (Ruedenauer et al. 2019; Vaudo et al. 2020). Therefore, the nutritional needs of pollinators need to be understood and improved. Plant communities should be designed to improve the diversity of specific species (Pollinator Partnership 2022).

10.8 Impact of Climate Change on Plant–Pollinator Interaction

Climate is an aggregation of weather conditions that occur in an area (Beillouin et al. 2020). Human-mediated activities, such as rapid industrialization and deforestation, along with sudden and hazardous climate changes, are known contributors to these changes (Baer and Singer 2018). Given that pollinators have lifecycles synchronized with climatic patterns, these rapid changes can negatively impact both plants and pollinators (Renner and Zohner 2018). Different components of climate change, such as drought and heat stress, shifts in rainfall patterns, and elevated CO_2 levels, can disrupt plant–pollinator interactions.

10.8.1 Reduction in Pollination Due to Drought Stress

Drought alters plant–pollinator interactions by modifying various factors such as plant signals or cues to insect visitors, such as plant height, floral display, number of flowers per plant, size, color, and shape of flowers, compound and composition of floral odors, and floral rewards for pollinators, such as nectar volume, pollen quantity, total sugar, and chemical concentration and composition (Descamps et al. 2021a, b). Floral guides, which are signal patterns that direct pollinators to floral rewards upon finding the flower, enhance foraging efficiency and pollen transmission (Harrap et al. 2020).

Plants under drought-stress conditions produce fewer flowers and have less nectar in each flower, but the pollen quantity remains the same in certain conditions. Although the overall sugar content remains unchanged in some cases, nectar from drought-stressed plants contains a reduced percentage of sucrose (Rering et al. 2020). Water stress has the worst impact on floral development, resulting in fewer flowers with smaller flower sizes and decreasing floral rewards for pollinators (Lambrecht et al. 2017; Descamps et al. 2018). Water unavailability decreases the number of floral units in all plant species, regardless of their functional drought content, thereby decreasing the number of floral units in all plant communities (Phillips et al. 2018). Pollinators are attracted by various volatile compounds secreted by flowers; however, when plants are faced with drought stress, the chemical composition of these compounds' changes (Campbell et al. 2019; Boncan et al. 2020; Höfer et al. 2021). These changes in plant attraction and decreased floral rewards lead to a decline in insect visitation and pollination. Several studies have reported a decline in pollinator populations when sampling is done from drought-inflicted plants, indicating that the insect population suffers losses during drought (Walter 2020; Hung et al. 2021).

10.8.2 Reduction in Pollination Due to Heat Stress

Due to annual increases in temperature, the nectar secretion of flowering species is expected to decrease, limiting the number of nectar supplies available to pollinators (Takkis et al. 2018). Heat stress is altering the flowering phenology of plants, leading to smaller floral displays and lower-quality floral rewards for pollinators. This increases the foraging time and energy demands of bees and alters their flower preferences (Keeler et al. 2021). Temperature influences flowering phenology and flower output by modulating the effects of water and nitrogen availability on vegetative and floral characteristics (Akter and Klečka 2020). Temperature heat shock not only affects flower attributes but also has a negative impact on pollinator biology; it damages sperm DNA and its viability in many species (Martinet et al. 2021). Less pollinator visitation has been reported in plants exposed to heat stress compared to plants not experiencing heat stress (Descamps et al. 2021a, b).

10.8.3 Impact of Changing Rainfall

Similar to drought and heat stress, sudden changes in rainfall patterns also significantly affect pollinators and their relationships with plants. Changes in rainfall patterns negatively affect phenological phase timing, nectar dilution, and pollen degradation. Pollinators are harmed by mechanical and energy restrictions, disruptions in foraging habitats, and sensory inputs (Lawson and Rands 2019). Lower flower visitation and pollinator abundance have been linked to increased precipitation (Antiqueira et al. 2020).

10.8.4 Impact of Elevated CO_2

It is hypothesized that increased levels of atmospheric CO_2 may have an impact on plant growth, biochemistry, physiology, and phenology, among other aspects of plant–pollinator interactions. However, there have not been many analyses of how atmospheric CO_2 might affect pollination relationships. This is a crucial information gap, as any unfavorable effects exerted by atmospheric CO_2 on these interactions could have a negative impact on plant and pollinator populations (Pringle 2016). The combination of this stressor with other stressors, such as habitat loss and fragmentation, agrochemicals, diseases, and invasive species, could result in a decline in pollinator numbers. High levels of CO_2 in the environment cause flowers to produce fewer volatile compounds and mature early, before the emergence of bees (LeBhun and Luna 2021; Crowley et al. 2021). In addition to other climate change factors, atmospheric CO_2 is anticipated to directly impact pollinator interactions; however, few studies have assessed this phenomenon.

10.9 Range Shifts of Some Wild Bees in Response to Climate Change

Studies on species distribution indicate that as temperatures increase, species distribution shifts toward polar regions and higher latitudes (Hegland et al. 2009). For example, cold-adapted bumblebee species have adapted to high elevations and mountain environments over the past 100 years (Kerr et al. 2015). Due to the impact of climate change, the dispersal range of five species of bumblebees has shrunk over the past 26 years. According to Imbach et al. (2017), there will be an 88% decrease in coffee-friendly locations by 2055. However, some studies have shown that a temperature rise expands bee-friendly habitats. For example, Dew et al. (2019) found that Australian species of carpenter bees, *Xylocopa aerata* (Hymenoptera: Apidae), are more suitable because of climate change. Additionally, the overlap between horticultural and wild bee-friendly habitats is decreasing owing to climate change (Polce et al. 2014). For instance, Bezerra et al. (2019) claimed that climate change has reduced the areas where the two species naturally overlap because of changes in the native range of *Xylocopa* bees (Apidae) and areas favorable for the cultivation of passion fruit. For the solitary ground-nesting bee, *Melitta leporina* (Hymenoptera: Melittidae), the outcome differs; as the temperature rises, the potential range of this species' distribution is expected to increase (Rahimi et al. 2021). Dew et al. (2019) estimated that the range of suitable areas (the Australian Continent) for the Australian small carpenter bee, *Ceratina australensis*, under climate change scenarios in 2070 would increase. A study conducted in New Jersey and eastern Pennsylvania, USA, showed that under the worst climate change scenario, honey bee pollination would decrease and native bee pollination would increase by 4.5% by 2099. This is because of the behavioral changes of the bees due to changes in temperatures (Rader et al. 2013).

10.10 Climate Change and Pollinator Taxa

10.10.1 Bees

Bees are highly affected by climate change (IPBES 2016). These abrupt climatic changes significantly impact bees by affecting their physiology, distribution, and behavior. In addition to bees, climate change also negatively affects the feeding sources necessary for development and survival. Beekeeping management practices are important for preserving honey bee colonies, overwintering, and bee productivity (Varikou et al. 2020). Bumble bees are anticipated to shift their habitat, making them among the species affected by climate change (Pyke et al. 2016).

10.10.2 Flies

Hoverflies are known to demonstrate strong resilience and a variety of responses to the effects of climate change. Species that inhabit mountainous regions are expected to decline. Additionally, widely distributed species are thought to be less affected by climate change than species with restricted ranges (Miličić et al. 2018). It is proposed that the anticipated shift of the 25 hoverfly species analyzed to lower altitudes, the majority of which are chronologically classified as widespread or Mediterranean species found on Mount Olympus (Greece), is a favorable response to climate change due to their adaptability (Minachilis et al. 2021). In contrast, mountainous bumblebees, which are cold-adapted, are anticipated to shift to higher altitudes because of climate change, and this should be more noticeable in southern Europe (Marshall et al. 2020).

The Syrphidae, Nemestrinidae, and Bombyliidae families inhabit all habitat types except for those with closed-canopy forests, which are connected to open, dry, and semi-arid regions, respectively (Larson et al. 2001). These families are found in areas with less than 400 mm of annual rainfall. Fly ranges may be altered, populations may change, and some species may become extinct as a result of climate change. Dipterans adapt to changing environmental conditions or reposition their geographical distribution to more advantageous locations in response to environmental changes. A decline in the diversity of dipteran species will lead to a decline in the flowering plant population, which will diminish yield (Orford et al. 2015; Nath et al. 2023).

10.10.3 Butterflies

Climatic changes also affect butterfly species. An increase in temperature leads to an increase in butterfly species, whereas a decrease in temperature results in a decrease in species (Hussain et al. 2011). Rainfall and high temperatures can affect butterfly abundance and species richness. Similarly, fires also affect the population present in a habitat (Saraf and Vijaykumar 2021). Butterflies are a perfect species to research the consequences of climate change since they are poikilothermic creatures, meaning that temperature affects their life cycle, activity, distribution, and abundance (Hill et al. 1999; Roy and Sparks 2000). Precipitation and moderate or rainy winters were found to have a negative impact on butterfly abundance. Long-term findings in British Lepidoptera are consistent with the detrimental impacts of winter weather (Dennis and Sparks 2007). The main possible detrimental consequences for the Apollo butterfly, *Parnassius smintheus* Doubleday, are habitat loss brought on by forest expansion and variations in snow cover throughout the overwintering season (Matter et al. 2011) (Fig. 10.2).

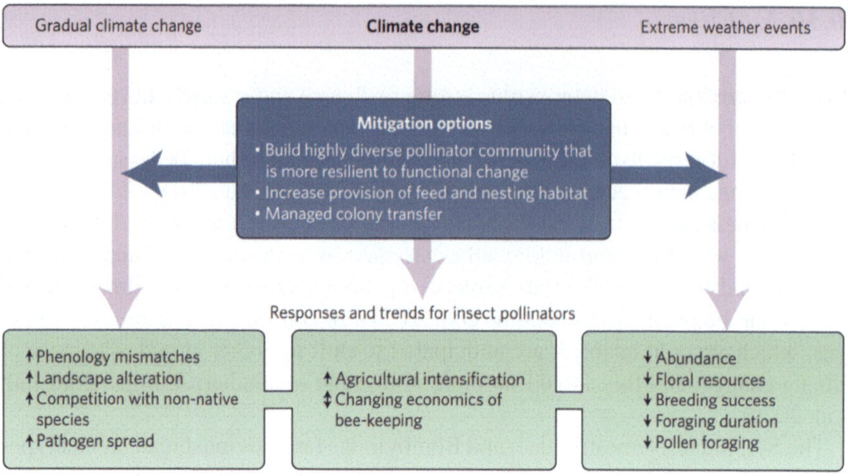

Fig. 10.2 Impact of climate change on pollinators. *Source* Settele et al. (2016)

10.11 Best Practices for Integrating Agriculture and Pollinator Conservation in a Changing Environment

The quality, structure, and management practices of surrounding landscapes significantly contribute to the richness and abundance of wild pollinators (Cole et al. 2022). Floral resources are limited in agricultural landscapes. These are the ultimate requirements for wild pollinators, which are mostly provided in semi-natural habitats. An increase in the distance of floral resources from bee habitats can lead to a decline in visitation rates and the abundance of wild pollinators (Lázaro and Alomar 2019). Even farm management practices are unable to preserve pollinator communities in isolated agrosystems (Monasterolo et al. 2022). Implementing such management practices is important because it helps increase the population of pollinators in agroecosystems to maintain food security (McKerchar et al. 2020).

10.11.1 Role of Farmlands

Farmlands play a significant role in preserving pollinator biodiversity (Shi et al. 2021; Toivonen et al. 2022). In agroecosystems affected by climate change, more effort is required to establish and sustain linkages for pollinators (Schatz et al. 2021).

10.11.1.1 Effects of Plant Characteristics on Conservation

When choosing which non-crop species to grow for pollinator conservation, plant characteristics that are significant for pollinators are often considered (Match and Potter 2018; Phillips et al. 2020). Plants that are native (can easily cope with the climate), economical for farmers, easily available, and good for pollinators can be cultivated (Bayle 2019). It is also possible to use flowering plants for many purposes, such as harvestable crops (such as sunflowers), materials, and nesting platforms for bees, windbreaks, and flowering crops that increase soil nitrogen and act as cover crops (Bentrup et al. 2019; Lee-Mäder et al. 2020). Diverse pollinator populations can benefit from various flower species exhibiting a range of complementary features during the growing season (Laha et al. 2020). An attempt should be made to use a variety of native plants that serve as equally common, abundant pollinators to create and maintain diversified pollination systems worldwide that are resilient to environmental changes (Grass et al. 2016; Samways et al. 2020).

10.11.1.2 Providing Nesting Conditions for Wild Pollinators

Wild pollinators require a variety of nesting substrates and flowering supplies offered by plant species to live and reproduce (Kremen et al. 2019). For instance, many solitary bee species make their nests on plant stems, excavated twigs, or above the ground in tunnels made of wood-boring beetles. To build nests, they need specific materials, such as tree resins, mud, leaves, and pebbles (Delaplane 2021). Many researchers have successfully created artificial nests to manage cavity-nesting bees. However, the conditions required for other wild pollinators, including ground-nesting bees, remain unknown (Antoine et al. 2021; Neumüller et al. 2022). The contribution of non-bee floral visitors, such as flies, butterflies, and beetles, should also be considered, even though bees are generally regarded as the most significant pollinators (Grass et al. 2016; Cusser et al. 2021). When bees do not visit flowers, non-bee pollinators play their role. Therefore, measures that take these creatures into account must also be considered (Cusser et al. 2021; Muñoz et al. 2021).

10.11.2 Building Strategies for Improving Pollinator Population

Healthy pollinator populations under uncertain environmental conditions can be overcome by including strategies that consider a diverse group of pollinators and interannual pollinator variability (Armstrong et al. 2021). Even though a few species of abundant wild pollinators can provide enough pollination services, initiatives to maintain diverse pollinator groups on farmlands are essential if we want to avoid dependence on a few wild species for the majority of pollination. Wild pollinator groups are adaptive in time and space, even when plants consistently produce floral supplies (Timberlake et al. 2019). However, it is very complicated to understand the dominant pollinator taxa among different years (Jachula et al. 2018).

10.11.3 Cost and Benefits

Stable and diverse pollinator communities could be of significant value, especially in long-term environmental and economic studies, together with other ecosystem service advantages for habitat enhancement (such as soil stability, pest management, and nutrient cycling) (Wratten et al. 2012). The costs and benefits of these studies can be very challenging to quantify, and there are all academic disciplines devoted to valuing things from economic and ecological perspectives (Cardinale et al. 2012).

10.11.4 Provision of Different Incentives

For better seed and fruit production, the main crops require insect pollination, and some incentives might be provided, such as subsidies by the government (Jacob et al. 2020). Planting wildflowers is one of the first stages in achieving this goal, which is adaptable, affordable, and practical compared to the price of hedgerows (Rey Benayas et al. 2020). The enhancement of nesting habitats is also an important factor (MacIvor 2016). The government must provide incentives and compensate growers for taking measures to conserve pollinators. If incentives are reduced, conservation practices may not be implemented (Jacob et al. 2020). Therefore, other strategies to promote pollinator-friendly strategies may prove to be longer-lasting and more effective. For instance, participating in pollinator conservation initiatives for purposes other than providing pollination services may be financially more advantageous for growers (Machatschek 2022).

10.12 Bees and Trees: Key Elements in the Solution to Climate Change

Climate change has affected ecosystems worldwide. As a result of climate change, the environment is becoming less healthy. Bees need a natural and healthy environment, free from pollution, to thrive. Climate change negatively impacts beekeepers' businesses, as they have to adapt to changes in deforestation and agricultural production. To promote sustainable production and reduce biodiversity loss, the planting of diverse forest species is encouraged worldwide (Pătruică and Simiz 2021).

Research has shown that incidences of forest fires can be reduced through beekeeping, highlighting the significant role of honey bees in restoring degraded habitats. To improve the performance of the apicultural sector, policies should be implemented. Honey and beeswax are harvested through beekeeping systems in forests and exported from Africa. In forest beekeeping systems, honey is produced in large quantities. In Zambia, research has also shown that harvesting approximately 1000 tons of honey from 15,000 km of forest can help reduce bushfire incidences (Lowore 2021).

Due to the changing climate, pollinators face many threats that impact their services. Changes in phenology can alter the relationships between plants and pollinators (Giannini et al. 2020). Loss of output, an increase in pests and diseases, and a greater fragility of ecological structures are the effects of this disturbing equilibrium on agroecosystems. It is important to establish a system that is not vulnerable to climate change and helps to conserve the diversity of pollinators. Climate-related events have a major impact on bees and can occasionally lead to the death of colonies. Fortunately, floods and fires affect small areas. Droughts, extremely cold temperatures, high temperatures, and continuous rain are additional climate variables that can affect bee populations (Faux and Kane 2021).

To assist beekeepers in adapting to these changing production conditions, specialized training should be made available. Insect-pollinated plants are at great risk due to the decline in bee populations. This means that food resources are also at risk, and as a result, the livelihood and well-being of humans are also affected (Shackleton et al. 2019). Climate change, the use of pesticides, and variations in temperature are causes of bee losses, which negatively impact nutritious crops such as vegetables and fruits. When bees affected by these conditions visit flowers, both the yield and quality of the fruit or seeds are reduced. This can lead to the deterioration of human health. Therefore, it is crucial to protect honey bees to maintain a clean and healthy environment. Maintaining bees can help fight hunger, maintain health conditions, and prevent malnutrition (Austin et al. 2019).

Beekeeping not only allows us to obtain honey but also improves the pollination of wild and commercial crop species. By maintaining commercially viable beekeeping practices, we can ensure both the pollination of various crops and the preservation of plant biodiversity (Kassa and Regasa 2020). Beekeeping, in all of its varied forms, contributes favorably to enhancing the environment. This includes growing trees to provide more bees with food, encouraging the planting of more trees, preventing tree cutting, and installing fire guards. Proper training and tree seedlings are provided as part of the intervention method to help recover damaged areas (Kumsa et al. 2021).

Manuka trees, which are crucial to New Zealand's beekeeping economy, were preserved and regenerated by the beekeeping sector. Beekeepers have been educating the public about the importance of forests as a source of food resources (nectar and pollen) for bees, emphasizing the need to take action to protect our ecosystems (Blignaut 2020). Beekeepers also play a role in the protection of pollinators other than bees by raising awareness of the safe use of agricultural practices, such as the use of pesticides in Australia and New Zealand. Beekeepers should collaborate with the government to work on protective measures to save bees and ensure food security (Rush and Obolonkin 2020).

10.13 Establishing Sustainable Supply Chains Through Pollinators

Sustainable pollinator management can be obtained by creating a roadmap that addresses the actions, policies, and implications of proper supply chain management strategies. There are different steps in strengthening a pollinator supply chain (Stout and Dicks 2022) (Fig. 10.3):

- The presence of crops dependent on pollinators must be assessed, and their exposure to pollinator loss must be assessed.
- Therefore, the extent to which crops rely on pollinators should be identified. After that, an assessment should be made on whether the crops could be substituted.
- Previous studies have been conducted on managing pollination risks, and the recommended measurements must be carefully reviewed.
- Further required actions must be identified, and a gap analysis must be conducted.
- Monitoring of pollinators should be done, and necessary actions must be implemented (Dicks et al. in prep).

Pesticide regulations must be improved to minimize pollinator exposure and ensure proper use. In addition, integrated pest management (IPM) strategies should be implemented to reduce the overreliance on pesticides. The standards for genetically modified crops should be elevated globally, and appropriate disease management systems should be developed. To support the preservation of pollinators, the implementation of pollinator-friendly practices should be encouraged, such as the use of agroforestry, organic farming, intercropping, control of invasive species, crop rotation, and the cultivation of flowering crops that support pollinators. Furthermore, incentives such as insurance for farmers should be provided to promote the adoption of these practices. The natural habitats of pollinators should be conserved and restored by providing food and nesting resources. Long-term monitoring of pollinator populations should be established to better understand their status, any declines, and the effectiveness of conservation efforts. The relationship between wild pollinators and their impact on crop yields should be studied through well-funded research projects. Effective pollinator conservation will require collaboration between NGOs, local communities, farmers, the general public, and stakeholders in relevant countries (Dicks et al. 2016; Potts et al. 2016b, a).

Fig. 10.3 Building pollinators supply chain

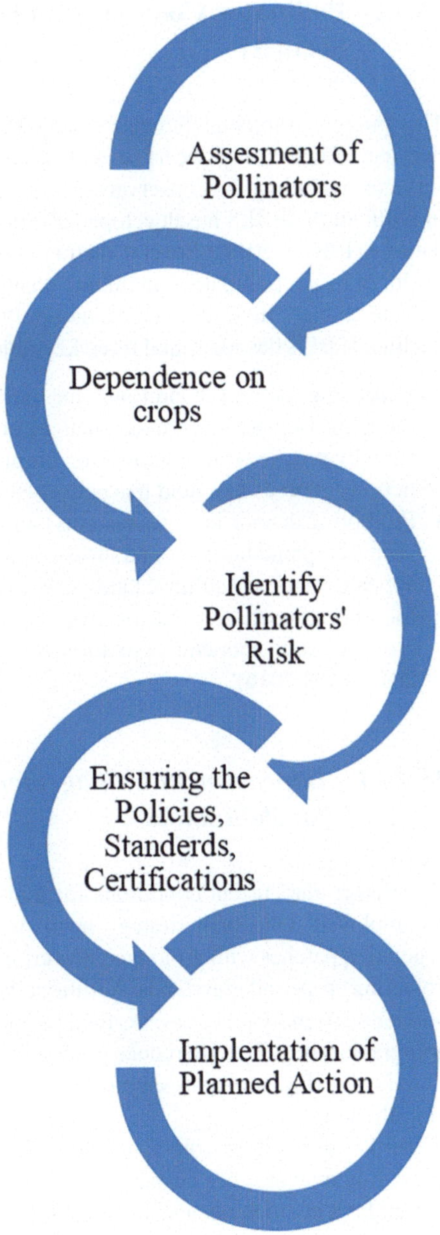

10.14 Pollinator Conservation Policies for Sustainable Food Security

The Intergovernmental Science-Policy Platform on Biodiversity and Ecosystem Services (IPBES) is concerned with issues related to biodiversity and ecosystem services, including the conservation of insect pollinators. In response to the decline of pollinators, IPBES has developed a variety of strategies, policies, and recommendations (IPBES 2019). Several factors contribute to the diversity, abundance, and health of insect pollinators, including changes in land use, pesticide use, pathogens, climate change, invasive species, and pollution. To mitigate the risks of pollinator decline, IPBES has identified three key objectives and associated responses:

- Improving current conditions for pollinators and preserving pollination by reducing immediate threats, such as encouraging farmers to adopt pollinator-friendly practices and establishing un-grazed vegetation with extended flowering periods in areas like field margins.
- Promoting diverse farming systems, supporting existing diverse farming systems, such as organic farming, and investing in ecological infrastructure to enhance the appearance of agricultural landscapes, such as restoring natural habitats.
- Incorporating diverse perspectives and values into management and fostering cross-sectoral, cooperative relationships between humans and pollinators (IPBES 2016; FRB 2016).

10.14.1 Improvement Management Conditions for Pollinators

To enhance the current conditions for managing pollinators, several strategies can be employed. One such strategy involves managing immediate risks by creating vegetation patches with prolonged flowering periods and effectively managing mass-flowering crops and grasslands. Additionally, farmers should be incentivized to adopt pollinator-friendly practices. Pesticide standards can be raised, and innovative technologies can be utilized to reduce pollinator exposure to pesticides and mitigate associated risks. IPM strategies should also be introduced to ensure the safe management of pollinators.

Furthermore, it is crucial to capitalize on immediate opportunities, such as improving managed bee husbandry and developing alternative-managed pollinators. Vacant land in cities can also be utilized to support pollinators. Adopting pollinator-friendly practices not only promotes biodiversity preservation but also enhances productivity and profitability for local farmers. Such practices include reducing the use of synthetic inputs and conserving and maintaining semi-natural ecosystems on farms and landscapes.

In an ideal scenario, there should be an adequate number of pollinators to provide crops with diverse and abundant pollination services. By taking appropriate measures at the right time and conserving pollinators, the world's economy could be positively impacted, as pollinators play a significant role in global food security (Marshman and Knezevic 2021).

10.14.2 Transformation of Agricultural Landscapes

Enhancing farming systems to support organic farming, diversified agriculture, and food security is crucial. To achieve this, implementing agricultural and food policies is necessary. Strengthening biocultural diversity involves leveraging local knowledge and governance that supports pollinators. Additionally, investments should safeguard heritage sites, enhance connectivity between patches, and support large-scale traditional practices for managing biocultural diversity.

10.14.3 Transforming the Relationship of Society with Nature

It is imperative to integrate people's knowledge and values into pollinator management. Agricultural practices should incorporate pollinator research. The exchange of knowledge among local knowledge holders, stakeholders, and scientists is crucial for enhancing the available information on pollinator conservation. Innovative pollinator activities should be supported by engaging stakeholders with diverse socio-cultural values related to pollination (IPBES 2022).

10.15 Conclusion

Insects, primarily from the orders Hymenoptera, Diptera, Lepidoptera, and Coleoptera, are the most efficient pollinators. Bees, the most abundant and effective flower visitors, are globally recognized for their essential role in pollination and honey production. Wild bees, dipterans, and lepidopterans also play crucial pollination roles. The decline of pollinators impacts not only crop yields but also the availability of micronutrients (e.g., vitamins A and C, folic acid, calcium, and fortified flour), as they rely on pollination. Approximately 80% of pollination services are provided by insects, and without them, the production of nuts, fruits, and seed crops could decrease by up to 90%. Major factors contributing to pollinator decline include habitat loss, landscape degradation, excessive agrochemical use, climate change, air pollution, insect predation, invasive species, pesticides, and diseases. Drought and water stress can also deter pollinators from visiting plants and disrupt their relationships. To address these challenges, it is possible to conserve insect pollinators by implementing strategies that support various pollinator species with floral and nesting resources. Collaborative efforts involving governments, researchers, stakeholders, and farmers can further enhance pollinator conservation through the development of pollinator-friendly initiatives and policies. Climate change has a significant impact on pollinators and their food sources, affecting their distribution, physiology, and behavior. Bees, hoverflies, and butterflies are particularly sensitive to sudden climatic changes. Butterfly populations decrease with dropping temperatures and increase with rising temperatures. In this context, the SDGs and regulations from the IPBES can greatly support pollinator conservation efforts. Adapting to changing environmental conditions necessitates the development of new conservation strategies to ensure the well-being of insect pollinator populations.

References

Abrol DP, Gorka AK, Ansari MJ, Al-Ghamdi A, Al-Kahtani S (2019) Impact of insect pollinators on yield and fruit quality of strawberry. Saudi J Biol Sci 26(3):524–530

Akter A, Klečka J (2020) Interactive effects of temperature, water, and nitrogen availability on the growth, floral traits, and pollination of white mustard, Sinapis alba

Allsopp MH, De Lange WJ, Veldtman R (2008) Valuing insect pollination services with cost of replacement. PLoS ONE 3(9):e3128

An J, Huang J, Shao Y, Zhang S, Wang B, Liu X, Williams PH (2014) The bumblebees of North China (Apidae, Bombus Latreille). Zootaxa 3830(1):1–89

Antiqueira PA, de Omena PM, Goncalves-Souza T, Vieira C, Migliorini GH, Kersch-Becker MF, Romero GQ (2020) Precipitation and predation risk alter the diversity and behavior of pollinators and reduce plant fitness. Oecologia 192(3):745–753

Antoine CM, Forrest JR (2021) Nesting habitat of ground-nesting bees: a review. Ecol Entomol 46(2):143–159

Armstrong A, Brown L, Davies G, Whyatt JD, Potts SG (2021) Honeybee pollination benefits could inform solar park business cases, planning decisions and environmental sustainability targets. Biol Cons 263:109332

Austin AJ, Lawson-Handley L, Gilbert JD (2019) Plugging the hunger gap: organic farming supports more abundant nutritional resources for bees at critical periods. bioRxiv, 837625

Austin AJ, Gilbert JDJ (2021) Solitary bee larvae prioritize carbohydrate over protein in parentally provided pollen. Funct Ecol 35:1069–1080. https://doi.org/10.1111/1365-2435.13746Baracchi

Baer HA, Singer M (2018) The anthropology of climate change: an integrated critical perspective. Routledge

Bailes EJ, Ollerton J, Pattrick JG, Glover BJ (2015) How can an understanding of plant–pollinator interactions contribute to global food security? Curr Opin Plant Biol 26:72–79

Bayle GK (2019) Ecological and social impacts of eucalyptus tree plantation on the environment. J Biodiversity Conservat Bioresource Managem 5(1):93–104

Beillouin D, Schauberger B, Bastos A, Ciais P, Makowski D (2020) Impact of extreme weather conditions on European crop production in 2018. Philos Trans R Soc B 375(1810):20190510

Bentrup G, Hopwood J, Adamson NL, Vaughan M (2019) Temperate agroforestry systems and insect pollinators: a review. Forests 10(11):981

Bergamini BAR, da Silva Elias MA, Franceschinelli EV (2013) Native bees pollinate tomato flowers and increase fruit production. J Pollinat Ecol 11:41–45

Bernhardt P (2000) Convergent evolution and adaptive radiation of beetle-pollinated angiosperms. Plant Syst Evol 222:293–320

Blignaut K (2020) Where is the profitability in the Manuka honey production in Northland? Kellogg Rural Leadership Programme Course 41

Boncan DAT, Tsang SS, Li C, Lee IH, Lam HM, Chan TF, Hui JH (2020) Terpenes and terpenoids in plants: interactions with environment and insects. Int J Mol Sci 21(19):7382

Campbell DR, Sosenski P, Raguso RA (2019) Phenotypic plasticity of floral volatiles in response to increasing drought stress. Ann Bot 123(4):601–610

Cardinale BJ, Duffy JE, Gonzalez A et al (2012) Biodiversity loss and its impact on humanity. Nature 486:59–67

Cassils JA (2004) Overpopulation, sustainable development, and security: Developing an integrated strategy. Popul Environ 25(3):171–194

Chabert S, Mallinger RE, Sénéchal C, Fougeroux A, Geist O, Guillemard V, Vaissière BE (2022) Importance of maternal resources in pollen limitation studies with pollinator gradients: a case study with sunflower. Agr Ecosyst Environ 330:107887

Chaplin-Kramer R, Dombeck E, Gerber J, Knuth KA, Mueller ND, Mueller M, Klein AM (2014) Global malnutrition overlaps with pollinator-dependent micronutrient production. Proc Royal Soc b: Biol Sci 281(1794):20141799

Chaplin-Kramer R, Dombeck E, Gerber J, Knuth KA, Mueller ND, Mueller M, Klein AM (2014) Global malnutrition overlaps with pollinator-dependent micronutrient production. Proc Royal Soc B: Biol Sci 281:20141799

Cole LJ, Baddeley JA, Robertson D, Topp CF, Walker RL, Watson CA (2022) Supporting wild pollinators in agricultural landscapes through targeted legume mixtures. Agr Ecosyst Environ 323:107648

Cook DF, Voss SC, Finch JT, Rader RC, Cook JM, Spurr CJ (2020) The role of flies as pollinators of horticultural crops: an Australian case study with worldwide relevance. Insects 11(6):341

Crowley LM, Sadler JP, Pritchard J, Hayward SA (2021) Elevated CO_2 impacts on plant-pollinator interactions: a systematic review and free air carbon enrichment field study. Insects 12(6):512

Cusser S, Haddad NM, Jha S (2021) Unexpected functional complementarity from non-bee pollinators enhances cotton yield. Agr Ecosyst Environ 314:107415

Decourtye A, Alaux C, Le Conte Y, Henry M (2019) Toward the protection of bees and pollination under global change: present and future perspectives in a challenging applied science. Current Opinion in Insect Sci 35:123–131

Delaplane KS (2021) Crop pollination by bees, Volume 1: Evolution, Ecology, Conservation, and Management. CABI

Dennis RLH, Sparks TH (2007) Climate signals are reflected in an 89-year series of British Lepidoptera records. Eur J Entomol 104:763–767

Descamps C, Quinet M, Baijot A, Jacquemart AL (2018) Temperature and water stress affect plant–pollinator interactions in Borago officinalis (Boraginaceae). Ecol Evol 8(6):3443–3456

Descamps C, Jambrek A, Quinet M, Jacquemart AL (2021) Warm temperatures reduce flower attractiveness and bumblebee foraging. Insects 12(6):493

Descamps C, Quinet M, Jacquemart AL (2021) The effects of drought on plant–pollinator interactions: what to expect? Environ Exp Bot 182:104297

Dicks LV, Viana B, Bommarco R, Brosi B, del Coro Arizmendi M, Cunningham SA et al. (2016) Ten policies for pollinators. Science 354(6315):975–976. Retrieved from https://doi.org/10.1126/science.aai9226

Doni F, Gasperini A, Soares JT (2020). What is the SDG 13?. In: SDG13–climate action: combating climate change and its impacts. Emerald Publishing Limited

Eeraerts M, Vanderhaegen R, Smagghe G, Meeus I (2020) Pollination efficiency and foraging behaviour of honey bees and non-Apis bees to sweet cherry. Agric for Entomol 22(1):75–82

Eilers EJ, Kremen C, Smith Greenleaf S, Garber AK, Klein AM (2011) Contribution of pollinator-mediated crops to nutrients in the human food supply. PLoS ONE 6(6):e21363

Ellis AM, Myers SS, Ricketts TH (2015) Do pollinators contribute to nutritional health? PLoS ONE 10(1):e114805

Estravis Barcala MC, Palottini F, Farina WM (2019) Honey bee and native solitary bee foraging behavior in a crop with dimorphic parental lines. PLoS ONE 14(10):e0223865
Everaars J, Settele J, Dormann CF (2018) Fragmentation of nest and foraging habitat affects time budgets of solitary bees, their fitness and pollination services, depending on traits: results from an individual-based model. PLoS ONE 13(2):e0188269
FAO (2018) Why bees matter? Food and agriculture organization of the United Nations
Faux CM, Kane TR (2021) Honey bees: disaster preparedness and response. Veterinary Clinics: Food Animal Pract 37(3):559–567
Filipiak M, Kuszewska K, Asselman M, Denisow B, Stawiarz E, Woyciechowski M, Weiner J (2017) Ecological stoichiometry of the honeybee: Pollen diversity and adequate species composition are needed to mitigate limitations imposed on the growth and development of bees by pollen quality. PLoS ONE 12:e0183236
Filipiak M, Weiner J (2017) Plant-insect interactions: the role of ecological stoichiometry. Acta Agrobot 70
Forister ML, McCall AC, Sanders NJ, Fordyce JA, Thorne JH, O'Brien J, Waetjen DP, Shapiro AM (2010) Compounded effects of climate change and habitat alteration shift patterns of butterfly diversity. Proc Natl Acad Sci USA 107:2088–2092
FRB (Fondation pour la Recherche sur la Biodiversité) (2016) Les acteurs français se mobilisent pour les pollinisateurs – plus de 250 initiatives en faveur des pollinisateurs mises en regard de l'Evaluation de l'IPBES (v2). In: Série FRB, Expertise et synthèse
Gallai N, Salles JM, Settele J, Vaissière BE (2009) Economic valuation of the vulnerability of world agriculture confronted with pollinator decline. Ecol Econ 68(3):810–821
Garratt MP, Breeze TD, Jenner N, Polce C, Biesmeijer JC, Potts SG (2014) Avoiding a bad apple: insect pollination enhances fruit quality and economic value. Agr Ecosyst Environ 184:34–40
Giannini TC, Costa WF, Borges RC, Miranda L, da Costa CPW, Saraiva AM, Imperatriz Fonseca VL (2020) Climate change in the Eastern Amazon: crop-pollinator and occurrence-restricted bees are potentially more affected. Reg Environ Change 20(1):1–12
Gill RJ, Baldock KC, Brown MJ, Cresswell JE, Dicks LV, Fountain MT, Potts SG (2016) Protecting an ecosystem service: approaches to understanding and mitigating threats to wild insect pollinators. In: Advances in ecological research. vol 54. Academic Press, pp 135–206
Grass I, Albrecht J, Jauker F, Diek€ otter T, Warzecha D, Wolters V, Farwig N (2016) Much more than bees—Wildflower plantings support highly diverse flower-visitor communities from complex to structurally simple agricultural landscapes. Agricul Ecosyst Environ 225:45–53
Greenleaf SS, Kremen C (2006) Wild bees enhance honey bees' pollination of hybrid sunflower
Hahn M, Brühl CA (2016) The secret pollinators: an overview of moth pollination with a focus on Europe and North America. Arthropod-Plant Interact 10:21–28
Harmon-Threatt A (2020) Influence of nesting characteristics on health of wild bee communities. Annu Rev Entomol 65:39–56
Harrap MJ, Hempel de Ibarra N, Whitney HM, Rands SA (2020) Floral temperature patterns can function as floral guides. Arthropod-Plant Interact 14(2):193–206
Hegland SJ, Nielsen A, Lázaro A, Bjerknes AL, Totland Ø (2009) How does climate warming affect plant-pollinator interactions? Ecol Lett 12(2):184–195
Hill JK, Thomas CD, Huntley B (1999) Climate and habitat availability determine 20th century changes in a butterfly's range margin. Proc Royal Soc London B 266:1197–1206
Höfer RJ, Ayasse M, Kuppler J (2021) Bumblebee behavior on flowers, but not initial attraction, is altered by short-term drought stress. Front Plant Sci 11:2255
Høye TT, Post E, Schmidt NM, Trøjelsgaard K, Forchhammer MC (2013) Shorter flowering seasons and declining abundance of flower visitors in a warmer Arctic. Nat Clim Chang 3:759–763
Hung KLJ, Sandoval SS, Ascher JS, Holway DA (2021) Joint impacts of drought and habitat fragmentation on native bee assemblages in a California biodiversity hotspot. Insects 12(2):135
Hussain KJ, Ramesh T, Satpathy KK, Selvanayagam M (2011) Seasonal dynamics of butterfly population in DAE Campus, Kalpakkam, Tamil Nadu India. J Threatened Taxa 3(1):1401–1414

IPBES (2016) Summary for policymakers of the assessment report of the Intergovernmental science-policy platform on biodiversity and ecosystem services on pollinators, pollination and food production. In: Potts SG, Imperatriz-Fonseca VL, Ngo HT, Biesmeijer JC, Breeze TD, Dicks LV, Garibaldi LA, Hill R, Settele J, Vanbergen AJ, Aizen MA, Cunningham SA, Eardley C, Freitas BM, Gallai N, Kevan PG, KovácsHostyánszki A, Kwapong PK, Li J, Li X, Martins DJ, NatesParra G, Pettis JS, Rader R, Viana BF (eds) Secretariat of the intergovernmental science-policy platform on biodiversity and ecosystem services, Bonn, Germany, pp 36

IPBES (2016) Summary for policymakers of the assessment report of the intergovernmental science-policy platform on biodiversity and ecosystem services on pollinators, pollination and food production. In: Potts SG, Imperatriz-Fonseca VL, Ngo HT, Biesmeijer JC et al. (eds), pp 1–30

Jachuła J, Wrzesień M, Strzałkowska-Abramek M, Denisow B (2018) The impact of spatio-temporal changes in flora attributes and pollen availability on insect visitors in Lamiaceae species. Acta Bot Croat 77(2):161–171

Jacob CT, Parida A, Kumar NK (2020) Conservation of India's agrobiodiversity towards increasing food, nutritional and livelihood security. Curr Sci 119(4):607

Jordan A, Patch HM, Grozinger CM, Khanna V (2021) Economic dependence and vulnerability of United States agricultural sector on insect-mediated pollination service. Environ Sci Technol 55(4):2243–2253

Kassa Degu T, Regasa Megerssa G (2020) Role of beekeeping in the community forest conservation: evidence from Ethiopia. Bee World 97(4):98–104

Keeler AM, Rose-Person A, Rafferty NE (2021) From the ground up: building predictions for how climate change will affect belowground mutualisms, floral traits, and bee behavior. Climate Change Ecol 1:100013

Kgaphola RN (2021) Quantifying the pollination ecosystem service provided by non-Apis bees in tomato production. Doctoral dissertation, North-West University (South Africa)

Klatt BK, Holzschuh A, Westphal C, YannClough IS, Pawelzik E, Tscharntke T (2014) Bee pollination improves crop quality, shelf life, and commercial value. Proc Royal Soc b: Biol Sci 281(1775):20132440

Kocher SD, Paxton RJ (2014) Comparative methods offer powerful insights into social evolution in bees. Apidologie 45(3):289–305

Kraemer ME, Favi FD, Niedziela CE (2014) Nesting and pollen preference of Osmia lignaria lignaria (Hymenoptera: Megachilidae) in Virginia and North Carolina Orchards. Environ Entomol 43:932–941

Kremen C, Albrecht M, Ponisio L (2019) Restoring pollinator communities and pollination services in hedgerows in intensively managed agricultural landscapes. In: The ecology of hedgerows and field margins. Routledge, pp 163–185

Kumsa T, Bareke T, Addi A, Roba K (2021) Beekeeping promotes the traditional home-garden conservation in Ethiopia. J Agricul Environ Int Developm (JAEID) 115(2):23–37

Laha S, Chatterjee S, Das A, Smith B, Basu P (2020) Exploring the importance of floral resources and functional trait compatibility for maintaining bee fauna in tropical agricultural landscapes. J Insect Conserv 24(3):431–443

Lambrecht SC, Morrow A, Hussey R (2017) Variation in and adaptive plasticity of flower size and drought-coping traits. Plant Ecol 647–660

Larson BMH, Kevan PG, Inouye DW (2001) Flies and flowers: taxonomic diversity of anthophiles and pollinators. Can Entomol 133(4):439–465

Lawson DA, Rands SA (2019) The effects of rainfall on plant–pollinator interactions. Arthropod-Plant Interact 13(4):561–569

Lawson SP, Kennedy KB, Rehan SM (2021) Pollen composition significantly impacts the development and survival of the native small carpenter bee Ceratina Calcarata. Ecol Entomol 46(2):232–239

Lázaro A, Alomar D (2019) Landscape heterogeneity increases the spatial stability of pollination services to almond trees through the stability of pollinator visits. Agr Ecosyst Environ 279:149–155

Lebuhn G, Droege S, Connor EF, Gemmill-Herren B, Potts SG, Minckley RL, Parker F (2013) Detecting insect pollinator declines on regional and global scales. Conserv Biol 27(1):113–120
LeBuhn G, Luna JV (2021) Pollinator decline: what do we know about the drivers of solitary bee declines? Current Opinion in Insect Sci 46:106–111
Lee-Mäder E, Vaughan M, Goldenetz-Dollar J (2020) Agroforestry and cover cropping for pollinators. In: Towards sustainable crop pollination services: measures at field, farm and landscape scales, pp 105–126
Lei M, Zhou Y, Luo Q (2019) Enhanced metaheuristic optimization: wind-driven flower pollination algorithm. IEEE Access 7:111439–111465
Lim SS, Vos T, Flaxman AD, Danaei G, Shibuya K, Adair-Rohani H, Pelizzari PM (2012) A comparative risk assessment of burden of disease and injury attributable to 67 risk factors and risk factor clusters in 21 regions, 1990–2010: a systematic analysis for the global burden of disease study 2010. The Lancet 380(9859):2224–2260
Mach BM, Potter DA (2018) Quantifying bee assemblages and attractiveness of flowering woody landscape plants for urban pollinator conservation. PLoS ONE 13(12):e0208428
Machatschek J (2022) Urban transition towards the inclusion of pollinators. Doctoral dissertation, Lund University
MacIvor JS (2016) Cavity-nest boxes for solitary bees: a century of design and research. Apidologie 1–17M
Mahmood L (2014) The metabolic processes of folic acid and Vitamin B12 deficiency. J Health Res Rev 1(1):112–125
Marshall L, Perdijk F, Dendoncker N, Kunin W, Roberts S, Biesmeijer JC (2020) Bumblebees moving up: shifts in elevation ranges in the Pyrenees over 115 years. Proc R Soc B 287(1938):20202201
Marshman J, Knezevic I (2021) What's in a name? Challenging the commodification of pollination through the diverse economies of'Bee Cities'. J Polit Ecol 28(1)
Martinet B, Zambra E, Przybyla K, Lecocq T, Anselmo A, Nonclercq D, Hennebert E (2021) Mating under climate change: impact of simulated heatwaves on the reproduction of model pollinators. Funct Ecol 35(3):739–752
McLaughlin JF, Hellmann JJ, Boggs CL, Ehrlich PR (2002) Climate change hastens population extinctions. Proc Natl Acad Sci USA 99:6070–6074
Michener CD (2007) The bees of the world, 2nd, Revised. Johns Hopkins University Press
Miličić M, Vujić A, Cardoso P (2018) Effects of climate change on the distribution of hoverfly species (Diptera: Syrphidae) in Southeast Europe. Biodivers Conserv 27:1173–1187
Minachilis K, Kougioumoutzis K, Petanidou T (2021) Climate change effects on multi-taxa pollinator diversity and distribution along the elevation gradient of Mount Olympus Greece. Ecol Indicat 132:108335
Monasterolo M, Poggio SL, Medan D, Devoto M (2022) High flower richness and abundance decrease pollen transfer on individual plants in road verges but increase it in adjacent fields in intensively managed agroecosystems. Agr Ecosyst Environ 333:107952
Monasterolo M, Musicante ML, Valladares GR, Salvo A (2015) Soybean crops may benefit from forest pollinators. Agricul Ecosyst Environ 202:217–222
Mpondo FT, Ndakidemi PA, Treydte AC (2021) Balancing bees and livestock: pastoralist knowledge, perceptions and implications for pollinator conservation in Rangelands Northern Tanzania. Tropical Conservat Sci 14:19400829211028130
Mugisha B (2019) Beekeeping around Bwindi. Scaling-up production of honey and other bee products through training in beekeeping best practices and the introduction of low-cost transitional hives. Beekeeping around Bwindi. Scaling-up production of honey and other bee products through training in beekeeping best practices and the introduction of low-cost transitional hives
Muñoz AE, Plantegenest M, Amouroux P, Zaviezo T (2021) Native flower strips increase visitation by non-bee insects to avocado flowers and promote yield. Basic Appl Ecol 56:369–378
Munyuli MBT (2011) Pollinator biodiversity in Uganda and in Sub Sahara Africa: landscape and habitat management strategies for its conservation. Int J Biodiv Cons 3:551–609

Nath R, Singh H, Mukherjee S (2023) Insect pollinators decline: an emerging concern of Anthropocene epoch. J Apic Res 62(1):23–38
Nayak RK, Rana K, Bairwa VK, Singh P, Bharthi VD (2020) A review on role of bumblebee pollination in fruits and vegetables. Phytopathology 9(3):1328–1334
Naz S, Fatima Z, Iqbal P, Khan A, Zakir I, Ullah H, Ahmad S (2022) An introduction to climate change phenomenon. In: Building climate resilience in agriculture. Springer, Cham, pp 3–16
Nepi M, von Aderkas P, Wagner R, Mugnaini S, Coulter A, Pacini E (2009) Nectar and pollination drops: how different are they? Ann Bot 104(2):205–219
Neumüller U, Burger H, Mayr AV, Hopfenmüller S, Krausch S, Herwig N, Ayasse M (2022) Artificial nesting hills promote wild bees in agricultural landscapes. Insects 13(8):726
Nicole W (2015).Pollinator power: nutrition security benefits of an ecosystem service
Nicolson SW, Wright GA (2017) Plant–pollinator interactions and threats to pollination: perspectives from the flower to the landscape
Odoux JF, Feuillet D, Aupinel P, Loublier Y, Tasei JN, Mateescu C (2012) Territorial biodiversity and consequences on physico-chemical characteristics of pollen collected by honey bee colonies. Apidologie 43(5):561–575
Ollerton J (2017) Pollinator diversity: distribution, ecological function, and conservation. Annu Rev Ecol Evol Syst 48:353–376
Ollerton J, Johnson SD, Cranmer L, Kellie S (2003) The pollination ecology of an assemblage of grassland Asclepiads in South Africa. Ann Bot 92:807–834
Oprita VA, Gavat C (2016) The behaviour of some apricot cultivars with late blooming in south-eastern Romania. In: III International symposium on horticulture in Europe-SHE2016 1242, October, pp 385–388
Orford KA, Vaughan IP, Memmott J (2015) The forgotten flies: the importance of non-syrphid Diptera as pollinators. Proc Royal Soc b: Biol Sci 282(1805):20142934
Osterman J, Theodorou P, Radzevičiūtė R, Schnitker P, Paxton RJ (2021) Apple pollination is ensured by wild bees when honey bees are drawn away from orchards by a mass co-flowering crop, oilseed rape. Agr Ecosyst Environ 315:107383
Panța ND (2021) Operationalizing farm sustainability, particularities from the beekeeping sector. Revista Economica 73
Pollinator Partnership (2022) Planting Guides Pollinator.org (2022). Available online at: https://www.pollinator.org/guides (Accessed 22 Feb 2022)
Pătruică S, Peț I, Simiz E (2021) Beekeeping in the context of climate change. Scientific Papers; Series D; Animal Science: Bucharest, Romania 64:2393-2260
Phillips BB, Shaw RF, Holland MJ, Fry EL, Bardgett RD, Bullock JM, Osborne JL (2018) Drought reduces floral resources for pollinators. Glob Change Biol 24(7):3226–3235
Phillips BB, Wallace C, Roberts BR, Whitehouse AT, Gaston KJ, Bullock JM, Osborne JL (2020) Enhancing road verges to aid pollinator conservation: a review. Biol Cons 250:108687
Pinke M (2013) How far is world agricultural production likely to be threatened by pollinator declines? J Sustain 1(1):9–9
Poletti S, Gruissem W, Sautter C (2004) The nutritional fortification of cereals. Curr Opin Biotechnol 15(2):162–165
Poole N, Echavez C, Rowland D (2018) Are agriculture and nutrition policies and practice coherent? Stakeholder evidence from Afghanistan. Food Secur 10(6):1577–1601
Potts SG, Imperatriz-Fonseca V, Ngo HT, Aizen MA, Biesmeijer JC, Breeze TD et al. (2016) Safeguarding pollinators and their values to human well-being. Nature 540(7632):220–229. Retrieved from https://doi.org/10.1038/nature20588
Pringle EG (2016) Integrating plant carbon dynamics with mutualism ecology. New Phytol 210:71–75
Pyke GH, Thomson JD, Inouye DW, Miller TJ (2016) Effects of climate change on phenologies and distributions of bumble bees and the plants they visit. Ecosphere 7(3):e01267
Quigley TP, Amdam GV, Harwood GH (2019) Honey bees as bioindicators of changing global agricultural landscapes. Current Opinion in Insect Sci 35:132–137

Rader R, Reilly J, Bartomeus I, Winfree R (2013) Native bees buffer the negative impact of climate warming on honey bee pollination of watermelon crops. Glob Change Biol 19(10):3103–3110

Rader R, Bartomeus I, Garibaldi LA, Garratt MPD, Howlett BG, Winfree R, Cunningham SA et al. (2016) Non-bee insects are important contributors to global crop pollination. Proc National Acad Sci 113(1):146–151

Rafique MK, Mahmood R, Stephen E, Irshad M, Sarwar G (2016) Pollination deficit in mango orchards at Multan, Pakistan. Pakistan J Zool 48(1)

Rahimi E, Barghjelveh S, Dong P (2021) Estimating potential range shift of some wild bees in response to climate change scenarios in North western regions of Iran. J Ecol Environ 45(1):1–13

Renner SS, Zohner CM (2018) Climate change and phenological mismatch in trophic interactions among plants, insects, and vertebrates. Annu Rev Ecol Evol Syst 49:165–182

Rering CC, Franco JG, Yeater KM, Mallinger RE (2020) Drought stress alters floral volatiles and reduces floral rewards, pollinator activity, and seed set in a global plant. Ecosphere 11(9):e03254

Rey Benayas JM, Altamirano A, Miranda A, Catalán G, Prado M, Lisón F, Bullock JM (2020) Landscape restoration in a mixed agricultural-forest catchment: Planning a buffer strip and hedgerow network in a Chilean biodiversity hotspot. Ambio 49(1):310–323

Roberts DP, Mattoo AK (2019) Sustainable crop production systems and human nutrition. Front Sustain Food Syst 3:72

Rodger JG, Bennett JM, Razanajatovo M, Knight TM, van Kleunen M, Ashman TL, Ellis AG (2021) Widespread vulnerability of flowering plant seed production to pollinator declines. Sci Adv 7(42):eabd3524

Roubik DW (ed) Pollination of cultivated plants: a compendium for practitioners. Food and Agriculture Organization of the United Nations

Roy DB, Sparks TH (2000) Phenology of British butterflies and climate change. Glob Change Biol 6:407–416

Ruedenauer FA, Spaethe J, Van der Kooi CJ, Leonhardt SD (2019) Pollinator or pedigree: which factors determine the evolution of pollen nutrients? Oecologia 191(2):349–358

Rush E, Obolonkin V (2020) Food exports and imports of New Zealand in relation to the food-based dietary guidelines. Eur J Clin Nutr 74(2):307–313

Samways MJ, Barton PS, Birkhofer K, Chichorro F, Deacon C, Fartmann T, Cardoso P (2020) Solutions for humanity on how to conserve insects. Biol Cons 242:108427

Saraf KK, Vijaykumar K (2021) Effect of climate change on the population of butterfly families-species richness, abundance and species composition across the different seasons of the year in Kalaburagi, Karnataka, India. World News of Natural Sci 34

Schatz B, Maxime D, Mickael H, Benoît G, Fabrice A, Colette S, Denis M (2021) Pollinator conservation in the context of global changes with a focus on France and Belgium. Acta Oecologica 112:103765

Senapathi D, Biesmeijer JC, Breeze TD, Kleijn D, Potts SG, Carvalheiro LG (2015) Pollinator conservation—the difference between managing for pollination services and preserving pollinator diversity. Current Opinion in Insect Sci 12:93–101

Settele J, Bishop J, Potts SG (2016) Climate change impacts on pollination. Nature Plants 2(7):1–3

Sgolastra F, Arnan X, Cabbri R, Isani G, Medrzycki P, Teper D, Bosch J (2018) Combined exposure to Sublethal concentrations of an insecticide and a fungicide affect feeding, ovary development and longevity in a solitary bee. Proc R Soc B 285(1885):20180887

Shackleton RT, Shackleton CM, Kull CA (2019) The role of invasive alien species in shaping local livelihoods and human well-being: a review. J Environ Manage 229:145–157

Shi X, Xiao H, Luo S, Hodgson JA, Bianchi FJ, He H, Zou Y (2021) Can landscape level semi-natural habitat compensate for pollinator biodiversity loss due to farmland consolidation? Agriculture. Ecosyst Environ 319:107519

Shivanna KR, Tandon R, Koul M (2020) 'Global pollinator crisis' and its impact on crop productivity and sustenance of plant diversity. In: Reproductive ecology of flowering plants: patterns and processes. Springer, Singapore, pp 395–413

Smith MR, Singh GM, Mozaffarian D, Myers SS (2015) Effects of decreases of animal pollinators on human nutrition and global health: a modelling analysis. The Lancet 386(10007):1964–1972
Stern N (2016) Current climate models are grossly misleading: Nicholas Stern calls on scientists, engineers and economists to help policymakers by better modelling the immense risks to future generations, and the potential for action. Nature 530:407–410
Stout JC, Dicks LV (2022) From science to society: implementing effective strategies to improve wild pollinator health. Philos Trans R Soc B 377(1853):20210165
Stout JC, Nombre I, de Bruijn B, Delaney A, Doke DA, Gyimah T, Kamano F et al. (2018) Insect pollination improves yield of shea (Vitellaria Paradoxa Subsp. Paradoxa) on the agroforestry parklands of West Africa. J Pollinat Ecol 22
Takkis K, Tscheulin T, Petanidou T (2018) Differential effects of climate warming on the nectar secretion of early-and late-flowering Mediterranean plants. Front Plant Sci 9:874
Timberlake TP, Vaughan IP, Memmott J (2019) Phenology of farmland floral resources reveals seasonal gaps in nectar availability for bumblebees. J Appl Ecol 56(7):1585–1596
Toivonen M, Huusela E, Hyvönen T, Marjamäki P, Järvinen A, Kuussaari M (2022) Effects of crop type and production method on arable biodiversity in boreal farmland. Agr Ecosyst Environ 337:108061
Trinkl M, Kaluza BF, Wallace H, Heard TA, Keller A, Leonhardt SD (2020) Floral species richness correlates with changes in the nutritional quality of larval diets in a stingless bee. Insects 11(2):125
Upadhyay RK (2014) Flower visiting, nectar feeding and oviposition behavioral responses in lepidopteran, hymenopteran and dipteran pollen collecting insect vectors. Int J Chem Biochem Sci 5:26–37
van der Sluijs JP, Vaage NS (2016) Pollinators and global food security: the need for holistic global stewardship. Food Ethics 1(1):75–91
Van Zandt PA, Johnson DD, Hartley C, LeCroy KA, Shew HW, Davis BT, Lehnert MS (2020) Which moths might be pollinators? Approaches in the search for the flower-visiting needles in the Lepidopteran haystack. Ecol Entomol 45(1):13–25
Varikou K, Kasiotis KM, Bempelou E, Manea-Karga E, Anagnostopoulos C, Charalampous A, Machera K (2020) A pesticide residues insight on honeybees, bumblebees and olive oil after pesticidal applications against the olive fruit fly Bactrocera oleae (Diptera: Tephritidae). Insects 11(12):855
Vasiliev D, Greenwood S (2020) Pollinator biodiversity and crop pollination in temperate ecosystems, implications for national pollinator conservation strategies: mini review. Sci Total Environ 744:140880
Vasiliev D, Greenwood S (2021) The role of climate change in pollinator decline across the Northern hemisphere is underestimated. Sci Total Environ 775:145788
Vaudo AD, Tooker JF, Patch HM, Biddinger DJ, Coccia M, Crone MK, Grozinger CM (2020) Pollen protein: lipid macronutrient ratios may guide broad patterns of bee species floral preferences. Insects 11(2):132
Vaudo A, Leonard A (2021) Pollen nutrition structures bee-wildflower community interactions in the Eastern Sierra. In: Pollinator nutritional research: from collecting and characterizing floral resource provisions to the inference of ecological and evolutionary consequences (Denver, CO)
Walter J (2020) Dryness, wetness and temporary flooding reduce floral resources of plant communities with adverse consequences for pollinator attraction. J Ecol 108(4):1453–1464
Wardhaugh CW (2015) How many species of arthropods visit flowers? Arthropod-Plant Interact 9:547–565
Willmer PG, Cunnold H, Ballantyne G (2017) Insights from measuring pollen deposition: quantifying the pre-eminence of bees as flower visitors and effective pollinators. Arthropod-Plant Interact 11(3):411–425
Woodcock BA, Edwards M, Redhead J, Meek WR, Nuttall P, Falk S, Pywell RF (2013) Crop flower visitation by honeybees, bumblebees and solitary bees: Behavioural differences and diversity responses to landscape. Agr Ecosyst Environ 171:1–8

Wratten SD, Gillespie M, Decourtye A, Mader E, Desneux N (2012) Pollinator habitat enhancement: benefits to other ecosystem services. Agr Ecosyst Environ 159:112–122

Wray JC, Elle E (2015) Flowering phenology and nesting resources influence pollinator community composition in a fragmented ecosystem. Landscape Ecol 30:261–272

Bezerra ADM, Pacheco Filho AJ, Bomfim IG, Smagghe G, Freitas BM (2019) Agricultural area losses and pollinator mismatch due to climate changes endanger passion fruit production in the Neotropics. Agric Syst 169:49–57

Brittain C, Kremen C, Garber A, Klein AM (2014) Pollination and plant resources change the nutritional quality of almonds for human health. PLoS One 9:e90082

Dew RM, Silva DP, Rehan SM (2019) Range expansion of an already widespread bee under climate change. Glob Ecol Conserv 17:e00584

Garibaldi LA, Steffan-Dewenter I, Winfree R, Aizen MA, Bommarco R, Cunningham SA, Klein AM (2013) Wild pollinators enhance fruit set of crops regardless of honey bee abundance. Science 339:1608–1611

Imbach P, Fung E, Hannah L, Navarro-Racines CE, Roubik DW, Ricketts TH, Harvey CA, Donatti CI, Läderach P, Locatelli B, Roehrdanz PR (2017) Coupling of pollination services and coffee suitability under climate change. Proc Natl Acad Sci USA 114:10438–10442

IPBES (2019) Global assessment report on biodiversity and ecosystem services of the Intergovernmental Science-Policy Platform on Biodiversity and Ecosystem Services. IPBES Secretariat

Kerr JT, Pindar A, Galpern P, Packer L, Potts SG, Roberts SM, Rasmont P, Schweiger O, Colla SR, Richardson LL, Wagner DL (2015) Climate change impacts on bumblebees converge across continents. Science 349:177–180

Kleijn D, Winfree R, Bartomeus I, Carvalheiro LG, Henry M, Isaacs R, Potts SG (2015) Delivery of crop pollination services is an insufficient argument for wild pollinator conservation. Nat Commun 6:7414

Leach ME, Drummond F (2018) A review of native wild bee nutritional health. Int J Ecol 2018:9607246

Lowore J (2021) Forest Beekeeping in Zambia: Analysing the Nexus of Sustainable Forest Management and Commercial Honey Trade. Univ Huddersfield PhD Thesis

MacInnis G, Forrest JR (2019) Pollination by wild bees yields larger strawberries than pollination by honey bees. J Appl Ecol 56:824–832

Mallinger RE, Gratton C (2015) Species richness of wild bees, but not the use of managed honeybees, increases fruit set of a pollinator-dependent crop. J Appl Ecol 52:323–330

Matter SF, Doyle A, Illerbrun K, Wheeler J, Roland J (2011) An assessment of direct and indirect effects of climate change for populations of the Rocky Mountain Apollo butterfly (*Parnassius smintheus* Doubleday). Insect Sci 18:385–392

McKerchar M, Potts SG, Fountain MT, Garratt MP, Westbury DB (2020) The potential for wildflower interventions to enhance natural enemies and pollinators in commercial apple orchards is limited by other management practices. Agr Ecosyst Environ 301:107034

Ollerton J, Winfree R, Tarrant S (2011) How many flowering plants are pollinated by animals? Oikos 120:321–326

Pervez M, Manzoor F (2023) Honey bee losses and pesticides threat: an Asian perspective. J Apic Res 62:64–75

Polce C, Garratt MP, Termansen M, Ramirez-Villegas J, Challinor AJ, Lappage MG, Biesmeijer JC (2014) Climatedriven spatial mismatches between British orchards and their pollinators: increased risks of pollination deficits. Glob Change Biol 20:2815–2828

Ruedenauer FA, Raubenheimer D, Kessner-Beierlein D, Grund-Mueller N, Noack L, Spaethe J, Leonhardt SD (2020) Best be (e) on low fat: linking nutrient perception, regulation and fitness. Ecol Lett 23:545–554

Mudassar Ali is a pollinator ecologist and Assistant Professor at MNS University of Agriculture, Multan, Pakistan. He has authored over 55 peer-reviewed publications focusing on pollinator ecology, insect pollination, and agroecosystem sustainability. His research highlights the ecological role of native insect pollinators, particularly solitary bees, and their contributions to crop production. Dr. Ali's work on the impact of landscape complexity on bee nesting behavior and post-harvest crop characteristics has enhanced the understanding of pollinator conservation.

Fawad Zafar Ahmad Khan is an insect ecologist and Assistant Professor of Entomology at MNS University of Agriculture, Multan, Pakistan. He has authored 41 peer-reviewed publications on insect behavior, ecology, and biological control. He earned his PhD in Entomology from the University of Georgia, USA, where he worked on understanding the role of arthropod predators in managing fall armyworm (*Spodoptera frugiperda*) populations in turfgrass ecosystems. Dr. Khan has worked on projects related to integrated pest management, pollination biology, and the conservation of beneficial arthropods. His research interests include plant-insect interactions and their application in pest management programs. His notable works include studies on native insect pollinators, the impact of landscape complexity on solitary bee nesting behavior, and predator-prey dynamics under water-limited conditions. Dr. Khan teaches courses including Introductory Entomology, Applied Entomology, Insect Morphology, and Insect Physiology.

Muhammad Awais Ahmad is a master's student at the Institute of Plant Protection, MNS University of Agriculture, Multan.

Chapter 11
Integrated Pest and Pollinator Management Under the Changing Climate

Fawad Zafar Ahmad Khan, Mudssar Ali, and Danyal Haider Khan

Abstract The concept of Integrated Pest and Pollinator Management (IPPM) focuses on managing pests while ensuring the safety and protection of insect pollinators within agricultural landscapes. IPPM represents a futuristic version of integrated pest management, considering the importance of pollinators while making decisions to manage insect pests in crops. This chapter presents a comprehensive overview of the significance of pollinators and the adaptation of integrated pest management into the IPPM approach, aiming to enhance the biodiversity of pollinators. The steps of integrated pest management include landscape modifications, crop field management, pest sampling, and thresholds, as well as the management of natural enemies. The chapter also provides examples of IPPM implementation in different crops and highlights the pivotal role of pollinators in shaping pest management strategies. IPPM represents the future of pest management, combining strategies to protect both crops and pollinator diversity.

Keywords Biodiversity management · IPPM · Pesticides · Pollinators · Climate change

11.1 Introduction

Agriculture is pivotal in securing food security for the growing global human population (Pradhan et al. 2015). To support the ever-rising population, it is essential to increase agricultural yield. However, due to several biotic and abiotic factors, achieving increased yield has become extremely challenging. The abiotic factors

F. Z. A. Khan
Institute of Plant Protection and Department of Outreach and Continuing Education, Muhammad Nawaz Shareef University of Agriculture Multan, Multan 60000, Pakistan
e-mail: fawad.zafar@mnsuam.edu.pk

M. Ali (✉) · D. H. Khan
Institute of Plant Protection, MNS University of Agriculture, Multan 60000, Pakistan
e-mail: mudssar.ali@mnsuam.edu.pk

M. Behnassi et al. (eds.), *Food Systems and Biodiversity in the Context of Environmental and Climate Risks*, https://doi.org/10.1007/978-3-031-89167-0_11

affecting agriculture yield are declining arable land, water scarcity, fertilizer use, and abrupt climate change. Pest infestations and the scarcity of pollinators are the primary biotic challenges to agricultural production. The first biotic factor limiting crop yield is the attack from arthropods. Arthropod pests cause an estimated US$470 billion in damage annually (Sharma et al. 2017). While the second factor that is limiting agriculture yield is declining pollination services by invertebrates, specifically bees (Cock et al. 2012; Reilly et al. 2020). The injudicious application of broad-spectrum pesticides harms the environment, killing beneficial non-target organisms, fostering pest resistance, and contaminating ecosystems with toxic residues. Pollination, an important ecological service, supports environmental sustainability and agricultural commerce with an estimated value ranging from 255 to 537 billion dollars annually. Pollination plays a crucial role in facilitating the production of around 35% of the global edible crop yield, as well as contributing to the reproductive success of around 75% of angiosperm species (Semeraro et al. 2018).

11.2 History of Pest Control

Agroecosystems have become less suited to many beneficial insects and are more friendly to pest species due to agricultural intensification, strict monoculture, and less frequent crop rotations. The struggle to manage agricultural pests has been continuing for thousands of years (Kumar and Kumar 2019). Crop pests, including arthropods, weeds, microbes, rodents, and nematodes, decrease crop quality and yield (Karar et al. 2021). Multiple pest management methods, including pest removal, selective plant breeding, and pesticide applications, have been used to manage these pests (Dent and Blinks 2020).

Since World War II, the use of chemical pesticides has increased in the agroecosystems, turning them from small-scale, reasonably diversified natural communities into strict monocultures (Katayama et al. 2015). Both inorganic substances, such as elemental sulfur and different arsenicals, as well as plant-derived chemicals, were used as early insecticides (Mateos Fernández et al. 2022). Beginning in the first decade of the 1940s, the developments in synthetic organic chemistry opened a new era of pest management. Since then, there has been a tremendous increase in the use of pesticides to manage pests, and consequently, crop yields have increased significantly. However, according to another school of thought, pesticides have not been highly successful in managing the crop losses caused by pests (Okere 2006).

During the early development of synthetic organic pesticides, a lack of environmental consciousness and government regulation, together with relatively affordable insecticides such as DDT, led farmers to use high amounts of insecticides without concern for safeguarding humans or non-target organisms from exposure (Atreya 2008). Even when the targeted pests weren't around, pesticides were often administered on a calendar-based schedule to various crops. Concern over potential pesticide-related dangers to human health and the environment started to grow (Jallow et al. 2017). Many countries have implemented strict legislation about the manufacturing, distribution, and utilization of pesticides, intending to protect human health and the environment. However, the enforcement of these regulations stays inconsistent in multiple regions (Pelaez et al. 2013).

In addition to inflicting a threat to humans and other non-target species, extensive use of pesticides also gives a financial burden to the growers (Mužinić and Želježić 2018; Pimentel and Burgress 2014). Moreover, repeated pesticide treatments may reduce the efficacy of a pesticide as the target pest population develops resistance against certain chemical groups (Huang et al. 2020). For example, the western corn rootworm, *Diabrotica virgifera virgifera* (LeConte), developed resistance against organophosphate and carbamates in maize-growing regions of the USA due to subsequent application of these insecticides (Meinke et al. 2021).

The adverse effects of pesticides include loss of important floral resources (due to spraying of herbicides). They also have negative impacts on the reproduction, navigation, and memory of pollinators, especially bees. Pesticide exposure can cause negative impacts of other stresses on pollinator populations, such as habitat loss, exposure to infections, and illness. There is widespread pesticide pollution. In agricultural landscapes, more than 90% of pollen samples are from bee hives and more than 90% of stream samples include chemical contamination. Honey bees and other pollinators, which are crucial to all fruit and most vegetable crops, are at risk from chemical exposure, reducing agricultural productivity (Ara and Haque 2021). Moreover, certain insecticide groups may not work under certain conditions. For example, under certain weather conditions or the complex feeding behavior of some insect pests. In such cases, pesticide use often exceeds normal levels, and consequently, the risk to the non-target species also rises.

11.3 Roles of Integrated Pest Management in Pest Control

An effective method for reducing pest populations while safeguarding the environment includes Integrated Pest Management (IPM). IPM programs use extensive information about the pest life cycle and environment. This knowledge, combined with conventional pest management strategies, helps manage pest damage cost-effectively while minimizing harm to non-target species and the environment. IPM is a pest management system and is not solely a pest control method.

The first phase of IPM involves determining the action threshold, i.e. the point at which pest numbers or environmental variables need pest control measures. Because not all weeds, insects, and other living things need management, the second phase is pest monitoring and identification. Many insect species are harmless and even useful to humans. IPM systems monitor and identify pests to minimize them via action thresholds. The third step emphasizes that preventing pest attacks is a priority. Rotating crops, selecting pest-resistant cultivars, and planting pest-free rootstock are examples of cultural methods to prevent pest infestation. The final step is insect pest control when IPM determines that pest control is necessary and preventative tactics are no longer effective. This includes all the possible options that limit the pest multiplication and spread.

Many agricultural producers use integrated pest management (IPM) to balance pest control with avoiding or minimizing the negative impacts of intensive pesticide use. In the 1970s, in response to rising understanding of pesticide side effects, the term "Integrated Pest Management" was first used in agriculture (Ehler 2006). This strategy puts an emphasis on combining cultural techniques and pest biology to manage insect pests in crops. IPM is a method for managing pests that is both environmentally and economically safe (Dara 2019). It combines a variety of instruments and techniques, including, insecticides, natural enemies, cultural practices, and host-plant resistance. The method aims to conserve and improve natural pest control mechanisms and is often knowledge-intensive rather than product-intensive. Although applying pesticides is a common part of the IPM program, the pesticides used in these programs are often narrow-spectrum and less dangerous for people and other non-target organisms than other pesticides (Gullino and Tavella 2020). IPM strategies frequently modify pesticide spraying procedures to lower exposure levels for people and the environment. This approach is also cost-effective for growers.

11.4 Biodiversity of Pollinators in Ecosystems

Many agricultural and nonagricultural crops depend on pollinators for successful pollination and subsequent commercial viability. Angiosperms require either biotic or abiotic agents for pollination because of the asymmetry between their anthers and stigmas. Pollination is a two-way interaction in which both participants receive benefits: the pollinating agent receives pollen, nectar, or both as a nutritive source, while the reproductive success of a plant is ensured. Depending on their taxonomic categorization and productivity, pollinators can be divided into various categories and subgroups, primarily mammals and arthropods (Traveset and Richardson 2014; Geeta and Berry 2020).

Birds are essential pollinators in the tropics and southern temperate zones because ornithophilous flowers are present in approximately one-third of 300 families of flowering plants (Dellinger et al. 2022). In contrast, over 2000 bird species from at least 50 families use flowers daily to consume nectar and pollen. Similarly, bats also pollinate more than 500 plant species, including mangoes, bananas, durians, guavas, and agaves (Ramírez-Fráncel et al. 2022). Some bats have evolved specialized mouthparts, such as long tongues and tube-shaped lips, to access nectar in the depths of the large bell-shaped flowers that bloom at night. Bats pollinate these flowers to produce flowers and fruit for feeding (Dutta et al. 2019).

Arthropods such as some spiders and many insects are primary pollinating agents of flowering plants. Insects belonging to the orders Hymenoptera, Diptera, Coleoptera, and Lepidoptera are the most effective pollinators (Mahfouz et al. 2012). Bees are the most numerous and productive floral visitors. Honey bee colonies are managed for their commercial use globally (Genersch et al. 2010), and colonies of bumblebees are also maintained in greenhouses because of their effect on boosting floral pollination and their non-stinging nature (Zhang et al. 2022). Out of the 20,000 bee species present worldwide, more than 90% are solitary (Sgolastra et al. 2019). These bees are skilled in foraging, and many of them are selective in the pollen they choose, visiting only particular flowers. The second category within Hymenoptera is wasps, whose adults feed on pollen and contribute to pollination (Summer et al. 2018). Although ants, another group within Hymenoptera, frequently visit flowers, they are not effective at transporting pollen (Eckols et al. 2021).

Several flies from 71 families of the order Diptera are flower visitors and constitute the second most abundant group of pollinators. In the order Diptera, the family Syrphidae has the most efficient pollinators that prefer flowers with bright colors and pungent odors. In many regions with marginally high temperatures, syrphids outperform bees as pollinating agents (Dunn et al. 2020). A vast diversity of lepidopteran butterflies and moths are floral visitors, but not all of them are pollinators. Butterflies, which are diurnal, prefer red-colored flowers. Moths, on the other hand, are nocturnal and visit flowers at night (Mertens et al. 2021).

11.5 Modification of IPM into IPPM for the Benefit of Pollinators

IPM indirectly aids pollinators by minimizing the use of harmful pesticides, but it does not fully aim towards a conservation approach. Merging pollinator management and IPM to form integrated pest and pollinator management (IPPM) can boost agricultural production by increasing pollination along with pest management (Egan et al. 2020). Although IPPM first concentrated on modifying chemical treatment to preserve pollinators while ignoring more extensive options for co-managing pest management and pollination, it can be compared to the original broad IPM strategy. By highlighting proactive landscape and agricultural field management as a fundamental IPPM foundation that could be complemented with reactive human-based inputs like pesticides, targets for pollinator conservation could be achieved (Lundin et al. 2021).

With arthropods serving as the primary group of organisms in charge of each function, an expanded IPPM paradigm comprises the management of agroecosystem functions driven by pests, natural enemies, and pollinators (Martin et al. 2019a,b; Cutler et al. 2022). Despite being typically treated separately, co-managing crop pollination and pest management have several benefits (Jung 2021). Co-management has pros and cons because of the overlapping life cycle and ecological distribution of arthropod pests, natural enemies, and pollinators. The three functional groups share a lot of environmental and human factors; one of these, pests, provides negative services (Kremen and Chaplin-Kramer 2007). Local and landscape habitat complexity may benefit natural enemies, like pollinators, which use numerous habitats and resources throughout their life cycles (Vialatte et al. 2017). Because the same species can

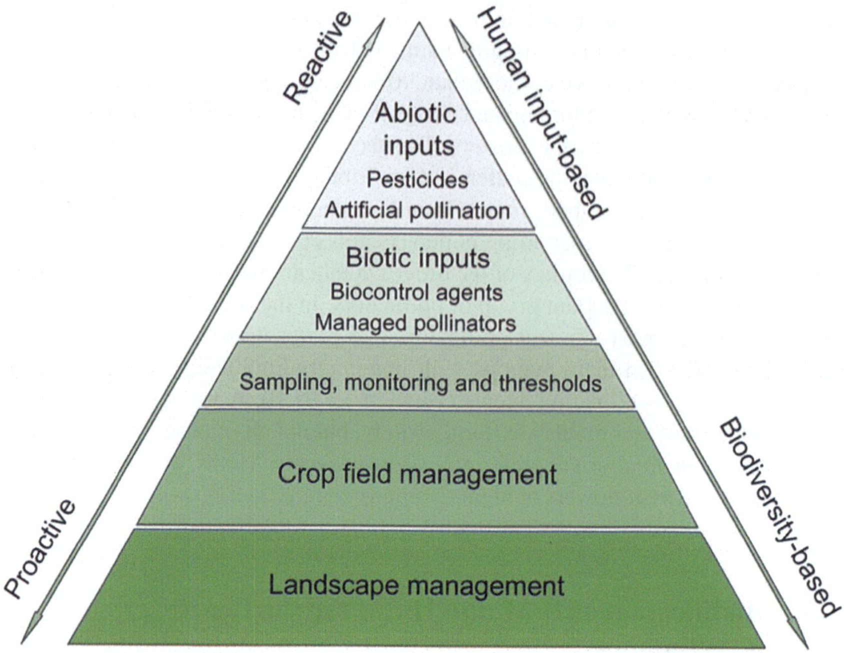

Fig. 11.1 Integrated pest and pollinator management framework. *Source* Lundin et al. (2021)

be involved in herbivory, predation, and pollination, co-management is especially crucial. For instance, larval hoverflies (family Syrphidae) are predators, but adult hoverflies serve as pollinators (Wotton et al. 2019). Some bee species, like the stingless bee, *Trigona spinipes* (Apidae), can act as beneficial pollinators or damaging pests depending on the crop type (Merle et al. 2021). IPPM faces challenges that are rather obvious, especially when it comes to the application of pesticides that can kill valuable non-target arthropods. IPPM methods employ strategies that balance the benefits of pest management with pollinator conservation (Fig. 11.1).

11.6 Steps of Integrated Pest and Pollinator Management

11.6.1 Landscape Modification

Management of pests at the landscape level is one of the initial steps of a wider IPPM strategy because this step directly suppresses damage caused by pests while fostering diverse groups of natural enemies that facilitate biological control of pests (Martin et al. 2019a, b). Activities performed at the landscape level can also help pollinators and pollination services. The core of the IPPM strategy is to increase the variety of wild and cultivated plants, protect or restore natural or semi-natural habitat, and maintain resources for pollinators and natural enemies. In certain circumstances, the presence of abundant semi-natural habitat might help with pest management by augmenting natural enemies that live there (Holland et al. 2016).

There are many studies conducted to investigate and elaborate on the consequences of landscape modification on pollinator diversity. Findings of some studies are discussed here. Pond management and restoration have been demonstrated to greatly increase floral supplies for pollinating insects by establishing a variety of unique microhabitats both inside and outside of ponds (Sayer et al. 2013). Therefore, ponds may provide adult pollinators that are foraging with both edge and center-of-field food supplies (Walton et al. 2021). Likewise, in agriculture, it is necessary to maintain mosaics where crops and untilled lands coexist. The vegetation on these untilled lands provides refuge and food resources to bees (Lentini et al. 2012). Planting a variety of native flowering plants on conservation strips in both urban and agricultural settings has proven beneficial for bee diversity (Sutter et al. 2017).

The impact of a semi-natural environment on the reaction of natural enemies is very different, and as pests can also benefit from it (semi-natural habitat), the results of pest control efforts vary (Haan et al. 2020). Likewise, pollinators often benefit from semi-natural environments on agricultural fields (Raderschall et al. 2021). Since there are more functionally significant species involved in determining pest management results than pollination outcomes, pest control may exhibit a more idiosyncratic response as a result of higher tri-trophic complexity and taxonomic variety.

11.6.1.1 Role of Semi-Natural Habitats in Pollinator Management

A semi-natural habitat is a biome that remains unaltered by human activity, preserving its biodiversity and ecological processes. Increasing the quality and quantity of semi-natural habitats is vital to maintain pollinators because they offer food (flowers, nectar, and pollen), shelter, nesting, and hibernating sites (Cole et al. 2017). Road verges are one such habitat, and their ubiquity means they have a particularly high potential to sustain pollinators (Monasterolo et al. 2020). These are the patches of ground that are located between the road's surface and its edge, such as a fence or hedgerow. Roadside vegetation is essentially a remnant of the environments that formerly surrounded them but were lost due to changes in land use, such as increased agricultural production. As a result, road verges can represent a variety of diverse habitat types, such as grassland, forest, hedgerows, and water bodies, such as banks, ditches, and reed beds. According to studies, roadside verges maintain pollinator populations that are comparable to those found in nearby environments in terms of both quantity and species (Phillips et al. 2020).

11.6.1.2 Landscape Diversification

Both pest and pollinator management are influenced by crop type, diversification, and landscape design. For instance, the number of pests with a host-specific niche can be efficiently reduced by interrupting the continuous planting of their host crops (Rusch et al. 2013), while an increase in crop diversity encourages the use of naturally existing predators for biological pest management (Crowder et al. 2010). While mass-flowering crops (but not highly managed cereals) enhance pollinator populations, the impacts of crop diversification on pollinators may be more complex and dependent on crop identities and their management (Riedinger et al. 2014). Thus, to achieve IPPM benefits, a strategic selection of crop arrangements is necessary, beyond merely increasing crop diversity. Pollination and pest control are benefited by landscape arrangements that include pockets of semi-natural habitat and tiny, asymmetrical fields (Aguilera et al. 2020).

A greater knowledge of the ecology of pests, predators, and pollinators is necessary to advance IPPM at the landscape scale. While decoupling resources for pests, the aim at the landscape scale should be to provide continuity of resources for pollinators and predators. By maintaining resource continuity in time and place at scales appropriate for beneficial organisms but not for species that are harmful to crops (Schneider et al. 2015), co-management possibilities may be created by taking use of variances in movement among organism groups (Schellhorn et al. 2015). Utilizing the different host-plant and environment preferences of pests and pollinators provides another possibility; for example, pests' access to nutrients can be disrupted by replacing host crop cover with alternative non-host plants, while the replaced crops still support pollinators and natural enemies (Parker et al. 2013). This is feasible on the landscape scale for a variety of crops that are largely susceptible to specialist pests but benefit from generalist pollinators and natural enemies.

11.6.2 Crop Management

To reduce pest outbreaks, several different field-based strategies known as 'cultural control' techniques have been developed. This includes increasing the variety of plants within a field, planting pest-resistant cultivars, and modifying soil tillage practices and agronomic inputs, such as fertilizer application and irrigation (Parry et al. 2020; Alyokhin et al. 2020). Extensive research on the effects of these techniques on pollinators is necessary before effectively adopting IPPM.

The addition of non-cropped vegetation, such as flower strips and hedgerows, intercropping (Alarcón-Segura et al. 2022), cover cropping (Bowers et al. 2020), and reducing weed control (Bowers et al. 2020), are ways to increase plant diversity in the cropped field or along its borders to promote pest control. Reduced weed control, along with flower strips and hedgerows, can also encourage pollinators and crop pollination (Albrecht et al. 2020). Intercropping and cover cropping are promising when the cropping system includes a variety of flowering herbs and shrubs, although their impacts on pollinators are still little understood. For both pest management and pollination, cropping systems with increased plant diversity should be created separately, by carefully selecting plants and considering how they affect pests, predators, and pollinators (Qian et al. 2021); it may therefore be possible to use diversification tactics that assist predators and pollinators while minimizing the benefits of crop pests and weed-related issues. To fully realize the promise of plant diversification at the field level for use as an IPPM tool, it is necessary to advance the overall understanding of functionally significant arthropods' resource usage.

11.6.2.1 Cover Crops

Research and conservation efforts have concentrated on integrating flowering plants into agricultural landscapes by planting on a farm's unused land and by cultivating flowering crops as cover crops. These measures aim to mitigate food scarcity for pollinators and to provide breeding sites for predators (Qian et al. 2021). Even though cover crops offer a variety of benefits to agroecosystems, their role in supplying floral resources to pollinators has received little attention. A study found that buckwheat cover crop treatments offered the highest summer-long floral cover, surpassing annual native wildflowers, and this better floral cover was associated with higher overall bee visitation rates. The floral preference of bees also varies among species; for example, honey bees and bumble bees were most attracted to mixes containing phacelia, whereas large-bodied solitary bees preferred mixes with cultivated sunflowers and wildflowers (Mallinger et al. 2019).

In winter, due to the lack of other pollen sources, many pollinators, including honey bees and native bees, can benefit from the presence of mass-flowering, melliferous crops (Ellis and Barbercheck 2015). The winter oilseed crops such as rapeseed (*Brassica napus* L.), pennycress (*Thlaspi arvense* L.), and camelina (*Camelina sativa* L.) provide flowers that attract pollinators and are useful as winter cover crops and sources of seeds for commercial crops (Eberle et al. 2015). These crops are green winter cover crops, which provide another important ecosystem service in addition to serving as a vital early feed supply for pollinators (Eberle et al. 2015). To preserve pollinator diversity, floral resources must be accessible throughout the landscape.

Similarly to pollinators, cover crops support predators by providing breeding sites and alternate food resources when prey is scarce (Duyck et al. 2011). A study was conducted to understand the impact of different cover crops that were grown adjacent to cotton on the biological control of cotton pests on a phenological basis. During the study, a gradual increase in predators' populations was observed over the years in the cotton fields that were adjacent to cover crops. This strategy was so effective that after the second year, there was only one pest that breached their economic threshold in experimental cotton fields, which was the heliothine *Heliothis virescens* (Fabricius) that only occurred in control cotton, which was not in direct contact with cover crops of crimson clover or rye cotton. After the second year of the experiment, heliothine predators and aphidophagous lady beetles were observed on cotton and cover crops. The three species of heliothine predators that were more prevalent were *Geocoris punctipes* (Say), *Orius insidiosus* (Say), and *Solentopsis invicta* (Buren). Compared to control cotton fields, the density of *G. punctipes* was greater in cotton fields that had previously been planted with crimson clover. The transmission of *G. punctipes* onto cotton in these crimson clover fields most likely resulted from intercropping cotton in live strips of the cover crop. Lady beetles appeared to transfer from cotton to cover crops. The greater density of red imported fire ants found in all conservation tillage cotton fields compared to control cotton fields was likely caused by conservation of the fire ant habitat during planting. Reduced *heliothine* levels in conservation tillage cotton with these cover crops compared to conventional tillage cotton without cover crops were found as a result of the buildup of predaceous fire ants and *G. punctipes* in these cover crops, according to a reduction in the number of times economic thresholds for heliothines were exceeded in crimson clover and rye compared with control fields (Tillman et al. 2004).

11.6.2.2 Development of Plant Cultivars Suited for IPPM

The use of pest-resistant agricultural cultivars is a crucial part of pest control (Dara 2019). A tactic similar to the use of pest-resistant cultivars for pest management is the selection or breeding of cultivars with pollinator-friendly traits. The simultaneous mapping and assessment of several crop attributes, such as pest resistance, pollinator reliance and attractiveness, and crop production, would present a significant challenge for breeding and cultivar selection from an IPPM perspective. Crop pest management also involves changing agronomic inputs like fertilizer and irrigation. For instance, organic fertilizers can enhance bottom-up control by strengthening crop defense against pests and top-down control by benefiting predators (Latha et al. 2019) while also enhancing pollinator visits to crop flowers (Cardoza et al. 2012). In general, little is known about how pests, predators, and pollinators are affected by water availability. Predators can be disturbed by irrigation, particularly floods, but crop pests and crop damage from pests can be affected by water availability in both favorable and unfavorable ways (Asiimwe et al. 2014). Flooding decreases floral attraction towards pollinators, causing food shortages for them (Walter 2020). Appropriate soil moisture management, avoiding under-, and over-irrigation, can increase nectar production and encourage pollinator visits (Dia et al. 2022).

11.6.2.3 Cultural Practices for Pest and Pollinator Management

The physical management of insect pests, such as exclusion netting and row coverings (cloth mesh to cover plant rows), has gotten more attention recently in addition to cultural control methods. Many of these treatments might involve integrating pollinators when the crop is pollinator-dependent (Mukherjee et al. 2019). With some research suggesting that benefits are larger in simple than in complex or cleared landscapes, the effectiveness of IPPM techniques at the field size (e.g. installation of flower strips) on pollination and pest control relies on the landscape context. When field activities are used over wide regions and scaled up to become practices at the landscape level, field and landscape management actions might occasionally overlap.

11.6.3 Sampling, Monitoring, and Thresholds

In order to choose which control measures and/or method to be used, monitoring and sampling are essential parts of pest management. The economic injury level (EIL) is the lowest pest density at which financial loss results from crop harm exceeds the cost of the management intervention. It is defined as the pest density at which action must be taken to prevent economic losses from exceeding the costs of management interventions, known as the economic threshold (ET) (Zhang and Swinton 2009). In practice, ETs for particular pests and cropping systems often lack availability, are outdated, or are not supported by scientific evidence. This is partly because a significant quantity of data is required for their assessment. Similarly, the establishment of a pollinator economic impact level (PEIL) is necessary, mirroring the EIL, to keep pollinator populations above a certain threshold and prevent economic losses caused by inadequate pollination (Flöhr et al. 2020).

The PEIL is potentially a helpful tool to assess the justification of pollinator management interventions, but just like EILs, the quantification of crop-specific PEILs requires a significant quantity of ecological and economic data. To make the PEIL idea more effective, further in-depth understanding of how to quickly assess using field sample data whether crops are pollen-restricted would be required. For example, while some pest management procedures, such as application of pesticides, can be implemented swiftly, pollinator control measures may face logistical challenges in rapid deployments, such as the dependency on the availability of surplus bee hives for managed bees. The joint economic effect level (jEIL), which incorporates crop production restrictions related to the activities and availability of both pests and pollinators, can also combine EIL and PEIL into a single decision indicator (Flöhr et al. 2020).

11.6.4 Release of Biocontrol Agents

Although biological control methods are more expensive than chemical controls, they offer a safer and more eco-friendly approach. Invertebrates and microorganisms that have been raised and released as biological control agents are frequently employed as biological control agents. Exotic biological control agents have traditionally been introduced with the intention of achieving long-term pest control (i.e., classical biological control), but augmentative releases that enhance existing natural enemy populations in the field (i.e., augmentative biological control) is a strategy that is becoming more and more popular (Ridgway 2013).

Predators, parasitoids, microorganisms, or nematodes are some examples of natural enemies of pests (Hajek and Eilenberg 2018). Predators quickly kill their victims by consuming them. The four primary types of predators are ants, mites, spiders, and certain true bugs. When a parasitoid attacks a pest as an adult, the pest is eventually eliminated by the parasitoid's larva. Parasitoid larvae slowly grow inside or externally while slowly eating a host. Based on the life stage that the larvae develop on, parasitoids are further divided into subgroups. To add biological control agents or to increase mortality from existing natural enemies, there are three methods: classical, conservation, and augmentation (Zhou et al. 2014).

Classical biological management entails locating the native geographic ranges of the natural enemies of alien pests. In the new areas where the pest has been introduced, those natural enemies are subsequently imported and released (Kenis et al. 2017). Conservation biological management is needed to sustain or grow the populations of endemic natural enemies or traditional biological pest control agents. The compatibility of insecticides with natural enemies and providing habitat or food for natural enemies to enhance populations locally are the two key study topics in biological control for conservation. The foundation of biological control for conservation is the established link between a natural enemy's ability to survive and ongoing predation or parasitism throughout the natural enemy's lifespan. Basically, offering food or other hosts draws in, sustains, or lengthens the lives of parasitoids or predators, allowing them to exert greater control over pest populations (Wyckhuys et al. 2013).

The third method of biological regulation is augmentation. Similar to traditional biological control, augmentation also releases native or endemic natural enemies that can be purchased from suppliers of beneficial species. (Ridgway 2013). Most businesses that provide natural enemies will advise their clients on the best product to buy and provide release assistance, or you can get advice from one of the generic biological control books. Either inoculation or inundation can lead to augmentation. The purpose of inoculation of natural enemies is not the instantaneous eradication of the pest population. The release of natural enemies with the intention of quick suppression of pests is known as inundation. Inundation employs predatory insects or mites because they instantly destroy their prey and are essential for quick impacts (Eilenberg et al. 2001). The method of releasing insects in the field varies from species to species. Some species are released by hand from pouches or bags, while others are introduced in the field by stapling cards on the leaves. These cards contain eggs of a biocontrol agent. Likewise, there are many species that are released with the use of drones.

Managed bees are commonly used as biological agents for crop pollination, similar to how biocontrol agents are used to manage pests on a commercial scale. The European honey bee (*A. mellifera* L.) pollinates most agricultural crops globally, although a number of bumblebee (*Bombus* spp.) and solitary bee species are also used in different regions. The value of managed honey bees to production varies substantially across different crops and geographical areas. Many studies have been conducted to determine the viability of utilizing species besides honey bees for agricultural pollination and to establish the ideal stocking levels of controlled pollinators (Artz and Nault 2011).

The management of pests and pollinators, which both depend on biological regulations, are generally independent of one another. The entomo-vectoring strategy, which uses insects (usually bees) as carriers of microbial control agents against diseases and insect pests, is a significant exception. Introducing microbial control agents against plant diseases and insect pests into crop blossoms through the use of pollinators as vectors is a novel, ecologically friendly control technique (Mommaerts and Smagghe 2011). In this scenario, pollinators serve a dual role, not only for pollination but also as vectors for disease and pest control. The use of biotic inputs for pest and pollinator control can work in tandem with efforts to develop fields and landscapes with healthy populations of wild beneficial arthropods. However, success depends on the harmonious interactions of the right vector, formulation, control agent, and dispenser. Additionally, it must be safe for the environment and human health as honey bees are the primary vector whose honey is consumed by people (Menzler-Hokkanen and Hokkanen 2017).

The limited variety of vector species is due to their widespread commercial availability. Due to the availability of bumble bees and honey bees throughout the year, pollination of crops may be timed to coincide with their flowering season. The adoption of the entomovector technique arose from the ineffectiveness of traditional biopesticide spraying methods. Because the agents may be delivered directly to the target area, in this case, the flowers, entomovector technology is thought to function better than spraying a biocontrol agent (Smagghe et al. 2020).

Landscapes containing a variety of floral resources help control pollinators, much as complex, resource-rich landscapes favor released biological control agents (Aguilera et al. 2020). However, introducing biotic IPPM agents into landscapes with limited resources may result in increased competition between wild and managed pollinators or antagonistic interactions between naturally existing and imported natural enemies.

11.7 IPPM in Different Crops Around the World

11.7.1 IPPM in Tropical Fruits

Tropical fruits are a broad category of fruits native to tropical climates. They are represented by members of a variety of families, such as Anacardiaceae (e.g. hog plum, imbu, and mango), Annonaceae (e.g. soursop, custard apple, and sugar apple), Bromeliaceae (e.g. pineapple), Passifloraceae (e.g. passion fruit), and Sapindaceae (e.g. rambutan, lychee, and longan). Tropical fruits are traditionally consumed close to the areas where they are farmed, either in a subsistence-style agricultural framework or foraged from the wild. In mango orchards, as farmland expands, fields may become increasingly isolated from pollinators, which lowers pollinator-dependent agricultural output. Planting small patches of native flowers within large farms, paired with the preservation of remnant natural habitat, can boost fruit productivity by attracting pollinators (Carvalheiro et al. 2012) (Table 11.1). Pesticides negatively impact mango pollinators. Consequently, using natural enemy weaver ants as a replacement is known to control thrips effectively in many crops, especially mangoes. In a specific study of mango orchards, it was observed that trees with a large population of weaver ants produced most of the flower panicles because they suppress thrip populations that cause damage to tree shoots (Peng et al. 2004).

Imbu, *Spondias tuberosa* (Arruda), whose fruit is the edible part, is highly dependent on pollinators for its production. *A. mellifera* is a major contributor to imbu pollination (Almeida et al. 2011); thus, the introduction of *A. mellifera* hives to imbu plants could enhance fruit production. Growing flowering strips near the plantations of avocado (*Persea americana* Mill.) and rambutans (*Nephelium lappaceum* L.) could result in increased visitation from syrphid flies, which are major pollinators and biological control agents (Mala and Reddy 2020).

Table 11.1 Integrated pest and pollinator management strategies for various crops

Crop	IPPM strategy	Reference
Mango (*Mangifera indica* L.)	Conservation strips of flowering plants	Carvalheiro et al. (2012)
Mango	Biocontrol agent	Peng et al. (2004)
Imbu (*Spondias tuberosa* Arruda)	Supplementary pollination by managed bees	Almeida et al. (2011)
Avocado (*Persea americana* Mill.)	Flowering conservation strips	Mala and Reddy (2020)
Rambutans (*Nephelium lappaceum* L.)	Flowering conservation strips	Mala and Reddy (2020)
Apricot (*Prunus armeniaca* Scop)	Pollinator-friendly cultivars	Audergon et al. (1997)
Peaches (*Prunus persica* L.)	Supplementary pollination by managed bees	Calzoni and Speranza (1995)
Plums (*Prunus simonii* Carrière)	Year-round flowering plants near plantations	Pratap et al. (2000)
Strawberries (*Fragaria* x *ananassa*)	Entomo-vectoring	Kaiser et al. (2016)
Raspberries (*Rubus idaeus*)	Entomo-vectoring	Kaiser et al. (2016)
Blueberries (*Vaccinium uliginosum* L.)	Installation of *Osmia* brood	Fortuin and Gandhi (2020)
Cereal crops	Strips of mass-flowering plants	Haenke et al. (2014)
Oil seed crops	Application of bio-pesticides	Briar et al. (2018)

11.7.2 IPPM in Stone Fruits

Stone fruit trees are deciduous species native to temperate regions. According to the pomological classification, the categories of stone fruits include apricot, plum, sweet and sour cherry, peach and nectarine, and cornelian cherry. All stone fruit varieties, except for the cornelian cherry, are members of the Rosaceae family of *Prunus* L. The *Cornus* genus of the Cornaceae family includes the Cornelian cherry (Milošević and Milošević 2020).

Plantation of such cultivars of apricot that are pollinator-friendly will not only provide more nectar sources to pollinating agents but also increase fruit yield (Audergon et al. 1997). Colonies of honey bees and bumblebees have a positive impact on the fruit yield of peaches and plums (Calzoni and Speranza 1995). The Asiatic honey bee, *Apis cerena* F., also plays an important role in pollination, but its population faces a shortage of floral resources before and after flowering periods. Providing it with year-round flowering plants will indirectly improve pollination of peach and plum trees (Pratap et al. 2000). Several entomopathogens are used for the suppression of pests such as aphids and plum curculio in stone fruit species that not only protect these fruits from harmful pesticide residues but also protect pollinators from toxins (Shapiro-ilan et al. 2016).

11.7.3 IPPM in Berry Fruits

Fruit berries, such as currants (*Ribes* spp.), blueberries (*Vaccinium* spp.), strawberries (*Fragaria* × *ananassa*), and raspberries (*Rubus idaeus* L.), are among the most well-liked and nutrient-dense fruits worldwide (Beattie et al. 2005). Entomo-vectoring is the best strategy to provide dual services of pollination and biological control in strawberries and raspberries, in which dispensers are installed on the colonies of honey bees and bumble bees that dispense entomopathogen powder onto bees (Kaiser et al. 2016). These bees visit flowers by transferring entomopathogens to and carrying pollen from them, ultimately pollinating and offering biocontrol simultaneously (Smagghe 2020). Different managed solitary bees, especially those of the genus *Osmia*, are deployed in the fields of raspberries and blueberries because of their excellent pollination (Fortuin and Gandhi 2020). Bee hotels are also sometimes installed in berry fields to provide nesting habitats for bees (McCallum et al. 2018).

11.7.4 IPPM in Cereals and Oil Seed Crops

Cereal crops comprise nine species within the Gramineae family, which are eaten as staple crops worldwide. These crops include barley (*Hordeum vulgare* L.), corn (*Zea mays* L.), wheat (*Triticum aestivum* L.), oat (*Avena sativa* L.), rice (*Oryza sativa*

L.), rye (*Secale cereale* L.), millet (*Pennisetum glaucum* L.), sorghum (*Sorghum bicolor* L. Moench), and triticale (× *Triticosecale* Wittmack), a man-made cereal from wheat and rye hybridization (Jauhar 2006). Nearly all of these cereal crops are self-pollinated and do not require pollinators, but they are heavily attacked by hemipteran pests, including aphids and whiteflies. The larvae of both syrphid flies and green lacewings are the best natural enemies for the pest complex of cereal crops, but their adults feed on pollen and nectar that is nonexistent in cereal crops (Bugg et al. 2008). Incorporating mass-flowering cover crops into cereal farming systems provides food resources and breeding sites for these natural enemies (Haenke et al. 2014).

Oil is a component of our diet and is an essential source of fatty acids (Calder et al. 2020). Oilseed crops are primarily used to produce oil. The two main commercial oilseed crops are *B. napus* and *B. campestris* (turnip). Other oilseed crops that produce high-quality edible oil include sesame, groundnut, sunflower, soybean, and safflower (Rashmi 2018). Castors and linseeds are non-edible oil-producing species. Most of these oilseed crops are cross-pollinated and attract a large diversity of pollinators. Providing supplementary pollination by managed bees and applying bio-pesticides are standard IPPM strategies for oilseed crops.

11.8 Conclusion

Owing to sudden and unpredictable global climate changes caused by rapid industrialization and deforestation, pest control is becoming increasingly difficult and expensive. The abrupt change in climate is causing shifts in the pest populations. Consequently, the increased pesticide use is not only contributing to a decline in pollinator diversity but also developing resistance in the pest populations. Therefore, IPPM offers a way to manage pests while supporting pollinator health. This approach enhances agricultural yield and supports biodiversity.

References

Aguilera G, Roslin T, Miller K, Tamburini G, Birkhofer K, Caballero-Lopez B, Lindström SAM, Öckinger E, Rundlöf M, Rusch A, Smith HG (2020) Crop diversity benefits carabid and pollinator communities in landscapes with semi-natural habitats. J Appl Ecol 57(11):2170–2179

Alarcón-Segura V, Grass I, Breustedt G, Rohlfs M, Tscharntke T (2022) Strip intercropping of wheat and oilseed rape enhances biodiversity and biological pest control in a conventionally managed farm scenario. J Appl Ecol

Albrecht M, Kleijn D, Williams NM, Tschumi M, Blaauw BR, Bommarco R, Campbell AJ, Dainese M, Drummond FA, Entling MH, Ganser D (2020) The effectiveness of flower strips and hedgerows on pest control, pollination services and crop yield: a quantitative synthesis. Ecol Lett 23(10):1488–1498

Almeida ALS, Albuquerque UP, Castro CC (2011) Reproductive biology of Spondias tuberosa Arruda (Anacardiaceae), an endemic fructiferous species of the caatinga (dry forest), under different management conditions in northeastern Brazil. J Arid Environ 75(4):330–337

Alyokhin A, Nault B, Brown B (2020) Soil conservation practices for insect pest management in highly disturbed agroecosystems–a review. Entomol Exp Appl 168(1):7–27

Ara ZG, Haque AR (2021) A comprehensive review on synthetic insecticides: toxicity to pollinators, associated risk to food security, and management approaches. J Biosyst Eng 46(3):254–272

Artz DR, Nault BA (2011) Performance of Apis mellifera, Bombus impatiens, and Peponapis pruinosa (Hymenoptera: Apidae) as pollinators of pumpkin. J Econ Entomol 104(4):1153–1161

Asiimwe P, Naranjo SE, Ellsworth PC (2014) Effects of irrigation levels on interactions among Lygus hesperus (Hemiptera: Miridae), insecticides, and predators in cotton. Environ Entomol 43(2):263–273

Atreya K (2008) Probabilistic assessment of acute health symptoms related to pesticide use under intensified Nepalese agriculture. Int J Environ Health Res 18(3):187–208

Audergon JM, Duffillol JM, Pinet C, Blanc P (1997) Pollination of three main apricot cultivars in France. In: XI International symposium on apricot culture, May, vol 488. pp 327–334

Beattie J, Crozier A, Duthie GG (2005) Potential health benefits of berries. Curr Nutr Food Sci 1(1):71–86

Bowers C, Toews M, Liu Y, Schmidt JM (2020) Cover crops improve early season natural enemy recruitment and pest management in cotton production. Biol Control 141:104149

Briar SS, Antwi F, Shrestha G, Sharma A, Reddy GV (2018) Potential biopesticides for crucifer flea beetle, Phyllotreta cruciferae (Coleoptera: Chrysomelidae) management under dryland canola production in Montana. Phytoparasitica 46(2):247–254

Bugg RL, Colfer RG, Chaney WE, Smith HA, Cannon J (2008) Flower flies (Syrphidae) and other biological control agents for aphids in vegetable crops

Calder PC, Waitzberg DL, Klek S, Martindale RG (2020) Lipids in parenteral nutrition: biological aspects. J Parenter Enter Nutr 44:S21–S27

Calzoni GL, Speranza A (1995) Pear and plum pollination: honey bees, bumble bees of both?. In: II Workshop on pollination, March, vol 423. pp 83–90

Cardoza YJ, Harris GK, Grozinger CM (2012) Effects of soil quality enhancement on pollinator-plant interactions. Psyche

Carvalheiro LG, Seymour CL, Nicolson SW, Veldtman R (2012) Creating patches of native flowers facilitates crop pollination in large agricultural fields: mango as a case study. J Appl Ecol 49(6):1373–1383

Cock MJ, Biesmeijer JC, Cannon RJ, Gerard PJ, Gillespie D, Jimenez JJ, Lavelle P, Raina SK (2012) The positive contribution of invertebrates to sustainable agriculture and food security

Cole LJ, Brocklehurst S, Robertson D, Harrison W, McCracken DI (2017) Exploring the interactions between resource availability and the utilisation of semi-natural habitats by insect pollinators in an intensive agricultural landscape. Agr Ecosyst Environ 246:157–167

Crowder DW, Northfield TD, Strand MR, Snyder WE (2010) Organic agriculture promotes evenness and natural pest control. Nature 466(7302):109–112

Cutler GC, Amichot M, Benelli G, Guedes RNC, Qu Y, Rix RR, Ullah F, Desneux N (2022) Hormesis and insects: effects and interactions in agroecosystems. Sci Total Environ 153899

Dai W, Yang Y, Patch HM, Grozinger CM, Mu J (2022) Soil moisture affects plant–pollinator interactions in an annual flowering plant. Philos Trans R Soc B 377(1853):20210423

Dara SK (2019) The new integrated pest management paradigm for the modern age. J Integrat Pest Managem 10(1):12

Dellinger A, Hamilton AM, Wessinger CA, Smith SD (2022) Flipped elevational pattern of pollination mode in tropical versus temperate Americas. bioRxiv

Dent D, Binks RH (2020) Insect pest management. Cabi

Dunn L, Lequerica M, Reid CR, Latty T (2020) Dual ecosystem services of syrphid flies (Diptera: Syrphidae): pollinators and biological control agents. Pest Manag Sci 76(6):1973–1979

Dutta J, Choudhury MM, Environmentalist G, Sen MT, Banerjee A, Tiwari C, Nagar EHK, Rawat DS (2019) Voice of environment newsletter
Duyck PF, Lavigne A, Vinatier F, Achard R, Okolle JN, Tixier P (2011) Addition of a new resource in agroecosystems: do cover crops alter the trophic positions of generalist predators? Basic Appl Ecol 12(1):47–55
Eberle CA, Thom MD, Nemec KT, Forcella F, Lundgren JG, Gesch RW, Riedell WE, Papiernik SK, Wagner A, Peterson DH, Eklund JJ (2015) Using pennycress, camelina, and canola cash cover crops to provision pollinators. Ind Crops Prod 75:20–25
Eckols T, Roberton B, Clark B, Rebar D (2021) Exploring the potential role of ants as pollinators in a tallgrass prairie following varied prescribed burns. Trans Kans Acad Sci 124(3–4):155–164
Egan PA, Dicks LV, Hokkanen HM, Stenberg JA (2020) Delivering integrated pest and pollinator management (IPPM). Trends Plant Sci 25(6):577–589
Ehler LE (2006) Integrated pest management (IPM): definition, historical development and implementation, and the other IPM. Pest Manag Sci 62(9):787–789
Eilenberg J, Hajek A, Lomer C (2001) Suggestions for unifying the terminology in biological control. Biocontrol 46(4):387–400
Ellis KE, Barbercheck ME (2015) Management of overwintering cover crops influences floral resources and visitation by native bees. Environ Entomol 44(4):999–1010
Flöhr A, Stenberg JA, Egan PA (2020) The joint economic impact level (jEIL): a decision metric for integrated pest and pollinator management. In: Integrative biological control. Springer, Cham, pp 17–38
Fortuin CC, Gandhi KJ (2020) Osmia lignaria (Hymenoptera: Megachilidae) produce larger and heavier blueberries than honey bees (Hymenoptera: Apidae). bioRxiv
Geeta R, Berry E (2020) Floral symmetry–what it is, how it forms, and why it varies. In: Reproductive ecology of flowering plants: patterns and processes. Springer, Singapore, pp 131–155
Genersch E, Von Der Ohe W, Kaatz H, Schroeder A, Otten C, Büchler R, Berg S, Ritter W, Mühlen W, Gisder S, Meixner M (2010) The German bee monitoring project: a long term study to understand periodically high winter losses of honey bee colonies. Apidologie 41(3):332–352
Gullino ML, Tavella L (2020) Chemical and natural pesticides in IPM: side-effects and application. In: Integrated pest and disease management in greenhouse crops. Springer, Cham, pp 441–454
Haan NL, Zhang Y, Landis DA (2020) Predicting landscape configuration effects on agricultural pest suppression. Trends Ecol Evol 35(2):175–186
Haenke S, Kovács-Hostyánszki A, Fründ J, Batáry P, Jauker B, Tscharntke T, Holzschuh A (2014) Landscape configuration of crops and hedgerows drives local syrphid fly abundance. J Appl Ecol 51(2):505–513
Hajek AE, Eilenberg J (2018) Natural enemies: an introduction to biological control. Cambridge University Press
Holland JM, Bianchi FJ, Entling MH, Moonen AC, Smith BM, Jeanneret P (2016) Structure, function and management of semi-natural habitats for conservation biological control: a review of European studies. Pest Manag Sci 72(9):1638–1651
Huang Y, Luo X, Tang L, Yu W (2020) The power of habit: does production experience lead to pesticide overuse? Environ Sci Pollut Res 27(20):25287–25296
Jallow MF, Awadh DG, Albaho MS, Devi VY, Thomas BM (2017) Pesticide risk behaviors and factors influencing pesticide use among farmers in Kuwait. Sci Total Environ 574:490–498
Jauhar PP (2006) Cytogenetic architecture of cereal crops and their manipulation to fit human needs: opportunities and challenges. Genetic Resources, Chromosome Eng Crop Improvem 2:1–25
Jung C (2021) Integrated pollinator-pest management (IPPM) strategy as future apple IPM. Korean J Appl Entomol 60(1):145–154
Karar ME, Alsunaydi F, Albusaymi S, Alotaibi S (2021) A new mobile application of agricultural pests recognition using deep learning in cloud computing system. Alex Eng J 60(5):4423–4432
Karise R, Dreyersdorff G, Jahani M, Veromann E, Runno-Paurson E, Kaart T, Smagghe G, Mänd M (2016) Reliability of the entomovector technology using Prestop-Mix and *Bombus terrestris* L. as a fungal disease biocontrol method in open field. Scientif Reports 6(1):1–8

Katayama N, Baba YG, Kusumoto Y, Tanaka K (2015) A review of post-war changes in rice farming and biodiversity in Japan. Agric Syst 132:73–84

Kenis M, Hurley BP, Hajek AE, Cock MJ (2017) Classical biological control of insect pests of trees: facts and figures. Biol Invasions 19(11):3401–3417

Kremen C, Chaplin-Kramer R (2007) Insects as providers of ecosystem services: crop pollination and pest control. In: Insect conservation biology: proceedings of the royal entomological society's 23rd symposium, June, CABI Publishing, pp 349–382

Kumar V, Kumar P (2019) Pesticides in agriculture and environment: Impacts on human health. Contaminants in Agricul Environ: Health Risks and Remediat 1:76–95

Latha ES, Rajan SJ, Reddy MN, Madhuri KS, Rao CS (2019) Conservation biological control: effect of bio fertilizers and bio pesticides in organic ecological engineering field of okra (Abelmoschus esculentus (L.) Moench). Int J Curr Microbiol App Sci 8(1):548–555

Lentini PE, Martin TG, Gibbons P, Fischer J, Cunningham SA (2012) Supporting wild pollinators in a temperate agricultural landscape: maintaining mosaics of natural features and production. Biol Conserv 149:84–92

Lundin O, Rundlöf M, Jonsson M, Bommarco R, Williams NM (2021) Integrated pest and pollinator management–expanding the concept. Front Ecol Environ 19(5):283–291

Mahfouz HM, Kamel SM, Belal A, Said M (2012) Pollinators visiting sesame (Sesamum indicum L.) seed crop with reference to foraging activity of some bee species. Cercetari Agronomice in Moldova 45(2):49–55

Mala BR, Reddy PV (2020) Diversity and abundance of insect pollinators on avocado (Persea americana Mill.) and rambutan (Nephelium lappaceum L.) and their contribution to fruit set. Pest Managem Horticul Ecosyst 26(1):76–82

Mallinger RE, Franco JG, Prischmann-Voldseth DA, Prasifka JR (2019) Annual cover crops for managed and wild bees: optimal plant mixtures depend on pollinator enhancement goals. Agr Ecosyst Environ 273:107–116

Martin EA, Dainese M, Clough Y, Báldi A, Bommarco R, Gagic V, Garratt MP, Holzschuh A, Kleijn D, Kovács-Hostyánszki A, Marini L (2019) The interplay of landscape composition and configuration: new pathways to manage functional biodiversity and agroecosystem services across Europe. Ecol Lett 22(7):1083–1094

Martin EA, Feit B, Requier F, Friberg H, Jonsson M (2019b) Assessing the resilience of biodiversity-driven functions in agroecosystems under environmental change. In: Advances in ecological research. vol 60. Academic Press, pp 59–123

Mateos Fernández R, Petek M, Gerasymenko I, Juteršek M, Baebler Š, Kallam K, Moreno Giménez E, Gondolf J, Nordmann A, Gruden K, Orzaez D (2022) Insect pest management in the age of synthetic biology. Plant Biotechnol J 20(1):25–36

McCallum RS, McLean NL, Cutler GC (2018) An assessment of artificial nests for cavity-nesting bees (Hymenoptera: Megachilidae) in lowbush blueberry (Ericaceae). Can Entomol 150(6):802–812

Meinke LJ, Souza D, Siegfried BD (2021) The use of insecticides to manage the western corn rootworm, Diabrotica virgifera virgifera, LeConte: history, field-evolved resistance, and associated mechanisms. Insects 12(2):112

Menzler-Hokkanen I, Hokkanen HM (2017) Entomovectoring: an agroecological practice of using bees for biocontrol. In: Agroecological practices for sustainable agriculture: principles, applications, and making the transition. pp 183–199

Merle I, Hipólito J, Requier F (2021) Towards integrated pest and pollinator management in tropical crops. Current Opinion in Insect Sci

Mertens JE, Brisson L, Janeček Š, Klomberg Y, Maicher V, Sáfián S, Delabye S, Potocký P, Kobe IN, Pyrcz T, Tropek R (2021) Elevational and seasonal patterns of butterflies and hawkmoths in plant-pollinator networks in tropical rainforests of Mount Cameroon. Sci Rep 11(1):1–12

Milošević T, Milošević N (2020) Soil fertility: plant nutrition vis-à-vis fruit yield and quality of stone fruits. Fruit Crops, pp 583–606

Mommaerts V, Smagghe G (2011) Entomovectoring in plant protection. Arthropod-Plant Interact 5(2):81–95

Monasterolo M, Poggio SL, Medan D, Devoto M (2020) Wider road verges sustain higher plant species richness and pollinator abundance in intensively managed agroecosystems. Agr Ecosyst Environ 302:107084

Mukherjee A, Knoch S, Chouinard G, Tavares JR, Dumont MJ (2019) Use of bio-based polymers in agricultural exclusion nets: a perspective. Biosys Eng 180:121–145

Mužinić V, Želježić D (2018) Non-target toxicity of novel insecticides. Arh Hig Rada Toksikol 69(2):86–101

Oerke EC (2006) Crop losses to pests. J Agric Sci 144(1):31–43

Parker JE, Snyder WE, Hamilton GC, Rodriguez-Saona C (2013) Companion planting and insect pest control. Weed and Pest Control-Convent New Challenges 10:55044

Parry H, Kalyebi A, Bianchi F, Sseruwagi P, Colvin J, Schellhorn N, Macfadyen S (2020) Evaluation of cultural control and resistance-breeding strategies for suppression of whitefly infestation of cassava at the landscape scale: a simulation modeling approach. Pest Manag Sci 76(8):2699–2710

Partap U, Shukla AN, Verma LR (2000) Pollination of peach and plum by *Apis cerana*. In: Asian bees and beekeeping: progress of research and development. Oxford and IBH, New Delhi, pp 171–173

Pelaez V, da Silva LR, Araujo EB (2013) Regulation of pesticides: a comparative analysis. Sci Public Policy 40(5):644–656

Peng RK, Christian K (2004) The weaver ant, Oecophylla smaragdina (Hymenoptera: Formicidae), an effective biological control agent of the red-banded thrips, Selenothrips rubrocinctus (Thysanoptera: Thripidae) in mango crops in the Northern Territory of Australia. Int J Pest Managem 50(2):107–114

Phillips BB, Wallace C, Roberts BR, Whitehouse AT, Gaston KJ, Bullock JM, Dicks LV, Osborne JL (2020) Enhancing road verges to aid pollinator conservation: a review. Biol Cons 250:108687

Pimentel D, Burgess M (2014) Environmental and economic costs of the application of pesticides primarily in the United States. In: Integrated pest management. Springer, Dordrecht, pp 47–71

Pradhan P, Fischer G, van Velthuizen H, Reusser DE, Kropp JP (2015) Closing yield gaps: how sustainable can we be? PLoS ONE 10(6):e0129487

Qian P, Bai Y, Zhou W, Yu H, Zhu Z, Wang G, Quais MK, Li F, Chen Y, Tan Y, Shi X (2021) Diversified bund vegetation coupled with flowering plants enhances predator population and early-season pest control. Environ Entomol 50(4):842–851

Raderschall CA, Bommarco R, Lindström SA, Lundin O (2021) Landscape crop diversity and semi-natural habitat affect crop pollinators, pollination benefit and yield. Agr Ecosyst Environ 306:107189

Ramírez-Fráncel LA, García-Herrera LV, Losada-Prado S, Reinoso-Flórez G, Sánchez-Hernández A, Estrada-Villegas S, Lim BK, Guevara G (2022) Bats and their vital ecosystem services: a global review. Integrat Zool 17(1):2–23

Rashmi S (2018) Dietary fatty acids and biochemistry of coconut, sesame, mustard and groundnut oils: comparative healthy aspects. J Food Tech Food Chem 1(105):1–3

Reilly JR, Artz DR, Biddinger D, Bobiwash K, Boyle NK, Brittain C, Brokaw J, Campbell JW, Daniels J, Elle E, Ellis JD (2020) Crop production in the USA is frequently limited by a lack of pollinators. Proc R Soc B 287(1931):20200922

Ridgway R (ed) (2013) In: Biological control by augmentation of natural enemies: insect and mite control with parasites and predators. vol 11. Springer Science and Business Media

Riedinger V, Renner M, Rundlöf M, Steffan-Dewenter I, Holzschuh A (2014) Early mass-flowering crops mitigate pollinator dilution in late-flowering crops. Landscape Ecol 29(3):425–435

Rusch A, Bommarco R, Jonsson M, Smith HG, Ekbom B (2013) Flow and stability of natural pest control services depend on complexity and crop rotation at the landscape scale. J Appl Ecol 50(2):345–354

Sayer J, Sunderland T, Ghazoul J, Pfund JL, Sheil D, Meijaard E, Buck LE (2013) Ten principles for a landscape approach to reconciling agriculture, conservation, and other competing land uses. Proc Natl Acad Sci USA 110:8349–8356

Schellhorn NA, Gagic V, Bommarco R (2015) Time will tell: resource continuity bolsters ecosystem services. Trends Ecol Evol 30(9):524–530

Schneider G, Krauss J, Riedinger V, Holzschuh A, Steffan-Dewenter I (2015) Biological pest control and yields depend on spatial and temporal crop cover dynamics. J Appl Ecol 52(5):1283–1292

Semeraro T, Pomes A, Del Giudice C, Negro D, Aretano R (2018) Planning ground-based utility scale solar energy as green infrastructure to enhance ecosystem services. Energy Policy 117:218–227

Sgolastra F, Hinarejos S, Pitts-Singer TL, Boyle NK, Joseph T, Lūckmann J, Raine NE, Singh R, Williams NM, Bosch J (2019) Pesticide exposure assessment paradigm for solitary bees. Environ Entomol 48(1):22–35

Shapiro-Ilan DI, Cottrell TE, Mizell RF, Horton DL (2016) Curative control of the peachtree borer using entomopathogenic nematodes. J Nematol 48(3):28–33

Sharma S, Kooner R, Arora R (2017) Insect pests and crop losses. In: Breeding insect resistant crops for sustainable agriculture. Springer, Singapore, pp 45–66

Smagghe G, Boecking O, Maccagnani B, Mänd M, Kevan PG (eds) 2020. Entomovectoring for precision biocontrol and enhanced pollination of crops. Springer Nature

Smagghe G (2020) Case studies on Entomovectoring in the greenhouse and open field. In: Entomovectoring for precision biocontrol and enhanced pollination of crops. Springer, Cham, pp 123–136

Sumner S, Law G, Cini A (2018) Why we love bees and hate wasps. Ecol Entomol 43(6):836–845

Sutter L, Jeanneret P, Bartual AM, Bocci G, Albrecht M (2017) Enhancing plant diversity in agricultural landscapes promotes both rare bees and dominant crop-pollinating bees through complementary increase in key floral resources. J Appl Ecol 54:1856–1864

Tillman G, Schomberg H, Phatak S, Mullinix B, Lachnicht S, Timper P, Olson D (2004) Influence of cover crops on insect pests and predators in conservation tillage cotton. J Econ Entomol 97(4):1217–1232

Traveset A, Richardson DM (2014) Mutualistic interactions and biological invasions

Vialatte A, Tsafack N, Hassan DA, Duflot R, Plantegenest M, Ouin A, Villenave-Chasset J, Ernoult A (2017) Landscape potential for pollen provisioning for beneficial insects favours biological control in crop fields. Landscape Ecol 32(3):465–480

Walter J (2020) Dryness, wetness and temporary flooding reduce floral resources of plant communities with adverse consequences for pollinator attraction. J Ecol 108(4):1453–1464

Walton RE, Sayer CD, Bennion H, Axmacher JC (2021) Improving the pollinator pantry: restoration and management of open farmland ponds enhances the complexity of plant-pollinator networks. Agric Ecosyst Environ 320:107611

Wotton KR, Gao B, Menz MH, Morris RK, Ball SG, Lim KS, Reynolds DR, Hu G, Chapman JW (2019) Mass seasonal migrations of hoverflies provide extensive pollination and crop protection services. Curr Biol 29(13):2167–2173

Wyckhuys KA, Lu Y, Morales H, Vazquez LL, Legaspi JC, Eliopoulos PA, Hernandez LM (2013) Current status and potential of conservation biological control for agriculture in the developing world. Biol Control 65(1):152–167

Zhang W, Swinton SM (2009) Incorporating natural enemies in an economic threshold for dynamically optimal pest management. Ecol Model 220(9–10):1315–1324

Zhang H, Han C, Breeze TD, Li M, Mashilingi SK, Hua J, Zhang W, Zhang X, Zhang S, An J (2022) Bumblebee pollination enhances yield and flavor of tomato in gobi desert greenhouses. Agriculture 12(6):795

Zhou H, Yu Y, Tan X, Chen A, Feng J (2014) Biological control of insect pests in apple orchards in China. Biol Control 68:47–56

Fawad Zafar Ahmad Khan is an insect ecologist and Assistant Professor of Entomology at MNS University of Agriculture, Multan, Pakistan. He has authored 41 peer-reviewed publications on insect behavior, ecology, and biological control. He earned his PhD in Entomology from the University of Georgia, USA, where he researched predator roles in managing fall armyworm (*Spodoptera frugiperda*) populations in turfgrass ecosystems. Dr. Khan has worked on projects related to integrated pest management, pollination biology, and the conservation of beneficial arthropods. His research interests include plant-insect interactions and their application in pest management programs. His notable works include studies on native insect pollinators, the impact of landscape complexity on solitary bee nesting behavior, and predator-prey dynamics under water-limited conditions. Dr. Khan teaches courses such as Introductory Entomology, Applied Entomology, Insect Morphology, and Insect Physiology.

Mudssar Ali is a pollinator ecologist and Assistant Professor at MNS University of Agriculture, Multan, Pakistan. He has authored over 55 peer-reviewed publications focusing on pollinator ecology, insect pollination, and agroecosystem sustainability. His research highlights the ecological role of native insect pollinators, particularly solitary bees, and their contributions to crop production. Dr. Ali's work on the impact of landscape complexity on bee nesting behavior and post-harvest crop characteristics has enhanced the understanding of pollinator conservation.

Danyal Haider Khan is a postgraduate student at the Institute of Plant Protection, MNS University of Agriculture, Multan. His research focuses on the conservation of solitary bees, studying nesting preferences and the impact of environmental factors such as nesting substrate, orientation, and landscape. He has developed artificial nesting sites to support solitary bee populations, contributing to pollinator conservation efforts.

Chapter 12
Are Desert Locust Outbreaks Linked to Changing Climates?

Rachid Sabbahi

Abstract Desert locust, *Schistocerca gregaria*, is a major pest of many crops and pastures across nearly 30 million km^2, covering Africa, north of the equator, the Middle East, and the Arab and Indo-Pakistani peninsulas. Like other locusts, *S. gregaria* can move from a solitary phase with low-density populations during recessions to a gregarious phase with high-density populations during invasions, when its hopper bands and swarms are devastating to agriculture. The importance of this pest has necessitated the establishment of a prevention strategy operated through numerous national monitoring and control centers located in the most affected countries. These systems are coordinated at the international level by the Food and Agriculture Organization, which has an information service responsible for monitoring the situation throughout the insect's range and forecasting the risk of outbreaks and invasions. This warning and risk prevention system has reduced the extent of invasions in most major agricultural areas. Current and future climate change may modify the locust risk to a greater or lesser extent. It is indeed in the presence of favorable climatic conditions (heavy rains and warmer temperatures) that invasions start from a smaller and more desert-like range. As a result, it is necessary to better assess the likely evolution of this agricultural pest in order to anticipate the desirable modifications of the current prevention system. Recent research has indicated a strong correlation between climate change and the frequency and intensity of desert locust outbreaks. The warming climate has led to more extreme weather events, such as heavy rainfall and cyclones, which create ideal breeding conditions for locusts. For instance, the Horn of Africa has experienced unprecedented locust invasions following unusual climatic patterns, with significant implications for regional food security and ecosystems. This book chapter aims to synthesize the current state of knowledge on the relationship between desert locusts and climate change and to identify the knowledge gaps and research needs for locust management and adaptation.

Keywords Climatic changes · Desert Locust · Food Security · Invasion · Outbreak · Plague · *Schistocerca gregaria* · Swarm

R. Sabbahi (✉)
Research Team in Science and Technology, Higher School of Technology of Laayoune, Ibn Zohr University, 3007, Laayoune, Morocco
e-mail: r.sabbahi@uiz.ac.ma

M. Behnassi et al. (eds.), *Food Systems and Biodiversity in the Context of Environmental and Climate Risks*, https://doi.org/10.1007/978-3-031-89167-0_12

12.1 Introduction

Desert locusts (DL), *Schistocerca gregaria* (Forskål) (Orthoptera: Acrididae), are short-horned grasshoppers that have been a symbol of agricultural devastation for centuries, with their ability to form massive swarms and consume vast quantities of vegetation (e.g. crops, grass, shrubs, and trees) (Salih et al. 2020). These swarms are not just a natural phenomenon but also a harbinger of the broader impacts of climate change on ecosystems and human livelihoods (UNEP 2021). The recent upsurges in locust outbreaks, particularly in Africa and the Middle East, have raised concerns about the role of a changing climate in exacerbating these events.

Climate change is altering the frequency, intensity, and distribution of rainfall patterns around the world, creating new opportunities and challenges for DL populations. The relationship between locust activity and climate has been well-documented, with locusts thriving in conditions of heavy rainfall, warm temperatures, tropical cyclones, floods, and droughts—conditions that are becoming more frequent due to global warming (Liu et al. 2024). The last five years have seen some of the hottest temperatures on record since the Industrial Revolution, and this has coincided with some of the most destructive locust swarms in recent history (UNEP 2021). Climate change can affect DLs in several ways, such as increasing the occurrence of extreme weather events that create favorable conditions for locust breeding and survival, altering the distribution and availability of suitable habitats for locusts, and interacting with other factors that influence locust dynamics and control efforts (Guan et al. 2021; Kimathi et al. 2020).

Schistocerca gregaria outbreaks can have severe and widespread impacts on agriculture and food security, affecting millions of people across several countries (FAO 2021). As the world grapples with the challenge of feeding a growing population, the threat of locusts to food security becomes even more significant. Innovative strategies to combat locust invasions are therefore urgently needed, including ecosystem-based approaches and improved early warning systems (UNEP 2021). These strategies are crucial in mitigating the impacts of locust invasions, which can lead to famine and economic hardship in affected regions. Controlling *S. gregaria* outbreaks requires a swift and coordinated response at local, national, regional, and international levels, but there are many challenges and limitations that hamper effective control efforts. Remote sensing, geographic information systems (GIS), and artificial intelligence (AI) can enhance the monitoring and forecasting of locust activity, and support the decision-making and planning of locust management (Klein et al. 2023; Kumar and Abdul Rahman 2021; Latchininsky and Sivanpillai 2010). Moreover, resilience and adaptation can help human communities cope with the impacts of locust outbreaks.

This chapter will explore (i) the impacts of *S. gregaria* outbreaks on agriculture and food security, (ii) the resilience and adaptation strategies of human communities to locust outbreaks, and (iii) the intricate link between DL outbreaks and climate change, examining the evidence from recent studies and the implications for future pest management and climate adaptation strategies. It will also delve into the mechanisms by which climate change may be influencing locust dynamics and discuss the measures being taken by nations and international organizations to address this growing threat.

12.2 Characteristics, Behavioral Phase Change, and Swarming of Desert Locusts

The Acrididae family, comprising over 6,700 species of grasshoppers, includes more than 100 species recognized as significant agricultural pests (Song et al. 2018). These species can cause substantial economic losses, particularly during periodic outbreaks that escalate from localized infestations to large-scale plagues (Gross 2021; Swaminathan and Swaminathan 2019). Notably, locusts—certain grasshoppers within this family—exhibit remarkable adaptability. They undergo behavioral and morphological transformations, altering their color and form, as a survival mechanism triggered by fluctuating population densities (Symmons and Cressman 2001). This adaptability makes locusts some of the most formidable insect pests globally.

The DL is a polyphagous, destructive species found throughout the world. Each locust is 7–8 cm long, weighs around 2 g, and undergoes a molting process after emerging from egg pods deposited at 5 to 10 cm beneath the soil surface (Showler 2008). Sandy soils are preferred for their ideal temperature and moisture levels, which are conducive to egg health. Its life cycle comprises three stages: egg (140–400 eggs per female, hatching after two weeks), nymph (with 5–6 instars, maturing into an adult in 4–6 weeks), and adult (capable of laying eggs after at least three weeks) (FAO and WMO 2016). Locusts can develop three to five generations annually, influenced by soil temperature and humidity (Showler 2008). The immature phase is particularly perilous due to the locusts' voracious appetite for plant matter, high mobility, and tendency to fly and migrate.

Locusts have a high reproductive capacity, and their breeding areas shift seasonally (Kimathi et al. 2020). Spring breeding occurs in Northwest Africa, East Africa, Saudi Arabia, Iran, and Pakistan. Summer breeding is concentrated in the Sahel regions, West Africa, Sudan, Eritrea, Ethiopia, and the Indo-Pakistani border (FAO 2020; Symmons and Cressman 2001). Winter breeding is confined to areas adjacent to the Red Sea and the Gulf of Aden, including Somalia.

In nature, *S. gregaria* exhibits two primary forms: solitary and gregarious, with intermediate levels known as 'transiens congregans'. The species exhibits distinct morphological, ecological, anatomical, biological, physiological, and ethological traits depending on the phase (Maeno and Tanaka 2011). For example, changes in color and form accompany behavioral shifts. The terms 'gregaricolor' and 'solitaricolor' refer to their coloration, while 'gregariform' and 'solitariform' describe their shape. Although these physical attributes can reflect past behavior, they are not reliable predictors of future behavior, making the monitoring of actual behavior essential for locust control strategies (Joshi et al. 2023).

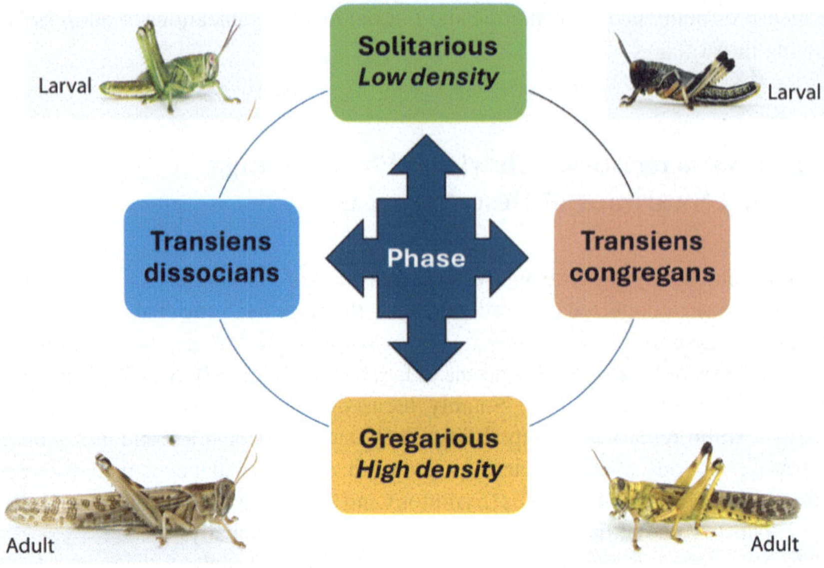

Fig. 12.1 Phase transformation cycle in the desert locust

The population dynamics of *S. gregaria* locusts are marked by shifts between solitary and gregarious phases (Fig. 12.1), punctuated by periods of remission, resurgence, upsurge, and invasion. DLs, typically found in the arid and semi-arid regions of Africa, the Middle East, and South-West Asia, usually exist as solitary insects (Landmann et al. 2023). The transformation from a solitary to a gregarious phase is characterized by an increase in population density, a shift to group behavior, and congregation (Symmons and Cressman 2001). The reproductive and movement potential of *S. gregaria* is triggered by favorable meteorological and ecological conditions, facilitating the transition from a harmless solitary phase to a destructive gregarious phase. When rainfall results in moist soil and abundant vegetation, locusts transition to a gregarious phase. If unchecked, the locust gregarization can lead to an outbreak characterized by the emergence of dense bands of nymphs and swarms of adults that can migrate over vast distances in search of food (Landmann et al. 2023).

12.3 Desert Locusts: A Plague on Crops

The DL has historically been a formidable adversary to agriculture, with its recurrent invasions inflicting extensive damage on crops and grazing lands (Le Gall et al. 2019). These swarms can cover an area of 29 million km^2—almost 20% of the Earth's landmass—threatening the livelihoods of people in more than 60 countries in Africa and Asia (Lecoq 2022). Locusts are known for their seasonal movements in search of favorable conditions for their development. The affected regions extend from Africa's North Atlantic shores to the Indo-Pakistani border and from the Mediterranean to the equator (Fig. 12.2).

Locust swarms, which can contain billions of locusts, have the potential to cover hundreds of square kilometers and consume massive amounts of vegetation, equivalent to the daily food intake of tens of thousands of people (Vosshall 2020). Should these outbreaks persist without control measures and if heavy rains continue to fall, breeding can intensify over several seasons, leading to more extensive hopper bands and adult swarms. This escalation is referred to as an upsurge, which usually impacts an entire region. In the absence of control and under continued favorable breeding conditions, locust numbers can grow exponentially, with infestations predominantly occurring as bands and swarms. The situation may intensify into a plague, a status declared when multiple regions are concurrently impacted (Gross 2021). Such a plague has the potential to diminish within a span of six months (Roffey and Magor 2003). While outbreaks are relatively common, only a minority progress to upsurges, and even fewer develop into plagues.

The threat posed by these insects is significant due to their ability to consume vast quantities of vegetation. In its gregarious form, a single locust can eat an amount equivalent to its weight (~ 2 g) each day (Villarreal, 2022), and with swarms containing up to 80 million individuals per square kilometer, they can collectively consume the same amount of food in one day as 35,000 people (FAO 2024). The most severe damage occurs when locusts occupy an area for an extended period of time. Notably, the last significant plague occurred between 1987–1989, and the most recent major upsurge was in 2003–2005 (Brader et al. 2006a). It's important to note

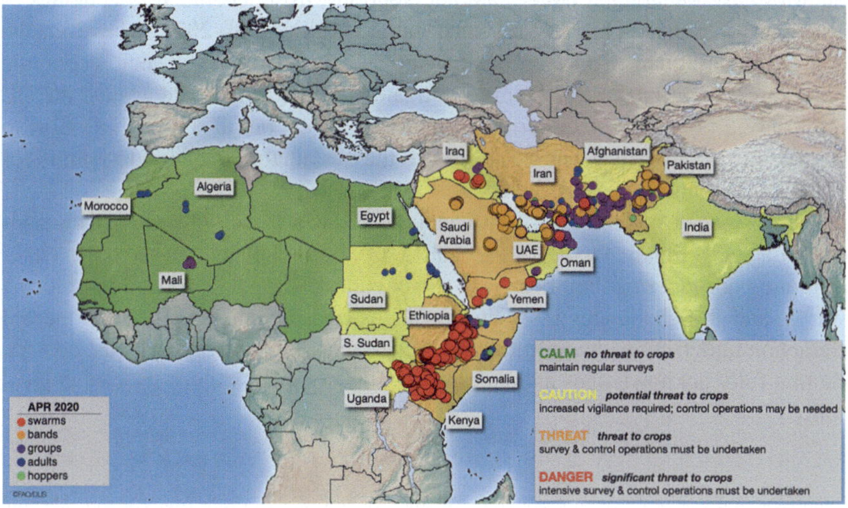

Fig. 12.2 Mapping *Schistocerca gregaria* threat dynamics during spring breeding in East Africa, the Arabian Peninsula, and Southwest Asia. *Source* FAO (2020)

that upsurges and plagues develop over many months, highlighting the importance of proactive locust management to prevent these events (Showler 2019).

The 2004 DL invasion led to considerable agricultural devastation in several nations, potentially precipitating food scarcity or even famine in parts of the Sahel. In Burkina Faso, the devastation was assessed at 80% of the anticipated cereal yield within the surveyed regions. Mali experienced similar destruction, with a 90% loss in cereal production. Mauritania faced the most severe impact, with cereal losses ranging from 90 to 100%. The invasion also heavily affected staple leguminous crops, such as cowpea and groundnut, with an estimated 85 to 90% loss in the expected yield for 2004 across the three nations. Pasturelands were not spared, with a third of them lost in Mali and Burkina Faso. In Mauritania, the situation was particularly dire, with fodder production plummeting by 85% (Brader et al. 2006a).

The impact goes beyond the loss of agriculture, leading to significant socio-economic challenges: it jeopardizes food security in vulnerable areas, increases the fragility of households, and triggers rural exodus (Sokame et al. 2024). Economically, the damage is dual: it reduces the primary income source for some nations and results in substantial financial burdens from extensive anti-locust campaigns. Historical invasions, such as those from 1987–1989 and 2003–2005, necessitated the mobilization of significant funds—700 and 400 million US dollars, respectively—across the affected countries (Brader et al. 2006a). However, these figures do not account for the indirect costs: decapitalization of farms, indebtedness of affected populations, market disruptions, and environmental impacts, which are difficult to quantify.

These compounded issues have the potential to destabilize the political landscape, particularly in the most impoverished nations. To counter these effects, affected countries have undertaken large-scale interventions, using chemical insecticides as the primary rapid response to the locust threat (Mullié et al. 2023). However, the ecological consequences of these actions are as concerning as the socio-economic and political disturbances they aim to mitigate. The ongoing battle against the DL is not only a fight to protect current agricultural assets but also a crucial effort to preserve the ecological balance and ensure the sustainability of food systems for future generations (Barasa et al. 2023).

The existence of *S. gregaria* is intricately linked to a tapestry of ecological, environmental, and climatic factors (Tang et al. 2023). Each element plays a role in either facilitating or hindering their life cycle, thereby influencing the probability of locust swarms and the challenges associated with their management. This multifaceted interaction underscores the need for an all-encompassing strategy in locust control, one that fully acknowledges and incorporates the complex web of their ecological reality.

12.4 Role of Climate Change in Desert Locust Outbreaks

DLs are masters of survival, deftly navigating a complex web of ecological and environmental factors that dictate their life cycle (Klein et al. 2023). These factors are far from passive; they actively shape the locusts' journey from hatching to maturity. The intricate dance between the locusts and their surroundings is marked by a series of ecological, habitat, and meteorological influences that imprint upon their developmental stages and migratory behaviors. Demonstrating remarkable adaptability, DLs flourish in the unpredictable expanses of arid landscapes, adjusting their growth, survival, and migration in response to environmental signals (Le Gall et al. 2019; Wang et al. 2021).

This complex interplay is pivotal, necessitating the identification of key ecological parameters that can be tracked, particularly through satellite imagery, to comprehend and forecast locust activity. In the context of climate change, research has established significant links with DL outbreaks. Climate change modifies various environmental elements, such as temperature, precipitation, and vegetation, which in turn substantially affect DL populations (Liu et al. 2024). Extreme weather events, for example, can foster conditions favorable for locust swarms, while shifts in habitats may introduce infestation risks to new regions (Guan et al. 2021). Additionally, climate-driven alterations in precipitation and vegetation patterns influence locust development and fecundity (Kimathi et al. 2020). The dynamics of locust populations are further affected by the interplay between climate change and other factors, including land use changes, ethnic conflict, political instability, and pest management strategies (Kassegn and Endris 2021; Salih et al. 2020; Showler 2003; Tang et al. 2023). Collectively, these elements, alongside climate change, shape the resilience of communities against the impact of locust invasions on agriculture and food security.

12.4.1 Rainfall Variability and Extreme Events

The ecological dynamics of DL depend on a series of environmental factors that are essential to its spread. Paramount among these is water availability, which serves as a catalyst for both reproduction and the growth of vegetation. Rainfall prediction is a common method for identifying potential locust breeding sites (Hielkema and Snijders, 1994); however, the precision of such predictions can vary widely, from 20 to 70%, particularly in arid regions, often resulting in an overestimation of actual rainfall (Dinku et al. 2010). The onset of rainfall marks the beginning of the breeding season, providing the necessary moist soil for egg deposition and supporting the vegetation that feeds the burgeoning locust populations. Additionally, heavy rainfall promotes robust plant growth, offering an abundance of crops and other plant life for the nourishment of juvenile locusts, which subsequently mature and disperse in swarms (Gómez et al. 2020).

Rainfall is the primary environmental factor influencing DL outbreaks, as it directly affects the availability of vegetation and the conditions necessary for egg-laying. The timing, duration, intensity, and spatial distribution of rainfall are crucial, as they determine the availability and quality of vegetation for feeding and egg-laying (FAO and WMO 2016). DLs typically breed in regions with annual precipitation below 200 mm (FAO 2024). While short, intense rainfall may be insufficient for egg development, extended periods of rain can sustain multiple locust generations (Wang et al. 2021). However, when rainfall significantly exceeds normal levels or when extreme weather events occur, these can lead to lush conditions that favor rapid locust multiplication.

Climate change is anticipated to exacerbate the variability and extremity of rainfall, as well as the occurrence of extreme weather events, which are key factors in DL outbreaks. Such changes are expected to disrupt the usual patterns of rainfall, potentially altering the timing and locations of locust swarms. Typically, locusts are restricted to certain desert areas during dry periods, but climate change could modify these habitats, leading to wetter or drier conditions (Guan et al. 2021; Kimathi et al. 2020; Salih et al. 2020). Research highlights the profound influence of heavy rains and strong winds on catalyzing DL epidemics (Chen et al. 2020; Kimathi et al. 2020). The potential 25% increase in the geographical range of locusts due to climate change underscores the need for urgent action (Liu et al. 2024). Irregular rainfall and warmer temperatures create the perfect breeding grounds for locust swarms, which can devastate multiple regions simultaneously. These swarms are driven by regional weather patterns that promote vegetation growth, aiding their dispersal. Conversely, when food is scarce, locusts become more mobile and form gregarious groups in search of new feeding and breeding grounds.

Floods, a consequence of heavy rainfall, can create new breeding grounds for locusts by inundating lands and promoting vegetation growth (Zhang et al. 2008). Notable flooding events in 2017 and 2018, fueled by abundant rainfall during critical growth periods, diminished suitable habitats by approximately 54 and 29%, respectively. This habitat reduction resulted in elevated locust densities, triggering a locust plague and areas of high-density locust populations (Zhao et al. 2023). In contrast, droughts have a dual effect on locust populations (Guan et al. 2021). They can either exhaust resources, causing locust dispersion or mortality, or they can safeguard locust eggs, which remain dormant for extended periods until favorable conditions return (Wang et al. 2023). The ensuing droughts and the resulting scarcity of vegetation compel the insects to congregate, precipitating the gregarious phase, during which locusts form bands and evolve into highly mobile swarms of flying adults (Despland et al. 2000; Georgiou et al. 2021).

12.4.2 Temperature Changes and Vegetation Dynamics

Temperature conditions significantly affect the development stages of DL eggs. According to Bennett (1976), these stages can last up to 65 days, but soil warmth can reduce this incubation period, potentially leading to larger swarms (Symmons and Cressman 2001). Higher temperatures are known to accelerate the metabolic, growth, and reproductive rates of locusts (Miller et al. 2009), although extreme heat can also result in stress and mortality. The ambient air temperature is crucial in determining the locusts' daily activities, such as feeding and movement, which are essential for swarm formation (FAO and WMO 2016).

Temperature plays a pivotal role in DL outbreaks, influencing their metabolism, growth, and geographical distribution. The optimal annual mean temperature for solitary DL development is 15–35 °C, conducive to their life cycle (Guan et al. 2021). While warmer conditions can speed up development, cooler temperatures may slow or even halt their progress. The thermal environment also affects the availability of vegetation, which is vital for locust sustenance and habitat. With climate change expected to increase average temperatures and the frequency of heatwaves, there could be alterations in vegetation patterns, potentially affecting locust habitats. Furthermore, temperature regulates locust development and behavior; higher temperatures shorten the time required for locusts to reach maturity, thus reducing the generational interval (Miller et al. 2009). However, adults usually take flight when temperatures exceed the minimum threshold for flight (~ 22 °C); at cooler temperatures, locusts in low bushes tend to move into the middle of the bush upon which they are roosting in order to hide (Maeno et al. 2019). Whereas, the anticipated rise in temperatures due to climate change could shorten the maturation and incubation periods in the spring, potentially allowing for an additional breeding cycle in regions such as North-West Africa, the Arabian Peninsula, and South-West Asia, increasing the risk of plagues (UNEP 2021).

The Intergovernmental Panel on Climate Change (IPCC) projects a global mean surface temperature increase of 1.5 °C by 2030, relative to pre-industrial levels, contingent on emission scenarios (Betts et al. 2023). Sub-Saharan Africa is predicted to see greater warming than the rest of the world (Umeh and Gil-Alana 2024). Such a rise could abbreviate locust life cycles and boost their reproductive rates in certain areas (Peng et al. 2020). Studies have consistently found that warmer sea surface temperatures, particularly in the Indian Ocean, are associated with increased locust activity in East Africa. Notable research by Stige et al. (2007) and Cressman et al. (2019) supports this correlation. Predictive ecological models suggest that by 2090, global warming could enlarge locust habitats by up to 25%, affecting new regions and altering locust behavior due to shifts in plant nutrition.

12.4.3 Wind Patterns and Tropical Cyclones

Atmospheric dynamics, particularly wind patterns, are crucial in guiding the migration of DLs. These winds enable locusts to traverse vast distances, initiate infestations in previously unaffected areas, and disperse widely (Symmons and Cressman 2001). Such wind currents are instrumental in the occurrence and spread of DL swarms, facilitating their far-reaching journeys. Locusts can traverse up to 150 km each day

(Villarreal, 2022), propelled by winds influenced by atmospheric phenomena like monsoons and El Niño-Southern Oscillation (ENSO, a recurring climate pattern across the tropical Pacific) (Liu et al. 2024). These phenomena, subject to the whims of climate change, can alter wind patterns, thus affecting locust migration and possibly increasing tropical cyclone frequency, which further aids their dispersal (UNEP 2021). Additionally, climate-driven changes in El Niño and La Niña events (the warm and cool phases of ENSO, respectively) could affect locust breeding in Africa. Although the precise impact of climate change on wind patterns is uncertain, any variations could allow locusts to invade new territories, contingent upon local ecological and climatic conditions (FAO and WMO 2016). Hence, climate change's role in shaping weather patterns and cyclones is becoming more evident as a catalyst for locust outbreaks.

The recent East African locust invasion illustrates how climate change intensifies extreme weather events such as tropical cyclones, creating environments (e.g. desert lakes and fertile areas) favorable for locust breeding (Salih et al. 2020). The 2018–2020 locust outbreak was sparked by two atypical cyclones, Mekunu and Luban, which caused intense precipitation events in the Arabian Peninsula, which provided the necessary moisture for locust breeding, multiplication, and gregarization (Wang et al. 2021). This was followed by severe rains in Yemen, Iran, and the India-Pakistan, where locusts usually travel (FAO 2024). By the end of 2019, the winds of yet another tropical cyclone, Pawan, were instrumental in transporting the locusts to East Africa, resulting in the most severe infestation in many years (Salih et al. 2020). A lack of preparedness, chronic political instability, and limited capacity made the invasion the worst in a quarter of a century for most countries and the worst in 70 years for Kenya and Uganda (Kimathi et al. 2020). Finally, just when the weather was drying out, the Horn of Africa experienced a cyclone in December 2019, which allowed locusts to stay and breed for two generations. This tipped the scale into an upsurge (FAO 2024).

Anticipated increases in extreme weather events due to climate change, including droughts, floods, and cyclones, could cause more frequent and intense locust swarms, particularly in west-central Asia, posing challenges to locust management efforts (UNEP 2021). Understanding these dynamics is critical for developing sophisticated strategies to thwart locust invasions. As climate change reshapes ecosystems, grasping its influence on locust behavior is imperative for integrating this knowledge into locust management and adaptation plans, with the aim of protecting agricultural and food resources.

12.5 Challenges and Opportunities for Desert Locust Control Under Climate Change

The population dynamics and risk management of *S. gregaria* are shaped by a complex interplay of factors. Climate change, rainfall patterns, vegetation availability, and the efficacy of monitoring and control measures are all significant contributors to this dynamic (Magor et al. 2008). These factors are deeply interconnected, presenting a spectrum of challenges and opportunities for managing locust risks effectively. At the heart of this is climate change, which exerts a profound influence on DL risk management (Liu et al. 2024; Meynard et al. 2017). It alters the effectiveness and sustainability of current monitoring systems, early warning mechanisms, and control strategies. Furthermore, climate change necessitates the adaptation of risk management practices and fosters innovation, ensuring that strategies evolve in response to changing environmental conditions (Meynard et al. 2020).

12.5.1 Desert Locust Monitoring and Early Warning Systems

Monitoring and early warning systems play a critical role in preventing and controlling *S. gregaria* outbreaks. These systems provide timely and accurate information about locust situations and forecasts. Current strategies for managing locust outbreaks focus on monitoring population dynamics, predicting movements, and intervening to reduce numbers and prevent swarm formation (Retkute et al. 2021).

The fundamental strategy to mitigate DL-related risks centers on preemptively eliminating larval bands during their solitary phase, which is generally harmless. This action is crucial to prevent locusts from advancing to the gregarious phase—a stage that poses a significant threat to agriculture and silvopastoral resources (Showler et al. 2021). Success in anti-locust measures relies heavily on accurate and up-to-date information, particularly meteorological and environmental data that influence locust development throughout their lifecycle (Table 12.1). Initiatives involve creating and validating environmental parameters critical to *S. gregaria* growth and population dynamics. These parameters include vegetation cover, rainfall, temperatures, and soil moisture levels, among other factors (Piou et al. 2019).

However, a major obstacle to locust management is the lack of predictive analytics capable of early crisis detection. Such analytics are essential for the timely implementation of control strategies and the efficient allocation of financial resources (Gómez et al. 2021, 2018; Klein et al. 2022). To address this, a comprehensive surveillance system covering all territories at risk of locust invasions is necessary. Continuous monitoring and assessment of essential parameters are vital components of this system.

Emerging technologies like GIS and AI offer new capabilities for monitoring and forecasting DL outbreaks. They provide detailed information on locust distribution and environmental conditions, facilitating the identification of potential outbreaks.

Table 12.1 Essential meteorological data for forecasting and managing desert locust populations

	Total	Actual	Forecast	Use	Stage
Precipitation and Humidity	Min/max	Daily Dekadal Monthly	+ 1 day + 10 day + 30 day Seasonal	Green vegetation Locust presence & egg-laying Breeding Migration	Outbreaks Plagues Recessions
Temperature	Data	Daily Dekadal Monthly	+ 1 day + 10 day + 30 day Seasonal	Egg & hopper development Adult takeoff/ migration	Upsurges Plagues Outbreaks
Wind	Direction Speed Height	12 h		Migration Invasion	Plagues

Source Cressman (2001); FAO and WMO (2016)

Innovations such as eLocust3 (a rugged tablet connected to a satellite antenna), the h3-hexagonal hierarchical geospatial indexing system, and the Doppler weather radar toolkits demonstrate the effectiveness of integrating satellite imagery, the global positioning system (GPS), and AI for real-time data analysis, significantly improving locust management (Anjita and Indu 2023; Klein et al. 2023; Qayyum et al. 2023). Risk models link *S. gregaria* population dynamics with satellite data on variables affecting multiplication and spread (Despland et al. 2004; Ellenburg et al. 2021; Gómez et al. 2019). The primary goal is to produce practical risk maps that stakeholders, especially those engaged in locust monitoring and control, can use. Providing these maps to policymakers and field operatives supports informed decision-making for ground surveys and enables ongoing preventive actions at the national level. However, challenges remain, including data quality, system compatibility, and specialized skill requirements (Gómez et al. 2021).

The World Meteorological Organization (WMO) plays a crucial role in mitigating locust outbreaks by providing timely weather forecasts that can signal potential exacerbations of locust conditions. This vital information enables the implementation of proactive measures. In a collaborative effort, the Food and Agriculture Organization (FAO) works alongside governments and key organizations, including the World Health Organization (WHO) and the UN Environment Programme (UNEP), to develop and refine standard operating protocols. These protocols guide the planning and execution of DL control campaigns, drawing upon the FAO's DL Guidelines, which are continually updated to embody current best practices. Moreover, the FAO is committed to monitoring locust activity closely and issuing alerts and forecasts to nations facing the threat of invasion. Central to this surveillance is the FAO's global Desert Locust Information Service (DLIS), which is instrumental in collecting, analyzing, and disseminating data. The DLIS integrates information from a variety of sources, such as national locust units, field surveys, satellite imagery, and weather data, ensuring a comprehensive approach to locust management (Cressman 2016).

However, despite these concerted efforts, the monitoring and early warning systems currently in place are not without their challenges. They face limitations and gaps, particularly in the context of climate change, which can affect their performance and reliability. These challenges include the need for improved data collection and processing capabilities, as well as the integration of new technologies and methodologies to enhance predictive accuracy and response times. Specifically, there is a notable deficiency in resources, funding, equipment, personnel, training, and coordination among national locust units, particularly in areas affected by conflict or those that are remote. This shortfall is documented in studies by Retkute et al. (2021), Roussi (2020), and Salih et al. (2020). Furthermore, the collection of timely and accurate data is often hindered by logistical, technical, or security challenges, as highlighted by Landmann et al. (2023). The availability of high-resolution and near-real-time satellite imagery, weather data, vegetation indices, drones, and other remote sensing products is insufficient for detecting locust habitats and movements, a gap identified by Anjita and Indu (2023), Enns et al. (2022), and Klein et al. (2021). The absence of accurate predictive models and dynamic forecasting tools that integrate multiple data sources and account for climate variability adds to the uncertainty in forecasting locust outbreaks, as noted by Sokame et al. (2024). Moreover, there is a lack of effective communication channels for disseminating locust information and warnings to stakeholders and communities at national, regional, and international levels, as Showler et al. (2021) have pointed out.

Addressing these limitations is essential for enhancing the capacity to combat *S. gregaria* invasions, thereby safeguarding food security and livelihoods. Collaborative efforts and advancements in technology and data integration are vital for overcoming these obstacles and establishing a more resilient approach to locust population management.

12.5.2 Strategies for Desert Locust Control and Challenges

The choice of control method and its intensity depend on the specific goals and the stage of the DL outbreak (Retkute et al. 2021). The primary method for controlling DLs involves the application of insecticides (Mullié et al. 2023). These are sprayed via aerial or ground methods at locust breeding and gathering sites. Using ultra-low volume (ULV) spraying formulations for pesticide application remains a prominent control method (Cressman 2016). Three types of sprayer platforms are used, each with distinct control-related characteristics: portable ULV sprayers (these have a slow work rate and can target areas of up to 10 ha), vehicle-mounted sprayers (they operate at a medium work rate and can cover areas ranging from 1 to 100 ha), and aircraft-mounted sprayers (these have a fast work rate and are effective for treating areas over 25 ha) (Dobson 2001). However, this approach has limitations. Organophosphates (e.g. fenitrothion and malathion), pyrethroids (e.g. deltamethrin), and carbamates (e.g. isophthalic acid), the main chemicals used, can harm non-target species (Mullié et al. 2023), including beneficial insects like pollinators that play an important role in agriculture, biodiversity conservation, ecosystem protection, and climate change adaptation (Sabbahi 2022a, 2022b). Additionally, traditional pesticide use poses risks to the environment and food security ecosystems without addressing the root causes of locust outbreaks, often linked to climate and land use/cover changes (Zhao et al. 2020).

Current research continues to explore non-chemical control methods, emphasizing entomopathogens like fungi, bacteria, and viruses (Matthews 2021). These biological agents, while environmentally friendly, face challenges such as slow action, inconsistent effectiveness, high production costs, and limited availability (Sabbahi et al. 2022). Specifically, *Metarhizium acridum*, a biopesticide targeting the Acrididae family, is effective against DLs (Githae and Kuria, 2021) but is not widely used due to its slow action, variable mortality rates among locusts, and difficulties in production and storage (Dobson 2001). Notably, its sole application during the 2019–2021 locust upsurge in Somalia showed promising outcomes. Entomopathogenic bacteria studied were from the species of *Serratia, Pseudomonas,* and *Bacillus.* These bacteria could be potential biocontrol agents for controlling DL nymphs (Bundey et al. 2003; Mashtoly et al. 2019; Reda et al. 2018). Understanding the dynamics between insect pathogens and their hosts is key to leveraging insect pathogens for sustainable biological pest control, extending beyond their use as mere bioinsecticides (Sabbahi et al. 2009b, 2009a). Future research should consider climate variability as a pivotal element in evaluating the efficacy of entomopathogenic fungi against *S. gregaria* (Kamga et al. 2022). Despite the presence of natural predators (e.g. *Dolichus halensis* Schaller and *Pheropsophus jessoensis* Morawitz) and parasitoids (e.g. *Blaesoxipha lineata* Fallén and *Telenomus* sp.), their impact on locust control is negligible because locusts can rapidly evade them (Li et al. 2022; Segoli et al. 2020). Human and avian predation does occur but does not significantly reduce locust numbers over large areas. Another alternative is the use of mechanical or physical barriers, such as nets, traps, or trenches, to capture or prevent DL movement (Babar 2023; Pandey et al. 2021). While cost-effective and easy to implement, these barriers have limitations like low efficiency, high maintenance, and interference with agricultural activities (Li et al. 2022).

The current approach to managing *S. gregaria* infestations is compromised by the environmental and financial drawbacks of chemical pesticides, the limited range of insecticides effective against all locust life stages, and the difficulty of maintaining their effectiveness and managing resistance under varying climatic conditions (Banik et al. 2022; Brader et al. 2006b; Khan et al. 2023; Khan and Ahmad 2019; UNEP 2021). Furthermore, there is a pressing need for enhanced integration and coordination of various control tactics, including biological, mechanical, cultural, and behavioral methods. Recognizing the significant economic, social, and environmental risks posed by DL outbreaks, the global community has highlighted the importance of preventive control as a sustainable strategy (Herbillon et al. 2024). Despite the efficacy of such measures, constant vigilance is essential to preemptively tackle locust upsurges that have the potential to become regional or global emergencies.

12.5.3 *Potential Adaptation and Innovation Options for Improving Desert Locust Risk Management*

Efforts to manage DLs are increasingly focused on sustainability and eco-friendly measures. These include enhancing early warning systems, improving locust surveillance, fostering regional collaboration, and advancing research on locust behavior and ecology (Showler 2019; Showler et al. 2021). Integrated pest management (IPM) practices, combining biological control, habitat management, and cautious insecticide use, aim for holistic and sustainable locust population management (Li et al. 2022; Pandey et al. 2021; Showler et al. 2022). Table 12.2 outlines the principal challenges, potential opportunities, and prospective strategies for managing DL populations.

Ultimately, the goal is to reduce the frequency and severity of *S. gregaria* plagues by combining advanced monitoring, targeted interventions, and sustainable land management that can help maintain the ecological balance and prevent the conditions that lead to large-scale locust upsurges. Strategic recommendations include enhancing agricultural system adaptation and resilience to climate change and variability, preparing local communities for locust emergencies, promoting stakeholder

Table 12.2 Main challenges, potential opportunities, and strategic recommendations for *Schistocerca gregaria* management

Challenges and barriers	Opportunities for improvement and innovation	Recommendations and future strategies
Vast and remote breeding areas that are often inaccessible or insecure	Use of new technologies for monitoring and control, like satellite imagery, drones, and mobile apps	Develop infrastructure and protocols for remote area access and security to facilitate timely interventions
High mobility and unpredictability of locust swarms	Development of alternative control methods that are environmentally friendly, like biopesticides and pheromones	Invest in research for predictive modeling of swarm movements to enhance response times and effectiveness
Limited resources and capacities of affected countries	Integration with sustainable agricultural development, such as crop diversification and soil conservation	Strengthen international cooperation and funding mechanisms to support affected regions during outbreaks
Environmental and health risks of insecticide use	Enhancement of national and regional capacities and cooperation through training and early warning systems	Promote the adoption of IPM practices to minimize reliance on chemical insecticides
Social and economic impacts on rural communities	Involvement of local communities in control activities, through awareness and participation	Implement community-based programs that empower local populations to actively participate in locust control efforts

collaboration, and supporting research and innovation in locust monitoring and control (Liu et al. 2024).

The escalating threat of climate change is compelling a transformation in *S. gregaria* risk management, driving the integration of cutting-edge technologies and novel strategies to enhance resilience. The deployment of eLocust3 devices, alongside the utilization of satellite imagery and AI, is revolutionizing the landscape of data collection and predictive analytics for locust activity (Qayyum et al. 2023). Concurrently, the advent of new insecticide formulas, such as biopesticides, and innovative delivery systems like drones is not only augmenting the safety and efficiency of control measures but also aligning with environmental sustainability goals (Fenibo et al. 2021; Matthews 2021). The shift towards alternative control methods, including the use of biological agents and mechanical barriers, reflects a growing commitment to reducing chemical dependency and fostering ecosystem balance (Babar 2023; El Bashir 1997). Policy enhancements are playing a pivotal role in strengthening collaborative efforts among stakeholders and ensuring a cohesive response to locust threats (Bennett 2020). Furthermore, the adoption of resilient agricultural practices and technologies is proving instrumental in safeguarding food security and supporting the livelihoods of communities vulnerable to locust devastation (Aich et al. 2022; Negera et al. 2023). These collective efforts underscore a proactive approach to locust risk management, emphasizing preparedness and adaptability in the face of climatic uncertainties.

UNEP plays a pivotal role in disseminating cutting-edge climate research, shaping multi-sectoral policies, and fostering sectoral resilience. Simultaneously, the FAO contributes to community resilience by introducing innovative technologies like solar dryers, which serve a dual purpose as climate action tools. These solar dryers empower communities to harvest crops earlier and preserve them more effectively, thereby mitigating the impact of locust incursions. By adopting climate-responsive measures—such as providing solar dryers to those involved in the agricultural supply chain—income can increase significantly, sometimes up to 30 times. Beyond safeguarding against unpredictable events like locust plagues, the production of these solar dryers can also stimulate new economic opportunities. Initiatives like these are critical for bolstering the climate resilience of Africa's most vulnerable groups.

Strengthening community resilience and adaptation remains paramount. Building resilient systems and empowering communities to resist and recover from locust invasions while adapting to new threats is a comprehensive strategy. This dual approach combines technological advancements with community empowerment, addressing the challenges posed by climate change. Research into community resilience and adaptation to locust outbreaks employs diverse methodologies. Quantitative studies, such as those by FAO (2021) and Sokame et al. (2024), quantify the impact of locust outbreaks on household resilience and vulnerability, revealing significant economic and food security consequences. Complementing these quantitative approaches are qualitative studies that delve into the strategies and factors influencing community responses to locust invasions, highlighting adaptive measures ranging from crop diversification to migration (Baraka et al. 2023a, 2023b; Barasa et al. 2023; Enns et al. 2022). Knowledge, resources, and policy play pivotal roles in shaping effective adaptation. However, there remains a need for more comprehensive approaches that capture the full spectrum of resilience and adaptation processes. Integrative research, accounting for the ecological and socio-economic complexities of locust outbreaks, aims to enhance our understanding and response to this global challenge (Sokame et al. 2024).

Climate change not only expands the potential habitat for locusts but also interacts with land use, human conflict, and pest management, complicating control efforts (Roussi 2020; Salih et al. 2020). Political instability and resource limitations in affected regions can hinder effective responses to locust outbreaks. Therefore, a comprehensive approach—one that considers these multifaceted challenges—is essential for managing the evolving threat of DLs. As the DL transcends borders and impacts vast regions, the international community's commitment to preventive measures underscores the need for a cohesive, proactive stance (Herbillon et al. 2024). Our overarching objective is clear: to protect agriculture—the bedrock of human sustenance—from the relentless attacks of these age-old pests. By doing so, we not only secure our current food sources but also ensure the stability of global food systems for the future. This ongoing battle demands unwavering vigilance, innovative solutions, and international collaboration, all aimed at maintaining the delicate equilibrium between humanity and the natural world.

12.6 Conclusion

Desert locust infestations present a significant challenge that necessitates a comprehensive and coordinated response. The escalation of these outbreaks, exacerbated by climate change, directly threatens agricultural productivity and food security. It is imperative to closely monitor climate change's impact on locust behavior and devise strategies that are both adaptive and resilient in order to curb the risk of invasions. Additionally, tackling the root causes of vulnerability and poverty is crucial, as these factors increase certain communities' susceptibility to the devastation caused by locust swarms. Strengthening collaboration among local, national, regional, and global entities can lead to a diminution of the DL threat, thereby safeguarding the livelihoods and welfare of countless individuals.

The increasing frequency and severity of *S. gregaria* attacks, fueled by climatic changes, create optimal conditions for their propagation and spread, posing a grave risk to food security and the sustenance of vulnerable populations. To combat this peril, proactive measures such as vigilant monitoring, forecasting of locust activities, outbreak prevention, and community resilience enhancement are essential. Addressing the fundamental issues of vulnerability and impoverishment is also vital, as these contribute to the heightened impact of locust crises on affected populations. A unified effort spanning various governance levels is required to mitigate the DL menace and ensure the protection of millions.

The economic ramifications of locust plagues are profound, affecting livelihoods, food availability, economic growth, and environmental health. This chapter underscores the potential influence of climate change on *S. gregaria* outbreaks and points out the particular vulnerability of Africa, the Middle East, and parts of Asia, regions already impacted by climate variability and socio-economic challenges. Even with a global temperature increase capped at 1.5 °C, the incidence of conducive conditions for locust surges could double compared to pre-industrial times, indicating a possible rise in future infestations. Forecasting and managing locust outbreaks remain formidable tasks for nations. A multi-sectoral approach is urgently needed, with shared responsibility across governmental levels. Building economic resilience to better withstand such threats while concurrently addressing the contributing factors, including climate change, is of paramount importance. Looking ahead, fostering intergovernmental partnerships and international cooperation for resource mobilization, coupled with cross-disciplinary leadership, will be key to minimizing socio-economic and environmental impacts and confronting the root causes of these challenges.

References

Aich A, Dey D, Roy A (2022) Climate change resilient agricultural practices: a learning experience from indigenous communities over India. PLOS Sustain Transform 1:e0000022. https://doi.org/10.1371/journal.pstr.0000022

Anjita NA, Indu J (2023) Leveraging weather radars for desert locust monitoring. Remote Sens Appl: Soc Environ 31:100983. https://doi.org/10.1016/j.rsase.2023.100983

Babar TK (2023) Risks of deserts locust and its mitigation. In: Ahmed M, Ahmad S (eds) Disaster risk reduction in agriculture. Springer Nature, Singapore, pp 361–392. https://doi.org/10.1007/978-981-99-1763-1_17

Banik A, Mondal MdF, Khan MdMR, Ahmed SR, Hasan MdM (2022) Screening and potent applicability analysis of commonly used pesticides against Schistocerca gregaria and Locusta migratoria: an integrative computational approach. Int J Trop Insect Sci 42:1971–1986. https://doi.org/10.1007/s42690-021-00726-x

Baraka GE, Sirera AM, Ong'amo GO (2023a) Limited preparedness: challenges during desert locust management in Kenya. Plant Protect 7:519–533. https://doi.org/10.33804/pp.007.03.4880

Baraka GE, Sirera AM, Ongamo GO, Baraka GE, Sirera AM, Ongamo GO (2023b) The dilemma of balancing between benefits and risks: desert locust management best practices. Path of Sci 9:8001–8008. https://doi.org/10.22178/pos.97-15

Barasa B, Watsusi CW, Gudoyi PM, Lutaaya N, Gumisiriza LT, Magaya JP (2023) Desert locust invasion in Uganda: effects on household food consumption and effective control interventions. Sustainability 15:14496. https://doi.org/10.3390/su151914496

El Bashir S (1997) Alternative strategy and tactics for the management of the desert locust, Schistocerca gregaria (Forsk.). In: Krall S, Peveling R, Ba Diallo D (eds), New strategies in locust control. Birkhäuser, Basel, pp 305–312. https://doi.org/10.1007/978-3-0348-9202-5_46

Bennett LV (1976) The development and termination of the 1968 plague of the desert locust, Schistocerca gregaria (Forskål) (Orthoptera, Acrididae). Bull Entomol Res 66:511–552. https://doi.org/10.1017/S000748530000691X

Bennett M (2020) Desert locust outbreak highlights gaps in risk governance | Good Governance Africa. https://gga.org/desert-locust-outbreak-highlights-gaps-in-risk-governance/ (Accessed 4 Aug 24)
Betts RA, Belcher SE, Hermanson L, Klein Tank A, Lowe JA, Jones CD, Morice CP, Rayner NA, Scaife AA, Stott PA (2023) Approaching 1.5 °C: how will we know we've reached this crucial warming mark? Nature 624:33–35. https://doi.org/10.1038/d41586-023-03775-z
Brader L, Djibo H, Faya FG, Ghaout S, Lazar M, Luzietoso PN, Ould-Babah MA (2006a) Towards a more effective response to desert locusts and their impacts on food security, livelihood and poverty multilateral evaluation of the 2003–05 Desert Locust Campaign, Food and Agriculture Organisation, Rome, Italy, pp 1–42
Brader L, Djibo H, Faye FG, Ghaout S, Lazar M, Luzietoso PN, Ould Babah MA (2006b) Towards a more effective response to desert locusts and their impacts on food security, livelihoods and poverty. Multilateral evaluation of the 2003–05 Desert locust campaign, Food and Agriculture Organisation, Rome
Bundey S, Raymond S, Dean P, Roberts SK, Dillon RJ, Charnley AK (2003) Eicosanoid involvement in the regulation of behavioral fever in the desert locust, Schistocerca gregaria. Arch Insect Biochem Physiol 52:183–192.https://doi.org/10.1002/arch.10081
Chen C, Qian J, Chen X, Hu Z, Sun J, Wei S, Xu K (2020) Geographic distribution of desert locusts in Africa, Asia and Europe using multiple sources of remote-sensing data. Remote Sens 12:3593. https://doi.org/10.3390/rs12213593
Cressman K (2001) Desert locust guidelines 3. Information and forecasting. FAO, Rome
Cressman K (2016) Chapter 4.2—desert locust. In: Shroder JF, Sivanpillai R (eds), Biological and environmental hazards, risks, and disasters. Academic Press, Boston, pp 87–105. https://doi.org/10.1016/B978-0-12-394847-2.00006-1
Despland E, Collett M, Simpson SJ (2000) Small-scale processes in desert locust swarm formation: how vegetation patterns influence gregarization. Oikos 88:652–662. https://doi.org/10.1034/j.1600-0706.2000.880322.x
Despland E, Rosenberg J, Simpson SJ (2004) Landscape structure and locust swarming: a satellite's eye view. Ecography 27:381–391. https://doi.org/10.1111/j.0906-7590.2004.03779.x
Dinku T, Ceccato P, Cressman K, Connor SJ (2010) Evaluating detection skills of satellite rainfall estimates over desert locust recession regions. J Appl Meteorol Climatol 49:1322–1332. https://doi.org/10.1175/2010JAMC2281.1
Dobson HM (2001) Desert locust guidelines 4. Control; FAO, Rome, Italy
Ellenburg WL, Mishra V, Roberts JB, Limaye AS, Case JL, Blankenship CB, Cressman K (2021) Detecting desert locust breeding grounds: a satellite-assisted modeling approach. Remote Sens 13:1276. https://doi.org/10.3390/rs13071276
Enns C, Bersaglio B, Karmushu R (2022) Disaster management takes to the skies: how new technologies are reconfiguring spatialities of power in desert locust management. Polit Geogr 98:102732. https://doi.org/10.1016/j.polgeo.2022.102732
FAO (2020) Desert Locust risk maps—2020 (WWW Document). URL https://www.fao.org/ag/locusts/en/archives/1340/2517/2518/index.html (Accessed 23 Mar 24)
FAO (2021) The impact of disasters and crises on agriculture and food security: FAO, Rome, Italy.https://doi.org/10.4060/cb3673en
FAO (2024) Desert locust (WWW Document). Food and Agriculture Organization of the United Nations. http://www.fao.org/locusts/en/ (Accessed 4 Feb 24)
FAO, WMO (2016) Weather and desert locusts (Technical No. 1175). World Meteorological Organization and Food and Agriculture Organization, Geneva
Fenibo EO, Ijoma GN, Matambo T (2021) Biopesticides in sustainable agriculture: a critical sustainable development driver governed by green chemistry principles. Front Sustain Food Syst 5.https://doi.org/10.3389/fsufs.2021.619058
Le Gall M, Overson R, Cease A (2019) A global review on locusts (Orthoptera: Acrididae) and their interactions with livestock grazing practices. Front Ecol Evol 7.https://doi.org/10.3389/fevo.2019.00263

Georgiou F, Buhl J, Green JEF, Lamichhane B, Thamwattana N (2021) Modelling locust foraging: how and why food affects group formation. PLoS Comput Biol 17:e1008353. https://doi.org/10.1371/journal.pcbi.1008353

Githae EW, Kuria EK (2021) Biological control of desert locust (Schistocerca gregaria Forskål). CABI Rev.https://doi.org/10.1079/PAVSNNR202116013

Gómez D, Salvador P, Sanz J, Casanova C, Taratiel D, Casanova JL (2018) Machine learning approach to locate desert locust breeding areas based on ESA CCI soil moisture. JARS 12:036011. https://doi.org/10.1117/1.JRS.12.036011

Gómez D, Salvador P, Sanz J, Casanova C, Taratiel D, Casanova JL (2019) Desert locust detection using earth observation satellite data in Mauritania. J Arid Environ 164:29–37. https://doi.org/10.1016/j.jaridenv.2019.02.005

Gómez D, Salvador P, Sanz J, Casanova JL (2020) Modelling desert locust presences using 32-year soil moisture data on a large-scale. Ecol Ind 117:106655. https://doi.org/10.1016/j.ecolind.2020.106655

Gómez D, Salvador P, Sanz J, Rodrigo JF, Gil J, Casanova JL (2021) Prediction of desert locust breeding areas using machine learning methods and SMOS (MIR_SMNRT2) near real time product. J Arid Environ 194:104599. https://doi.org/10.1016/j.jaridenv.2021.104599

Gross M (2021) How locusts become a plague. Curr Biol 31:R459–R461. https://doi.org/10.1016/j.cub.2021.05.007

Guan J, Li M, Ju X, Lin J, Wu J, Zheng J (2021) The potential habitat of desert locusts is contracting: predictions under climate change scenarios. PeerJ 9:e12311. https://doi.org/10.7717/peerj.12311

Herbillon F, Piou C, Meynard CN (2024) An increase in management actions has compensated for past climate change effects on desert locust gregarization in western Africa. Heliyon e29231. https://doi.org/10.1016/j.heliyon.2024.e29231

Hielkema JU, Snijders FL (1994) Operational use of environmental satellite remote sensing and satellite communications technology for global food security and locust control by FAO: the ARTEMIS and DIANA systems. Acta Astronaut 32:603–616. https://doi.org/10.1016/0094-5765(94)90071-X

Joshi S, Musne P, Munjale A, Munde O, Mundhe S, Murkute Y, Khan M (2023) Agriculture monitoring system using locust intrusion detection method. In: 2023 2nd International conference on edge computing and applications (ICECAA). Presented at the 2023 2nd International conference on edge computing and applications (ICECAA), pp 249–251. https://doi.org/10.1109/ICECAA58104.2023.10212375

Kamga SF, Ndjomatchoua FT, Guimapi RA, Klingen I, Tchawoua C, Hjelkrem A-GR, Thunes KH, Kakmeni FM (2022) The effect of climate variability in the efficacy of the entomopathogenic fungus Metarhizium acridum against the desert locust Schistocerca gregaria. Sci Rep 12:7535. https://doi.org/10.1038/s41598-022-11424-0

Kassegn A, Endris E (2021) Review on socio-economic impacts of 'Triple Threats' of COVID-19, desert locusts, and floods in East Africa: evidence from Ethiopia. Cogent Soc Sci 7:1885122. https://doi.org/10.1080/23311886.2021.1885122

Khan MA, Ahmad W (2019) Synthetic chemical insecticides: environmental and agro contaminants. In: Khan MA, Ahmad W (eds), Microbes for sustainable insect pest management : an eco-friendly approach. vol 1. Springer International Publishing, Cham, pp 1–22. https://doi.org/10.1007/978-3-030-23045-6_1

Khan BA, Nadeem MA, Nawaz H, Amin MM, Abbasi GH, Nadeem M, Ali M, Ameen M, Javaid MM, Maqbool R, Ikram M, Ayub MA (2023) Pesticides: impacts on agriculture productivity, environment, and management strategies. In: Aftab T (ed) Emerging contaminants and plants: interactions, adaptations and remediation technologies. Springer International Publishing, Cham, pp 109–134. https://doi.org/10.1007/978-3-031-22269-6_5

Kimathi E, Tonnang HEZ, Subramanian S, Cressman K, Abdel-Rahman EM, Tesfayohannes M, Niassy S, Torto B, Dubois T, Tanga CM, Kassie M, Ekesi S, Mwangi D, Kelemu S (2020)

Prediction of breeding regions for the desert locust Schistocerca gregaria in East Africa. Sci Rep 10:11937. https://doi.org/10.1038/s41598-020-68895-2

Klein I, Oppelt N, Kuenzer C (2021) Application of remote sensing data for locust research and management—A review. Insects 12:233. https://doi.org/10.3390/insects12030233

Klein I, van der Woude S, Schwarzenbacher F, Muratova N, Slagter B, Malakhov D, Oppelt N, Kuenzer C (2022) Predicting suitable breeding areas for different locust species—A multi-scale approach accounting for environmental conditions and current land cover situation. Int J Appl Earth Obs Geoinf 107:102672. https://doi.org/10.1016/j.jag.2021.102672

Klein I, Uereyen S, Eisfelder C, Pankov V, Oppelt N, Kuenzer C (2023) Application of geospatial and remote sensing data to support locust management. Int J Appl Earth Obs Geoinf 117:103212. https://doi.org/10.1016/j.jag.2023.103212

Kumar KS, Abdul Rahman A (2021) Early detection of locust swarms using deep learning. In: Patnaik S, Yang X-S, Sethi IK (eds) Advances in machine learning and computational intelligence. Springer, Singapore, pp 303–310. https://doi.org/10.1007/978-981-15-5243-4_27

Landmann T, Agboka KM, Klein I, Abdel-Rahman EM, Kimathi E, Mudereri BT, Malenge B, Mohamed MM, Tonnang HEZ (2023) Towards early response to desert locust swarming in eastern Africa by estimating timing of hatching. Ecol Model 484:110476. https://doi.org/10.1016/j.ecolmodel.2023.110476

Latchininsky AV, Sivanpillai R (2010) Locust habitat monitoring and risk assessment using remote sensing and GIS technologies. In: Ciancio A, Mukerji KG (eds), Integrated management of arthropod pests and insect borne diseases. Springer Netherlands, Dordrecht, pp 163–188. https://doi.org/10.1007/978-90-481-8606-8_7

Lecoq M (2022) Schistocerca gregaria (desert locust). CABI Compendium CABI Compendium, pp 49833.https://doi.org/10.1079/cabicompendium.49833

Li S, Feng S, Ullah H, Tu X, Zhang Z (2022) IPM—Biological and integrated management of desert locust. J Integr Agric 21:3467–3487. https://doi.org/10.1016/j.jia.2022.09.017

Liu X, Zhang D, He X (2024) Unveiling the role of climate in spatially synchronized locust outbreak risks. Sci Adv 10:eadj1164. https://doi.org/10.1126/sciadv.adj1164

Maeno KO, Ould Ely S, Ould Mohamed S, Jaavar MEH, Nakamura S, Ould Babah Ebbe MA (2019) Defence tactics cycle with diel microhabitat choice and body temperature in the desert locust, Schistocerca gregaria. Ethology 125:250–261. https://doi.org/10.1111/eth.12845

Maeno K, Tanaka S (2011) Phase-specific responses to different qualities of food in the desert locust, *Schistocerca gregaria*: developmental, morphological and reproductive characteristics. J Insect Physiol 57:514–520. https://doi.org/10.1016/j.jinsphys.2011.02.004

Magor JI, Lecoq M, Hunter DM (2008) Preventive control and desert locust plagues. Crop Prot 27:1527–1533. https://doi.org/10.1016/j.cropro.2008.08.006

Mashtoly TA, El-Zemaity MS, Abolmaaty A, Abdelatef GM, Marzouk AA, Reda M (2019) Phylogenetic characteristics of novel Bacillus weihenstephanensis and Pseudomonas sp. to desert locust, Schistocerca gregaria Forskål (Orthoptera: Acrididae). Egypt J Biol Pest Control 29:85. https://doi.org/10.1186/s41938-019-0188-7

Matthews GA (2021) New technology for desert locust control. Agronomy 11:1052. https://doi.org/10.3390/agronomy11061052

Meynard CN, Gay P-E, Lecoq M, Foucart A, Piou C, Chapuis M-P (2017) Climate-driven geographic distribution of the desert locust during recession periods: subspecies' niche differentiation and relative risks under scenarios of climate change. Glob Change Biol 23:4739–4749. https://doi.org/10.1111/gcb.13739

Meynard CN, Lecoq M, Chapuis M-P, Piou C (2020) On the relative role of climate change and management in the current desert locust outbreak in East Africa. Glob Change Biol 26:3753–3755. https://doi.org/10.1111/gcb.15137

Miller GA, Clissold FJ, Mayntz D, Simpson SJ (2009) Speed over efficiency: locusts select body temperatures that favour growth rate over efficient nutrient utilization. Proc Royal Soc B: Biol Sci 276:3581–3589. https://doi.org/10.1098/rspb.2009.1030

Mullié WC, Prakash A, Müller A, Lazutkaite E (2023) Insecticide use against desert locust in the horn of Africa 2019–2021 reveals a pressing need for change. Agronomy 13:819. https://doi.org/10.3390/agronomy13030819

Negera M, Alemu T, Hagos F, Haileslassie A (2023) Impacts of climate-smart agricultural practices on farm households' climate resilience and vulnerability in Bale-Eco region, Ethiopia. Environ Dev Sustain. https://doi.org/10.1007/s10668-023-03962-y

Pandey M, Suwal B, Kayastha P, Suwal G, Khanal D (2021) Desert locust invasion in Nepal and possible management strategies: a review. J Agricul Food Res 5:100166. https://doi.org/10.1016/j.jafr.2021.100166

Peng W, Ma NL, Zhang D, Zhou Q, Yue X, Khoo SC, Yang H, Guan R, Chen H, Zhang X, Wang Y, Wei Z, Suo C, Peng Y, Yang Y, Lam SS, Sonne C (2020) A review of historical and recent locust outbreaks: Links to global warming, food security and mitigation strategies. Environ Res 191:110046. https://doi.org/10.1016/j.envres.2020.110046

Piou C, Gay P-E, Benahi AS, Babah Ebbe MAO, Chihrane J, Ghaout S, Cisse S, Diakite F, Lazar M, Cressman K, Merlin O, Escorihuela M-J (2019) Soil moisture from remote sensing to forecast desert locust presence. J Appl Ecol 56:966–975. https://doi.org/10.1111/1365-2664.13323

Qayyum MA, Yasin M, Wakil W, Hunter D, Ghazanfar MU, Wajid M, Saeed S, Ashfaq M (2023) Advanced technologies for monitoring and management of locusts. In: Locust outbreaks. Apple Academic Press

Reda M, Mashtoly TA, El-Zemaity ME-S, Abolmaaty A, Abdelatef GM (2018) Phylogenic and antagonistic characteristics of novel Bacillus cereus isolates against desert locust, Schistocerca gregaria Forskal (Orthoptera: Acrididae). Egypt J Biol Pest Control 28:51. https://doi.org/10.1186/s41938-018-0056-x

Retkute R, Hinton RGK, Cressman K, Gilligan CA (2021) Regional differences in control operations during the 2019–2021 desert locust upsurge. Agronomy 11:2529. https://doi.org/10.3390/agronomy11122529

Roffey J, Magor JI (2003) Desert locust population dynamics parameters. Technical Series-Desert Locust Field Research Stations (FAO)

Roussi A (2020) Why gigantic locust swarms are challenging governments and researchers. Nature 579:330–330. https://doi.org/10.1038/d41586-020-00725-x

Sabbahi R (2022) Economic value of insect pollination of major crops in Morocco. Int J Trop Insect Sci 42:1275–1284. https://doi.org/10.1007/s42690-021-00645-x

Sabbahi R, Merzouki A, Guertin C (2009) Potential effect of *Beauveria bassiana* (Hypocreales: Clavicipitaceae) on *Anthonomus signatus* (Coleoptera: Curculionidae) in strawberries. Biocontrol Sci Tech 19:729–741. https://doi.org/10.1080/09583150903100815

Sabbahi R, Hock V, Azzaoui K, Saoiabi S, Hammouti B (2022) A global perspective of entomopathogens as microbial biocontrol agents of insect pests. J Agricul Food Res 10:100376. https://doi.org/10.1016/j.jafr.2022.100376

Sabbahi R, Lavallée R, Merzouki A, Guertin C (2009a) Differentiation of entomopathogenic fungus Beauveria bassiana (Ascomycetes: Hypocreales) isolates by PCR-RFLP. Phyto 90:49–56. https://doi.org/10.7202/044022ar

Sabbahi R (2022a) Effects of climate change on insect pollinators and implications for food security—Evidence and recommended actions. In: Behnassi M, Gupta H, Barjees Baig M, Noorka IR (eds) The food security, biodiversity, and climate nexus. Springer International Publishing, Cham, pp 143–163. https://doi.org/10.1007/978-3-031-12586-7_8

Salih AAM, Baraibar M, Mwangi KK, Artan G (2020) Climate change and locust outbreak in East Africa. Nat Clim Chang 10:584–585. https://doi.org/10.1038/s41558-020-0835-8

Segoli M, Kishinevsky M, Rozenberg T, Hoffmann I (2020) Parasitoid abundance and community composition in desert vineyards and their adjacent natural habitats. Insects 11:580. https://doi.org/10.3390/insects11090580

Showler AT (2019) Desert locust control: the effectiveness of proactive interventions and the goal of outbreak prevention. Am Entomol 65:180–191. https://doi.org/10.1093/ae/tmz020

Showler AT, Ould Babah Ebbe MA, Lecoq M, Maeno KO (2021) Early intervention against desert locusts: current proactive approach and the prospect of sustainable outbreak prevention. Agronomy 11:312.https://doi.org/10.3390/agronomy11020312

Showler AT, Shah S, Sulaiman Khan S, Ullah S, Degola F (2022) Desert locust episode in Pakistan, 2018–2021, and the current status of integrated desert locust management. J Integrat Pest Managem 13:1. https://doi.org/10.1093/jipm/pmab036

Showler AT (2003) The importance of armed conflict to Desert Locust control, 1986–2002. Orth 12:127–133. https://doi.org/10.1665/1082-6467(2003)012(0127:TIOACT)2.0.CO;2

Showler AT (2008) Desert Locust, Schistocerca gregaria Forskål (Orthoptera: Acrididae) plagues. In: Capinera JL (ed) Encyclopedia of entomology. Springer Netherlands, Dordrecht, pp 1181–1186. https://doi.org/10.1007/978-1-4020-6359-6_885

Sokame BM, Agboka KM, Kimathi E, Mudereri BT, Abdel-Rahman EM, Landmann T, Rwaheru MM, Abdalla O, Mafabi MM, Lubango LM, Tonnang HEZ (2024) An integrated assessment approach for socio-economic implications of the desert locust in Eastern Africa. Earth's Future 12:e2023EF003841. https://doi.org/10.1029/2023EF003841

Song H, Mariño-Pérez R, Woller DA, Cigliano MM (2018) Evolution, diversification, and biogeography of grasshoppers (Orthoptera: Acrididae). Insect Systemat and Divers 2:3. https://doi.org/10.1093/isd/ixy008

Swaminathan R, Swaminathan T (2019) Taxonomy of Orthoptera with emphasis on Acrididae. In: Indian insects. CRC Press

Symmons P, Cressman K (2001) Desert locust guidelines 1. biology and behavior. FAO, Rome

Tang Q, Feng J, Zong D, Zhou J, Hu X, Wang B, Wang T (2023) Potential spread of desert locust Schistocerca gregagia (Orthoptera: Acrididae) under climate change scenarios. Diversity 15:1038. https://doi.org/10.3390/d15101038

Umeh SC, Gil-Alana LA (2024) Trends in temperatures in Sub-Saharan Africa. Evidence of global warming. J Afri Earth Sci 213:105228. https://doi.org/10.1016/j.jafrearsci.2024.105228

UNEP (2021) Desert locusts' upsurges: a harbinger of emerging climate change induced crises?—Foresight Brief No. 022 January 2021. United Nations Environment Programme, Briefs, Summaries, Policies and Strategies

Villarreal M (2022) Desert locusts: can mathematical models help to control them?. In: Emmer M, Abate M (eds) Imagine math 8: dreaming venice. Springer International Publishing, Cham, pp 405–417. https://doi.org/10.1007/978-3-030-92690-8_26

Vosshall LB (2020) Catching plague locusts with their own scent. Nature 584:528–530. https://doi.org/10.1038/d41586-020-02264-x

Wang L, Zhuo W, Pei Z, Tong X, Han W, Fang S (2021) Using long-term earth observation data to reveal the factors contributing to the early 2020 desert locust upsurge and the resulting vegetation loss. Remote Sensing 13:680. https://doi.org/10.3390/rs13040680

Wang X, Li G, Wang S, Feng C, Xu W, Nie Q, Liu Q (2023) The effect of environmental changes on locust outbreak dynamics in the downstream area of the Yellow River during the Ming and Qing dynasties. Sci Total Environ 877:162921. https://doi.org/10.1016/j.scitotenv.2023.162921

Zhang Z, Cazelles B, Tian H, Christian Stige L, Bräuning A, Stenseth NC (2008) Periodic temperature-associated drought/flood drives locust plagues in China. Proc Royal Soc B: Biol Sci 276:823–831. https://doi.org/10.1098/rspb.2008.1284

Zhao L, Huang W, Chen J, Dong Y, Ren B, Geng Y (2020) Land use/cover changes in the Oriental migratory locust area of China: implications for ecological control and monitoring of locust area. Agr Ecosyst Environ 303:107110. https://doi.org/10.1016/j.agee.2020.107110

Zhao L, Li H, Huang W, Dong Y, Geng Y, Ma H, Chen J (2023) Outbreak mechanism of locust plagues under dynamic drought and flood environments based on time series remote sensing data: implication for identifying potential high-risk locust areas. Remote Sensing 15:5206. https://doi.org/10.3390/rs15215206

Rachid Sabbahi is a Professor at the Higher School of Technology—Laayoune, Ibn Zohr University, Morocco, since 2019. He completed a B.Sc. in Animal Biology at University of Hassan II (Casablanca, Morocco), a M.Sc. (2003) and a Ph.D. (2008) in Biological Sciences at Quebec University (Montreal, Canada). Dr. Sabbahi specializes in biological control of insect pests and plant diseases, plant-microbe interaction, environmental microbial biotechnology, and insect pollination and crop production. In addition, Dr. Sabbahi has co-managed many research projects both nationally and internationally, published several scientific papers and participated in many international conferences.

Chapter 13
Food Security Face to Invasive Species and Climate Change in Tunisia

Ben Jamâa Mohamed Lahbib, Ahlem Belhaj, Besma Mrabet, Essia Limem, Imed Jaouadi, Hichem Asses, Lassaad Medhioub, Mohamed Chaouachi, Mouna Mhafdhi, Nadia Debbabi, and Noura Touati

Abstract This chapter is an overview of the Tunisian food security and the national Integrated Pest Management (IPM) strategies adopted to face invasive alien species (IAS) pathways of entry, to control pests and to enhance export and domestic demand. In Tunisia, the agricultural sector involves 12.6% of gross domestic product (GDP) and employs almost one quarter of the country's labor force. However, this sector is

B. J. M. Lahbib (✉)
General Directorate of Plant Health and Control of Agriculture Inputs; Ministry of Agriculture, Water Resources and Fisheries, Eaux et Forêts (INRGREF), University of Carthage, Institut National de Recherches en Génie Rural, Tunis, Tunisia
e-mail: benjamaaml@gmail.com

A. Belhaj · B. Mrabet · E. Limem · I. Jaouadi · H. Asses · L. Medhioub · M. Chaouachi · M. Mhafdhi · N. Debbabi · N. Touati
General Directorate of Plant Health and Control of Agricultural Inputs, Ministry of Agriculture, Water Resources and Fisheries, Tunis, Tunisia
e-mail: ahlemgueddich@gmail.com

B. Mrabet
e-mail: besma.mrabetsamaali@gmail.com

E. Limem
e-mail: essialimem@yahoo.fr

I. Jaouadi
e-mail: jaouadi_i@yahoo.fr

H. Asses
e-mail: Hichem.Asses@fao.org

L. Medhioub
e-mail: medhioublassaad@gmail.com

M. Chaouachi
e-mail: mohamed.chawachi@gmail.com

M. Mhafdhi
e-mail: mounamhafdhi@gmail.com

N. Debbabi
e-mail: nadiadabbabi@gmail.com

M. Behnassi et al. (eds.), *Food Systems and Biodiversity in the Context of Environmental and Climate Risks*, https://doi.org/10.1007/978-3-031-89167-0_13

threatened by various IAS introduced in the twentieth century (such as *Phyllocnistis citrella* in 1994, *Tuta absoluta* in 2008, *Rhynchophorus ferrugineus* in 2011, *Erwinia amylovora* in 2012, *Phyllosticita citricarpa* in 2019, and *Citrus tristeza virus* and *Dactylopius opuntiae* in 2021). These IAS are subject to a national control strategy to eradicate the disease from contaminated areas, and prevent its introduction into free areas. This program also contributes preventing the introduction of other pests (such as *Xylella fastidiosa, Spodoptera frugiperda,* and *Palmaspis phoenicis*) recently detected in neighboring Mediterranean countries and can dramatically induce a disaster to domestic arable lands. In addition to pest control, training and communication programs are conducted to develop National Phytosanitary Capacity, share information, and take measures against IAS. Tunisia has also implemented certification programs to preserve agricultural sectors and to replant damaged areas with healthy and resistant varieties.

Keywords Citrus · Cactus · Date · Olive strategy · Tunisia

13.1 Agricultural Tunisian Sector Facing the Natural Disasters

The 1992 Rio Summit and the United Nations Framework Convention on Climate Change set a global framework for the intergovernmental effort to meet the challenge posed by climate change. Indeed, the climate system is an unstable resource constantly affected by greenhouse gas emissions. Most countries are affected by global warming and climate-induced events, especially drought and flooding. At the national level, the assessment of climate change impacts on the economy and society should be developed for a better adaptation strategy. In the case of Tunisia, agriculture and natural resources, namely hydraulic resources, environmental protection, marine coastal system, livestock, etc. are the most concerned sectors (WB 2020). In this perspective, future agricultural policies should consider a system of protection of natural resources against plant diseases and invasive alien species (IAS) (OSS 2020).

The Tunisian Republic covers 163,610 km^2 with a population of over 12 million. The country is situated on the North African Mediterranean Sea coast bordering Algeria to the west, and Libya to the south-east. In the north the climate is Mediterranean with mild and rainy winters, hot and dry summers. Forty per cent of the country is comprised of the Sahara Desert, with much of the remainder consisting of particularly fertile soil, with easily accessible coasts. The north is mountainous, which, moving south, gives way to a hot, dry central plain. The south is semiarid, and merges into the Sahara. The lowest point is Chott el Gharsa, at − 17 m, and the highest is Jebel ech Chambi, at 1544 m.

N. Touati
e-mail: tnoura@gmail.com

In Tunisia, the agricultural sector employs nearly 14.5% of the total working population and contributes 12% of the GDP. It also contributes to the commercial balance, accounting for 10% of exports. The agricultural activity is conducted over 10 million hectares throughout 516,000 farmlands. The income for a large part of the population depends on agriculture and the use of natural resources. The sector, highly important, is evolving in a fragile natural environment, with a highly sensitive dependence on weather conditions. Furthermore, experts' forecasts for the upcoming years are not optimistic, with several regions, especially in the center and in the south, expected to sustain weather conditions most unfavorable to agriculture: temperature variations, heavier rainfall, and frequent droughts. Other experts are even more pessimistic, predicting that under the effect of extreme phenomena, some activities are likely to disappear in the future. Most widespread risks are droughts, floods, windstorms, hail, snow, frost, forest fires, diseases, insects, and faunal pests. According to Atlas Magazine (2019), the Tunisian agricultural sector suffers from some constraints among which:

- About 50% of farmlands are below five hectares in size, belonging to small farmers who sustain important losses in recurrent fashion;
- Natural resources are limited, lands are hardly arable with severe erosion;
- Poor average rainfall with irregular rain in space and time;
- Increasing frequency and severity of extreme phenomena such as droughts and floods: hail and torrential rain are the most dreaded phenomena by farmers;
- Rising temperatures by 2.1 °C and decrease in rain by − 10 to − 30% by 2050.

Natural disasters have caused 172.3 million USD in losses for the Tunisian agricultural sector during the period 2011–2016. Moreover, the season 2017–2018 was estimated at 33 million USD in damage, and the State covered only 6% of the total losses (1.9 million USD). The rest was at the expense of farmers.

Despite the socio-economic importance of agriculture in Tunisia, this sector is also threatened by various IAS introduced in the twentieth century as *Phyllocnistis citrella* in 1994, *Tuta absoluta* in 2008, *Rhynchophorus ferrugineus* in 2011, *Erwinia amylovora* in 2012, *Phyllosticita citricarpa* in 2019, and *Citrus tristeza virus* and *Dactylopius opuntiae* in 2021. For a protection purpose, a national strategy has been established to: (i) face IAS pathways of entry; (ii) control pests; and (iii) enhance export and domestic demand. All these IAS are being subject to a national control strategy to reach the same goal, eradicate the disease from contaminated areas, and prevent its introduction into free (healthy) areas as is the case of the quarantine bacterium *Xylella fastidiosa.*

13.2 The Main Crops Facing Invasive Pests

13.2.1 Case of the Date

Socio-economic Importance of the Date Sector in Tunisia

In Tunisia, the date sector is one of the pillars of the agricultural sector with olive oil and citrus (Fig. 13.1). The palm date area covers more than 58,000 ha, means 13% of all arboreal area (DGPA 2024). In fact, Tunisia has 5.5 million palm trees of which 66% (3.5 million palm trees) of Deglet Nour. Palm trees are distributed in 4 governorates: Gafsa (5.2%), Gabes (11.6%), Tozeur (16.7%), and Kebili (66.5%).

Dates represent 19% of Tunisian agricultural exports. In 2022, dates production was 368,000 tons and export 125,000 tons. Tunisia export dates (Fig. 13.2), especially Deglet Nour (GID 2024), to 85 countries. In addition to its economic importance, the dates sector also has a social importance. Tunisia has 60,000 farmers with 2 million working days per year, 400 collect centers, 115 packing stations, and 260 exporting companies. A large workforce is necessary for the entire production and packaging chain in the dates sector (GIF Tunisia 2021). During the last 30 years (1990–2020), the dates production increases three times (Fig. 13.3).

Tunisia is the fifth producer of dates in the world (GID 2024), with 9% quantity of production, and the first in the world with 20% quality of the dates produced (GIF Tunisia 2020). However, the palm date is threatened by various pests, mainly Palm green scale (*Palmaspis phoenicis*), Date moth (*Ectomyeloïs ceratoniae*), and

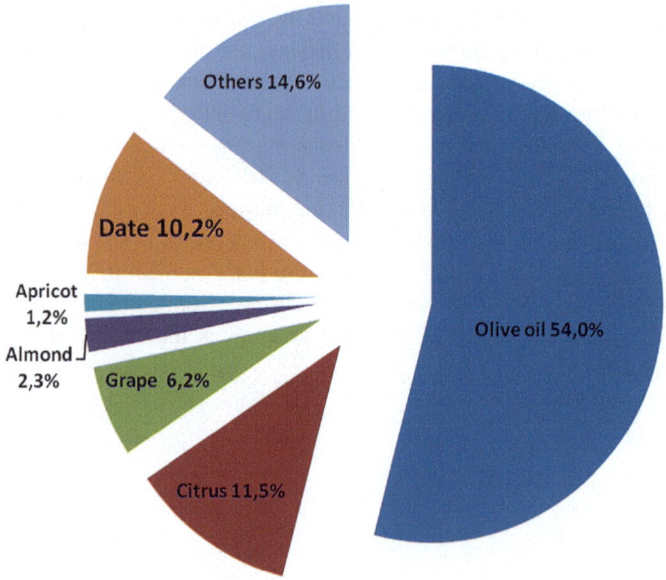

Fig. 13.1 Annual tree production, 2018. *Source* agridata.tn, Ministry of agriculture of Tunisia

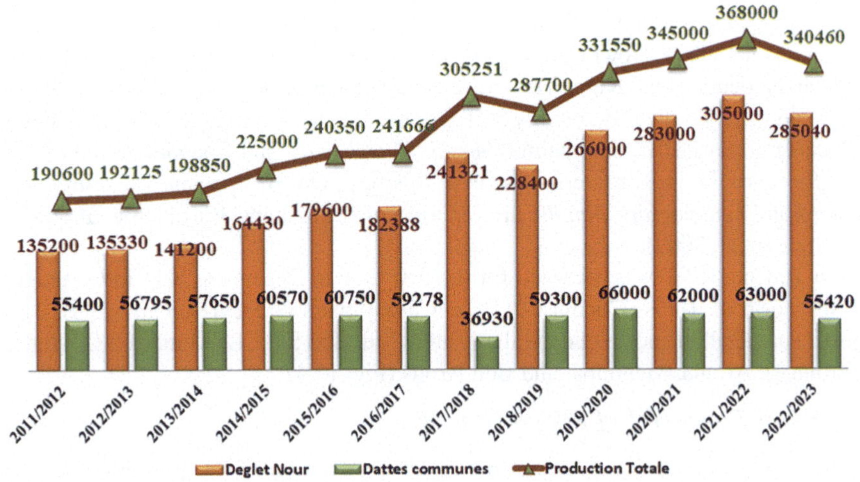

Fig. 13.2 Evolution of dates production per variety

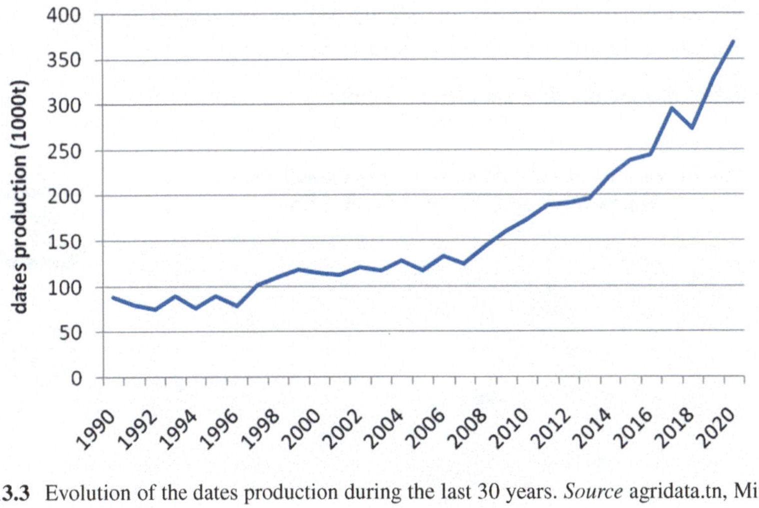

Fig. 13.3 Evolution of the dates production during the last 30 years. *Source* agridata.tn, Ministry of agriculture of Tunisia

Date palm mite (*Oligonychus afrasiaticus*). Red Palm Weevil (*Rhynchophorus ferrugineus*) was observed on Canary palm.

The Red Palm Weevil: Rhynchophorus ferrugineus

The Red Palm Weevil (RPW), *Rhynchophorus ferrugineus,* attacks palms in many countries of the world: India (the original home), Southeast Asian countries, the Arabian Gulf, the Middle East, most of the Mediterranean countries, California, USA, and others. The insect has three to four generations per year and develops four stages (egg, larva, pupa, and adult), with the larval stage is the harmful one. The insect spends all its growth phases inside the tree trunk causing damages that can induce the death of the palm. The entire development cycle of the RPW lasts about 4 months. The female lays roughly 300 eggs near the base of the crown. Canary palm (*Phoenix canariensis*), date palm (*Phoenix dactylifera*), and Coconut palm (*Cocos nucifera*) are the most preferred hosts of this pest. On the opposite, *Washingtonia filifera* and *Chamaerops humilis* are reported to be resistant (Barranco et al. 2000; Dembilio et al. 2009).

In Tunisia, RPW was recorded for the first time in December 2011 in Carthage, in the north of Tunis (Chebbi 2011) on Canary palm (*Phoenix canariensis*). The insect spread all the north of capital city area (Sidi Bou Said, La Marsa, La Goulette, Carthage), Ariana, Manouba, and Ben Arous (Fig. 13.4).

Legislative Framework of RPW in Tunisia

RPW appears in the list of quarantine organisms. The entry was prohibited for all plants by an order issued by the Minister of Agriculture on May 31st, 2012 and a list of quarantine organisms was established. On October 24th, 2012, another order was issued to control and eradicate the pest. On June 26th, 2015, a decree was issued and reported that the import of plants, plant products, and any object in whole or in part of the Palmaceae family is prohibited from all countries.

Action Plan and Strategies Adopted in Tunisia

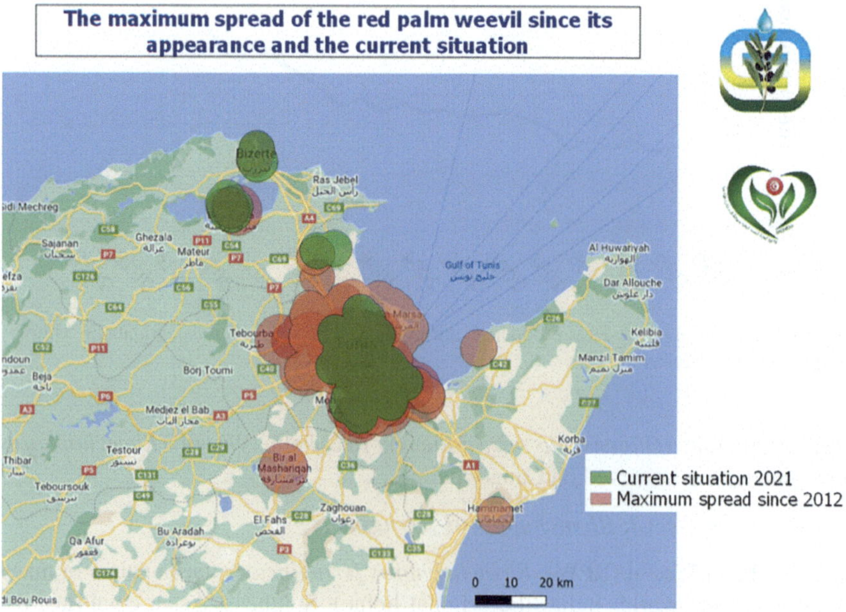

Fig. 13.4 Infestation area: maximum spread in red and current situation in 2021 spread

- **Period 2012–2014**: The FAO TCP/RAB/3307 project was launched in June 2012 for 2 years. It consists of supporting efforts of technical interventions to control the RPW. The main technical interventions were as follows: (i) localization of palm trees using GIS to control infestation; (ii) cutting of infested palm fronds and treatments of palms; (iii) foliar spray of palms using chemical products in infested area and endo-therapy injection every 6 to 12 months.
- **Period 2016–2019**: A national strategy to control and eradicate the RPW was conducted in the infested area and to prevent introduction and dissemination of RPW in the area of date palm tree production. This strategy consists of encirclement of the affected areas, limitation of their scope and preparation of the removal process.
- **Period 2020–2022**: During this period, the national strategy to control and eradicate the RPW was extended to cover all the area in the north and the dates production areas in the south of the country.

RPW has only been detected on *Phoenix canariensis* in urban areas in the north in 2011. The total number of georeferenced Canary palm was 52,000 trees with 9000 infested ones. Overall, 18% of Canary palms were eliminated. The national strategy to control and eradicate the RPW in the north of Tunisia adopted in 2016 to limit its propagation consists of several actions: (i) field control in affected areas; (ii) preventive measures to avoid spread of the RPW to free areas and to dates production areas; (iii) capacity building by training activities, awareness, guidance; and (iv) scientific research. The control of the RPW was conducted over the year and consists of:

- Treatment of the palm fronds 48 h before cutting, then cutting the tops of the affected palm fronds and sorting.
- A monthly foliar treatment of palm tree.
- An annual endotherapy treatment (in the beginning of the spring).
- Mass trapping network consists of 800 traps distributed over all the infested areas.
    - After the first assessment, indicators of the RPW monitoring were approved (Fig. 13.5): chemical control by monthly foliar treatment of all palms (threshold ~ 42.3%); and the endotherapy treatment of all palms once a year from March to May (threshold ~ 71.5%). The mechanical treatment (apical cut of infested palm trees) and the mass trapping are fundamental to monitor the population dynamics of the RPW.
- The national strategy for the prevention of the RPW in the dates production areas in the south of Tunisia consists of the establishment of rules and regulations. Moreover, an observatory to monitor the health status of oases and prevent invasion of pests was created. Furthermore, agricultural research, training, and guidance are thoroughly solicited. This proactive plan allows a rapid intervention against RPW.

The Date Palm Green Pit Scale: Palmaspis phoenicis

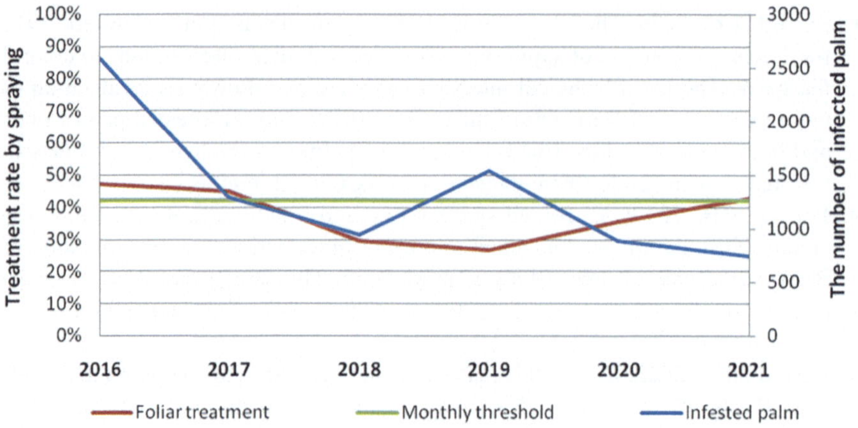

Fig. 13.5 Assessment of treatment of infested palms from 2016 to 2021

Palm green scale insect, *Palmaspis phoenicis,* is an invasive pest detected in the Middle-East, Near-East, North Africa, and USA. Tunisia is free from *Palmaspis phoenicis*, although the risk of its introduction is very high since the pest exists in Libya. *Palmaspis phoenicis* has two short generations in spring and summer, and one long generation in autumn and winter (Ali 1989), overwintering as an immature female and mating in the following May. The female development cycle lasts 85–95 days in spring and summer, and 15–180 days in autumn and winter. Contrariwise, the male development takes only 50–60 days. Adult flight is observed in May–October and female lays between 2 and 22 eggs. Further, males represent only 25% of the total *Palmaspis* population on the palms.

The cochineal sucks the plant sap causing dryness and death of the infested palm fronds. Furthermore, it covers the date clusters and prevents plant respiration and photosynthesis process. Nymphs and females attack all the green palm parts and the fruits of the host with preference to young palms, especially those in shaded, moist areas away from direct sunlight.

Symptoms of palm infection are conspicuous; the green fruits turn yellow and dry out before maturity. On palms, the color of the fronds and leaves turn from green to complete yellowing, and if the infection is severe on the fronds, the palm may eventually die. The attack of this pest may affect the productivity of the tree and leads to losses in production from a single palm, up to 50 and 80% of the production. The management of this pest consists of: cleaning the top of the palm from the affected leaves and the armpits from the remnants of the fruits, and burning all; and getting rid of neglected palms and dead palms by cutting them into small pieces to be grinded (Ahmed et al. 2013).

13.2.2 Case of the Citrus

Socio-economic Importance of the Citrus Sector in Tunisia

The citrus industry occupies an important position in Tunisian agriculture as it covers 28,000 Ha (5% of the tree area) and the annual expansion rate of about 500 ha over the last decade (Fig. 13.6). It provides a source of income for 12,000 farmers, the majority of whom are small farmers, and 11 packaging units. It represents a source of living for over 25 thousand families and an estimated seasonal labor force of about 3 million working days per year (GIFRUIT 2022).

The citrus production area is located in the northeastern Tunisia in Cap Bon region, which represents approximately 75% of the total area, and the governorate of Nabeul alone accounts for 69% of this area. Other production sites can be found in the Governorates of Bizerte, Jendouba, and Sidi Bouzid.

Due to the diversity of cultivated varieties and the staggering of production, each year, this sector supplies the local market with fruits for 8 months from October until May. Several varieties of these citrus fruits are required in 12 foreign markets such as Valencia Late and Lemon, but the country-specific Maltese orange remains by far the most popular variety. Tunisia is, in fact, the world's leading exporter of half-blooded Maltese Orange with France since it receives more than 90% of the volume exported. The value of citrus fruits exports represents 14% of the total value of agricultural exports by providing currency income of 15 million dinars per year. The total value of citrus production is estimated to be about 125 million dinars, which is not less than 3% of the agricultural domestic output (GIRUIT 2022).

The citrus production increased from about 80,000 tons in the season 2008/2009 to more than 100,000 tons in the season 2019/2020 (Fig. 13.7).

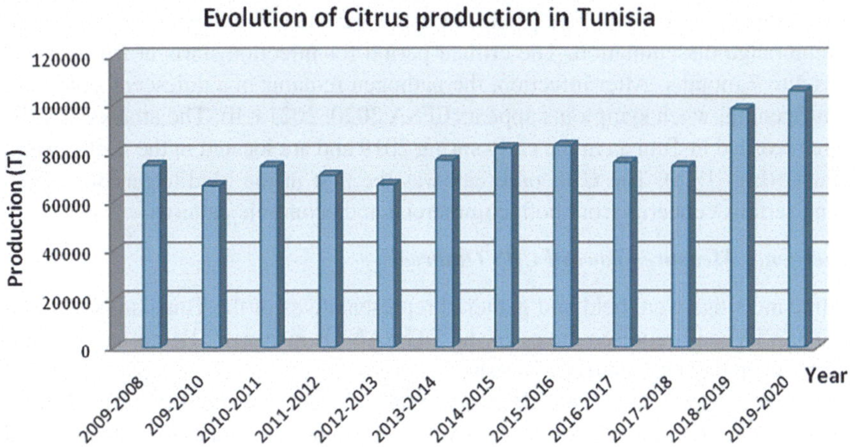

Fig. 13.6 Evolution of the citrus production area in Tunisia. 2020 *Source* GIF ()

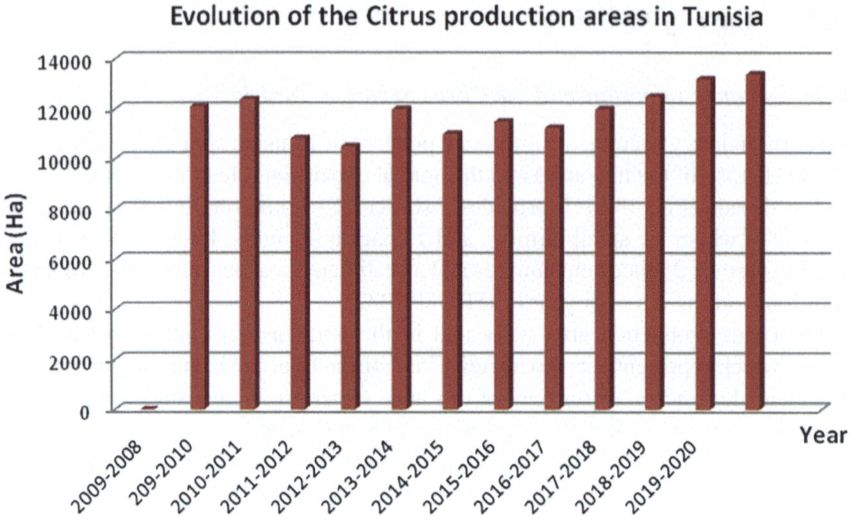

Fig. 13.7 Evolution of citrus production in Tunisia. 2020 *Source* GIF ()

During the 2018–2019, the citrus production sector faces up to many challenges in Tunisia, mainly the emergence of new pests/diseases such as *Citrus Tristeza virus* (CTV) recorded since 2018 (Najar et al. 2021), and Citrus Black Spot (CBS) caused by *Phyllosticta citricarpa,* occurred in 2019.

The Citrus Black Spot Phyllosticta citricarpa

Phyllosticta citricarpa, produces asexual spores (*pycnidiospores*) on citrus fruit lesions, twigs and leaf litter from which they can disperse on other plant parts. Sexual spores (*ascospores*) are produced in the leaf litter, and they are important for longue range dissemination. The critical period for infection starts at fruit set and lasts 4 to 7 months. After infection, the pathogen remains in a quiescent state until fruit ripening, when symptoms appear (EFSA 2020, 2021a; b). The attacks of CBC were recorded in Tunisia in the early spring 2019 and are located in the northeastern Tunisia (Fig. 13.8). The CBS outbreak was the first in the Mediterranean region, rising serious concerns from both commercial and economic patterns.

Emergency Measures Toward CBS Outbreak

Online meetings were held and gathered representatives of the Tunisian NPPO (G/D DGSVIA) and representatives of the DG SANTE in April 2019 to have a clear overview on this pest status in Tunisia.

Emergency measures were discussed with relevant stakeholders (GIF Tunisia 2020) for the: (i) intensification of surveys (7 groups) with sample collection and GPS coordinates. All stakeholders were involved and were introduced to the SOP to perform surveillance as well as relevant documents; (ii) establishment of laboratory for analysis according to ISPMs methods (purchase of necessary equipment and

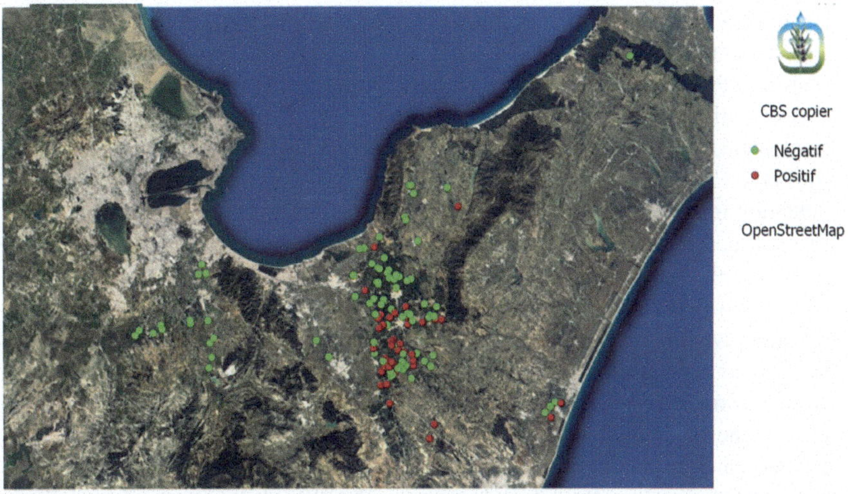

Fig. 13.8 CBS distribution in Tunisia by November 2019: red spots = infected orchards-(G/D plant health and control of agricultural products)

reagents); (iii) implementation of a fungal treatment schedule with 50% subsidy on fungicides intended to control CBS over an area of 2000 ha of infected orchards; and (iv) information days to raise awareness on the impact of the disease and the control methods—international trade fair for agriculture, agricultural machinery and fisheries (SIAMAP Tunisia, September 2019); the G/D of DGSVIA was also invited to the "CBS survey toolkit and contingency planning" (Lisbon-Portugal, November 2019) to share the Tunisian experience toward CBS detection and management published in the EFSA supporting publication. These activities cost 393,000 Dinars (121,255.13 Euros) and successfully achieved their goals.

In October 2019, the quarantine laboratory (G/D DGSVIA) was able to perform more reliable testing using molecular tools (Real-time PCR) to detect *P. citricarpa* on citrus fruits, and accurate mapping of the disease has been developed, so that citrus producers could benefit from 50% subsidy on chemical treatment to control the disease. Besides, an export certification scheme for citrus was set: exported fruits must come from certified orchards free from CBS, followed by a thorough inspection in the packaging house. Tunisia exported more than 10,467 T of citrus fruits during the year 2020, including 7543 T exported to France (GIF Tunisia 2021).

The National Strategy to Control CBS

A mid-term (2020–2022) contingency plan was developed (the National Strategy to Control Citrus Black Spot). This strategy involved the highest decision-making sphere in the country, including the ministerial council, to approve funds (more than 8 million dinars over 3 years). The strategy focuses essentially on six axes:

- Phytosanitary surveys were extended to all citrus production areas in the country (12 groups), with sampling and mapping.

- Laboratory tests performed according to the international standards. The quarantine laboratory purchased all necessary reagents and equipment to perform real-time PCR to detect CBS on samples coming from orchards or from packing houses willing to export. Moreover, the quarantine laboratory was also invited to participate to a proficiency test organized by ANSES (the French National Agency for Food, Environmental and Occupational Health Safety) regarding *P. citricarpa* detection in 2020 with 100% of conformity.
- Promoting scientific research on CBS biology, control and post-harvest measures, including the publication of the first report of Citrus Black Spot disease caused by *Phyllosticta citricarpa* on *Citrus limon* and *C. sinensis* in Tunisia. Further research works are still ongoing with focus on determining the biology of the fungus (including the existence of teleomorph *Guignardia citricarpa*), its epidemiology, the means to control the disease in post-harvest, as well as the effectiveness of fungicides in controlling the disease under Tunisian conditions. In fact, early research output concerning fungicide treatments have allowed to put in place a treatment schedule to control CBS while waiting for pesticide registration on citrus crops.
- To ensure the control of CBS through the correct application of the treatment schedule, especially since it requires 5–6 fungicide sprays, which increases the production cost especially for small farmers, the Ministry of Agriculture, Water Resources and Fisheries, and as part of the National Strategy to control CBS, allowed citrus producers in infested areas to receive a 50% subsidy on fungicides intended to control the disease mainly Mancozeb and Azoxystrobine.
- Establishment of an export certification scheme for citrus with a thorough inspection in packaging house according to related ISPMs. This export certification scheme continued to be adopted with exported fruits originating from production sites that are subject to treatments and cultivation techniques to control *P. citricarpa,* as well as official inspections on the production sites during the vegetative season, confirming the absence of the fungus.
- Rising awareness of producers, as well as owners of packing houses were mentored and supervised, and benefited from information days with presentations on CBS biology and control, activities undertaken in the national strategy and other quarantine fungi on citrus crops.

Citrus tristeza virus (CTV)

The CTV (EPPO A2 List) is classified as a quarantine organism according to Tunisian phytosanitary legislation officially since 2018, only in some area subject to official control.

The CTV is naturally transmitted in a semipersistent mode by several aphid species. The brown citrus aphid (*Toxoptera citricida*) is the most efficient vector because of its great abundance on citrus trees and its high efficacy of vectoring highly pathogenic isolates.

The symptoms of CTV depend on the nature of the invading isolate and the sensitivity of the citrus host. The most dramatic symptom referred to as “quick decline” results from rapid death of trees grafted onto sour orange rootstocks.

Other CTV-related diseases such as “stem pitting” of grapefruit or sweet orange trees are not related to a specific rootstock and do not result in death, but are often associated with stunting and poor productivity. Symptomless CTV isolates are used to provide protection (“cross-protection”) against stem pitting, but not against the quick decline (Bar-Joseph and Dawson 2008).

National Control Program of CTV

This organism is the subject of a national control strategy to eradicate the disease from contaminated areas and preventing its introduction into free areas. This strategy includes: prospecting and eradication, vector processing, nursery control, scientific research, training, and extension. In fact, surveys with sampling and laboratory tests have been carried out in different production regions mainly affected, such as the region of Nabeul, in order to limit the spread of the disease (Fig. 13.9). Surveys have highlighted the outbreaks of infestations in the areas of the governorate of Nabeul: Menzel Bouzelfa, Beni Khalled, and Soliman (Table 13.1). The *Citrus tristeza virus* control strategy monitoring committee regularly meets to assess the progress of the various components of the strategy. The main constraint encountered is mainly linked to the abstention of the majority of farmers from the application of compulsory control measures, which widened the initial source of contamination. Faced with this situation, the decision to ban the use of the traditional rootstock (sour orange: *Citrus aurantium*) was taken as well as its replacement by rootstocks torrents to the disease.

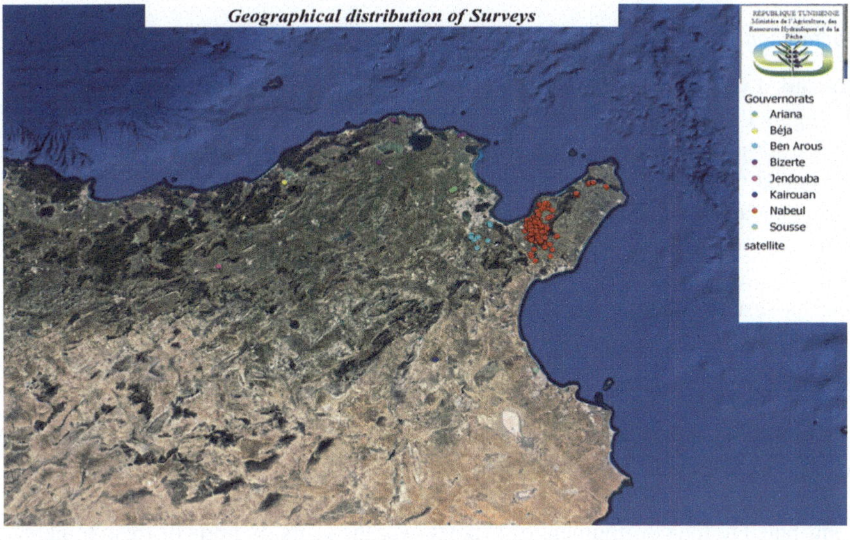

Fig. 13.9 Geographic areas of surveys

Table 13.1 Outbreaks of CTV in the north east citrus areas of Tunisia

Region	Zone	Area (Ha)	Sample count	Number of infections	Contamination rate
Ariana	Ariana	12	23	0	–
Ben Arous	Ben Arous	99	45	0	–
Sousse	Sousse	0	5	0	–
Kairouan	Kairouan	4	23	0	–
Bizerte	Bizerte	144	18	0	–
Béja	Béja	3	19	0	–
Nabeul	Mel. Bouzelfa	478	10,818	1936	17.9%
	Béni Khalled	330	9885	911	9.2%
	Bou Argoub	27	399	0	–
	Takelsa	73	62	0	–
	Soliman	194	4162	589	14.2%
	Grombalia	9	8	0	–
	Hammamet	7	4	0	–
	Nabeul	9	38	0	–
	Haouaria	0	0	0	–
	Total Nabeul	1127	25,376	3436	13.5%
Total		1388	25,509	3436	13.5%

13.2.3 The Case of Vegetable Crops

Socio-economic Importance of the Tomato Sector in Tunisia

Tomato, *Lycopersicon esculentum Mill* is a vegetable crop of large importance throughout the world. Its annual production accounts for 107 million metric tons with fresh market tomato representing 72% of the total (FAO 2002). It is the most consumed fruit in Africa, both in its raw and processed forms. Its production has a particular socio-economic significance, as it particularly offers employment opportunities to women, who contribute to over 60% of the labor force. It is the first horticultural crop in Tunisia with a production area of 29,000 hectares/year and a total harvest of 1.2 million metric tons (GIL 2016) of which nearly 70% are processed (Tomatonews 2011). Tomatoes are grown both under plastic covered greenhouses and in open field.

Tunisian tomato exports come mainly from fresh tomatoes, geothermal crops, and processed tomatoes (tomato concentrate, dried tomato, etc.). The geothermal tomato is characterized by a specific taste quality due in particular to the mineral composition of the water and the microclimate of southern Tunisia. This quality has contributed to the development of exports over the years. Exports of fresh tomatoes increased from 2481 tons during the 2004/2005 campaign to 13,981 tons during the 2013/2014 campaign (GIL 2016). However, several pests oppose this production, mainly *Tuta absoluta.*

The tomato leaf miner: Tuta absoluta

Regulation

In September 2004, the EPPO Council decided to add *Tuta absoluta* at A1 list (EPPO Reporting Service no. 09-2004 Num. article: 2004/129) and recommends its member countries to regulate it as quarantine pest (A1 pests are absent from the EPPO region). In September 2009, the EPPO Council approved the addition of *Tuta absoluta* to the EPPO A1 and A2 Lists of pests recommended for regulation. In order to draw the attention of NPPOs to these newly added pests, they have also been placed on the EPPO Action List. The later displays management measures which are recommended to the EPPO member countries and brief descriptions of the areas endangered by these pests. As some A1 pests are now found with a limited distribution in the EPPO region, they were transferred to the A2 List (EPPO Reporting Service no. 09-2009 Num. article: 2009/169).

Detection of *Tuta absoluta* in Tunisia

In October 2008, leaf mines were observed on field tomato crops during the phytosanitary inspections carried out in the region of Akouda (Tunisian Sahel coast). Leaf samples were collected from the affected plants and the causal agent was identified by the Tunisian quarantine laboratory as *Tuta absoluta* (EPPO A1 List). Further surveys carried out in other tomato-producing regions showed that the pest was also present in the regions of Kairouan (central part of Tunisia) and Bizerte (north of Tunisia).

In order to prevent any further spread of *T. absoluta*, phytosanitary measures were applied, including in particular the use of pheromone traps and the development of integrated control strategies combined use of biological control agents and chemical insecticides, information campaigns for the growers, nurserymen, and other stakeholders (EPPO 2009a; b).

Control of *Tuta Absoluta* in Tunisia: A Success Story

The female of *T. absoluta* deposits eggs on the underside of leaves, stems and to a lesser extent on fruits. After hatching, young larvae penetrate into tomato fruits, leaves/stems to feed and develop inducing mines and galleries (EPPO 2005). Tomato plants can be attacked at any developmental stage, from seedlings to mature stage (Braham and Hajji 2012).

At the beginning of the invasion of *T. absoluta*, the pest disturbed existing pest control programs in greenhouses and open fields, forcing Tunisian growers to use more chemical insecticides to reduce its impact (Abbes et al. 2012a, b).

Several approaches have been used for monitoring and controlling *Tuta absoluta*, in both the native and exotic ranges of the pest.

The methods currently being used in Africa, and mostly in Tunisia as identified by the household survey conducted in 2018, include pesticides, pheromone traps, destroying infected plants, staking, organic pesticide, and crushing larvae (CABI 2019). Furthermore, a national program was adopted to control *T. absoluta*, including cultural practices, the use of pheromones, the setting up of insect-proof screens in greenhouses, and the use of organic insecticides including *Bacillus thuringiensis kurstaki* (Btk) and plant extract-based products (Abbes et al. 2012a, b). In addition, experiments in biological control, using the predatory Mirid *Nesidiocoris tenuis* Reuter (Heteroptera: Miridae) and the parasitoid wasp *Trichogramma cacoeciae* Marchal (Hymenoptera: Trichogrammatidae) were conducted in greenhouses and open-field crops (producing tomatoes for fresh consumption as well as for processing). The parasitoid, *T. cacoeciae* is very efficient to reduce the infestation caused by larvae of *T. absoluta* with an optimal dose recommended to Tunisian farmers around 20 Trichogramma/plant in protected and open filed crops (Cherif 2018).

The mass trapping strategy demonstrate the need to be applied over an isolated field, in the whole area or under greenhouse conditions to minimize the influence of adult migration (Braham 2014).

The combined use of mass trapping and insecticides gave encouraging results given that problems of resistance were not found. In fact, the optimal density of sex pheromone traps to install in the greenhouses was fixed to two sex pheromone traps/greenhouse with the application of insecticides is sufficient for the control of *T. absoluta* in protected crops (Cherif 2018).

The total netting of the greenhouses is not economically rewarding since the netting of openings can give the same level of protection against *T. absoluta*. This netting setup can be ameliorated by using insecticide-treated nets to enhance the role of the physical barrier (Harbi et al. 2015).

Two strategies against *T. absoluta* were conducted (Abbes et al. (2012a). The first used Integrated Pest Management (IPM) through mass trapping was the release of *Nesidiocoris tenuis* and rational use of compatible insecticides. The second strategy was based on conventional use of synthetic insecticides (9 treatments). This experiment revealed at the end of the culture a 20% infestation rate of leaves in the IPM cropping system versus 98% in the conventional cropping system. The infestation rate of fruit was 18.2% in the IPM cropping system versus 46.8% with the conventional one.

The national program based on IPM strategy against *T. absoluta* has been funded by the Tunisian state until 2016 with the collaboration of the DGSVIA, agronomic research, and other stakeholders. This strategy comprised six major components summarized in Table 13.2. The Tunisian tomato farmers know how to control *T. absoluta*, that reached the equilibrium in its environment, and the pest no longer presents a major problem for the country.

13.2.4 The Case of Cereal Crops (Cereal, Maize)

Socio-economic Importance of the Cereal Crops Sector in Tunisia

In Tunisia, the cereal production represents 13% of the added agricultural value. Its contribution to the GDP remains very minimal around 1.44% on average over the last 10 years. Cereal farming occupies nearly 33% of the useful agricultural area (UAA) and mobilizes 250,000 farms, of which only 11,000 are over 50 ha (40% of surface

Table 13.2 Components of the national strategy against *T. absoluta*

Components	For open field	For green houses
1. Monitoring	2 sex pheromone traps/ha or per green house	
2. Threshold	30 oths/trap/day	
3. Cultural practices	Weed removal (host of *T. absoluta*)	
	Destroying/Removal of Infested leaf and fruits	
	Plow the soil and remove plant debris at the end of cultivation	
4. Physical methods	• Mass tapping: 30 sex pheromone traps/ha	• Mass tapping: 2 sex pheromone traps/ Greenhouse
		• Double airlock netting
5. Treatment of Chemical pesticides/Biological pesticides	Many experiments have been conducted and many active substances belonging to different chemical families have been registered	
6. Biological control	Predator*: Nesidiocoris tenuis* Reuter (Heteroptera: Miridae) Parasitoid *Trichogramma cacoeciae* Marchal (Hymenoptera: Trichogrammatidae)	

area). It assures an important part of livestock feed and supply the food chain (Khaldi and Bouali 2017). According to the Statista (2024) in the 2022/2023 crop year, the total land area used for cereal farming in Tunisia reached 1.28 million hectares. Most of these lands were under barley and durum wheat production, each exceeding 570,000 ha. On the other hand, a significantly smaller area of agricultural land was devoted to soft wheat and triticale farming. In 2021, the total cereals produced in Tunisia amounted to over 1.67 million metric tons. However, this crop is threatened by various pests, mainly the fall armyworm, maize stalk borers, *Sorghum midge Stenodiplosis sorghicola* / *Contarinia sorghicola* and Sorghum stem borers *Busseola fusca, Chilo partellus, Sesamia calamistis* (CABI 2019).

The Fall armyworm Spodoptera frugiperda

Regulation

The fall armyworm (FAW), *Spodoptera frugiperda,* is a transboundary invasive pest, native to tropical and subtropical regions of the American country. In 2016, it was recorded in the west of Africa, but since then it has spread to many countries in the region including Egypt, Sudan and Yemen. In Tunisia, the regulatory status of this pest is on the list of quarantine organisms as was indicated in the order of the Minister of Agriculture on May 31st, 2012. Until today, Tunisia is a free country from FAW.

It is a voracious and polyphagous insect feeding on more than 80 plant species belonging to 27 families. The pest has a preference to the Poaceae family, mainly maize, sorghum, and wheat and can cause devastating economic losses. The adult female moth is a strong flyer, traveling up to 100 km in a single night. Several natural enemies regulate the FAW population levels, including generalist predators (ants, earwigs, etc.), specialized parasitoids (small wasps that kill eggs or larvae), and pathogens (bacteria, virus, and fungus).

Female lays eggs in masses of 50–200 eggs. Larvae generally emerge 3 to 5 days following oviposition and migrate to the whorl and pass through 6 instars larvae (FAO 2017).

The invasive nature of this pest is underscored by its way of spread and its versatility in terms of ecological requirements and biological behavior.

Action Plan in Tunisia

As part of monitoring spread of the FAW in the region and its recent high spread in Africa, Tunisia has taken preventives measures against the introduction of this pest by mobilizing a working committee including members of the DGSVIA and the Research Institutes belonging to the Institution de la Recherche et de l'Enseignement Supérieur Agricole (IRESA). This committee established an action plan based essentially on 3 axes:

- A monitoring system (early detection) based on visual prospecting and trapping (network of pheromone traps).
- Workshops and awareness program (leaflets, posters, information days, etc.).
- Update of the phytosanitary legislation by issuing a circular establishing emergency measures intended to prevent the introduction of this pest.

Since 2020 until 2023, a network of the FAW traps was established in different regions of Tunisia, from the north to the south, especially in the governorates of Ariana, Manouba, Bizerte, Nabeul, Jendouba, Sidi Bouzid, and Gabes. This network of pheromone traps was established on Maize, sorghum and artichoke crops to detect the occurrence of this insect. Results of the monitoring system of this pest ensured by trapping and prospection plots of the main host plants prove that until now Tunisia is a country free from *Spodoptera frugiperda.* To enhance the efforts of the Ministry of Agriculture to mitigate the impact and risk of FAW in Tunisia, a technical cooperation program was implemented in the country as in the other North African countries.

13.2.5 The Case of Grape and Olive

Socio-economic Importance of Grape and Olive Sectors in Tunisia

Grape Sector

Tunisia is a country with a history and a civilization linked to vines and wine for several millennia. Already, the art of cultivating vines dates back to antiquity. The vine and the wine are indeed part of the local cultural heritage. Tunisian vineyards cover 25.500 ha and 75% are located in the governorates of Nabeul, Ben Arous, and Bizerte. About 11.500 ha are dedicated to table vines and 9.250 ha to wine vine. The wine exports include the Europe (France, Belgium, Luxembourg, and The

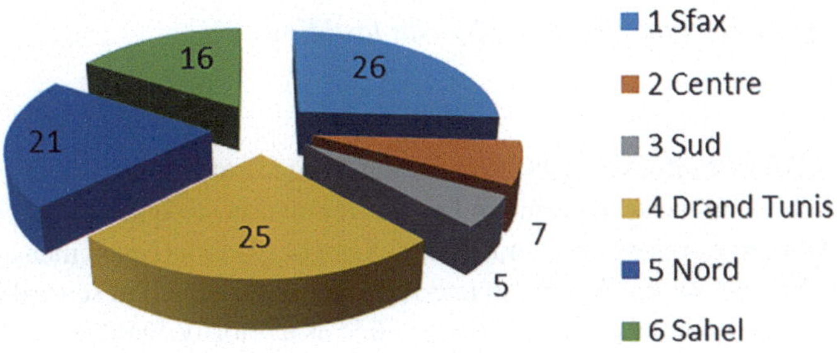

Fig. 13.10 Regional distribution of exporters. *Source* OLIVAE (2017)

Netherlands) and Japan. Despite the importance of this crop, the country is still dependent on foreigners for grape plants to reach its national needs.

Olive Sector

Olive growing orchards is the Tunisia's main agricultural activity and its socio-economic role is of paramount importance since it constitutes a pillar of the country's cultural heritage. Tunisian olive groves are spread over about a third of the cultivated areas in the country, i.e., 1.9 million hectares are devoted to this crop and more than 310.000 producers (65% of all farmers). Most of the olive oil exporters are based in the southern part of the country in Sfax (Fig. 13.10) due to the vast areas planted with olive trees in this region, as well as by the very large number of oil mills that are installed there, i.e., 342.

The cultivation of about 95% of olive tree is carried out in dry conditions in varied climatic conditions. The number of olive trees in Tunisia is estimated at around 60 million including 30% in the North, 45% in the Center, and 25% in the South. Olive trees are most often planted in monoculture and sometimes intercropped with other fruit trees.

Tunisia is the best known Southern Mediterranean country in the field of olive cultivation and the export of olive oil to the European Union. Revenues from olive oil exports increased from 33.9% in 2021 to 40.6% in 2022. The National Observatory of Agriculture (ONAGRI) announced that the export earnings of olive oil have reached 1.941 million dinars (MD). In fact, the first export agreements were concluded with the European Union in the 1980s. Since then, this market has become the traditional customer for Tunisian olive oil, with more than 80% of exports. During the 1980s, Italy was Tunisia's first customer with 58.2% of total exports and mainly imports olive oil to supply its refineries. France was the second European purchaser and imports the quality oil for direct consumption. Today, the European Union is still the priority market for Tunisian olive oil and widened other destinations, mainly Spain and Portugal. However, olive tree faced several pests mainly *Xylella fastidiosa.*

The Olive Quick Decline Syndrome Xylella fastidiosa

Regulation

Xylella fastidiosa (Wells et al. 1987) is a xylem-limited pathogenic bacterium, listed in EPPO A2 list of pests recommended for regulation as quarantine pests, and Chap. 1 of Directive 2000/29/EC and is the subject of EU emergency measures. This bacterium is present on the American and Asian continents, and in several countries of the European community where it is classified as a priority agent according to the implemented regulation 2016/2031.

The pathogenesis of *X. fastidiosa* is similar to that *of Xanthomonas campestris* with a translucent colony composed of extracellular polysaccharides (EPS); the bacterium forms aggregate leading to xylem occlusion, which leads to symptoms of water stress, then desiccation followed by plant death (Rechka and Lefort 2017). This pest attacks the conductive vessels of many hosts and specific cultivated or wild plants; it is reported in the literature about 595 plant species, 275 genera, and 85 families (EFSA 2021a; b). It is transmitted by xylem-sucking insects and by methods of propagation of plant material. Genetically diverse, this species is divided into six subspecies (*X. fastidiosa fastidiosa, X. fastidiosa pauca, X. fastidiosa multiplex, X. fastidiosa sandyi, X. fastidiosa morus, X. fastidiosa tashke*), the four-first being the most damaging and now being found in numerous countries worldwide. It is considered as a real threat for the whole Mediterranean basin.

The intensive monitoring is an effective means of early detection of the bacterium and its vectors in an area or a country. The sampling procedure on the plant, the sampling type and period depend on the host species, the biological cycle of insect vectors and on the occurrence of suspect symptoms (Valentini et al. 2017).

The bacterium *X. fastidiosa* is passively delivered into the xylem sap by spittlebugs vectors. These small hemipterans mostly range from 7 to 9 mm in length (Almeida and Purcell 2006). In Tunisia, four spittlebug species were recorded, namely: *Philaenus tesselatus, Neophilaenus campestris, Neophilaenus lineatus,* and *Philaenus maghresignus*. The two first species seem to be potential vectors of the pest. Consequently, the risk of spreading the bacteria is significant because of the international trade (Boukhris-Bouhachem et al. 2023).

Tunisian Action Plan

Since its first report on olive trees in southern Italy/Puglia region in 2013, and its spread to other countries of the EU (France, Spain, and Portugal), Tunisia has applied emergency measures including strict procedures quarantine and phytosanitary control. The bacterium is well-positioned at the gates of the Near East and North Africa countries (NENA). Thus, countries/regions where the bacterium has not yet been reported should strengthen preventive measures and apply quarantine and phytosanitary control procedures, as well as implement surveillance programs for this devastating pathogen. The NPPOs should adopt urgent measures and prepare a contingency plan/national strategy against this bacterium in case of its entrance.

National Monitoring in Tunisia

A large-scale production vegetation with intensive agronomic management (open fields and nurseries) are the main plant hosts targeted for this monitoring namely, Olive tree (*Olea europaea*); Stone fruit trees of *Prunus* spp. (Almond *Prunus dulcis*, Plum *Prunus domestica* and Peach *Prunus persica*); vines (*Vitis vinifera*); Citrus (*Citrus* spp.); and Pistachio (*Pistacia vera*). Likewise, host plants in natural or semi-natural areas, with particular interest for ornamental, culinary, medicinal, shrubs, and herbaceous vegetation were also monitored namely, Rosemary, Oleander, Lavender, Geranium, and Hibiscus. Likewise, forest trees mainly cork oak (*Quercus suber*) and shade trees, present in forest patches in parks or along roads were subjected to this monitoring.

Regulatory Status of *Xylella Fastidiosa* in Tunisia

All trade decisions relating to the *X. fastidiosa* pest are governed by the Ministerial Orders: (i) the decree of the Ministry of Agriculture of May 31st, 2012 fixing the list of quarantine organisms; (ii) order of the Ministry of Agriculture of June 26th, 2015 fixing the list of prohibited plants and plant products whose entry into Tunisian territory is prohibited; (iii) order of the Ministry of Agriculture of February 19th, 2016, setting the phytosanitary requirements as well as the methods of control of plants and plant products imported into Tunisia.

In 2016, the NPPO submitted a declaration informing the Commission of the EU of the "Free Status" of this pest in the country. This declaration was made public on the website of the DGSVCIA. As part of the application of strict phytosanitary measures and preventing contamination by this quarantine pest, the Tunisian NPPO issued a circular prohibiting imports from countries contaminated by the bacterium under No. 307 dated on December 27th, 2016. The latter was recently repealed by a new circular No. 208 dated on November 09th, 2020.

National Strategy

In 2019, a national strategy/contingency plan was set by the competent services of the DGSVCIA, approved in November 2019 by a Ministerial Council entitled "The National Strategy for Preventing the Intrusion and the Spread of *Xylella fastidiosa* into the Tunisian territory." This national contingency plan includes: (i) an annual national survey programs (territory monitoring); (ii) a consignment check procedures in the exploration and importations of goods; (iii) a strengthening of the capacity of the plant quarantine laboratory for diagnostic analyzes, in addition to developing the border controller's capacities; (iv) a multi-lateral collaboration with stakeholders (research, professionals, regional delegates for agricultural development, sub-directorate of control and protection and National Oil Office); (v) issuance and update of legislative texts as regards the national phytosanitary measures implemented.

Official Survey Organization in Tunisia-DGSVCIA-Tunisian ONPV

Official surveillance in Tunisia for the specified quarantine pest *X. fastidiosa* is directed by the NPPO, the DGSVCIA under the supervision of the Ministry of Agriculture, Fisheries, and Hydraulic Resources.

The DGSVCIA has developed a national strategy 2020–2024 as a part of the prevention against the introduction of the pest into Tunisian territory and has designated responsibility for official analyzes (bacteria and vector insects) to the quarantine laboratory. This latter is responsible for issuing final report of sample analyzes received. The confirmation of the species is done in collaboration with the research laboratories involved in the national strategy.

In consultation with all stakeholders involved in the execution of the official national surveillance plan, and particularly the national strategy, ad hoc official monitoring surveys were carried out in the governorates and delegations of the country as indicated in the cards below, during the period extending from 2020 to 2022. The samples were analyzed in the quarantine laboratory. The areas (governorates and localities) surveyed were chosen on the basis of the geo-local distribution of the species targeted by the monitoring procedure (geo-distribution of strategic crops, mapping, and zoning). In all plant protection sub-directorates located in the North, the Center and the South of the country (Fig. 13.11).

Control under the DGSVCIA ensured the coordination of prospecting and sample collection with the respective cellules territoriales de vulgarization (CTVs) and the Commissariat Régional au développement Agricole (CRDAs). More than 17,000 samples were collected from different regions and on different species (olives, grapes, citrus, Prunus, ornamental plants). In addition, about 11,000 samples were collected from 128 nurseries (olives, grapes, citrus, Prunus, ornamental plants, Pistachio).

13.2.6 The Case of the Cactus

Socio-economic Importance of the Cactus Sector in Tunisia

Originally from Central America, the cactus is one of the most important crops in Tunisia. It covers 600.000 ha distributed in the center (55%) mainly in Kasserine, Siliana, Sidi Bou Zid, Kairouan and in the South (25%) in Gafsa. To promote this sector, the Office d'Elevage et des Pâturages (OEP) has opted for plantation projects with approximately 145.860 ha of spineless cactus since 1990. The country is among the most important country in terms of culture with an annual national production estimated at 550.000 tons.

Initially used for making hedges and fences, virtues of this culture have been diversified, bringing together animal and human food, cosmetics and phytomedicine, which provides Tunisia with a large domestic market allowing wealth and development of poor and deprived regions by creating new jobs. At the same time, on the world market, exporting prickly pear fruit allows Tunisia to occupy the 4th position with 60% of exportation designated for the Gulf countries. Despite the importance of this agroforestry species, it is threatened by several pests mainly *Dactylopius opuntiae.*

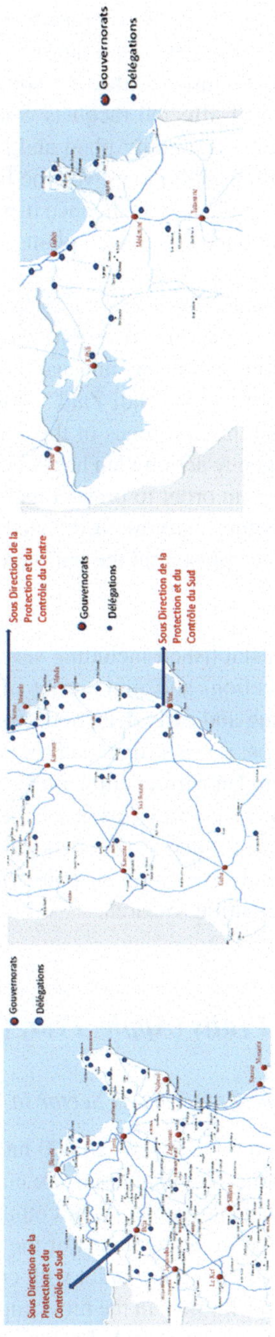

Fig. 13.11 Geo-distribution of the sampled crops

Dactylopius opuntiae

The hemipteran insect, *Dactylopius Opuntiae* is considered as one of the invasive alien species and classified among the list of quarantine pests in many countries. It is one of the most important pest's insects infecting *Opuntia* species. Symptoms are manifested by the desiccation of affected racquets causing their death in case of an outbreak. Although, the cactus is a hardy plant and knows few enemies, it is very important to be vigilant on the health plan given the high spread of the cactus mealybug already present in the region and which, once installed, it become difficult to control with only uprooting and burying infested plants.

Emergency Plan of Tunisia

At the end of the surveillance and prospecting operations carried out by the competent services of the Tunisian Agriculture Ministry, unusual symptoms of infection on the cactus plantations had been noticed in "Sidi Zaid" (governorate of Mahdia) in August 2021, which is likely to be the first focus of the cactus mealybug. Faced to this suspicious situation, an emergency action plan has been established to eradicate the first detected focus of the insect in order to prevent its spread. The plan is based on mechanical intervention (uprooting, burying of infested hedges of cactus) and a chemical treatment of cactus hedges present in the buffer zone.

Action Plan to Control

A national control program was established including several points: (i) establishment of exploration system, detection, and monitoring of mealybug *Dactylopius opuntiae* in Tunisia; (ii) prospecting and early detection; (iii) scientific research; (iv) training and awareness (workshops, posters, flyers, etc.); (v) ongoing development of a PRA on this harmful organism (in partnership with CABI).

Currently, a national strategy to control this insect is being established and including five tasks: (1) phytosanitary survey; (2) chemical treatment in zones with low infestation level; (3) mechanical destruction in zones with high infestation level; (4) promoting training; and (5) scientific research.

13.2.7 The Case of Pome Fruits (Apple, Pear, Quinces)

Socio-economic Importance of the Pome Fruits Sector in Tunisia

Apple and pear plantations cover a total area of 24.000 ha, including 18,356 ha of apples and 3100 ha of pears. Apple tree plantations are concentrated in the north-western Tunisia (Siliana and Kef) and the center of the country (Kasserine). Pear tree plantations are mainly concentrated in the northeast of the country (Manouba, Ben Arous, and Bizerte). This sector supplies the market with crops throughout the year and is the main source of income for farmers in the most important production areas. Despite its importance, several pests were signaled namely the Fire Blight, *Erwinia amylovora*.

National Program to Control the Fire Blight, Erwinia amylovora

Regulation

According to Tunisian phytosanitary legislation, *Erwinia amylovora* (EPPO A2 List) causative agent of fire blight is classified as a quarantine organism. Its official status in Tunisia is present only in restricted area subject to official control. This organism is the subject of a national control strategy that differs according to the status of the region.

Action Plan of Tunisia

The Tunisian Agriculture Ministry works to surround the disease in affected areas by reducing infections and bacterial inoculum to the lowest level and protecting free areas from disease transmission. The parts of this strategy are as follows: (i) prospecting and eradication; (ii) nursery control; (iii) scientific research; and (iv) training and extension.

Prospecting and Eradication

Following the detection of the disease, a vast prospecting campaign was carried out with sampling and analysis in all production areas. The results of these surveys enabled the detection of the disease in the pear production areas in the north of the country. The apple production areas are free from the disease.

Subsidies for uprooting and replanting in contaminated areas have been distributed to farmers. They are around TND1400/ha for uprooting (USD470) and TND600/ha for replanting (USD200). This action allowed the uprooting of 4150 ha of contaminated areas between 2013 and 2017.

Preventive measures have been taken to maintain the disease-free status of apple production areas such as the control of nurseries, an extensive awareness campaign. For pears, a national program for the revival of the sector has been established by the introduction of new varieties tolerant to the disease.

Evaluation

The fire blight control strategy monitoring committee regularly meets to assess the progress of the various components of the strategy.

The pear tree areas were significantly affected by the disease. The introduction of new varieties tolerant to the bacteria is in the final phase of testing and evaluation. As for apple-growing areas, there are increased efforts to preserve their disease-free status. A national nursery for the production of apple tree seedlings in the Kasserine region has been set up in order to remedy the problems of illegal nurseries.

13.3 Certified Plants

13.3.1 Importance of Stone Fruits Plants Certification in the World

With the intensification of fruit production after World War II, nurserymen and fruit farmers became aware that a main prerequisite for high yields of quality fruit would be the planting of healthy, efficient fruit trees. In Germany, a program started for the careful selection of healthy, efficient mother trees for rootstock and scion cultivars, producing high yields of quality fruit. Moreover, viruses were eliminated by heat treatment and/or meristematic tissue culture. From 1978 onwards the usage of virus-free planting material was made compulsory. In Great Britain in 1953, the nuclear stock association formed to provide a method of propagating and distributing healthy pest and disease-free planting stock of commercially grown fruit varieties to the fruit industry. Today, the NSA works in cooperation with The Animal and Plant Health Agency (APHA) to promote the production and the use of certified healthy proven planting material from high quality parent stock. In France, an agreement was signed in 1952 between INRA and Centre d'Experimentation de la Pépinière (CEP) in order to foster the propagation of the best clone of peach and plum to guarantee their trueness to type. In 1962, the Centre Inter-Professional des Fruits et Legumes (CTIFL) joined this activity and the propagation of rootstocks began in 1963, while the freedom from infectious disease was included in 1967, and only in 1980 that the certification of fruit crops was implemented by a preliminary technical regulation. Italy has developed a voluntary certification scheme adopted by the national nursery and was extended through the decree in 1987 by the Ministry of agriculture. This decree was amended two times in 1989 in order to establish the national technique committee for the implementation and management of certification service, and in 1991 to define the interconnections between the Ministry, local boards, institutions, and inter-professional organizations. In 1997, a ministerial decree was published to specify new techniques for the production of the propagation plant material of the stone fruit and their rootstocks.

13.3.2 History of Citrus, Olive, Stone Fruits Plants Certification Program in Tunisia

In Tunisia, the National program of the production of certified propagation material has been set up since 1991 with the collaboration among INRAT/GIAF/DGPA/FAO in the framework of the project RAB/88/025. This project aims to the sanitation of the main varieties of citrus and vines in Tunisia and provides nursery with planting material conforms to the ecotype and free from virus and diseases.

The Certification of Citrus Propagation Material

The program started by the: (i) elimination of several viral diseases using thermotherapy and viral remediation (meristem micrografting); (ii) creation of foundation block of all the commercial varieties in Tunisia; (iii) introduction of new varieties from INRA Corsica; (iv) creation of mother block in GIF; (v) production of certified plants.

The program faces many challenges since the laboratory of virus diagnosis was ready only in 2002. The number of certified nurseries increased from one nursery to three, and currently there are seven certified nurseries in Tunisia.

The Certification of Grapevine Propagation Material

The certification program started by the clonal and the sanitary selection of four rootstocks and various varieties of grapevine, then the creation of pre-multiplication center. During the decade 2002/2012, there was no improvement in the national program of grapevine certification because of the non-involvement of the inter profession to this program and the certified plants were imported from France and Italy. It was only in 2007 that the clonal and the sanitary selection of stone fruit and olive plants started trough the Tunisian-Italian project "*Support of fruit and vegetable production in Tunisia.*"

Because of the presence of *Xyllela fastidiosa* in many European countries, the importation of plant material was restricted, and then the national program of certified propagation material was taken back.

The Certification of Olive Propagation Material

The program started by the: (**i**) clonal selection and the viral sanitation of main varieties of olive in Tunisia by the National Institute for Agronomic Research (INRAT); (**ii**) creation of foundation block in Oil National Office (ONH); (**iii**) creation of mother block in 2009 with basic material introduced from Spain (Arbequina- Arbosana-Koroneiki); and (**iv**) production of basic material by ONH.

The Certification of Stone Fruit Propagation Material

The program started by the: (i) clonal selection and the viral sanitation of main varieties of stone fruit in Tunisia by National Institute for Agronomic Research (INRAT); (ii) creation of premultiplication center in GIF; (iii) transfer of the of premultiplication center to Citrus Technical Center (CTA); (iv) production of pre-basic material by CTA; and (v) creation of multiplication center in GOVPF.

The certification scheme of different species was published in decree n°2002-101 dated January 18, 2000 and decree n°2007-1985 published July 30, 2007.

The certification scheme was revised in 2019 to reach all the EPPO standards, from the techniques of production of the propagation material until the techniques used in diagnosis laboratory, and the *Xyllela fastidiosa* and *Guignardia citricarpa* had been introduced to the list of the pest controlled during the certification process.

13.3.3 Tunisian Certification Schemes

In Tunisia, certification schemes for plants and seeds are in perfect harmonization with those of EPPO (1992, 1993). In fact, Tunisian Certification Schemes describe the steps to be followed for the production of vegetative propagated planting material of a particular cultivated plant, whose health status is attested by an official certificate. In a certification scheme, the certified material is descended by a fixed number of steps from individual plants tested and free from pests, and is then maintained and propagated under rigorous conditions excluding any recontamination. Hereby, we gave the examples of citrus and olive certification schemes (Fig. 13.12).

Before giving an official certification, a rigorous control is carried out by the competent departments of the DGSVIA which consists of technical control (in-situ) and a sanitary control (in-vivo). The technical control is the verification of varietal

Fig. 13.12 Scheme of citrus and olive certification schemes in Tunisia

Technical control (On-site) *Sanitary control (laboratory)*

Fig. 13.13 Technical and sanitary controls of plants

authenticity, the health status and the compliance with technical multiplication standards. The sanitary control is the verification of the absence of pests listed as prohibited pathogens in the production of certified plants using international standardized methods and techniques (Fig. 13.13).

Qualified nurseries produce about 10,000 of certified citrus plants, 20,000 of certified olive plants, and 3 million of certified rootstocks.

13.4 Conclusion

In Tunisia, the agricultural sector plays a crucial role to ensure food security. However, since the twentieth century various transboundary plant pests and diseases threaten this sector namely, *Phyllocnistis citrella*, *Tuta absoluta*, *Rhynchophorus ferrugineus*, *Erwinia amylovora*, *Phyllosticita citricarpa*, *Citrus tristeza virus*, and *Dactylopius opuntiae*. Hopefully, the country is free from *Xylella fastidiosa*, *Spodoptera frugiperda*, and *Palmaspis phoenicis* recently detected in neighbor Mediterranean countries and can dramatically induce a disaster to our arable land.

The Tunisian Government established a national control strategy to eradicate the plant pests and diseases from contaminated areas and to prevent the introduction of other pests. Aware of the danger that could destroy Tunisian agriculture, the Government spent more than 40 million dollars during the last decade, intending for the pest control, training, development of National Phytosanitary Capacity, information sharing and measures against IAS. Tunisia has also implemented certification programs to preserve agricultural sectors and to replant damaged areas with healthy and resistant varieties.

References

Abbes K, Harbi A, Chermiti B (2012a) Comparative study of 2 protection strategies against Tuta absoluta (Meyrick) in late open field tomato crops in Tunisia. EPPO, Bull 42(2):167–344

Abbes K, Harbi A, Chermiti B (2012b) The tomato leafminer *Tuta absoluta* (Meyrick) in Tunisia: current status and management strategies. EPPO Bull 42(2):167–344

Ahmed MA, Abdelbagi AO, Elshafi e, AFH, Fageer EA, Abass IA (2013) Efficacy of imidacloprid (Confi dor 200 SL) and improved cultural practices in the control of the green date palm pit scale insect (Asterolecanium phoenicis Rao) (Palmaspis phoenicis) (Homoptera: Asterolecaniidae) in Northern Sudan. Sci Res Essays 8:1752–1758

Ali AA (1989) Studies on *Asterolecanium phoenicis* (Rao) A date palm scale insect in El-Golid area. MSc Thesis, Faculty of Agriculture University of Khartoum, Sudan

Almeida RPP, Purcell AH (2006) Patterns of *Xylella fastidiosa* Colonization on the Precibarium of Sharpshooter Vectors Relative to Transmission to Plants Annals of the Entomological Society of America, vol 99

Atlas Magazine (160) (2019) The national strategy to control citrus black spot in Tunisia

Barranco P, de la Peña JA, Martín MM, Cabello T (2000) Host rank for *Rhynchophorus ferrugineus* (Olivier, 1790) (Coleoptera: Curculionidae) and host diameter. Bol Sanid Vegetal, Plagas 26:73–78

Bar-Joseph M, Dawson WO (2008) Citrus tristeza virus. In: Encyclopedia of virology, 3rd edn, pp 520–525

Boukhris-Bouhachem S, Souissi R, Abou Kubaa R, El Moujabber M, Gnezdilov V (2023) Aphrophoridae as potential vectors of *Xylella fastidiosa* in Tunisia. Insects 14(2):119. https://doi.org/10.3390/insects14020119

Braham M, Hajji L (2012) Management of Tuta absoluta (Lepidoptera, Gelechiidae) with Insecticides on Tomatoes; in book: insecticides—pest engineering

Braham M (2014) Is mass trapping technique useful for the control of the tomato Leafminer, Tuta absoluta (Lepidoptera: Gelechiidae)? Greener J 1–19

Centre for Agricultural Bioscience International (CABI) (2019) Tomato leaf miner (Tuta absoluta) impacts & coping strategies for Africa. Evidence Note. Consorsium (CABI). https://www.invasive-species.org/wp-content/uploads/sites/2/2019/04/Tuta-Evidence-Note_FINAL.pdf

Chebbi H (2011) First Record of *Rhynchophorus ferrugineus* on *Phoenix canariensis* in Tunisia. Tunisian J Plant Prot 6:149–153

Cherif A (2018) These de doctorat en Sciences Agronomiques: "La mineuse de la tomate *Tuta absoluta* (Lepidoptera: Gelechiidae) : Paramètres biologiques, écologiques et alternatives de lutte» 225p.

Dembilio Ó, Jacas JA, Llácer E (2009) Are the palms *Washingtonia filifera* and *Chamaerops humilis* suitable hosts for the red palm weevil, Rhynchophorus ferrugineus (Col. Curculionidae)? J Appl Entomol 133(7):565–567

EFSA Supporting Publication (2020) Workshops report for cooperation in crisis preparedness for *Phyllosticta citricarpa* in the European Union, pp 25–33

European Food Safety Authority (EFSA) (2021a) Update of the Xylella spp. host plant database-Decembre 2020-approved Mai 2021

European Food Safety Authority (EFSA) (2021b) Pest survey card on *Phyllosticta citricarpa*. EFSA Supporting publication 2021: EN-7105

European and Mediterranean plant Protection Organization EPPO (2004) Reporting Service no. 09-2004 Num. article: 2004/129) European and Mediterranean plant Protection Organization EPPO (2005). Bulletin OEPP/EPPO Bull 35:434, 434–435

European and Mediterranean plant Protection Organization EPPO (2009a) Reporting Service no. 09-2009 Num. article: 2009/169)

European and Mediterranean plant Protection Organization EPPO (2009b) Reporting Service no. 03-2009 Num. article: 2009/042 Num. article: 2009/042

European and Mediterranean plant Protection Organization EPPO (2013) Reporting Service no. 05-2013 Num. article: 2013/095 : First report of *Erwinia amylovora* in Tunisia

European and Mediterranean plant Protection Organiztion EPPO Reporting Service no. 05-2021 Num. article: 2021/107 : First report of *citrus tristeza virus* in Tunisia

Food and Agriculture Organization FAO (2002) Production yearbook, 2000, vol 54. FAO, Italy

Food and Agriculture Organization FAO (2017) Fall ArmyWorm (*Spodoptera frugiperda*): Identification, biology and ecology. For more information on Fall Armyworm please visit: www.fao.org/Africa, www.plantwise.org/fallarmyworm

Groupement Interprofessionnel des dates (2024). https://www.nefzawa.net/blog/66d790d131d8aaf5b5db753c

Groupement Interprofessionnel des Fruits Tunisia (GIF) (2020). http://gifruits.com/actualites.php?category=rapport2020

Groupement Interprofessionnel des Fruits Tunisia (GIF) (2021). http://gifruits.com/actualites.php?category=rapport2021

Groupement Interprofessionnel des Fruits Tunisia (GIF) (2022). http://gifruits.com/actualites.php?category=rapport2022

Groupement Interprofessionnel des Légumes (GIL) (2016) Annual statistic, https://www.gil.com.tn/fr/

Harbi A, Abbes K, Dridi-Almohandes B, Chermiti B (2015) Efficacy of insect-proof nets used in Tunisian tomato greenhouses against Tuta absoluta (Meyrick) (Lepidoptera: Gelechiidae) and potential impact on plant growth and fruit quality. J Entomol Acarol Res 2012; 44: e-J Entomol Acarol Res 2015; 47:5256

ONAGRI (Observatoire National de l'Agriculture) (2020) Indicateurs clés des filières agricoles en Tunisie 2020, pp 12–13

IPPC website. Official pest reports—Tunisia (TUN-02/4 of 2021-03-29) Premier signalement du virus de la tristeza ou *Citrus tristeza virus* (CTV) en Tunisie

Khaldi R, Bouali S (2017) Analyse de la filière céréalière en Tunisie et identification des principaux points de dysfonctionnement à l'origine des pertes. FAO, Projet GCP/RNE/004/ITA: réduction des pertes et gaspillages alimentaires à travers le développement des chaînes de valeur pour la sécurité alimentaire en Egypte et en Tunisie, pp 10–11

Najar A, Hamdi I, Mahmoud S, Medhioub L, Jaouadi I, Varsani A, Jemmazi A (2021) First report of citrus tristeza virus in commercial citrus orchards in Tunisia. J Plant Pathol 103:1051–1052. https://doi.org/10.1007/s42161-021-00857-7

Observatoire du Sahara et du Sahel, OSS (2020) Espèces Exotiques Envahissantes Terrestres au Maghreb Arabe. OSS, p 148

OLIVAE (2017) Journal Officiel du COI–Conseil Oléicole International N°24

Rechka Y, Lefort F (2017) Une bactérie inquiétante pour l'agriculture marocaine Groupe plantes et pathogènes, Institut Terre Nature Environnement. https://www.agri-mag.com/2017/06/21/xtlella-fastidiosa/

Tomatonews (2011). https://www.tomatonews.com/en/italy-climate-change-to-hit-processing-tomato-production_2_2288.html

Valentini F, Cavallo G, Maria D'Onghia A (2017) Sampling procedures of plant material for the survey of *Xylella fastidiosa* in Puglia Region, Italy

Wells JM, Boligala CR, Hsueh-Yun H, William GW, Linda MP, Don JB (1987) *Xylella fastidiosa* Gram-Negative, Xylem-Limited, Fastidious plant bacteria related to *Xanthomonas* spp. Int J Syst Evol Microbiol, vol 37

World Bank (2020) Convergence: five critical steps toward integrating lagging and leading areas in the Middle East and North Africa. © World Bank, Washington, DC. http://hdl.handle.net/10986/33187 License: CC BY 3.0 IGO

Ben Jamâa Mohamed Lahbib From 2017 to 2023, Pr. Ben Jamâa leads the General Directorate of Plant Health and Control of Agricultural Inputs (Ministry of Agriculture, Water Resources and Fisheries, Tunisia) and manages more than 10 projects. His main responsibilities include: building programs on phytosanitary measures by controlling pests and diseases of main national crop systems production; monitoring the progress of quarantine pests; implement control programs and limit their spread; and establishing strategies for sustainable management of Transboundary Plant Pests and Diseases. Pr. Ben Jamâa leads the Elaboration and Execution of National Strategies of Control of the Quarantine and Emerging Pests and the Study of the effects of pesticides on human health and the environment in Tunisia. Since 2023, he has been working as Executive Director of the Near East Plant Protection Organization (NEPPO). He is responsible of the management of the Organization and ensures coordination among member countries of the region to develop and implement a regional strategy to control Transboundary Plant Pests and standards of plant protection. He coordinates international and regional cooperation to: support plant protection and enhance capabilities; facilitate and encourage regional and international trade exchange; and prevent the entry & spread of pests; Pest control in appropriate ways. Mr. Ben Jamâa also is a Professor in Forest Entomology at the Institut National de Recherches en Génie Rural, Eaux et Forêts (INRGREF). He is experienced in Plant Health with more than 30 years in Invasive Pest Species, methods of Monitoring and Integrated Control. He published more than 150 scientific papers and chapter in books and supervised more than 30 theses and master's degrees in pest and pest control, and is a reviewer in various scientific journals. Pr. BEN JAMÂA obtained his PhD in Plant Protection from the National Institute of Agronomy of Tunisia (INAT), the Agricultural Engineering degree in Forest Entomology in 1992 from the Institute of Agronomy and Veterinary Hassan II (Morocco), and the Engineer degree in 1989 from INAT. He joined the INRGREF in 1992 as a Researcher in Forest Entomology and leads for 15 years (from 2003 to 2017) at the Laboratory of Management and Valorization of Forest Resources. During this period, he managed about 30 National and International projects in plant health, coordinated the research program, moderated workshops and meetings, and executed budget. He was a member of committee of theses/masters; Member of the university accreditation committee of researcher; and member of committee of Recruitment and Promotion of Researcher.

Ms. Ben Hadj Ali Since March 2016, she is a Chief Engineer in the Quarantine Laboratory in the General Directorate of Plant Health and Control of Agricultural Inputs (DGVPCIA), Ministry of Agriculture, Water Resources and Fisheries, Tunisia. Since 2012, she has been the technical manager of Nematology and Mycology analyses and quality manager of the Quarantine Laboratory. Ms. Ben Hadj Ali gets her Master's degree in "Biological Agriculture" in 2007 and a National Engineering Diploma in "Agronomy of Semi-Arid Regions" in 2003 from the National Institute of Agronomy of Tunisia (INAT).

Dr. Besma Mrabet is a General Engineer and the Head of the Virological Analysis Laboratory in the General Directorate of Plant Health and Agricultural Input Control (DGSVCIA), Ministry of Agriculture, Water Resources and Fisheries, Tunisia. She gets an extensive experience in the field of certification of seeds and plants in Tunisia and, in the detection of viral diseases (viruses, viroids and similar viruses) transmitted by seeds and plants using molecular tools (conventional and real-time PCR). Since 2024, Dr. Mrabet is the Quality management manager (RMQ)-ISO 17025 for the official laboratories of the DGSVCIA (Plant Quarantine Laboratory, Seed and Plant Analysis Laboratory; Virological Analysis Laboratory and Pesticide Analysis Laboratory). In 2019, she leads the Virological Analysis Laboratory (DGSVCIA), and in 2010 she leads the Legislation and Quality Department in the DGSVCIA in charge of the Mornag Virological Control Station. From 2002 to 2009, she was an engineer in the Plant Quarantine Laboratory (Under Directorate of Phytosanitary Control at Crossing Points-DGSVCIA).

Dr. Essia Limem is a Tunisian agronomist, expert in plant protection, specializing in integrated pest management (IPM) and biocontrol, focusing on the biodiversity of beneficial insects in citrus groves in the General Directorate of Plant Health and Agricultural Input Control (DGSVCIA). With over a decade of experience, she has played key roles in pest management strategies for crops, mainly olive, palm, citrus, and vegetables. Dr. Limem headed the Entomology Department in DGSVCIA, where she developed biological and biotechnical control programs, supervised mass rearing of beneficial insects, implementing organic farming regulations and contributed to pesticide registration. She leads Tunisia's efforts in managing Fall Armyworm and participated in the FAO Focus Group on Climate Change and Phytosanitary Issues. Currently (since 2023), she is the Technical Coordinator of Plant Production and Protection for the FAO's BIOREST project in Tunisia.

Mr. Imed Jaouadi is a Chief Engineer, specialist in Plant Protection, in charge of phytosanitary control at crossing points in the General Directorate of Plant Health and Agricultural Input Control (DGSVCIA). He supervises the national strategies for the control and prevention of new quarantine organisms and assists in project management of the 'Projet d'Appui à la Compétitivité et à l'Exportation (PACE)'; the 'Programme d'Appui au Développement Rural (PADR)'; and international standards such as 'ISO 17020, ISO 19011, ISO 31000, ISO 17025'. Since 2020, he is a Member of the Board of Directors of the National Plant Protection Society and since 2019, he is the National coordinator of the phytosanitary component of the 'Support for official control services for animal and plant products (ASCO)' project funded by the European Union (EU) and managed by the Food and Agriculture Organization of the United Nations. In 2015, Mr. Jaouadi led the Department of Statistics, Data Bank and External Relations.

Mr. Hichem Asses Since 2013, he is a chief engineer, specialized in plant protection and pest control in the General Directorate for Plant Health and Control of Agricultural Inputs (DGSVCIA), Ministry of Agriculture, Water Resources and Fisheries, Tunisia. He is experienced in the development and implementation of national strategies for the control of quarantine pests, mainly the Red Palm Weevil (Rhynchophorus ferrugineus) and the cactus mealybug (Dactylopius opuntiae). Currently he is the national coordinator of the program of prevention of the introduction of the Fall armyworm (Spodoptera frugiperda) into Tunisia. Within this framework, he is working closely with various stakeholders to anticipate and manage this growing threat, by developing control strategies. In 2022, Mr. ASSES took up the post of junior technical expert in plant production and protection at the Food and Agriculture Organization's subregional office, where he has worked as an expert on several projects in the North Africa region.

Mr. Lassaad Medhioub Since 2013, he is a Chief Engineer in the General Directorate of Plant Health and Control of Agricultural Inputs Inputs (DGSVCIA), Ministry of Agriculture, Water Resources and Fisheries, Tunisia. He participated in the development of guides and procedures for phytosanitary control and territorial surveillance. As a Phytosanitary Controller at crossing Points throughout Tunisia, he has played since the beginning of his professional career, a key role in the development and monitoring of the implementation of national strategies aimed at combating quarantine organisms mainly fire blight, citrus tisteza virus, citrus black spot disease. Mr. Medhioub is experienced in managing prospecting teams, data storage, mapping, phytosanitary monitoring against emerging harmful organisms.

Mr. Mohamed Chaouachi is a Senior Engineer, specialized in crop defense in the Quarantine Service of the General Directorate of Plant Health and Agricultural Input Control (DGSVCIA), Ministry of Agriculture, Water Resources and Fisheries, Tunisia. He is the Project manager of the national Red Palm Weevil strategy (monitoring, training agents, seminar and meeting planning, and management) and a th member of the RPW monitoring and research committee.

Ms. Mouna Mhafdhi is a General Engineer and a Deputy Director in Plant Health Division in the General Directorate of Plant Health and Control of Agricultural Inputs (DGSVCIA), Ministry of Agriculture, Water Resources and Fisheries, Tunisia. She worked as an engineer specialized in Plant Protection since 2001. She is experienced in establishing plant health strategies based on IPM methods, evaluation of strategies of control of pests and diseases in crops, and preparation of training plan about methods of control of pests and diseases for different crops. She is performing several tasks as a member in the National Unit to control mosquitoes, the sub-committee of Think Tank Advisory Committee to advance the Citrus sector, the Dates sector and Board of Directors of National oil office. Ms. Mhafdhi has a key role in the update of the Tunisian Law of Plant health as part of "the support for official control services" project: ASCO project (Appui aux Services de Contrôle Officiel) discussing the Tunisian Law of Food Safety and participating in the editing and achievement of the specifications of the phytosanitary information system as part of the above-mentioned project. Ms. Mhafdhi is a focal point of Tunisia in (i) "Desert Locust Control Commission in the Occidental Region (CLCPRO)", and (ii) the phytosanitary aspects at the project "The Antimicrobial Resistance (AMR) Multi-Partner Trust Fund" (UNJP/TUN/047/UNJ project).

Ms. Nadia Debbabi is a Chief Engineer in Phythopathology, working as a Phytosanitary Controller in the International Airport of Tunisia -Carthage. She is experienced in Phytosanitary control at post borders, Diagnostic analyzsis of quarantine bacteria, Issuance and extensions of import restriction BANs (In the form of a circular or ministerial orders), Planification and carry out Trainings on the bacterium and the diseases it causes for the benefit of phytosanitary control agents, staff of crop protection offices and farmers, Carry out PRA for emergent organisms and Implementation of National Phytosanitary Measures to manage the national strategy on the frame of pests and crop diseases. Ms. Debbabi is the Responsible of The National Strategy of Xyllela fastidiosa, involved in the Establishment of national surveillance plans, the Tunisian phytosanitary measures and requirements, and the Management of the national committee for the national X. fastidiosa strategy.

Ms. Noura Touati is an Agronomic Engineer and the Head of the Seeds and Plants Certification Department in the General Directorate of Plant Health and Control of Agricultural Inputs (DGSVCIA). She is experienced in the field of nursery inspection and seed multipliers control for more than 12 years, in developing working tools (procedure manual, sampling protocol, database, etc.). She's a member of a sector expert working group on Fruit (including hop) and Vine for the RNQP Project part II, charged to review the current list of pest/host combinations established in the EU Fruit Marketing Directives, and confirm or adjust current measures as laid down in the EU Fruit Marketing Directives, including requirements for the new pest/host combinations proposed for classification as Regulated Non-Quarantine Pests by the European union. Ms. Touati participated in putting certified olive propagation material of local varieties on the market for the first time in Tunisia and contributed to the revision of several technical standards of production and certification of propagating material, particularly that of the Olea europaea, Malus and Citrus. She intends to effectively implement a certification system for grafted vine plants. She began her career as Plant Production Manager in an agricultural company; she was in charge of the management of a peach orchard, a vineyard, and a nursery for the production of grafted vine plants.

Chapter 14
The Impact of Climate Changes on Food Security and Phytopathogens

Sadraddin Gurbanov, Iqra Zaheer, Sadia Muneer, and Mirza Barjees Baig

Abstract The chapter provides crucial information on climate change and its association with phytopathogens and food security. Climate change directly or indirectly influences food safety and associated health risks. The climate has been changing rapidly globally, resulting in devastating consequences upon social and environmental determinants of health such as clean air, safe drinking water, food chains, and food safety for humans. Climate change has become an unprecedented challenge for the planet's ecosystem, biodiversity, human health, and global economy. Climate change has also been the most pressing problem in the field of plant health, and ultimately animal and human health. Moreover, climate change is adversely affecting agricultural production systems around the world, posing negative impacts on food security. It affects international trade flows of agricultural products and changes the invasion potential, exposure, and prevalence of pests around the world. Numerous studies carried out on grain or fruit crops demonstrate that the pest risk posed by pathogens, (viruses, bacteria, parasites, etc.) weeds, as well as other undesirable biotic and abiotic factors, tend to increase in agricultural systems due to climate change. Habitats of several plants and animal species are being negatively influenced by climate change due to changes in water availability and quality. This may lead to several terrestrial, freshwater, and marine species, to cause changes in geographic ranges, seasonal activity, migration patterns, population size, and inter-species interactions. However, the impact on the yield of most agricultural crops is quite severe. Numerous studies indicate that climate change affects biological systems at various

S. Gurbanov
Institute of Genetic Resources of the National Academy of Sciences of Azerbaijan, Baku, Azerbaijan

I. Zaheer (✉)
Livestock Production Research Institute, Okara, Pakistan
e-mail: dr.iqzaheer@gmail.com

S. Muneer
University of Agriculture, Faisalabad, Pakistan

M. B. Baig
Prince Sultan Institute for Environmental, Water and Desert Research, King Saud University, Riyadh, Kingdom of Saudi Arabia
e-mail: mbbaig@ksu.edu.sa

M. Behnassi et al. (eds.), *Food Systems and Biodiversity in the Context of Environmental and Climate Risks*, https://doi.org/10.1007/978-3-031-89167-0_14

levels, from genes to ecosystems. The biologists noted that climate change in association with anthropogenic factors negatively affects 82% of the main ecological processes; from genetic diversity to the functioning of ecosystems. Phytopathogens under climate change not only affect the decline in agricultural yields but also deteriorate their quality, thereby threatening food security. The chapter presents a detailed review and situational analyses on the impacts of climate change, biological processes, and diversity threatening food security. The purpose of this chapter is to assess the potential impacts of climate change on plant pests and on plants, animals, and human health.

Keywords Climate change · Global warming · Phytopathogens · Plant diseases · Food and nutrition insecurity · Environmental issues

14.1 Introduction

14.1.1 Sources of Food and Food Security

It has long been known that food sources generally fall into two groups: plants and animals and related foods. The main sources of traditional types of food from animals are procured from cattle, poultry as well as from plants including fruits, vegetables, cereals, and legumes (Pingali et al. 2023).

14.1.1.1 Animals as Food Source

Animal products are used as food directly or indirectly. Milk, eggs, and meat are important examples of food from animals. Milk of cows, buffaloes, sheep, goat, and camels are a great source of milk. Milk is an ideal food because it is rich in vitamins, minerals, proteins, carbohydrates and fats and cows, buffaloes, sheep, goat, and camels are a great source of milk. Eggs obtained from chickens, ducks, geese, and quails also are a rich source of protein, minerals, and vitamins. Meat is of two types—red meat and white meat. The meat of cow, goat, sheep, and pigs has a lot of fat and is called red meat. White meat contains less fat and is obtained from chicken and fish. White meat is healthier and can be easily digested as compared to the red meat. Meat is rich in proteins, vitamins, zinc, phosphorus, and iron.

The livestock farming is an important source of providing animal origin food in Azerbaijan and the animal husbandry practices in Azerbaijan trace back to ancient times; as the Terekemins (nomadic tribes) were famous for sheep breeding. Since 1930's, the cattle farming began to develop in the region. The most popular cattle breeds of Azerbaijan; Holstein–Friesian cattle and Kara-ala/Gara-ala (Indigenous white and black cattle breed) are brought from Russia, Ukraine, the Baltic States (Pingali et al. 2023; Kartvelishvili 2009).

Livestock raising is one of the leading and constantly developing branches of agriculture in the Republic of Azerbaijan. Since last two decades, the production of meat in live weight increased to 391.8 thousand tons (1.8%), and milk yield up to 1.6732 million tons (2.9%). In poultry farms, the production of eggs increased to 1.3456 billion pieces (4.3%). The number of cattle increased, reaching 2.7208 million heads (0.6%), and small ruminants reaching up to 8.7407 million heads. Since 2016, production growth in agriculture increased by 1.8%, including crop production by 1.2%, livestock production by 2.5%. In Azerbaijan, there are 1.25 head of cattle and 3.9 head of sheep for every farmer family. These indicators have significantly improved compared to 1990's due to privatization of farming units, improved veterinary and animal husbandry services and agricultural subsidiary programs by the government.

Climate change can have negative and sometimes detrimental effects on livestock health. It can affect the nature of diseases (the frequency, distribution, seasonality, and predictability of diseases in livestock and zoonosis and other parameters, which makes it difficult to properly plan the fight against disease outbreaks and increases the costs associated with them. This, in turn, leads to negative consequences in livestock production, to a decrease in livestock products.

The animal husbandry services can help conserve biodiversity and genetic resources for food and agriculture. More than a third of the world's ruminants are kept on pasture lands. When properly managed, livestock production can contribute to essential life functions, nutrient cycling, soil organic carbon sequestration, and conservation of agricultural land. At the same time, animal husbandry has negative factors of influence on climate change, as it is a direct source of greenhouse gases, intestinal fermentation of manure, as well as an indirect source in the production of feed (agricultural and forage crops). Livestock systems produce large volumes of manure and by-products, as well as leakage of nutrients into water and air (Kartvelishvili 2009).

14.1.1.2 Plants as a Food Source

Plants are a source of a wide variety of nutrients required to keep the human body in an impeccable functioning state. Humans consume fruits, flowers, stems of some plants, leaves and stem-like lettuce, celery, roots of some plants like carrots, beetroot, and seeds like wheat, rice, etc. All food comes from plants, even animals depend on plants. Hence, humans directly or indirectly consume plants as food.

Crop production holds a significant share in agriculture of Azerbaijan. Favorable natural land conditions and geographic characteristics of Azerbaijan has the potential to satisfy the population's need for the staple crops production, at specified times of the year. Since ancient times, agriculture has been well developed on the territory of Azerbaijan, since grain farming had been the main occupation of local people. As civilization developed, several occupations developed alongside agriculture and contributed to national economy. The Caucasus region has become a major center of farming grains, cotton, tobacco, and silk worm in Azerbaijan. Land owners and entrepreneurs have increased their attention to the production of a variety of agricultural products that caters quality and quantity agricultural food products to domestic and foreign markets, which are of strategic importance from the point of view of food security (ESCAP 2021).

In order to ensure food security of the population and to increase the export potential in the agricultural sector, modern agricultural parks and large farms are being introduced in the country.

Since grain production (wheat, corn, rice, barley, etc.) occupies an important place in Azerbaijan's agriculture among crop production since the population consumes 65% grain of the total production. The annual volume of wheat production in Azerbaijan is 1.7 million tons, the remaining 1.3 million tons are covered by purchases in Russia and Kazakhstan.

Despite some positive achievements in the development of the agro-industrial complex, the Republic of Azerbaijan faces numerous environmental problems such as: land degradation, climate change, and water pollution. Climate change has the potential to have a profound impact on food safety of both animal and plant origin. It basically means a change in precipitation patterns, an increase in the number of extreme weather events and an overall increase in average annual temperatures. Climate change has an impact on the spread and resistance of microorganisms (viruses, bacteria, fungi, etc.), parasites, harmful algae, disease vectors, as well as on the clinical picture of related foodborne diseases. The presence of pesticide residues in plant and animal products will depend on the varying activity of these chemicals and pests. Climate change increases the risk of food contamination with heavy metals and persistent organic pollutants from changing crop varieties and farming practices, and from changing soils (Sutherst et al. 2011).

14.1.1.3 Food Security

An important stage in the state agricultural policy was the "State program for the socio-economic development of the regions of the Republic of Azerbaijan in 2008–2015, which, among other things, aims to strengthen the food security of Azerbaijan. In parallel with it, the "State Program for the Reliable Provision of the Population with Food Products" is being implemented, which aims at reducing imports of food products and creating their own food reserves. Climate change, agriculture, food issues, nutritional quality, and different diets are all linked. Climate change leads to changes in temperature and rainfall, as well as the frequency and severity of weather events. Rising temperatures, heat waves, droughts, floods, cyclones, forest fires, and land degradation have a negative impact on agriculture. In those regions where there are serious food difficulties, such as: a drop in crop yields and livestock productivity, a decrease in the volume of fishery production, food security issues, they take a leading position. In these cases, the volume of food decreases, the quality deteriorates and safety fades into the background. As a result, food prices rise, which in turn significantly affects the availability and access to healthy diets. Climate change can also be a driver of nutritional changes in major crops, including declines in protein, some essential minerals and vitamins, and food insecurity (Change 2012).

14.1.2 Climate Change in Azerbaijan

General climate change can be observed in several countries, and Azerbaijan is no exception. In recent years, like other geographical locations, weather conditions in Azerbaijan are not typical for the seasons. For example, February, which is usually the harshest month of the winter, the temperature is higher than normal and some days in low-lying areas the temperature rises 15–20% of the normal range (Change 2012). It is known that among all sectors of the economy, agriculture is the most dependent on climatic conditions. Major land in Azerbaijan is dedicated to agriculture; especially grain crops. The lack of precipitation during the sowing season in regions where there is usually sufficient precipitation during the fall season has created serious difficulties. Due to insufficient soil moisture, most of the sown seeds are not able to germinate, or the germinated ones do not develop further (Geneti 2019). The existing conditions implicated by climatic changes are unfavorable for plants. More so, it has led to increase in infestations and pests such as rodents. The rodent population feeds on the crop as well as the stored grains, causing serious food and economic losses. Severe and erratic climatic changes cause decreases in yield and the quality of the produce.

According to experts, more than 80% of agriculture depends on irrigation. In this regard, the depletion of water resources under the influence of climate change also puts a toll on agriculture. The strategic road map for the production and processing of agricultural products in Azerbaijan provides preventive measures to reduce the impact of climate change on agriculture through adaptation and mitigation measures. The adaptation measures include state and private level mechanisms to adjust agricultural activities to climate change. In case of depleting water resources due to climate change, one must abandon traditional irrigation and apply modern irrigation methods. The use of modern irrigation methods such as drip irrigation, surface irrigation, sprinkler system not only allow to conserve water, but also prevent soil salinization (Sakellariou-Makrantonaki et al. 2002; Sengupta 1985). On the other hand, mitigation measures such as carbon sequestration and manure management address the causes of climate change.

14.1.2.1 Impact of Climate Change on Natural Resources Impacting Food Security

In 1995 Azerbaijan joined the UNFCCC (Framework Convention on Climate Change) and by 2020, the national legal framework for reducing greenhouse gas emissions was put into place through "National Development Concept" by the collaboration of State Program for Poverty Reduction and Sustainable Development, the

Green Economy Concept and the National Program for Reforestation and Expansion of Forests. By 2030, Azerbaijan plans to reduce greenhouse gas emissions by 35%. Climate change is expected to result in lower yields (especially rain-fed potatoes and cotton), reduced available water resources for irrigation, delayed planting, crop destruction, and soil erosion (Bayramli 2020; Abbasov et al. 2022).

14.1.2.2 Land Degradation

Even though Republic of Azerbaijan was able to cater the food requirements of the population with its progressive agriculture, about 36% of the land (29.888 km^2) in Azerbaijan is subject to degradation, including 34% of agricultural land and 8% of forest land. About 34% of irrigated land on mountain slopes suffers from erosion and loss of fertility. According to the Ministry of Agriculture, 41.8% (3610 km^2) of land in the country has been degraded, and 20.7% is in a state of extreme degradation. One million hectares out of 1.6 million hectares of arable land in Azerbaijan are susceptible to degradation. The largest land degradation zones are located in Ordubad (93.4%), Julfi (90%), Apsheron (86%), Khizi (80.4%), and Gobustan (76.5%) regions. The factors effecting the land degradation are unsustainable land use, including over tillage on mountain slopes, water and wind erosion, salinization (Table 14.1), overgrazing and mining operations. In order to stop rapid land degradation and its effective reversal, Azerbaijan is in the process of developing land degradation neutrality targets (NDB) (Gurbanov 2021; Abbasov el al. 2023; İsmayilov and Jabrayilov 2019; Bayramov 2013).

14.1.2.3 Water-Related Problems in Azerbaijan

Republic of Azerbaijan is located at the terminal of the river transit system of the South-Eastern Caucasus, therefore, polluted waters enter here from the territories of neighboring states (Gulıyev 2021). Fresh water supplies are rapidly depleting as well. The shortage of water detrimentally affects agriculture and food security. The available water resources are heavily polluted, and poor-quality drinking water leads to the spread of serious diseases and undermines people's health (Сембеков and Кукеева 2014). The shortage of water is subject to several reasons such as the reduction in precipitation, the water level in the rivers is low. The main rivers, the Kura and Araz, are fed from sources in Turkey and Georgia. These countries must receive heavy rainfall to fuel the distribution canals. A warm winter without heavy rainfall observed in Azerbaijan threatens the drying up of fresh water sources in the country (Gurbanov 2018).

According to the World Resources Institute (WRI), Azerbaijan "by 2040, one-fifth of the world's countries will have low water levels. Such that climate change disrupts the cycle of precipitation, and population growth increases the demand for water." According to a report by the WRI, Azerbaijan ranks 18th out of 33 among the high-risk countries to face drought. This does not mean that Azerbaijan can experience

Table 14.1 Phytopathogens and the diseases caused by them[a]

Diseases	Hosts/plants affected	Etiological agent	Disease characteristics	Distribution
Phytopathogens of wheat				
Smut and rust	Wheat and grass families; *Poaceae* and *Cyperaceae*	Fungus; ***Tilletia tritici***	Brownish-yellow to bright orange spots that form on leaves, filled with a powdery substance	Cosmopolitan specially in Great Britain, China, Canada, Russia, and Kazakhstan
Dust smut	Wheat	*Ustilago tritici*	Yield losses up to 40%	Cosmopolitan
Russian wheat mosaic	Wheat, Barley, rye	Russian wheat mosaic virus (Winter Wheat Russian Mosaic Virus (WWRMV)) from family *Rhabdoviridae*	Irregular leaf mottling (light and dark green or yellow patches or streaks). Leaves are commonly stunted, curled, or puckered; veins may be lighter than normal or banded with dark green or yellow	The territories of the Former USSR
Phytopathogens of rye				
Rust smut	*Rye*	*Urocystis occulta, Rabh-Tilletiaceae*	Lead grey longitudinal stripes and cracks on stems, loop shape appearance of stem	Non-Black Earth Zone, Western Siberia, north-western regions of Russian Federation

(continued)

Table 14.1 (continued)

Diseases	Hosts/plants affected	Etiological agent	Disease characteristics	Distribution
Head smut	Rye	*Sphacelotheca reiliana*	The primary source for head smut inoculum is the teliospores that overwinter in the soil	North Caucasus, the Transcaucasus (including Azerbaijan), Kuibyshev, Orenburg, Amur regions of the Russian Federation, Primorsky Territory (29)
Dusky smut		Fungus; *Urocystis occulta*	Linear sori occur	Cosmopolitan
Dust smut/Corn smut		Fungus; *Ustilago maydis*	Tumor-like galls that vary in size (1 cm to ≥ 30 cm diameter). All meristematic tissues are susceptible. Galls formation on; on ears, tassels, stalks, nodal shoots, and mid-ribs of leaves	Cosmopolitan
Root rot		Fungi; *Armillaria mellea, Clitocybe tabescens, Fusarium* spp., *Pythium* spp., *Phytophthora* spp.	Loss of vigor, become stunted and yellow, and may wilt or die back and drop some leaves	Cosmopolitan
Septoria		Ascomycete fungus; *Mycosphaerella graminicola*	STB is characterized by necrotic lesions on leaves and stems that develop after infected cells collapse, and is more prevalent during cool, wet weather	Europe, Asia, America
Damping-off		Fungi; *Rhizoctonia* spp., and *Fusarium* spp., *Pythium* spp.	Damping off pathogens can cause root rot or crown rot in mature plants	Cosmopolitan

(continued)

Table 14.1 (continued)

Diseases	Hosts/plants affected	Etiological agent	Disease characteristics	Distribution
Fusarium head blight/ Root and stem base rot of cereals		Mainly; *Fusarium graminearum* also *Fusarium avenaceum*, *Fusarium culmorum*, and *Fusarium poae*	Kernels are shriveled, chalky, and white-pinkish. Disease spreads upward and downward sequentially infecting spikelets. Bleaching of spikelets, occasional pink/Orange spore masses	Cosmopolitan
Ergot		Fungus; *Claviceps purpurea*	Yield reduction, the significance of the disease is primarily related to the toxic alkaloids present in the ergots (sclerotia)	Cosmopolitan
Powdery mildew	Rye, infects all cucurbits, including muskmelons, squash, cucumbers, gourds, watermelons, and pumpkins	Fungus; *Podosphaera xanthii*	Premature death of leaves, decreased yield and fruit quality	Cosmopolitan
Phytopathogens of potato				
Potatoes leaf roll (PLR)	Potato and night shade family Potato	Potato leaf roll virus (PLRV)	Upper leaves turn fawn, directed upward, Reddened leaf edges, curled, hard to the touch	Cosmopolitan
		PLRV Y	It causes a prominent upward rolling of the leaves, and the plants are stunted and have a stiff upright growth	

(continued)

Table 14.1 (continued)

Diseases	Hosts/plants affected	Etiological agent	Disease characteristics	Distribution
		PRLV X	Also causes a prominent upward rolling of the leaves, and the plants are stunted and have a stiff upright growth	
		PSTV	Also causes a prominent upward rolling of the leaves, and the plants are stunted and have a stiff upright growth	
Mosaic/Latent	Potato, tomato, tobacco, pepper	Potato virus X (PVX)	No pathological signs, Decreased yield (15%), severe cases yield losses (≥ 30%)	Cosmopolitan
Wrinkled mosaic	Potato, tomato, pepper, weeds	Potato virus Y (PVY)	Necrotic spots (the entire leaf blade may die off	Cosmopolitan
Potato spindle tuber viroid (PSTVd)	Potato, tomato	Potato spindle tuber viroid (PSTVd)	Yield is reduced by 30–90%, reduced number of elongated and erect stems, Leaves elongated, hard, yellowish, wrinkled, Delayed tuber formation, tubers are smaller, elongated, increased eyes and sharper eyebrows. The potato spindle tuber viroid (PSTV) which is often grouped with the virus diseases, has largely been eliminated through systematic testing over the past two decades	Cosmopolitan

(continued)

Table 14.1 (continued)

Diseases	Hosts/plants affected	Etiological agent	Disease characteristics	Distribution
Phytopathogens of barley				
Barley spots (Septoria)	*Barley*	*Septoria hordei, Septoria graminum, Septoria nodorum, Phaeosphaeria nodorum*	Premature death of leaves, decreased yield	Cosmopolitan
Dark brown spot		*Drechslera sorokiniana*	Brown or brownish streaks	Cosmopolitan
Striped spot (striped helminthosporium)		*Drechslera graminea* (*Helminthosporium gramineum*)	Leaves, stem, ear show pale yellow spots, which gradually lengthen into light brown stripes with a narrow pink border and black embryo symptom	Cosmopolitan
Net spotting		*Drechslera teres*	Separate oval brown spots appear with a white–pink border and mutually perpendicular black stripes that form a reticular pattern	Cosmopolitan
Grey green spots		*Rhynchosporium graminicola* (secalis)	Atery gray-green elongated spots with a dark brown border on the leaves, later turn gray-blue bloom of conidial sporulation. Leaves curl and dry. Grain show light brown spots with a dark brown halo	Cosmopolitan
Stem (linear) rust		*Puccinia graminis*	Shrunken grain, rusty-brown oblong uredinia in stems and leaves	Cosmopolitan
Yellow rust		*Puccinia striiformis*	Rounded lemon-yellow uredinia up to 0.5 mm in diameter Arranged in groups, in the form of longitudinal dotted lines, with a chlorotic border	Cosmopolitan
Barley dwarf rust		*Puccinia hordei*	Leaves and sheaths in the form of small chaotically located light yellow or light brown uredinia	Cosmopolitan

(continued)

Table 14.1 (continued)

Diseases	Hosts/plants affected	Etiological agent	Disease characteristics	Distribution
Phytopathogens of Sunflower				
Downy mildew	Sunflower	Fungus-like organism; *Peronospora manshurica*	Disease is superficial and causes no yield loss, but can cause defoliation of plants and reduced yields under rare conditions	Cosmopolitan; occurs during periods of high humidity and moderate temperatures
Alternaria		*Alternaria helianthi*	Premature death of the affected parts of the plants (fracture and drying of the stem). 40% decreased production (32)	South Africa
Sunflower gray rot		*Botrytis cinerea, Macrophomina phaseolina*	Affected parts turn brown and become covered with a gray bloom	Cosmopolitan
Phytopathogens of grapes				
Oidium (powdery mildew)	Grapes	*Erysiphe necator*	Massive damage to un ripened fruit and chlorophyll containing parts of plant, early wilting	Cosmopolitan diseases of grapevines in the world
Mildew (downy mildew)		*Plasmopara viticola*	Shrinkage and drying of the fruit, leaves, yellow spots on chlorophyll containing parts of plant with white fluff	North America, Western Australia
Black spot (Phomopsis, Escoriasis)		Fungus; *Phomopsis viticola*	Small dark leaf spots with yellow halos that appear during spring and black cane lesions on the lower internodes	–

(continued)

Table 14.1 (continued)

Diseases	Hosts/plants affected	Etiological agent	Disease characteristics	Distribution
Anthracnose (Grape pox, Bird's eye)		Fungus; *Elsinoe ampelina*	In chlorophyll-containing parts; a point lesion turns into ulcers, which can ring shoots and inflorescences, as a result of which the latter break or dry out	The disease is most destructive in warm, wet seasons in the world
Rubella		Fungus; Ascomycetes	Leaves in white grape—yellow or orange spots, leaves of red varieties—scarlet or reddish-brown. Affected leaves dry up and fall off. Sick berries and inflorescences rot and fall off	Cosmopolitan
Black rot		Fungus; *Guignardia bidwellii*	Striated necrosis (5–10 mm) is replaced by profuse ulceration and decay of large areas The green berries are covered with bulges that turn into depressed black spots. Gradually, diseased berries acquire a black-purple color, wrinkle and dry out	North America, Europe, South America, and Asia (during hot and humid weather)
White rot (white mouth/Grapevine white rot)		*Coniella diplodiella*	Gray-white, coloration the berries begin to rot, then dry up. At the end, the affected bunch falls off entirely	Cosmopolitan
Gray rot Botrytis (Gray rot or Noble rot)		*Botrytis cinerea*	Characteristic gray bloom, the berries become licky, sour, unpleasant moldy smell. On the green parts of the plant and shoots, gray rot takes the form of large brown spots with and without bloom	Cosmopolitan

(continued)

Table 14.1 (continued)

Diseases	Hosts/plants affected	Etiological agent	Disease characteristics	Distribution
Aspergillus rot (Aspergillosis)		*Aspergillus niger*	By a white bloom that covers all the berries or completely blackened rotten bunches (extreme case)	Cosmopolitan
Alternaria		*Alternaria alternata*	Point necrosis develops into rounded and shapeless spots with dark edges	Cosmopolitan
Phytopathogens of forages				
Root rot and Fusarium wilting	Sweet potatoes, tomatoes, legumes, melons, bananas	*Fusarium* spp.	Rot appears on the upper part of the main root or on the lateral ones, where dark brown dry or rotting spots are visible. The core of the root becomes rotten and the plant dies. Withering of plants occurs as a result of blockage of the vascular bundles of the root and base of the stem	Cosmopolitan
Rust	Clover	Fungus; *Uromyces fallens Kern*	Bright yellow etsidia develop on the underside of clover leaves. By the beginning of flowering, brown uredopustules with urediospores are formed. Spores re-infect other plants, and by the end of the growing season, dark brown teliopustules are formed	Cosmopolitan
Rust	Alfafa, Spurge	Fungus; *Uromyces striatus Schr*	Leaves grow poorly, dry up quickly and fall off. The stems are unproductive. Plants lose a lot of carbohydrates and protein	Cosmopolitan
Powdery mildew	Clover	*F. trifolii*	Rounded spots with leucorrhoea appear on the leaves, which, growing, form a continuous bloom. Over time, black cleistothecia form on it, which is why the plaque takes on a dirty color. The leaves become brittle and die off	Cosmopolitan
	Alfafa	*F. medi-caginis*		
	Sainfoin	*F. onobrichis*		

(continued)

Table 14.1 (continued)

Diseases	Hosts/plants affected	Etiological agent	Disease characteristics	Distribution
Clover anthracnose	Clover	*Colletotrichum trifolii*	Streaks and dark spots appear on the shoots, cotyledons and stems, the seedlings die	Cosmopolitan
Brown spot	Alfafa	*Pseudopeziza medicaginis Fckl*	Brown, rounded spots appear on the leaves, at first small, then increasing to 2–3 mm in diameter. The edges of the spots are jagged-edged. On stems and beans, spots are oblong	Cosmopolitan
	Clover	*Pseudopeziza trifolii Fckl*		
Downy mildew	Alfalfa	*Peronospora aestivaiis Syd*	Lesion can be diffuse and local, leaves are chlorotic, often curling downward, on the underside they are covered with a brownish bloom of fungal sporulation. Plants stunted, do not bloom and gradually dry out	Cosmopolitan
Dodder	Clover, Alfalfa, and Sainfoin	Parasitic flowering plant; 5 species: clover, European, southern, thyme and thin-stemmed	Affected plants grow poorly, often die off, and the quantity and quality of seeds decrease	Cosmopolitan

[a] Adapted from the following sources; Juroszek and Tiedemann (2013), Majumder et al. (2013), Vega et al. (2019), Cavicchioli et al. (2019), Raffel et al. (2013), Harvell et al. (2002), Altizer et al. (2013), Anderegg et al. (2013), Bakkeren et al. (2008), Scheffers et al. (2016), Hammond and Hammond (2003), Fones et al. (2017), Arogundade et al. (2021), Shahbazi (2013), Jafarova and Hajiyeva (2018), Duisenbekuly et al. (2024), Hasanov et al. (2021), Khatibi et al. (2021), Hashimova (2023)

drought due to low water levels. Although, the country has 7–8 times more water resources than it needs however, there is scarcity of fresh and clean water (due to high water loss) for drinking and irrigation (Ahouissoussi et al. 2014).

For several years, scarcity of irrigation water has been a serious concern for the farmers and agricultural production. The volume of irrigated lands in the country is 1.4 million hectares. Furthermore, in recent years Azerbaijan, Georgia, and Armenia, which use same water resource of Kura and Araz rivers are expanding irrigated arable land to produce more agricultural products such that their demand for water has increased; Azerbaijan (1.2 billion cubic meters), Georgia's (2 billion cubic meters), and Armenia (1 billion cubic meters). According to the statistics provided by the State Statistics Committee of Azerbaijan on water from natural resources, its consumption and loss, Azerbaijan ranks first in water loss with 28% among the countries of the Commonwealth of Independent States. Water loss in other CIS countries: 20% in Armenia, 13% in Kazakhstan, 12% in Ukraine, 10% in Russia, 7% in Moldova, and 5% in Belarus. Obsolete irrigation systems are the main cause of water losses. The main canals on the balance of the Amelioration and Water Resources Open Joint-Stock Company were also built during the Soviet era. Such that the Upper Karabakh canal and the Upper Shirvan canal were commissioned in 1958, the Samur-Absheron canal in 1940, the Bash-Mughan canal in 1960, and the Bash Mil canal in 1976 (Abduev et al. 2024).

Rainwater harvesting is not organized in the country. Existing reservoirs and lakes have been seized by oligarchic officials and used for other purposes. Farmers and villagers are not allowed to use the water collected in these lakes. In-farm reservoirs remaining from the Soviet era have become unusable because they have not been repaired for a long time (Altizer et al. 2013).

Despite the fact that the Takhtakorpu and Shamkirchay reservoirs, which were completed due to large-scale investments, were put into operation several years ago, there has been no positive change in the supply of irrigation water in agriculture over the years. On the contrary, water shortages have recently become more acute in the country. Despite the fact that more than 411,764,705.80 USD (700 million AZN) was spent on the construction of Shamkirchay reservoirs. The investment in the Takhtakorpu reservoir was 588,235,294 USD (1 billion AZN). About 50,000 ha of arable land were projected to be irrigated with these reservoirs (Новрузов 2023).

The problem of freshwater shortage has become a topical issue not only in the region but also in the world. International experience shows that some countries have managed to solve the problem of water shortage by preventing wasteful use of irrigation water. Another problem is the lack of centralized coordination of the country's water management. The government should take a number of urgent steps to address the existing problems in this area. (1) First of all, a special state program should be developed to improve the water management system. (2) In order to ensure the centralized coordination of the water management system, one of the Deputy Prime Ministers should be responsible for the sector. (3) In order to meet the needs of agriculture for irrigation water in a sustainable and loss-free manner, the transfer of water to the fields through metal and plastic pipes should be organized. (4) In order to prevent water loss in irrigated arable lands, it is necessary to start the rapid introduction of cost-effective irrigation techniques and advanced irrigation technologies in all farms (Abduev et al. 2024; Altizer et al. 2013).

14.1.3 Impact of Climate Change on Phytopathogens

14.1.3.1 Phytopathogens and Plant Diseases

The term "Phytopathogen" comes from the Greek words—"*phyto*" ..., "*patho*" ..., *and* "*gene*," an environmental factor that causes plant diseases. In general, **phytopathogens** can be viruses, bacteria, fungi, insects, as well as meteorological factors (sharp fluctuations in temperature or humidity), toxins and toxic substances, etc. (Dangl 2013). A **living phytopathogen** can be described as an organism harmful to plants (viruses, bacteria, fungi, parasitic animals, and plants, etc.) or a variety/biotype of plants and animals that are harmful to plants or plant products (2). Whereas, plant diseases can be described as "processes that occur in a plant under the influence of various reasons—pathogens and unfavorable environmental conditions, manifest themselves in dysfunction (photosynthesis, respiration, synthesis of plastic and growth substances, water flow, nutrients), the structure of the body and cause premature death of the plant or damage to its individual organs" (Garrett et al. 2006). More than 30 thousand different plant diseases are known till date.

Plant diseases are caused by countless phytopathogens/factors lead to huge crop losses all over the world, including in Azerbaijan, thereby threatening food security.

14.2 Impact of Climate Change on Phytopathogens and Food Security

The climate of our planet is indeed changing rapidly as a result of global warming, an increase in the concentration of CO_2 in the atmosphere and other anthropogenic factors, which is confirmed by numerous examples. This leads to a change in the biological characteristics of many living organisms, including plants, animals, insects, microorganisms, and phytopathogens (Schulze et al. 2017). Rapid climate change is making it difficult for species and ecosystems to adapt. And this leads to a change in the geographical distribution of many species of phytopathogens (Singh et al. 2023; Raza and Bebber 2022). For example, historically on the territory of the CIS, (in particular, on the territory of Russia), the causative agent of fusarium of grain crops (fungus; *Fusarium graminearum*) is localized in two zones: in the North Caucasus and in the Far East. Temperature can affect the function of the virulence genes of parasites and the resistance genes of plant varieties. When the ambient temperature changes, the dominance of the species can change. Apparently, when predicting the phytosanitary situation in connection with climate warming, one should pay special attention to thermally stable species. It is recommended to constantly monitor the emergence of new plant diseases, adjust protective measures against diseases, and create varieties with a broader ability to adapt to changing environmental conditions (Royo et al. 2009).

Climate change disrupts interactions between species, forces them to adapt to undesirable conditions, migrate, replacing some species in the community with others, or leads to their extinction (Scheben et al. 2016; Mooney et al. 2009; Upadhyay 2020). Climate change can increase the risk of disease by altering the adaptation of the host plant and the parasite (Muluneh 2021). Climate change affects the emergence and spread of diseases in marine and terrestrial biota (Marcogliese 2008) depending on various socio-economic factors, environmental factors, as well as pathogen-host relationships (Cavicchioli et al. 2019). Forest death caused by drought and heat stress can be exacerbated by the attack of pathogenic organisms. For crops, a number of interacting factors, including carbon dioxide levels, climate change, plant health, and species-specific interactions with pathogens, are important when it comes to crop responses to pathogens (Garrett et al. 2006). Many microorganisms can cause plant diseases (fungi, bacteria, viruses, viroids, oomycetes) and thus affect food production, cause hunger (for example, the oomycete *Phytophthora infestans* caused the potato famine in Ireland) and pose a threat to food security (Hutchins and Fu 2017). The spread of pathogens and the emergence of diseases are stimulated by the transportation of products and the introduction of new species, and are also influenced by weather factors and the environment.

Anthropogenic climate change has a negative impact on local life forms, which allows pathogens to cause disease more often. The impact on aquaculture, livestock, and crops threatens the global food supply for humans. Human activities such as population growth and transportation, combined with climate change, are increasing the antibiotic resistance of pathogens and stimulating the spread of pathogens, both transmissible and waterborne, thereby increasing the incidence of diseases in humans, other animals and plants.

Disasters continue to occur around the world, for example, in 2020, hordes of desert locusts invaded northern Kenya, and this pest infestation has become the worst for the country over the past 70 years. Locusts destroyed crops and threatened food security for 3 million people. The impact of the locust plague has also been devastating to some of the other poorest countries in Africa, the Middle East and South Asia, as governments and communities prepared for the COVID-19 pandemic. The scale and strength of this locust infestation is associated with climate change—powerful cyclones that hit the Arabian Peninsula, and heavy rains gave impetus to lush vegetation in usually arid locust breeding areas. Cyclonic vortices facilitated the flight of growing flocks of insects to East Africa. And just as Kenya and other countries had to grapple with the pandemic, climate change has created another risk factor—one that is expected to intensify in the future and hit the most vulnerable populations the hardest.

According to some estimates, in 2020 alone, the increase in the number of the extremely poor caused by the pandemic ranged from 88 to 115 million people.

Moreover, climate change could put up to 132 million more people in poverty by 2030. The world community will not be able to return to the previous rate of poverty reduction if it does not tackle the problem of climate change. It is our responsibility to help the poorest countries adapt and build resilience to shocks. Various aspects of the problem of global climate change, including its possible catastrophic impact on agriculture are discussed at all levels in recent years. Climate change can result in the extinction of 30–40% of plant and animal species, degradation of key ecosystems, reduced agricultural productivity and thus to the aggravation of the problem of food safety (Hutchins and Fu 2017; Cavicchioli et al. 2019).

Climate change disrupts interactions between species, forces them to adapt to undesirable conditions, migrate, replacing some species in the community with others, or leads to their extinction. An infinite number of examples can be cited about the influence of climate, phytopathogens, various biotic and abiotic factors on the plant organism, on food security. Since the domestication of crops, humans have been observing a variety of pests, phytopathogens and environmental factors that cause crop losses, which often lead to hunger, disease, and social tension. According to Savary et al., on average, globally, 10–28% of crops are lost due to pests during cultivation. And after the harvest, there are further losses, with the developing countries doing the worst. Moreover, in addition to crop losses, the presence of mycotoxins (toxins of fungal origin) aggravates the situation. Molds parasitize many types of food products such as cereals, dried fruits, nuts, and spices. Mold can occur both before and after harvest, during storage, and/or on finished food under conditions of favorable temperature and high humidity (Hoffmann and Sgrò 2011).

References

Abbasov RK, Allahverdiyev R, Zaynalov R, Habilov A, Aliyeva R (2023) Azerbaijan National Ecosystem Assessment. Government of Azerbaijan, RECC Azerbaijan, Baku, Azerbaijan

Abbasov R, Karimov R, Jafarova N (2022) The Caspian Sea and its values in Azerbaijan. Ecosystem services in Azerbaijan: value and losses. Springer International Publishing, Cham, pp 1–28

Abduev M, Asgarova M, Karimova A, Mammadova T, Sadiq M, Mammadova K (2024) Technological solutions for water resources management in Azerbaijan to achieve sustainable development: monitoring and improvement of management and forecasting. Rev Gestão Tecnologia 24(2):96–113

Ahouissoussi N, Neumann JE, Okan C, Boehlert B, Strzepek K (2014) Reducing the vulnerability of Azerbaijan's agricultural systems to climate change: impact assessment and adaptation options. World Bank Publications

Altizer S, Ostfeld RS, Johnson PT, Kutz S, Harvell CD (2013) Climate change and infectious diseases: from evidence to a predictive framework. Science 341(6145):514–519

Anderegg WR, Kane JM, Anderegg LD (2013) Consequences of widespread tree mortality triggered by drought and temperature stress. Nat Clim Chang 3(1):30–36

Arogundade O, Ajose T, Matthew JO, Osijo IA (2021) Current and emerging pests and diseases of cucumber (Cucumis sativus L.) in Africa. Cucumber Econ Values Its Cultivation Breed 179:473–479

Bakkeren G, Kämper J, Schirawski J (2008) Sex in smut fungi: structure, function and evolution of mating-type complexes. Fungal Genet Biol 45:S15–S21

Bayramli G (2020) The environmental problems of Azerbaijan and the search for solutions. Wseas Trans Environ Dev 16:423–433

Bayramov E (2013) Quantitative assessment of vegetation renaturation and soil degradation and their control by climate and ground factors along rights-of-way of petroleum/gas pipelines, Azerbaijan (Doctoral dissertation, Dresden, Technische Universität Dresden, Diss., 2013)

Cavicchioli R, Ripple WJ, Timmis KN, Azam F, Bakken LR, Baylis M, Behrenfeld MJ, Boetius A, Boyd PW, Classen AT, Crowther TW (2019) Scientists' warning to humanity: microorganisms and climate change. Nat Rev Microbiol 17(9):569–586

Change C (2012) The Republic of Azerbaijan

Сембеков НН, Кукеева ФТ (2014) Environmental risks in Central Asia and its impact on regional cooperation. Вестник КазНУ. Серия международные отношения и международное право 66(2)

Dangl JL (ed) (2013) Bacterial pathogenesis of plants and animals: molecular and cellular mechanisms, vol 192. Springer

Duisenbekuly AK, Kulbay B, Bigeldiyeva Z, Zhakipbekova D, Daurbayeva M (2024) Food security in Kazakhstan and Azerbaijan: challenges and strategies for economic and sustainable development. Int J Food Syst Dyn 15(3):267–277

ESCAP U (2021) Business process analysis: study of cereal exports from Kazakhstan to Azerbaijan

Fones HN, Fisher MC, Gurr SJ (2017) Emerging fungal threats to plants and animals challenge agriculture and ecosystem resilience. The fungal kingdom, pp 787–809

Garrett KA, Dendy SP, Frank EE, Rouse MN, Travers SE (2006) Climate change effects on plant disease: genomes to ecosystems. Annu Rev Phytopathol 44(1):489–509

Geneti TZ (2019) Review on the effect of moisture or rain fall on crop production. Civ Environ Res 11(2)

Gulıyev A (2021) Old and current logistics management in Azerbaijan and economic benefits of international logistic management in Azerbaijan (Master's thesis, İzmir Ekonomi Üniversitesi)

Gurbanov S (2021) Role of natural gas consumption in the reduction of CO_2 emissions: case of Azerbaijan. Energies 14(22):7695

Gurbanov A (2018) Hydrological situation in Azerbaijan and environmental aggression of Armenia against Azerbaijan. Threat World: Hydrol Crime Ecol Genocide 11:55

Hammond J, Hammond RW (2003) The complete nucleotide sequence of isolate BYMV-GDD of Bean yellow mosaic virus, and comparison to other potyviruses. Adv Virol 148:2461–2470

Hutchins DA, Fu F (2017) Microorganisms and ocean global change. Nat Microbiol 2(6):1–11

Harvell CD, Mitchell CE, Ward JR, Altizer S, Dobson AP, Ostfeld RS, Samuel MD (2002) Climate warming and disease risks for terrestrial and marine biota. Science 296(5576):2158–2162

Hasanov E, Garayusifova A, Tongren EJ, Geleishvili M (2021) Descriptive study of cattle and dog rabies cases in the Sheki-Zagatala region of Azerbaijan (2015–2016): knowledge, attitudes, and practices of people towards rabies. PLoS ONE 16(11):e0248572

Hashimova AC (2023) Main problems and solutions of agriculture in Azerbaijan. In: Conference book, p 31

Новрузов РЄ (2023) Peculiarities of the formation of innovative infrastructure in the territories of Azerbaijan limited from the occupation. Київський Економічний Науковий Журнал 1:53–59

Hoffmann AA, Sgrò CM (2011) Climate change and evolutionary adaptation. Nature 470(7335):479–485

МГЭИК О (2014) Изменение климата 2014 г.(Электронный ресурс) доклад of climate change on organisms harmful to plants. http://www.ipcc.ch/home_languages_main_russian.shtml

İsmayilov M, Jabrayilov E (2019) Protected areas in Azerbaijan: landscape-ecological diversity and sustainability. Ankara Üniversitesi Çevrebilimleri Dergisi 7(2):31–42

Jafarova F, Hajiyeva A (2018) Food security in state programs of Azerbaijan (for example of the major Caucasus). In: 55th international scientific conference on economic and social economic and social development (book of proceedings vol 2/4), p 257

Juroszek P, von Tiedemann A (2013) Climate change and potential future risks through wheat diseases: a review. Eur J Plant Pathol 136:21–33

Kartvelishvili T (2009) Cattle sector and dairy chain developments in Georgia, Azerbaijan and Armenia. In: The cattle sector in Central and Eastern Europe. Wageningen Academic, pp 133–152

Khatibi M, Abdulaliyev G, Azimov A, Ismailova R, Ibrahimov S, Shikhiyev M, Agalarov D, Seyidov T, Omarov A, Suleymanova C, Zeynalova S (2021) Working towards development of a sustainable brucellosis control programme, the Azerbaijan example. Res Vet Sci 137:252–261

Majumder D, Rajesh T, Suting EG, Debbarma A (2013) Detection of seed borne pathogens in wheat: recent trends. Aust J Crop Sci 7(4):500–507

Marcogliese DJ (2008) The impact of climate change on the parasites and infectious diseases of aquatic animals. Rev Sci Tech 27(2):467–484

Mooney H, Larigauderie A, Cesario M, Elmquist T, Hoegh-Guldberg O, Lavorel S, Mace GM, Palmer M, Scholes R, Yahara T (2009) Biodiversity, climate change, and ecosystem services. Curr Opin Environ Sustain 1(1):46–54

Muluneh MG (2021) Impact of climate change on biodiversity and food security: a global perspective—a review article. Agric Food Secur 10(1):1–25

Pingali P, Boiteau J, Choudhry A, Hall A (2023) Making meat and milk from plants: a review of plant-based food for human and planetary health. World Dev 170:106316

Raffel TR, Romansic JM, Halstead NT, McMahon TA, Venesky MD, Rohr JR (2013) Disease and thermal acclimation in a more variable and unpredictable climate. Nat Clim Chang 3(2):146–151

Raza MM, Bebber DP (2022) Climate change and plant pathogens. Curr Opin Microbiol 70:102233

Royo C, Elias EM, Manthey FA (2009) Durum wheat breeding. Cereals pp 199–226

Sakellariou-Makrantonaki M, Kalfountzos D, Vyrlas P, Kapetanos B (2002) Water saving using modern irrigation methods. In: 3rd International forum integrated water management: the key to sustainable water resources, Hydrorama, Athens, Greece, pp 96–102

Scheffers BR, De Meester L, Bridge TC, Hoffmann AA, Pandolfi JM, Corlett RT, Butchart SH, Pearce-Kelly P, Kovacs KM, Dudgeon D, Pacifici M (2016) The broad footprint of climate change from genes to biomes to people. Science 354(6313):aaf7671

Schulze A, Roberts R, Pietersen G (2017) First report of the detection of Bean yellow mosaic virus (BYMV) on Tropaeolum majus; Hippeastrum spp., and Liatris spp. in South Africa. Plant Dis 101(5):846

Scheben A, Yuan Y, Edwards D (2016) Advances in genomics for adapting crops to climate change. Curr Plant Biol 6:2–10

Sengupta N (1985) Irrigation: traditional vs modern. Econ Polit Wkly, pp 1919–1938

Shahbazi P (2013) Selection, characterization and evaluation of streptomyces for the control of anthracnose in chilli. University of Malaya (Malaysia)

Singh BK, Delgado-Baquerizo M, Egidi E, Guirado E, Leach JE, Liu H, Trivedi P (2023) Climate change impacts on plant pathogens, food security and paths forward. Nat Rev Microbiol 21(10):640–656

Sutherst RW, Constable F, Finlay KJ, Harrington R, Luck J, Zalucki MP (2011) Adapting to crop pest and pathogen risks under a changing climate. Wiley Interdisc Rev: Clim Change 2(2):220–237

Upadhyay RK (2020) Markers for global climate change and its impact on social, biological and ecological systems: a review. Am J Clim Chang 9(03):159

Vega D, Gally ME, Romero AM, Poggio SL (2019) Functional groups of plant pathogens in agroecosystems: a review. Eur J Plant Pathol 153:695–713

Dr. Sadradin Gurbanov has been a leading plant pathologist in Azerbaijan for over three decades. He has published extensively in academic journals and authored several books. Dr. Gurbanov worked with the Institute of Genetic Resources at the National Academy of Sciences of Azerbaijan until his retirement.

Dr. Iqra Zaheer is a specialist in veterinary pathology, holding a D.V.M., M.Phil., and Ph.D. in Pathology from the University of Agriculture, Faisalabad, Pakistan. With nearly 10 years of experience in animal disease diagnostics, Dr. Zaheer currently serves as the Incharge of the Health Control Division at the Livestock Production Research Institute, Okara, Pakistan. She has published several scientific articles and book chapters with both national and international publishers.

Dr. Sadia Muneer holds a D.V.M. and M.Phil. in Microbiology from the University of Agriculture, Faisalabad, Pakistan. She has published numerous book chapters and scientific articles. As an early career scientist, Dr. Muneer actively participates in high-impact academic projects related to microbiology and veterinary science.

Dr. Mirza Barjees Baig is a professor at the Prince Sultan Institute for Environmental, Water, and Desert Research, King Saud University, Saudi Arabia. He earned his MS degree in International Agricultural Extension in 1992 from Utah State University, Logan, Utah, USA, and was placed on the "Roll of Honor". He completed his Ph.D. in Extension for Natural Resource Management from the University of Idaho, USA, and was honored with the "1995 Outstanding Graduate Student Award". He has published extensively on the issues associated with natural resources in national and international journals. He has also made oral presentations about agriculture, natural resources, and the role of extension education at various international conferences. He serves on the editorial boards of many international journals and is a member of many international professional organizations. He has produced six books on climate change, food security, and climate-smart agriculture. Food waste, water management, degradation of natural resources, the deteriorating environment, and their relationship with society and the community are his areas of interest. He has attempted to develop strategies to conserve natural resources, promote the environment, and develop sustainable communities. He started his scientific career in 1983 as a researcher at the Pakistan Agricultural Research Council, Islamabad, Pakistan. He served at the University of Guelph, Ontario, Canada, as a Special Graduate Faculty member from 2000 to 2005. He served as a foreign professor at the Allama Iqbal Open University (AIOU), Pakistan, through the Higher Education Commission from 2005 to 2009.

Chapter 15
Biofortification and Sustainable Intensification of Soil: Perspectives on Rice Cultivation in India

Rohan Rama Krishnan, Sivakumar Bhadrapriya, J. Nayana, Nachiketha R. Rao, S. Nithya, Praveen Nagella, and Jameel M. Al-Khayri

Abstract Biofortified crops have inherently been known to acquire climate-smart traits and stress resilience. Climate-smart agriculture integrates climate change into the planning and implementation of sustainable agricultural strategies. Biofortification is a climate-smart concept that enhances crop nutrient quality and quantity through conventional breeding, agronomic practices, or genetic engineering. It will enrich food availability, stability, accessibility, and utilization and positively impact the health, livelihood, production, and distribution of food crops. The system of rice (*Oryza sativa* L.) intensification involves a set of agronomic principles to improve the structure and functioning of the soil system by fortifying it with organic matter and micronutrients. With the exceeding urbanization and population explosion, food security is a primary concern for policymakers all around the globe. Widespread zinc, iodine, iron, and selenium micronutrient malnutrition is a significant cause of numerous health problems in human populations where rice is part of the staple diet. Climate-smart biofortification is a durable and effective option to reach the vast numbers of malnourished populations scattered across the world sustainably. Approaches have been strategized worldwide under rice biofortification research projects for maintaining, increasing, and introducing new micronutrients in rice grain. Biofortification has been safely implemented as an environmentally friendly approach to produce higher yields at low costs without undesirable soil effects.

R. Rama Krishnan · S. Bhadrapriya · J. Nayana · N. R. Rao · S. Nithya · P. Nagella
Department of Life Sciences, School of Sciences, CHRIST (Deemed to be University), Bengaluru, India
e-mail: j.nayana@science.christuniversity.in

P. Nagella
e-mail: praveen.n@christuniversity.in

J. M. Al-Khayri (✉)
Department of Agricultural Biotechnology, College of Agriculture and Food Sciences, King Faisal University, Al-Ahsa, Saudi Arabia
e-mail: jkhayri@kfu.edu.sa

M. Behnassi et al. (eds.), *Food Systems and Biodiversity in the Context of Environmental and Climate Risks*, https://doi.org/10.1007/978-3-031-89167-0_15

Prospective advancements can be achieved by integrating mineral and organic fertilizers with superior germplasm, promoting improved nutrient uptake and localization in the consumed parts of the crop.

Keywords Agronomic practices · Biofortification · Climate-smart agriculture · Malnutrition · *Oryza sativa*

15.1 Introduction

Soil comprises organic matter, gases, liquids, minerals, and living organisms, collectively supporting life. It adapts to the complex biotic community in response to climatic conditions, native biota, time, topography, and parent material. It provides various ecological services helpful to humans, like providing food, fiber, wood, raw materials, and physical infrastructural support. Aside from supplying food for mankind, which adds to the near-immediate provisioning benefits, it also provides long-term, less tangible benefits. For instance, soil regulates flood mitigation, nutrient cycling, contaminant filtering, greenhouse gas reduction, carbon storage, waste detoxification, and recycling. Figure 15.1 depicts the soil food web, which is a network of complex relationships between the diverse groups of fauna and flora found in the soil, such as bacteria, fungi, protozoa, nematodes, microarthropods, and the larger plants and animals. Soil also checks the growth of pest and disease populations and acts as a cultural hub for recreation, aesthetics, and heritage values. With the significant rise in the human population, the soaring per capita calorie consumption, and changing diets, the demand for food and fiber is escalating swiftly (Kopittke et al. 2019). As a result, global soil management will suffer dire repercussions, especially for agronomic purposes.

Since the age of the first civilization, food production by man has been often acquainted with systematic agriculture, which is intrinsically associated with soils. About 98.8% of the daily calorific consumption by human beings is directly linked with soil (FAO 2021), with the remaining 1.2% accounted for by aquatic sources. Soils produce agricultural products by biological and chemical processes influenced by soil quality. Agricultural activities implemented by farmers will deplete, preserve, or expand the stock of soil quality. "Land degradation" refers to the deterioration of soil quality. More severely degraded lands reduce productivity, though the effects vary depending on the production conditions and processes employed. Lower productivity is caused by either declining yields or increased production costs due to decreased input efficiency. Land degradation may result in more yield variability and, as a result, higher costs for risk-averse farmers (Noel et al. 2015).

Soil contamination is a critical process in which activities instigated by man alter soil properties significantly. In this case, the repercussions can be matched to several soil-generating factors, and thus the consequences of soil contamination are far-reaching. Contaminants are frequently lethal to the soil microflora (and, in some cases, humans), causing disturbances in the native fauna and flora. Innate soil quality

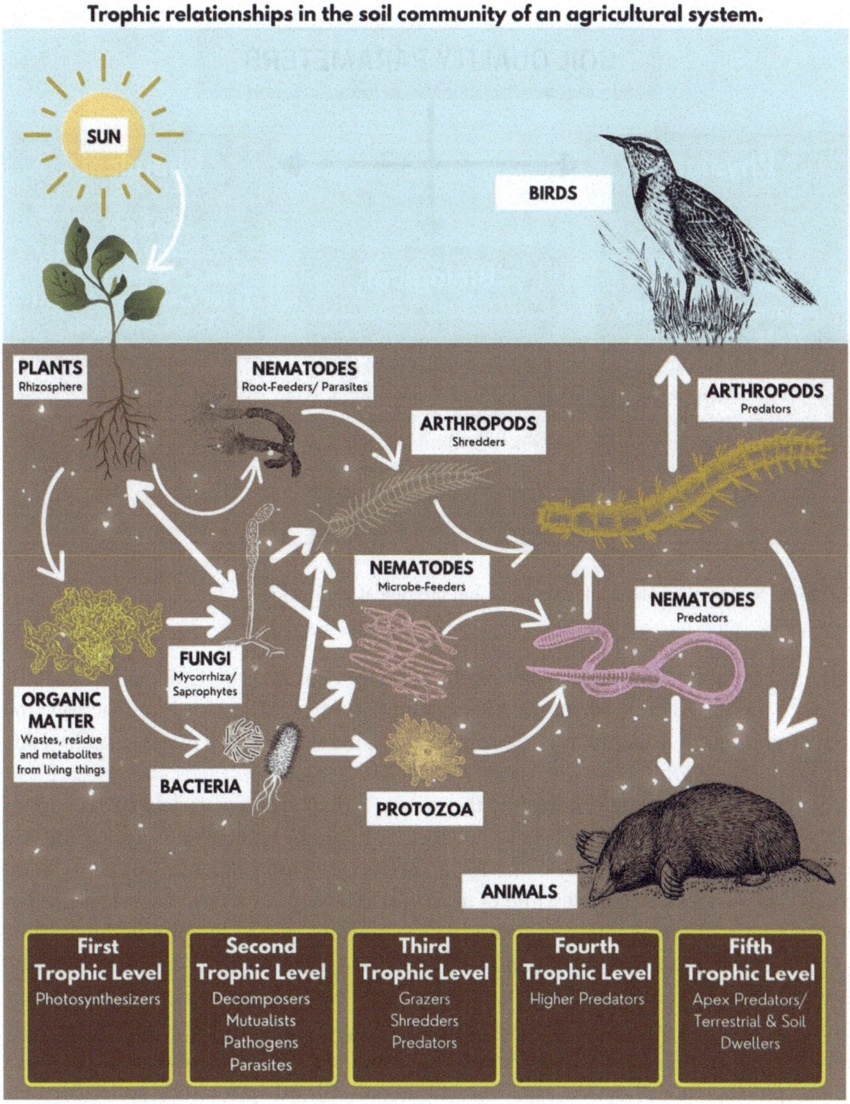

Fig. 15.1 Trophic inter-relationships in the soil biota of an agricultural system

parameters are depicted in Fig. 15.2. Pollutant levels can change soil composition, adsorption capacity, water repellency, soil pH, and redox properties. Contaminated "hot spots" may form due to contamination from a single source. The reduction of the soil organic carbon (SOC) level is another factor that is readily stimulated by anthropogenic activities. SOC is the world's largest carbon reservoir that impacts the physicochemical properties of soil and its availability. The conversion of natural soil to agroecosystems, plowing, wetland drainage, burning of biomass or its removal,

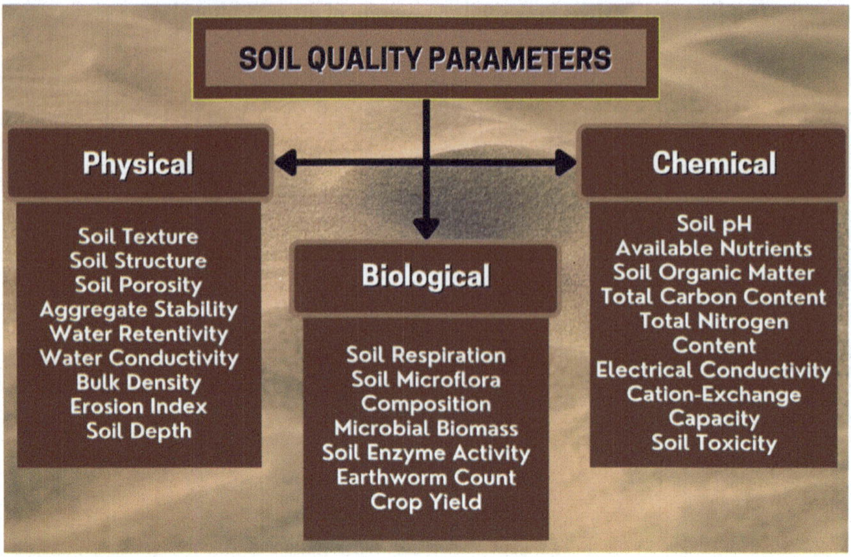

Fig. 15.2 Soil quality physical, chemical, and biological parameters

and most notably, climate change, are the primary causes of SOC depletion. In soils prone to increased erosion, nutrient loss, salinization, acidification, compaction, and contamination, SOC depletion is accelerated at a higher rate on a massive scale (FAO 2015).

Soil quality is thus determined by the grower's response to a variety of exogenous factors as well as the grower's socio-economic situation. Humans rely on natural and managed (agricultural) ecosystems for a variety of "environmental (or ecosystem) goods and services." (Kibblewhite et al. 2007). These products and services are by-products of biological activities. Natural processes that aid in the production of food and fiber, such as nutrient cycles and biological pest and disease management, as well as water flow and quality regulation and influence on the gaseous composition of the atmosphere, have consequences for global climate control. It should be highlighted, however, that these functions are dependent on a complex interplay with the soil's abiotic physical and chemical environment. In natural and agricultural soils, a diverse range of organisms thrives, leading to a wide range of soil-based benefits (Wall 2005).

In order to understand the impacts of urbanization on soil, it is prudent to recognize the fundamental role of soils in ecosystems. Brady and Weil have established five fundamental ecological functions of soils: supporting the growth of higher plants, regulating water within the hydrologic system, mediating nutrient cycles, providing habitat for living organisms, and serving as an engineering medium (Brady and Weil 2008). The intrinsic factors of soil under the threat of urbanization might be threefold: Soils have a physical structure, pH, and characteristic temperature that aid in supporting higher plants; Soils regulate the cycling of essential nutrients and micronutrients, water movement, and waste disposal; Soils contribute to biodiversity. As a result, these three characteristics can be used to assess changes in the soil. Urbanization has thus impacted soils, causing them to undergo compaction and crusting and have deteriorating properties (Marcotullio et al. 2008).

The two pillars of all societies are food security and environmental security. The United Nations' projection of 9.7 billion people by 2050 exemplifies the world's unending population growth (United Nations 2019). The global food demand is anticipated to rise by 70–100%, and to meet the additional food demand, yields of major food crops are expected to increase by 50% (Godfray et al. 2010). Furthermore, drastic climate change has significantly impacted global food security (Fahad et al. 2015). Land degradation is one of the utmost severe consequences of global climate change, and it is stirring at an alarming rate due to a variety of factors (Fig. 15.3), including vegetation loss, soil erosion, inadequate water management, excessive fertilizer use, soil and groundwater degradation, urbanization, and poor agricultural practices (Wingeyer et al. 2015).

Climate has an impact on soil structure, stability, and compaction. Climate change has been reported to cause a shift in the precipitation patterns and temperature regime. A climate change would result in a disturbance in the biochemical properties of the soil, as heat and water cause chemical variations in the soil. Climate change has an insurmountable impact on the weathering process of soil. Several studies have shown that soil carbon emissions increase as air temperature rises. The anticipated extreme temperature inhibits the population, structure, and composition of soil microorganisms. As a result, climate studies have predicted that soil microbial biomass will decrease as temperatures rise (GirijaVeni et al. 2020).

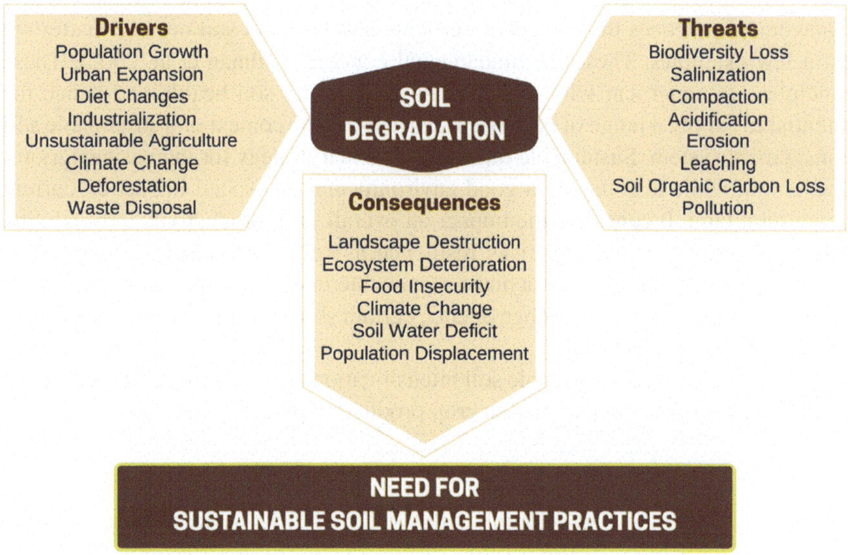

Fig. 15.3 Soil degradation by unsustainable agricultural practices

Sustainable development is the strategy of preserving productivity by replacing utilized resources with resources of equal or greater value without damaging or threatening natural biotic systems (Kahle and Gurel-Atay 2014). There is an imminent need for sustainable soil management approaches, as indicated in Fig. 15.3, due to massive soil degradation caused by the increasing population increase. Soil is crucial to achieving sustainable development goals. Several of the United Nations General Assembly's (UNGA) "Agenda 2030" resolution's Sustainable Development Goals (SDGs) (Fig. 15.4) are either directly or indirectly related to soil. Soil plays an important role in the supply of ecosystem services due to its biological, chemical, and physical qualities. A combination of manmade and natural causes influences these traits (Zambo et al. 2018). Agriculture has traditionally provided only one service: arable land for agricultural and livestock food production. Primary and secondary production relies on soil-based ecosystem services such as nutrient cycling, soil structure preservation, and biotic population regulation. Several industrialized economies have already shown a strong desire for this. The preservation of the whole range of ecosystem services for future generations must be a key goal of sustainable agriculture. As a result, agricultural soils must maintain their multifunctional capability (Kibblewhite et al. 2007).

A protocol for acceptable levels of soil-based ecosystem functions, particularly the balance between food production and other functions that support soil conservation, water flow and quality, crop, livestock, human health control, and greenhouse gas emissions, is required for sustainable soil health management. Organic matter inputs to meet the demand for carbon and energy supply to the soil biota, balanced with crop nutrient demand, and the development of integrated (i.e., organic plus inorganic) nutrient management systems in which inorganic fertilizers are used in precise dosages in combination with equally carefully designed organic matter management practices that conserve nutrients and levels of soil organic matter are just a few examples. These are fundamental concepts (Tilman et al. 2002). These principles are congruent with those required to ensure soil health, and hence the potential to supply a range of ecosystem services, in the context of a sustainable soil management system. Sustainable intensification is a strategy for increasing agricultural output while also having a good environmental and social impact on current agricultural land. It considers the impact on overall farm productivity, profitability, stability, production and market risks, resilience, as well as individual farmer interests and innovative capabilities. To achieve sustainable intensification goals in conjunction with farmers and other change agents, certain strategies and principles must be followed (Donovan 2020).

Biofortification and sustainable soil intensification are two prominent methods of sustainable intensification applied in crop production practices in India.

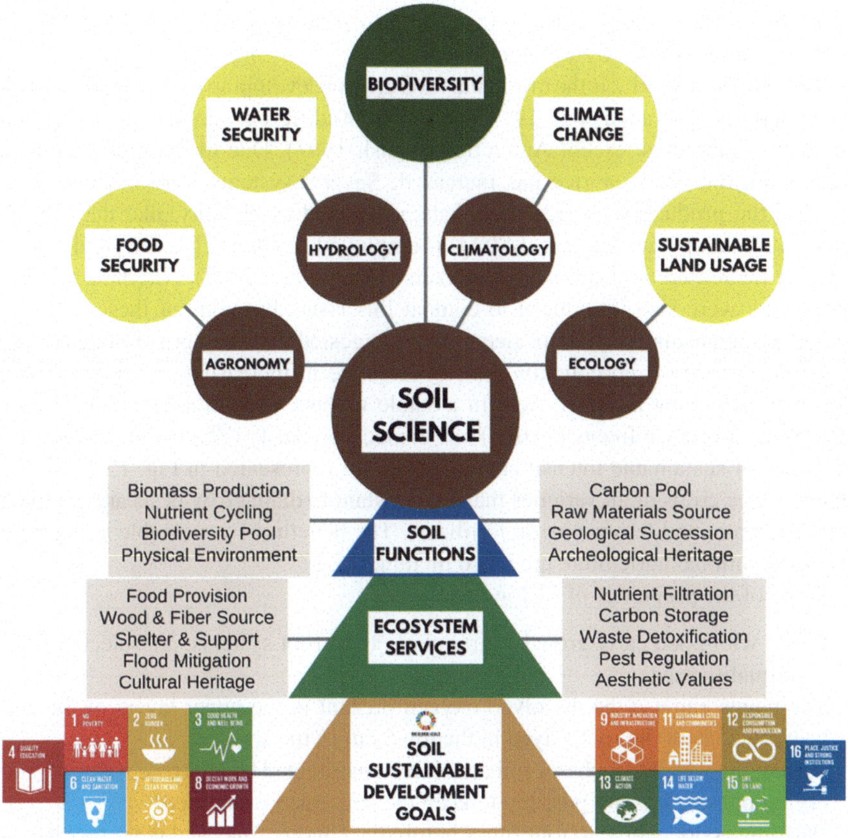

Fig. 15.4 Inter-relationships between the soil biome activities and the range of ecosystem services offered by soil

15.2 Soil Intensification—Demand for Soil

Soil is the most diverse natural ecosystem, serving several functions like food provisioning, providing fiber, energy, physical support for infrastructure, recycling, and detoxifying waste, greenhouse gas regulation, flood mitigation, carbon storage, regulation of pests and disease populations, heritage values, etc. (Tian and Niu 2015). It is the loose material that covers most of the land and consists of inorganic and organic matter. Due to rapid population growth, changing lifestyles, and increased consumption, the greatest pressure is placed on soil. It has become the need of the hour to convert it into a sustainable ecosystem to counter soil degradation. The properties of a natural ecosystem as compared to a sustainable ecosystem are given in Table 15.1. An estimate states that 98.8% of the calories in a human come from the soil, the remaining 1.2% from aquatic life. The world has seen an increase in crop yield per unit area of soil. It has been said that with the increase in population, there should be a 70% increase in food production from 2005 to 2050 (Kopittke et al. 2019). These problems can be solved by soil intensification.

Rice, *Oryza sativa* L., is the most vital food crop in developing nations, followed by wheat and maize (Ladha et al. 1993). Over 40,000 different varieties are cultivated on all the continents, except Antarctica (Khush 1997). Due to the rapid boom in population, the need for rice has increased. Several systems were introduced to increase the production of rice with decreasing land availability, like the SSCM. Site-specific crop management (SSCM) is employed to attain a greater yield of rice with the decreased application of fertilizers (Dobermann 2002). Alternate wetting and drying were also introduced to combat this issue, but none of these methods helped in increasing the yield of rice. It was suggested that crop growth rate (CGR) at the later stage of reproduction is most effective in increasing the yield of rice. However, achieving a higher yield in a stable manner is insufficient. This is how SRI (System of Rice Intensification) has gained popularity (Zhao et al. 2009). The strategies of sustainable intensification of soil are represented in Fig. 15.5. Higher yields of rice crops of all varieties that are resistant to damage by pests are obtained with less water and fewer mineral fertilizers. The benefits of sustainable agriculture over conventional agriculture are listed in Table 15.2.

Some of the properties of SRI include:

- Plants with good early growth, particularly their root systems, are selected from high-quality seeds.
- The plants can use the dissolved oxygen present in the upper layers of the soil because 75% of the root stayed in the top 6 cm of the soil.
- During the vegetative phase, soil aeration is maintained by alternate watering and drying phases. It is not possible to keep the soil always saturated.
- Weeding is done three to four times before the canopy shuts and the rotary hoe is used to remove it.
- Sparse plant populations in widely spaced squares instead of large populations planted in rows.
- To increase organic matter, compost should be used instead of chemical fertilizers.
- Single seedlings are planted in muddy, shallow, unflooded fields instead of planting 3–4 seedlings in flooded paddies (Randriamiharisoa et al. 2006).

Table 15.1 Properties of natural ecosystems versus sustainable ecosystems

Natural ecosystem	Sustainable ecosystem
High species diversity	Species diversity is moderate
Output stability is moderate	High output stability
Nutrient cycle is closed	Nutrient cycle is semi-closed
Trophic relationships are closed	Trophic relationships are intermediate
Resilience is high	Resilience is medium
Dependence on external input is low	Dependence on external input is medium

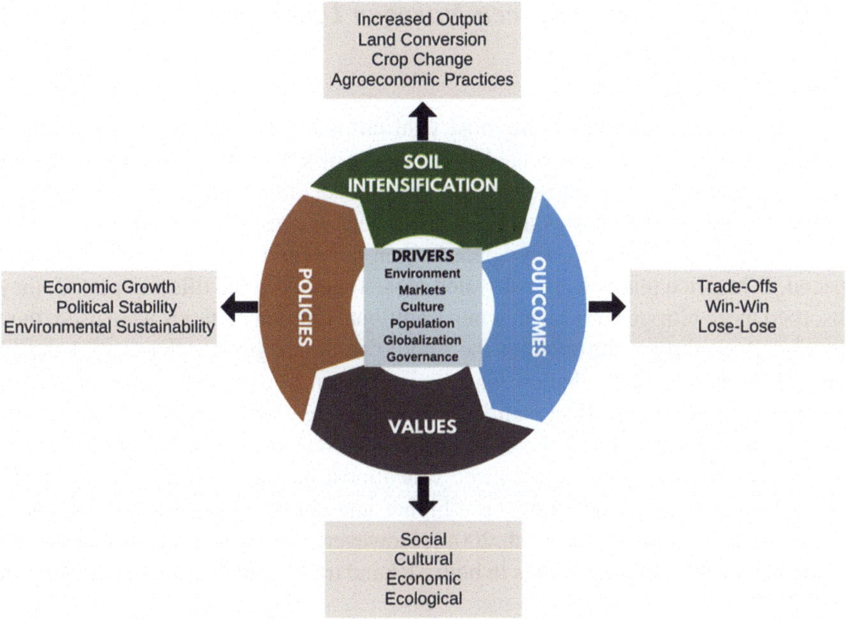

Fig. 15.5 Strategies involved in sustainable intensification of soil

Table 15.2 Differences between conventional and sustainable forms of agricultural intensification

Conventional agricultural intensification	Sustainable agricultural intensification
The primary goals of farming include increasing crop and livestock yield	Apart from increasing crop and livestock yield, the primary goals of farming include improving natural and social capital and improving knowledge
Knowledge about farming is solely experience-driven	Participatory research and scientific development lead to new technologies and farming practices
Ideas and techniques come to the farmers from public or private research	Apart from private/public research for farmers, ideas and techniques also come from peer-to-peer learning and participatory dissemination
The utilization of external inputs to replace supporting, regulating, and provisioning services obtained from agricultural landscapes is emphasized	Emphasizes the many ecological services supplied by agricultural landscapes and the need to understand the link between agricultural and non-agricultural landscape components

15.2.1 Relationship Between Nitrogen Uptake and Yield of Rice Crops

Plants use nitrogen as one of the most vital nutrients. The amount and pattern of nitrogen absorbed by rice plants are largely responsible for their output. Plant density was suggested to have an impact on nitrogen uptake (Takahashi et al. 1973). When a basal nitrogen fertilizer was applied, the levels of nitrogen were seen to reduce more in plants that were closely spaced when compared to plants that are widely spaced. When rice plants were cultivated close together in the tillering period, they absorbed more nitrogen, however, when traditional cultural methods were used, they absorbed less nitrogen throughout the middle and later growth phases (Takahashi et al. 1973). In rice fields, shallow water management resulted in a higher amount of exchangeable ammonium than traditional water management strategies. Although the nitrogen uptake was quicker in narrowly spaced plants, the total amount of nitrogen in both narrow and widely spaced rice was almost the same (Toriyama and Ando 2022). However, SRI plants take up a higher amount of nitrogen when compared to traditional flooding (Zhao et al. 2009). However, there was a linear relationship between nitrogen and grain yields in both SRI and traditional flooding practices (Lin et al. 2009).

15.2.2 Effect of Non-Flooded Soil Environment on Rice Crop Yield

Paddy fields are usually flooded, resulting in an oxidized surface layer and a reduced plow soil layer. Lowland rice has evolved to thrive in anoxic environments by developing aerenchyma, which aids with air diffusion (Jackson and Armstrong 1999). The rhizosphere is a good place for nitrification to occur. Even under submerged settings, NO_3 accounts for about 40% of rice plant absorption. Mid-season rains with adequate percolation, deep plowing, and high levels of organic compounds are required for a decent yield. When high nitrogen was employed, it was discovered that intermittent irrigation followed by submergence resulted in a 10% increase in nitrogen uptake and a 10% rise in spikelets compared to constantly flooded fields. In saturated soil conditions, rather than flooded soil conditions, there was an increase in root development (Toriyama and Ando 2022). As a result of SRI's success, various crops like finger millet (*Eleusine coracana*), wheat (*Triticum aestivum*), Teff (*Eragros tistef*), mustard (*Brassica rapa*), broad bean (*Vicia faba*), peas (*Pisum sativum*), and lentils (*Lens culinaris*) are being grown through sustainable intensification (Abraham et al. 2014). Soil intensification practices now in use erode soils in an unsustainable manner. Salinization, pollution, acidity, and compaction are thought to be responsible for 33% of soil degradation (FAO 2015). Soil deterioration is affecting 52% of the land.

15.2.3 Drawbacks of Soil Intensification

15.2.3.1 Release of Greenhouse Gases and Organic Matter Loss

Soil organic matter (SOM) is necessary for several functions, like maintenance of chemical, physical, and biological fertility, improving water infiltration, and slow-release reservoirs for nutrients like nitrogen and phosphorus. Agricultural production can lead to the mineralization of organic matter. It has been found that agriculture reduces the carbon content in organic matter by 30%, nitrogen by 42%, phosphorus by 27%, and sulfur by 33% (Kopittke et al. 2019). Because soils play such an important role in the carbon cycle, mineralization can also result in the creation of carbon dioxide, a greenhouse gas. As a result, a rise in agricultural production has a direct impact on climate change due to carbon dioxide emissions. Temperature changes, resulting from climate change, also cause the thawing of rich organic carbon soils. Increased microbial activity at high temperatures, oxygenation of the soil profile, and the loss of organic material could all contribute to this decrease. Organic matter also conserved soil fertility by retaining water, often known as green water, which accounts for 90% of global agricultural output. Cropping leads to a reduction of green water and reduces overall soil productivity. Therefore, appropriate sustainable methods have been designed to increase water and nutrition availability (Mekonnen and Hoekstra 2011).

15.2.3.2 Soil Erosion

Every year, 75 billion tonnes of soil is being eroded from the terrestrial ecosystem. It has impacted many terrestrial and forest ecosystems and is considered one of the most serious environmental issues. Water, wind energy, and loss of vegetation are the three major causes of soil erosion and are exceedingly high on slopes (Pimentel and Kounang 1998). Apart from the loss of nutrients, soil erosion causes a reduced volume of rooting media and reduced water content. Due to a reduction in green water, every 10 cm of soil lost results in a 4% reduction in yield. As a result, preventing soil erosion is critical to preserving soil fertility and productivity. Improved vegetation, conservation tillage, and restoration of marginal lands are some of the ways for reducing soil erosion.

15.2.3.3 Acidification

The decrease in soil pH is called soil acidification. The major causes of soil acidification are acid rain, use of fertilizers, etc. Plant growth is inhibited as a result. Aluminum at high concentrations is phytotoxic, meaning it inhibits plant root growth. Toxic metals accumulate in food due to soil acidification (Zhao et al. 2015). It is estimated that 30% of ice-free topsoils and 75% of subsoils are acidic (FAO 2015).

Topsoil acidity can be reduced by the addition of lime, although it is not economic. However, it is difficult to do so in the subsoil regions, given the difficulty in placing lime in the subsoil. Acidification can be reduced by improving nitrogen use efficiency. Improving air quality by the reduction of nitrogen and sulfur can reduce acid rains, which can prevent acidification. Improving agricultural systems and practices can prevent the deterioration of soil organic matter and soil quality. Careful matching of nitrogen supply and nitrogen demand of the plant must be made to prevent acidification of the soil. Removal or restoration of marginal soils from agriculture is another method, as it suffers high degradation and has low productivity. Improving the quality and schedule of irrigation water is another method to improve agricultural methods (Letey et al. 2011). It is necessary to limit or prohibit soil conservation from natural ecosystems to agricultural lands, as well as to promote ways that recognize the significance of soil ecosystem services.

15.3 Rice Cultivation in India: Current Scenario

Rice is one of the world's major food crops, sustaining over 65% of India's population and about 4 billion people, or half of the world's population. Rice is grown across the world, with India being the fourth largest producer (Table 15.3). Many Asian countries' cultures and heritages are dependent on the crop, making it critical to people's food and livelihood security and directly contributing to the achievement of the Sustainable Development Goals (SDG). As of the year 2020, rice is grown on 43.78 million ha in India, with a production level of 118.43 million tonnes and productivity of around 2705 kg/ha (Directorate of Economics and Statistics, Department of Agriculture 2020). About 4454.77 thousand tonnes (worth ₹31,026.33 crores) of Basmati and 5056.28 thousand tonnes (worth ₹14,400.32 crores) of non-Basmati rice are being exported by India (Directorate of Economics and Statistics, Department of Agriculture 2020). It is grown in a variety of soil types and climatic conditions. The Kharif season accounts for a sizable portion of rice production.

Rice production in India has increased nearly 5.75 times in the last 70 years, from 20.58 million tonnes in the year 1950 to nearly 118.43 million tonnes in 2019–20. (Directorate of Economics and Statistics—Department of Agriculture 2020). Rice productivity has increased dramatically in India in recent years. Rice production has risen because of the development of high-yielding rice cultivars that respond well to high fertilizer doses, as well as superior packing procedures devised by agricultural scientists for different regions. Despite a significant rise in rice output across the country, there are still some locations where rice productivity is extremely poor. Rice productivity varies greatly from location to region due to a multitude of factors, including soil type, fertility, rainfall patterns, floods, waterlogging, climatic conditions, and so on. To boost the country's production, improved technology and a variety of initiatives could be employed (Fahad et al. 2017).

Rice is farmed in India in altitudes and climates that vary greatly. As a result, rice-growing seasons vary based on temperature, rainfall, soil type, water availability,

Table 15.3 World's top ten rice-producing countries, harvested area, yield, production, varieties, and environmental conditions

Country	Harvested area (ha)	Yield (kg/ha)	Production (mt)	Rice varieties	Environmental conditions
Bangladesh	11,417,745	48,088	54,905,891	Chandina, Moya, Ghazi, Kiron, Dishari	Sub-tropical monsoon climate. Alluvial soil enriched with heavy silt
Brazil	1,677,705	66,108	11,091,011	Acrefino, Aimoré, Cuiabana, Rio Paraguay	Hot and humid in the north, semi-arid in the north-east, temperate in the south and tropical semi humid in the center west
China	30,341,784	70,402	213,610,729	Japonica, Indica, Zengcheng, Wannian, Panjin	High temperature and adequate rainfall in the southeast, low temperature and little rainfall in the north-east. Irrigated fields are seen here
India	45,000,000	39,623	178,305,000	Brown, Basmati, Mogra, Indrayani, Jasmine	Humid to sub-humid climate with clayey, loamy, alluvial soil
Indonesia	10,657,275	51,279	54,649,202	Pandan Wangi, Setra Ramos, Rojolele, Batang Lembang, Glutinous rice	Tropical climate with abundant rain and high temperature. Irrigated lowland fields are irrigated
Japan	1,462,000	66,390	9,706,250	Hatsuga-Genmai, Haigamai, Buzukimai, Musenmai, Mochigome	Humid temperate with adequate rainfall in the months of June and July

(continued)

Table 15.3 (continued)

Country	Harvested area (ha)	Yield (kg/ha)	Production (mt)	Rice varieties	Environmental conditions
Myanmar	6,655,811	37,711	25,100,000	Paw HsanHmwe, Mayan, Emata, Zeera, Ngasein	The central dry zone is irrigated, while the delta zone gets enough water and rich soils. Upland areas are cold in winter
Philippines	4,718,896	40,888	19,294,856	Makiling, Molawin, Liliw, Chico, Jaro	Tropical marine climate with relatively even rainfall throughout the year
Thailand	10,401,653	29,064	30,231,025	Long grain, Glutinous, Jasmine rice, Hom Mali, Sao Hai	An average of 27 °C and northeast monsoon during October and southwest monsoon during May
Vietnam	7,222,618	59,201	42,758,897	DoiMoi, MocTuyen, Tran Chau Lun, Nanh Chon, Than Nong	Humid tropical in southern lowlands, temperate in northern highlands with north-eastern winter and south-western summer monsoons

Source FAO (2020)

and other climatic factors across the country. Figure 15.6 depicts an overview of the agronomic practices used in rice growing in India. The average temperature in the country's eastern and southern regions is determined to be conducive to rice growing throughout the year. As a result, rice is farmed in two or three crops per year in the eastern and southern states. In the northern and western areas of the country, where rainfall is unusually abundant and winter temperatures are relatively low from May to November, only a single crop of rice is farmed. Rice is farmed in India during three seasons: autumn, winter, and summer (Table 15.4). These three seasons are called after the crop's harvesting season. Pre-Kharif rice is another name for autumn rice. Rice is sown in the pre-Kharif season, which runs from May to August. The timing of sowing, however, varies widely from state to state according to the weather and rainfall pattern. Between September and October, it is harvested. About 7% of the crop is produced this year. The bulk of rice varieties planted during this season is of short duration, ranging from 90 to 110 days. The country's main rice-growing season is Kharif or winter. It is called winter rice or Kharif rice, depending on when it is harvested. Rice is sown in June-July and harvested in November–December for the winter (Kharif) season. This season, about 84% of the country's rice harvest is grown, with medium- to long-duration types being the most common. Summer rice is referred to as Rabi rice. Summer rice is harvested from March to June after being seeded from November to February. Summer rice accounts for only 9% of overall rice output, with early maturing types being the most popular (Singh 2010).

Irrigated rice covers around 22.00 million hectares, accounting for approximately 49.5% of the total rice cultivation area in the country. The rainfed ecosystem includes both upland and lowland ecosystems. Assam, Bihar, Eastern M.P., Orissa, Eastern U.P., West Bengal, and the North-Eastern Hill region are all part of the eastern zone. There is no standing water in the rainfed upland rice field after a few hours of rain. The country's entire area under upland rainfed rice is roughly 6.00 million ha, accounting for 13.5% of the country's total rice crop area. Rice productivity in the uplands is extremely poor, at around 0.90 t/ha. Low-land rice covers around 14.4 million hectares, accounting for 32.4% of the country's total rice cultivation area, with rice productivity ranging from 1.0 to 1.2 t/ha (Singh 2010).

The country's major rice-growing areas, as shown in Fig. 15.7, can be divided into five regions based on climatic conditions: (a) North-Eastern Region (Assam and North-eastern states); (b) Eastern Region (Bihar, Chhattisgarh, Jharkhand, Madhya Pradesh, Orissa, Eastern Uttar Pradesh, and West Bengal); and (c) Northern Region (Haryana, Punjab, Western Uttar Pradesh, Uttarakhand, Himachal Pradesh Southern region (Andhra Pradesh, Karnataka, Kerala, and Tamil Nadu). Rainfed rice cultivation accounts for over 40% of India's rice-growing area, with eastern India accounting for more than 70%. Rainfed highland accounts for 23% of the total rainfed area, whereas rainfed lowland accounts for 77%. The entire rainfed upland and 52% of the rainfed lowlands are under threat of drought. Approximately 17% of rainfed lowlands are prone to flooding. India's rice output has improved tremendously over the years. However, environmental deterioration and climate change have posed enormous problems in recent years (Wassmann et al. 2009; Pathak and Aggarwal 2012). Due to low and uncertain income, a damaged natural resource base, expanding labor and energy shortages, and climate change risks, Indian agriculture is very susceptible and unsustainable. Another issue facing rice farmers in many locations, particularly in the country's northwestern region, is how to dispose of rice straw in an environmentally safe and economically feasible manner (Pathak and Aggarwal 2012). The fundamental aim of the Indian rice sector is to produce enough rice to feed an expanding population while minimizing the natural resource base.

The principal rice-growing areas are concentrated in the Eastern region, which receives heavy rainfall and severe flooding on an almost annual basis. The rice harvest has been severely damaged. Crop failure also occurs in upland areas as a result of excessive rainfall or drought. Rice crops are regularly subjected to soil moisture stress

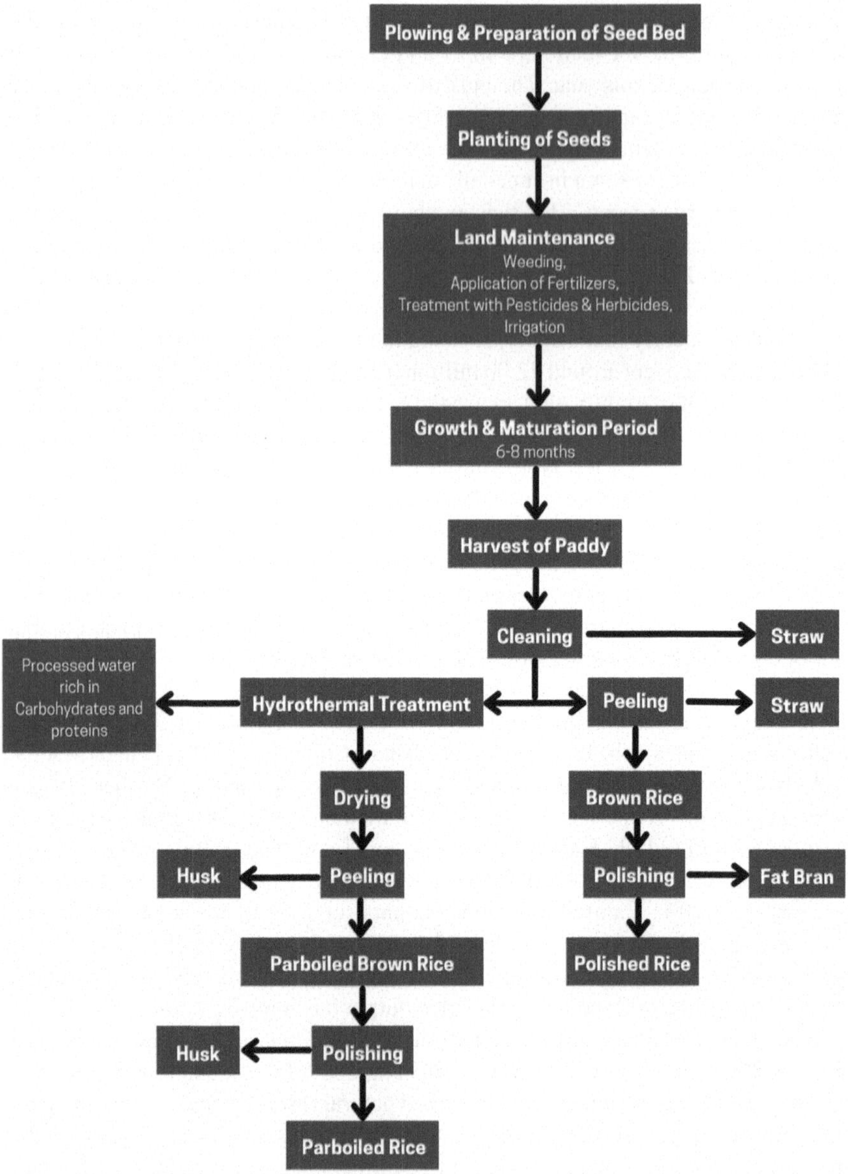

Fig. 15.6 Flowchart depicting agronomic practices involved in rice cultivation

Table 15.4 Rice cropping seasons in India (Directorate of Rice Development (DRD) 2020)

Crop season	Local name	Sowing	Harvesting	Production (%)	Area (%)
Autumn–Kharif	Aus/Kar	May–June	September–October	39.4	43.91
Winter–Rabi	Aman/ Sali/ Karthika	June–July	November–December	54.2	48.79
Summer–Zaid	Boro/ Dalua	November–December	March–April	6.4	7.24

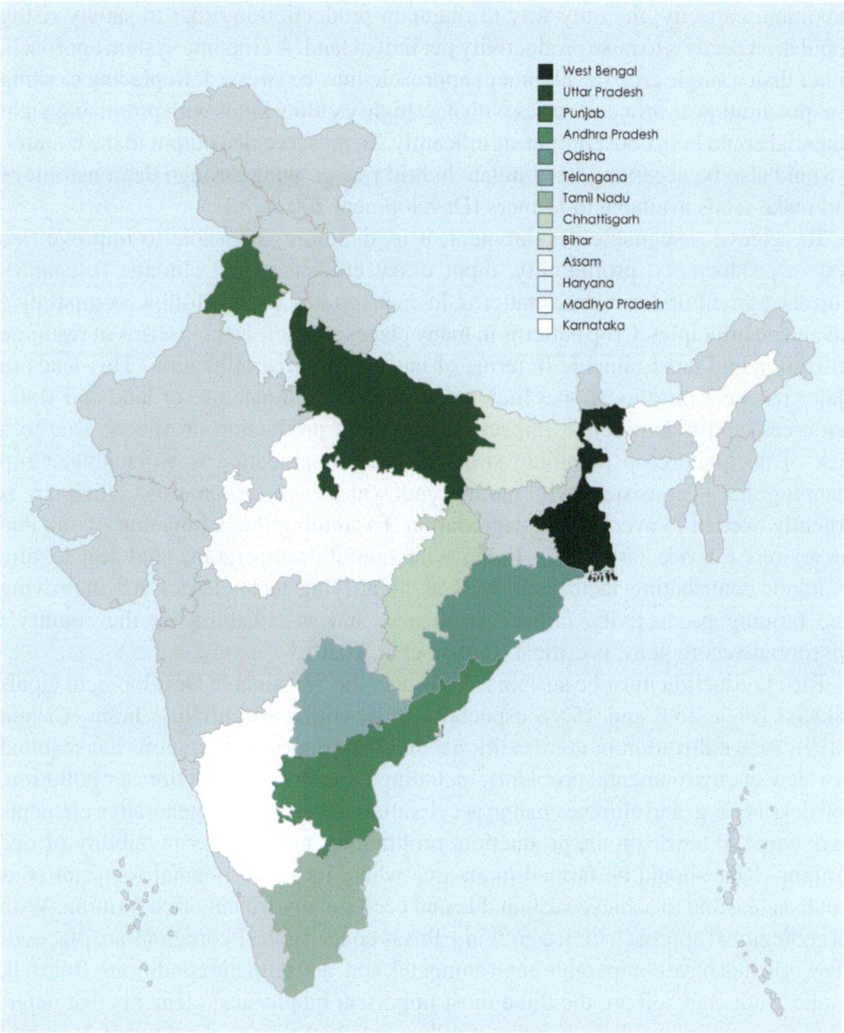

Fig. 15.7 Geographical map of India depicting major rice-growing regions

because of erratic and insufficient rainfall. Low and unpredictable rainfall as well as poor soil conditions cause intermittent soil moisture stress in Madhya Pradesh, Orissa, and parts of Uttar Pradesh. Poor drainage, low-lying physiography, and excessive rainfall in submergence-prone lowlands are causing flash floods, waterlogging, and submergence in Assam, North Bihar, West Bengal, and Eastern Uttar Pradesh. Rice crops are grown in the uplands under rainfed circumstances, and their growth is strongly reliant on the monsoon's whims. In years of scarce or unequal rainfall distribution, the crop fails to owe to drought, whereas in years of excessive rainfall, particularly during blooming and fruiting, grain setting is weak, and developed grains germinate on the panicles. The rice farming area has virtually reached its maximum capacity; the only way to maintain production in order to satisfy rising population needs is to raise productivity per unit of land. A cropping system approach, rather than a single crop development approach, may be stressed. Replacing existing low-potential/pest-prone varieties with new high-yielding kinds with promising yield potential could help boost output significantly. To preserve rice output in the country, it would also be necessary to stimulate hybrid rice growing through demonstrations and make seeds available to farmers (Development 2002).

To achieve sustainable development, it is, therefore, advisable to improve rice system productivity, profitability, input usage efficiency, and climatic resistance. Current agricultural land use patterns in many states do not follow comparative advantage principles. Crop patterns in many places are inefficient in terms of resource utilization and unsustainable in terms of natural resource utilization. This leads to major resource misallocation, efficiency loss, indiscriminate use of land and water resources, and a detrimental impact on long-term production prospects. Due to a lack of adequate crop planning, soil and water degradation is worsening. Crop planning that is consistent with natural endowment and resource use efficiency is urgently needed to avert future degradation. Examining the delineation of suitable ecoregions for rice farming in India, with rainfall, temperature, and soil texture as major contributing factors, as well as identifying technologies for improving rice farming productivity, climate resilience, and sustainability in the country's appropriate ecoregions, is critical (Pathak et al. 2020).

Rice production must be sustainable to meet the Sustainable Development Goals (SDGs) (Figs. 15.8 and 15.9), especially in emerging nations like India (Cropin 2019). Rice cultivation in non-traditional and inappropriate ecoregions has resulted in a slew of environmental problems, including groundwater depletion, air pollution, soil degradation, and climate change acceleration. All of these degenerative elements have wreaked havoc on the production, profitability, and long-term viability of rice farming. Rice should be farmed in an area where its environmental footprint is as small as feasible to achieve sustainable and ecologically friendly rice farming. With an ecoregional approach to rice growing, this is conceivable. Ecoregions are places of the world that have comparable environmental, soil, and climatic conditions. Rainfall, temperature, and soil are the three most important biophysical elements that determine rice ecoregions. Rice is a wet-weather crop that thrives. As a result, it should be planted in places with more than 1000 mm of annual rainfall. The temperature has a significant impact on rice growth, development, and yield. Rice, as a tropical

Fig. 15.8 Pillars of sustainable farming

and subtropical plant, requires temperatures ranging from 20 to 30 °C. Clayey loam soil (27–40% clay content) in monsoon land is considered the finest for rice growing due to its excellent water-retention ability (KPMG Advisory Services 2019; Pathak et al. 2020).

The "Green Revolution" was first implemented in the 1960s by the Consultative Group on International Agricultural Research (now CGIAR) (the International Rice Research Institute and the Center for Wheat and Maize Improvement merged), which was successful in increasing grain production through the development of high-yielding varieties (HYVs). HYV grains, on the other hand, have fewer nutrients; in the case of rice, polishing reduces nutrient content even further, particularly iron (Fe) and zinc (Zn). Following that, in 1991, the CGIAR initiated research on creating "micronutrient-dense" staple crops under the label of biofortification in response to the worldwide nutrition community's concern about micronutrient deficiency as a global problem (Brooks 2011). In 2003, CGIAR started the HarvestPlus Challenge Program as a global endeavor with the goal of breeding biofortified staple food crops like wheat, maize, rice, and cassava (Pfeiffer and McClafferty 2007; Bouis et al. 2000). Biofortification of rice grain with iron and zinc began in 1992 and 1995,

Fig. 15.9 Strategies of sustainable rice farming

respectively (Gregorio 2002). Nutritional improvement is also an objective of 12 of the UN's 17 Sustainable Development Goals (United Nations 2015).

15.4 Biofortification: A Sustainable Approach

Biofortification, also known as "biological fortification," refers to nutritionally enhanced food crops that are grown and manufactured using modern biotechnology techniques, conventional plant breeding, and agronomic approaches to promote bioavailability to humans. It is a brand-new, promising, cost-effective, and long-term approach to delivering micronutrients to people who do not have access to a wide range of foods or other micronutrient therapies. Biofortification has been widely used for crop improvement around the world (Fig. 15.10). Unfortunately, major food crops lack the micronutrients required for proper human development (Agarwal et al. 2018; Beal et al., 2017).

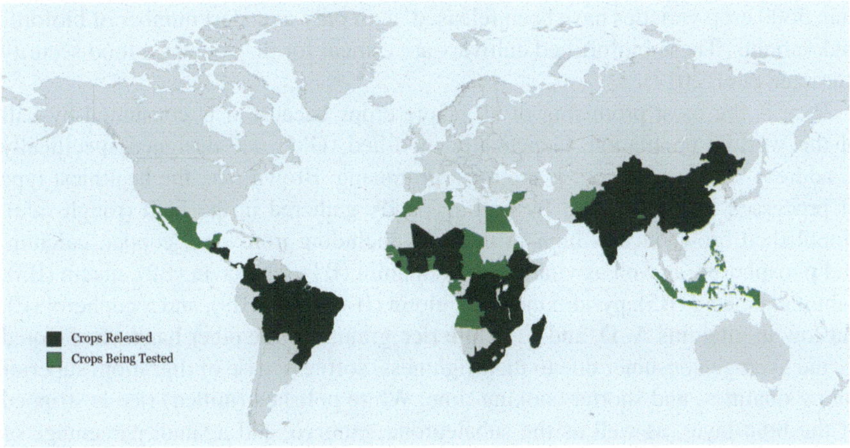

Fig. 15.10 Biofortified crop map—countries applying biofortification for crop development

Inadequate vitamin and mineral intake, also referred to as hidden hunger or micronutrient malnutrition, has a deleterious impact on human health. Hidden hunger affects over two billion people worldwide, with the largest prevalence in Africa and South Asia. Malnutrition has the greatest impact on women and children of reproductive age, including health issues, stunted growth, and a number of birth-related issues. Human diets are frequently lacking in seven minerals (zinc, iron, copper, magnesium, calcium, iodine (I), and selenium (Se), as well as a variety of vitamins, with zinc, iron, vitamin A, and vitamin B9 (folate) deficits among the most severe and widespread (Beal et al. 2017). Biofortification of the use of agricultural practices to increase the number of micronutrients in food crops is widely acknowledged as a critical strategy for reducing micronutrient malnutrition. Micronutrient deficiencies such as iron deficiency or iron deficiency anemia (IDA), zinc deficiency, and vitamin-A deficiency (VAD) are all frequent and can have serious consequences (Bouis et al. 2000). Every day, more than 24,000 people die because of "hidden hunger" and malnutrition around the world. Food must be supplemented with various biological and chemical addictions, as well as the food processing system, to overcome these deficiencies. Staple food crops that have been biofortified (including bioengineered) are a long-term solution that can be particularly beneficial to persons who have limited access to a variety of dietary options.

Food provides energy, protein, essential fats, vitamins, antioxidants, and minerals to support our daily metabolic needs. In addition, anti-nutritional substances included in edible food portions are harmful to human health. Billions of people throughout the world suffer from unbalanced food consumption, which leads to poor health and socioeconomic conditions (Dubock 2017). To date, the focus has primarily been on generating high-yielding cultivars to feed the world's ever-growing population. The Indian Council of Agricultural Research has improved the nutritional quality of high-yielding grains, pulses, oilseeds, vegetables, and fruits (ICAR). So far, more than 5600 crop varieties have been released, with only a modest number of biofortified variants. These biofortified cultivars are critical for the country's food security (Straeten et al. 2017).

Rice is the most promising of all staple crops because it is consumed by half of the world's population. Genetically modified (GM) rice designed specifically to address "hidden hunger" is the most promising. Brown rice, the healthiest type of processed rice, is created by hulling paddy gathered in the field (rough rice). Unpolished brown rice is high in minerals including iron, zinc, copper, calcium, and phosphorus, as well as vitamins like thiamin (B1), riboflavin (B2), niacin (B3), pantothenic acid (B5), pyridoxine (B6), biotin (B7), folate (B9), and tocopherol (E), but low in vitamins A, D, and C. White rice grains, on the other hand, are favored by the average consumer due to their lightness, softness, ease of digestion, superior eating qualities, and shorter cooking time. White polished (milled) rice is stripped of the bran layer, as well as the subaleurone, embryo, and a small percentage of the endosperm. The nutritional quality of polished (milled) rice is lower than that of brown rice, with iron content reduced by 2.14 times (from 8.8 to 4.1 ppm) to 4.75 (from 19 to 4 ppm), zinc content reduced by 1.83 times (33–18 ppm), and other minerals, vitamins, fats, proteins, and fibers reduced by 1.83 times (33–18 ppm) (Garg et al. 2018; Majumder et al., 2019; McGuire 2015). Mineral reduction amounts, on the other hand, may vary depending on rice varieties and grain milling procedures. Although education and awareness have increased brown rice consumption, the vast majority of rice consumers still choose white polished rice, encouraging scientists to consider developing nutritionally superior rice varieties through biofortification (Muthayya et al. 2013).

15.4.1 Challenges Toward Sustainable Biofortification

One of the most significant problems in developing fortified crops is the high cost of research and regulatory compliance, which is due to the strict regulation of transgenic crops. In the case of biofortified crops, when profit margins for private technology companies are limited, the lack of public funding exacerbates the problem. Intellectual property (IP) issues must be addressed because genetically modified technology is frequently patented. 16 patents and 72 intellectual property issues had to be handled in the process of making golden rice free to disadvantaged farmers. Farmers and consumers must adopt biofortified crops for them to be effective, which raises a variety of challenges. Public acceptance is also important, especially if the new trait changes the look of the crop, such as color (as in golden rice), taste, or dry matter content. Adoption will rely heavily on appropriate information activities. A solid market network and channels for the diffusion of agricultural information, as well as widespread technological dissemination, are also necessary for success. A lack of agricultural infrastructure in some poor countries, notably in Africa, is a significant impediment to the adoption of novel biofortified cultivars (Bouis et al. 2024).

15.4.2 Biofortified Rice: Limitations and Advantages

The process of making plant foods more nutritious as they grow is known as biofortification. Typical fortification nutrients are added to the food during processing (Nestel et al. 2006). Nearly a billion people rely on rice as their main diet. The aleurone layer protects the rice endosperm (the starchy and most edible part of the rice seed). Nutrients are found in the aleurone layer of the dehusked rice grain, but this layer is lost during polishing and milling. Rice that has not been processed or rice that has a foul taste goes rancid. Rice accounts for 30–50% of a person's daily calorie consumption. Rice agriculture is, for the most part, India's and the world's primary source of employment. Rice endosperm, embryo, and bran contain 32, 13, and 55% iron, respectively, and 57, 9, and 34% zinc, respectively. Approximately 20.7, 17.05, and 36.61% of Zn are lost during dehulling, polishing, and boiling, respectively (Fig. 15.11). Rice is typically planted in flooded soil, where zinc is less available to the crop due to sulfur and carbonate binding under reduced conditions. It is a genetically modified rice genome that's been improved with provitamin A (β-carotene). All golden rice credits go to the Rockefeller Foundation, the European Union, and the Swiss Federal Institute of Technology. Potrykus and Beyer are credited with creating β-carotene-rich golden rice (Beyer et al., 2002). They altered the genetic makeup of soil bacteria (*Agrobacterium tumefaciens*) using the crtl gene and the daffodil gene. Golden rice is not something that can be bred. Golden rice 1 (SGR1) has a modified promoter and contains 5–7 mg of α-carotene per gram of rice, but golden rice 2 contains 31 g of α-carotene per gram of rice and has the Pys daffodil gene substituted with the maize gene. IR 64, IR 36 (mega varieties with broad Asian coverage), BRRI Dhan 29 (the most popular boro rice variety in Bangladesh), DRR Dhan 48 (a zinc-rich variety released for the South Indian states), DRR Dhan 49 (High zinc-rich variety released for the states of Kerala, Maharashtra, and Gujarat), PSB Rc 82 (the most popular rice variety in the Philippines), OS 6561 (the most popular in Vietnam), Chehirang (the leading variety in Indonesia), and Swarna (the leading variety in India) are some of the popular carotene-enriched rice varieties (important in India).

Rice biofortification has a multitude of benefits, including increased nutritional value, lower mortality from micronutrient deficiencies in adults and children, fewer dietary deficiency syndromes, and a healthier population with more powerful and quick immune responses to infections. With these advantages, biofortification of rice has some disadvantages, such as high processing costs (machinery, technology, patenting, etc.), potentially negative interactions of biofortified rice on other plants/non-GMO rice crops, resulting in the extinction of wild rice varieties, and low significant equivalence (inability to provide high micronutrient and protein content compared to supplements).

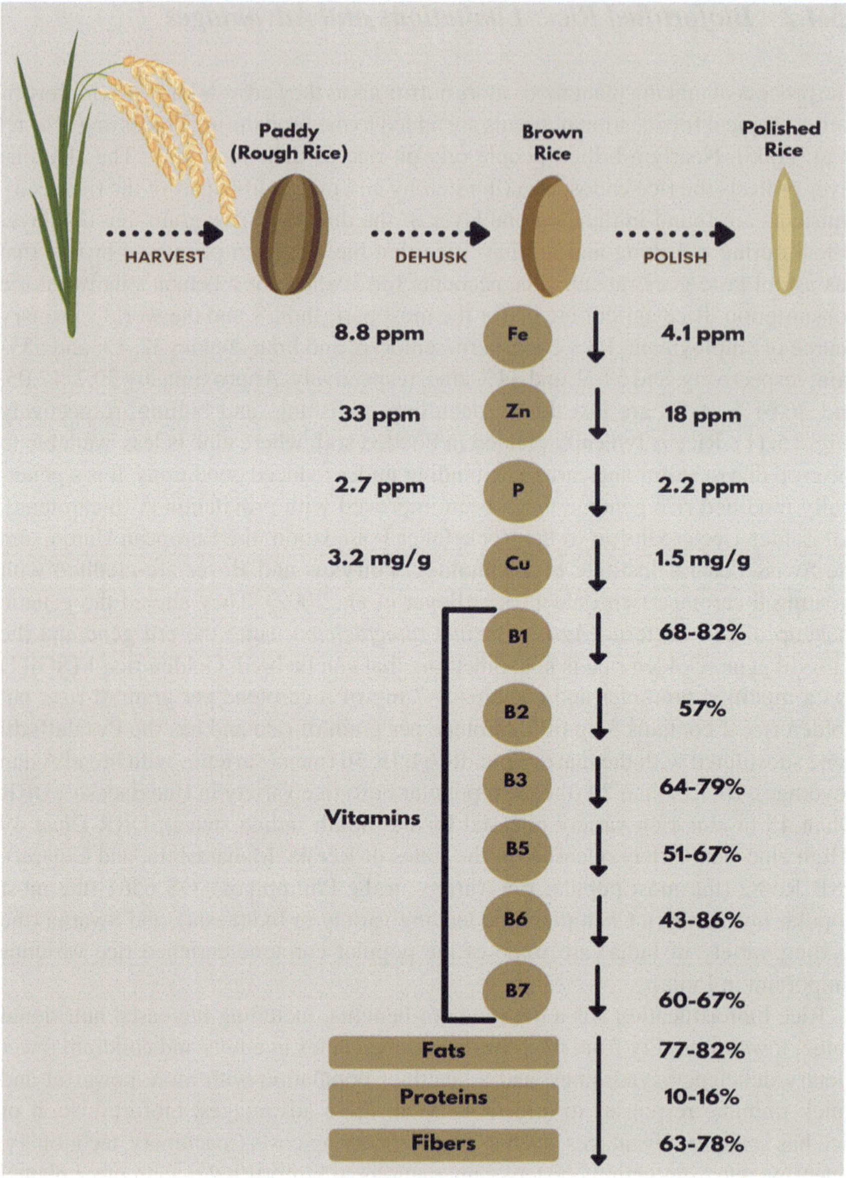

Fig. 15.11 Loss of nutrients from rice grain due to the milling process

15.5 Nutrition Availability and Evidence of Efficacy

15.5.1 Justification for Biofortification

Over the last 50 years, agricultural research for poor countries has improved the production and availability of calorie-dense staple crops, but not so much for micronutrient-dense non-staples like vegetables, pulses, and animal products. Biofortification is a supplement to current treatments like supplementation and industrial food fortification because no one intervention can completely eliminate micronutrient deficiencies (Sperotto et al. 2012). Biofortification, on the other hand, has two different benefits: long-term cost-effectiveness and the ability to reach underserved rural populations. Unlike supplementation and commercial fortification, which need continuous financial investments, an upfront investment in plant breeding provides farmers with micronutrient-rich biofortified planting material at virtually no marginal cost. After they've been generated, nutritionally enhanced crops can be tested and tweaked for different settings and geographies, increasing the return on the initial investment. Once the micronutrient trait has been mainstreamed into the principal breeding objectives of national and international crop development programs, agriculture research institutes' recurring costs for monitoring and maintenance are low. Biofortified crops are also a potential strategy to reach rural areas that may not have access to a diverse diet or other micronutrient therapies. Target micronutrient levels for biofortified crops are calculated based on current consumption trends to meet the particular dietary needs of women and children. After meeting the needs of the home, surplus biofortified crops are sold in rural and urban retail venues (Bouis and Saltzman 2017).

15.5.2 Nutritional Bioavailability and Efficacy Evidence

Those who consume a lot of high-carbohydrate foods, such as white rice, are more likely to be deficient in thiamin (vitamin B1). Genetic engineering has been used to try to increase the amount of thiamin in foods. The discovery of regulatory elements, as well as the identification of biosynthesis, transport, and salvage genes, has laid the framework for the first successful thiamin engineering attempts. However, many aspects of thiamin metabolism are unknown, hindering future advancement in thiamin biofortification (Straeten et al. 2017). Three of the world's top five staple crops (rice, wheat, and cassava) fail to meet the recommended daily need for vitamin B6 when consumed in large quantities. Unlike vitamin B1, genetic engineering has been utilized to raise vitamin B6 levels in cassava to target levels. Furthermore, increasing the correct B6 vitamin in plants in a controlled manner has the potential to improve stress tolerance and bioavailability (Strobbe and Straeten 2017). Vitamin B9 insufficiency is highly common among folates, according to research. Effective biofortification engineering in rice, tomato, Mexican common bean, and lettuce, as well as contemporary biofortification technologies in potato and other staple crops, have all come through the biosynthetic route. It has also been reported on the importance of preserving vitamin stability, which has been accomplished in rice, as well as the requirement to efficiently biofortify more staple crops with folate.

Many people in Western countries have chronic vitamin E deficiency symptoms, according to reports. On the other hand, the long-term consequences of chronic insufficiency in humans have yet to be discovered. The manufacturing process has been drawn out, and plant breeding and engineering strategies to increase vitamin E content are now being used. Researchers conclude that one of the most effective techniques for raising active vitamin E content in crop plants is through the conversion of different forms of tocopherols and that a better understanding of chlorophyll turnover will likely open up new avenues for vitamin E biofortification (Straeten et al. 2017). It has been proven that transgenic technologies can be used to supplement milled rice with considerable amounts of iron and zinc. Substituting this biofortified transgenic rice for non-biofortified rice on any given day would result in an additional 30% of the estimated average requirement (EAR) of iron and a 60% increase in the EAR of zinc in the diet (Trijatmiko 2016). High-iron bean and pearl millet have been generated through conventional breeding and have been shown to improve iron status in Rwandan young women and Indian school pupils. To acquire the correct amounts of iron in polished rice grains, however, genetic engineering was required. Another noteworthy example of how known genetic approaches may be successfully used species by species is warranted. Incorporating iodine-rich plant items into one's diet, as well as using iodized salt, may thus help to enhance a population's iodine status. No staple crops, however, have been successfully biofortified with iodine to date. It is consequently necessary to identify and modify the genetic features that regulate iodine uptake, mobilization, and retention in plants. Traditional breeding processes alone are unable to increase the micronutrient content of crops to nutritionally adequate levels in many cases due to the lack of natural diversity. Metabolic engineering, whether through genetic alteration or innovative genome editing and epigenetic approaches, is the most valuable choice for supplementing breeding technology or achieving an increase in numerous nutrients delivered by a single overexpression cassette carrying the different features (Straeten et al. 2017).

In addition to simultaneously boosting the amounts of multiple micronutrients, the issue of micronutrient long-term stability in stored agricultural products must be examined. This is especially essential for grain crops because poor farmers can keep seeds for up to a year before harvesting them. Furthermore, bio-accessibility and bioavailability can influence how efficiently various nutrients are carried through the blood, stored, and used, depending on the biochemical nature of the resource (Díaz-Gómez et al. 2017). To achieve long-term elimination of micronutrient deficiencies, biofortification approaches must be paired with diversification, nutrition education, and promotion of locally produced nutritious crops, with the added benefit of contributing to a lower carbon footprint. Golden rice has a high bio-conversion rate of -carotene, and the nutritional effects of GM biofortified crop intake are positive, with daily servings occasionally exceeding daily predicted average requirements, according to the clinical trial (Straeten et al. 2017).

15.6 Crop Development for Sustainable Food Systems

15.6.1 Introduction to Sustainable Food Systems Development

Food systems encompass all processes involved in the production, aggregation, processing, distribution, consumption, and disposal of food products derived or generated from agriculture, fisheries, forestry, and other natural contexts in which they are ingrained. Agriculture, waste management, and other subsystems of the food system interact with other important systems such as energy, health, and others.

A sustainable food system (SFS) is one that ensures everyone has access to food security and nutrition while simultaneously conserving the economic, social, and environmental underpinnings for future food security and nutrition. Economic (constant profitability), social, and environmental sustainability are all factors that SFS analyzes and aims to achieve (FAO 2015).

A sustainable food system includes the entire agri-value chain, from pre-production to production and supply networks to consumption. Long before the crop is grown, the construction of a sustainable food system begins. Genetic resources are kept and utilized in the pre-production stage to improve the quality and productivity of domesticated crops in the future. Agricultural input enhancements, such as crop protection agents, are included in pre-production and can help increase yield and profitability while preventing future land conversion for cultivation. Furthermore, agricultural technologies are considered in order for farmers to succeed in the face of climate change and other rapidly changing circumstances.

During the production stage, farmers must deal with problems such as pests and illnesses that affect the crops, unexpected weather, managing food loss, and coping with changing market conditions. Due to higher fertility and easier access to natural water supplies than uplands, a range of production alternatives is advocated, including the utilization of lowland valleys (in drought-stricken West African countries) as acceptable rice farming sites. Food security can be improved by diversifying food production locations that can make use of available water resources and manage crop output and long-term productivity sustainably. These areas are known as SMART-valleys, and they can increase water retention at a lesser cost while also reducing the risk of fertilizer loss due to flooding. This approach could be implemented with one growing season and benefit the farmers.

The stages involved in obtaining food from a farmer to a client are referred to as the supply chain stage. The items are converted into sellable form, processed, packed, and stored before being sent to market. To strive to reduce post-harvest losses, innovative supply chain programs are being deployed. Consumption stage is an integrated food system that considers not only how food is produced but also how it is prepared, consumed, and discarded by customers (Reynolds et al. 2019).

15.6.2 Changing Food Systems

When considering the food system, rapid population growth, urbanization, rising prosperity, changing consumption habits, globalization, climate change, and natural resource depletion must all be considered. Over the last three decades, there have been numerous positive developments in food systems, including the growth of food industries, which has resulted in an increase in off-farm employment opportunities, and also satisfying consumer preferences for taste, form, and quality by expanding food options beyond local staples. The rapid changes that have accompanied this process have created significant issues, with far-reaching implications for food security and nutrition. These include widely available and consumed highly processed, high-calorie, low-nutrition-value foods; limited access to viable markets for small-scale producers and agribusinesses; elevated levels of food loss and waste; the greater toll of food safety, animal and human health issues; and increased energy intensity and ecological footprint associated with the lengthening and industrial growth of food supply chains. As a result, a better understanding of how various food systems work will be crucial in ensuring that these systems evolve in such a way that their negative effects are minimized while their positive effects are maximized (FAO 2021).

15.6.3 Limitations of Current Approaches

Because of the complexity of food systems, a thorough and coordinated plan is required. Food security and nutrition are frequently complex concerns with divisive solutions that cut over disciplinary, institutional, and divisional boundaries. These issues posed by expanding global food systems are the result of interactions between different scales and levels, necessitating coordinated efforts from all stakeholders, including public and private actors, at the local, national, regional, and global levels, on a variety of fronts, including agriculture, trade, policy, health, the environment, gender norms, education, and infrastructure. Traditional food security measures have been production-based, with the goal of directly influencing food security through boosting food supplies. Focusing just on food production often overlooks other areas where underlying causes of underperformance in the food system and the potential for greater impact can be addressed. However, as the food system has evolved over the last few decades, it has developed intricate linkages and feedback loops that have impacted food security and nutrition in a variety of ways (FAO 2015).

15.6.4 Release of Biofortified Crops

Biofortification uses agronomic approaches such as traditional breeding, genetic engineering, and genome editing to improve the nutritional quality of food crops.

Biofortification of staple crops such as rice is carried out in poor nations with the goal of providing targeted micronutrients to those who do not have access to a varied diet or other existing therapies such as fortified foods and supplements. It is regarded as environmentally friendly, cost-effective, and food-based. In communities that are low in these components, iron, zinc, and vitamin A deficiency can cause major developmental and public health concerns ("Biofortification," September 10, 2021).

Vitamin A deficiency is an immunological deficiency caused by a lack of vitamin A in the diet. During the maturity of golden rice seeds, β-carotene is synthesized and accumulated. The human body can then convert β-carotene to vitamin A with efficiency (Dubock 2017). β-carotene is a vitamin A precursor. The endosperm contains a precursor for β-carotene called geranyl geranylpyrophosphate (GGPP) (Malik and Maqbool 2020). The phytoene synthase gene from daffodil was used to convert Japonica cultivar T309. A seed-specific Glutenin (Gt1) promoter was in charge of the gene. Phytoene is found in the endosperm of the altered plants. The genes *Erwinia uredovora* phytoene desaturase gene (crtI), daffodil phytoene synthase (psy1), and lycopene cyclase (lcy) from daffodil were used to increase β-carotene accumulation in rice endosperm. The CaMV 35S promoter was used to produce the crtI and Icy genes, whereas the Gt1 promoter was used to express the psy1 gene (Malik and Maqbool 2020). Because of the presence of β-carotene, rice supplemented with pro-vitamin A has a yellow/orange tint and is aptly referred to as golden rice.

15.6.5 High Protein, Nutrient-Rich Rice Varieties of NRRI

CR DHAN 310 (Protein-Rich Variety)

The ICAR-National Rice Research Institute in Cuttack developed CR Dhan 310, a high protein rice variety with an average of 10.3% protein in milled rice, by upgrading the popular high yielding variety Naveen. It has now been released by the Central Variety Release Committee. Rice is one of the most important staple meals, particularly for Asians, however, it contains the least grain protein of all cereals (6–8%). As a result, protein shortage among the poor, for whom rice is a staple diet, is a serious nutritional concern. The panicle is tall and thin, with medium grains. It is a compact semi-dwarf (110 cm) medium-early (123 days) semi-dwarf plant with good initial growth and tillering capabilities. The variety has a high head rice recovery (65%) and good cooking quality because of its alkali spreading value of 4 and intermediate amylose concentration (24.7%). Zinc is also present in milled rice in a moderate proportion (15 ppm). This type contains higher glutelin content and essential amino acids like lysine and threonine than its parent, Naveen, assuring greater nutritional value and intestinal function in humans. This variety would be a good source of nutrition for persons who rely on rice as their primary source of nutrition because it is not only high-yielding but also protein-rich. The average grain yield of this variety at the national level was 4483 kg/ha, surpassing the yield-check, Samba Mahsuri, by 6.81% in multi-locational testing. It also surpassed IR 64, another national check, as well as Naveen, a local check. Tolerance to leaf blast, brown spot, sheath rot, stem borer, gall midge biotype 1, and leaf folder was found to be moderate. This high-protein rice variety has been awarded a national identity number, IC 614804, by the NBPGR in New Delhi for release in Odisha, Uttar Pradesh, and Madhya Pradesh (Fig. 15.12).

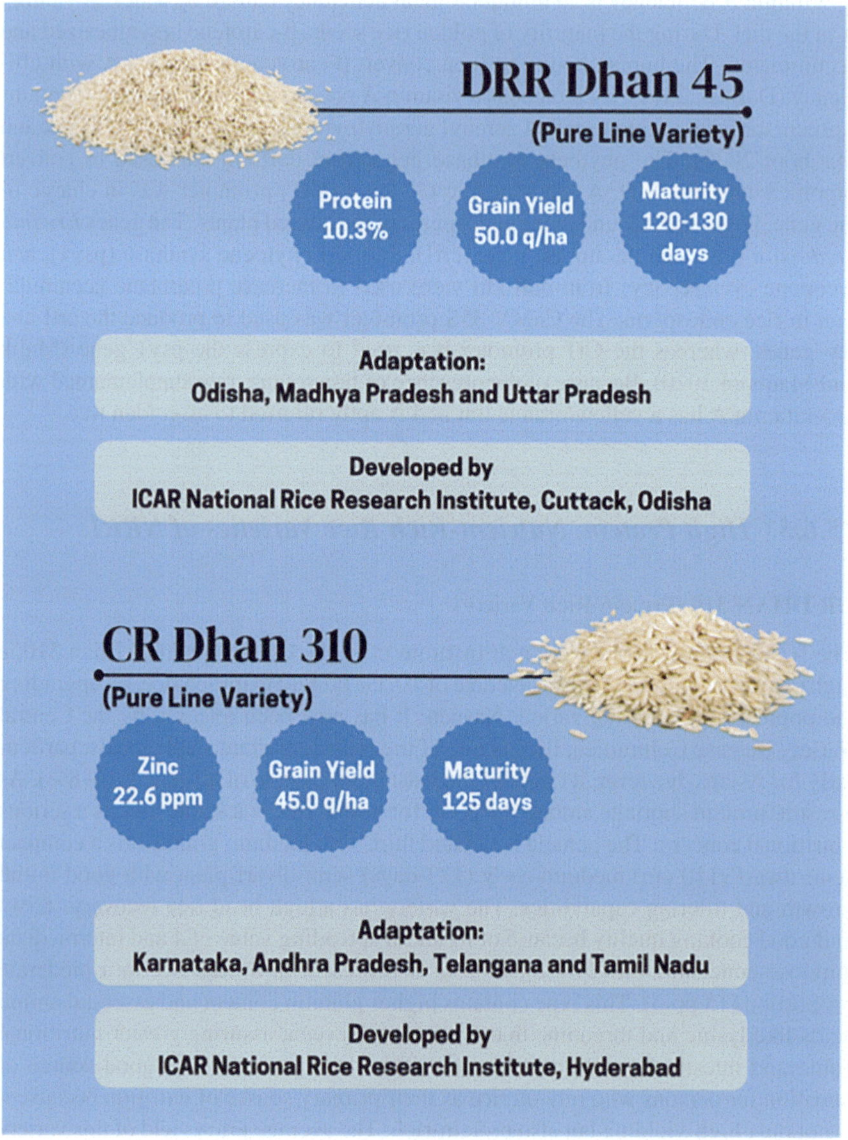

Fig. 15.12 Summary of the protein-rich and the zinc-rich rice variety

DRR DHAN 45 (Zinc-Rich Variety)

DRR DHAN 45 (IET 23832), India's first zinc-rich and high-yielding variety (Fig. 15.12), was created at ICAR-IIRR and launched in 2015. It is the result of a cross between IR 73707–45-3-2-3 and IR 77080-B-34-3. IET 23832 is a biofortified semi-dwarf cultivar with a non-lodging plant type and long-slender grains suitable for irrigated medium-duration situations (125 days) (Bandumula et al. 2019). It is the first high zinc rice variety to be recognized at the national level, with a total mean zinc content of 22.6 ppm (24.0 ppm in AP, KA, and TN) in polished rice. It was developed by traditional breeding without reducing yield using HarvestPlus material. Because of its high zinc content and yield performance of above 5 t/ha, it has been recommended for the states of Tamil Nadu, Andhra Pradesh, and Karnataka. It contains a high amount of amylose and is appropriate for cooking (20.7%). Resistance to blast, sheath rot, and rice tungro virus is moderate. The variety is a biofortification proof OF concept that can alleviate hidden hunger and mineral shortages, ultimately improving the country's nutritional security. Zinc insufficiency is a major health issue around the world. It causes lowered immunity, stunting, and possibly death. The genetic enrichment of rice with zinc is an efficient strategy to prevent malnutrition.

Drr Dhan 48

DRR Dhan 48, a high zinc variety (20.91 ppm), is a semi-dwarf, medium-slender variety with duration of 135–140 days and is suitable for irrigated ecology. It is resistant to BLB possessing genes of xa21 + xa13 + xa5. It has a yield potential of 5.0–5.5 t/ha and has been released for the states of Karnataka, Telangana, Kerala, and Tamil Nadu (Abbasi et al. 2024).

Drr Dhan 49

DRR DHAN 49, a high zinc (25.2 ppm), is a semi-dwarf variety with a duration of 130–135 days and a medium-slender grain type. It is suitable for irrigated ecologies. It has resistance with BLB resistance having genes of xa21 + xa13 and is tolerant to Blast to a certain degree. It was released in 2018 for the states of Kerala, Maharashtra, and Gujarat (Bandumula et al. 2019).

15.6.6 Impact of the Sustainable Food System

The food system is one of the primary causes of environmental effects, such as climate change. The 2019 EAT-Lancet report emphasizes the need for a nutritious diet derived from a sustainable food system, which can help mitigate climate change's effects (Fanzo et al. 2022). The food system must be sustainable in terms of its environmental impact, as well as the economic and social aspects of the enormous number of interconnected firms operating at various levels and scales (Fanzo et al. 2022). Closing nutrient loops, boosting the efficiency of the sustainable food system, producing higher-quality food, or bio-refining unavoidable losses (De Boer and Van

Ittersum 2018) are all examples of circular food systems that could provide both environmental and economic benefits (Fanzo et al. 2022).

15.7 Conclusions and Recommendations

Climate-smart agronomic techniques must be aimed at boosting soil carbon stocks in order to improve soil and water productivity and, as a result, reduce greenhouse gas emissions. A system-wide capacity development strategy can help to bring about a gradual transition to climate-smart agricultural production. Its goal is to generate more transformative and impactful effects by supporting long-term, endogenous development based on national empowerment. Because of its unique agronomic management practices and implications for global food security, this chapter focused on the long-term management of paddy rice agriculture. Climate-Smart Agriculture and Soil Management methods, on the other hand, can be useful in other crop production systems as well. When it comes to sustainable soil and land management methods, strategically restricting land-use change by minimizing the need to extend croplands can help to reduce methane emissions while also increasing the soil's carbon storage capacity. Increasing productivity allows us to get more out of the area we already have under cultivation. This eliminates the necessity for agricultural invasion and aids in the reduction of emissions caused by agricultural expansion.

Crop production has been found to increase when soil and water conservation practices such as shading, soil mulching, rainwater collection, and the use of fences or windbreaks to prevent evaporation are used. Achieving balanced nitrogen cycling across production systems and conserving the soil on the field are the most resourceful ways of soil intensification and crop management options. Direct seeding, in combination with sustainable agricultural residue management and an integrated soil fertility management system, protects the soil. Crop rotations that are effective for carbon sequestration can also keep a positive balance of nitrogen. Carbon sequestration is bolstered by increasing the complexity of crop rotations and incorporating legume crops. Rice cropping systems can be made more efficient by diversifying them by including an additional legume crop in the typical crop rotation method. Adding legumes to cropping systems to diversify them has proven to be a successful technique for increasing food and nutrition security while conserving land and water, hence enhancing agricultural sustainability and the environment. Actively growing roots produce exudates, and, notably in the case of legumes, it favors mycorrhizal associations.

In lowland/irrigated paddy rice farming, farmers have numerous water-saving options. No-tillage with mulching for soil cover; raised beds; land leveling; intermittent wetting and drying cycles; and aerobic rice are some of the choices (Thakur et al. 2022). When compared to traditional tillage techniques, zero tillage with residue retention (ZTR) has recently been introduced as a better crop establishment practice for cereal-cropping systems to improve productivity, nutrient concentration levels, nutrient uptake, and micronutrient biofortification. ZTR significantly resulted in a reduction of moisture stress on cereal crops in cropping systems developed in semi-arid locations. Micronutrient biofortification of rice grain from ZTR and cropping system expansion with a supplementary legume, particularly in poor countries, are potential management methods for reducing micronutrient deficiency and hidden hunger. We've established (1) that substituting traditional crop establishment procedures with zero tillage and residues left on the soil surface, as well as (2) integrating summer legumes in cropping systems, may easily boost rice yield and biofortification. Future research is needed to fully understand GHG emissions from all crops within each cropping system, as well as in-depth investigations on paddy rice in-season temporal fluctuations.

15.8 Future Prospects

With the proof-of-concept of various biofortified crops through genetic engineering, socio-economic studies have shown that biofortification by genetic engineering as a health policy intervention is not only well accepted by the majority of its key beneficiaries, with consumers' premium levels generally exceeding 20%, but is also cost-effective, especially when multiple micronutrients are targeted in one single crop (multi-biofortification) (Bashir et al. 2013).

With rising population and food demand, crop cultivars with higher yields are needed, which can lower micronutrient levels in edible plant components. As a result, there is a need to produce more productive cultivars that maintain acceptable micronutrient concentrations as a result of climate change (Table 15.5). Biofortification is a proven approach for combating malnutrition, especially in poor and developing nations where many people depend on staple food crops with low micronutrient concentrations. Many countries are implementing the biofortification of staple crops to eliminate micronutrient deficiencies and boost human health. Superior biofortified rice cultivars have resulted through improved micronutrient fertilization, traditional plant breeding, and genetic engineering. To increase micronutrient concentrations in staple crops, prospective research should focus on (i) the integration of both agronomic and genetic strategies to increase mineral mobilization to phloem-fed tissues and (ii) the discovery of mechanisms that affect mineral homeostasis in plants. A combination of methods (for example, improving crop efficiency in taking up mineral nutrients and/or improving vitamin/protein production using conventional breeding and genetic engineering technologies) should be examined, followed by increased fertilization of key nutrients. Because of its precision in editing the genome and generating transgenic crops with desirable traits, sequence-specific nuclease technology, including CRISPR/Cas9 and other gene-editing technologies created in recent decades, has become the method of choice. They were also expected to circumvent the limitations of transgenic plants developed through traditional methods.

Table 15.5 Genetically modified rice cultivars

Cultivar	Fold increase in Fe/Zn	References
Japonica cv. Taipei 309	6.3/1.6	Wirth et al. (2009)
Japonica cv. Tsukinohikari	3.4/2.31	Masuda et al. (2009)
Japonica cv. Tsukinohikari	1.0/1.0, 1.1/1.1, 1.4/1.3	Masuda et al. (2008)
Japonica cv. Nipponbare	2.2/1.4, 4.2, 2.2, 2.2/1.4	Johnson et al. (2011)
Japonica cv. Dongjin	3/2.7	Lee et al. (2011)
Japonica cv. Dongjin	/2.2	Lee et al. (2009)
Japonica cv. Xiushui 110	1.0/1.3	Zheng et al. (2010)
Japonica cv. Kitaake	3	Goto et al. (1999)
Japonica cv. Taipei 309	2	Lucca et al. (2001)
Japonica cv. Kitaake	3.0/1.1	Qu et al. (2005)
Japonica cv. TsukinoHikari	1.2/1.6	Nozoye et al. (2011)
Japonica cv. TsukinoHikari	4.4	Ishimaru et al. (2010)
Japonica cv. TsukinoHikari	6, 4/1.6	Masuda et al. (2012)
Japonica cv. Dongjin	1.1/1.1	Lee and An (2009)
Indica cv. Pusa Sugandh II	2.1/1.4	Paul et al. (2012)
Indica cv. IR68144	3.7/1.4	Vasconcelos et al. (2003)

In light of available biofortification technology and stakeholders that fund biofortification initiatives, it is necessary to refine planning, monitor, and evaluate biofortification programs. Even though the HarvestPlus alliance is doing a good job, indicators for evaluating the efficacy of biofortification activities and setting priorities are required. There is a need to establish communication and marketing strategies that consider ethical values when it comes to the production and use of biofortified staples. To make it acceptable and convince consumers to invest in micronutrient-rich food, the same strategy may not be practicable for different countries; consequently, governments should pursue promising strategies for their people. Food quality and safety verification are essential, as is recognizing the hazards of excessive consumption. With environmental changes in mind, further research into probable allergies and toxicities associated with increased micronutrient intake is still needed. Food content information and health claims must be highlighted according to international standards and national policies. When designing effective and open communication and implementation plans where GM foods are regarded as significant in reaching food security goals, governments should consider local food and agricultural contexts, as well as the unique context of each community (Straeten et al. 2017). Biofortification schemes should be tailored to a single country's regulatory framework. Although conventional breeding improvements have not been exposed to laws, the transgenic technique requires a legal framework that is suited for its application. Methodical installation efforts by public sector institutions and policies to promote biofortified crop varieties can successfully ensure a substantial increase in their adoption and acceptability.

Acknowledgements The authors express appreciation for the support provided by the Deanship of Scientific Research, Vice Presidency for Graduate Studies and Scientific Research, King Faisal University, Saudi Arabia (Grant No. KFU251925).

References

Abbasi A, Waheeb SA, Dawood AI, Zafar S, Hina A, Zaman QU, Ali HM, Jaremko M, Abdelsalam NR (2024) Biofortification of leafy vegetables: maximizing nutritional potential for well-being of humans. In: Harnessing crop biofortification for sustainable agriculture, pp 275–294

Abraham B, Araya H, Berhe T, Edwards S, Gujja B, Khadka R et al (2014) The system of crop intensification: reports from the field on improving agricultural production, food security, and resilience to climate change for multiple crops. Agri Food Secur 3(4). https://doi.org/10.1186/2048-7010-3-4

Agarwal S, Mangrauthia SK, Sarla N (2018) Genomic approaches for micronutrients biofortification of rice. In: Hossain MA, Kamiya T, Burritt DJ, Tran LP, Fujiwara T (eds) Plant micronutrient use efficiency. Academic Press, pp 245–260

Bandumula N, Neeraja CN, Waris A, Durbha S, Muthuraman P (2019) An economic analysis of biofortified rice varieties Oryza. Int J Rice 56:405–410

Bashir K, Takahashi R, Nakanishi H, Nishizawa NK (2013) The road to micronutrient biofortification of rice: progress and prospects. Front Plant Sci 4(15)

Beal T, Massiot E, Arsenault JE, Smith MR, Hijmans RJ (2017) Global trends in dietary micronutrient supplies and estimated prevalence of inadequate intakes. PLoS One 12(4):e0175554

Beyer P, Al-Babili S, Ye X, Lucca P, Schaub P, Welsch R, & Potrykus I (2002) Golden rice: introducing the beta-carotene biosynthesis pathway into rice endosperm by genetic engineering to defeat vitamin A deficiency. J Nutr 132(3):506S–510S

Bouis HE, Saltzman A (2017) Improving nutrition through biofortification: a review of evidence from HarvestPlus, 2003–2016. Glob Food Sec 49–58

Bouis H, Graham R, Welch R (2000) The consultative group on international agricultural research (CGIAR) micronutrients project: Justification and objectives. Food Nutr Bull 21(4):374–381

Bouis H, Foley J, Lividini K, Jumrani J, Reinke R, Van Der Straeten D, Zagado R, Boy E, Brown LR, Mudyahoto B, Alioma R, Hussain M, Pfeiffer WH (2024) Biofortification: future challenges for a newly emerging technology to improve nutrition security sustainably. Curr Dev Nutr 8(12):104478

Brady NC, Weil RR (2008) The nature and properties of soil, 14th edn. Pearson Prentice Hall, Upper Saddle River, New Jersey

Brooks S (2011) Is international agricultural research a global public good? The case of rice biofortification. J Peas Stud 38(1):67–80

Cropin (2019) Sustainable rice practices safeguarding the environment sustaining livelihoods. (Cropin). Retrieved from https://www.cropin.com/blogs/sustainable-rice-practices-safeguarding-the-environment-sustaining-livelihoods/

De Boer IJ, Van Ittersum MK (2018) Circularity in agricultural production. Wageningen University & Research

Development DO (2002). Rice in India: a status paper. Directorate of Rice Development, Ministry of Agriculture, Department of Agriculture, Co-Operation & Farmer Welfare, Government of India, Patna

Díaz-Gómez J, Twyman RM, Zhu C, Farré G, Serrano JS, Portero-Otin M, Christou P (2017) Biofortification of crops with nutrients: factors affecting utilization and storage. Curr Opin Biotech 115–123

Directorate of Rice Development (DRD) (2020) Rice in India: a status paper. Retrieved 25 Dec 2021, from https://drdpat.bih.nic.in/Status%20Paper%20-%2002.htm

Dobermann A (2002) A critical assessment of the system of rice intensification (SRI)
Donovan M (2020) What is sustainable intensification? CIMMYT. Retrieved 01 Jan 2022, from International Maize and Wheat Improvement. Center
Dubock A (2017) An overview of agriculture, nutrition and fortification, supplementation and biofortification: golden rice as an example for enhancing micronutrient intake. Agric Food Secur 6(1):13
Fahad S, Hussain S, Saud S, Tanveer M, Bajwa A, Hassan S, Huang J (2015) A biochar application protects rice pollen from high-temperature stress. J Plant Physio Biochem 96:281–287
Fahad S, Bajwa A, Nazir U, Anjum S, Farooq A, Zohaib A, Huang J (2017) Crop production under drought and heat stress: plant responses and management options. Front Plant Sci 29(8):1147. https://doi.org/10.3389/fpls.2017.01147
Fanzo J, Rudie C, Sigman I, Grinspoon S, Benton TG, Brown ME, Covic N, Fitch K, Golden CD, Grace D, Hivert MF, Huybers P, Jaacks LM, Masters WA, Nisbett N, Richardson RA, Singleton CR, Webb P, Willett WC (2022) Sustainable food systems and nutrition in the 21st century: a report from the 22nd annual Harvard Nutrition Obesity Symposium. Am J Clin Nutr 115(1):18–33
FAO (2015) Status of the world's soil resources (SWSR)—main report. Food and Agriculture Organization of the United Nations and Intergovernmental Technical Panel on Soils, Rome, Italy. Retrieved from http://www.fao.org/3/i5199e/I5199E.pdf
FAO (2021) World food and agriculture—statistical yearbook 2021. Food and agriculture organization, Rome. https://doi.org/10.4060/cb4477en
Food and Agriculture Organization of the United Nations (FAO) (2020) FAOSTAT: crops and livestock products. Retrieved 28 Dec 2021, from https://www.fao.org/faostat/en/#data/QCL/visualize
Garg M, Sharma N, Sharma S, Kapoor P, Kumar A, Chunduri V, Arora P (2018) Biofortified crops generated by breeding, agronomy, and transgenic approaches are improving lives of millions of people around the world. Front Nutr 5:12
GirijaVeni V, Srinivasarao C, Sammi Reddy K, Sharma K, Rai A (2020) Soil health and climate change. J Clim Change Soil Int 751–767. https://doi.org/10.1016/b978-0-12-818032-7.00026-6
Godfray HJ, Beddington JR, Crute I, Haddad L, Lawrence D, Muir J, Toulmin C (2010) Food security: the challenge of feeding 9 billion people. J Sci 327(5967):812–818. https://doi.org/10.1126/science.1185383
Goto F, Yoshihara T, Shigemoto N, Toki S, Takaiwa F (1999) Iron fortification of rice seed by the soybean ferritin gene. Nat Biotech 17(3):282–286
Gregorio G (2002) Progress in breeding for trace minerals in staple crops. In: Symposium: plant breeding: a new tool for fighting micronutrient malnutrition. J Nutr 132(3):500S–502S
Ishimaru Y, Masuda H, Bashir K, Inoue H, Tsukamoto T, Takahashi M, Nishizawa NK (2010) Rice metal nicotianamine transporter, OsYSL2, is required for the long-distance transport of iron and manganese. J Plant 62(3):379–390
Jackson M, Armstrong W (1999) Formation of aerenchyma and the processes of plant ventilation in relation to soil flooding and submergence. Plant Bio 1(3):274–287
Johnson AA, Kyriacou B, Callahan DL, Carruthers L, Stangoulis J, Lombi E, Tester M (2011) Constitutive overexpression of the OsNAS gene family reveals single-gene strategies for effective iron-and zinc-biofortification of rice endosperm. PLoS ONE 6(9):e24476
Kahle L, Gurel-Atay E (2014) Communicating sustainability for the green economy. Routledge, New York
Khush GS (1997) Origin, dispersal, cultivation and variation of rice. Plant Mol Biol 35(1/2):25–34
Kibblewhite M, Ritz K, Swift M (2007) Soil health in agricultural systems. Philos Trans R Soc Lond Ser B Biosci 363(1492):685–701. https://doi.org/10.1098/rstb.2007.2178
Kopittke PM, Menzies NW, Wang P, McKenna BA, Lombi E (2019) Soil and the intensification of agriculture for global food security. Environ Int 132:105078. https://doi.org/10.1016/j.envint.2019.105078

KPMG Advisory Services (2019) Rice industry review. Retrieved from https://assets.kpmg.com/content/dam/kpmg/ng/pdf/audit/rice-industry-review.pdf

Ladha J, Tirol-Padre A, Reddy K, Ventura W (1993) Prospects and problems of biological nitrogen fixation in rice production: a critical assessment. New Horiz Nitrogen Fixation 677–682

Lee S, An G (2009) Overexpression of OsIRT1 leads to increased iron and zinc accumulations in rice. Plant Cell Environ 32(4):408–416

Lee S, Jeon US, LJ, Kim YK, Persson DP, Husted S, An G (2009) Iron fortification of rice seeds through activation of the nicotianamine synthase gene. Proc Natl Acad Sci 106(51):22014–22019

Lee S, Persson DP, Hansen TH, Husted S, Schjoerring JK, Kim YS, An G (2011) Bioavailable zinc in rice seeds is increased by activation tagging of nicotianamine synthase. J Plant Biotech 9(8):865–873

Letey J, Hoffman G, Hopmans J, Grattan S, Suarez D, Corwin D et al (2011) Evaluation of soil salinity leaching requirement guidelines. Agri Water Manag 98(4):502–506

Lin X, Zhu D, Chen H, Zhang Y (2009) Effects of plant density and nitrogen application rate on grain yield and nitrogen uptake of super hybrid rice. Rice Sci 16(2):138–142

Lucca P, Hurrell R, Potrykus I (2001) Genetic engineering approaches to improve the bioavailability and the level of iron in rice grains. Theor Appl Genet 102(2):392–397

Majumder S, Datta K, Datta SK (2019) Rice biofortification: high iron, zinc, and vitamin-A to fight against "hidden hunger." Agronomy 9(12):803

Malik KA, Maqbool A (2020) Transgenic crops for biofortification. Front Sustain Food Syst 4:571402

Marcotullio P, Braimoh A, Onishi T (2008) The impact of urbanization on soils. In: Land use and soil resources. Springer Netherlands, pp 201–250. https://doi.org/10.1007/978-1-4020-6778-5_10

Masuda H, Suzuki M, Morikawa KC, Kobayashi T, Nakanishi H, Takahashi M, Nishizawa NK (2008) Increase in iron and zinc concentrations in rice grains via the introduction of barley genes involved in phytosiderophore synthesis. Rice 1(1):100–108

Masuda H, Usuda K, Kobayashi T, Ishimaru Y, Kakei Y, Takahashi M, Nishizawa NK (2009) Overexpression of the barley nicotianamine synthase gene HvNAS1 increases iron and zinc concentrations in rice grains. Rice 2(4):155–166

Masuda H, Ishimaru Y, Aung MS, Kobayashi T, Kakei Y, Takahashi M, Nishizawa NK (2012) Iron biofortification in rice by the introduction of multiple genes involved in iron nutrition. Sci Rep 2(1):1–7

McGuire S (2015) FAO, IFAD, and WFP. The state of food insecurity in the world 2015: meeting the 2015 international hunger targets: taking stock of uneven progress. Rome: FAO. Adv Nutr 6(5):623–624

Mekonnen M, Hoekstra A (2011) The green, blue and grey water footprint of crops and derived crop products. Hydrol Earth Syst Sci 15(5):1577–1600

Muthayya S, Rah JH, Sugimoto JD, Roos FF, Kraemer K, Black RE (2013) The global hidden hunger indices and maps: an advocacy tool for action. PLoS One 8(6):e67860

Nestel P, Bouis HE, Meenakshi JV, Pfeiffer P (2006) Biofortification of staple food crops. J Nutr 136(4):1064–1067

Noel S, Mikulcak F, Etter H, Stewart N (2015) Report for policy and decision makers: reaping economic and environmental benefits from sustainable land management. Economics of Land Degradation (ELD) Initiative, Bonn, Germany. Retrieved from https://hdl.handle.net/20.500.11766/4881

Nozoye T, Nagasaka S, Kobayashi T, Takahashi M, Sato Y, Sato Y, Nishizawa NK (2011) Phytosiderophore efflux transporters are crucial for iron acquisition in graminaceous plants. J Biol Chem 286(7):5446–5454

Pathak H, Aggarwal P (2012) Low carbon technologies for agriculture: a study on rice and wheat systems in the Indo-Gangetic plains. Indian Agricultural Research Institute, New Delhi

Pathak H, Tripathi R, Jambhulkar N, Bisen J, Panda B (2020) Eco-regional-based rice farming for enhancing productivity, profitability and sustainability. ICAR-National Rice Research Institute, ICAR, Cuttack, Odisha. https://doi.org/10.13140/RG.2.2.26486.42563

Paul S, Ali N, Gayen D, Datta SK, Datta K (2012) Molecular breeding of Osfer 2 gene to increase iron nutrition in rice grain. GM Crops Food 3(4):310–316

Pfeiffer WH, McClafferty B (2007) HarvestPlus: breeding crops for better nutrition. Crop Sci 47:S88–S100

Pimentel D, Kounang N (1998) Ecology of soil erosion in ecosystems. Ecosystems 1:416–426

Qu LQ, Yoshihara T, Ooyama A, Goto F, Takaiwa F (2005) Iron accumulation does not parallel the high expression level of ferritin in transgenic rice seeds. Planta 222(2):225–233

Randriamiharisoa R, Barison J, Uphoff N (2006) Soil biological contributions to the system of rice intensification. Bio Appr Sust Soil Syst 409–424

Reynolds C, Goucher L, Quested T, Bromley S, Gillick S, Wells VK, Evans D, Koh L, Carlsson-Kanyama A, Katzeff C, Svenfelt Å, Jackson P (2019) Review: consumption-stage food waste reduction interventions—what works and how to design better interventions. Food Policy 83:7–27. https://doi.org/10.1016/j.foodpol.2019.01.009

Singh M (2010) Rice productivity in India under variable climates. ICAR, New Delhi

Sperotto RA, Ricachenevsky FK, de Abreu Waldow V, Fett JP (2012) Iron biofortification in rice: it's a long way to the top. Plant Sci 24–39

Straeten DV, Fitzpatrick TB, Steur HD (2017) Editorial overview: biofortification of crops: achievements, future challenges, socio-economic, health and ethical aspects. Curr Opin Biotech 7–10

Strobbe S, Straeten DV (2017) Folate biofortification in food crops. Curr Opin Biotech 202–211

Takahashi J, Shoji S, Nogi T, Wada G (1973) The fate of fertilizer nitrogen applied to the paddy field and its absorption by rice plant 5. The fate of basal nitrogen in the paddy field and its absorption by rice plant under the various cultural conditions. Proc Crop Sci Soc Japan 42:91–96

Thakur V, Sharma A, Sharma P, Kumar P, Shilpa (2022) Biofortification of vegetable crops for vitamins, mineral and other quality traits. J Hortic Sci Biotechnol 97(4):417–428

Tian D, Niu S (2015) A global analysis of soil acidification caused by nitrogen addition. Environ Res Lett 10(2):024019

Tilman D, Cassman K, Matson P, Naylor R, Polasky S (2002) Agricultural sustainability and intensive production practices. Nature 418(6898):671–677

Toriyama K, Ando H (2022) Towards an understanding of the high productivity of rice with system of rice intensification (SRI) management from the perspectives of soil and plant physiological processes. Soil Sci Plant Nutr 57(5):636–649

Trijatmiko KD (2016) Biofortified indica rice attains iron and zinc nutrition dietary targets in the field. Sci Rep

United Nations (2015) Sustainable development goals. The United Nations (UN). Retrieved from https://www.un.org/sustainabledevelopment/sustainable-development-goals/

United Nations Do (2019) World population prospects 2019: highlights. United Nations, New York

Vasconcelos M, Datta K, Oliva N, Khalekuzzaman M, Torrizo L, Krishnan S, Datta SK (2003) Enhanced iron and zinc accumulation in transgenic rice with the ferritin gene. Plant Sci 164(3):371–378

Wall DH (2005) Sustaining biodiversity and ecosystem services in soils and sediments. Island Press, Washington, USA

Wassmann R, Jagadish S, Sumfleth K, Pathak H, Howell G, Ismail A, Heuer S (2009) Regional vulnerability of climate change impacts on Asian rice production and scope for adaptation. Adv Agro 91–133 (Chapter 3)

Wingeyer A, Amado T, Pérez-Bidegain M, Studdert G, Varela C, Garcia F, Karlen D (2015) Soil quality impacts of current South American agricultural practices. Sustainability 7(2):2213–2242. https://doi.org/10.3390/su7022213

Wirth J, Poletti S, Aeschlimann B, Yakandawala N, Drosse B, Osorio S, Tohge T, Fernie AR, Günther D, Gruissem W, Sautter C (2009) Rice endosperm iron biofortification by targeted and synergistic action of nicotianamine synthase and ferritin. J Plant Biotech 7(7):631–644. https://doi.org/10.1111/j.1467-7652.2009.00430.x

Zambon I, Benedetti A, Ferrara C, Salvati L (2018) Soil matters? A multivariate analysis of socioeconomic constraints to urban expansion in Mediterranean Europe. Ecolo Econom 146:173–183

Zhao L, Wu L, Li Y, Lu X, Zhu D, Uphoff N (2009) Influence of the system of rice intensification on rice yield and nitrogen and water use efficiency with different new application rates. Exp Agri 45(3):275–286

Zhao F, Ma Y, Zhu Y, Tang Z, McGrath S (2015) Soil contamination in China: current status and mitigation strategies. Environ Sci Tech 49(2):750–759

Zheng L, Cheng Z, Ai C, Jiang X, Bei X, Zheng Y, Shou H (2010) Nicotianamine, a novel enhancer of rice iron bioavailability to humans. PLoS ONE 5(4):e10190

Mr. Rohan Rama Krishnan has recently graduated from a master's program in Food Science (Biotechnology) at the University of Leeds, awarded with distinction, focusing on sustainable edible packaging of orange-fleshed sweet potatoes using chitosan-based coatings. He previously earned a master's in Botany and a triple major bachelor's in Chemistry, Botany, and Zoology from Christ University, Bangalore. His research interests include food microbiology and sustainable food systems and adds to his credit previous publications on similar themes.

Ms. Bhadrapriya Sivakumar is pursuing her Ph.D. in Neuroscience at the University of Saskatchewan, Canada. Her research focuses on therapies for peripheral nerve regeneration. Previously, she worked on gene therapy for ALS at Rajiv Gandhi Centre for Biotechnology in Kerala. She has published research papers on fungi from the Western Ghats and neurodegenerative therapies and has contributed to several book chapters on topics such as climate change impacts and CRISPR-Cas9 genome editing.

Ms. Nayana Jagannath is a research associate at Mynvax Private Ltd. in Bengaluru, India, specializing in analytical research and development. She holds a master's degree in Biotechnology from Christ University, Bangalore, and has previously worked at Bhabha Atomic Research Centre. Her research interests include drug development and biotechnological applications for pharmaceuticals.

Mr. Nachiketha R. Rao is pursuing a second master's degree in Neurobiology at the Rheinland-Palatinate Technical University, Germany. He previously earned a Master of Science in Biotechnology and a bachelor's degree in Chemistry, Botany, and Zoology from Christ University, Bangalore. His research interests include neurodegenerative diseases, particularly Alzheimer's, and he has published research on the fungi of the Western Ghats. His current research focuses on electrophysiological techniques in Alzheimer's disease studies.

Ms. Nithya Suresh is a research associate at Jubilant Biosys in Bengaluru, India, where she specializes in protein purification and drug discovery. She earned her master's degree in Biotechnology from Christ University, Bengaluru, and has contributed to academic publications on CRISPR-Cas9 genome editing and phytochemical extraction techniques. Her research interests include X-ray crystallography, surface plasmon resonance, and therapeutic development.

Dr. Praveen Nagella is an associate professor at CHRIST (Deemed to be University), Bangalore, India, specializing in plant tissue culture, medicinal plants, stress physiology, and secondary metabolites. He completed his Ph.D. at Karnatak University in 2010 and later worked as a post-doctoral researcher at Konkuk University, South Korea. Dr. Nagella has published over 100 articles and 25 book chapters and serves as a reviewer for several international journals.

Dr. Jameel M. Al-Khayri is a professor at the Department of Agricultural Biotechnology, King Faisal University, Saudi Arabia. He holds a B.S. in Biology from the University of Toledo (1984), an M.S. in Agronomy (1988), and a Ph.D. in Plant Science (1991) from the University of Arkansas. He is a prominent member of the International Society for Horticultural Science and serves as the national correspondent for the International Association of Plant Tissue Culture and Biotechnology. Dr. Al-Khayri has dedicated over three decades to date palm biotechnology research, authoring 130 research articles, 80 book chapters, and editing 30 reference books. In 2022, he was recognized among the top 2% of Scopus-cited scientists. He is the editor of three Springer book series and has organized numerous international scientific conferences. In addition to his research and teaching, Dr. Al-Khayri has held administrative roles, including assistant director of the Date Palm Research Center and vice dean for Development and Quality Assurance. He served as a member of Saudi Arabia's Majlis Ash Shura from 2009 to 2012 and focuses on using biotechnology to support agricultural sustainability and food security.

Chapter 16
Plants Response to Climate Change—Mitigation Strategies to Improve Crop Yield

Amtul Bari Tabinda, Rimsha Javed, Javairia Ansar, Adeel Mahmood, and Abdullah Yasar

Abstract Climate change significantly impacts agriculture production, affecting crops both directly through alterations in abiotic factors such as temperature, rainfall, and CO_2 levels and indirectly by modifying soil health, insect populations, and pest infestations. As a result of climate variations, both abiotic and biotic stresses have a detrimental impact on crop productivity. Climate change's complex effect compromises food security, reduces agricultural yields, and places immense pressure on economic development, particularly in the least developed countries. Food demand may rise by 85% as the population is projected to grow by approximately 900 million by 2050 placing immense stress on ecological integrity and food supply requirements. Plants respond swiftly to establish stress-specific signaling by producing plant hormones and antioxidants, and by undergoing physiological, morphological, and biochemical changes under stress conditions. To combat the drastic impacts of climate change, climate-smart agriculture (CSA) offers an effective approach. Crop rotations, intercropping, adjusting fertilizer rates, maintaining grain nutritional content, and improving pest, disease, and weed management through integrated pest and pathogen management could help mitigate the negative effects of climate change. General Circulation Models (GCMs) and Representative Concentration Pathways (RCPs) are widely used to quantify the negative effects of climate change on agriculture and to generate future weather projections. Environmental policies such as initiatives for sustainable farming, water treatment and reuse in agriculture, rainwater harvesting, and the restoration and preservation of ecosystems can support the adoption of CSA.

Keywords Climate change · Plant response · Mitigation · Food security · Climate-smart agriculture

A. B. Tabinda (✉) · R. Javed · J. Ansar · A. Yasar
Sustainable Development Study Centre, Government College University Lahore, Lahore, Pakistan
e-mail: amtulbaritabinda@gcu.edu.pk

A. Mahmood
Department of Environmental Sciences, GC Women University Sialkot, Sialkot, Pakistan

M. Behnassi et al. (eds.), *Food Systems and Biodiversity in the Context of Environmental and Climate Risks*, https://doi.org/10.1007/978-3-031-89167-0_16

16.1 Introduction

Climate change is recognized as a significant global issue with substantial adverse impacts on agriculture, water resources, forest ecosystems, biodiversity, and socio-economic well-being. Climate change is considered serious threat to food security with rapid climatic variation and pressure on agro-ecological environments affecting crop growth. Globally, crop production and food security are compromised due to the adverse impacts of climate change. Crop yield and climatic patterns are closely related; in some regions, climatic changes may render areas inappropriate for cultivation, while in others, it may enhance crop production (Chandio et al. 2024). Climate variations trigger both biotic and abiotic stresses which adversely impact agriculture production. Climate change has both direct and indirect implications for agriculture regions. Direct impact on agriculture arises from physical changes in agroecological conditions such as temperature, and rainfall patterns. Indirect impacts are caused by fluctuations in species distribution, competition, disease outbreaks, and the dominance of invasive species, all of which influence the demand for agricultural products in a region. Indirect impacts are challenging to assess due to the dispersed and interrelated nature of influencing parameters (Kumar 2016; Andrianarimanana et al. 2023).

Climate change has severe implications for underdeveloped nations, primarily due to inadequate access to resources and a lack of proper information about its devastating effects. Anthropogenic activities have significantly threatened ecological integrity, particularly, in impoverished communities worldwide. Increased greenhouse gas emissions have contributed to substantial global warming since the Industrial Revolution. As the Earth warms the frequency of extreme climatic events has increased, including abrupt disasters such as floods, heatwaves, droughts, and storms, all of which threaten food security and human livelihoods (Mulagha-Maganga et al. 2024; Syeda et al. 2022). Major climatic variations include intense heat waves, changes in rainfall patterns, temperature increases, rising sea levels, elevated atmospheric CO_2 concentrations, variations in microorganism abundance, and outbreaks of diseases, pests, and weeds (Ahmad et al. 2018).

According to FAOSTAT (2017), food demand possibly rise to 85% as the population is expected to grow approximately by 9 billion in 2050 that puts huge stress on ecological integrity and food requirements. Despite limited water availability, the increase in evapotranspiration due to longer growing seasons and higher temperatures might raise agricultural irrigation requirements by up to 5–20% in 2080 globally (Abou Seeda et al. 2021). Agricultural production is significantly impacted by environmental variations that negatively affect ecosystem health and natural systems. Environmental changes progressively impact crop productivity particularly through abiotic stresses. Predominately, abiotic stresses such as water shortages, soil infertility, intense temperature, and pollution are the main contributors determining crop yield (Shah et al. 2024).

The impact of climate change on agriculture sectors is primarily predicted by the frequency and intensity of extreme stress events, crop damage, and their effects on daily life. Climatic variations mostly affect underdeveloped nations due to poor environmental conditions that reduce crop productivity (Tong et al. 2024). However, scientists are focusing on new strategies to address the challenges posed by high temperatures and excessive CO_2 buildup to find new solutions. Abiotic stress significantly impacts plant development and crop yield. Rapid industrialization and modernization continuously alter the natural environment leading to extreme climatic events that disrupt ecosystem functioning globally (Kang et al. 2009; Nasir et al. 2020). The physical and chemical factors included in abiotic stresses are temperature, heat and cold waves, drought, water logging, ozone accumulation, salinity, and emission of gases. Fluctuations in various parameters considerably put food security under great threat. Temperature is a crucial parameter for plant growth and survival. Heavy rainfall, famines, temperature fluctuations, salinity, pest infestation, and pollution emissions are all anticipated pressures on crop yield. The frequency of global warming is ultimately amplified with the passage of time. All living biota have been impacted by the drastic environmental conditions either directly or indirectly (Ullah 2017).

The major challenge to food security is rapid rise in global population and reduction in productivity of required crops. There have been probabilities that the frequency of extreme climatic events would enhance in the near future. It is expected higher drought rates projected from 15.4 to 44.0% by 2100 and reduction in crop yield up to 50% by 2050. Water–plant relations are greatly affected by the higher evapotranspiration rate and physiological changes in plant as a response to high temperatures (Raza et al. 2019). According to plant species and developmental stage, a responsive mechanism to climate change is different depending upon plant type. Temperature mainly affects plants by reducing crop duration growth. Malhi et al. (2021) reported that a 2 °C rise in temperature potentially reduces the yield of maize, rice, and wheat in tropical regions. Tropical crops are more vulnerable to elevated temperatures as already have optimum temperature however more temperatures put huge stress on crop survival. Plant growth and many physiological processes are heavily influenced by temperature (Crous 2019; Malhi et al. 2021).

"Due to enormous scale and extreme sensitivity to climatic variables, the agriculture sector is more susceptible to the effects of climate change which has a significant economic impact. A slight change in weather parameters greatly influences the crop yield. The production of crops is substantially impacted by variations in rainfall patterns and temperature. Each parameter has a different impact on crops according to location, adaptability, irrigation practice, crop type, and magnitude of climatic fluctuations such as high temperature, decreased precipitation, and increased CO_2 concentration as observed in Iran. Additionally, production of crops expected in China such as corn, wheat, and rice, respectively 18.26 ± 12.13, 45.10 ± 11.55, and $36.25 \pm 10.75\%$ till 2100 as a continuous change in temperature perceived. When deviation in the external environment occurs, modification in the homeostatic mechanism results in plant adaptation to extreme environmental situations. At this stage, plants needed slight modification in genetic and metabolic mechanisms. These mechanisms involve the metabolic reprogramming in cellular systems to initiate biochemical functions of the extrinsic conditions. Plant microbiome gives superior assistance to plants in acquiring complements, opposing diseases, and enduring abiotic stresses (Shefali et al. 2021; Grymshi et al. 2022; Abbas 2022).

"To ensure food security, climate-smart crop cultivars are necessity of the time to tackle the limitations. Climate-smart agriculture technologies that significantly reduce the adverse impact of climate change on crop productivity are water-smart practices; laser land leveling, rainwater harvesting, micro-irrigation, crop diversification, raised-bed planting, direct-seeded rice, nutrient-smart practices; precision nutrient application, leaf color charts, crop residue management, weather-smart activities; stress-tolerant varieties, ICT based agrometeorological services, carbon-smart activities; zero tillage, legumes, crop residue management and knowledge-smart activities; agricultural extensions to enhance capacity-building (Malhi et al. 2021; Ogundeji et al. 2022).

16.2 Plant Adaptation to Extreme Climatic Stress

Extreme environmental conditions and climate variability increase stress on plants. The drastic implications of climate change on the productivity of crops varied according to region, some regions expected to see an increase in crop yields while others will experience a decrease. Temperature extremes, frequent droughts, and erratic flooding are key stress factors that can reduce cereal yields, with wheat production being particularly affected. It is estimated that wheat crop yield may decrease by 6% for every 1 °C increase in temperature. Furthermore, water deficit and temperature extremes can also have a significant impact on the reproductive phase of plant growth (Kalra and Kumar 2019). It has been observed that flower inflorescence and development in cereal crops are adversely affected by water scarcity.

Climate change has a complex impact on food security affecting agricultural yields and placing pressure on income growth and distribution, particularly in the least developed countries. In Asia, rice and wheat are the major contributors to food availability and security. Rising temperatures and rainfall variability have negatively affected crop growth and development. Reduction in the productivity of major crops such as rice and wheat have been observed to depend on regional climatic patterns (Gray and Brady 2016).

Meeting the growing demand for food will require a 60% increase in wheat production by 2050, posing a significant challenge. In regions with arid to semi-arid climates, crop yields have declined due to increased temperatures in low-latitude areas. It is reported that rice, wheat, and maize yields have declined in China by droughts and floods, and this trend is expected to intensify in the future. Rice is particularly vulnerable to steady nighttime temperature increases, with biomass and yield reductions ranging from 16 to 52% if temperatures exceed the critical threshold of 24 °C by 2 °C. In Asian arid to semi-arid regions already facing low productivity and drought stress. Climate change has negatively impacted wheat quality and grain production due to erratic rainfall, high-intensity rainfall, and rising temperatures. Climate variability has decreased wheat productivity by about 17.6% in the Egyptian North Nile

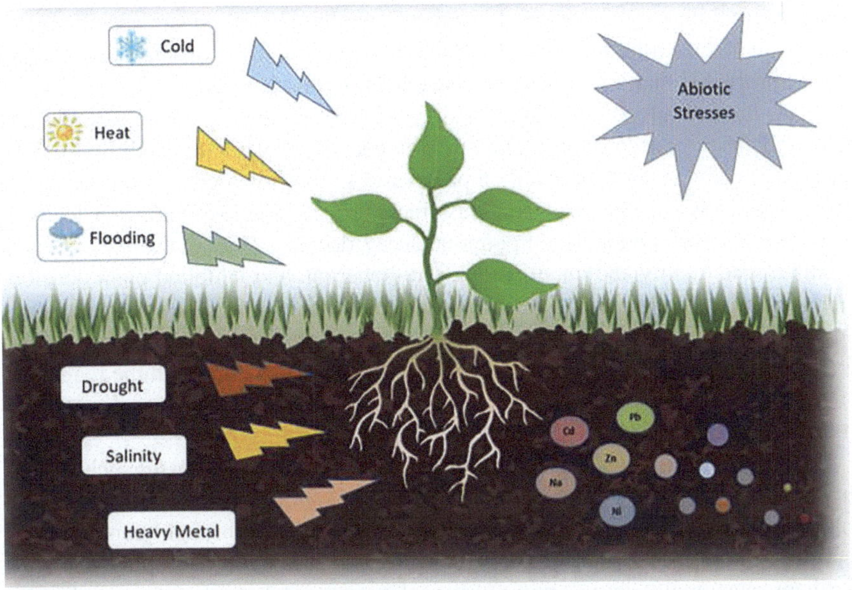

Fig. 16.1 Impact of abiotic stresses on plant growth. *Source* Raza et al. (2019)

Delta, as well as India and China, due to rising temperatures, increased precipitation, and frequent weed insect invasions (Habib-ur-Rahman et al. 2022). Figure 16.1 illustrated the impact of abiotic stresses on plant growth.

Climate change negatively affects crop production and viability by increasing pest outbreaks and their migration to main crops, respectively. The factors crucial for Pest population propagation are humidity and temperature. For instance, leaf miner and coffee nematode infestation are projected to develop as a result of a rise in monthly generations contrary to climatic circumstances of 1961–1990, which could adversely affect crop yields (Malhi et al. 2021; Wu et al. 2022). As the global earth temperature rises, it significantly modifies climate, leading to abiotic stress that poses a threat to crop species. Among the various environmental stressors, drought and heat are the most prevalent and significantly affect plant growth and development. Plants require an optimum temperature for normal growth and blooming, and temperature fluctuations can heavily influence plant physiology. Heat stress negatively impacts grain production and yield, while cold stress can lead to sterility, and drought stress affects plant morphology and physiology.

These climatic challenges result in significant changes, including molecular, biochemical, physiological, and morphological modifications, and are predicted to increase in frequency and intensity due to climate change. Therefore, it is crucial to develop stress-tolerant varieties as a mitigation strategy (Raza et al. 2019). Adapting to climate change primarily involves modifying farm practices that are influenced by social, political, and economic conditions and policy decisions that align with climatic variability and extremes. Precision agriculture based on site-specific information can reduce nitrogen application without compromising profitability, making it more profitable than managing the entire field (Crous 2019).

Plants have developed mechanisms to recognize previous exposure to stress and defend themselves accordingly. The initial impact of unfavorable conditions can be seen at the cellular level, which then progresses to physiological symptoms. Water stress negatively affects plant anatomy, including the photosystem, and leads to reduced root development, water potential decay, decreased size of the leaf, stomatal opening, seed quality and quantity, delayed flowering, and inhibited plant development and growth. However, plants have developed unique mechanisms to manage optimal water resources and regulate their growth when exposed to harsh conditions. Excess light leads to photooxidation, which can be managed through the regulation of enzymes and other biomolecules using ROS or reactive oxygen species (Shefali et al. 2021).

To mitigate the impact of climate change and reduce abiotic stresses such as heat, drought, flood, water logging, and nutrient-deficient soil, the development of stress-tolerant, higher-yielding, and early-maturing crop varieties through plant breeding and biotechnological approaches are crucial. Crop rotations, intercropping, adjusting fertilizer rates, maintaining grain nutritional content, and improving pest, disease, and weed management through integrated pest and pathogen management could also help mitigate the negative effects of climate change. Low-cost adaptation strategies such as optimizing fertilizer and irrigation management, adjusting sowing dates based on weather conditions, and upgrading seed varieties can help farmers increase crop yield (Syeda et al. 2022).

16.3 Factors Limit Plant Productivity

Agricultural productivity is limited by several factors, with biotic and abiotic stresses being the most critical. Continuous stress conditions lead to the generation of reactive oxygen species (ROS) in various cell divisions. ROS damages biomolecules and disrupts cellular activity, ultimately causing cell death. Microbes, which are the most prevalent living organisms worldwide, have intricate physiological and metabolic systems that allow them to deal with potentially reactive oxygen species that are produced as a result of environmental stressors. The plant microbiome is made up of numerous complex elements. Increasing environmental changes during constant stress, making an important mark, and finding out how plants and microbes work together to protect against environmental issues. Abiotic stress is the primary factor limiting crop production globally. Plants are susceptible to a variety of stress conditions. A plant's response to abiotic factors depends on their quantity or intensity. Plants require specific quantities of nonliving environmental elements for successful growth. Abiotic stress refers to any modification in the biochemical or natural environment that deviates from ideal environmental conditions. Abiotic stress, which affects plants in various ways, is the principal factor limiting global food production.

The degree or quantity of something affects a plant's abiotic factor (Müller et al. 2015).

16.3.1 Drought

Under drought conditions, high CO_2 levels in leaves cause an increased production of ROS magnifying stress levels. When stomata close, CO_2 cannot pass through the leaf, and increased oxygen levels favor ROS production under drought. ROS generation breaks down membranes, restricting plant growth, photosynthesis, and respiration. ROS limits cells to form carbohydrates, lipids, proteins, and nucleic acid when they are under stress from dehydration. Recent studies show that tomato plants produce osmo-protectants when exposed to combined heat and salt stress, but not to either stress alone (Xu et al. 2021). Plants are sensitive to water availability. Different plant species have diverse responses to drought because of how the SOD enzyme works, which plays a big role in antioxidant absorption. In bluegrass, drought stress not only slows down the migration of SOD, but the genes for FeSOD and Cu/ZnSOD are also uncontrolled. When Alfalfa is stressed by drought, CU/ZnSOD and FeSOD are upregulated. This indicates that defenses act differently in according to species and tissues.

A higher level of salts accumulation plants via topsoil, disturb natural signaling of ions inside cells and tissue that react in different ways to the stressors caused by salinity. When a plant receives a stress condition, the genetic expression may change due to harmful. Because there was more salt in the topsoil, its osmotic potential was reduced, which made the plants lose ions. This situation can have a dangerous effect on plant growth by slowing down the germination of seeds, growth of seedlings, and aging of plants. The amounts of amino acids like methionine, arginine, and cysteine are also restricted under salt stress. The combination of proline in the cells greatly mitigates salt stress. Plant responses as stressed by salt are the synthesis of nitric oxide, the start of antioxidants, hormonal control, and the mixing of glycine betaine. Moreover, limited water availability or high salinity disturbs plants from getting nutrients. As a consequence, plant tissues damages, and probably affects agricultural yield (Shefali et al. 2021).

16.3.2 Heat

According to the investigation, the mutual salt and heat impact diverse metabolomics profiles related to physiological and biomolecular data. ROS production exhibits a vital secondary signal for cellular metabolism to promote plant development (Raza et al. 2019). Heat stress is a significant issue in agriculture hindering plants' ability to modify their genetic makeup, biochemistry, and structural components, thereby affecting crop output. Elevated temperatures negatively impact plants at various stages of growth. It inhibits the germination of seeds, and photosynthesis, and reduces membrane permeability. Plants produce a moderate number of metabolites and phytohormones in response to heat stress, signaling the production of heat shock proteins, and ROS. Maintaining membrane stability, and activating calcium-dependent protein kinase and mitogen-activated protein kinase are plant's defensive mechanisms against heat stress. Scavenging of reactive oxygen species (ROS), chaperone signaling, transcriptional regulation, and antioxidant metabolites are plant defensive mechanisms (Nguyen and Scrimgeour 2022).

16.3.3 Water

Metabolomic studies have examined the activity of free amino acids, abscisic acid, and oxidative phases in barley under water scarcity. Barley cultivars were given less water so that the genomic differences in metabolomics during the grain-forming stage could studied. Inhibiting protein production is the first metabolic response to abiotic stressors. Abiotic stressors are also mostly responsible for changes and processing that occur after translation (Abbas 2022).

16.3.4 Plants and Microorganisms Relation Under Stress

Using pattern recognition receptor (PRR) proteins on the outside or inside of the cell, plants can identify pathogens by detecting extracellular particles known as pathogen-associated molecular patterns (PAMPs). For instance, Ef-TU proteins, peptidoglycans, lipopolysaccharides, and bacterial flagellin. There are four levels in a plant's immune system. PAMPs for microorganisms are identified and named at level 1. It can adhere to various PRRs based, on the surface of the cell which initiates the development of the plant's improved immunity (PTI) and makes it resistant, to tentative colonization and growth. Some viruses create effectors that make it more deadly at level 2. With PTI, the effectors wait and prepare the susceptibility to be activated. Leucine-rich replicate receptor proteins that bind nucleotides find the effector at level 3. The body becomes resistant to disease through a mechanism known as effector-triggered immunity (ETI). In level 4, natural selection has made new effectors that boost virulence until plants make new receptors, causing infections to control ETI (Kumar 2016).

16.4 Plant Defensive Response to Stress

Abiotic stressors significantly affect plant morphology and physiology. The optimal temperature range for plant growth and development is between 10 and 35 °C. The specified temperature increase will allow plants to produce surplus energy, but a bigger temperature increase will slow plant growth and cause photosynthesis to decline to lethal levels. The drought stress limits turgor pressure, which slows down cell formation. Photosynthesis is greatly affected under water-scarce conditions, which reduce enzyme activity and metabolic processes ultimately disrupting photosynthesis function. Environmental variations, such as increased CO_2 concentration slow plant respiration, and raise internal heat levels. When the temperature was increased from 15 to 40 °C, the plant's respiration rate increased, disrupting the morphological characteristics of particular crops. A slight increase in temperature triggers the production of xylulose-1,5-bisphosphate, an inhibitory enzyme that inactivates the functioning of the Rubisco enzyme (Kalra and Kumar 2019).

16.4.1 Plants Physiological Response

Growing stress-tolerant plants and understanding responses to various stress conditions are essential for global food security and efficient agriculture to meet population demand. Plants respond to climatic conditions according to metabolism, physiology, and gene expression. Although numerous investigations have been conducted, only a small number of putative sensors have been identified. Plants perceive any fluctuations from the immediate environmental signals. It is evident that plants' tissues and organs are greatly affected by diverse pressures and react accordingly. For instance, different root cells or tissues have varied transcriptional responses to various stresses. Cellular responses to salt stress, drought, and chemical contamination include stress-responsive protein production, enhanced associated solutes, and increased antioxidant ratios. The strains are thought of as primary strains, leading to secondary strains such as oxidative and osmotic stress (Syeda et al. 2022).

16.4.2 Hormonal Response

Plant hormones provide defense mechanisms against environmental challenges and are essential for plant growth and evolution. The most effective abiotic stress response pathways are abscisic acid-mediated, followed by other phytohormone-dependent defense pathways, such as ethylene (ET), jasmonic acid, and salicylic acid that deteriorate plants' responses to ecological stress. (Shefali et al. 2021). Hormones have a significant role in the regulation of numerous signaling pathways and responses to a variety of abiotic stressors, including salicylic acid (SA), abscisic acid (ABA), and ethylene. By interacting with some other hormones, ABA plays a crucial part in the control of stress responses. In particular, ABA is important for the functioning of stomata either closing or opening under stress, seed propagation, and inactivity during various stages of plant growth. It is understood that the signaling cascade formed by PYR/PYL/RCAR-PP2C-SnRK2 efficiently regulates seed dormancy and is produced by ABA. Drought conditions substantially slow down plant growth while also raising the ABA concentration in cell walls. ABA builds up when there is drought stress, which regulates transpiration and prevents stomatal disclosure. Moreover, ABA stimulates a variety of physiological processes in plants, including water stress, stomatal closure regulation, and the production of several stress-responsive genes (Ullah 2017).

Three components make up the signaling cascade: PYR/PYL/RCAR proteins, SnRK2/OST1 protein kinase, and PP2C protein phosphatases. By altering several signal transduction processes in response to climate variability, phytohormones play an essential role in the stress response. Ethylene is one of the most significant phytohormones. Under various abiotic and biotic climatic stressors, ethylene plays a role in controlling seed germination, ripening, leaf growth, and senescence. Ethylene is thought to function as a signaling channel between plant growth and changes in the weather. The factors that modulate the production of ethylene include abiotic stressors such as heat, frost, water scarcity or logging, contamination of heavy metals, and nutritional insufficiency. Plants' ethylene response factors (ERFs), which are members of a large transcription factor (TF) family, become active in response to a variety of physiological and environmental stressors (Raza et al. 2019).

Environmental stress results in biochemical changes. Many abiotic factors that limit crop output have an impact on plant development growth. Long-term salt exposure stress sets the stage for cell damage and osmotic balance disturbance. Controlling osmotic and ionic equilibrium in the cells requires tolerance to salinity stress. Plants protect their delicate plant skins from highly salty portions, ions coming from the roots, or ions remaining away from the cytoplasm to fend off salinity. Several plants demonstrated a method to deal with cold temperatures during icing conditions by enhancing defense response through the process of cold acclimation. When plants detect stress, they respond quickly and forcefully to establish signaling in accordance with stress by producing hormones and a group of flavonoids and phenolic acids in plants.

The main factor causing ROS generation is abiotic stress. Under usual conditions, the production and destruction of reactive oxygen species are in equilibrium. During stressful situations, it disrupts balance by increasing the production of ROS, which is extremely detrimental to the organism since it has a negative impact on the structure and function of the biomolecules. Plants create ROS in chloroplasts, peroxisomes, and mitochondria. The electron transport mechanism is less efficient, which causes mitochondria to produce oxygen radicals and hydrogen peroxide. Due to elevated oxygen levels and lower molecular oxygen in the electron transport chain as compared to others, chloroplasts are the main source of O_2 and H_2O_2. These super oxides are converted into hydrogen peroxide through the action of the enzyme SOD, also known as superoxide dismutase. Furthermore, hydrogen peroxide is responsible for the production of hydroxyl radicals. When xanthine oxidase is present in the peroxisomal matrix, it generates O· radicals that cause xanthine and hypoxanthine to corrode into uric acid (Gray and Brady 2016; Malhi et al. 2021).

16.5 Strategies to Combat Climate Change in Agriculture

Climate-smart agriculture (CSA) is an effective approach to combat the drastic impacts of climate change. Climatic fluctuations and increasing food demand, CSA seeks to revive the agriculture extension and climate accessibility to meet growth targets. To reduce or eliminate greenhouse gases (GHGs) and resilience, CSA projects must be prepared to handle trade-offs and interactions between crop yield, adaptation, and mitigation measures. It is necessary to adapt existing crop management techniques into more climate-resilient methods to mitigate effects (Al-Kilani 2024; Ren et al. 2023). To enhance agriculture productivity practices such as crop rotation, fertilization methods, seed selection, germination techniques, and irrigation systems must be incorporated. (Table 16.1) describes adaptations for the agricultural sector to combat climate change with their efficiency and income benefits.

To forecast climatic variability different crop models have been utilized according to limited regional scale, associated with remote sensing, relational layers of biophysical and socio-economic elements. It is now necessary to construct an effective operational platform of advanced yield forecasting, on a national scale, possibly on operational/integrated clusters on state levels (Kalra and Kumar 2019).

16.5.1 Cultural Techniques

Farmers use various effective strategies including adjusting planting and harvesting dates, selecting crops with shorter growing seasons, rotating crops, using irrigation methods, and modifying cropping plans. All of these strategies are highly advantageous for crop adaptability under climatic stress situations. Changes in sowing timing, the use of cultivars resistant to drought, and advanced crop development are significant approaches to mitigate climate-associated risks and enhancing crop capacity for adaptation to ensure the safety and security of the food supply. To combat weather pressures, it is crucial to choose the right sowing time, planting density, and watering practices. Moreover, fertilizers are essential for crop enhancement to cope with global warming and survival of plants. Fertilizers not only provide plants with essential nutrients and energy but also keep the soil fertile and productive boosting crop growth (Raza et al. 2019; Kumar 2016). The schematic representation of climate change's impact on crop production and their response is shown in Fig. 16.2.

Table 16.1 Climate-smart agricultural adaptation to combat climate change

CSA Technology	Country	Crop	Enhanced economic benefit	Efficiency improvement	References
Site-specific nutrient management	Philippines	Rice	106 US$/ha	The partial factor productivity of nitrogen-enhanced	Pampolino et al. (2007)
	India	Rice	168 US$/ha		
	Vietnam	Rice	34 US$/ha		
Zero tillage bed furrows	Pakistan	Rice–Wheat	–	Increased fertilizer use efficiency, higher water productivity, irrigation water saving	Latif et al. (2013)
Zero tillage	Haryana	Wheat	Higher net income	Lower cultivation cost and Enhanced production by 1.88%	Tripathi et al. (2013)
Drought tolerant varieties	Semi-arid tropics of India	Groundnut	Reduction in variable cost upto 17%	Lower variability in yield, Increase in yield by 23%, increased share of risk benefits in total benefits	Birthal et al. (2012)
Stress-tolerant crop varieties	Nyando basin of Kenya	Multiple crops and livestock	Increased in household income (83% from crops and 76% through livestock)	Improved livestock breeds, Increased household income leading to household asset accumulation and investment	Ogada et al. (2020)
Laser Land leveling	Pakistan	Wheat	INR 23,250/acre	Minimized irrigation time and saving of 21% irrigation water	Wagan et al. (2015)
Drip irrigation	India	Eggplant	Increased net return up to 54%	Reduction in use of water fertilizer and electricity	Narayanamoorthy et al. (2018)
Direct-seeded rice	India	Rice	16% increase in household income	Reduced irrigation and preparation costs	Mishra et al. (2017)

(continued)

Table 16.1 (continued)

CSA Technology	Country	Crop	Enhanced economic benefit	Efficiency improvement	References
Improved crop varieties	Indo-Gangetic plains of India	Cropping system of rice–wheat	INR 15,712/ha/year	Improved net return	Khatri-Chhetri et al. (2016)
Site-specific nutrient management	Plains of upper Gangetic	Wheat	INR 68,980/ha over FFP	Yield increased by 29%	Singh et al. (2015)
Drip irrigation	India	Okra	INR 72,711/acre	Less cultivation cost, saving of electricity and irrigated water	Narayanamoorthy and Devika (2018)

16.5.2 Conventional Techniques

Genetic divergence analysis, a key component of inbreeding is used to access polymorphism, inbreeding, and recombination to help achieve plant perfection. Genetic divergence studies significantly benefit the development of novel cultivars based on genomic distance and similarity. Landraces are an efficient source for determining genetic information. For instance, a wheat landrace stored in a data bank holds greater genetic diversity providing a valuable foundation for stress resistance due to its adaptability to diverse environmental conditions (Habib-ur-Rahman et al. 2022).

16.5.3 Genomic Strategies

16.5.3.1 Omics-Led Breeding and Marker-Assisted Selection (MAS) Technique

Breeding programs are integrated with genomic techniques in several crops to accomplish multi-trait assembly. Genetics and transcriptome analysis are utilized to identify phenotypes under various environmental variation connections. Genomics allows for the investigation of molecular pathways underlying resistance to abiotic stress. These methods support the creation of climate-smart crops for increased output and yield in response to changing climatic conditions. High throughput sequencing and phenotyping enable genomic-led breeding to identify various stresses that threaten crop yield. Despite different biotic and abiotic challenges present, molecular plant breeding is a crucial strategy for increasing agricultural yield and productivity. Marker-assisted selection (MAS) promotes rapid breeding and plays a critical role in improving agricultural productivity and crop traits (Abou Seeda et al. 2021).

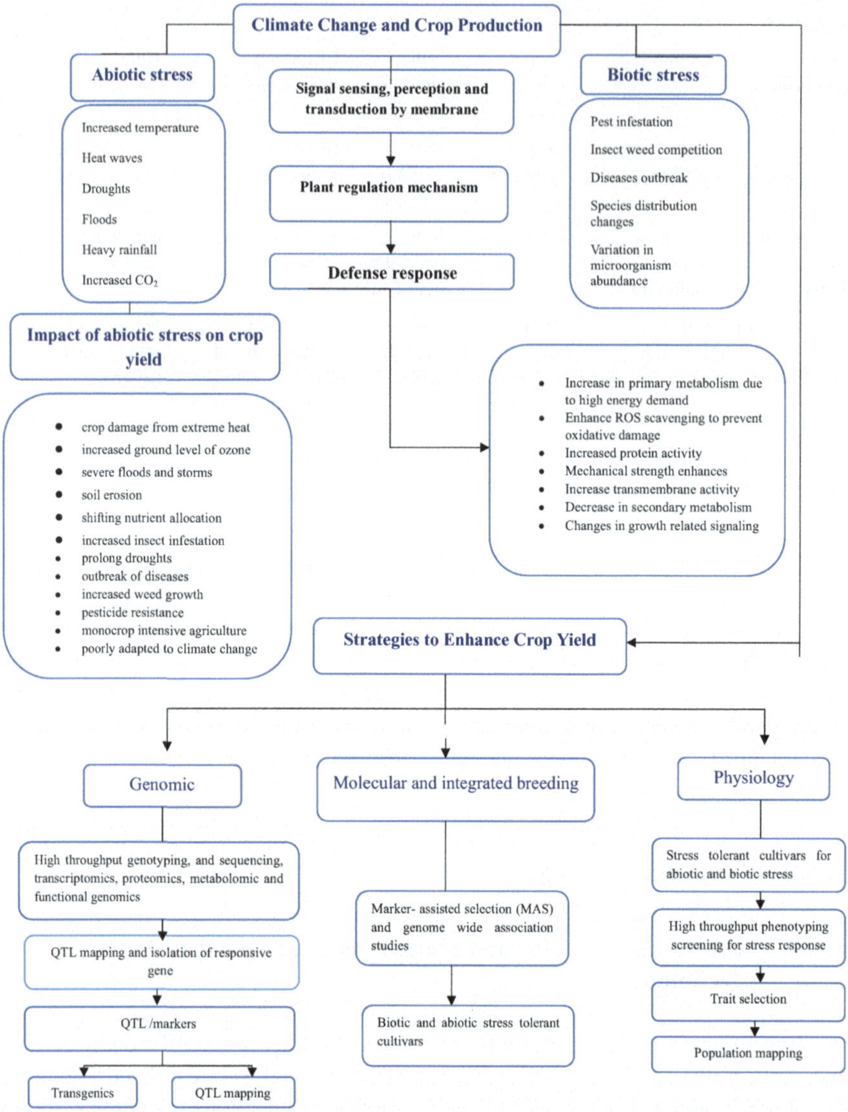

Fig. 16.2 Schematic representation of impact of climate on crop production; plant response to stress condition and strategies to combat adverse climatic effects on agriculture

The tremendous advancements in crop genomics have led to the discovery of DNA markers useful for marker-assisted breeding. The development of new sequencing techniques has reduced the difficulty of studying genetic variability and enabled DNA polymorphism recognition, notably single nucleotide polymorphism (SNP) markers. Exact QTL mapping precision increased generally after 10–30 cm to greater than 1 cm. To find the QTLs associated with drought stress, durum wheat (*Triticum turgidum* L.) and bread wheat (*Triticum aestivum* L.) were investigated for the marker selection method. Barley is grown on a variety of different types of land all over the world, but drought conditions have a significant impact on it globally. Two unique barley cultivars that responded to drought conditions quite differently were chosen for QTL mapping and a double haploid. QTL mapping study was conducted for malting features. A QTL study revealed that the MSA is particularly responsible for genomic regions controlling the malting feature, which can be useful (Müller et al. 2015).

16.5.3.2 For Stress Tolerance; Genome-Wide Association Studies (GWAS)

Genome-wide association studies (GWAS) are an effective method for identifying allelic variations related to specific traits by analyzing the full set of genetic variants in various crop varieties. GWAS is widely used in plants to address biotic and abiotic stressors. Heat tolerance, salt tolerance, and drought tolerance have all been studied using GWAS. Certain areas influencing proline accumulation were identified after linking SNPs from genotypic and phenotypic data were investigated. In Sorghum bicolor, number of stress-tolerant genes related to heat and cold stressors were investigated. Analyses of genotyping and phenotyping were done using GWAS. Thirty SNPs were discovered for genes involved in the development of anthocyanins and glucose absorption, that strongly correlated with frost stress during the growth phase of sorghum seedling. Similarly, 12 SNPs for heat stress were identified in seedlings, regulated by genes involved in sugar metabolism and ion transport mechanisms (Raza et al. 2019).

16.5.3.3 To Enhance Crop; Genome Selection (GS)

Genotyping sequencing and SNP along with diversity array technology (DArT) GBS, are the most efficient tools used in genome selection (GS). These studies demonstrate the effectiveness of GS in crop breeding. Genomic selection strategies have been widely developed in wheat to identify germplasms with the highest capacity for climate change adaptation (Nasir et al. 2020).

16.5.3.4 For Stress Tolerance; Genetic Engineered Plants

Research into stress-responsive transcription factors (TFs) can be used to develop agricultural cultivars resistant to stress. In genetically modified crops, these TFs can regulate the phenotypes of genes linked to diverse stressors. To combat biotic and abiotic stressors, genetic engineering has created a wide variety of transgenic plants. In comparison to natural plants, these genetically modified plants exhibit notable climate tolerance. Several transcription factors (TFs), including the AP2/ERFBP group, are acknowledged as being plant-specific TFs. AP2/ERFBP TFs control many plant growth pathways and play a role in physical and biological stress responses. In response to stressors, DREB TFs have been studied in a variety of plant species (wheat, barley, tomato, soybean, rice, maize, and Arabidopsis). Many studies have focused on the regulating mechanism of DREB1 in rice and Arabidopsis under cold stress, whereas DREB2 responds to drought, salinity, and high temperature stressors. As a result of biotic and abiotic challenges, NAC TFs play a crucial role in a variety of activities, including flower development, and plant regulation mechanisms regarding cell division and adaptation. As an extensive plant variety with genetic sequencing such as rice (151), Arabidopsis (117), maize (152), and soybean (152), numerous NAC TFs were found (Raza et al. 2019; Kumar 2016; Kang et al. 2009).

16.5.3.5 Genome Editing Strategies

By using sequence-specific nucleases, Genomic Editing (GE) is the most effective method for modifying plant genomes. GE holds unmatched potential for crop development, enhancing food security, and promoting climate-smart agricultural practices. The GE technologies have had a significant impact on plant breeding techniques. Researchers are now looking into new methods for precise, quick modifications of crop genomes to increase crop production and protect crops from various challenges. The site-specific endonucleases utilized in genome editing technology include CRISPR-Cas9, transcription activator-like effector nucleases (TALENs), and zinc-finger nucleases (ZFNs). In the genome editing tools, the CRISPR/Cas9 is an evolving and advanced GE method compared to TALENs and ZENs due to great precision, reasonable and multiple site-specific editing inside the genome (Shefali et al. 2021). Agricultural Improvement Using the CRISPR/Cas9 System Modern genome editing techniques like CRISPR/Cas9 are based on the prokaryotic defense system that is activated by type II RNA organization and protects prokaryotes against invading viruses. CRISPR-based approaches were extensively used in plant genomes in contrast to other genome editing tools, i.e., TALENs/ZFNs. The CRISPR/Cas9 tool is evolving into an ecologically benign method for genome editing, offering non-transgenic solutions to enhance food security and combat climatic extremes (Ogundeji et al. 2022).

16.6 Development of Crop Model to Sustain Food Security

Climate change poses a significant threat to regional food availability due to the complex interactions within agricultural systems. Therefore, it is need of the time to develop crop simulation models to monitor climate and take relative strategic plans to improve crop production. General Circulation Models (GCMs) and Representative Concentration Pathways (RCPs) are extensively utilized to quantify the negative effects of climate change on agriculture and generate future weather data (Habib-ur-Rahman et al. 2022). Crop models can predict crop production based on changes in CO_2 levels and heat intensity at different growth stages.

Currently, advanced technologies for evaluating the effect of climate change on crop and livestock production are limited to modeling approaches. Modeling tools are being widely used across the globe to assess the effects of climate change on agriculture, including crop and livestock production. To understand the complete agricultural system and the impact of future climate on farmers' livelihoods, we adopted the RAPs methodology, which combines crop, economic, and climate modeling. After comparing different methods, the Agricultural Model Inter-comparison and Improvement Project (AgMIP) was the most suitable method among all (Crous 2019).

Crop modeling is an essential tool for achieving food security as it allows for a dynamic simulation of crop growth by integrating various processes using computers. These models provide a transparent surrogate for real crops and help in understanding the factors affecting climate-linked with crop yield. By simulating how crops respond to changes in environmental factors such as CO_2 levels, temperature, and crop management practices, crop models can forecast future farm-level productivity and evaluate the gap in actual and predicted yield estimation. It also evaluated the different management substitutions and environmental impacts, respectively. Simulation models provide a non-destructive and time-efficient method for estimating natural resource status and testing hypotheses about plant growth processes (Kumar 2016; Malhi et al. 2021; Abbas 2022).

16.7 Conclusion

Anthropogenic activities compromise the natural environment as consequence extreme climatic events disturb ecosystem functioning worldwide. Agricultural productivity is limited by a number of parameters, the vital are biotic and abiotic stresses. Major climatic variation includes intense heat waves, modification in rainfall patterns, temperature, increase in sea level, rising atmospheric CO_2 concentration, variation in microorganism abundance, outbreak of diseases, and pest and weed infestation. Meeting the growing demand for food will require a 60% increase in wheat production by 2050, posing a significant challenge. There have been probabilities that the frequency of extreme climatic events would be enhanced in the near future. The development of advanced mitigation and adaptative mechanisms is necessary to create climate-smart, resilient agriculture that enhances crop productivity. Crop modeling can also contribute to crop improvement programs by identifying suitable future management practices, understanding genotype-environment interactions, and measuring climate change threats to crop production.

References

Abbas S (2022) Climate change and major crop production: evidence from Pakistan. Environ Sci Pollut Res 29(4):5406–5414

Abou Seeda MA, Abou El-Nour EAA, Yassen AA, Zaghloul SM (2021) Climatic changes and their effects on physiological and nutritional status of crop plants: importance and strategies. A review. Middle East J Agric Res 10(3):866–936

Ahmad Z, Anjum S, Waraich EA, Ayub MA, Ahmad T, Tariq RMS, Ahmad R, Iqbal MA (2018) Growth, physiology, and biochemical activities of plant responses with foliar potassium application under drought stress: a review. J Plant Nutr 41:1734–1743

Al-Kilani MR (2024) Agricultural land measures for climate change adaptation in arid regions: can the farmers do it alone? J Arid Agric 10:82–93

Andrianarimanana MH, Qin Z, Tanteliniaina MFR (2023) The impact of environmental policy on the global consumption of climate-friendly crops: evidence from crops reducing carbon emissions. Environ Dev 45(100774):1–14

Birthal PS, Nigam SN, Narayanan AV, Kareem KA (2012) Potential economic benefits from adoption of improved drought-tolerant groundnut in India. Agric Econ Res Rev 25(1):1–14

Chandio AA, Zhang H, Akram W, Sethi N, Ahmad F (2024) Climate change and crop production nexus: assessing the role of technological development for sustainable agriculture in Vietnam. Int J Clim Change Strat Manag 16(2):177–200. https://doi.org/10.1108/IJCCSM-11-2022-0138

Crous KY (2019) Plant responses to climate warming: physiological adjustments and implications for plant functioning in a future, warmer world. Am J Bot 106(8):1–3

FAO, FAOSTAT (2017) Food agriculture. Organization. United Nations

Gray SB, Brady SM (2016) Plant developmental responses to climate change. Dev Biol 419:64–77

Grymshi D, Crespo-Cebada E, Elghannam A, Mesías FJ, Díaz-Caro C (2022) Understanding consumer attitudes towards ecolabeled food products: a latent class analysis regarding their purchasing motivations. Agribusiness 38(1):93–107

Habib-ur-Rahman M, Ahmad A, Raza A, Hasnain MU, Alharby HF, Alzahrani YM, Bamagoos AA, Hakeem KR, Ahmad S, Nasim W, Ali S, Mansour F, Sabagh ELA (2022) Impact of climate change on agricultural production; issues, challenges, and opportunities in Asia. Front Plant Sci 13(925548):1–22

Kalra N, Kumar M (2019) Simulating the impact of climate change and its variability on agriculture. In: Syed SM (ed) Climate change and agriculture in india: impact and adaptation, pp 21–28 (Chapter 3)

Kang Y, Khan S, Ma X (2009) Climate change impacts on crop yield, crop water productivity and food security: a review. Prog Nat Sci 19:1665–1674

Khatri-Chhetri A, Aryal JP, Sapkota TB, Khurana R (2016) Economic benefits of climate-smart agricultural practices to smallholder farmers in the Indo-Gangetic plains of India. Curr Sci 110:251–1256

Kumar M (2016) Impact of climate change on crop yield and role of model for achieving food security. Environ Monit Assess 188(465):1–14

Latif A, Shakir AS, Rashid MU (2013) Appraisal of economic impact of zero tillage, laser land levelling and bed-furrow interventions in Punjab, Pakistan. Pak J Eng Appl Sci 13:65–81

Malhi GS, Kaur M, Kaushik P (2021) Impact of climate change on agriculture and its mitigation strategies: a review. Sustainability 13(1318):1–21

Mishra AK, Khanal AR, Pede VO (2017) Economic and resource conservation perspectives of direct seeded rice planting methods: evidence from India. In: Proceedings of the agricultural and applied economics association's 2017 AAEA annual meeting, Chicago, IL, USA

Mulagha-Maganga A, Chiwaula LS, Kambewa P, Ngaiwi ME (2024) A Ricardian analysis of the economic impacts of climate change on agricultural production in the low-income agrarian economy: estimates from Malawi's 2010–2019 LSMS longitudinal data. J Agric Food Res 15:100995. https://doi.org/10.1016/j.jafr.2024.100995

Müller C, Elliott J, Chryssanthacopoulos J, Deryng D, Folberth C, Pugh TAM, Schmid E (2015) Implications of climate mitigation for future agricultural production. Environ Res Lett 10(12):1–12
Narayanamoorthy A, Devika N (2018) Economic and resource impacts of drip method of irrigation on okra cultivation: an analysis of field survey data. J Land Rural Stud 6(1):15–33
Narayanamoorthy A, Bhattarai M, Jothi P (2018) An assessment of the economic impact of drip irrigation in vegetable production in India. Agric Econ Res Rev 31(1):105–112
Nasir IR, Rasul F, Ahmad A (2020) Climate change impacts and adaptations for fine, coarse, and hybrid rice using CERES-rice. Environ Sci Pollut Res 27:9454–9464
Nguyen CT, Scrimgeour F (2022) Measuring the impact of climate change on agriculture in Vietnam: a panel Ricardian analysis. Agric Econ 53(1):37–51
Ogada MJ, Rao EJ, Radeny M, Recha JW, Solomon D (2020) Climate-smart agriculture, household income and asset accumulation among smallholder farmers in the Nyando basin of Kenya. World Dev Perspect 18:100203
Ogundeji AA, Danso-Abbeam G, Jooste J (2022) Climate information pathways and farmers' adaptive capacity: insights from South Africa. Environ Dev 44(100743):1–18
Pampolino MF, Manguiat IJ, Ramanathan S, Gines HC, Tan PS, Chi TTN, Buresh RJ et al (2007) Environmental impact and economic benefits of site-specific nutrient management (SSNM) in irrigated rice systems. Agric Syst 93(1–3):1–24
Raza A, Razzaq A, Mehmood SS, Zou X, Zhang X, Lv Y, Xu J (2019) Impact of climate change on crops adaptation and strategies to tackle its outcome: a review. Plants 8(34):1–29
Ren X, He J, Huang Z (2023) Innovation, natural resources abundance, climate change and green growth in agriculture. Resour Policy 85:103970. https://doi.org/10.1016/j.resourpol.2023.103970
Shah WUH, Lu Y, Liu J, Rehman A, Yasmeen R (2024) The impact of climate change and production technology heterogeneity on China's agricultural total factor productivity and production efficiency. Sci Total Environ 907:168027
Shefali, Sankhla MS, Kumar R, Sonone SS (2021) Phytomicrobiome studies for combating the abiotic stress. Biointerface Res Appl Chem 11(3):10493–10509
Singh VK, Shukla AK, Singh MP, Majumdar K, Mishra RP, Rani M, Singh SK (2015) Effect of site-specific nutrient management on yield, profit and apparent nutrient balance under pre-dominant cropping systems of Upper Gangetic Plains. Indian J Agric Sci 85(3):335–343
Syeda A, Raza T, Bhatti TT, Eash NS (2022) Climate impacts on the agricultural sector of Pakistan: risks and solutions. Environ Chall 6(100433):1–5
Tong Q, Yuan X, Zhang L, Zhang J, Li W (2024) The impact of livelihood capitals on farmers' adoption of climate-smart agriculture practices: evidence from rice production in the Jianghan Plain, China. Clim Risk Manag 43:100583. https://doi.org/10.1016/j.crm.2023.100583
Tripathi RS, Raju R, Thimmappa K (2013) Impact of zero tillage on economics of wheat production in Haryana. Agric Econ Res Rev 26(1):101–108
Ullah S (2017) Climate change impact on agriculture of Pakistan-A leading agent to food security. Int J Environ Sci 6:76–79
Wagan SA, Memon QUA, Wagan TA, Memon IH, Wagan ZA (2015) Economic analysis of laser land leveling technology water use efficiency and crop productivity of wheat crop in Sindh, Pakistan. J Environ Earth Sci 5:21–25
Wu G, Xie Y, Li H, Riaz N (2022) Agricultural ecological efficiency under the carbon emissions trading system in China: a spatial difference-in-difference approach. Sustainability 14(8):470
Xu Y, Chu C, Yao S (2021) The impact of high-temperature stress on rice: challenges and solutions. Crop J 9:963–976

Dr. Amtul Bari Tabinda is a tenured professor at the Sustainable Development Study Center, Government College University, Lahore. She holds a Ph.D. in Chemistry from the University of the Punjab and has over 36 years of teaching and research experience. Dr. Tabinda has completed more than 10 research projects and published over 137 articles in high-impact journals. Her work focuses on biogas production, water quality, and wastewater treatment, and she has made significant contributions to the field of environmental sustainability.

Ms. Rimsha Javed is an M.Phil. in Environmental Science from Government College University, Lahore, where she received a gold medal for securing first position. Her research focuses on the health risks of heavy metals in polystyrene containers. She has published over seven papers in international journals and worked as a research associate at GCU Lahore.

Ms. Javairia Ansar is an M.Phil. in Environmental Science from the Sustainable Development Study Center, Government College University, Lahore where she has secured second position. Her thesis research explores biodiesel production from fish waste oil using organic waste catalysts. She holds a bachelor's degree in Environmental Science from the same institution and currently works as a research associate on a Pakistan Science Foundation (PSF) funded project. She has published over four research papers in international journals.

Dr. Adeel Mahmood is an associate professor and Chairperson of the Department of Environmental Science at Government College Women University, Sialkot. With a post-doctorate in environmental biology, Dr. Mehmood has published over 113 articles in international refereed journals, focusing on ecotoxicology, micropollutants, microplastics, and heavy metals. His research portfolio includes significant contributions to understanding environmental pollutants and their impact on ecosystems.

Dr. Abdullah Yasar is an esteemed academic with over 30 years of experience in environmental science and engineering. He holds a Ph.D. in Environmental Science from the University of the Punjab, where he received the prestigious HEC Merit Scholarship. He is currently a professor and campus engineer at the Sustainable Development Study Center (SDSC), Government College University, Lahore. Dr. Yasar has published over 130 papers in international refereed journals and conference proceedings. His expertise includes industrial wastewater treatment, ozone-based treatment plants, environmental degradation, and sustainable energy solutions. Dr. Yasar has collaborated on numerous high-impact projects funded by organizations such as HEC, GIZ, and the Pakistan Science Foundation.

Chapter 17
Plant Diversity and Habitat Composition of AlUla Oasis: Insights from Northwest Saudi Arabia

Monier M. Abd El-Ghani, Faten Zubair Filimban, Zaheer-ud-din Khan, Mirza Barjees Baig, and Abdulmalek A. Al Shaikh

Abstract AlUla oasis in northwestern Saudi Arabia is a historic and cultural landmark known for its unique floristic diversity, rich cultural heritage, and environmental significance. For thousands of years, it has served as a critical hub along the ancient Incense Road, linking civilizations, and supporting thriving ecosystems. This review assesses AlUla's floristic composition and habitat diversity, highlighting its distinctive sandstone mountains, volcanic formations, and lush oases that support an array of plant and animal species, some endangered. Traditional agricultural practices, such as date palm cultivation and ancient qanat irrigation, reveal the resourcefulness of its inhabitants and underscore the oasis's ecological importance. Our analysis identifies 58 plant species across 30 families, primarily in desert ecosystems, with Fabaceae, Asteraceae, and Poaceae as the most represented families. Conservation initiatives by the Royal Commission for AlUla (RCU) aim to balance the development of sustainable tourism and

M. M. Abd El-Ghani
Cairo University Herbarium, Department of Botany and Microbiology, Faculty of Science, Cairo University, Giza, Egypt
e-mail: mmounir@sci.cu.edu.eg

F. Z. Filimban
Plant Taxonomy of the Arabian Peninsula, Division of Botany, Department of Biological Sciences, College of Sciences, King Abdulaziz University, Jeddah, Saudi Arabia
e-mail: ffilimban@kau.edu.sa

Z. Khan
Department of Botany, GC University Lahore, Lahore, Pakistan
e-mail: zaheeruddin@gcu.edu.pk

M. B. Baig (✉) · A.A. Al Shaikh
Prince Sultan Institute for Environmental, Water and Desert Research, King Saud University, Riyadh, Kingdom of Saudi Arabia
e-mail: mbbaig@ksu.edu.sa

A. A. Al Shaikh
e-mail: aasheikh@ksu.edu.sa

M. Behnassi et al. (eds.), *Food Systems and Biodiversity in the Context of Environmental and Climate Risks*, https://doi.org/10.1007/978-3-031-89167-0_17

agriculture with biodiversity preservation. By comparing AlUla's ecosystems to similar arid regions, this review emphasizes the need for continued conservation and sustainable land use to protect this unique desert oasis, fostering a balance between cultural preservation and environmental stewardship.

Keywords AlUla Oasis · Northwestern Saudi Arabia · Floristic composition · Habitat diversity · Desert ecosystems · Biodiversity conservation · Endangered species · Sustainable tourism · Royal commission for AlUla (RCU) · Cultural heritage

17.1 Historical and Cultural Background

Saudi Arabia occupies about 80% of the Arabian Peninsula's surface area. However, most of this area is barren desert, except for the southwestern highlands, which receive ample rainfall and have a mild climate that supports plant life (Masrahi 2012). Known for its beautiful terrain, Saudi Arabia's northwestern area is home to the historic and cultural oasis of AlUla. It has a long history of human habitation and was a significant trading hub for goods traveling throughout the Arabian Peninsula. Situated at the crossroads of several continents, AlUla served as a meeting place for civilizations that have left behind a rich legacy (Burckhardt 2013). It was a market city due to its location along the Incense Road, a network of routes that facilitated the trade of silk, spices, and other luxury goods via the kingdoms of Axum, Arabia, Egypt, and India (Engler 1895).

AlUla region is well-known for its traditional methods of water management, irrigation, and land usage, all of which have contributed to the development of a lush and fruitful area that testifies to the ingenuity of the people who have lived there for ages (Burckhardt 2013). It is home to many natural and cultural treasures, including the Mada'in Salih UNESCO World Heritage Site (Engler 1895). Hegra (Mada'in Salih), where the Nabataeans created rock-cut tombs with imposing facades similar to those at the Jordanian site of Petra, was their principal center in AlUla.

Hegra, the first UNESCO World Heritage Site in Saudi Arabia, is the most well-known and renowned location in AlUla (Osman et al. 2014). An ancient city covering 52 ha, Hegra served as the main southern metropolis of the Nabataean Kingdom. It has almost 100 well-preserved tombs with elaborate facades carved into sandstone outcrops. After the Romans defeated the Nabataeans in 106 CE, Hegra is thought to have been their southernmost outpost (Sawicki 2016).

AlUla was established in the sixth century BCE as an oasis in a desert valley with fertile soil and abundant water. It was located along the Incense Road, facilitating the trade of luxury goods like silk, spices, and other items through the kingdoms of Arabia, Egypt, India, and Axum. AlUla stands on the site of the biblical city of Dedan but was founded with the ancient North Arabian kingdom of Lihyan, which ruled from the 5th to the second century BCE (Rashad 2024). The oasis's earlier history includes different eras. The Dedanite kingdom spanned the seventh and sixth centuries BCE. Dedan is mentioned in the Harran Inscriptions, which describe how Nabonidus, the Babylonian king, conquered Tayma, Dedan, and Yathrib (Medina), three cities in northern Arabia, in 552 BCE or possibly a little later (Ghazanfar and Fisher 1998). The kingdom is believed to have been hereditary from the beginning of the fifth century BCE. The following 400 years, up until around 100 BCE, were ruled by the Kingdom of Lihyan. The Nabataeans then ruled the area at least until the Romans invaded their capital Petra in 106 CE. Hegra, now known as Mada'in Salih, served as the Nabataeans' secondary capital. As a result, Hegra, located 22 km north of AlUla, became the new center of power in the area.

The Royal Commission for AlUla (RCU) was established in July 2017 to protect and develop AlUla, a region of outstanding natural and cultural significance in northwestern Saudi Arabia (Engler 1895). In order to maintain ecological balance and preserve the region's biodiversity, it is crucial to protect the natural environment. The commission also creates plans for heritage conservation and preservation (Alam et al. 2017). The preservation of AlUla region is essential for achieving the Saudi Green Initiative, which aims to restore the land and halt desertification in the area (Zohary 1982). To accomplish this, it is essential to protect the area's natural environment.

The RCU has undertaken various development projects in the region, including five districts planned for preservation priorities under Saudi Arabia's Vision 2030 program. These districts include AlUla Old Town, Dadan, Jabal Ikmah, Nabataean Horizon, and Hegra Historical City, which are part of a larger development plan for AlUla that aims to transform the city into a world-class tourist destination while preserving its cultural and historical heritage (High Commission for the Development of Arriyadh 2014).

A variety of desert ecosystems, including sandstone mountains, canyons, valleys, and oases, can be found in AlUla region. A variety of plants and animals, including several rare and endangered species, occur there (Rashad 2024).

The French priests Antonin Jaussen and Raphael Savignac, who visited the area three times in 1907, 1908, and 1910, conducted one of the most thorough examinations of the area. They gathered numerous Lihyanite, Minaean, Thamudic, and Nabataean inscriptions while studying the ruins at Hegra and Dedan. Their work became the foundation for all subsequent analysis and investigation into the region's history (Mandaville 2013).

There are several plant and animal species in AlUla. Date palms, acacia trees, and tamarisks can all be found in AlUla Valley. Numerous bird species, such as the Arabian partridge and the desert owl, are also found in the area. The northernmost range of the Arabian partridge is found in the recently established Harrat Uwayrid Biosphere Reserve (HUBR) by the RCU in northwest Saudi Arabia (Alam et al. 2017). The restriction of hunting and the creation of conservation and reintroduction plans for threatened and endemic species, like the Arabian partridge, are part of the HUBR mandate. Although it has been presumed that the number of Arabian partridges is steady across their range (Migahid 1978), accurate data on the population trends and conservation status are not accessible because neither it nor many other endemic species in this area have a monitoring program.

AlUla is currently undergoing a significant development project with the goal of making the region a top tourist destination. New hotels, resorts, and tourist attractions will be built as part of AlUla Vision project, which also includes the preservation and restoration of the area's numerous historic sites.

This review aims to shed light on the floristic diversity and habitat composition of AlUla oasis in comparison with other comparable oases in Saudi Arabia and neighboring countries within the arid region.

17.2 Environmental Features

17.2.1 Location and Landforms

AlUla region covers an area of around 23,301 km^2 and is situated between 26° 37′ N and 37° 55′ E, with an elevation range of approximately 678.5 to 694.9 m above sea level (Fig. 17.1). The region is distinguished by a large, arid environment with sand dunes and dry mountains (Burckhardt 2013; Engler 1895; Ghazanfar and Fisher 1998; Rashad 2024). There are five key formations that make up the region's landforms. The most recognizable are the sandstone mountains, composed of red and pink sandstone that has been weathered over millions of years to generate stunning formations like cliffs and canyons. AlUla's sandstone formations bear witness to the area's geological past.

The sandstone was produced over 485 million years ago, during the Cambrian Period (Ghazanfar and Fisher 1998). Its remarkable geology has been further shaped by tectonic activity, which has produced volcanic landscapes, basalt

Fig. 17.1 Location map of AlUla region

plateaus, and sandstone massifs (Ghazanfar and Fisher 1998). Due to these geological features, the valley acts as a water catchment basin with a surface area of around 700 km^2, allowing it to flow southward into the villages of AlUla (Ghazanfar and Fisher 1998). Jabal Al-Fil (Elephant Rock) and Jabal Ikmah (Heritage Mountain), both located in the Mada'in Salih archaeological site, are two of the most well-known sandstone mountains in AlUla. The UNESCO World Heritage Site of Mada'in Salih is located on the sandstone mountain known as Jabal Mada'in Salih. This ancient city was once a significant trading center for the Nabataeans (Osman et al. 2014).

Ancient lava flows produced the volcanic rocks. They stand out clearly from the nearby sandstone and are often black or dark crimson in color. Harrat Khaybar, located south of AlUla, is home to some of the most noticeable volcanic rocks. The oases are desert-enclosed regions with rich flora. These oases, which receive their water from underground reservoirs, serve as essential habitats for both plants and animals. The Mada'in Salih oasis, situated in the heart of the Hegra archaeological site, is the most well-known oasis in AlUla. Deep valleys known as canyons were formed by water erosion. These canyons, which are frequently narrow and steep, create a striking backdrop for the desert surroundings. The As-Siq and Jabal Faynan are two of the most renowned canyons. Large areas are occupied by sand and gravel deserts devoid of vegetation. The Nafud Desert, situated north of AlUla, is the most well-known desert in the region. Extreme geological diversity is present in the inland desert plains, mountain ranges, and valleys (Rashad 2024).

17.2.2 Climate

Arid and dry conditions typical of a desert region characterize the climate in AlUla region (Osman et al. 2014; Rashad 2024). Summers are long, hot, dry, and clear, with average daily high temperatures exceeding 34.4 °C. August is the hottest month of the year in AlUla, with average high temperatures of 37.7 °C and lows of 21.7 °C (Burckhardt 2013). Winters are brief, chilly, dry, and primarily clear, with average daily high temperatures below 23.9 °C. In AlUla, January is the coldest month, with average lows of 3.9 °C and highs of 21.1 °C (Burckhardt 2013). The district averages 28.6 °C throughout the year, which is 1.55% warmer than the Saudi Arabian average (Rashad 2024). Annual rainfall amounts to roughly 355.6 mm, and the average humidity is 47% (Engler 1895).

17.2.3 Land Use and Human Activities

In AlUla, tourism dominates land use. Numerous historic archaeological sites can be found in the area, notably Mada'in Salih, a UNESCO World Heritage Site. The Royal Commission for AlUla (RCU) is promoting the area as a top travel destination and has significant plans for infrastructure investment in the tourism industry.

Agriculture is also a significant land use in AlUla. The region is experiencing a growing water shortage, and there is a limited quantity of land suitable for agriculture. There are several oases in the area where water is available for irrigation. Dates are a staple food in the region and are grown for local consumption and export (Burckhardt 2013; Ghazanfar and Fisher 1998). The mango cultivation area in AlUla covers around 255 ha (630 acres), producing 1600 tons annually (Burckhardt 2013). Farmers in AlUla are growing Arabic Moringa, a crop that is in high demand (Osman et al. 2014; Rashad 2024). The crop is being farmed extensively, and farmers are being taught the mechanisms of growing it (Osman et al. 2014; Rashad 2024). AlUla's fertile agricultural land is suitable for various crops across the seasons, especially citrus fruits (Ghazanfar and Fisher 1998). In the past, barley and wheat were grown and consumed in AlUla, as well as lentils and peas (Zohary 1982).

Livestock rearing (sheep, goats, and camels) is a significant land use in AlUla, serving as an important source of food and revenue for the local population. However, due to the growing scarcity of water and grazing pasture, the number of animals is decreasing. Industrial land use is relatively insignificant in AlUla.

In addition to these traditional activities, there are a number of new and emerging human activities in AlUla:

1. **Sustainable Tourism**: The RCU is committed to developing AlUla as a sustainable tourist destination. This means that the development of tourism will be done in a way that protects the region's natural and cultural heritage.
2. **Renewable Energy**: AlUla has a number of renewable energy resources, such as solar and wind power. The RCU is developing these resources to help meet the region's energy needs and reduce its reliance on fossil fuels.
3. **Research and Development**: The RCU is investing in research and development to support the sustainable development of AlUla. This includes research into new ways to conserve water, improve agricultural productivity, and reduce the impact of tourism on the environment.

The human activities in AlUla are constantly evolving. The RCU is committed to developing the region in a sustainable way that benefits the local community and protects the environment.

17.3 Botanical Exploration and Ecosystem Types

17.3.1 History of Botanical Exploration

The earliest recorded botanical exploration of AlUla was conducted by the French explorer Jean-Louis Burckhardt in the early nineteenth century. Burckhardt (2013) was a keen botanist and made a detailed record of the plants he observed in AlUla.

Another early explorer was the German botanist Eduard Glaser. Engler (1895) was particularly interested in the region's desert plants and made extensive botanical collections in AlUla. The twentieth century saw several other botanists explore AlUla, including the British botanists David Ingram and Michael Zohary. Zohary (1982) focused on the region's endemic plants, while Zohary (1982) was interested in the historical uses of plants in the region. Their work has helped to further our understanding of the flora of AlUla.

In recent years, there has been a renewed interest in the botanical exploration of AlUla, due in part to the increasing recognition of the region's importance as a biodiversity hotspot. Gandreau et al. (2022) published an account on the native species for AlUla landscaping, produced for internal RCU purposes to provide resources for landscaping projects. Fifty-eight native plants were included for landscaping and wilderness regeneration. These have been selected based on several criteria to meet the objectives of landscaping and restoration. Together they represent about 20% of the potential plants of AlUla, with a wealth of forms and striking habits embodying the different natural environments (deserts, wadis) and the plants historically grown in the area or needed to regenerate degraded areas in the reserves.

In 2021, AlUla Oasis was designated a UNESCO World Heritage Site, which has helped to raise awareness of the region's unique natural heritage. Today, there are a number of ongoing botanical research projects in AlUla aimed at documenting the region's plant diversity, understanding its ecology, and developing strategies for its conservation.

17.3.2 Ecosystem Types

There are two main ecosystems can be recognized in AlUla region: the desert and the oasis. A detailed description of each ecosystem is given below, and a full floristic list is displayed in Table 17.1. The diverse habitats provide shelter, food, and breeding grounds for a diverse array of species, making AlUla an ecological hotspot.

17.3.2.1 The Desert Ecosystem

The desert ecosystem is characterized by a vast landscape with sparse vegetation, arid mountains, and sand dunes (Burckhardt 2013; Engler 1895; Ghazanfar and Fisher 1998). The main types of sand dunes found in AlUla region are barchan (crescent-shaped), transverse (linear), star, and parabolic (U-shaped). The first two types are the most common, while the latter two are less so. The type of sand dune found in a particular area depends on factors such as wind direction, the amount of sand available, and the slope of the ground. The sand dunes in AlUla are constantly changing shape and size due to the wind, making them a dynamic

Table 17.1 Floristic composition of trees of AlUla Oasis, and their traditional uses

Family	Species	Arabic name (translation)	Arabic name (arabic script)	Traditional uses	References
Arecaceae	*Hyphaene thebaica* (L.) Mart.	Doum Palm (Doum)	الدُوم	**Edible fruit** used for making sweet beverages; **seeds** used in crafts and as prayer beads; **leaves** woven into mats, baskets, and ropes; **wood** used for construction and fuel	(El Amin 1990; Zohary 1982)
	Phoenix dactylifera L.	Date Palm (Nakheel Al-Balah)	نخيل البلح	**Staple food** source; dates consumed fresh or dried; **leaves** used for weaving mats, baskets, and roofing; **trunk** used in construction; **fibers** used for ropes; **sap** sometimes used to make syrup or vinegar	(Barreveld 1993; Zohary 1982)
Capparaceae	*Maerua crassifolia* Forssk.	Bushveld Shepherd's Tree (Al-Hamr)	السَرح	**Fodder** for livestock; **leaves and roots** used in traditional medicine for digestive issues and as a diuretic; **wood** used for fuel	(Ghazanfar 1994)
Fabaceae	*Vachellia gerrardi* var. *najdensis* (Chaudhary) Ragup, Seigler, Ebinger & Maslin	Najd Acacia (As-Sunt Al-Najdi)	السنط النجدي	**Shade provider** in arid areas; **pods and leaves** used as **fodder**; **wood** used for fuel and making tools; **gum** extracted for traditional medicinal uses	(Neuwinger 1996; Wickens 1995)
	Vachellia tortilis (Forssk.) Galasso & Banfi	Umbrella Thorn Acacia (Al-Sammar)	السَمر	**Wood** used for fuel and charcoal; **leaves and pods** as **fodder**; **gum arabic** harvested; **bark and leaves** used in traditional medicine for treating colds and skin ailments	(Fagg and Allison 2004; Wickens 1995)

(continued)

Table 17.1 (continued)

Family	Species	Arabic name (translation)	Arabic name (arabic script)	Traditional uses	References
	Vachellia tortilis subsp. *raddiana* (Savi) Kyal. & Boatwr.	Raddiana Acacia (Al-Sayyal)	السيال	Similar uses as above; **wood** for fuel; **fodder** for livestock; **gum** extraction; **environmental** use in soil stabilization and desertification control	(Fagg and Allison 2004; Wickens 1995)
Moraceae	*Ficus palmata* Forssk. subsp. *palmata*	Fig (Al-Teen Al-Barri)	التين البري	**Edible fruits** consumed fresh; **leaves and latex** used in traditional medicine for treating skin conditions; **shade** tree near dwellings and farms	(Duke 2002; Orwa et al. 2009)
	Ficus salicifolia Vahl	Willow-leaf Fig (Al-Athab)	الأثب	**Edible fruits**; **ornamental** planting; **leaves** may be used in traditional remedies; **shade** provider in agricultural areas	(Orwa et al. 2009; Shah et al. 2018)
Moringaceae	*Moringa peregrina* (Forssk.) Fiori	Moringa (Al-Baan Al-Arabi)	البان العربي	**Seeds** yield **edible oil** (Ben oil) used in cooking and cosmetics; **leaves and pods** consumed as food rich in nutrients; **traditional medicine** uses for various ailments; **wood** used as fuel	(Anwar et al. 2007; Lalas and Tsaknis 2002)
Rhamnaceae	*Ziziphus spina-christi* (L.) Desf.	Christ's Thorn Jujube (Al-Sidr)	السدر	**Edible fruits** eaten fresh or dried; **leaves** used to make **herbal shampoos** and skin treatments; **honey production** from flowers; **wood** used in carpentry and as fuel; **medicinal uses** for coughs and digestive issues	(Abdellaoui et al. 2014; Kirtikar and Basu 1918)

(continued)

Table 17.1 (continued)

Family	Species	Arabic name (translation)	Arabic name (arabic script)	Traditional uses	References
Salvadoraceae	*Salvadora persica* L.	Toothbrush Tree (Al-Araak)	الأراك	**Twigs** used as natural **toothbrushes (Miswak)** for oral hygiene; **leaves and roots** used medicinally for digestive and respiratory conditions; **berries** edible; **oil** extracted from seeds	(Almas 2001; Seshathri and Thiyagarajan 2011)
Sapindaceae	*Dodonaea viscosa* subsp. *angustifolia* (L.f.) J. G. West	Sand Olive (Al-Shath)	الشَث	**Traditional medicine** uses for treating fevers and skin conditions; **leaves** used as antiseptic; **wood** used for tool handles and as fuel; **ornamental** planting for hedges and windbreaks	(Ahmed et al. 2014; Chopra 1956)
Tamaricaceae	*Tamarix nilotica* (Ehrenb.) Bunge	Nile Tamarisk (Al-Athal)	الأثل	**Wood** used in construction and as fuel; **planted** for **windbreaks** and **soil stabilization**; **bark and leaves** used in traditional remedies for digestive issues; **shade** provider in deserts	(Boulos 1983; Danin 1991)

and interesting landscape. This ecosystem is characterized by several species such as *Acacia gerrardii* subsp. *negevensis*, *Acacia raddiana*, *Ziziphus spina-christi*, *Maerua crassifolia*, *Moringa peregrina*, *Tamarix nilotica*, *Haloxylon persicum*, *Lycium shawii*, *Retama raetam*, and *Capparis cartilaginea* (Photos 17.1, 17.2, 17.3, 17.4, 17.5, 17.6, 17.7 and 17.8).

The Sharaan Nature Reserve in AlUla is dedicated to protecting the region's ecosystem and wildlife (Burckhardt 2013). Additionally, the National Tropical Botanical Garden is involved in conservation efforts to protect *Brighamia insignis*, an endangered endemic species to Kaua'i (Engler 1895; Ghazanfar and Fisher 1998; Zohary 1982).

17.3.2.2 The Oasis Ecosystem

An oasis is a green patch excavated in the sterile desert. AlUla oasis has the typical structure of an oasis where vast areas of date-palm groves (*Phoenix dactylifera* L.) occupy the cultivated lands surrounded by escarpments. Agriculture and vegetation in these areas are mainly groundwater-dependent (Abd El-Ghani 1998;

Photo 17.1 *Ziziphus spina-christi*

Photo 17.2 *Phoenix dactylifera*

Photo 17.3 *Capparis cartilaginea*

Photo 17.4 *Hyoscamus muticus*

Photo 17.5 *Cleome droserifolia*

Photo 17.6 *Vachellia tortilis* subsp. *raddiana*

Photo 17.7 *Tamarix nilotica*

Photo 17.8 *Blepharis edulis*

Bornkamm and Kehl 1990). The major source of irrigation water used for cultivation is underground (fossil) water, which appears on the soil surface in the form of naturally flowing springs (some are hot) or is pumped from wells.

A complex irrigation system (qanat in Arabic, or canal) consisting of a network of canals and wells has been in use for centuries in AlUla to provide water for irrigation and daily use. Ain Tidal was one of the most important water springs on which the irrigation system relied in the old town of AlUla (Burckhardt 2013). The wells in AlUla (10–20 m deep) are dug to access the water table (Osman et al. 2014). The wells are wide and can store large amounts of water for long periods. Water is drawn from the wells using animal-powered pulleys or pumps (Osman et al. 2014). The springs or wells were used to irrigate the date palm groves and other crops in the region. The network of canals distributes water from the wells to the date palm gardens (Burckhardt, 2013). The canals are lined with clay and designed to be gravity-fed, with water flowing downhill from the wells to the gardens (Engler 1895).

The qanat system in AlUla has been affected by the destruction and demolition taking place in the region, as modern pumping systems have replaced the traditional qanats. The irrigation system in AlUla is an important part of the region's cultural heritage and has been recognized by UNESCO as an important element of the Arab world's iconic date palm culture (Zohary 1982). Similar patterns of irrigation systems have been described from Feiran Oasis, Sinai Peninsula, Egypt (Abd El-Ghani and Fahmy 1998), and the oases of the Western Desert of Egypt (Abd El-Ghani and Fawzy 2006).

17.3.3 Floristic Composition and Biological Spectrum

Table 17.1 summarizes the flora of AlUla oasis, excluding the floristic composition of date-palm orchards, which needs special future attention. A total of 58 species (49 perennials and 9 annuals) from 30 families of vascular plants were recorded mainly from the desert ecosystem. In terms of species richness, Fabaceae was the largest family with the highest number of species (7 species), followed by Asteraceae (6 species), Poaceae (4 species), and Amaranthaceae, Apocynaceae, Brassicaceae, and Capparaceae (3 species each) (Tables 17.2, 17.3, 17.4 and 17.5).

The dominance of Fabaceae, Asteraceae, and Poaceae reflects a typical desert flora, as has been reported from similar investigations in arid regions, for example, (Al Shaye et al. 2020) in Riyadh region, Saudi Arabia; Al-Mutairi et al. (2016) in the Tabuk region of Saudi Arabia; Salama et al. (2018) from Kharga Oasis in the Western Desert of Egypt; Salama et al. (2018) in the wadi ecosystem of central Eastern Desert, Egypt; Enright et al. (2005) in Kirthar National Park, Sindh, Pakistan; Abd El-Ghani and Fahmy (1998) in Feiran Oasis, Sinai Peninsula, Egypt; and Abd El-Ghani (1998) in a transect between Makkah and Medina in Saudi Arabia. Four genera have more than one species, including *Vachellia* (3 species), and *Ficus*, *Capparis*, *Lavandula*, and *Artemisia* (2 species each).

Table 17.2 Floristic composition of shrubs and subshrubs of AlUla Oasis, and their traditional uses

Family	Species	Arabic name (transliteration)	Arabic name (arabic script)	Traditional uses	References
Amaranthaceae	*Haloxylon persicum* Bunge	White Saxaul (Al-Ghadha)	الغضى	**Wood** used as high-quality fuel and charcoal; **branches** used for fencing and construction; **important** for **soil stabilization** and combating desertification; **fodder** for camels during drought	(Read 1990; Remesh et al. 2023)
	Aerva javanica (Burm.f.) Juss. ex Schult.	Kapok Bush (Al-Tarfh)	الطَرف	**Fibers** from the plant used for stuffing cushions and mattresses; **traditional medicine** uses for treating wounds and infections; **grazed** by livestock; **soil** binder in erosion control	(Skalli et al. 2019)
Anacardiaceae	*Searsia tripartita* (Ucria) Moffett	Sumac (Al-Artan)	العرتن	**Leaves and fruits** used as **spice (sumac)** for culinary purposes; **medicinal uses** include treating digestive issues; **tannins** extracted for use in leather processing; **dye** production from plant parts	(Le Floc'h 1983; Skalli et al. 2019)
Asteraceae	*Artemisia monosperma* Delile	White Wormwood (Al-Ather)	العاذر	**Traditional medicine** for treating stomach ailments, colds, and as an antiseptic; **aromatic** plant used in **incense** and **perfumes**; **grazed** by animals when other forage is scarce	(Hudaib and Aburjai 2006; Sreelatha et al. 2009)

(continued)

Table 17.2 (continued)

Family	Species	Arabic name (transliteration)	Arabic name (arabic script)	Traditional uses	References
Fabaceae	*Retama raetam* (Forssk.) Webb & Berthel.	White Broom (Ar-Ratam)	الرتم	**Roots and stems** used in traditional medicine for skin diseases; **flowers** used in perfumery; **shade** provider in deserts; **nitrogen-fixing** plant improving soil fertility	(Batanouny et al. 1999)
	Senna italica Mill.	Italian Senna (Al-Eshreq)	العشرق	**Leaves and pods** used as a **laxative** in traditional medicine; **treatment** for skin conditions; **cultivated** for medicinal purposes; **grazed** by livestock in some regions	(Akhtar et al. 2000; Wyk et al. 2002)
Polygonaceae	*Calligonum comosum* L'Hér	Arta (Al-Arta)	الأرطى	**Wood** used as fuel; **roots and stems** used in traditional medicine for digestive issues; **important** for **sand dune stabilization**; **provides** shelter and food for desert wildlife	(El-Bana and Al-Mathnani 2009; Hegazy and Amer 2001)
Resedaceae	*Ochradenus baccatus* Delile	Sweet Mignonette (Al-Ghardaq)	القرضي	**Berries** consumed by birds and sometimes humans; **traditional medicine** uses for anti-inflammatory purposes; **roots** used in remedies; **grazed** by livestock; **soil** stabilizer in erosion control	(Al-Qura'n 2008; Lev and Amar 2000)

(continued)

Table 17.2 (continued)

Family	Species	Arabic name (transliteration)	Arabic name (arabic script)	Traditional uses	References
Rhamnaceae	*Ziziphus nummularia* (Burm.f.) Wight & Arn.	Jujube (Al-Sider Al-Barri)	السدر البري	**Fruits** are edible; **leaves** used in traditional medicine for skin diseases; **branches** used for fencing due to thorns; **leaves** used as fodder; **honey production** from flowers	(Kirtikar and Basu 1918; Pareek 1996)
Solanaceae	*Lycium shawii* Roem. and Schult	Desert Thorn (Al-Awsaj)	العوسج	**Fruits** are edible and have medicinal properties; **used** in traditional medicine for diabetes and hypertension; **branches** used for fencing; **grazed** by camels and goats	(Batanouny et al. 1999)
	Withania somnifera (L.) Dunal	Ashwagandha (Al-Abaab)	العبعب	**Roots and leaves** used in traditional medicine as a tonic and for stress relief; known as **"Ashwagandha"**, widely used in herbal remedies; **berries** sometimes used in treatments for inflammation	(Dar et al. 2015; Mishra et al. 2000)

Table 17.3 Floristic composition of vines and lianas of AlUla Oasis, and their traditional uses

Family	Species	Arabic name (translation)	Arabic name (arabic script)	Traditional uses	References
Apocynaceae	*Pergularia tomentosa* L.	Hairy Periploca (Al-Ghalkah)	الغلقة	Latex used in traditional medicine for skin diseases and wound healing; leaves used as poultices; fibers from stems used for making ropes; grazed by livestock with caution due to potential toxicity	(Aati et al. 2019; Uphof 1968)
Capparaceae	*Capparis cartilaginea* Decne.	Caper Bush (Al-Shaflah Al-Jabali)	الشفلح الجبلي	**Buds and fruits** pickled and used as a **condiment**; **medicinal uses** for rheumatism and as diuretic; **roots and bark** used in traditional remedies; **ornamental** planting due to attractive flowers	(Migahid 1978; Rivera and Obón 1995)
	Capparis spinosa L.	Spiny Caper (Al-Shaflah)	الشفلح	Similar uses as above; **capers** widely used in cuisine; **medicinal properties** for various ailments; **stabilizes** soil on rocky slopes; **flowers** attract pollinators	(Inocencio et al. 2000; Rivera and Obón 1995)
Cucurbitaceae	*Cucumis prophetarum* L.	Wild Cucumber (Al-Shari)	الشري	**Fruits** used in traditional medicine as a purgative; **leaves and roots** used for treating skin conditions; **caution** advised due to potential toxicity; **grazed** minimally by wildlife	(Dymock et al. 1891; Ghazali et al. 1996)

Table 17.4 Floristic composition of perennial herbs of AlUla Oasis, and their traditional uses

Family	Species	Arabic name (translation)	Arabic name (arabic script)	Traditional uses	References
Aizoaceae	*Aizoon canariense* L	Canary Aizoon (Al-Daa'a)	الدعاع	**Leaves** sometimes used as a **vegetable** in salads; **traditional medicine** uses for cooling and soothing properties; **ground cover** in erosion control; **grazed** by livestock	(Students' Flora of Egypt 1974; Sukhorukov et al. 2023)
Apiaceae	*Ferula sinaica* Boiss.	Sinai Ferula (Al-Kalakh)	الكلخ	**Resin** extracted used in traditional medicine for digestive issues and as an anti-inflammatory; **used** to make **"Asafoetida"**, a spice and remedy; **caution** due to strong odor and potential toxicity in large amounts	(Bellakhdar 1997; Parsa and Maleki 1978)
Apocynaceae	*Gomphocarpus sinaicus* Boiss.	Sinai Milkweed (Al-Ushr AlSinai)	العشر السيناوي	**Latex** used in traditional medicine for skin ailments; **fibers** from seed pods used for stuffing and textiles; **toxic** if ingested; **used** in **folk remedies** with caution	(Ghazanfar 1994)
	Solenostemma arghel (Delile) Hayne	Argel (Al-Hargal)	الحرجل	**Leaves and stems** used in herbal teas for treating kidney and liver ailments; **medicinal uses** for respiratory issues; **requires careful dosage** due to potential toxicity; **dried** parts used in incense	(Abdel-Mogib et al. 2002; El-Kamali and Khalid 1996)

(continued)

Table 17.4 (continued)

Family	Species	Arabic name (translation)	Arabic name (arabic script)	Traditional uses	References
Asphodelaceae	*Asphodelus fistulosus* L.	Hollow Asphodel (Al-Brooq)	البروق	**Bulbs and roots** used in traditional medicine for skin conditions; **ornamental** plant due to its flowers; **grazed** by livestock when other forage is scarce; **caution** due to potential mild toxicity	(Gibbs 1974; Moerman 2009)
Asteraceae	*Asteriscus graveolens* (Forssk.) Less.	Arabian Starwort (Al-Niqidan)	النقيدان	**Aromatic herb** used in traditional medicine for digestive issues and colds; **infusions** made from leaves and flowers; **used** in **perfumes** and **incense**; **grazed** by animals	(Batanouny et al. 1999; Hammiche and Maiza 2006)
	Pulicaria incisa (Lam.) DC.	Incised Fleabane (Al-Rabbal)	الرَبل	**Traditional medicine** uses as an anti-inflammatory and for respiratory ailments; **aromatic** plant used in **incense**; **flowers** attract pollinators; **grazed** by livestock	(Ali et al. 2017; Mothana and Lindequist 2005)
Boraginaceae	*Moltkiopsis ciliata* (Forssk.) I. M. Johnst	Hairy Moltkiopsis (Al-Halamahl)	الحلَمة	**Notable for** its **bitter fruits**; **used** in traditional medicine as a purgative; **caution** due to high toxicity; **handled** carefully; historically used in treatments for parasites	(Bailey 1949; Frohne 2005)

(continued)

Table 17.4 (continued)

Family	Species	Arabic name (translation)	Arabic name (arabic script)	Traditional uses	References
Brassicaceae	*Eremobium aegyptiacum* (Spreng.) Asch. ex Boiss.	Egyptian Desert Cress (Al-Tarrbah)	التَّربه	**Leaves and seeds** used in traditional medicine for digestive issues; **pungent taste** similar to cress; grazed by livestock; seeds may be used as a spice in some regions	(Boulos 1999; El-Gamal 2001)
	Farsetia burtoniae Oliv.	Burton's Farsetia (Al-Maqar)	الحَفري	**Traditional uses** limited; **plant** may have medicinal properties but not widely documented; **grazed** by livestock; **adapted** to arid conditions	(Alfarhan 2000; Zohary 1982)
Convolvulaceae	*Convolvulus spicatus* Peter ex Hallier f.	Spicate Bindweed (Al-Rakhama)	رَخامى	**Roots and leaves** used in traditional medicine as a laxative; **flowers** attract pollinators; **grazed** by animals cautiously due to potential mild toxicity; **ornamental** uses for ground cover	(Austin 1998; Dalziel 1937)
Euphorbiaceae	*Chrozophora tinctoria* (L.) A. Juss.	Dyer's Croton (Al-Tanoom)	التنوم	**Leaves and stems** yield a blue dye used historically for textiles; **traditional medicine** uses include treatments for skin conditions; **caution** due to potential toxicity; **grazed** minimally by livestock	(Evans 2009; Maurya 2018)

(continued)

Table 17.4 (continued)

Family	Species	Arabic name (translation)	Arabic name (arabic script)	Traditional uses	References
Fabaceae	*Tephrosia purpurea* subsp. *apollinea* (Delile) Hosni & El Karemy	Purple Tephrosia (Al-Tiqtaq)	الطقطاق	**Used** in traditional medicine for respiratory ailments and as an anti-inflammatory; **nitrogen-fixing** plant improving soil fertility; **grazed** by livestock; **flowers** attract bees	(Singh and Kumar 2019; Youssef et al. 2024)
Lamiaceae	*Lavandula coronopifolia* Poir.	Lavender (Al-Wathahah)	الوذحة	**Aromatic flowers** used in **perfumes**, **incense**, and traditional medicine for relaxation and treating headaches; **infusions** used as herbal teas; **ornamental** planting; **attracts** pollinators	(Basch et al. 2004; Rivera and Obón 1995)
	Lavandula pubescens Decne.	Downy Lavender (Al-Khuzama)	الخزامى	Similar uses as above; **essential oils** extracted for therapeutic uses; **traditional remedies** for respiratory issues; **ornamental** uses; **supports** bee populations	(Basch et al. 2004; Rivera and Obón 1995)
Malvaceae	*Abutilon fruticosum* Guill. & Perr.	Shrubby Indian Mallow (Al-Hatmi)	الحتمي	**Leaves and flowers** used in traditional medicine for coughs and colds; **fibers** from stems used in rope making; **ornamental** uses due to attractive flowers; **grazed** by livestock	(Gupta 2004; Khare 2008)

(continued)

Table 17.4 (continued)

Family	Species	Arabic name (translation)	Arabic name (arabic script)	Traditional uses	References
Plantaginaceae	*Kickxia pseudoscoparia* V. W. Sm. & D. A. Sutton	False Scoparia (Safeeirah)	صفيرا	**Traditional uses** limited; **may** have medicinal properties but not well documented; **grazed** by livestock; **found** in sandy and rocky habitats	(Migahid 1996; Van Slageren 2001)
Poaceae	*Cenchrus ciliaris* L.	Buffel Grass (Al-Sabat)	السبط	**Important forage** grass for livestock; **used** in **pasture improvement** and **soil conservation**; **drought-resistant**; **planted** to prevent desertification	(Humphreys 1997; White 2000)
	Hyparrhenia hirta (L.) Stapf	Common Thatching Grass (Al-Sanfi)	الصنفي	**Used** for **thatching roofs** and making brooms; **grazed** by livestock; **soil binder** in erosion control; **traditional uses** in crafts	(Osman et al. 2014; Perveen and Qaiser 2012)
	Cenchrus divisus (J. F. Gmel.) Verloove, Govaerts & Buttler	Desert Grass (Al-Halfa)	الحلفا	**Used** for **fodder**; **soil stabilization**; **ornamental grass** in landscaping; **tolerant** of arid conditions; **grazed** by livestock	(Chaudhary and Saudi Arabia National 1989; Shaltout and Mady 1996)
	Stipagrostis ciliata (Desf.) De Winter	Feather Grass (Al-Saliyan)	صليان	**Grazed** by livestock; **used** in **sand dune stabilization**; **adapted** to desert environments; **limited traditional uses**; **potential** for use in rehabilitation of degraded lands	(Breman and Kessler 2012; El-Keblawy and Ksiksi 2005)

(continued)

Table 17.5 Floristic composition of annual herbs of AlUla Oasis, and their traditional uses

Family	Species	Arabic name (Translation)	Arabic name (Arabic script)	Traditional uses	References
Amaranthaceae	*Bassia eriophora* (Schrad.) Asch.	Woolly Bassia (Al-Qattinah)	القطينة	**Grazed** by livestock; **used** in traditional medicine for skin conditions; **important** in salt-affected soils; **soil stabilization** properties; **limited human uses** due to potential toxicity	(Reddy et al. 2014; Zayed and Wink 2005)
Asteraceae	*Anthemis deserti* Boiss.	Desert Chamomile (Oqhwan)	أقحوان	**Aromatic herb** used in **herbal teas** for digestive issues and relaxation; **similar** to chamomile; **traditional medicine** uses for colds and fevers; **flowers** attract pollinators	(Leporatti and Lattanzi 1994; Shahat et al. 2017)
Asteraceae	*Artemisia scoparia* Waldst. & Kit.	Red stem Wormwood (Ash-Sheeh Al-Ahmar)	الشيح الأحمر	**Used** in traditional medicine for digestive issues and as an antiseptic; **aromatic** plant used in **incense**; **caution** advised due to potential toxicity in large amounts; **grazed** by livestock sparingly	(Huang et al. 2021; Mamedov and Craker 2001)
	Calendula triptero-carpa Rupr.	Marigold (Al-Hanwah)	الحنوة	**Flowers** used in traditional medicine for skin treatments and anti-inflammatory purposes; **ornamental** uses due to bright flowers; **used** in making dyes; **attracts** beneficial insects	(Preethi et al. 2009; Singh and Kumar 2019)
Boraginaceae	*Arnebia hispidissima* (Sieber ex Lehm.) A. DC.	Prophet's Flower (Al-Mulleh)	المُلَيِّح	**Produces** a red dye from roots used historically in textiles and cosmetics; **traditional medicine** uses for skin ailments; **handled** with care due to potential skin irritation; **flowers** attract pollinators	(Ali et al. 2017; Ghorbani 2005)
Brassicaceae	*Cakile ara-bica* Velen.	Arabian Sea Rocket (Al-Salih)	السليح	**Leaves** sometimes consumed as a **salad green**; **tolerant** of saline conditions; **used** in traditional remedies for digestive issues; **grazed** by livestock; **soil stabilization** along coasts and deserts	(Davy et al. 2006; Rodman 1974)

(continued)

Table 17.5 (continued)

Family	Species	Arabic name (Translation)	Arabic name (Arabic script)	Traditional uses	References
Fabaceae	*Trigonella stellata* Forssk.	Star Fenugreek (Al-Naffl)	النفل	**Seeds and leaves** used in traditional medicine for digestive health; **similar** to fenugreek in uses; **aromatic** herb used in cooking; **grazed** by livestock; **nitrogen-fixing** properties improve soil fertility	(Saeidnia et al. 2014; Wani and Kumar 2018)
Polygonaceae	*Rumex vesicarius* L.	Bladder Dock (Al-Humayd)	الحُميِّض	Leaves consumed as a leafy vegetable with sour taste; used in salads and soups; traditional medicine uses for detoxification and digestive aid; rich in vitamins and minerals; grazed by livestock	(Gupta 2004; Khan et al. 2014)
Zygophyllaceae	*Tribulus terrestris* L.	Puncture Vine (Al-Qataf)	القطف	**Fruits and roots** used in traditional medicine for urinary and reproductive health; **handled** with care due to sharp spines; **used** as a tonic; **grazed** by livestock cautiously; **weed** in some agricultural areas	(Chopra 1956; Kokoska et al. 2002)

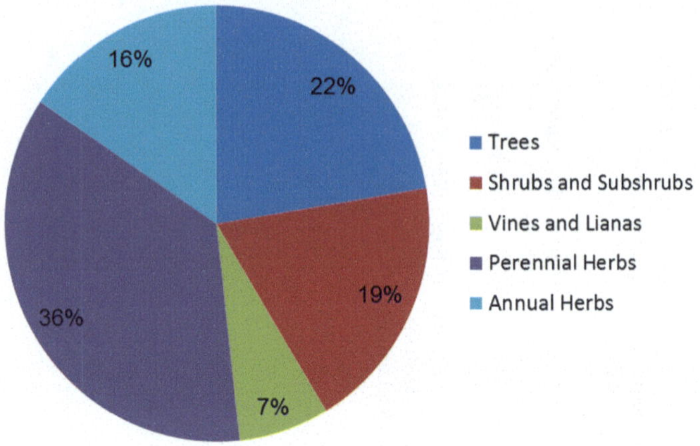

Fig. 17.2 Biological spectrum showing the percentages of growth forms in the flora of AlUla oasis

Trees, shrubs and subshrubs, vines and lianas, perennial herbs, and annual herbs constitute the biological spectrum of the flora (Fig. 17.2). The contributions of the perennial herbs (36%) were the highest, followed by trees and shrubs, while vines and lianas were the least represented (7%).

17.4 Conclusions

AlUla oasis stands as a remarkable example of ecological, historical, and cultural resilience within an arid desert landscape. With a distinctive blend of sandstone mountains, fertile oases, and historical trade routes, AlUla embodies an ecological richness that has flourished alongside human ingenuity. The presence of rare and endangered species highlights AlUla's role as an ecological hotspot, necessitating a sustainable approach to land use. While current conservation efforts by the RCU are promising, they must evolve to address the ongoing challenges posed by tourism expansion, agricultural demands, and climate change.

Recommendations

- **Expand Conservation Areas**: Increase protected zones, such as Sharaan Nature Reserve, to safeguard AlUla's endemic flora and fauna and mitigate the environmental impacts of expanding tourism.
- **Enhance Sustainable Water Management**: Improve irrigation methods, preserving traditional qanat systems while integrating modern technologies to reduce water consumption and address shortages in agriculture.

- **Promote Native Landscaping**: Encourage the use of native plants in landscaping for tourism and local development projects, which can support biodiversity, enhance resilience, and reduce water usage.
- **Implement Continuous Monitoring**: Establish monitoring programs for both plant and animal species to track ecological changes and measure the effectiveness of conservation efforts, particularly for endangered species like the Arabian partridge.
- **Foster Community Involvement**: Engage local communities in conservation and sustainable land-use initiatives to ensure that tourism and agricultural practices align with cultural preservation and ecological health.

Acknowledgements The authors are grateful to Dr. Muhammad Waheed, Dr. Memuna, and Dr. Sohaib Muhammad, Department of Botany, GC University, Lahore, Pakistan for critically going through the earlier drafts, making valuable suggestions to bring manuscript in the final shape. The authors also feel highly obliged and indebted to the entire research team of Dr. Zaheer-ud-din Khan (Distinguished Professor), including his post-graduate and doctoral students for their academic support and cooperation.

References

Aati H, El-Gamal A, Shaheen H, Kayser O (2019) Traditional use of ethnomedicinal native plants in the Kingdom of Saudi Arabia. J Ethnobiol Ethnomed 15:1–9

Abd El-Ghani MM (1998) Environmental correlates of species distribution in arid desert ecosystems of eastern Egypt. J Arid Environ 38(2):297–313

Abd El-Ghani MM, Fahmy AG (1998) Composition of and changes in the spontaneous flora of Feiran Oasis, S Sinai, Egypt, in the last 60 years. Willdenowia 28(1/2):123–134

Abd El-Ghani MM, Fawzy AM (2006) Plant diversity around springs and wells in five oases of the Western Desert, Egypt. Int J Agric Biol 8(2):249–255

Abdellaoui R, Yahyaoui F, Neffati M (2014) Population structure and genetic diversity of a medicinal plant species. Pak J Biol Sci 17(2):182–189

Abdel-Mogib M, Albar H, Batterjee S (2002) Chemistry of the genus Plectranthus. Molecules 7(2):271–301

Ahmed N, Mahmood A, Tahir S, Bano A, Malik RN, Hassan S, Ashraf A (2014) Ethnomedicinal knowledge and relative importance of indigenous medicinal plants of Cholistan desert, Punjab Province. Pak J Ethnopharmacol 155(2):1263–1275

Akhtar MS, Iqbal Z, Khan MN, Lateef M (2000) Anthelmintic activity of medicinal plants with particular reference to their use in animals in the Indo-Pakistan subcontinent. Small Rumin Res 38:99–107

Al Shaye NA, Masrahi YS, Thomas J (2020) Ecological significance of floristic composition and life forms of Riyadh region, Central Saudi Arabia. Saudi J Biol Sci 27(1):35–40

Alam H, Khattak JZK, Ppoyil SBT, Kurup SS, Ksiksi TS (2017) Landscaping with native plants in the UAE: a review. Emir J Food Agric 29(10):729–741

Alfarhan AH (2000) A floristic account on Raudhat Khuraim Central Province Saudi Arabia. Saudi J Biol Sci 8(1)

Ali NAA, Chhetri BK, Dosoky NS, Shari K, Al-Fahad AJ, Wessjohann L, Setzer WN (2017) Antimicrobial, antioxidant, and cytotoxic activities of *Ocimum forskolei* and *Teucrium yemense* (Lamiaceae) essential oils. Medicines 4(2):17

Almas D (2001) The anti micro bial effects of seven different types of Asian chewing sticks

Al-Mutairi K, Al-Shaami S, Khorshid Z, Moawed M (2016) Floristic diversity of Tabuk province, north Saudi Arabia. JAPS J An Plant Sci 26(4)

Al-Qura'n S (2008) Taxonomical and pharmacological survey of therapeutic plants in Jordan
Anwar F, Latif S, Ashraf M, Gilani AH (2007) Moringa oleifera: a food plant with multiple medicinal uses. Phytoth Res Int J Dev Pharmacol Toxicol Eval Nat Prod Deriv 21(1):17–25
Austin DF (1998) Convolvulaceae morning glory family. J Ariz Nevada Acad Sci 61–83
Bailey LH (1949) Manual of cultivated plants most commonly grown in the continental United States and Canada
Barreveld W (1993) Date palm products. Bulletin 101
Basch E, Foppa I, Liebowitz R, Nelson J, Smith M, Sollars D, Ulbricht C (2004) Lavender (*Lavandula angustifolia* miller). J Herb Pharmacother 4(2):63–78
Batanouny K, Aboutabl E, Shabana M, Soliman F (1999) Wild medicinal plants in Egypt, vol 154. The Palm Press, Cairo
Bellakhdar J (1997) Contribution à l'étude de la pharmacopée traditionnelle au Maroc: la situation actuelle, les produits, les sources du savoir (enquête ethnopharmacologique de terrain réalisée de 1969 à 1992) Université Paul Verlaine-Metz
Bornkamm R, Kehl H (1990) The plant communities of the Western Desert of Egypt
Boulos L (1983) Medicinal plants of North Africa
Boulos L (1999) Flora of egypt
Breman H, Kessler J-J (2012) Woody plants in agro-ecosystems of semi-arid regions: with an emphasis on the Sahelian countries, vol 23. Springer Science & Business Media
Burckhardt JL (2013) Travels in Arabia: comprehending an account of those territories in Hedjaz which the Mohammedans regard as sacred. Routledge
Chaudhary SA, Saudi Arabia National Herbarium (1989) Grasses of Saudi Arabia. National Herbarium, National Agriculture and Water Research Center, Ministry of Agriculture and Water
Chopra R (1956) Glossary of Indian medicinal plants. CSIR
Dalziel JM (1937) The useful plants of west tropical Africa
Danin A (1991) Plant adaptations in desert dunes. J Arid Environ 21(2):193–212
Dar NJ, Hamid A, Ahmad M (2015) Pharmacologic overview of Withania somnifera, the Indian Ginseng. Cell Mol Life Sci 72:4445–4460
Davy AJ, Scott R, Cordazzo CV (2006) Biological flora of the British isles: *Cakile maritima* Scop. J Ecol 94(3):695–711
Duke J (2002) Handbook of medicinal herbs. CRC Press
Dymock W, Warden CJH, Warden CJH, Hooper D (1891) Pharmacographia Indica: a history of the principal drugs of vegetable origin, met with in British India, vol 2. K. Paul, Trench, Trübner & Company, Ltd
El Amin H (1990) Trees and shrubs of the Sudan. Ithaca Press
El-Bana MI, Al-Mathnani A (2009) Vegetation-soil relationships in the Wadi Al-Hayat area of the Libyan Sahara. Aust J Basic Appl Sci 3(2):740–747
El-Gamal AA (2001) Sesquiterpene lactones from *Smyrnium olusatrum*. Phytochemistry 57(8):1197–1200
El-Kamali H, Khalid SA (1996) The most common herbal remedies in Central Sudan
El-Keblawy A, Ksiksi T (2005) Artificial forests as conservation sites for the native flora of the UAE. For Ecol Manage 213(1–3):288–296
Engler A (1895) "Die" Pflanzenwelt Ost-Afrikas und der Nachbargebiete, vol 5. Geographische Verlagshandlung Dietrich Reimer
Enright NJ, Miller BP, Akhter R (2005) Desert vegetation and vegetation-environment relationships in Kirthar National Park, Sindh, Pakistan. J Arid Environ 61(3):397–418
Evans WC (2009) Trease and evans' pharmacognosy. Elsevier Health Sciences
Fagg CW, Allison G (2004) Acacia senegal and the gum Arabic trade: monograph and annotated bibliography
Frohne D (2005) Poisonous plants: a handbook for pharmacists, doctors, toxicologists, biologists and veterinarians, 2nd edn. Taylor & Francis. https://books.google.com.pk/books?id=gC91F_SwVEUC

Gandreau D, Moriset S, Arleo L, Hajmirbaba M, Henous M, Munoz NS, Saoudi A, Venton C (2022) AlUla Saudi Arabia: final report 2019–2022. CRAterre

Ghazali GE, Tohami MSE, Elegami A, Abdalla WE, Mohammed MG (1996) Medicinal plants of the Sudan, part IV, Medicinal Plants of Northern Kordofan. National Centre for Research, Khartoum, Sudan

Ghazanfar A (1994) Arabian medicinal plants. Library of Congress Cataloging in the USA, USA

Ghazanfar SA, Fisher M (1998) Vegetation of the Arabian peninsula. Springer Science & Business Media

Ghorbani A (2005) Studies on pharmaceutical ethnobotany in the region of Turkmen Sahra, north of Iran: (part 1): general results. J Ethnopharmacol 102(1):58–68

Gibbs RD (1974) McGill-Queen's University Press. https://doi.org/10.1515/9780773592889

Gupta ANT (2004) Reviews on Indian medicinal plants, vol 3. Are-Azi

Hammiche V, Maiza K (2006) Traditional medicine in Central Sahara: pharmacopoeia of Tassili N'ajjer. J Ethnopharmacol 105(3):358–367

Hegazy A, Amer W (2001) Altitudinal and latitudinal diversity of the flora on the Eastern and Western sides of the Red Sea. In: 3rd IUPAC–international conference on biodiversity. Antalya, Turkey, 3–8 Nov 2001

High Commission for the Development of Arriyadh (2014). Landscape plants for Arriyadh region, 1st edn. Projects & Planning Centre, Riyadh, Saudi Arabia

Huang X, Liu T, Zhou C, Huang Y, Liu X, Yuan H (2021) Antifungal activity of essential oils from three Artemisia species against *Colletotrichum gloeosporioides* of Mango. Antibiotics 10(11):1331. https://www.mdpi.com/2079-6382/10/11/1331

Hudaib MM, Aburjai TA (2006) Composition of the essential oil from Artemisia herba-alba grown in Jordan. J Essent Oil Res 18(3):301–304

Humphreys LR (1997) The evolving science of grassland improvement

Inocencio C, Rivera D, Alcaraz F, Tomás-Barberán FA (2000) Flavonoid content of commercial capers (*Capparis spinosa*, *C. sicula* and *C. orientalis*) produced in mediterranean countries. Eur Food Res Technol 212:70–74

Khan TH, Ganaie MA, Siddiqui NA, Alam A, Ansari MN (2014) Antioxidant potential of *Rumex vesicarius* L.: in vitro approach. Asian Pac J Trop Biomed 4(7):538–544

Khare CP (2008) Indian medicinal plants: an illustrated dictionary. Springer Science & Business Media

Kirtikar KR, Basu BD (1918) Indian medicinal plants, vol 2. Publisher not identified Basu, Bhuwaneśwari Âśrama

Kokoska L, Polesny Z, Rada V, Nepovim A, Vanek T (2002) Screening of some Siberian medicinal plants for antimicrobial activity. J Ethnopharmacol 82(1):51–53

Lalas S, Tsaknis J (2002) Characterization of *Moringa oleifera* seed oil variety "Periyakulam 1." J Food Compos Anal 15(1):65–77

Le Floc'h E (1983) Contribution à une étude ethnobotanique de la flore tunisienne. Ministère de l'Enseignement Supériur et de la Recherche Scientifique

Leporatti M, Lattanzi E (1994) Traditional phytotherapy on coastal areas of Makran (Southern Pakistan)

Lev E, Amar Z (2000) Ethnopharmacological survey of traditional drugs sold in Israel at the end of the 20th century. J Ethnopharmacol 72(1–2):191–205

Mamedov N, Craker LE (2001) Medicinal plants used for the treatment of bronchial asthma in Russia and Central Asia. J Herbs Spices Med Plants 8(2–3):91–117

Mandaville JP (2013) Flora of eastern Saudi Arabia. Routledge

Masrahi Y (2012) A brief illustrate to wild plants in Jizan region. King Fahad Library, Jeddah 302

Maurya H (2018) An updates on *Chrozophora tinctoria* for Its medicinal values. J Curr Pharma Res 9(1):2589–2603

Migahid AM (1978) Migahid and Hammouda's flora of Saudi Arabia

Migahid AM (1996) Flora of Saudi Arabia, vols. 1, 2 and 3, 4th edn. King Saud University, Riyadh

Mishra L-C, Singh BB, Dagenais S (2000) Scientific basis for the therapeutic use of *Withania somnifera* (ashwagandha): a review. Altern Med Rev 5(4):334–346

Moerman DE (2009) Native American medicinal plants: an ethnobotanical dictionary. Timber Press Portland, OR

Mothana RA, Lindequist U (2005) Antimicrobial activity of some medicinal plants of the island Soqotra. J Ethnopharmacol 96(1–2):177–181

Neuwinger HD (1996) African ethnobotany: poisons and drugs: chemistry, pharmacology, toxicology. CRC Press

Orwa C, Mutua A, Kindt R, Jamnadass R, Anthony S (2009). Agroforestree Database: a tree reference and selection guide version 4.0 (http://www.worldagroforestry.org/sites/treedbs/treedatabases.asp). Diakses Tanggal 10(09)

Osman AK, Al-Ghamdi F, Bawadekji A (2014) Floristic diversity and vegetation analysis of Wadi Arar: a typical desert Wadi of the Northern Border region of Saudi Arabia. Saudi J Biol Sci 21(6):554–565

Pareek S (1996) Medicinal plants in India: present status and future prospects. In: Prospects of medicinal plants, p 14

Parsa A, Maleki Z (1978) Flora of Iran. Offset Press. https://books.google.com.pk/books?id=M81GAAAAYAAJ

Perveen A, Qaiser M (2012) Pollen flora of Pakistan-LXIX. Poaceae. Pak J Bot 44(2):747–756

Preethi KC, Kuttan G, Kuttan R (2009) Anti-inflammatory activity of flower extract of *Calendula officinalis* Linn. and its possible mechanism of action. Ind J Exp Biol 47(2):113–120

Rashad J (2024) AlUla: renaissance of an Oasis. UNESCO Retrieved from https://www.unesco.org/en/alula/fellowships?hub=158596

Read M (1990) Plants of Dhofar (the Southern Region of Oman, traditional economic and medicinal uses) Anthony G. Miller and Miranda Morris the office of the adviser for conservation of the environment, Diwan of Royal Court, Sultanate of Oman. 1988, 361pp., HB. In UK, available from Holmes McDougall Ltd., Allander House, 137–141 Leith Walk, Edinburgh EH6 8NS.£ 37.50 inc postage. Oryx 24(4):232–232

Reddy VV, Vishal Gupta N, Raghunandan H, Nitin Kashyap U (2014) Quality risk management in pharmaceutical industry: a review. Int J Pharm Tech Res 6(3):908–914

Remesh M, Al Faify EA, Alfaifi MM, Al Abboud MA, Ismail KS, Al-Namazi AA, Masrahi YS (2023) Ethnobotanical survey of medicinal plants native to the mountains of Jazan, southwestern Saudi Arabia. Int J Adv Appl Sci 10(9):218–227

Rivera D, Obón C (1995) The ethnopharmacology of Madeira and Porto Santo Islands, a review. J Ethnopharmacol 46(2):73–93

Rodman JE (1974) Systematics and evolution of the genus *Cakile* (Cruciferae). Contrib Gray Herb Harv Univ 205:3–146

Saeidnia S, Manayi A, Gohari AR, Abdollahi M (2014) The story of beta-sitosterol—a review

Salama M, Ataman-Millhouse L, Sobral F, Terrado G, Scarella A, Bourlon MT, Adiga SK, Udupa KS, Mahajan N, Patil M (2018) Barriers and opportunities of oncofertility practice in nine developing countries and the emerging oncofertility professional engagement network. J Glob Oncol 4:1–6

Sawicki P (2016) Les tombeaux nabatéens de Hégra

Seshathri K, Thiyagarajan T (2011) Antimicrobial activity of chewing sticks of Jimma-Ethiopia against Streptococcus pyogens. J Phytol 3(8):34–37

Shah A, Sarvat R, Shoaib S, Ayodele A, Nadeem M, Qureshi T, Ishtiaq M, Abbas A (2018) An ethnobotanical survey of medicinal plants used for the treatment of snakebite and scorpion sting among the people of Namal Valley, Mianwali District, Punjab, Pakistan. Appl Ecol Environ Res 16(1)

Shahat EA, Bakr RO, Eldahshan OA, Ayoub NA (2017) Chemical composition and biological activities of the essential oil from leaves and flowers of *Pulicaria incisa* sub. *candolleana* (Family Asteraceae). Chem Biodiv 14(4):e1600156

Shaltout K, Mady M (1996) Analysis of raudhas vegetation in central Saudi Arabia. J Arid Environ 34(4):441–454

Singh B, Kumar D (2019) Ethnobotanical aspects of Nagaur District, Rajasthan. Lulu.com.

Skalli S, Hassikou R, Arahou M (2019) An ethnobotanical survey of medicinal plants used for diabetes treatment in Rabat, Morocco. Heliyon 5(3)

Sreelatha S, Padma P, Umadevi M (2009) Protective effects of *Coriandrum sativum* extracts on carbon tetrachloride-induced hepatotoxicity in rats. Food Chem Toxicol 47(4):702–708

Students' Flora of Egypt (1974) Cairo University. https://books.google.com.pk/books?id=qYsLAQAAIAAJ

Sukhorukov AP, Nilova MV, Kushunina M, Mazei Y, Klak C (2023) Evolution of seed characters and of dispersal modes in Aizoaceae. Front Plant Sci 14:1140069

Uphof JCT (1968) Dictionary of economic plants

Van Slageren M (2001) Wildflowers of Saudi Arabia. Kew Bull 56(4):1021

Wani SA, Kumar P (2018) Fenugreek: a review on its nutraceutical properties and utilization in various food products. J Saudi Soc Agric Sci 17(2):97–106

White R (2000) Grassland ecosystems. World Resources Institute, p 81

Wickens GE (1995) Role of Acacia species in the rural economy of dry Africa and the Near East. Food & Agriculture Org

Wyk B-EV, Heerden FV, Oudtshoorn BV (2002) Poisonous plants of South Africa. Briza Publications, Pretoria

Youssef AM, Doaa AM, Gaber Y (2024) Chemical compounds investigation and profiling of antimicrobial and antiviral constituents of *Tephrosia purpurea* subsp. apollinea. J Biol Res Boll Soci Ital Biol Sper

Zayed R, Wink M (2005) β-carboline and quinoline alkaloids in root cultures and intact plants of *Peganum harmala*. Z Naturforsch C 60(5–6):451–458

Zohary M (1982) Plants of the bible: a complete handbook to all the plants with 200 full-color plates taken in the natural habitat. Cambridge University Press

Prof. Dr. Monier Mohamed Abd El-Ghani is a distinguished Egyptian botanist and taxonomist whose academic journey began with a B.Sc. in Botany from the Faculty of Science, Cairo University, in 1976. He pursued advanced studies in the same institution, earning his M.Sc. (1981) and Ph.D. (1985) in Plant Taxonomy and Flora, focusing on vegetation structure and biodiversity in agro-ecosystems, particularly alien plant species in the Western Desert oases and the Faiyum region. His research interests span biodiversity in arid regions, encompassing Egypt and Saudi Arabia. His extensive work includes studies on alien plant communities, environmental influences on plant distribution, island biogeography, desert vegetation dynamics, host parasite relationships, and desiccation-tolerant plants. These efforts have culminated in the publication of seven books and over 100 scientific papers, with 75 appearing in international peer-reviewed journals and 25 in national and regional outlets. In 1992, Prof. Abd El-Ghani co-authored *The Plant Red Data Book of Egypt, Part I: Woody Perennials*, a seminal contribution to the documentation and conservation of Egypt's threatened flora. His expertise has been instrumental in laying the groundwork for subsequent Red Data Books on Egyptian plant species. From 1997 to 1998, he received a prestigious research fellowship from the Alexander von Humboldt Foundation, awarded to distinguished scholars through global competition. At the Institute of Ecology, Technical University of Berlin, he advanced his studies on biodiversity, biogeography, and vegetation environment interactions in Egypt's principal ecosystems, including the oases of the Western Desert, southern Sinai, and the Eastern Desert. This period enriched his scientific capabilities and provided access to modern techniques and literature, enhancing his expertise in ecological studies. In 2000, Prof. Abd El-Ghani became the Director and Keeper of the Cairo University Herbarium. He initiated the Cairo University Database Project, a venture informed by his visits to the Royal Botanic Gardens and the British Museum (1980–1984) in England and the Botanischer Museum in Berlin (1997–1998). His research in the Western Desert has increasingly focused on its phytogeography, with collaborative efforts alongside German scientists to address critical issues in plant distribution patterns and analysis. Prof. Abd El-Ghani's achievements have been recognized with numerous awards. He received the Egyptian State Encouragement Prize in Biological Sciences in 2000 and 2003 for his contributions to both theoretical and applied sciences. In 2015, he was awarded a D.Sc. (Doctor of Science) degree in Plant Taxonomy and Flora from Cairo University, becoming the first to earn this distinction in his field. The same year, he received the State Prize for Excellence in Basic Sciences for 2014 and the Georg Forster Prize for Scientific Research (Climatic Changes) from the Alexander von Humboldt Foundation in Germany. Prof. Abd El-Ghani's legacy is a testament to his dedication to advancing botanical science, conservation, and education in Egypt and beyond.

Dr. Faten Zubair Filimban is a distinguished associate professor at King Abdulaziz University (KAU) in Jeddah, specializing in Botany. In her academic capacity, she is responsible for educating and mentoring both undergraduate and postgraduate students, including the supervision of M.Sc. and Ph.D. candidates from Saudi Arabia and beyond, thus contributing to the cultivation of future leaders in the botanical sciences. Dr. Filimban's academic background is marked by significant international experience. She completed her PhD and second master's degree at the University of Edinburgh in collaboration with the Centre for Middle Eastern Plants (CMEP) at the Royal Botanic Garden Edinburgh (RBGE), Scotland, where she engaged with leading botanical experts and participated in diverse research initiatives. From 2019 to 2021, Dr. Filimban led a postdoctoral project at the Royal Botanic Garden Edinburgh, funded by KAU, aimed at creating an electronic platform for the identification of "Urban Plant Species" in Saudi Arabia. This initiative provides an accessible, systematic tool for both researchers and the public. In 2021, she served as a visiting researcher at the Royal Botanic Gardens, Kew, London, focusing on the botanical diversity of the Arabian Peninsula. This position enabled her to expand her research efforts and benefit from the expertise of leading global scientists. Since 2023, Dr. Filimban has been serving as a consultant at the Deanship for Scientific Research at KAU. Her expertise in the flora of the Arabian Peninsula and Socotra has facilitated collaborations with scientists from multiple disciplines, particularly in the study of the medicinal properties of native plants. Her work has shown potential in addressing specific cancers and other diseases. In addition, she serves as a consultant to the Department of Gardens and Parks Management at Jeddah Municipality, where she develops lists of cultivated and native plants suited to the city's environment, thereby supporting sustainable urban landscaping. In 2024, she spearheaded a conservation initiative in Jeddah, reporting directly to the Governor, further underscoring her dedication to environmental conservation. Dr. Filimban is an active participant in the academic community, volunteering as a peer reviewer for various prestigious journals and organizing workshops and scientific events to promote dialogue and knowledge in the field of Botany.

Dr. Zaheer-ud-din Khan serves as a distinguished professor at the Department of Botany, GC University Lahore Pakistan. He did his master degree with distinction from the University of the Punjab in 1974 and started teaching and supervising graduate and postgraduate students of Plant Diversity in GC University Lahore. He did his Ph.D. in 1990 from Kings College, University of London in the field of Food Microscopy. Dr. Khan has a strong research background in Plant Systematics, Advance Plant Anatomy, and Ethnobotany. Some of his notable research works include Ethnobotanical Studies; (He has conducted extensive research on the ethnobotanical aspects of various plant species in Pakistan, including studies on the useful shrubs of district Kotli, Azad Jammu & Kashmir.) DNA Barcoding (including the identification of the "Lahore District Flora" using rbcl and matK markers); Taxonomic Reaffirmation. (He has contributed to the taxonomic reaffirmation of some members of family Cannabaceae, Moraceae, Rhamnaceae, Rosaceae, and Urticaceae of order rosales using DNA barcoding markers and Anatomical studies.) He worked out the microstructural/Pharmacognostic and Phytochemical evaluation of adulteration or admixturing in the processed food and herbal products using simple and electron microscopy and chemotaxonomic evaluation techniques including DNA barcoding. He successfully completed HEC research projects on the "Conservation Status of Flora of Punjab" and Ethnofloristic Study of Kotli, AJK". Dr. Khan has also helped WWF-Pakistan, World Bank, Pakistan WAPDA Tarbela Dam, EPA Punjab, and Punjab Small Industries Corporation in completing their projects. His research has been published in various journals, including the Pakistan Journal of Botany and Journal of Biodiversity and Environmental Sciences. He has published 134 research articles in National and International Journals, with an impact factor of 83.23 to his credit. He has a significant number of citations, with over 1767 and an h-index of 21. He is also the author of text books of BSc level. Dr Khan won HEC cash awards like: Best University Teacher Award, 2009; Productivity Allowance for Publications of Research articles, 2009–2011; and Productive Science Award, 2012 by PCST. Recently, he is bestowed with "Life Time Achievement Award" by his Alma Mater, GC University Lahore. He is the member, life member, and life fellow of various professional organizations and societies of Pakistan, such as Pakistan Botanical Society, Biological Society of Pakistan, and Pakistan Association for the Advancement of Science. Dr. Khan has represented Pakistan at many International fora and has attended several academic conferences and meetings as a resource person.

Dr. Mirza Barjees Baig is a professor at the Prince Sultan Institute for Environmental, Water, and Desert Research, King Saud University, Saudi Arabia. He earned his M.S. degree in International Agricultural Extension in 1992 from Utah State University, Logan, Utah, USA, and was placed on the "Roll of Honor". He completed his Ph.D. in Extension for Natural Resource Management from the University of Idaho, USA, and was honored with the "1995 Outstanding Graduate Student Award". He has published extensively on the issues associated with natural resources in national and international journals. He has also made oral presentations about agriculture, natural resources, and the role of extension education at various international conferences. He serves on the editorial boards of many international journals and is a member of many international professional organizations. He has produced six books on climate change, food security, and climate-smart agriculture. Food waste, water management, degradation of natural resources, the deteriorating environment, and their relationship with society and the community are his areas of interest. He has attempted to develop strategies to conserve natural resources, promote the environment, and develop sustainable communities. He started his scientific career in 1983 as a researcher at the Pakistan Agricultural Research Council, Islamabad, Pakistan. He served at the University of Guelph, Ontario, Canada, as a Special Graduate Faculty member from 2000 to 2005. He served as a foreign professor at the Allama Iqbal Open University (AIOU), Pakistan, through the Higher Education Commission from 2005 to 2009.

Dr. Abdul-Malik Bin Abdul-Rahman Al-Al shaikh is the director of the Prince Sultan Institute for Environmental, Water, and Desert Research at the King Saud University (KSU) in Riyadh, Saudi Arabia. He earned his Ph.D. in Arid Land Studies from the University of Arizona, USA, in 1983. He also serves as the General Secretary of the Prince Sultan bin Abdulaziz International Prize for Water (PSIPW). The Prize is awarded bi-annually at the United Nations Headquarters. Being the Director and Research Chair, he oversees the Research Programs at the Institute. He is the team leader of many research projects concerning the environment, desertification, climate change, and water harvesting in Saudi Arabia including the King Fahd Project for Water Harvesting and Storage in Saudi Arabia, Prince Sultan Project for Villages and Hamlets Rehabilitation in Saudi Arabia, the Space Images Atlas of Saudi Arabia, and the Environmental Atlas of Saudi Arabia. Being the chairman of the international conferences on water resources and arid environments, he successfully organized them at the King Saud University. He has brought world fame water scientists to Saudi Arabia and published 9 proceedings of organized international conferences. He has published extensively in journals of international repute.

Chapter 18
Agriculture, Pesticide Pollution, and Climate Change—Interconnected Challenges

Javaid Iqbal, Sajid Ali, Adnan Zahid, Ammara Fatima, Aisha Zahra, Kashaf Siddiqi, and Mukhtar Ahmad

Abstract The complex nexus of agriculture, pesticide pollution, and climate change presents interconnected challenges to environmental sustainability, human health, and global food security. While pesticides play a crucial role in food security and crop yield, their widespread use has led to pollution and health concerns, promoting exploration of alternative pest management strategies to mitigate ecological and human impacts. Addressing the imperative need for advanced agricultural practices to meet projected food demand in the coming years, this chapter scrutinizes the escalating threat of climate change, pesticide use, and risks associated to food security. Fluctuating rainfall patterns, rising temperatures, and high greenhouse gases (GHGs) emission pose severe risk to global food production, leading to a surge in chronically undernourished populations. Moreover, climate change intensifies the impact of pesticide pollution through altered weather patterns, increased pest pressures, and ecosystem shifts, further escalating agricultural challenges. Drawing from empirical literature, the comprehensive and detrimental effects of synthetic pesticides across various dimensions are underscored, encompassing environmental pollution, air quality, GHGs, and biodiversity decline, notably affecting insects. The

J. Iqbal (✉)
Department of Plant Protection, College of Food and Agriculture Sciences, King Saud University, Riyadh, Saudi Arabia
e-mail: jiqbal@ksu.edu.sa

S. Ali · A. Zahid · A. Zahra · K. Siddiqi
Department of Agronomy, University of the Punjab, Lahore, Pakistan
e-mail: sajid.agronomy@pu.edu.pk

A. Zahid
e-mail: adnan.iags@pu.edu.pk

A. Fatima
Department of Environmental Science, Lahore College for Women, University, Lahore, Pakistan
e-mail: ammara.fatima@lcwu.edu.pk

M. Ahmad
Department of Agronomy, Pir Mehr Ali Shah-Arid Agriculture University, Rawalpindi, Pakistan
e-mail: ahmadmukhtar@uaar.edu.pk

M. Behnassi et al. (eds.), *Food Systems and Biodiversity in the Context of Environmental and Climate Risks*, https://doi.org/10.1007/978-3-031-89167-0_18

pesticide bioaccumulation in food chains jeopardizes mammals, non-target animals, and ecosystem balance. Moreover, the persistence of certain pesticide residues in air, water, and soil underscores their potential risks, especially carcinogenic effects. Given the escalating climate-related challenges impacting global crop productivity and food security, this chapter advocates for the necessity of adaptive plans, integrated sustainable approaches, precision farming technologies, policy interventions, and regulatory framework to promote eco-friendly pest management strategies, aimed at improving agricultural systems. Furthermore, it offers a comprehensive insight into the multifaceted challenges facing modern agriculture, highlighting the persistent risks to ecosystems and human health. Interdisciplinary research, collaboration, and transformative actions are imperative to promote resilient agricultural systems, protect environmental integrity, and ensure food security amid evolving global challenges.

Keywords Pesticide pollution · Climate change · Sustainability · Food security · Crop productivity

18.1 Introduction

With an estimated 60% increase in food production required in the coming years, agricultural practices must advance, particularly to prevent what Beddington referred to as the 'perfect storm' (Bilal et al. 2019). Global warming in the twenty-first century, currently at 1.09 °C, may surpass 1.5 and 2 °C as projected by the Intergovernmental Panel on Climate Change (IPCC) unless greenhouse gases (GHGs) emission are significantly curtailed and reduced to 'net-zero' by 2050 or later. Rodenticides, fungicides, herbicides, molluscicides, nematicides, insecticides, and herbicides are collectively referred to as pesticides (Bernardes et al. 2015). Pesticides are widely recognized for their crucial role in fostering agricultural development by reducing crop losses and enhancing both the quantity and quality of food produced at a reasonable cost (Aktar et al. 2009; Fenik et al. 2011; Strassemeyer et al. 2017). During the World War II, the imperative to enhance food production and control diseases spread by insects led to a significant increase in pesticide development. Furthermore, from the 1940s onwards, the increased use of synthetic crop protection agents further enhanced food production (Bernardes et al. 2015).

Measuring agricultural competitiveness is essential due to the agriculture sector's susceptibility to climate change, which poses one of the greatest threats to the planet (Ullah et al. 2021). Natural resource constraints from agricultural output underscore the need for productivity improvements to ensure food security and supply for expanding population. The consequences of global warming are especially dangerous to agriculture and food security, especially for small-scale farmers (Janni et al. 2024; Morton 2007; Schlenker and Lobell 2010).

In many regions of the world, food productivity and the food supply chain have been significantly impacted by climate change (Lobell et al. 2011; Praveena and Malaisamy 2024; Tchonkouang et al. 2024). While the increased crop yield and expansion of agricultural areas have indeed been closely associated with the widespread usage of pesticides and the management of vector-borne diseases, the outcomes also include unintended consequences including pesticide residues exposure (Singh and Kumar 2024), the resurgence of vector-borne illnesses like malaria, and the development of resistance in insects.

The adverse impacts of pesticides on the environment and human health are manifold, and their extensive use has become painfully obvious (Eddleston et al. 2002; Garrido et al. 2004; Rajmohan et al. 2020). In agriculture, insecticides primarily contribute to increased food yield. Despite their effectiveness, pesticides are used more frequently due to additional advantages, such as reduced crop damage, increased crop yield, and vector disease control, which indirectly improve food quality (Arias-Estévez et al. 2008). However, the negative effects of overusing and improperly applying synthetic pesticides have necessitated alternative approaches to pest management (Mahmood et al. 2016). The overuse of these hazardous pesticides affects the plants and animals that constitute the entire ecological framework, with profound effect on human and other systems.

18.2 Climate Change Effects on Agriculture

Climatic change over extended periods, involving variations in humidity, precipitation, and temperature, is a well-understood concept. Global food production is seriously threatened by rising temperatures, climate extremes, increased levels of CO_2 and other GHGs, and altered precipitation patterns. According to a recent FAO assessment of global food security, the population experiencing undernourishment increased from 777 million in 2015 to 815 million in 2016 (FAO 2017). The number of undernourished people remained at 733 million in 2023, with projections indicating that approximately 581 million people could still face hunger by 2030, demonstrating that the world is significantly off track to achieve zero hunger by 2030 (FAO et al. 2024). The most notable degradations have been observed in areas experiencing armed conflict and in regions where war coexists with natural calamities like floods or droughts. Due to their geographic location, the majority of developing nations face serious climate-related challenges that affect agricultural productivity. Most developing nations are located in tropical and semi-desert regions, making them highly vulnerable to rising temperatures and water scarcity (Fonta et al. 2011). Extreme meteorological conditions, such as floods, droughts, variations in precipitation and temperature, and an increased CO_2 concentrations are key features of climate change that affect agriculture (Zewdie 2014). The IPCC's report predicts that by the last two decades of the twenty-first century, temperatures in Africa are projected to increase by more than 2 °C under medium scenarios. According to the SRES (Special Report on Emission Scenarios), rainfall is predicted to decrease in northern Africa and in the southwestern regions of southern Africa. Forecast for sub-Saharan Africa's in the middle and end of the twenty-first century is unclear (IPCC 2022). Numerous studies show that sub-Saharan Africa is significantly affected by climate change in terms of agricultural yield and food security, as much of the region's population relies on climate-sensitive agro-economic practices for their livelihoods (Challinor et al. 2007; Knox et al. 2012; Kotir 2011; Thompson et al. 2010). Under future climate scenarios, one of the biggest concerns may disease outbreaks (Garrett et al. 2006), which could compromise the productivity and sustainability of food crops.

18.2.1 Pesticide Pollution and Impacts

Chronic human illnesses have been linked to components of synthetic pesticides through ingestion, occupational exposure (or home) use, eating pesticide-contaminated foods, or inhaling (or contact) pesticide-contaminated air (Damalas and Koutroubas 2016; Kumari et al. 2014). Because most synthetic pesticides accumulate in the environment and degrade slowly, they contaminate soil and groundwater, contributing to ozone layer thinning (Sande et al. 2011; Wimalawansa and Wimalawansa 2014).

18.2.2 Airborne Pesticide Drift and Atmospheric Implications

It is common to categorize pesticide drift as either primary or secondary migration. When pesticides are applied and drift off-target, this is referred to as primary movement (Jones et al. 2019). Drift occurs when air flow during the application causes the active ingredient in a pesticide to travel beyond of the targeted area after passing through the applicator's spray nozzle (Combellack 1982). Following the use of herbicides, secondary migration happens (Jones et al. 2019; Mueller 2015). Compared to primary movement, variables related to secondary movement can be more challenging to quantify and are much harder to manage. When a pesticide is applied, it may drift as liquid droplets, vapor, or dust particles. Pesticide drift is not restricted to the period immediately following or during an application, but it can also happen hours or even days later. Since receptors that are not target may be acutely exposed, there is a significant chance that they will experience negative consequences during and immediately after pesticide spray (Felsot et al. 2010). Additionally, pesticide drift residues can concentrate in stable air masses or inversions, allowing them to travel long distances (Hernández et al. 2008). Thus, the degree of exposure and the chemical's toxicity to the impacted non-target receptor will determine the risk chance of a negative outcome.

18.2.3 Impact on Air Quality

Pesticide contamination is a significant source of pollution that endangers plants, human, and wildlife health, as well as air quality. The majority of pesticide residues in the air originate from pesticide volatilization from the plants or soil. Pesticides used in agriculture frequently result in spray drift that disperses into the air (Langenbach et al. 2017). Every technique of pesticides application has the potential to cause air pollution and chemical exposure. Pesticides lost through drift can account for 2–25% of the total applied (Pan et al. 2020).

This process pollutes the surrounding environment, negatively impacting global ecosystems (Kim et al. 2017). For instance, pesticides used in southern U.S. fields, such as chlordane, toxaphene, and hexachlorocyclohexanes, were volatilized, transported by atmospheric processes, and deposited on Canada's Great Lakes from the atmosphere through condensation (Sultana et al. 2014). Quantifying the extent of pollution caused by pesticides is challenging, and up to 30–50% of applied pesticides may evaporate into the atmosphere (Van den Berg et al. 1999), making them a potential cause of atmospheric organic pollution (Samsonov et al. 1998). Outdoor air pollution mainly consisting of particulate matter (Bloomfield et al. 2006) contributes to annually 3.3 million premature deaths worldwide, predominantly in Asia with estimated 2.5 million deaths. If emissions keep rising at the current rate, this number might double by 2050 (Lelieveld et al. 2015). Approximately 15–40% of applied pesticides diffuse into the atmosphere during application, primarily through volatilization or spray drift (Sinfort et al. 2009). Most commonly used pesticides have poor volatility, which causes them to be adsorbed on the surface of airborne particles (Coscollà et al. 2014). The transformation and transportation processes they undergo could produce secondary chemicals that are much more dangerous than the original pesticides that are released into the environment. The changes in air quality, atmospheric composition, and related climate change centered on these factors (Woodrow et al. 2019).

New scientific methods and technologies, as well as practical steps like prohibiting pesticides of high risks, creating a national implementation plan (NIP), and implementing integrated pest management (IPM) can collectively mitigate the harmful effects of pesticide contamination on the environment and non-target organisms.

18.2.4 Greenhouse Gas (GHG) Emissions

Agricultural practices, particularly the manufacturing of synthetic insecticides, emit more GHGs than synthetic fertilizer production due to their energy-intensive nature. These emissions significantly contribute to the alteration of the atmosphere (Gallardo 2024). GHG emissions from the manufacturing and application of pesticides accounted for 51% of total GHG emissions in intensive apple production, 37% in viticulture, and 12% in sugar beet cultivation (Cech et al. 2022). Many studies have examined how pesticides effect non-target biodiversity (Astaykina et al. 2020; Baćmaga et al. 2018). Nonetheless, there are few studies that establish a link between GHG emissions, pesticide use, and climate change. Although synthetic nitrogen fertilizer production requires a significant amount of energy, its average GHG emissions are significantly lower than those from pesticide manufacturing (Lal 2004). Pesticides significantly modify the quantity of gases released into the atmosphere and the activity of soil microorganisms. The gradual increase in N_2O and CO_2 gas emissions over time indicates that these pesticides disrupt soil microbiological processes, destabilizing the environmental equilibrium and leading to the accumulation of gaseous byproducts of microbial activity (Jezierska-Tys et al. 2021). The invasion or discovery of new insects also impact the agriculture and pest management strategies, as witnessed by the discovery of soybean aphid in North America in 2000, resulting in a significant increase in insecticide application (Hanson et al. 2017; Ragsdale et al. 2011). The emission of 10.6 kg of CO_2 greenhouse gases equivalent per hectare of treated soybeans is reported due to production, transportation, and insecticides application against soybean aphids (Heimpel et al. 2013). Without a threshold for aphid density beyond which farmers were advised not to spray, emissions would have increased.

Although pesticides have a lower GHG emission footprint than fertilizers, nonetheless, there are still opportunities to save emissions, particularly for crops that require many pesticides. Alternative crop protection strategies should be employed more frequently, particularly preventive strategies like cover crops, mixed crops, and varied crop rotations that can improve biodiversity, mitigate the effects of climate change, and increase soil carbon stores (Duong et al. 2019). Solutions must address both concerns simultaneously since biodiversity loss and climate change are intertwined and cannot be seen as distinct challenges.

18.2.5 Pesticides Persistence, Bioaccumulation, and Long-Term Impacts on Environment

Anthropogenic activities are primarily responsible for the continual increase in environmental contamination, as the most widely used chemicals and their byproducts infiltrate into soil and water sources (Bilal et al. 2019; Pullagurala et al. 2018). Agricultural products carry residual pesticides, make their way to the food chain, progressively reaching to higher trophic levels of the food web, and endanger the animal and human health. To prevent the buildup of pesticides in the ecosystem and their detrimental effects on living organisms, stringent regulations are imperative. Initial reports from 1960s have described the pesticides' harmful effects on people, wildlife, and the environment (Carson 2002; Davis 2014). Plant uptake is a primary route through which pesticides enter the food chain, causing bioaccumulation (or bio-magnification) in humans and other animals (Alengebawy et al. 2021; Gupta and Gupta 2020; Paterson et al. 1990).

An organochlorine pesticide, endosulfan which is known for its broad-spectrum activity, is primarily used in pest control systems to protect crops (Rani et al. 2019). However, due to its persistence, endosulfan residues bioaccumulate and bio-magnify in the food chain, contributing to environmental pollution (Milesi et al. 2020). Persistent organic pollutants (POPs) found in pesticides, such as dichloro-diphenyl-trichloroethane (DDT), dieldrin, and hexachlorobenzene, are environmental contaminants known for their long-lasting persistence (Yadav et al. 2016). The conflict between pollution and production is evident through various types of toxins found in surface water, closely linked to farming practices (Fernandes et al. 2019; Zafar et al. 2016). Pesticide residues can enter water sources through percolation, soil erosion, and surface runoff. These losses caused by pesticide residues can be attributed to both nutrients runoff (Zafar et al. 2016) and to poor no-tillage management practices (Tiecher et al. 2017).

A pressing concern in modern agriculture is the emergence of pesticide resistance, posing significant challenges to pest management strategies (Takahashi et al. 2017). The second half of the twentieth century witnessed an exponential rise in pesticide resistance cases worldwide, promoting industries to explore novel chemicals with distinct modes of action to overcome resistance. This led to the establishment of initiatives such the Analysis of Global Pesticide Resistance in Arthropods and the International Insecticide Resistance Action Committee (IRAC), primarily focusing on resistance management, and developing a resistance classification system based on pesticide mode of action (Sparks et al. 2021; Whalon et al. 2008).

The widespread dispersion of pesticide residues across the environment and agricultural landscapes results in catastrophic contamination of ecosystems affecting dust, sediments, water, soils, and air as well as the adulteration of human diets (Aamir et al. 2018; Kumar et al. 2018; Li 2018). Aquatic systems, aquaculture, marine food, and fisheries supplies are also jeopardized by pesticide residue pollution, and studies have validated the neurotoxic nature of pesticides (Pereira et al. 2015); consequently, broad pesticide exposure poses major health risks and can cause death through direct or indirect means (Simonelli et al. 2007).

The hazardous consequences of pesticide residues on people and non-human biota in aquatic and terrestrial environments are escalating, necessitating the production of food with fewer impurities and higher-quality food in large quantities. Training programs, such as organized by Centre for Agriculture and Biosciences International (CABI) and educational programs, should be implemented to educate farmers on the appropriate use of these chemicals, aiming to reduce the adverse effects of pesticides.

18.2.6 Pesticide Impact on Biodiversity

Growing evidence suggests that pesticides have negative ecological consequences on the diversity of organisms in agricultural environments, despite their widespread use. Agricultural intensification has been criticized for its detrimental effects on biodiversity and ecosystems (Emmerson et al. 2016; Habel et al. 2019; Hallmann et al. 2017; Ickowitz et al. 2019; Sánchez-Bayo and Wyckhuys 2019). Pesticide use has been associated with reduced biodiversity in aquatic systems (Beketov et al. 2013; Brühl and Zaller 2019; Mineau and Whiteside 2013). The widespread use of neonicotinoid pesticides has raised concerns about potential harm to non-target organisms. The widespread application of neonicotinoids led to substantial declines in bird biodiversity between 2008 and 2014 (Li et al. 2020). A worldwide long-term monitoring program was initiated during 2012 to evaluate the effects of unintended agricultural practices, especially pesticides, on biodiversity. Focusing on several taxonomic groups, including birds, plants, insects, and earthworms, the program aimed to address unintended effects on non-target species. Known as 500 ENI, this monitoring program seeks formal, legal, and regulatory inclusion in the biological survey (Andrade et al. 2021).

Research indicates that insects face survival challenges, with an association between the decline in insect populations, pesticide use (Basset and Lamarre 2019; van der Sluijs 2020), climate change (Bodlah et al. 2023), and the global extinction of insect species (Sánchez-Bayo and Wyckhuys 2019). According to Mineau and Whiteside (2013), pesticide use correlates with a decline in grassland bird populations in the United States. However, worries persist that the reported statistics may inadequately reflect the true impact of pesticides usage, despite the fact that extensive research highlights the complex mechanism underlying declines in population (Arpaia et al. 2021). The issue of pesticide fraud is particularly important in less developed nations, where sale of fake pesticides can reach 20–30%, in contrast to 5–7% in affluent nations (Płonka et al. 2016). Long-lasting pesticides and spray drift, either individually or in combination, can negatively impact biodiversity in protected areas (Martín-López et al. 2011).

Despite concerted efforts in Europe to mitigate habitat loss and poaching pressure, a significant investment in agro-environment programs is designated to increase biodiversity. The decline in biodiversity in modern agricultural landscapes has been documented through the collaborative efforts of the European Academies Science Advisory Council (EASAC) for Ecosystems Services and Biodiversity in Europe, the Intergovernmental Science-Policy Platform on Biodiversity and Ecosystem Services (IPBES), and the International Union for Conservation of Nature (IUCN) (Ruckelshaus et al. 2020). Biodiversity, which plays a crucial role in ecosystem stability and human survival, is declining due to human activity (Linhares et al. 2023). The IPBES highlighted a growing trend of global biodiversity decline and the benefits derived by people from nature (Ruckelshaus et al. 2020). It is imperative to address these concerns through science-driven policy aimed to mitigate their impacts. The detrimental effects of pesticides on non-target organisms have significant economic implications, particularly in the global decline of pollinators and biodiversity (Goulson et al. 2015; Mamy et al. 2023). To restore biodiversity and improve opportunities for agriculture reliant on biodiversity-based ecosystem services, such as biological pest control, a significant shift toward farming with limited or no pesticide usage across large regions is essential. This approach is exemplified by the European Union's farm to fork strategy, which aims to half the use and risk of chemical pesticides by 2030 (EU 2020, 2022).

18.2.7 Interactive Link Between Pesticide Application and Climate Change

Food security in the twenty-first century may be threatened by climate change, which can alter insect distribution patterns and increase pests' resistance to pesticides. Pest losses to crops are exacerbated by anthropogenic climate change, which alters insect distribution patterns and increases pesticide resistance, posing a significant threat to global food security and safety (Carvalho 2006; Deutsch et al. 2018; Palumbi 2001). In particular, climate change expands the geographic range of many pests, particularly in historically low-risk areas (Bebber et al. 2013). In some regions, pesticide use and resistance have increased simultaneously (Gress and Zalom 2019).

Global change factors affect species traits, their interactions with competitors, natural enemies, resources, and the locations of ideal temperatures and habitats.

The greatest threat to global biodiversity comes from habitat changes caused directly by human land use, with different mechanisms affection various biomes (Newbold et al. 2015).

In temperate landscapes, insects face different pressures from agricultural intensification. The rate at which insects in tropical forest are being lost due to cultivation and fragmented is concerning. An increase in the relative abundance of insects is associated with nutrient-enriched plant communities and dense vegetation cover (Betzholtz et al. 2013; Carvalheiro et al. 2020; Ubach et al. 2020; Wölfling et al. 2019).

Although global insect population have not consistently declined over the past 50 years, they have shown clear changes in variety, abundance, and distribution that align with the widespread impacts of climate change and land use (Lenoir et al. 2020). Insects are powerful indicators of environmental risks due to their rapid responses to environmental change and eco-evolutionary adaptations to global change, highlighting the importance of conservation efforts and legislative measures to enhance environmental quality and biodiversity (Outhwaite et al. 2020; Singer and Parmesan 2020; Thomas et al. 2009).

18.2.8 Synergistic Effect on Ecosystem Health

Globally, pesticide consumption notably increased from 2 to 3.5 million tons annually between 2019 and 2020, with herbicides accounting for 47.5%, insecticides 29.5%, fungicides 17.5%, and other pesticides 5.5% (Fig. 18.1) (De et al. 2014; FAO 2022b; Sharma et al. 2019).

The global use of pesticides varies significantly by continent as illustrated in Figs. 18.2 and 18.3. In Africa, pesticide use in agriculture accounted to 209,418.48 tons, with use of 0.7 kg ha^{-1} in cropland area, 0.2 kg per capita, and 0.5 g/Int$ based on agricultural production value. In America, pesticide use in agriculture amounted to 1,887,222.99 tons, with use of 5.1 kg ha^{-1} in cropland area, 1.8 kg per capita, and 2.0 g/Int$ based on agricultural production value. In Asia, pesticide use in agriculture

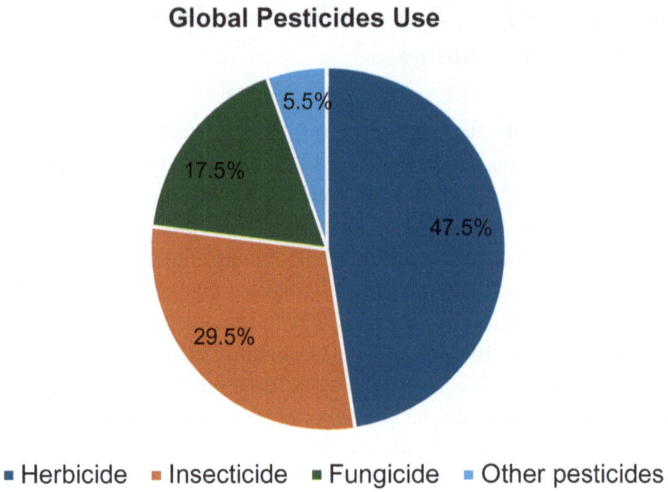

Fig. 18.1 Global pesticide usage data. *Source* Modified after De et al. (2014), FAO (2022b), and Sharma et al. (2019)

accounted to 1,053,930.32 tons, 1.9 kg ha^{-1} in cropland area, 0.2 kg per capita, and 0.5 g/Int$ in terms of agricultural production value. In Europe, 479,552.47 tons pesticides were used with 1.7 kg ha^{-1} in cropland area, 0.7 kg per capita, and 0.9 g/Int$ in terms of agricultural production value. In Oceania, pesticide use in agriculture amounted to 67,372.8 tons, where 2.0 kg ha^{-1} pesticides were used in cropland area, 1.5 kg per capita and 1.0 g/Int$ in terms of agricultural production value.

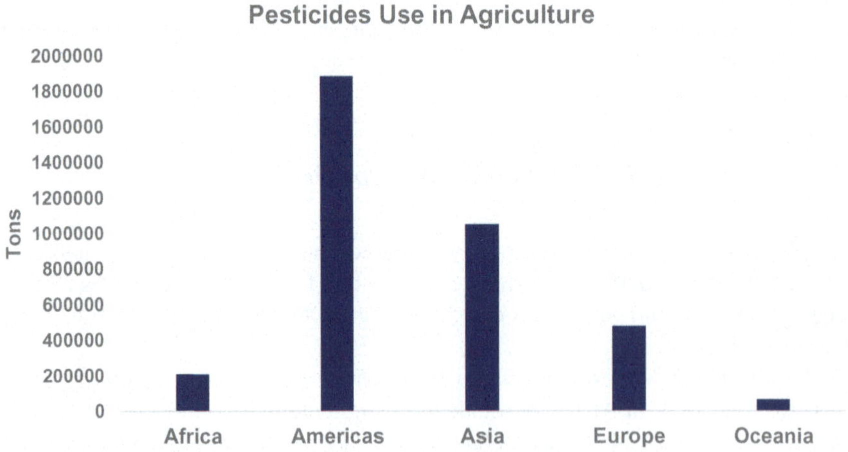

Fig. 18.2 Pesticide use in agriculture (tons) across different continents. *Source* FAO (2022a)

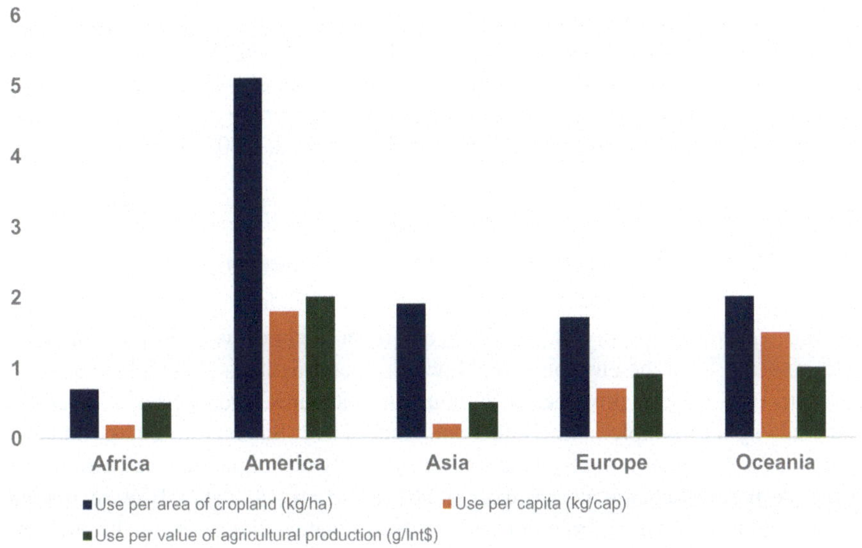

Fig. 18.3 Pesticide use across different continents in 2022. *Source* FAO (2022a)

Systemic insecticides and their derivatives pose relatively low risk to beneficial biota including beneficial insects, when encountered during feeding on plant tissues (Alengebawy et al. 2021). In contrast, systemic pesticides have been found to cause high mortality in social pollinators (such as honey bees) and nectar-feeding parasitoids by contaminating floral and extra floral nectars (El-Wakeil et al. 2013; Stapel et al. 2000). Pesticides, including neonicotinoid and pyrethroids, interfere with the olfactory cognitive functions of honey bees (Abuagla et al. 2024a, b; Iqbal et al. 2019), disrupting their pollinating activities (Naharki and Regmi 2020; Suhail et al. 2001). Availability of high-quality, uncontaminated food (pollen and nectar) enhances the bee immunity, boosting their resilience to environmental stresses, including pesticides (Iqbal 2009; Iqbal et al. 2024). Many pesticides, including dieldrin, endrin, chlordane, thiobencarb, and hexachlorobenzene, resist decomposition and remain in the atmosphere for extended periods. Additionally, pesticide residues can bioaccumulate, with concentrations increasing by up to 70,000 times (Zaka et al. 2019). Residues of pesticides can be found in a wide range of common food items, water, and human breast milk (Kaur et al. 2024; Nicolopoulou-Stamati et al. 2016). Pesticide residues directly and indirectly adversely affect the agricultural soil's physicochemical, biological, and faunal qualities. These residues also have the ability to suppress soil microbial communities and lower enzyme activity. Moreover, pesticides can induce oxidative stress potentially resulting in chlorosis, necrosis, twisting of the leaves, and disruption in photosynthesis. Pesticides have been also found to restrict nitrogen metabolism, affect enzyme activity, cause developmental abnormities in fruits and grains, and alter leaf pigmentation (Alengebawy et al. 2021; Shahid et al. 2023).

Environmental stressors significantly increases the toxicity of specific pesticides, as indicated by a meta-analysis (Liess et al. 2001). Examples in meta-analysis comprise food stress, UVB radiation, and competition, showing that, depending on the severity of environmental stress, organisms can become up to 100 times more sensitive to toxicants (Beketov and Liess 2005; Liess et al. 2001, 2016).

18.3 Bioremediation of Pesticides—An Emerging Trend

Previously, landfills, pyrolysis, recycling, and other methods were utilized for pesticides removal from the environment. However, these methods also yielded adverse environmental effects, produced hazardous intermediates, and proved to be costly and challenging to implement (Hattab and Ghaly 2012; Saleh et al. 2020; Siddiqua et al. 2022). Bioremediation is a natural, eco-friendly technique that can eliminate a variety of harmful chemicals (Porto et al. 2011; Yadav et al. 2017). Bioremediation uses the natural capacity of microorganisms to eliminate pollutants from the environment, making it an effective approach. It is economical, eco-friendly, and adaptable (Finley et al. 2010; Jain et al. 2005).

Microorganisms, such as fungi, algae, and bacteria, can metabolize chlorinated and non-chlorinated pesticides, heavy metals, organic compounds, and other chemical substances, facilitating the removal of organic and inorganic pollutants from the soil environment (Megharaj and Naidu 2017). During biodegradation, microorganisms break down harmful substances by oxidizing parent compounds to produce H_2O, CO_2, or less harmful molecules (Doolotkeldieva et al. 2018). Pesticide bioremediation relies on enzymes that are produced by soil microorganisms during various metabolic processes. The effectiveness of pesticide bioremediation is influenced by the type of microbe—such as bacteria, cyanobacteria, and fungi—as well as the type of pesticide and environmental conditions, all of which play a collective role in degradation processes (Ortiz-Hernández et al. 2013).

Enzymes derived from biological sources break down pesticides and purify polluted agricultural soils and groundwater. Enzymes have capability to restore the polluted environments by effectively neutralizing a variety of pesticides, including organochlorine, carbamide, organophosphate, and DDT pesticides (Kaur et al. 2023; Rehmat 2012; Scott et al. 2008). Microbially assisted pesticide breakdown is the best, most effective, and environmentally friendly method. Microbial communities are crucial role in nutrient recycling, biodegradation of pesticides, and the restoration of contaminated habitats (Kaur et al. 2023; Pande et al. 2020). Microbial communities' integrity and diversity can be altered by pesticides in the soil and water. However, microbes are able to tolerate the presence of environmental pollutants, because to their flexible genetic makeup and adaptive metabolism.

To develop effective treatment methods that incorporate field investigations, a comprehensive understanding of bioremediation of diverse pollutants is essential. Pesticides age in the field and sequester into soil nanopores, where they become inaccessible to microorganisms for degradation. Therefore, research should concentrate

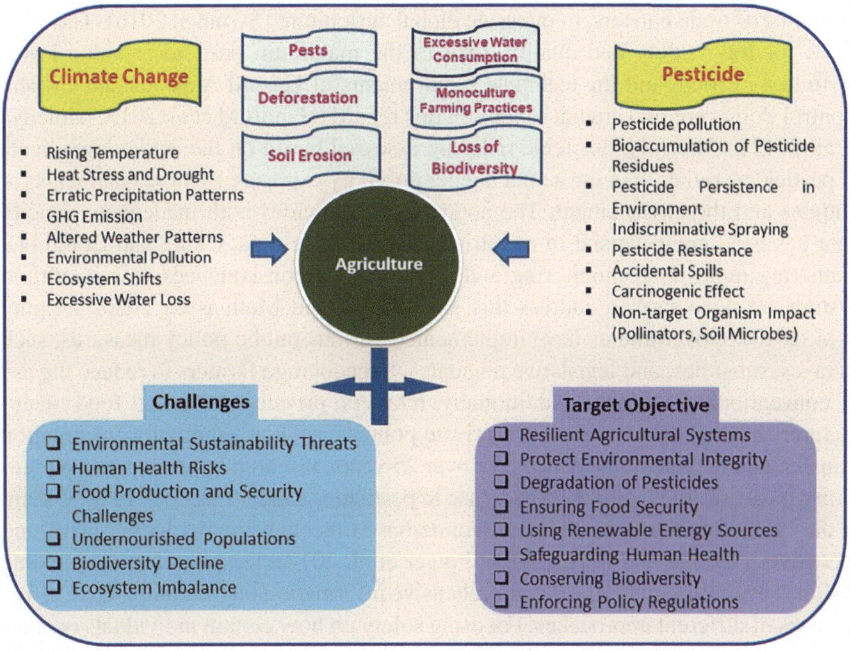

Fig. 18.4 Summary of interconnected challenges in agriculture, pesticide pollution, and climate change

on improving bound pesticide residues' bioavailability and that of their metabolites through an integrated strategy. Numerous studies have demonstrated the microbial density including bacteria, fungi, and algae involved in bioremediation processes of various pesticides (Chen et al. 2012; Cycoń et al. 2009, 2013; Doolotkeldieva et al. 2018; Lopes et al. 2022; Pant et al. 2013; Porto et al. 2011; Qu et al. 2015; Raju and Bidlan 2017; Randika et al. 2022; Reddy et al. 2024; Rigas et al. 2007; Yang et al. 2018). Figure 18.4 summarizes the interconnected challenges in agriculture, pesticide pollution, and climate change.

18.4 Policy Regulations

Mitigating the dangers associated with pesticide use is a prominent topic of discussion among global stakeholders in food value chains and agricultural policies (Lai 2017; Osteen and Fernandez-Cornejo 2013). Strategies to reduce pesticide-associated risks include implementing more sustainable farming systems, developing new technologies and agricultural inputs, and introducing food-labeling measures. Europe is leading in developing export regulations and pesticide laws, which are also known as non-tariff trade barriers, to integrate global agriculture (Swinnen 2016). The standards for food safety and quality, such as the maximum pesticide residue levels permitted in food, and the technical requirements of Hazard Analysis and Critical Control Points are two crucial aspects in this regard (Handford et al. 2015; Nimenya et al. 2012). Currently, pesticide risks are assessed based on the active ingredients in pesticides, rather than the actual danger posed by exposure to these chemicals for humans and the environment. The approval of pesticides with inadequate toxicity data has been controversial in countries, such as Argentina, Chile, and Colombia. Gathering more data, monitoring and modeling the consequences on agricultural system are necessary to address this gap (Kudsk and Mathiassen 2020; Topping et al. 2020). Governments have implemented various public policy measures, such as taxes, subsidies, and legislative mandates, to encourage farmers to reduce the use of conventional pesticides. Additionally, retailers, private actors, and food manufacturers have also been employed private policies, such as premiums, information transfer, and technical assistance. For over 25 years, research have focused on regulatory measures to encourage a decrease in pesticides usage (Falconer 1998; Oskam et al. 1998), often emphasizing economic-based mechanisms including taxes and incentives (Böcker and Finger 2016; Finger et al. 2017). Despite offering valuable insights, available studies lack comprehensive information on the comparative effectiveness of different approaches. Focusing solely on how certain individuals or actor groups are targeted by policies should not be the main concern, as targeted behaviors frequently depend on broader contextual factors in addition to individual actions and decisions (Schoonhoven and Runhaar 2018). Recognizing the influence of groups, such as interest groups and green NGOs on the social context, as well as understanding how interactions with knowledge institutions shape innovation, provides insights into this phenomenon. The agri-food industry is transitioning toward sustainability as its ultimate goal (Squatrito et al. 2020). NGOs can play a significant role in implementing 'The International Code of Conduct on Pesticide Management' to promote sustainable pesticide use, particularly in developing countries (FAO & WHO 2023). NGOs are promoting sustainable use of pesticide, such as prohibiting toxic pesticides, ending illegal fumigations, and enforcing local regulations (Laohaudomchok et al. 2020; Pan et al. 2020). However, concerns have been raised about NGOs' involvement in promoting 'greenwashing' practices in certain regions, and reports have highlighted their contribution to the persistence of outdated pesticides (Loha et al. 2018; Partzsch et al. 2019).

In sustainable agriculture, it is crucial to consider pesticide licensing requirements and their sustainable application. Certain principles for sustainable pesticide use have proposed with the aim of addressing both the social and environmental dimensions of pesticide management. These include compensating for unavoidable consequences, identifying and communicating risk, reducing pesticide usage, optimizing risk management, and internalizing external cost (Colin et al. 2020; Frische et al. 2018; Lechenet et al. 2017; Samways et al. 2020). Figure 18.5 demonstrates the integrated strategies for tackling agricultural challenges, pesticide pollution, and climate change.

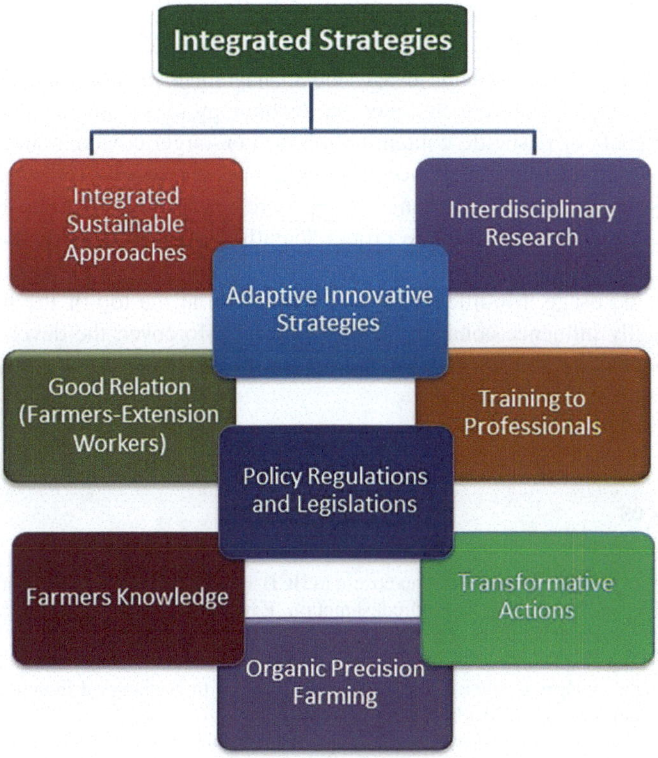

Fig. 18.5 Strategies for addressing challenges and objectives in agriculture, pesticide pollution, and climate change

18.5 Conclusion

Consideration must be given to the dynamics of pesticides in the environment, including their degradation and transfer, when applying them to plants. Improper handling, usage, and behavior with pesticides can result in both environmental degradation and contaminated food. Furthermore, environmental and socioeconomic factors are influenced by climate change. Pesticide degradation, including chemical and microbiological breakdown as well as photolysis, is accelerated by climate change, along with volatilization, runoff, and leaching. Consequently, because of climate change, exposure and risks to human health from pesticide pollution are increased by the growing usage of insecticides and other pesticides. Several factors affect pesticide use. First, pesticide manufacturers should strive to provide high-quality products, particularly in agriculture, where pesticide active components must be developed as rain-fast formulations. Seasonal precipitation, temperature, and other environmental conditions significantly impact farmers' decision-making regarding pesticide use. Secondly, research should concentrate on occupational and environmental exposures, as well as the health hazards associated with pesticide to improve our comprehension of pesticide usage and management in the future. The adoption of innovative scientific techniques and technology is essential to minimize the harmful impacts of pesticide contamination on non-target creatures and the environment, along with implementation of integrated pest management practices. It is crucial to share the scientific findings of exposure, occupational and environmental health risk assessments to provide proper scientific training for pesticide application, encourage safety for applicators, and prevent harmful health impacts arising from pesticide usage. Modified pesticide application at the top of the food chain will eventually influence consumer exposure level. Moreover, the development of bio-pesticides must be prioritized to reduce overall pesticide contamination in the environment.

References

Aamir M, Khan S, Li G (2018) Dietary exposure to HCH and DDT congeners and their associated cancer risk based on Pakistani food consumption. Environ Sci Pollut Res 25(9):8465–8474. https://doi.org/10.1007/s11356-017-1129-1

Abuagla MIB, Iqbal J, Raweh HSA, Alqarni AS (2024a) Insight into olfactory learning, memory, and mortality of *Apis mellifera jemenitica* after exposure to acetamiprid insecticide. Insects 15(7):473. https://doi.org/10.3390/insects15070473

Abuagla MIB, Iqbal J, Raweh HSA, Alqarni AS (2024b) Olfactory learning behavior and mortality of the honey bee *Apis mellifera jemenitica* in response to pyrethroid insecticide (Deltamethrin). Toxics 12(1):25. https://doi.org/10.3390/toxics12010025

Aktar MW, Paramasivam M, Sengupta D, Purkait S, Ganguly M, Banerjee S (2009) Impact assessment of pesticide residues in fish of Ganga river around Kolkata in West Bengal. Environ Monit Assess 157(1–4):97–104. https://doi.org/10.1007/s10661-008-0518-9

Alengebawy A, Abdelkhalek ST, Qureshi SR, Wang MQ (2021) Heavy metals and pesticides toxicity in agricultural soil and plants: ecological risks and human health implications. Toxics 9(3). https://doi.org/10.3390/toxics9030042

Andrade C, Villers A, Balent G, Bar-Hen A, Chadoeuf J, Cylly D, Monestiez P (2021) A real-world implementation of a nationwide, long-term monitoring program to assess the impact of agrochemicals and agricultural practices on biodiversity. Ecol Evol 11(9):3771–3793. https://doi.org/10.1002/ece3.6459

Arias-Estévez M, López-Periago E, Martínez-Carballo E, Simal-Gándara J, Mejuto J-C, García-Río L (2008) The mobility and degradation of pesticides in soils and the pollution of groundwater resources. Agr Ecosyst Environ 123(4):247–260. https://doi.org/10.1016/j.agee.2007.07.011

Arpaia S, Smagghe G, Sweet JB (2021) Biosafety of bee pollinators in genetically modified agro-ecosystems: current approach and further development in the EU. Pest Manag Sci 77(6):2659–2666. https://doi.org/10.1002/ps.6287

Astaykina A, Streletskii R, Maslov M, Belov A, Gorbatov V, Stepanov A (2020) The impact of pesticides on the microbial community of agrosoddy-podzolic soil. Eurasian Soil Sci 53:696–706. https://doi.org/10.1134/S1064229320050038

Baćmaga M, Wyszkowska J, Kucharski J (2018) The influence of chlorothalonil on the activity of soil microorganisms and enzymes. Ecotoxicology 27(9):1188–1202. https://doi.org/10.1007/s10646-018-1968-7

Basset Y, Lamarre GP (2019) Toward a world that values insects. Science 364(6447):1230–1231. https://doi.org/10.1126/science.aaw7071

Bebber DP, Ramotowski MA, Gurr SJ (2013) Crop pests and pathogens move polewards in a warming world. Nat Clim Chang 3(11):985–988. https://doi.org/10.1038/nclimate1990

Beketov MA, Kefford BJ, Schäfer RB, Liess M (2013) Pesticides reduce regional biodiversity of stream invertebrates. Proc Natl Acad Sci 110(27):11039–11043. https://doi.org/10.1073/pnas.1305618110

Beketov MA, Liess M (2005) Acute contamination with esfenvalerate and food limitation: chronic effects on the mayfly, *Cloeon dipterum*. Environ Toxicol Chem 24(5):1281–1286. https://doi.org/10.1897/04-256R1.1

Bernardes MFF, Pazin M, Pereira LC, Dorta DJ (2015) Impact of pesticides on environmental and human health. In: Andreazza AC, Scola G (eds) Toxicology studies-cells, drugs and environment. InTechOpen Limited, UK, pp 195–233. https://doi.org/10.5772/59710

Betzholtz P-E, Pettersson LB, Ryrholm N, Franzén M (2013) With that diet, you will go far: trait-based analysis reveals a link between rapid range expansion and a nitrogen-favoured diet. Proc R Soc b: Biol Sci 280(1750):20122305. https://doi.org/10.1098/rspb.2012.2305

Bilal M, Iqbal HM, Barceló D (2019) Persistence of pesticides-based contaminants in the environment and their effective degradation using laccase-assisted biocatalytic systems. Sci Total Environ 695:133896. https://doi.org/10.1016/j.scitotenv.2019.133896

Bloomfield J, Williams R, Gooddy D, Cape J, Guha P (2006) Impacts of climate change on the fate and behaviour of pesticides in surface and groundwater—a UK perspective. Sci Total Environ 369(1–3):163–177. https://doi.org/10.1016/j.scitotenv.2006.05.019

Böcker T, Finger R (2016) European pesticide tax schemes in comparison: an analysis of experiences and developments. Sustainability 8(4):378. https://doi.org/10.3390/su8040378

Bodlah MA, Iqbal J, Ashiq A, Bodlah I, Jiang S, Mudassir MA, Fareen AGE (2023) Insect behavioral restraint and adaptation strategies under heat stress: an inclusive review. J Saudi Soc Agric Sci. https://doi.org/10.1016/j.jssas.2023.02.004

Brühl CA, Zaller JG (2019) Biodiversity decline as a consequence of an inappropriate environmental risk assessment of pesticides. Front Environ Sci 7:177. https://doi.org/10.3389/fenvs.2019.00177

Carson R (2002) Silent spring. Houghton Mifflin, Boston, United States. https://search.library.wisc.edu/catalog/999966520502121

Carvalheiro LG, Biesmeijer JC, Franzén M, Aguirre-Gutiérrez J, Garibaldi LA, Helm A, Schweiger O (2020) Soil eutrophication shaped the composition of pollinator assemblages during the past century. Ecography 43(2):209–221. https://doi.org/10.1111/ecog.04656

Carvalho FP (2006) Agriculture, pesticides, food security and food safety. Environ Sci Policy 9(7–8):685–692. https://doi.org/10.1016/j.envsci.2006.08.002

Cech R, Leisch F, Zaller JG (2022) Pesticide use and associated greenhouse gas emissions in sugar beet, apples, and viticulture in Austria from 2000 to 2019. Agriculture 12(6):879. https://doi.org/10.3390/agriculture12060879

Challinor A, Wheeler T, Garforth C, Craufurd P, Kassam A (2007) Assessing the vulnerability of food crop systems in Africa to climate change. Clim Change 83:381–399. https://doi.org/10.1007/s10584-007-9249-0

Chen S, Geng P, Xiao Y, Hu M (2012) Bioremediation of β-cypermethrin and 3-phenoxybenzaldehyde contaminated soils using Streptomyces aureus HP-S-01. Appl Microbiol Biotechnol 94:505–515. https://doi.org/10.1007/s00253-011-3640-5

Colin T, Monchanin C, Lihoreau M, Barron AB (2020) Pesticide dosing must be guided by ecological principles. Nat Ecol Evol 4(12):1575–1577. https://doi.org/10.1038/s41559-020-01302-1

Combellack J (1982) Loss of herbicides from ground sprayers. Weed Res 22(4):193–204. https://doi.org/10.1111/j.1365-3180.1982.tb00164.x

Coscollà C, Munoz A, Borrás E, Vera T, Ródenas M, Yusa V (2014) Particle size distributions of currently used pesticides in ambient air of an agricultural Mediterranean area. Atmos Environ 95:29–35. https://doi.org/10.1016/j.atmosenv.2014.06.022

Cycoń M, Żmijowska A, Wójcik M, Piotrowska-Seget Z (2013) Biodegradation and bioremediation potential of diazinon-degrading *Serratia marcescens* to remove other organophosphorus pesticides from soils. J Environ Manage 117:7–16. https://doi.org/10.1016/j.jenvman.2012.12.031

Cycoń M, Wójcik M, Piotrowska-Seget Z (2009) Biodegradation of the organophosphorus insecticide diazinon by *Serratia* sp. and *Pseudomonas* sp. and their use in bioremediation of contaminated soil. Chemosphere 76(4):494–501. https://doi.org/10.1016/j.chemosphere.2009.03.023

Damalas CA, Koutroubas SD (2016) Farmers' exposure to pesticides: toxicity types and ways of prevention. Toxics 4(1):1. https://doi.org/10.3390/toxics4010001

Davis FR (2014) Banned: a history of pesticides and the science of toxicology. Yale University Press. https://doi.org/10.2307/j.ctt13x1tbs

De A, Bose R, Kumar A, Mozumdar S (2014) Targeted delivery of pesticides using biodegradable polymeric nanoparticles. Springer. https://doi.org/10.1007/978-81-322-1689-6

Deutsch CA, Tewksbury JJ, Tigchelaar M, Battisti DS, Merrill SC, Huey RB, Naylor RL (2018) Increase in crop losses to insect pests in a warming climate. Science 361(6405):916–919. https://doi.org/10.1126/science.aat346

Doolotkeldieva T, Konurbaeva M, Bobusheva S (2018) Microbial communities in pesticide-contaminated soils in Kyrgyzstan and bioremediation possibilities. Environ Sci Pollut Res 25:31848–31862. https://doi.org/10.1007/s11356-017-0048-5

Duong TT, Brewer T, Luck J, Zander K (2019) A global review of farmers' perceptions of agricultural risks and risk management strategies. Agriculture 9(1):10. https://doi.org/10.3390/agriculture9010010

Eddleston M, Karalliedde L, Buckley N, Fernando R, Hutchinson G, Isbister G, Senanayake N (2002) Pesticide poisoning in the developing world—a minimum pesticides list. Lancet 360(9340):1163–1167. https://doi.org/10.1016/s0140-6736(02)11204-9

El-Wakeil N, Gaafar N, Sallam A, Volkmar C (2013) Side effects of insecticides on natural enemies and possibility of their integration in plant protection strategies. In Trdan S (ed) Insecticides—development of safer and more effective technologies. Intech, Rijeka, Croatia, pp 1–54. https://doi.org/10.5772/54199

Emmerson M, Morales MB, Oñate JJ, Batáry P, Berendse F, Liira J, Bengtsson J (2016) Chapter Two—How agricultural intensification affects biodiversity and ecosystem services. In: Dumbrell AJ, Kordas RL, Woodward G (eds) Advances in ecological research, vol 55. Academic Press, pp. 43–97. https://doi.org/10.1016/bs.aecr.2016.08.005

EU (2020) Farm to fork strategy: for a fair, healthy and environmentally-friendly food system. https://food.ec.europa.eu/horizontal-topics/farm-fork-strategy_en

EU (2022, August 29, 2024) Farm to fork: new rules to reduce the risk and use of pesticides in the EU. https://ec.europa.eu/commission/presscorner/detail/en/QANDA_22_3694

Falconer K (1998) Managing diffuse environmental contamination from agricultural pesticides: an economic perspective on issues and policy options, with particular reference to Europe. Agr Ecosyst Environ 69(1):37–54. https://doi.org/10.1016/S0167-8809(98)00095-4

FAO & WHO (2023) International code of conduct on pesticide management-guidance on the monitoring and observance of implementation of the code of conduct. https://www.who.int/publications/i/item/9789240069800

FAO (2017) The state of food security and nutrition in the world: building resilience for peace and food security (9251098883). https://www.fao.org/3/I7695e/I7695e.pdf

FAO (2022a) FAOSTAT statistical database: pesticides use. https://www.fao.org/faostat/en/#data/RP

FAO (2022b) Pesticides use, pesticides trade and pesticides indicators—global, regional and country trends, 1990–2020. FAOSTAT Analytical Briefs 46. Food and Agriculture Organization of the United Nations. https://doi.org/10.4060/cc0918en

FAO, IFAD, UNICEF, WFP, WHO (2024) The State of Food Security and Nutrition in the World 2024—Financing to end hunger, food insecurity and malnutrition in all its forms. Rome, Italy. https://doi.org/10.4060/cd1254en

Felsot AS, Unsworth JB, Linders JB, Roberts G, Rautman D, Harris C, Carazo E (2010) Agrochemical spray drift; assessment and mitigation—a review. J Environ Sci Health B 46(1):1–23. https://doi.org/10.1080/03601234.2010.515161

Fenik J, Tankiewicz M, Biziuk M (2011) Properties and determination of pesticides in fruits and vegetables. TrAC, Trends Anal Chem 30(6):814–826. https://doi.org/10.1016/j.trac.2011.02.008

Fernandes G, Aparicio VC, Bastos MC, De Gerónimo E, Labanowski J, Prestes OD, Dos Santos DR (2019) Indiscriminate use of glyphosate impregnates river epilithic biofilms in southern Brazil. Sci Total Environ 651:1377–1387. https://doi.org/10.1016/j.scitotenv.2018.09.292

Finger R, Möhring N, Dalhaus T, Böcker T (2017) Revisiting pesticide taxation schemes. Ecol Econ 134:263–266. https://doi.org/10.1016/j.ecolecon.2016.12.001

Finley SD, Broadbelt LJ, Hatzimanikatis V (2010) In silico feasibility of novel biodegradation pathways for 1, 2, 4-trichlorobenzene. BMC Syst Biol 4(1):1–14. https://doi.org/10.1186/1752-0509-4-7

Fonta W, Edame G, Anam BE, Duru E (2011) Climate change, food security and agricultural productivity in Africa: issues and policy directions. Int J Humanit Soc Sci 1(21):205–223

Frische T, Egerer S, Matezki S, Pickl C, Wogram J (2018) 5-point programme for sustainable plant protection. Environ Sci Eur 30:1–17. https://doi.org/10.1186/s12302-018-0136-2

Gallardo RK (2024) The environmental impacts of agriculture: a review. Int Rev Environ Resour Econ 18(1–2):165–235. https://doi.org/10.1561/101.00000166

Garrett KA, Dendy SP, Frank EE, Rouse MN, Travers SE (2006) Climate change effects on plant disease: genomes to ecosystems. Annu Rev Phytopathol 44:489–509. https://doi.org/10.1146/annurev.phyto.44.070505.143420

Garrido EM, Delerue-Matos C, Lima J, Brett AO (2004) Electrochemical methods in pesticides control. Anal Lett 37(9):1755–1791. https://doi.org/10.1081/AL-120039425

Goulson D, Nicholls E, Botías C, Rotheray EL (2015) Bee declines driven by combined stress from parasites, pesticides, and lack of flowers. Science 347(6229):1255957. https://doi.org/10.1126/science.1255957

Gress BE, Zalom FG (2019) Identification and risk assessment of spinosad resistance in a California population of *Drosophila suzukii*. Pest Manag Sci 75(5):1270–1276. https://doi.org/10.1002/ps.5240

Gupta S, Gupta K (2020) Bioaccumulation of pesticides and its impact on biological systems. In: Pesticides in crop production: physiological and biochemical action, pp 55–67. https://doi.org/10.1002/9781119432241.ch4

Habel JC, Ulrich W, Biburger N, Seibold S, Schmitt T (2019) Agricultural intensification drives butterfly decline. Insect Conserv Divers 12(4):289–295. https://doi.org/10.1111/icad.12343

Hallmann CA, Sorg M, Jongejans E, Siepel H, Hofland N, Schwan H, de Kroon H (2017) More than 75% decline over 27 years in total flying insect biomass in protected areas. PLoS One 12(10):e0185809. https://doi.org/10.1371/journal.pone.0185809

Handford CE, Elliott CT, Campbell K (2015) A review of the global pesticide legislation and the scale of challenge in reaching the global harmonization of food safety standards. Integr Environ Assess Manag 11(4):525–536. https://doi.org/10.1002/ieam.1635

Hanson AA, Menger-Anderson J, Silverstein C, Potter BD, MacRae IV, Hodgson EW, Koch RL (2017) Evidence for Soybean Aphid (Hemiptera: Aphididae) resistance to Pyrethroid Insecticides in the Upper Midwestern United States. J Econ Entomol 110(5):2235–2246. https://doi.org/10.1093/jee/tox235

Hattab M, Ghaly A (2012) Disposal and treatment methods for pesticide containing wastewaters: critical review and comparative analysis. J Environ Prot 03:431–453. https://doi.org/10.4236/jep.2012.35054

Heimpel GE, Yang Y, Hill JD, Ragsdale DW (2013) Environmental consequences of invasive species: greenhouse gas emissions of insecticide use and the role of biological control in reducing emissions. PLoS ONE 8(8):e72293. https://doi.org/10.1371/journal.pone.0072293

Hernández F, Marín JM, Pozo ÓJ, Sancho JV, López FJ, Morell I (2008) Pesticide residues and transformation products in groundwater from a Spanish agricultural region on the Mediterranean Coast. Int J Environ Anal Chem 88(6):409–424. https://doi.org/10.1080/03067310701724772

Ickowitz A, Powell B, Rowland D, Jones A, Sunderland T (2019) Agricultural intensification, dietary diversity, and markets in the global food security narrative. Glob Food Sec 20:9–16. https://doi.org/10.1016/j.gfs.2018.11.002

IPCC (2022) Global Warming of 1.5°C. An IPCC special report on impacts of global warming of 1.5°C above preindustrial levels and related global greenhouse gas emission pathways, in the context of strengthening the global response to the threat of climate change, sustainable development, and efforts to eradicate poverty. Cambridge University Press, Cambridge, UK and New York, NY, USA. https://doi.org/10.1017/9781009157940

Iqbal J, Alqarni AS, Raweh HSA (2019) Effect of sub-lethal doses of imidacloprid on learning and memory formation of indigenous Arabian bee (*Apis mellifera jemenitica* Ruttner) adult foragers. Neotrop Entomol 48(3):373–380. https://doi.org/10.1007/s13744-018-0651-2

Iqbal J, Manzoor M, Boruah T, Wani S, Wachkoo A (2024) Sex and caste determination and life cycle of honey bees. In: Wachkoo AA, Nayik GA, Uddin J, Ansari MJ (eds) Honey bees, beekeeping and bee products, 1 edn. CRC Press. https://doi.org/10.1201/9781003354116-3

Iqbal J (2009) Stress activated protein kinase: central mediator of stress-and infection-induced changes in sensory processing, learning and memory in honeybee (Apis mellifera) (Ph.D. Thesis, Universität des Saarlandes, Saarbrücken, Germany). https://publikationen.sulb.uni-saarland.de/handle/20.500.11880/22642

Jain RK, Kapur M, Labana S, Lal B, Sarma PM, Bhattacharya D, Thakur IS (2005) Microbial diversity: application of microorganisms for the biodegradation of xenobiotics. Curr Sci 89(1):101–112

Janni M, Maestri E, Gullì M, Marmiroli M, Marmiroli N (2024) Plant responses to climate change, how global warming may impact on food security: a critical review (review). Front Plant Sci 14. https://doi.org/10.3389/fpls.2023.1297569

Jezierska-Tys S, Joniec J, Bednarz J, Kwiatkowska E (2021) Microbiological nitrogen transformations in soil treated with pesticides and their impact on soil greenhouse gas emissions. Agriculture 11(8):787. https://doi.org/10.3390/agriculture11080787

Jones GT, Norsworthy JK, Barber T (2019) Off-target movement of diglycolamine dicamba to non-dicamba soybean using practices to minimize primary drift. Weed Technol 33(1):24–40. https://doi.org/10.1017/wet.2018.90

Kaur R, Singh D, Kumari A, Sharma G, Rajput S, Arora S, Kaur R (2023) Pesticide residues degradation strategies in soil and water: a review. Int J Environ Sci Technol 20(3):3537–3560. https://doi.org/10.1007/s13762-021-03696-2

Kaur R, Choudhary D, Bali S, Bandral SS, Singh V, Ahmad MA, Chandrasekaran B (2024) Pesticides: an alarming detrimental to health and environment. Sci Total Environ 915:170113. https://doi.org/10.1016/j.scitotenv.2024.170113

Kim K-H, Kabir E, Jahan SA (2017) Exposure to pesticides and the associated human health effects. Sci Total Environ 575:525–535. https://doi.org/10.1016/j.scitotenv.2016.09.009

Knox J, Hess T, Daccache A, Wheeler T (2012) Climate change impacts on crop productivity in Africa and South Asia. Environ Res Lett 7(3):034032. https://doi.org/10.1088/1748-9326/7/3/034032

Kotir JH (2011) Climate change and variability in Sub-Saharan Africa: a review of current and future trends and impacts on agriculture and food security. Environ Dev Sustain 13:587–605. https://doi.org/10.1007/s10668-010-9278-0

Kudsk P, Mathiassen SK (2020) Pesticide regulation in the European Union and the glyphosate controversy. Weed Sci 68(3):214–222. https://doi.org/10.1017/wsc.2019.59

Kumar B, Mishra M, Verma V, Rai P, Kumar S (2018) Organochlorines in urban soils from Central India: probabilistic health hazard and risk implications to human population. Environ Geochem Health 40:2465–2480. https://doi.org/10.1007/s10653-018-0112-1

Kumari K, Kumar KR, Rao CN (2014) Adverse effects of chemical fertilizers and pesticides on human health and environment. J Chem Pharm Sci 3:150–151

Lai W (2017) Pesticide use and health outcomes: evidence from agricultural water pollution in China. J Environ Econ Manag 86:93–120. https://doi.org/10.1016/j.jeem.2017.05.006

Lal R (2004) Carbon emission from farm operations. Environ Int 30(7):981–990. https://doi.org/10.1016/j.envint.2004.03.005

Langenbach T, Mano D, Campos MM, Cunha AL, De Campos TM (2017) Pesticide dispersion by spraying under tropical conditions. J Environ Sci Health B 52(12):843–849. https://doi.org/10.1080/03601234.2017.1359040

Laohaudomchok W, Nankongnab N, Siriruttanapruk S, Klaimala P, Lianchamroon W, Ousap P, Siriwong W (2020) Pesticide use in Thailand: current situation, health risks, and gaps in research and policy. Hum Ecol Risk Assess 27(5):1147–1169. https://doi.org/10.1080/10807039.2020.1808777

Lechenet M, Dessaint F, Py G, Makowski D, Munier-Jolain N (2017) Reducing pesticide use while preserving crop productivity and profitability on arable farms. Nature Plants 3(3):1–6. https://doi.org/10.1038/nplants.2017.8

Lelieveld J, Evans JS, Fnais M, Giannadaki D, Pozzer A (2015) The contribution of outdoor air pollution sources to premature mortality on a global scale. Nature 525(7569):367–371. https://doi.org/10.1038/nature15371

Lenoir J, Bertrand R, Comte L, Bourgeaud L, Hattab T, Murienne J, Grenouillet G (2020) Species better track climate warming in the oceans than on land. Nat Ecol Evol 4:1044–1059. https://doi.org/10.1038/s41559-020-1198-2

Li Z (2018) Health risk characterization of maximum legal exposures for persistent organic pollutant (POP) pesticides in residential soil: an analysis. J Environ Manage 205:163–173. https://doi.org/10.1016/j.jenvman.2017.09.070

Li Y, Miao R, Khanna M (2020) Neonicotinoids and decline in bird biodiversity in the United States. Nat Sustain 3(12):1027–1035. https://doi.org/10.1038/s41893-020-0582-x

Liess M, Champeau O, Riddle M, Schulz R, Duquesne S (2001) Combined effects of ultraviolet-B radiation and food shortage on the sensitivity of the Antarctic amphipod Paramoera walkeri to copper. Environ Toxicol Chem 20(9):2088–2092. https://doi.org/10.1002/etc.5620200931

Liess M, Foit K, Knillmann S, Schäfer RB, Liess H-D (2016) Predicting the synergy of multiple stress effects. Sci Rep 6(1):32965. https://doi.org/10.1038/srep32965

Linhares Y, Kaganski A, Agyare C, Kurnaz IA, Neergheen V, Kolodziejczyk B, Flores Bueso Y (2023) Biodiversity: the overlooked source of human health. Trends Mol Med 29(3):173–187. https://doi.org/10.1016/j.molmed.2022.12.002

Lobell DB, Schlenker W, Costa-Roberts J (2011) Climate trends and global crop production since 1980. Science 333(6042):616–620. https://doi.org/10.1126/science.1204531

Loha KM, Lamoree M, Weiss JM, de Boer J (2018) Import, disposal, and health impacts of pesticides in the East Africa Rift(EAR) zone: a review on management and policy analysis. Crop Prot 112:322–331. https://doi.org/10.1016/j.cropro.2018.06.014

Lopes PRM, Cruz VH, de Menezes AB, Gadanhoto BP, de Almeida Moreira BR, Mendes CR, Montagnolli RN (2022) Microbial bioremediation of pesticides in agricultural soils: an integrative review on natural attenuation, bioaugmentation and biostimulation. Rev Environ Sci Biotechnol 21(4):851–876. https://doi.org/10.1007/s11157-022-09637-w

Mahmood I, Imadi SR, Shazadi K, Gul A, Hakeem KR (2016) Effects of pesticides on environment. In Hakeem K, Akhtar M, Abdullah S (eds) Plant, soil and microbes, vol 1. Springer, Cham, pp 253–269. https://doi.org/10.1007/978-3-319-27455-3_13

Mamy L, Pesce S, Sanchez W, Aviron S, Bedos C, Berny P, Leenhardt S (2023) Impacts of neonicotinoids on biodiversity: a critical review. Environ Sci Pollut Res Int. https://doi.org/10.1007/s11356-023-31032-3

Martín-López B, García-Llorente M, Palomo I, Montes C (2011) The conservation against development paradigm in protected areas: valuation of ecosystem services in the Doñana social–ecological system (southwestern Spain). Ecol Econ 70(8):1481–1491. https://doi.org/10.1016/j.ecolecon.2011.03.009

Megharaj M, Naidu R (2017) Soil and brownfield bioremediation. Microb Biotechnol 10(5):1244–1249. https://doi.org/10.1111/1751-7915.12840

Milesi MM, Durando M, Lorenz V, Gastiazoro MP, Varayoud J (2020) Postnatal exposure to endosulfan affects uterine development and fertility. Mol Cell Endocrinol 511:110855. https://doi.org/10.1016/j.mce.2020.110855

Mineau P, Whiteside M (2013) Pesticide acute toxicity is a better correlate of US grassland bird declines than agricultural intensification. PLoS ONE 8(2):e57457. https://doi.org/10.1371/journal.pone.0057457

Morton JF (2007) The impact of climate change on smallholder and subsistence agriculture. Proc Natl Acad Sci 104(50):19680–19685. https://doi.org/10.1073/pnas.0701855104

Mueller TC (2015) Methods to measure herbicide volatility. Weed Sci 63(SP1):116–120. https://doi.org/10.1614/WS-D-13-00127.1

Naharki K, Regmi S (2020) Risk assessment of pesticidal toxicity and threats on pollinators: a review on honey bee. Turk J Agric—Food Sci Technol 8(12):2556–2561. https://doi.org/10.24925/turjaf.v8i12.2556-2561.3681

Newbold T, Hudson LN, Hill SL, Contu S, Lysenko I, Senior RA, Collen B (2015) Global effects of land use on local terrestrial biodiversity. Nature 520(7545):45–50. https://doi.org/10.1038/nature14324

Nicolopoulou-Stamati P, Maipas S, Kotampasi C, Stamatis P, Hens L (2016) Chemical pesticides and human health: the urgent need for a new concept in agriculture. Front Public Health 4:148. https://doi.org/10.3389/fpubh.2016.00148

Nimenya N, Ndimira PF, de Frahan BH (2012) Tariff equivalents of nontariff measures: the case of European horticultural and fish imports from African countries. Agric Econ 43(6):635–653. https://doi.org/10.1111/j.1574-0862.2012.00609.x

Ortiz-Hernández ML, Sánchez-Salinas E, Dantán-González E, Castrejón-Godínez ML (2013) Pesticide biodegradation: mechanisms, genetics and strategies to enhance the process. In: Chamy R, Rosenkranz F (eds) Biodegradation—life of science, vol 10. InTech, pp 251–287. https://doi.org/10.5772/56098

Oskam A, Vijftigschild R, Graveland C, Van Dam Y (1998) Additional EU policy instruments for plant protection products: a report within the second phase of the programme: possibilities for future EC environmental policy on plant protection products (9074134556). Wageningen Pers, Netherlands

Osteen CD, Fernandez-Cornejo J (2013) Economic and policy issues of US agricultural pesticide use trends. Pest Manag Sci 69(9):1001–1025. https://doi.org/10.1002/ps.3529

Outhwaite CL, Gregory RD, Chandler RE, Collen B, Isaac NJ (2020) Complex long-term biodiversity change among invertebrates, bryophytes and lichens. Nat Ecol Evol 4(3):384–392. https://doi.org/10.1038/s41559-020-1111-z

Palumbi SR (2001) Humans as the world's greatest evolutionary force. Science 293(5536):1786–1790. https://doi.org/10.1126/science.293.5536.1786

Pan D, He M, Kong F (2020) Risk attitude, risk perception, and farmers' pesticide application behavior in China: a moderation and mediation model. J Clean Prod 276:124241. https://doi.org/10.1016/j.jclepro.2020.124241

Pande V, Pandey SC, Sati D, Pande V, Samant M (2020) Bioremediation: an emerging effective approach towards environment restoration. Environ Sustain 3(1):91–103. https://doi.org/10.1007/s42398-020-00099-w

Pant G, Mistry SK, Sibi G (2013) Isolation, identification and characterization of p, p′-DDT degrading bacteria. J Environ Sci Technol 6(3):130–137. https://doi.org/10.3923/jest.2013.130.137

Partzsch L, Zander M, Robinson H (2019) Cotton certification in Sub-Saharan Africa: Promotion of environmental sustainability or greenwashing? Glob Environ Chang 57:101924. https://doi.org/10.1016/j.gloenvcha.2019.05.008

Paterson S, Mackay D, Tam D, Shiu WY (1990) Uptake of organic chemicals by plants: A review of processes, correlations and models. Chemosphere 21(3):297–331. https://doi.org/10.1016/0045-6535(90)90002-B

Pereira LC, de Souza AO, Bernardes MFF, Pazin M, Tasso MJ, Pereira PH, Dorta DJ (2015) A perspective on the potential risks of emerging contaminants to human and environmental health. Environ Sci Pollut Res 22:13800–13823. https://doi.org/10.1007/s11356-015-4896-6

Płonka M, Walorczyk S, Miszczyk M (2016) Chromatographic methods for the determination of active substances and characterization of their impurities in pesticide formulations. TrAC, Trends Anal Chem 85:67–80. https://doi.org/10.1016/j.trac.2016.03.011

Porto ALM, Melgar GZ n, Kasemodel MC, Nitschke M (2011) Biodegradation of pesticides. In: Margarita S (ed) Pesticides in the modern world. IntechOpen. https://doi.org/10.5772/17686

Praveena K, Malaisamy A (2024) Climatic shifts and agricultural strategies: a thorough review on impact of climate change on food security and crop productivity. Int J Environ Clim Change 14(1):817–831. https://doi.org/10.9734/ijecc/2024/v14i13900

Pullagurala VLR, Rawat S, Adisa IO, Hernandez-Viezcas JA, Peralta-Videa JR, Gardea-Torresdey JL (2018) Plant uptake and translocation of contaminants of emerging concern in soil. Sci Total Environ 636:1585–1596. https://doi.org/10.1016/j.scitotenv.2018.04.375

Qu J, Xu Y, Ai G-M, Liu Y, Liu Z-P (2015) Novel *Chryseobacterium* sp. PYR2 degrades various organochlorine pesticides (OCPs) and achieves enhancing removal and complete degradation of DDT in highly contaminated soil. J Environ Manag 161:350–357. https://doi.org/10.1016/j.jenvman.2015.07.025

Ragsdale DW, Landis DA, Brodeur J, Heimpel GE, Desneux N (2011) Ecology and management of the soybean aphid in North America. Annu Rev Entomol 56(1):375–399. https://doi.org/10.1146/annurev-ento-120709-144755

Rajmohan KS, Chandrasekaran R, Varjani S (2020) A review on occurrence of pesticides in environment and current technologies for their remediation and management. Indian J Microbiol 60(2):125–138. https://doi.org/10.1007/s12088-019-00841-x

Raju SM, Bidlan R (2017) 16S metagenomic analysis and taxonomic distribution of enriched microbial consortia capable of simultaneous biodegradation of organochlorines by illumina platform. Biotechnol Commun 10:697–703. https://doi.org/10.21786/bbrc/10.4/13

Randika JLPC, Bandara PKGSS, Soysa HSM, Ruwandeepika HAD, Gunatilake SK (2022) Bioremediation of pesticide-contaminated soil: a review on indispensable role of soil bacteria. J Agric Sci—Sri Lanka. https://doi.org/10.4038/jas.v17i1.9609

Rani R, Kumar V, Usmani Z, Gupta P, Chandra A (2019) Influence of plant growth promoting rhizobacterial strains *Paenibacillus* sp. IITISM08, *Bacillus* sp. PRB77 and *Bacillus* sp. PRB101 using *Helianthus annuus* on degradation of endosulfan from contaminated soil. Chemosphere 225:479–489. https://doi.org/10.1016/j.chemosphere.2019.03.037

Reddy K, Jose S, Fayaz T, Renuka N, Ratha SK, Kumari S, Bux F (2024) Microbe-assisted bioremediation of pesticides from contaminated habitats. In: Malik A, Garg VK (eds) Bioremediation for sustainable environmental cleanup, 1st edn. CRC Press, Boca Raton, pp 109. https://doi.org/10.1201/9781003277941

Rehmat H (2012) Enzyme based removal of pesticides from soil and water: a review

Rigas F, Papadopoulou K, Dritsa V, Doulia D (2007) Bioremediation of a soil contaminated by lindane utilizing the fungus Ganoderma australe via response surface methodology. J Hazard Mater 140(1–2):325–332. https://doi.org/10.1016/j.jhazmat.2006.09.035

Ruckelshaus MH, Jackson ST, Mooney HA, Jacobs KL, Kassam KS, Arroyo MTK, Ouyang Z (2020) The IPBES global assessment: pathways to action. Trends Ecol Evol 35(5):407–414. https://doi.org/10.1016/j.tree.2020.01.009

Saleh IA, Zouari N, Al-Ghouti MA (2020) Removal of pesticides from water and wastewater: chemical, physical and biological treatment approaches. Environ Technol Innov 19:101026. https://doi.org/10.1016/j.eti.2020.101026

Samsonov YN, Makarov VI, Koutsenogii KP (1998) Physicochemical model and kinetics of pesticide constituent evaporation out of multi-ingredient polydisperse aerosols. Pestic Sci 52(3):292–302. https://doi.org/10.1002/(SICI)1096-9063

Samways MJ, Barton PS, Birkhofer K, Chichorro F, Deacon C, Fartmann T, Hallmann CA (2020) Solutions for humanity on how to conserve insects. Biol Conserv 242:108427. https://doi.org/10.1016/j.biocon.2020.108427

Sánchez-Bayo F, Wyckhuys KA (2019) Worldwide decline of the entomofauna: a review of its drivers. Biol Cons 232:8–27. https://doi.org/10.1016/j.biocon.2019.01.020

Sande D, Mullen J, Wetzstein M, Houston J (2011) Environmental impacts from pesticide use: a case study of soil fumigation in Florida tomato production. Int J Environ Res Public Health 8(12):4649–4661. https://doi.org/10.3390/ijerph8124649

Schlenker W, Lobell DB (2010) Robust negative impacts of climate change on African agriculture. Environ Res Lett 5(1):014010. https://doi.org/10.1088/1748-9326/5/1/014010

Schoonhoven Y, Runhaar H (2018) Conditions for the adoption of agro-ecological farming practices: a holistic framework illustrated with the case of almond farming in Andalusia. Int J Agric Sustain 16(6):442–454. https://doi.org/10.1080/14735903.2018.1537664

Scott C, Pandey G, Hartley CJ, Jackson CJ, Cheesman MJ, Taylor MC, Oakeshott JG (2008) The enzymatic basis for pesticide bioremediation. Indian J Microbiol 48(1):65–79. https://doi.org/10.1007/s12088-008-0007-4

Shahid M, Singh UB, Khan MS (2023) Metabolomics-based mechanistic insights into revealing the adverse effects of pesticides on plants: an interactive review. Metabolites 13(2). https://doi.org/10.3390/metabo13020246

Sharma A, Kumar V, Shahzad B, Tanveer M, Sidhu GPS, Handa N, Parihar RD (2019) Worldwide pesticide usage and its impacts on ecosystem. SN Appl Sci 1:1–16. https://doi.org/10.1007/s42452-019-1485-1

Siddiqua A, Hahladakis JN, Al-Attiya W (2022) An overview of the environmental pollution and health effects associated with waste landfilling and open dumping. Environ Sci Pollut Res Int 29(39):58514–58536. https://doi.org/10.1007/s11356-022-21578-z

Simonelli A, Basilicata P, Miraglia N, Castiglia L, Guadagni R, Acampora A, Sannolo N (2007) Analytical method validation for the evaluation of cutaneous occupational exposure to different chemical classes of pesticides. J Chromatogr B 860(1):26–33. https://doi.org/10.1016/j.jchromb.2007.10.001

Sinfort C, Cotteux E, Bonicelli B, Ruelle B, Douchin M, Berenger M, De Rudnicki V (2009) Influence des conditions et matériels de pulvérisation sur les pertes de pesticides au sol et dans l'air en viticulture Languedocienne. Colloque National du Groupe Français d'études et d'applications des pesticides, Toulouse, France

Singer MC, Parmesan C (2020) Colonizations cause diversification of host preferences: a mechanism explaining increased generalization at range boundaries expanding under climate change. Glob Change Biol 27:3505–3518. https://doi.org/10.1111/gcb.15656

Singh M, Kumar A (2024) Pesticide pressure on insect and human population: a review. Arch Curr Res Int 24(4):188–202. https://doi.org/10.9734/ACRI/2024/v24i4679

Sparks TC, Storer N, Porter A, Slater R, Nauen R (2021) Insecticide resistance management and industry: the origins and evolution of the Insecticide Resistance Action Committee (IRAC) and the mode of action classification scheme. Pest Manag Sci 77(6):2609–2619. https://doi.org/10.1002/ps.6254

Squatrito S, Arena E, Palmeri R, Fallico B (2020) Public and private standards in crop production: their role in ensuring safety and sustainability. Sustainability 12(2):606. https://doi.org/10.3390/su12020606

Stapel JO, Cortesero AM, Lewis WJ (2000) Disruptive sublethal effects of insecticides on biological control: altered foraging ability and life span of a parasitoid after feeding on extrafloral nectar of cotton treated with systemic insecticides. Biol Control 17(3):243–249. https://doi.org/10.1006/bcon.1999.0795

Strassemeyer J, Daehmlow D, Dominic A, Lorenz S, Golla B (2017) SYNOPS-WEB, an online tool for environmental risk assessment to evaluate pesticide strategies on field level. Crop Prot 97:28–44. https://doi.org/10.1016/j.cropro.2016.11.036

Suhail A, Ahmad M, Fareed T, Iqbal J, Iqbal M (2001) Insecticidal mortality and pollination role of honeybee (*Apis mellifera* L.) on some Solanaceous crop. Pak J Biol Sci 4(Suppl. 6):565–566

Sultana J, Syed JH, Mahmood A, Ali U, Rehman MYA, Malik RN, Zhang G (2014) Investigation of organochlorine pesticides from the Indus Basin, Pakistan: sources, air–soil exchange fluxes and risk assessment. Sci Total Environ 497:113–122. https://doi.org/10.1016/j.scitotenv.2014.07.066

Swinnen J (2016) Economics and politics of food standards, trade, and development. Agric Econ 47(S1):7–19. https://doi.org/10.1111/agec.12316

Takahashi D, Yamanaka T, Sudo M, Andow DA (2017) Is a larger refuge always better? Dispersal and dose in pesticide resistance evolution. Evolution 71(6):1494–1503. https://doi.org/10.1111/evo.13255

Tchonkouang RD, Onyeaka H, Nkoutchou H (2024) Assessing the vulnerability of food supply chains to climate change-induced disruptions. Sci Total Environ 920:171047. https://doi.org/10.1016/j.scitotenv.2024.171047

Thomas JA, Simcox D, Clarke RT (2009) Successful conservation of a threatened Maculinea butterfly. Science 325(5936):80–83. https://doi.org/10.1126/science.1175726

Thompson HE, Berrang-Ford L, Ford JD (2010) Climate change and food security in sub-Saharan Africa: a systematic literature review. Sustainability 2(8):2719–2733. https://doi.org/10.3390/su2082719

Tiecher T, Caner L, Minella JPG, Pellegrini A, Capoane V, Rasche JWA, dos Santos Rheinheimer D (2017) Tracing sediment sources in two paired agricultural catchments with different riparian forest and wetland proportion in southern Brazil. Geoderma 285:225–239. https://doi.org/10.1016/j.geoderma.2016.10.008

Topping CJ, Aldrich A, Berny P (2020) Overhaul environmental risk assessment for pesticides. Science 367(6476):360–363. https://doi.org/10.1126/science.aay1144

Ubach A, Páramo F, Gutiérrez C, Stefanescu C (2020) Vegetation encroachment drives changes in the composition of butterfly assemblages and species loss in Mediterranean ecosystems. Insect Conserv Divers 13(2):151–161. https://doi.org/10.1111/icad.12397

Ullah F, Shah SAA, Saqib SE, Yaseen M, Haider MS (2021) Households' flood vulnerability and adaptation: empirical evidence from mountainous regions of Pakistan. Int J Disaster Risk Reduction 52:101967. https://doi.org/10.1016/j.ijdrr.2020.101967

Van den Berg F, Kubiak R, Benjey W, Majewski M, Yates S, Reeves G, Van der Linden A (1999) Emission of pesticides into the air. Water, Air, Soil Pollut 115:195–218. https://doi.org/10.1023/A:1005234329622

van der Sluijs JP (2020) Insect decline, an emerging global environmental risk. Curr Opin Environ Sustain 46:39–42. https://doi.org/10.1016/j.cosust.2020.08.012

Whalon ME, Mota-Sanchez D, Hollingworth RM (2008) Global pesticide resistance in arthropods. CABI Publishing. https://doi.org/10.1079/9781845933531.0000

Wimalawansa SA, Wimalawansa SJ (2014) Agrochemical-related environmental pollution: effects on human health. Glob J Biol, Agric Health Sci 3(3):72–83

Wölfling M, Uhl B, Fiedler K (2019) Multi-decadal surveys in a Mediterranean forest reserve–do succession and isolation drive moth species richness? Nat Conservation 35:25–40. https://doi.org/10.3897/natureconservation.35.32934

Woodrow JE, Gibson KA, Seiber JN (2019) Pesticides and related toxicants in the atmosphere. Rev Environ Contam Toxicol 247:147–196. https://doi.org/10.1007/398_2018_19

Yadav IC, Devi NL, Li J, Zhang G, Shakya PR (2016) Occurrence, profile and spatial distribution of organochlorines pesticides in soil of Nepal: implication for source apportionment and health risk assessment. Sci Total Environ 573:1598–1606. https://doi.org/10.1016/j.scitotenv.2016.09.133

Yadav KK, Gupta N, Kumar V, Singh JK (2017) Bioremediation of heavy metals from contaminated sites using potential species: a review. Indian J Environ Prot 37(1):65

Yang X, Wei H, Zhu C, Geng B (2018) Biodegradation of atrazine by the novel *Citricoccus* sp. strain TT3. Ecotoxicol Environ Saf 147:144–150. https://doi.org/10.1016/j.ecoenv.2017.08.046

Zafar M, Tiecher T, de Castro Lima JAM, Schaefer GL, Santanna MA, Dos Santos DR (2016) Phosphorus seasonal sorption-desorption kinetics in suspended sediment in response to land use and management in the Guapore catchment, southern Brazil. Environ Monit Assess 188:1–20. https://doi.org/10.1007/s10661-016-5650-3

Zaka SM, Iqbal N, Saeed Q, Akrem A, Batool M, Khan AA, Bibi R (2019) Toxic effects of some insecticides, herbicides, and plant essential oils against *Tribolium confusum* Jacquelin du val (Insecta: Coleoptera: Tenebrionidae). Saudi J Biol Sci 26(7):1767–1771. https://doi.org/10.1016/j.sjbs.2018.05.012

Zewdie A (2014) Impacts of climate change on food security: a literature review in Sub Saharan Africa. J Earth Sci Clim Change 5(8):225. https://doi.org/10.4172/2157-7617.1000225

Dr. Javaid Iqbal is an Associate Professor of Plant Protection at King Saud University, Saudi Arabia, with a PhD in Insect Neurobiology from Saarland University, Germany. His research focuses on sensory processing in bees, insect neuroethology, pollinator health, and insect chemical ecology. Dr. Iqbal has over 15 years of academic experience, and his work has been published in renowned international journals. He has also led several research projects and has contributed to insect pest control and apiculture research.

Dr. Sajid Ali is an esteemed academician in Agronomy, specializing in Medicinal Plants and Conservation Agronomy. He earned his PhD from Justus Liebig University in Germany and currently serves as Assistant Professor at the University of the Punjab, Lahore. With over 50 publications in prestigious international journals and contributions to eight book chapters, Dr. Ali has made substantial contributions to the academic and scientific community. He has supervised 3 PhD and over 30 MSc students, focusing on sustainable agricultural practices and advancing knowledge in medicinal plants.

Dr. Adnan Zahid is an Assistant Professor at the Department of Agronomy, University of the Punjab, Lahore. He earned his PhD from the same institution and was awarded the CLIFF-GRADS fellowship supported by the CGIAR Research Program on Climate Change, Agriculture, and Food Security (CCAFS). As part of his fellowship, Dr. Zahid worked at USDA's Agricultural Research Service in Oregon, USA. He has published 15 research articles and 5 book chapters. His research focuses on carbon modeling, soil health, and climate-smart agriculture.

Dr. Ammara Fatima is an Associate Professor at the Department of Environmental Science, Lahore College for Women University (LCWU), where she specializes in environmental sustainability. She earned her PhD from Justus Liebig University Giessen, Germany. With over 25 publications in international journals and contributions to four book chapters, Dr. Fatima has also supervised 3 PhD and 40 M.S. students, advancing research in environmental science, with a focus on sustainable practices and ecological preservation.

Ms. Aisha Zahra completed her MSc (Hons.) in Agronomy from the University of the Punjab, Lahore, where she was a gold medalist. She worked as a research assistant in the Carbon Sequestration Laboratory at the University of the Punjab. Ms. Zahra's research focuses on soil health, sustainable agriculture, and climate-smart practices.

Ms. Kashaf Siddiqi holds an MSc (Hons.) in Agronomy from the University of the Punjab, Lahore, where she worked as a research assistant in the Soil Science laboratory. Her research focuses on soil health and precision agriculture.

Dr. Ahmad Mukhtar is an Assistant Professor at the Department of Agronomy, PMAS-Arid Agriculture University, Rawalpindi. He holds a PhD in Agronomy and completed postdoctoral research at Washington State University, USA, and the Swedish University of Agricultural Sciences. Dr. Mukhtar has published over 150 research articles and 25 book chapters. His research interests include climate change impact assessment, agroecosystem modeling, and climate adaptation strategies in agriculture.

Chapter 19
The *Agdal* Management System in the High Atlas, Morocco, Between the Sustainability Imperative and the Tragedy of the Commons

Aziz Radi, Elhoucine Chougrani, and Mohamed Behnassi

Abstract The Agdal Management System, a centuries-old tradition among the *Amazigh* communities in the Moroccan High Atlas, represents a sophisticated form of communal land management. This system regulates the use of high-altitude pastures through shared access and temporal restrictions, ensuring the sustainable utilization of natural resources. The *Agdal* operates within a framework of customary law, with decisions traditionally made by local assemblies known as "*jmaâ*". This system has historically provided a balance between resource conservation and the livelihood needs of pastoral communities, allowing for periods of rest and regeneration of pastures. In recent decades, the *Agdal* system has faced significant challenges due to increased State intervention and the influx of private capital. These changes have often clashed with traditional practices, leading to a crisis of legitimacy. State projects and policies have sometimes failed to adequately incorporate the local knowledge and practices that are central to the *Agdal's* effective functioning. Additionally, the pressure from private actors seeking to exploit these communal resources for profit has further complicated the management of the *Agdal*, often at the expense of communal interests. Amidst these challenges, local NGOs have emerged as crucial stakeholders, attempting to mediate between traditional practices

A. Radi
English for Economics, University Cadi Ayyad, Marrakesh, Morocco
e-mail: a.radi@uca.ac.ma

E. Chougrani
International Law, University Cadi Ayyad, Marrakesh, Morocco
e-mail: e.chougrani@uca.ac.ma

M. Behnassi (✉)
Senior Environmentalist Expert, Economic Social and Environmental Council (ESEC), Center for Environment, Human Security and Governance (CERES), Ministry of Higher Education and Innovation, Rabat, Morocco
e-mail: behnassi@gmail.com

Center for Environment, Human Security and Governance (CERES), Rabat, Morocco

Ministry of Higher Education and Innovation, Rabat, Morocco

M. Behnassi et al. (eds.), *Food Systems and Biodiversity in the Context of Environmental and Climate Risks*, https://doi.org/10.1007/978-3-031-89167-0_19

and modern governance structures. These organizations aim to empower local populations, revitalize traditional management practices, and advocate for policies that recognize and integrate the *Agdal's* ecological and cultural values. Despite these efforts, the future of the *Agdal* remains uncertain. The system's resilience hinges on the populations' ability to adapt to changing socioeconomic and ecological conditions while preserving the communal ethos that underpins the *Agdal*. This chapter delves into the complex interactions between State policies, private interests, and traditional community governance within the *Agdal* system. It highlights the delicate balance required to manage these commons in ways that respect both environmental sustainability and the socioeconomic needs of local populations. The research underscores the importance of inclusive and adaptive management strategies that can reconcile the diverse and sometimes conflicting interests at play. Through this analysis, the chapter contributes to broader discussions on the governance of common-pool resources, the role of traditional knowledge systems, and the implications of modernization on traditional local practices.

Keywords *Agdal* · Commons · Resource sustainability · Local governance · Stakeholders

19.1 Introduction

The commons represent multidimensional forms of collective management of tangible and intangible resources that communities seek not only to exploit in order to guarantee their survival, but also to protect for the sustainability of their resources. The research on the commons since the mid-1980s has shown the complexity and multidisciplinary dimensions of this subject area (Coward 1980; De los Reyes 1980; MacKenzie 1979; Wittfogel 1957). A brief review of the literature reveals a shift from traditional commons relating to natural resources to new commons around digital resources whose collective management is supposed to ensure fair access to everyone.[1] By virtue of their diversity and richness, the existing studies, both theoretical and empirical, provide a better understanding of the link between the management of collective resources and the building of communities. They also make it

[1] "James Boyle draws an analogy between the "enclosures" of the era of primitive accumulation and these "new enclosures" of intellectual property. The privatization of land was defended in its time in the name of agricultural productivity, the increase of which was supposed to eradicate famines and shortages—even at the cost of terrible urban misery. Today, we are witnessing a "new wave of enclosures," justified in turn by the race for innovation or the urgencies of global food supply. However, the differences between the two phenomena are not insignificant. While the use of land is mutually exclusive (what one person appropriates, another cannot use), the use of knowledge and information is "non-rivalrous": The good does not diminish in the use of a genetic sequence, software, or a digitized image. This is why, from the monk scribe to email, through traditional printing and photocopying, the cost of reproduction has continually decreased. It is also why private appropriation is now justified by the stimulation of research, rather than the use of the product" (Bensaid 2007: 40-41).

possible to grasp this community ties beyond the binary alternative between private and public domains. Indeed, they highlight millenary practices of equal access, an attachment to the system of sharing and collective decision-making rather than to individual property within traditional communities.

This chapter attempts to shed light on such system through the ancestral institution[2] of *Agdal*.[3] It focuses on the traditional management of the commons by the *Amazigh* communities, within the institutional framework of the *Agdal* in the Moroccan High Atlas. While so doing, our major concern will be the relevance rather than the exhaustiveness of our research. We based the latter on a previous body of research in this subject area, both in the literature on the commons as well as that related to the study of *Agdal* per se.

In this analysis, we proceed as follows: First, we briefly define the key concepts used throughout the chapter, including the concept of *Agdal* as a "common" underlying its institutional framework. Then, we examine the three perspectives typically used to approach the concept of commons: the State, the private actors, and the community. Next, we explore the crisis of legitimacy faced by this ancestral institution when confronted with the State and private actors. The differing perspectives of various stakeholders and their influence on the community management of the *Agdal* will be presented. Following this, we describe the conflict of interest between the State's top-down approach and the traditional management of *Agdal*, which relies on trust among stakeholders. Finally, we discuss how the withdrawal of the State, the rise of the private operator, and the weakening of traditional communities have paved the way for a new actor, the NGO, to emerge. In conclusion, we attempt to address whether the commons, exemplified by the *Agdal* institution, represent a continuing or declining ancestral form of collective commons practice.

---

[2] "Institutions are the rules of the game in a society or, more formally, are the humanly devised constraints that shape human interaction. In consequence, they structure incentives in human exchange, whether political, social, or economic. (...) Institutions include any form of constraint that human beings devise to shape human interaction. Are institutions formal or informal? They can be either... Institutional constraints include both what individuals are prohibited from doing and, sometimes, under what conditions some individuals are permitted to undertake certain activities. As defined here, they therefore are the framework within which human interaction takes place (...) they consist of formal written rules as well as typically unwritten codes of conduct that underlie and supplement formal rules" (North 1990:3–4).

[3] "The high-altitude pastures used for transhumance, known as Agdal, find their place within this overall agro-pastoral system, and their use is embedded in the traditional mode of utilizing tribal lands. These transhumance pastures are located, indeed, as extensions of the Douar territories or at most within a half-day's walk from the usual places of residence. The use of the Agdal logically fits into a general pattern of pastoral land use, alternating between stays in the Douars and stays in the high-altitude pastures".

19.1.1 Definition of Key Concepts

- **Commons**: Both the terms "commons" and "global public goods" (GPGs) are widely used in political discourses on how to tackle challenges, such as climate change, food security, and transmission of knowledge. On the one hand, in academic writings and social movements, the commons have come to represent an "alternative model of social organization" going beyond the "market-State" dichotomy (Menatti 2017; Boonen and Brando 2016). The commons have been often claimed as a new institutional and organizational framework—both local and global—in defining a non-capitalist society (Menatti 2017).
- **Common good**: The "common good" refers to a good proper to and attainable only by the entire community. It is not a good external to the group but a good that must be collectively shared while promoting the personal benefit of its members. The common good takes precedence over the individual goods of each member[4] (Cochran 1980). Yet, this does not override the right of its individuals from enjoying it (Dupré 1993).
- **Self-interest**: The "prisoner's dilemma" focuses on rational users and self-interest. This theory predicts that as long as benefits outweigh costs, rational users will ignore the possibility of future gains and continue to exploit resources (Clancy 1998). The economic problem of inefficient resource allocation does not come from the interdependence of resources but rather from the tendency of self-interested individuals to engage in free riding—the "hope to snatch some selfish benefit", in Samuelson's words (Frischmann et al. 2019).
- **Collective action**: As Kutz (2000) puts it, the challenge of collective action is "bridging the gap between the statements true of the group and the statements true of its members".
- **Free rider**: This term refers to a member of a group who obtains benefits from group membership but does not bear a proportional share of the costs of providing the benefits (Olson 1965; Albanese and van Fleet 1985). In *The Logic of Collective Action*, Mansur Olson describes a paradox underlying this same action. He shows that a group of individuals who would benefit from mobilizing and are aware of it may choose to do nothing. The rational individual surely has an advantage in collective actions but has no interest in participating in them. By choosing to do so, the individual is adopting a free-rider behavior. "Indeed, unless the number of individuals in a group is quite small, or unless there is coercion or some other special device to make individuals act in their common interest, rational, self-interested individuals will not act to achieve their common or group interests" (Olson 1965:2, author's emphasis). The classic example of the free-rider problem

[4] "Some scholars (Sgard 2010:5) distinguish between the common good (*singular*) and common goods (*plural*). Common goods could be considered as common-pool resources (Earth's ecological resources: water, air, etc.). Common good, instead, means, sensu lato, something belonging to the community and endowed with a political appeal. Namely, the common good is something of general interest; it is related to the debate on contemporary forms of democracy and values (Olwig 2003:15), sustainable development and ethical ecological issues, and concerns future generations" (Menatti 2017:649).

is the "Tragedy of the Commons" (Hardin 1968), in which individuals acting in their own self-interest end up being worse off than if they had considered the interest of the community.

19.1.2 The Agdal *Concept*

The *Agdal* concept (plur. *igudlan, igdalen*) has a complex etymology and history in the Moroccan *Amazigh* setting. To simplify, the term *Agdal* refers, in the High Atlas, to a time-honored community management tradition. The latter includes the protection of forest and pastoral resources in a well-identified and enclosed territory. "In fact, the word *Agdal* is very often used to refer to the territory itself. However, the *Agdal* encompasses both the land and the rules associated with it". It is the seasonal provisional defense of a space or resource for the sake of ensuring a minimal recovery for vegetation and contributing to the replenishment of plant species.

In this sense, the *Agdal* encompasses here its pastoral, fruit, forage, and forest dimensions. This ancestral management practice promotes the regeneration of "natural resources" and the preservation of biodiversity. Its primary objective is to allow the utilization of these assets at crucial times of the year. In sum, the *Agdal* is a three-dimensional concept which includes an institution, a community, and (in-) tangible resources. The Agdal's management system of the commons involves at least three fundamental generic properties: a shared vision (i.e., a clear identification of users and rules); a common objective (i.e., resource conservation); and a shared course of action (i.e., a temporary respite from use).[5]

The decisions to open and close this space[6] are debated within the customary assembly "*Jmaâ*".[7] The closing and opening dates are announced 8 days in advance

[5] "The '*Agdal'* concept also delineates two orders of fundamental symbolic value in the *Amazigh* tradition, namely honor and some form of ceremonial blessing called 'baraka'. The first regulates exchanges and relationships inside traditional communities, while the second guides the ones between the customary assembly and God for the protection of *Agdals* (Auclair et al. 2011). "The saint ensured the fertility and prosperity of the livestock; in return, the herders offered the saint various gifts and sacrifices. Gellner (1969; 1981) demonstrated, through their physical location (and that of their sanctuary) at the boundaries between tribes, how the saints played the role of arbitrators in conflicts, guaranteed agreements, and assisted merchants by ensuring the safety of passage".

[6] Regarding his study of the *Agdal* of Yagour, Mahdi argues that, through the combined power of belief and legend alone, "The regulation of the Agdal, with its fixed opening and closing dates, is thus associated with the figure of the saint, giving it a sacred character that makes it serious and perilous". On a related note, Tozy et al. (1990:207–208) consider that "The uniqueness of this Agdal lies in it being placed under the patronage of a local saint (Sidi Fares). The Agdal is then called 'agdal n'sidi fares', as it belongs to the Saint and is considered a sacred space. Several rituals are performed at the opening of the pasture and during the stay of the transhumant herders to honor the memory of the Saint and thank him for establishing the very principles of pastoral organization and instituting the order that governs the social organization of the Agdal".

[7] "Regarding the concept of '*Jmaâ'*, Paul Pascon (1965:186) notes that "The example of the Jmaâ is enlightening. The village assembly in the irrigated plain of Haouz is not the same as in the

by a herald in the *souks* and mosques. The opening date is not set in stone. It can, during a drought, be renegotiated, when the demand for fodder from breeders in the spring is high. The customary assembly management of resources aims at regulating fair access within the *Amazigh* communities. This evenhanded regulation of access is a guarantee of the sustainability of resources. To this end, the community enacts a variety of rules for the allocation and use of the common assets. In fact, according to Giraud (2020 cited in Bonnin 2021:1), the concept of the commons itself includes "the resources, whether symbolic or material, that a community chooses to manage by establishing rules that are themselves subject to deliberation. What defines the common is not the nature of the resource, but the political act of a collective that continuously subjects its own practices to communal discernment in the protection and promotion of what it values". It also helps thwart possible conflicts between groups holding the rights to access the coveted resources.

Historically, the community management of *Agdals*, particularly pastoralist ones, is justified by a double concern. It is supposed to manage the inner rivalry between group members by ensuring them equal access to pasture. Concomitantly, it should secure, to its members, the *Agdal*'s grazing resources to the exclusion of other groups of neighboring[8] herders (Barrière 2012:229). Apparently, such a concern could imply that the preservation of resources, through the above-mentioned vegetative recovery of the *Agdal*, would be more vital than the management of competition. Actually, the stakes are much higher. They lie in a subtle game of priorities[9] to be established between both dimensions, as the *Agdal*'s legitimacy crisis discussed below would show.[10]

---

neighboring Chleuh mountains, nor certainly as it was in the past. Neither Jmaâ, nor community assembly, nor communal assembly, nor ban; should we call it today a regular neighborhood group, or as Gurvitch might say, a 'supra-functional, reduced, durable (or permanent) group, occasionally assembled, formed in fact, partially organized, etc.'? The final solution, then, is to describe a kind of Jmaâ containing subgenres, species, and varieties similar to biology and attempt to establish the affiliations (…) The jmaa appears as the basic political institution of human groups constituted as a sort of primary economic and social cell, the qbila (or taqbilt). The Jmaâ imposes itself on the agnatic lineages and also emerges from them. A kind of balance between the bonds of blood and those of neighborhood seems to gradually give preeminence to territoriality".

[8] Regarding the regression of alliance systems between pastoral communities, Carrière notes: "The reduction of the Béni M'tir grazing routes resulted in the disruption of a very fragile balance, which had been maintained by a pastoral agreement between them and the Béni M'guild. Since they could no longer access the same grazing routes as before among the Béni M'tir, the Béni M'guild refused to offer them the summer grazing areas they had previously provided in return" (Pascon 1974 in Carriere 1995:19).

[9] "It is therefore a highly unequal system, as each person brings all the animals they can to the pasture and tries by all means (transported water tanks, high-altitude camps, annexation of pastureland) to extract the maximum resources. There is no cooperative spirit in the modern sense of the term, as the rights holder claims a right for themselves, which they reluctantly share with others. Under these conditions, "the principle of management is not the joint development of resources, but the control of competition for their individual use" (Bourbouze 2009:31).

[10] Ironically, the *Agdal* regime does not seem to extend to pastures with enough available herbaceous resources. It is as if the abundance of resources put on hold competition from pastoralists. Conversely,

The clash between the imperative of preservation and the competition set in motion by different community members in order to benefit from the *Agdal*'s resources has not always been a cooperative game.[11] In the past, under the *Agdal*'s system, offenses were punished by a levy in kind proportional to the offense committed. This sanction was administered by a community chief (*amghar n'ugdal)*, himself a breeder, who is legally and legitimately recognized, by the customary assembly "*jmaa*" for his honesty, probity, and availability. In line with the above-mentioned three-dimensions of the Agdal, three criteria define the regimen of the *Agdal*:

- A subject of law: a lineage or territorial community (lineage, *village,* or tribe).
- An object of law: a resource space (forest, cultivated land, arboreal area).
- A legal privilege: exclusive "collective" enjoyment over resources available in a well-defined spot (Barrière 2012:209).

One could add a fourth decisive feature, namely that it is also a governance structure designated to oversee access to resources (*jmaa*, *naib*, *amghar*,) and set penalties for offenders. The same customary assembly, which designates an *amghar n-ugdal* or *n'tuga* (grass guard) to enforce the *Agdal*'s closure and opening dates, reports any violations of the customary rules and sets penalties for offenders (Ilahiane 1999).

As a hybrid concept, the *Agdal* covers two legal entities, both national and local. This *Agdal* system is conceived less in terms of a "collective property" than of a "common heritage". It is the land regime in which the *Agdal* is embedded that will define its legal status (Barrière 2012:210).[12] The analysis here will focus more on common resources managed collectively. The *Agdal* is made up of an interweaving of collective material and immaterial resources, a knowledge-linked customary practice, which cannot be appropriated by any individual. It is a heritage intended for intergenerational transmission, every generation needs to pass the Earth and our natural and cultural resources on in at least as good condition as we received them.[13] As

---

the Commons' destruction is triggered because they are free or by virtue of their abundance or a bit of both (Hardin 1968).

[11] "While the technique of short-term grazing prohibition has proven effective in a number of pastoral projects, controlling the grazing load remains a challenge on communal lands where the rights holders, as a matter of principle, do not accept limitations on the number of animals. Therefore, the solutions adopted usually focus on limiting the duration of grazing rather than the number of animals, which is an indirect way of limiting the load. Attempts to reduce the number of animals by authoritative measures or to charge for grazing generally result in failure" (Bourbouze 2009:25).

[12] The reference to the perception of the land by the traditional communities of the *Agdal* is a three-dimensional reality: (a) a technical dimension: land as a physical substance on which the community works in order to survive; (b) a social dimension: the land as an entity on which the fundamental social relations are woven; and (c) a sacred dimension: the land as the dwelling place of the ancestors.

[13] Edith Brown Weiss, Climate Change, Intergenerational Equity, and International Law, Vermont Journal of Environmental Law, Vol. 9, No. 3 (Spring 2008), ???616.

an incarnation of "commons", its durability is the end result of its permanent reconfiguration over time. In fact, it is a continuous process rather than as an intrinsic property.[14]

19.2 The Commons: Three Angles of Approach

The collective management of resources, materialized by the *Agdal* tradition, is based on the mechanism of collective action. Such system itself depends on the different frameworks and/or institutional arrangements put in place between the members of the community. For it to be successful, this management should overcome the problem of overexploitation of common resources. The misuse of these resources was analyzed by several authors (Warming 1911; Gordon 1957) before it was approached by a biologist (Hardin 1968), in an article published in *Science* titled *The Tragedy of the Commons.*

For the sake of describing this tension between individual and collective interest in the collective management of resources, Hardin cites the famous example of communal grazing. In fact, being driven by his individual interest, each breeder would be inclined to increase the number of livestock since the pasture does not belong to him. Thus, he would maximize his individual gains while increasing the cost for the community as a whole. Hardin concludes that the combination of individual interests inevitably leads to the overuse or even the destruction of the resource. As a solution, he advocates for the privatization of common resources. He proposes to convert common resources into private properties within which the owners would be incited to develop a more rational management stance. An alternative solution in the literature, often associated with Pigou (1920), would be to entrust their management to the State.

Between the market and/or the State, Ostrom (1990) proposes a third way for the management of common resources. In *Governing the commons*, she suggests letting the users themselves create their own governance systems. The uniqueness of this approach, where communities themselves govern common resources, would demonstrate their capacity for self-governance and thus prevent the overuse of resources.

Through an impressive quantity, diversity, as well as analytical subtlety of case studies, Ostrom attempted to identify salient points emanating from the various institutional arrangements ingeniously cobbled together by communities in order to achieve a successful management of the commons. Thus, through various cases worldwide, Ostrom showed that many communities managed to avoid the tragedy of the commons, while the ownership of the resources at their disposal is neither private nor public, but collective. In this regard, it is interesting to note that, "(...) Regardless of the conditions for accessing collective resources, their use is carried

[14] As we will see below, local NGOs constitute some form of more or less rehabilitation of the commons in renovated forms adapted to the current context.

out individually, and participation in their management does not lead to the existence of strong institutions. The body that manages the collective is, in principle, the *jmaà*, which lacks legal status. It is worth noting that the term refers to a group of people bound by common interests, and therefore, the *jmaà* is not always the same in a given area. There can be a *jmaa* of a tribe, a fraction, a village, a neighborhood, or a lineage, depending on the type of issue being addressed". Actually, by creating small-scale institutions adapted to local conditions, these communities manage their resources sustainably.[15]

Conversely, there are also just as many cases where, despite the arrangements put in place by these same communities, they fail to curb the immoderate exploitation of common resources. The strength of Ostrom's commons management model lies not only in its success stories, but also in its ability not to sweep failures under the rug. Bourbouze confirms this duality in practices and uses when he infers from "The analysis of these traditional organizations evokes a dual sentiment: one of coherence and balance, which can be suggested by this "community production of law" serving a cooperative, flexible management closely adapted to a complex environment; and another, more tumultuous, reflecting the conflicts and individualistic practices that herders develop to appropriate space". The result is a vast research project initiated by Ostrom. It is connected to the very nature of the institutional mechanisms most likely to promote an efficient management of the commons. In this sense, her contribution, at the crossroads of several academic disciplines, is also a huge research project on the tools that could provide the bases for a comprehensive theory of management of the commons.

The focus of collective action lies, for Ostrom, in the resolution of social dilemmas pertaining to the management of these commons. In this respect, perhaps the major point, in the management of the commons, would relate to the communication potential between individuals. According to her, it would be essential to consider the actors' leeway for deliberating among themselves in order to jointly develop their own operating rules. Communication between the members of a given community is, in her view, a critical factor in the development of trust relationships within the group. In addition, the role of communication, from Ostrom's perspective, is crucial in successful collective management, since individuals would be unable to calculate a full battery of strategies regardless of the diversity of situations they would face. Indeed, few real-life situations offer the necessary and complete information that may serve as a basis for every possible course of action that individuals might undertake.

While Ostrom recognizes the importance of the rules implemented by a community, she nonetheless stresses that they would be ineffective without the involvement of stakeholders in the development and monitoring of different actors with these

[15] These 8 principles are: Boundaries of users and resource are clear; congruence between benefits and costs; users had procedures for making own rules; regular monitoring of users and resource conditions; graduated sanctions; conflict resolution mechanisms; minimal recognition of rights by government nested enterprises (Ostrom 1990:90 and also their discussion in pp 91–102).

rules. In this sense, the monitoring activities in force within the communities themselves would be regarded as an integral part of the commons. The involvement of actors in monitoring activities combined with their communication ingenuity seems to take precedence over State intervention and perhaps may even sometimes transcend property rights over the common resources as stated above.

19.3 The *Agdal*'s Legitimacy Crisis

As already mentioned, the collective management of common resources is part of an extensive history of the Moroccan peasantry. Structured into tribes, customary assemblies enjoyed during the pre-colonial period a relative autonomy in decision-making from the central State. However, these traditional communities will subsequently undergo a triple transformation that will significantly affect both their status and roles.[16] They will be placed under the respective administration of the Protectorate, the Independent State, and the private investment. As a result, the community goods have become subject to new statuses, thereby jeopardizing and eventually breaking the bond between the customary assemblies and their resources. With the emergence of new actors, it has become particularly problematic to speak of commons per se in the case of the *Agdal*. Admittedly, rural communities have relinquished their long-held rights in the local management of collective resources, such as water, pastures, or certain forests, in favor of the State and the private sector.

The appropriation mode of space initiated by the administration of the Protectorate as well as by the 1919 *Dahir* would promote the individual appropriation of the commons. The traditional governance system has experienced a sharp decline in favor of individual management systems. Thus, the traditional modes of regulation of the commons embodied by the *Agdal* have been to a large extent weakened. Accordingly, the *Agdal* has experienced a double commons' legitimacy crisis, both ecological and institutional.

The ecological crisis, linked to the degradation of ecosystems—and recently climate change—has been exacerbated by an institutional crisis due to the destabilization of the *Agdal*'s traditional management systems (Auclair et al. 2012). Put under the supervision of State authority, the *Agdal* seems to have lost much of its initial flexibility and efficiency.[17] To make matters worse, the pressure on resources has increased as a result of population growth and land divisions. This has occurred at

[16] Terres collectives au Maroc, dahir du 20 janvier 1919.

[17] "What should be retained from traditional practices is indeed the least easily transmittable: flexibility and the art of compromise in execution and implementation. However, where these organizations persist, it is essential to support them and recognize their strengths, which lie in a subtle blend of competitiveness and solidarity among users, flexible rules with apparent leniency, the unscrupulous clientelism of local leaders alongside proclaimed egalitarianism, the friendliness of local arrangements paired with the intensity of formal demands, and the contrast between simple appearances and the complexity of details".

the expense of the flexibility and fluidity that was once the trademark of the customary regulatory system (Romagny et al. 2018:59).

Hence, the rules established by the community for the harvesting of construction timber, for example, have gradually given way to direct negotiations between villagers and foresters, marginalizing *in fine* customary assemblies (Romagny et al. 2018:60). Bereft of reciprocity and trust, the *Agdal* institution has been confronting not only an illegitimacy crisis but also an unprecedented existential predicament. Until a recent past, the sustainable management of natural resources has been part of a shared framework of renewed horizontal commitment by traditional community members for the sake of preserving the common good. Today, however, the new top-down institutional approach has transformed the *Agdal* into territories for economic projects and profit instead (Ilbert 2011:64).

19.4 Clashing Visions

From what precedes, emerge two visions of the *commons* which apparently everything seems to oppose. First, there is the vertical State approach, deemed by its critics, as being at the root of the *Agdal*'s loss of momentum. This approach is characterized by the establishment of new State property rights in a space formerly governed, through customary assemblies, by customary law (*ourf*) over millennia. The second approach considers that the tribal institution of the *Agdal* is no longer suited to the current realities. Critics of the collective decree consider it to carry the seeds of destruction of the commons. Any communal use of a given resource implies competition between users, which will ultimately push them to exploit it into extinction (Hardin 1968). Bourbouze succinctly sums up this dilemma by stating that "It is true that the status of collective ownership prohibits economic collateral for accessing credit. It prevents secure access by granting only usufruct rights, thus hindering or discouraging land development and the intensification of its use. The critics of collective ownership argue that it carries risks of resource destruction and environmental degradation. Since there are many entitled users, the number of operators is abnormally high, leading to resource overexploitation. Moreover, common land use implies competition among users, which drives them to overexploit the resources (…). A sharing arrangement, if not implying equality, at least requires some equity, close monitoring, guidance, and control over the investments made (such as wells dug and the methods of land development adopted). The authorities find it difficult to resist this strong trend, which challenges the foundations of the collective pastoral system and promotes, whenever possible, individual entrepreneurial exploitation" (Bourbouze 2009:31–32).

Which of the two opposing visions, that of the State or that of customary assemblies, is better oriented toward achieving the objective of a sustainable management of forest resources described by Ostrom (2010) while avoiding the scenario of a tragedy of common goods raised by Hardin (1968)? Clearly, in both cases, the vision seems extreme. Neither vision has been able to warrant any sustainability of the commons. They have not resulted in a blameless conservation due to the abusive exploitation of forest resources by logging companies. Simultaneously, they have not contributed to improving the living conditions of populations dependent on these forest resources for their survival activities.[18]

In light of the facts, the rationale underlying the sharing of the commons is hardly compatible with simplistic arguments. The exploitation of resources is not necessarily conditioned by their status: both State-managed property and traditional commons invariably show multiple examples of good and inappropriate management alike. Although verifiable, the opposition between traditional practices and the public policies pertaining to the management of the commons remains sterile and counterproductive. What is more, it leaves the following fundamental question unanswered. Should the responsibility for deforestation be attributed to certain local actors or rather to an inefficient management system unsuited to the local realities?[19]

Whatever vision is adopted, it should take into account the local particularities of the territory concerned and the needs of local populations. In essence, environmental policies cannot be dissociated from poverty reduction policies. The interdependence of these policies is critical. Positive action in the economic or social sphere will affect the environmental sphere.

Theoretically, a territory is endowed with a system of "good management" when, at the level of behavior, interdependence and coordination facilitate links between agents. This is about the implementation of moral and cultural patterns which translate into "behavior models". Obviously, a good coordination of activities is at the basis of optimality. Institutional efficiency is reflected in the existence of established rules to manage these links through the organization of individual behaviors.

This would amount to saying that the actions of the various agents should be consistent with the objectives imposed by "good management". If that is not the case, those in charge should exert a constraint for the respect of these rules. Sanctions are taken in the event that these rules are transgressed. Only the integration of these norms into the behavior of each actor will determine their strategies with regard to the environment. Concretely, either in the case of taxation policies through payments for the exploitation of a resource, such as wood, or in that of conservation through the establishment of protected areas for eco-tourism, the number of free riders remains important. Oddly, "the intervention of the forester in the field of pastoral restoration thus grants him a special standing among rural populations; his role shifts from the difficult task of protection to the sphere of wealth-creating activity" (Aubert 2013:311).

[18] "(...) Establishing a nature park solely aimed at preserving the ecosystem and protecting wildlife in a region where the population has a very low standard of living and demands attention would be socially unacceptable. It would face such hostility that, in the long run, the remedy would be worse than the problem" (Bourbouze 2009:29).

[19] "The Agdal is both a physical and socio-cultural space. It is primarily an annual meeting and convergence area for herders from different clans or tribes. The sustainability of this structure hinges on the answer to a central question: How can collective discipline be established to manage these spaces and ensure their shared use by herders from different tribes and clans? How can conflicts be prevented and competition among these different users be regulated?".

In a country marked by extreme poverty, awareness-raising policies stumble against this harsh reality, which is as unavoidable as it is vicious. At best, a proper functioning of a system begins with a similar collective representation in the mind of every community member. At worst, we would envisage a minimum agreement on fundamental benchmarks for the management of the commons.

19.5 Who is in Charge?

The management of the commons raises the question of trust between national and local stakeholders. If the confidence of the population is to be recovered by foresters, will the administration, for its part, be able to trust the population? The answer to this question depends on an adequate understanding of the reasons behind this mutual suspicion.

First, the idea of entering into a discussion with the population does not appeal to everyone. Historically, the forester has never been in a position to ask locals anything. Likewise, the local populations cherish a strong feeling of being free in the land of their ancestors. For fear of losing their authority, forest officers defend their exclusive right to make decisions. Ironically, the administration does not intend to relinquish its powers in favor of communities whose cooperation it desperately seeks. Some foresters consider the participation of populations as a constraint with which they should deal. Therefore, they are wary of any opportunity that might be used by the *Agdal* communities to reclaim their own land. As Aubert relevantly point out "The negotiation with rural populations does not aim to integrate their needs or management modes into the administration's management methods but to exclude them from these forests in a different manner" (Aubert 2013:310).

Second, the individualization of the relationship with the natural resource may be at the origin of its depletion and perhaps even behind the emergence of potential conflicts. The individualization itself is generated and accelerated by the mechanisms resulting from public policies, definitely too often exclusive and disempowering. Then, the market imperatives have come to surf on such policies.[20] These same concerns have largely contributed to the exclusion and downgrading of the long-established practice of sharing and solidarity between rights holders (Romagny et al. 2018:65). Through the decay of the *Agdal*'s collective system, the notion of the common good is therefore losing momentum in rural Morocco (Romagny et al. 2018:67).

Living in extreme precariousness, some community members adopt a short-term thinking. The socioeconomic pressures of "here and now" override the medium- and long-term environmental future commitments. The breeders, for instance, are too destitute to resist the temptation to lead their herds to young plantations. However, should we consider poverty as necessarily a cause behind the destruction of the

[20] Michael Sandel. What Money Can't Buy: The Moral Limits of Markets, Farrar, Straus and Giroux, Year: 2012.

commons? If the answer to this question is affirmative, how can one justify that such a traditional governance mode has lasted for centuries? This vision perceives the population as an endogenous socioeconomic constraint that must be controlled. According to its tenets, the curse of the *Agdal*'s resources lies with the vulnerable inhabitants that populate it. However, are rural populations themselves able to play such a role given their high level of illiteracy and attachment to beliefs and practices often at odds with development objectives?

According to Chauveau (1998:157), "every project, whether participatory or not, triggers a hidden participation by the populations that differs from the one authoritatively assigned (in the case of non-participatory projects) or encouraged (in the case of participatory projects) by the project. Project evaluations conducted with a deep understanding of the local context reveal several common trends, regardless of whether the projects are participatory:

- Populations reinterpret projects in their own way, according to the social dynamics and local divisions that preexist the projects;
- They inevitably engage in "unplanned participation" in relation to the projects, either to share the new resources generated by the projects or to secure or strengthen their position in relation to specific local "issues" regulated or disputed by different groups of actors;
- Populations have the ability to create a "margin of maneuver" in relation to the objectives and means of the projects and to develop a whole range of active strategies that anticipate the effects of the projects on local dynamics, either to protect, strengthen, or transform them;
- They integrate the new resources and constraints brought by the projects into their "off-project" strategies;
- Populations have the capacity to influence the trajectory of the projects, either by neutralizing them, marginalizing them, selectively breaking them down, or adopting them while diverting them from their objectives.

In the same vein, the State, centralizing and authoritarian, does not always have the means of its top-down policies. The image of the ranger[21] as a policeman did not have the expected deterrent effect. On the contrary, it has made, among the rural population, the fear of sanctions more critical than the destruction of an entire ecosystem of which the *Agdal* is the symbol. Instead of a genuine participatory forest management, the system of fines becomes the only connection point between the rural population and the forest ranger. Such a system is ineffective since infringements proliferate. On the contrary, these penalties only encourage informal arrangements at the basis of opportunistic behavior (North 1990). Moreover, the degree of corruption remains an obstacle to the credibility of sanctions. Additionally, close social relations make it difficult to impose sanctions within the group. Their validation by an authority

[21] "If you want to succeed in the forest, you must succeed beside it. And we need people to believe in us. We've been in the forest for 100 years, and they've never believed in us. For them, the forester is an 'enemy,' in quotation marks. To improve this perception, it takes time. And here, we brought people into the forest who are not foresters, to create a sort of intermediary. That was the project" (A regional manager, former project leader, Meknes, December 2008, cited in Aubert 2013:307).

outside the group implies a coercive power. The latter will seek to legitimize the resort to institutions outside the tribe for the application of the established rules (Barrière 2012:231).

19.6 Local NGOs

The emergence of local NGOs is taking place in a context that is not quite favorable to the development of practices around the commons. It is characterized in particular by the partial disengagement of the State coupled with a tendency to disintegrate collective action. Of course, the socio-spatial structures of the *Agdal* have changed. The eruption of the private operator and the development of individualism, mentioned earlier, have resulted in a significant dissolution of traditional communities. This dislocation resulted in a restructuring of the very concept of *Agdal* and gave rise to new forms of management of the commons. Thus, with the associative actor, a new form of local governance is initiated in connection with the structural changes undergone both by the central State and traditional communities.

Since there seems to be no trickledown effect of national State policies on local populations, NGOs emerged as a new actor on the management front of the commons. To compensate both for the inadequacies in the State's national policies and the weakness or even disappearance of customary communities, the State has designed a policy for the promotion of local NGOs. In short, the State intends to establish the vision of a regime that will include the *Agdal* in a collaborative local governance system.

Undeniably, the sustainable management of natural resources requires a solid consensual base. It would have to constantly deal with the dilemma of the commons. It must guarantee immediate equitable access to shared common resources without undermining long-term environmental biodiversity commitments. NGOs are meant to reshape the link to be established between local populations and the *Agdal* as an ecological space. The bond does not consist in a vertical power relationship. Rather, it would serve as a vehicle for a horizontal transmission of an obligation, to the rising generations, of both a natural and cultural heritage on which the future of the whole community revolves. Thus, the common heritage is founded on a relationship of commonly shared obligations. It defines the perimeter of a link in terms of rights and duties. It is neither a status nor a power relationship. It is a commitment of an entire community with regard to the future of the commons (Barrière 2012:236).

The choice of NGOs as a necessary governance framework for the foundation of local collective action should, in principle at least, respond to this imperative. It is called upon to succeed where the traditional institution of *Agdal* has lost impetus and, by extension, to compensate for the insufficiency of public policies thereof. With this objective in mind, the forestry administration encourages pastoralists to organize themselves into pastoral local NGOs. Its approach aims at finding an interlocutor within the population, which has gone astray with the disappearance of the *jmaâ* and the rise of individualism. Yet, the target communities perceive this change, hence their fear, as a way for the administration to (re)assert its power rather than an authentic opportunity for the local populations to regain their territory. Such an apprehension partly finds its justification in the attitude of some administration officials who believe there is no need to give local communities additional usufruct rights. This conflict of visions takes the air of a test of legitimacy between central power and local counter-power (Bnidir 2010:98).

Local NGOs must make it possible to reconnect with a collective and not individual vision of the commons, where the offender finds himself sanctioned, not through fines, but because everyone comes to dine at his place. The social control of behavior and practices explains the adoption of regulations. Its purpose is not the sanction but the disapproval of a social conduct. The various conventions are the expression of a model of behavior to be followed without which a pecuniary penalty is mandatory (Barrière 2012:228).

Nevertheless, the downside is that these associations are active in a marginalized space. Their demands lie outside the centers of power likely to shape the local organization of power by means of the associative action (Cefai 1996). According to Bnidir (2010:104), it is a race for leadership between two apparently antagonistic groups: On the one hand, association activists, either deriving from notable families or development agents, are outsiders in the local political arena; conversely, the local elected representatives stand for the typical ideal figure of the insider seeking to maintain their communal leadership and maintain at all costs the status quo. Regardless of whomever line prevails, one of the major difficulties of these local NGOs stems above all from the lack of political and entrepreneurial experience of their members insofar as it comes to the management of the *Agdal*.

In the end, the main issue remains always related to the diverging interpretations relating to "public good" and "local collective interest", as well as the designation of those who are their actual legitimate custodians. As Chauveau (1998:168) points out "(…) Development interventions fundamentally fall within the broader field of development, which involves the confrontation of different logics and cultures and, above all, unequal power relations. This means that the actors involved (the State, projects, NGOs) accumulate the capacity to intervene and financial resources, and they explicitly or implicitly attempt to impose a model that is inevitably misaligned with local social and economic realities (and sometimes even technical realities)".

19.7 The Agdal, a Mutating Institution[22]

Let us admit, like Chauveau, that development interventions all inevitably fit into the overall framework of development. It follows that the different types of stakeholders, the State, the private operator or the NGOs, end up creating a kind of gap with local social and economic realities. Within this context of clash between different priorities and mindsets, it can be argued that the main feature of this traditional form

[22] See Dominguez (2012:286–292) and Bourbouze.

of organization can be viewed in two ways. The first view relates to a balanced and invariable customary institution at the service of a flexible management intimately adjusted to a complex and changing environment. Considering the failures of State development projects for decades, this institution has been able to promote an elastic and participatory management of the commons. The ecological, technical, and social knowledge associated with the *Agdal* has made it possible to shape a unique mutual reinforcement between the agricultural, pastoral, and forestry dimensions. Such an experiential knowledge could inspire future development projects embedded in local realities.

Yet, no matter how much vital, the success of these concepts of "flexibility" and "participation" remains dependent on how they are practiced. Indeed, this second view suggests a more turbulent image of the *Agdal* insofar as individualistic postures are likely to oppose various herders seeking to appropriate the mountainous space (Bourbouze 2009:30). In fact, the traditional forms of collective exploitation of pastoral resources have even declined. There is the weight of demographic pressure, a powerful rise in individualism, and massive interventions in the process of transformation of the rural space by the State. Accordingly, communities have been engaged in a competition and an inordinate utilization of the commons, which have provided legitimacy for the strong pressures to privatize them.

However, irrespective of the cheerful or grim outlook adopted, it is fundamental to envision the *Agdal* by recognizing the particular organic link between the ecosystem and the populations that manage it. More often than not, State and private actors find it difficult to integrate into their legislation and representations and project this essential dimension of co-dependence and co-construction between nature and traditional communities. Thus, the approaches in terms of profit, performance for market valuation, naturalist conservation, culturalist, and social rural development methods have reasonably little to do with representations and knowledge premises forged in domestic and community spaces. The implication is not that the various top-down projects do not take into account local knowledge. On the contrary, they do so but only partially and selectively while disqualifying whatever issues located beyond their immediate concerns.

On the one hand, this dissociation is amplified by the strong sectorization of projects (forest, arboriculture, culture, and ecosystem). Local actors are, therefore, left with little room for the involvement and implementation of their ancestral knowledge. Local knowledge is designed to manage the integration/inter-operability of the functions within the same space. The *Agdal* institution is a local laboratory for the practice of democratic governance. Consequently, governance, which gives full place to local institutions, is thus marginalized. Political and economic considerations reinforce the ongoing weakening of local knowledge, on the other hand. Local communities, having cohabited for centuries with their ecosystem, morph, as we said above, from an ally into a threat that should be reprimanded and combated.

The fate of the Agdals depends on the perception and discourse of change developed about them.

19.8 Conclusion

Throughout this chapter, we have tried to briefly define the key concepts used namely *Agdal* as a "common", underlying the *Agdal* as an institution. We have delineated the three different angles from which the concept of "commons" is usually approached, that is the State, the private, or the community. Then, we have tackled the crisis of legitimacy of this ancestral institution in its confrontation with the State and private actors.

We have succinctly portrayed the competing visions of the different protagonists/stakeholders and their impact on the community management of the *Agdal*. Subsequently, we have tried to describe the conflict of interest between the vertical State approach and that of the traditional management of *Agdal* based on the trust factor between stakeholders. Finally, we have argued how the withdrawal of the State, the rise of the private operator, and the weakening of traditional communities allowed a new actor, the NGO, to emerge. Finally, we have raised the question of whether the commons, of which the *Agdal* institution is the symbol, constitute an ongoing or declining ancestral collective form of practice of commons.

Either the *Agdal*'s perpetuation or disappearance depends on community practices, as stated earlier, that decides to self-organize in order to benefit from its resources as a common good. Throughout the centuries, customary assemblies have been responsible for managing the conditions of access in order to protect these resources against various threats. The risks may emanate from within communities themselves through individual free-riding behaviors but also in particular from the appropriation by private actors. A common can take the form of a material natural resource (forest, pasture, rangelands) or immaterial (values). The pursuit of a financial goal is never the main motivation of an individual who engages in a community so to speak. In other words, social values cannot be reduced to economic ones. Within the *Agdal* institution, the commons represent more than a mere market value.

While private property is based on exclusion, the commons, on the contrary, are based on the principle of inclusion. The values of solidarity, flexibility, and tolerance have, over many centuries, formed fortifications[23] against the principle of market access to these collective goods. The resilience of the ancestral institution of *Agdal* relies on the ability of communities to maintain the integrity of their commons at all times. In his attempt to explain the strength of resistance of this traditional High Atlas institution in a hostile context, Bourbouze highlights the values of "flexibility" and "tolerance" when he asserts, with relevance, that "the first quality of the organization is simplicity, which relies on a few principles: defined territories and identified rights holders, restrictions (rather than prohibitions) on the rights to build shelters, cultivate land, take animals together, and watering rights. It also involves

[23] Geographic isolation also contributed to the resistance of the Agdal: "The [isolation] has certainly protected these systems from at least two highly destabilizing factors: strict State control and the influx of urban capital. State projects for rangeland management that have emerged over the past thirty years have generally demonstrated great clumsiness and have shown an inability to consider customary management where it still existed".

the establishment of *Agdal* (a traditional system of communal land management), which requires a decision-making body, a system of guardianship, and sanctions". As pointed out above, demographic pressure, the rise of individualism and market values constantly challenge community resistance. Collective tenacity remains conditioned by community vigilance regarding the emergence of possible enclosures.

Curiously, this difficulty in tangibly measuring the value of the commons strikes them with the seal of economic invisibility. The engagement of actors in a traditional community cannot be approached algorithmically. It seems hardly conceivable if not impossible, in the logic of the market, to understand that non-market practices produce value. Admittedly, a model of remuneration for the actors of the commons struggles to exist. Yet, thinking or re-visiting the commons, like Ostrom, makes the search for new paths crucial. The commons, as we saw above in the angles of approach, are a sort of third economic path lying between the State and the market.

We consider this research as an attempt to synthesize a large body of general literature related to the study of the commons, and to specific studies of the *Agdal* in particular. Obviously, such an endeavor would be ineffective unless it is followed by further relevant theoretical and empirical examination. By way of conclusion, we will say, in line with Bensaid's insight (2007:44), that "just as a traveler does not own the road on which he is traveling, likewise, a farmer does not own the field on which he is sowing. All capital, whether material or intellectual, being a collective work, therefore forms collective property [...] All accumulated capital being social property, no one can have exclusive ownership of it".

References

Albanese, van Fleet (1985) Rational behavior in groups: the free-riding tendency. Acad ManagemenI Rev 10(2):244–255. https://www.researchgate.net/publication/256528388_The_free_riding_tendency_in_organizations/link/5be314ff299bf1124fc2d31f/download

Aubert PM, Romagny B (2012) De l'économie néo-institutionnelle et patrimoniale à la sociologie de l'action organisée. In: Auclair L, Alifriqui M (eds) Agdal. Patrimoine socio-écologique de l'Atlas marocain, IRCAM—IRD éditions, 649 p, pp 245–282. http://horizon.documentation.ird.fr/exl-doc/pleins_textes/divers13-07/010059469.pdf.

Aubert PM (2013) Les évolutions de la politique forestière au Maroc: entre réappropriation du modèle forestier français et idéalisation de la tribu. http://documents.irevues.inist.fr/bitstream/handle/2042/53627/RFF_2013_4_305_Aubert.pdf

Auclair L, Baudot P, Genin D, Romagny B, Simenel R (2011) Patrimony for resilience: evidence from the forest Agdal in the Moroccan High Atlas Mountains. Ecol Soc 16(4):24. 10.5751IES-04429-160424

Auclair L, Alifriqui M (dir.) (2012) Agdal: patrimoine socio-écologique de l'Atlas marocain. Rabat (MAR); Marseille: IRCAM ; IRD, 647 p. (Publications de l'Institut Royal de la Culture Amazighe.Centre d'Etudes Historiques et Environnementales : Série Colloques et Séminaires ; 29). Colloque sur la Gestion de la Biodiversité des Agdals, Marrakech (MAR), 2007/05/10-12. Auclair et al., 2013

Auclair L (1996) L'appropriation communautaire des forêts dans le Haut Atlas marocain. Les cahiers de Sciences Humaines (1):177–194

Auclair L (2012) Un patrimoine socio-écologique à l'épreuve des transformations du monde rural. In: Auclair L, Alifriqui M (eds) Agdal. Patrimoine socio-écologique de l'Atlas marocain, IRCAM—IRD éditions, 649 p, pp 23–72. http://horizon.documentation.ird.fr/exl-doc/pleins_textes/divers13-07/010059469.pdf

Barrière O (2012) Nature juridique de l'Agdal. De la propriété collective au patrimoine commun. In: Agdal. Patrimoine socio-écologique de l'Atlas marocain, IRCAM—IRD éditions, pp 209–244. http://horizon.documentation.ird.fr/exl-doc/pleins_textes/divers13-07/010059469.pdf

Bensaïd D (2007) Les dépossédés. Karl Marx, les voleurs de bois et le droit des pauvres, La Fabrique, 82 p

Bnidir M (2010) Leaders associatifs et élus locaux au Maroc : épreuves de face-à-face et controverses dans les arènes du développement, Karthala, Politique africaine, 2010/4 N° 120, pp 87–104. https://www.cairn.info/revue-politique-africaine-2010-4-page-87.htm

Bonnin C et al (2021) Adaptation et hybridation des communs en territoire Aït Oucheg, Haut Atlas, Maroc. In: Revue de géographie alpine N°109–1, https://horizon.documentation.ird.fr/exl-doc/pleins_textes/2021-07/010081528.pdf

Boonen C, Brando N (2016) Revisiting the common ownership of the earth: a democratic critique of global distributive justice theories. Glob Justice Theory Pract Rhetor 9(2):134–154. https://doi.org/10.21248/gjn.9.2.115

Bourbouze A (1999) Gestion de la mobilité et résistance des organisations pastorales des éleveurs du Haut Atlas marocain face aux transformations du contexte pastoral maghrébin. In: Naimir Fuller M (ed) Managing mobility in african rangeland: the legitimization of transhumance. IT publications: FAO and Beijer Inter. Instit. Ecolog. Economics, London, pp 236–265. https://ter ramed.iamm.fr/images/publications/6_gestion_mobilite_resistance.pdf

Carriere M (1995) Impacts des systèmes d'élevage pastoraux sur l'environnement en Afrique et en Asie tropicale et subtropicale aride et sur-aride, Centre de Coopération Internationale en Recherche Agronomique pour le Développement/ Département d'Elevage et de Médecine vétérinaire (CIRAD-EMVT). https://agritrop.cirad.fr/520761/1/ID520761.pdf

Cefaï D (1996) La construction des problèmes publics. Définitions de situations dans des arènes publiques, pp 43–66. https://www.persee.fr/doc/reso_0751-7971_1996_num_14_75_3684

Chauveau (1998) Communiquer dans l'affrontement: la 'participation cachée' dans les projets participatifs ciblés sur les groupes ruraux défavorisés. https://horizon.documentation.ird.fr/exl-doc/pleins_textes/divers16-11/010068525.pdf

Clancy EA (1998) The tragedy of the global commons the tragedy of the global commons. Indiana J Glob Leg Indiana J Glob Leg Stud. Indiana University School of Law 5(2). https://www.repository.law.indiana.edu/cgi/viewcontent.cgi?article=1136&context=ijgls

Cochran CE, Simon, YR (1978) The common good": a note on the concept. Ethics 88(3):229–239

Coward EW (1980) Principles of social organization in an indigenous irrigation system. Hum Organ 38:28–36. https://doi.org/10.17730/humo.38.1.764782138jh83416

De los Reyes RP (1980) Communal gravity systems: organizational profiles. Institute of Philippine Culture, Quezon City

Dominguez Gregorio P (2012) Vers l'éco-anthropologie. Une approche multidisciplinaire de l'agdal pastoral du Yagour (Haut Atlas de Marrakech), 428 p

Domínguez Mohamed Mahdi y Pablo, Regard anthropologique sur transhumance et modernité au Maroc. https://www.redalyc.org/pdf/296/29611869003.pdf

Dupré L (1993) The common good and the open society. Rev Polit 55(4):687–712

Frischmann BM, Marciano A, Ramello GB (2019) Tragedy of the commons after 50 years. J Econ Perspect 33(4), Fall 2019:211–228

Giraud G (2020) La recherche en économie et les défis posés par le désastre environnemental sont aujourd'hui des périphéries. In: Revue internationale et stratégique, 2020/3 119:6–18

Gordon H (1957) The economic theory of a common property resource: the fishery. J Polit Econ 62:124–142

Hardin G (1968) The tragedy of the commons. Science 162:1243–1248. https://doi.org/10.1080/19390450903037302

Herzenni A (2012) L'*agdal* à l'épreuve des stratégies familiales et de l'action collective, le cas du douar Ifrane dans la vallée des Ayt Bouguemmez. In: Agdal. Patrimoine socio-écologique de l'Atlas marocain, IRCAM—IRD éditions, pp 375–410. https://www.researchgate.net/profile/Ramou-Hassan/publication/338121567_Les_agdals_d%27Imilchil_Modalites_de_gestion_et_mutations/links/5e00c19ca6fdcc2837389635/Les-agdals-dImilchil-Modalites-de-gestion-et-mutations.pdf

Ilahiane H (1999) The Berber agdal institution: indigenous range management in the Atlas mountain. Ethnology 38(1):21–45. https://doi.org/10.2307/3774085

Ilbert H (2011) Les indications géographiques, outils de construction des terroirs et de la biodiversité à l'aune des règles internationales du commerce :quelles tendances en Méditerranée ? », dans Delfosse C. (dir.), La mode du terroir et les produits alimentaires, Paris, Indes savantes, pp 301–321

Kutz (2000) Acting together. Philos Phenomenol Res 61(1):1–31. http://pacherie.free.fr/COURS/MSC/Kutz-Acting-togethe-PPR-2000.pdf

Lahssini S, Vanuxem S (2016) Les Paiements pour Services Écologiques ou la reconquête de la Mâamora par ses habitants? Enquête sur une forêt emblématique du littoral méditerranéen marocain. Développement durable & territoires 7(1), avril 2016. https://doi.org/10.4000/developpementdurable.11165

Lahssini S, Vanuxem S (n.d.) Les Paiements pour services Écologiques ou la reconquête de la Mâamora par ses habitants? https://journals.openedition.org/developpementdurable/11165

MacKenzie WC (1979) Rational fishery management in a depressed region: the Atlantic Ground fishery. J Fish Res Board Can 36:811–826. https://doi.org/10.1139/f79-119

McAfee K (2012) The contradictory logic of global ecosystem services markets. Dev Chang 43(1):105–131. https://doi.org/10.1111/j.1467-7660.2011.01745.x

Menatti L (2017) Landscape: from common good to human right. Int J Commons 11(2):641–683. https://doi.org/10.18352/ijc.738

North DC (1990) Institutions, institutional change and economic performance. Cambridge University Press

Olson M (1965) The logic of collective action. Public Good and the Theory of Groups, Harvard University Press

Ostrom E (1990) Governing the Commons: The Evolution of Institutions for Collective Action. Cambridge University Press, New York

Ostrom E (2010) Beyond markets and states: polycentric governance of complex economic systems. Am Econ Rev 100(3):641–672

Pascon P (1965) Désuétude de la jmaa dans le Haouz de Marrakech. Les Cahiers de Sociologie édités par l'Institut de sociologie de l'Université Mohammed V septembre - octobre - novembre 1965, no 1, 185–194

Pigou AC (1920) The economics of welfare. Macmillan, London

Romagny B, Aderghal M, Auclair L, Ilbert H, Lemeilleur S (2018) Communs en crise Agdals, terres collectives, forêts et terroirs au Maroc. Revue internationale des études du développement 2018/1 (No 233), pp 53–73. https://www.cairn.info/revue-internationale-des-etudes-du-developpement-2018-1-page-53.htm

Tozy M, Mahdi M (1990) Aspects du droit communautaire dans l'Atlas marocain. Droit et Société, Année 15, pp 203–210. https://www.persee.fr/doc/dreso_0769-3362_1990_num_15_1_1077

Warming J (1911) On the rent of fishing grounds. Hist Polit Econ 15:391–396

Weiss Edith Brown, Climate Change, Intergenerational Equity, and International Law, Vermont Journal of Environmental Law, Vol. 9, No. 3 (Spring 2008)

Wittfogel KA (1957) Oriental despotism. A comparative study of total power. Yale University Press, New Haven

Aziz Radi Dr. Radi received his PhD (1993) in Pragmatics from the *Marc Bloch Humanities University* in Strasbourg, France. While preparing his doctoral dissertation, he started a teaching career in France. He authored two English language manuals: *Speaking Already* (2005) and *First Steps into English* (2007). He worked on the perception of North African immigrants in the European media. In this regard, he was invited as a guest lecturer to the *Center of Refugees Studies, Oxford University*, UK. He gained a Fellowship in 2010 on Civic Education, Citizenship and Leadership at Maxwell School of Citizenship, Syracuse University, N.Y. His research mainly focuses on issues such as citizenship, civic engagement, protest movements, and cultural mediation. He is currently an *English for Economics* lecturer at the *Faculty of Law* in the *Economics Department, Cadi Ayyad University* in Marrakesh, Morocco.

Elhoucine Chougrani Dr. Chougrani is a professor of International Law and International Relations at Cadi Ayyad University, Faculty of Law, Economics and Social Sciences, Marrakech. He is the director of Arab Journal of International Law (ARJIL) and a member of the International Center for Crisis Management Research. His research focuses mainly on international environmental law, with recent publications on water, climate justice, and the rights of future generations. He was born on May 1, 1971, in Morocco. His doctoral research is entitled "The European Investment Bank and Environmental Protection in the Mediterranean Region" (2003). His latest publication in Arabic is entitled "The Rights of Future Generations in the Light of Current Developments in the Arab World" (Arab Center for Political Research and Studies (Lebanon, 2018)), and "Contradictions in International Law. A Preliminary Analysis" (Center for Arab Unity Studies (Lebanon, 2019)). Dr. Chougrani El Houcine also has extensive professional experience as an accountant general at the Ministry of Economy and Finance (2002–2014).

Chapter 20
The Role of Exchange Collectives in the Agroecological Transition in Morocco

Mouad Chari and Larbi Aziz

Abstract The transition to sustainable agriculture is now a primary focus of agricultural policy concerns. In this context, agroecology emerges as a promising solution that simultaneously ensures food security, resilience to climate change, and especially the preservation of biodiversity, which is considered one of the principles for sustainable agriculture. The adoption of agroecological practices that respect the biological diversity is encouraged by groups of farmers and/or consumers, including RIAM—Network of Agroecological Initiatives in Morocco. This article analyzes through an empirical study, the cognitive factors influencing this transition within RIAM. It aims, firstly, to understand how the adhesion of agroecological farmers to a collective improves their practices and supports them in their transition, and secondly, to identify how knowledge is generated and shared within this collective. Ultimately, the study's findings could guide future strategies for fostering agroecological practices and strengthening community-based initiatives in sustainable agriculture.

Keywords Agroecology · Biodiversity · Collective · Knowledge · Morocco

20.1 Introduction

It is widely acknowledged that intensive agriculture has led to the homogenization and standardization of production methods (shortening of rotations and reduction in the number of cultivated and raised species). This approach aimed at achieving economies of scale and simplify work has led to the specialization of production systems and territories based on their comparative advantages (Meynard et al. 2013). Intensive agriculture is responsible for negative effects on the environment

M. Chari (✉) · L. Aziz
Department of Rural Development Engineering, National School of Agriculture of Meknes, Meknes, Morocco
e-mail: mchari@enameknes.ac.ma

L. Aziz
e-mail: laziz@enameknes.ac.ma

M. Behnassi et al. (eds.), *Food Systems and Biodiversity in the Context of Environmental and Climate Risks*, https://doi.org/10.1007/978-3-031-89167-0_20

(Touzard and Fournier 2014), leading to the exploitation of renewable resources (land, soil fertility, water, and biodiversity), surpassing the recovery capacity of many ecosystems and resources (Gliessman 2015; Altieri 2018). This situation has been exacerbated by the increased variability in precipitation due to climate change.

To address this situation, many authors suggest that implementing agroecological practices and diversifying crops is a viable alternative. These practices represent a paradigm shift as they aim not only to reduce environmental impacts but also to enhance biological diversity and associated ecosystem services (Scherr and McNeely 2008). Furthermore, by focusing on biodiversity, crop rotation, and polyculture practices, agroecology enhances the resilience of agroecosystems to climate change (Wezel et al. 2009).

However, the implementation of ecological processes in agriculture faces cognitive challenges. These include a lack of knowledge about these processes and their interactions with other elements of the cropping system, and the sensitivity of these processes to specific context. Additionally, there is a need to develop local references, and the necessity for farmers to reconnect with observational skills and knowledge in order to assess these processes and responds accordingly (De Tourdonnet et al. 2013).

In this regard, the formation of collectives is a major challenge in the agroecological transition, enabling the construction of new agricultural knowledge. These collectives, where knowledge and local expertise are shared, offer a valuable insight into the evolution of agricultural practices (Vergote et al. 2019). These collectives also serve to build relationships and knowledge exchange among farmers, making them important learning spaces in the pursuit of new forms of modernity (Deléage 2012). Similarly, these collectives are considered as drivers of change (Ansaloni and Fouilleux 2006), generating knowledge (Darré 1994, 1996; Compagnone 2014) and providing references for technical knowledge, thereby mitigating the risks and uncertainties perceived by farmers when considering changes in practices (Nguyen et al. 2013).

The main objective of this article is to demonstrate, through a case study, the central role of exchange collectives in promoting agroecology by the production and dissemination of knowledge. Specifically, it aims to examine how joining an exchange collective, such as the Network of Agroecological Initiatives in Morocco (RIAM),[1] influences the cognitive capital of its members and facilitates the agroecological transition in Morocco. The Rabat-Salé-Kénitra region has been chosen because it has a significant number of agroecological farmers and the RIAM is more represented and dynamic there.

[1] A practice exchange collective in agroecology, founded in January 2013 with the aim to facilitate exchanges, training, and resource sharing in the field of ecological agriculture.

20.2 Methodology

20.2.1 Presentation of the RIAM Collective

The Network of RIAM was founded in January 2013 with the aim of facilitating exchanges, training, and resource sharing in the field of ecological agriculture. After operating informally for three years, RIAM officially became an association in June 2015, led by a Strategic Steering Committee comprises of various stakeholders committed to ecology and the development of sustainable agriculture.

RIAM is structured into several divisions to promote different aspects of ecological agriculture. The marketing division facilitates exchanges between farmers and consumers by establishing markets and promoting the SPG label.[2] The agritourism division encourages ecotourism initiatives that integrate agroecology. The urban agriculture division focuses on developing urban farms for environmentally friendly sustainable food production. The training division offers courses and educational materials to enhance knowledge in ecological agriculture. Additionally, RIAM participates in international discussions on healthy food and sustainable agriculture. The research and experimentation division conducts projects based on advanced scientific knowledge in agroecology (RIAM 2018).

RIAM has achieved various significant milestones, including the establishment of a specification document for plant production, developed in collaboration with stakeholders in the sector. This document is regularly updated to ensure crop quality. It comprises guidelines on various aspects, including crop selection, soil care, disease and pest prevention, seed selection, agricultural equipment, water, energy, and waste management, and the well-being of farm workers, along with provisions for ongoing training (RIAM 2018). Similarly, RIAM has implemented a Participatory Guarantee System (SPG) in accordance with the International Federation of Organic Agriculture Movements (IFOAM). This system ensures the quality of products from ecological agriculture through the active participation of stakeholders, emphasizing trust, networking, and knowledge sharing. It provides an alternative to third-party certification, promoting local markets and short supply chains.

20.2.2 Presentation of the Study Area

The Rabat-Salé-Kénitra region covers a total area of 17,570 km^2, representing approximately 2.5% of the national territory. According to the RGPH[3] of 2014, it has a population of 4.6 million inhabitants.

This region's topography is characterized by two major watersheds: the Bouregreg and Chaouia basin and the Sebou basin, which is supplied by several sub-basins.

[2] See below.

[3] General Population and Housing Census.

The hydrographic network includes major rivers, such as the Bouregreg, Sebou, Beht, Ouergha, and Rdat. It is also home to important dams, including the Sidi Mohammed Ben Abdellah and El Kansera dams. Furthermore, the region has significant groundwater aquifers, including the Gharb aquifer, Maâmora aquifer, Témara aquifer, Shoul aquifer, and Tanoubart aquifer (SIREDD 2023).

The region has a semi-arid Mediterranean climate shaped by both maritime and continental influences. Winters are mild and rainy, while summers are humid and temperate. Occasionally, the dry wind known as Chergui can affect the summer conditions. The uneven terrain causes significant local climate variations, with apparent fluctuations in temperatures (minimum temperatures can drop to approximately 4 °C, and maximum temperatures may reach up to 40 °C). During summer, maximum temperatures range from 16 to 26 °C, with occasional peaks reaching 38–40 °C, although such extremes are rare (SIREDD 2023).

The region's fertile land, abundant water resources, and developed agro-industrial infrastructure attract public and private investments in the agricultural sector, playing a crucial role in the economy of the Rabat-Salé-Kénitra region. In fact, the agriculture

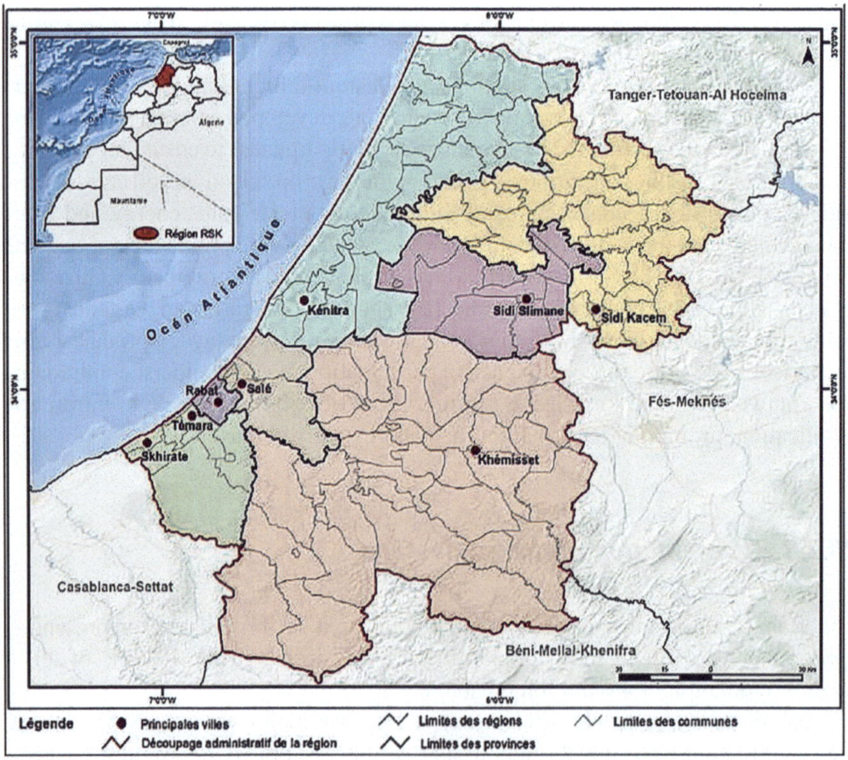

Fig. 20.1 Geographic situation and administrative boundaries of Provinces of Rabat-Salé-Kénitra Region (RGPH 2014)

sector contributes nearly 18% of the national agricultural GDP, with utilized agricultural land representing 11% of the national territory and 12.5% of irrigated areas. The region contributes over 15% of the country's cereals production, while intensive livestock farming and beekeeping are also significant economic pillars. Additionally, the region is conducive to a wide variety of crops, including citrus fruits, sugar crops, red fruits, rice, avocados, bananas, industrial tomatoes, and various other fruits and vegetables. This agricultural diversity has contributed to the development of agroecology, which has gained significant attraction in the region due to significant initiatives from RIAM.

20.2.3 Methodological Aspects

The study was conducted using a survey with two investigative tools: a questionnaire for agroecological farmers and an interview guide for RIAM board members. The questionnaire was specifically designed for agroecological farmers and included both quantitative and qualitative components, addressing the following areas: farmers' profile, their journey to agroecology, their practices before and after adopting agroecology, techniques used, knowledge exchange with peers and within RIAM, and their experience with the Participatory Guarantee System. A non-representative random sampling method, specially snowball sampling, was utilized, resulting in fifteen surveys conducted with farmers in the region and semi-structured interviews with seven managers of the collective. Data analysis was carried out using a descriptive method.

20.3 Reasons for Joining an Exchange Collective

20.3.1 Profiles of the Farmers Surveyed

Among the 15 farmers interviewed, six had adopted agroecology from the outset of their farming activities. This observation highlights a trend among agricultural farmers to choose more ecological methods from the beginning, reflecting a growing awareness of environmental issues and a strong commitment to agricultural sustainability. Notably, many of these farmers are newcomers to the field, such as retired executives and professionals, who possess a strong ecological inclination and concern for their own health as well as that of consumers.

For farmers who did not initially adopt agroecology, the transition to this sustainable mode of production was gradual. These conventional farmers progressively increased the area dedicated to agroecological practices on their farms. They have slowly introduced methods, such as using organic seeds, eliminating chemical products, applying mulching, and using drip irrigation.

The adoption of agroecological practices has allowed these farmers with greater autonomy in managing their farms, notably through the production of their own seeds. However, they have noted that this shift has led to increased responsibilities beyond production, such as in transport, marketing, and sales. This process of change also extends to their dietary and consumption habits, encouraging them to become engaged farmers and consumers who advocate for agroecology.

20.3.2 Farmers' Reasons for Joining RIAM

The reasons cited by the surveyed farmers for joining RIAM are diverse. Six farmers stated that they joined the collective to gain new knowledge, share their experiences, and enhance their skills. This is achieved through participating in training sessions where they have the opportunity to interact with experts working in the field. This highlights the importance of information exchange and continuous learning within this agricultural community. As noted by Deléage (2012), collectives facilitate relationship building and knowledge exchange among farmers, making them essential for learning and exploring new forms of modernity. They also serve as interactive spaces, where members can mutually benefit from each other's experiences and skills, as outlined by (De Silva and Ranasinghe 2017) , who found that farmers often learn more effectively from peers than from traditional information sources.

Five farmers highlighted the importance of community engagement and recognition as motivating factors for their membership in RIAM. For them, belonging to a collective allows a sense of connection to a community that shares common values and goals related to sustainable agriculture. This means that collectives enhance farmers' social networks, fostering collaboration and mutual support. In this context, collectives act as a social reference for engaged farmers (Demeulenaere and Goulet 2012). Furthermore, Prost et al. (2019) argue that the formation of strong social ties within collectives contributes to a greater resilience of production units.

Three farmers joined RIAM for commercial purpose. For these individuals, RIAM represents an opportunity to expand their professional networks and access new markets. This includes advantages in purchasing and gaining consumer trust through the SPG label. Indeed, the RIAM collective specifically contributes in establishing commercial circuits and enhancing product value. This highlights the importance of the collective in promoting and supporting the commercial activities of farmers engaged in agroecology.

One another farmer mentioned a combination of commercial opportunities, knowledge sharing, and skill enhancement as reasons for their membership in RIAM.

20.3.3 Reasons for Joining the RIAM According to Collective Managers

Farmers join RIAM for various reasons, primarily to gain new knowledge, share experiences, and enhance skills through training and peer interactions, which fosters continuous learning and community engagement. Additionally, membership provides access to professional networks and new markets, promoting collaboration, recognition, and resilience in sustainable agriculture practices. These various reasons for joining the collective were also highlighted by RIAM managers, who emphasized their motivation to promote and disseminate agroecology on both national and African scales. To achieve this, RIAM implements various initiatives aimed at training and supporting farmers in adopting agroecological practices. According to managers, farmers gain access to valuable resources and information, including farmer seeds, traditional varieties, and in-depth knowledge of agroecological practices through training and exchanges within the collective. The SPG certification also provides benefits that allow farmers to engage in agroecology and learn new sustainable practices during SPG visits. The training sessions organized by the RIAM encourage farmers to question their current methods and adopt more environmentally friendly approaches. This demonstrates that exchange collectives provide technical knowledge and references, helping to mitigate the risks and uncertainties that farmers may perceive when considering changes in practices (Nguyen et al. 2013).

On the other hand, RIAM managers observe that the collective serves as both a platform for exchanges with other agroecology stakeholders and a means to expand farmers' commercial opportunities. Indeed, networking and the complementarity within the network are considered highly beneficial, offering farmers access to a wider network of contacts, thereby facilitating the exchange of products and knowledge to enhance their practices. Additionally, SPG certification allows farmers to adopt new sustainable practices, build consumer trust, market their agroecological products more effectively, and ultimately improve their incomes.

20.4 Knowledge Production and Appropriation Within the Collective

20.4.1 The Modes of Knowledge Production and Sharing Within the Collective

The collective managers indicated that mechanisms for sharing and valuing information and knowledge have been established to foster communication and collaboration among members. Firstly, the RIAM website, its Facebook page, and WhatsApp groups are used to publish important information, such as invitations to training sessions and SPG visits, allowing members to easily access the latest updates and share their knowledge within the group. Additionally, email communication serves as an essential means for sharing more detailed and official information within the network.

Regular meetings organized by the collective and its committees provide essential opportunities for exchanging information and knowledge among members. During these in-person meetings, discussions address ongoing projects, advancements in agroecology, and the sharing of best practices and practical experiences. These exchanges intensify during the training sessions organized by the collective for its members and during the implementation of the specifications and SPG label developed by RIAM. These sessions involve more detailed and precise discussions, particularly when implementing rules related to crop selection, soil care, disease and pest prevention, seed choice, agricultural equipment, water, energy, and waste management.

Furthermore, RIAM organizes farm visits and seed exchange fairs, and seed production workshops. These annual fairs, held in partnership with 10–15 farmers, provides theoretical and practical training on seed selection and reproduction. This event has led to the establishment of seed banks and nurseries for growing vegetable, aromatic, and medicinal plants that are at risk of disappearing in Morocco. The goal is to enable farmers to gain independence and autonomy while raising awareness about the importance of local seeds, which are better adapted to local environmental conditions.

Overall, these mechanisms for sharing and valuing knowledge within the collective facilitate communication, information sharing, and pooling of knowledge among members. By utilizing these diverse communication channels, the collective can strengthen cohesion, promotes mutual learning, and stimulates innovation in the field of agroecology.

20.4.2 Farmers' Use/Appropriation of Knowledge Generated Within the Collective

Our results show that the farmers who are members of RIAM utilize the knowledge produced within the collective in various ways. Firstly, they adapt this knowledge to their own specific context, recognizing the need for a personalized approach to their agricultural practices. Each farmer tailors the knowledge to fit their environment and specific needs, seeking solutions that best align with their individual circumstances. The content disseminated by RIAM plays an inspiring role in guiding farmers through their improvement process. Thus, the knowledge produced by the collective is used flexibly and personally by the farmers. This highlights the role of collectives in co-constructing solutions tailored to the specific challenges of each agricultural region (Snapp et al. 2010).

To facilitate the appropriation of knowledge and agroecological practices, the interviewed managers stated that the adoption process unfolds through training sessions implemented using an andragogical approach that considers the context of the members. This approach includes different stages: initiation, awareness, conviction, and finally, ultimately the adoption of agroecological practices. Members gradually understand the importance of these sustainable practices and integrate them into their agricultural activities.

20.5 Building Trust and Managing Conflict: Two Key Elements for Collective Cohesion

20.5.1 Establishing and Maintaining Trust Between Collective Members

Our results indicate that trust within the RIAM collective is considered a key element for the success and sustainability of the network by both managers and ordinary members. This trust is established through several actions and opportunities. Firstly, collective members trust the network and its actors due to shared values of agroecology and a commitment to associative work. The strengthening of this trust between farmers and consumers, members of the network, is ensured by the SPG certification. The SPG label guarantees the quality of ecologically produced products through the active participation of all stakeholders. During SPG visits, consumers are part of the evaluation teams, allowing them to verify compliance to the specifications.

The training sessions organized by RIAM also provide an opportunity to build and strengthen trust among farmers, providing opportunities for exchange, learning, and meeting with other collective members. However, some interviewed farmers remain skeptical about "fraudulent" practices by certain other farmers and raising doubt about the origin of the products sold by them in eco-solidarity markets. These concerns highlight the need for greater control measures to maintain trust within the network and ensure the integrity of the products offered in these markets.

20.5.2 Conflict Management Within the Collective

The RIAM has implemented mechanisms that our respondents consider effective in managing potential conflicts and reinforcing trust among members. The collective has established a mediation committee, which intervenes in conflict situations by conducting investigations and dialogues with the concerned parties. This allows for a comprehensive evaluation of the situation and providing a resolution that is respectful to all concerned parties.

While the collective addresses conflicts reactively, RIAM has adopted preventive measures to avoid market conflicts by regularly alternating the locations of product stands. This practice ensures fairness in customer access and reducing tensions among farmers. Transparency, according to our respondents, is a fundamental principle within the collective, which fosters conflict prevention. Open discussions and communication are encouraged and considered essential tools for the proper functioning of the network.

RIAM's internal regulations are crucial with clear directives on how to handle conflicts. Article 8 of these regulations stipulates conditions under which membership may revoked, such as non-compliance with statutes, regulations, or Charter, or when a member damages the collective's image and credibility. Article 8–2 specifies the procedure for expulsion including issuance of summon to the member to be delivered by registered letter to explain their action before the Strategic Orientation Committee (COS). The COS then takes necessary measures if fault is proven. For serious misconduct, expulsion of member may be pronounced for a maximum of three years, with the obligation to return any material and prohibiting the individual from communicating on behalf of the collective. The decision and measures are communicated to the concerned member within 15 days via registered letter after the COS meeting.

Overall, these measures reflect RIAM's commitment to maintaining a climate of trust and cooperation among its members. These conflict management mechanisms play a crucial role in strengthening the collective and supporting its mission to promote agroecology.

20.6 Discussion

The implementation of ecological processes in agriculture faces cognitive challenges, including a lack of knowledge about these processes and their interactions with other elements of the cultivation system, the sensitivity of these processes to context, the need to develop local references, and the necessity for farmers to reconnect with knowledge and observational skills to evaluate processes and act accordingly (De Tourdonnet et al. 2013). In this context, we focused on analyzing the cognitive levers that shape the agroecological transition within the RIAM collective. Our results highlight the role of RIAM as a catalyst for promoting agroecology in Morocco. Indeed, a convergence of interests between the farmers and the RIAM managers has been demonstrated, underscoring the shared importance placed on raising awareness about agroecology and sharing knowledge within the collective.

The motivations of the farmers, as well as the concrete benefits they derive from their membership in this collective, confirm the effectiveness of the network in creating a dynamic community focused on learning, community engagement, and economic success. These motivations and benefits are diverse, with the most frequently cited being the exchange of knowledge and sharing of experiences within the collective. The collectives foster the building of relationships and the exchange of knowledge among farmers (de Deléage 2012). This makes learning a social process that occurs within communities of practice, where individuals share experiences, ideas, and problems to address common challenges (Wenger 1998). Such learning is often more effective because farmers get benefit from the experience of their peers rather than relying solely on traditional sources of information (De Silva and Ranasinghe 2017).

Moreover, the major changes observed among farmers following the adoption of agroecology demonstrate the positive impacts both economically and in the introduction of sustainable practices, thereby reinforcing the relevance of collective engagement. According to the RIAM managers, membership in the collective has significantly influenced the agricultural practices of the member farmers. Through access to resources and information, farmers gain a deep understanding of agroecological practices via training and exchanges within the collective. They benefit from local seeds and traditional varieties. Thus, exchange collectives provide references and technical knowledge, mitigating the risks and uncertainties perceived by farmers when considering changes in practices (Nguyen et al. 2013). This dimension is crucial because the adoption of agroecological practices can reduce yields and increase costs related to labor, certification, monitoring, and sometimes inputs (Côte et al. 2022). Therefore, this exchange of knowledge and experiences through strong social connections within collectives could enhance resilience of farms (Prost et al. 2019).

Being a part of RIAM allows farmers to feel connected to a community that shares similar values and goals regarding agroecology. This community engagement is seen by farmers as a benefit of their membership in an exchange collective, serving as a social reference point (Demeulenaere and Goulet 2012) from them.

In summary, the benefits of membership in RIAM, such as knowledge acquisition, network strengthening, and commercial gains, positively contribute to the evolution of agricultural practices. The essential role of the collective in providing technical support, promoting commercial circuits, and enhancing agroecological products is evident, demonstrating RIAM's positive influence on the agricultural activities of its members.

Moreover, farmers within RIAM adapt the knowledge produced within the collective to their specific context, recognizing the need for a tailored approach to their agricultural practices. Each individual appropriates the knowledge based on their environment and specific needs. The content disseminated by RIAM plays an inspiring role in this process, guiding farmers in their improvement and appropriation efforts. This indicates that the co-construction of solutions tailored to the specific challenges of each agricultural region can be significantly facilitated by collectives (Snapp et al. 2010). This adaptability is crucial because agroecology relies on context-specific knowledge and does not offer universal solutions (FAO 2018). To achieve this adaptability, farmers combine their own knowledge with the experiences of their peers and the external technical knowledge gained from the RIAM's training. These findings align with those of Vergote et al. (2019) which have shown that farmers combine experiential knowledge with scientific knowledge. In addition, the study identifies several weaknesses experienced by RIAM. According to the collective's board members, these weaknesses include the lack of institutional recognition, insufficient financial resources, challenges related to management, regulation, volunteer-based model, and resistance to change within the collective. They find that management poses a challenge for the collective, as decision-making and coordinating activities become difficult with the limited resources. Furthermore, they believe that the volunteer-based operational model has contributed to these management difficulties. Additionally, the collective's internal regulations require constant adjustments to meet evolving needs. While the collective aims to grow or diversify, this can trigger reluctance due to the uncertainty and risk associated with changing the collective's economic or operational model. This resistance is manifested by the limited ambitions among some 'activist' members who do not wish to advance the collective to a more mature stage. They fear that professionalizing the collective, characterized by adopting standards, recruiting human resources, and implementing professional management practices may compromise the friendly and supportive atmosphere they have experienced. For these members, the volunteer-based model represents a form of freedom in management. This resistance may be related to the complexity of coordinating with various stakeholders, as suggested by Dumont et al. (2014), who found that the management and coordination of collectives can be complex, especially when engaging diverse actors in agriculture. Furthermore, the volunteer-based operational model can create organizational issues and influence the work environment within the collective. Decision-making and participation can be challenging, particularly as the number of members increases and geographic distances widen. This can negatively impact the member engagement and the long-term sustainability of the collectives, as highlighted by Lamine et al. (2019).

The diversity of RIAM members, while potentially enriching, can also present challenges in decision-making and aligning goals. The collective comprises both agroecological farmers and consumers, who may have different objectives. In this regard, Birkin et al. (2018) found that the participant's diversity can complicate decision-making and goal alignment. This complexity is tied to the challenges of coordination and management, as divergent aspirations can make decision-making more difficult (Dumont et al. 2014). The predominance of elite collective's members raises questions about the inclusive impact, and potentially limiting its reach if it fails to integrate other categories important to the agroecological transition, such as small-scale farmers. Moreover, succession represents a crucial challenge for the collective. This involves the gradual transfer of the collective's responsibilities, skills, and leadership from current members to newly recruited members. This process is essential to ensure the continuity and sustainability of the collective, enabling an effective transition of power and knowledge.

20.7 Conclusion

The RIAM model demonstrates the importance of collective commitment, knowledge sharing, and trust in promoting more sustainable farming practices. Membership of this exchange collective has enabled its members to move positively and progressively toward environmentally friendly farming practices. Moreover, the flexible use of the knowledge produced by the collective demonstrates the adaptability and customization required for the successful adoption of agroecological practices. In addition, the collective has put in place conflict management mechanisms that have helped to maintain an environment conducive to its survival. However, to develop further at national level, RIAM needs to promote agroecology among small-scale producers, involve stakeholders concerned with this type of agriculture, and improve the communication and management skills of its leaders. In addition, it is asked to support circular farming practices, organize training for young people and small producers, and develop a strategy for promoting agroecological products to consumers.

References

Altieri MA (2018) Agroecology, the science of sustainable agriculture, 2nd edn. CRC Press, Taylor and Francis Group, Boca Raton London New York

Ansaloni M, Fouilleux E (2006) Changement de pratiques agricoles. Acteurs et modalités d'hybridation technique des exploitations laitières bretonnes. Économie rurale. Agricultures, Alimentations, Territoires 292:3–17

Birkin F, Martin G, Gibson RB (2018) How local networks facilitate the adoption of agroecological farming practices: a case study from New Zealand. Agric Hum Values 35(1):1–14

Compagnone (2014) Les viticulteurs bourguignons et le respect de l'environnement. Réseaux de dialogues professionnels et dynamique de changement. Revue française sociol 55(2):319–358

Côte FX et al (2022) Levers for the agroecological transition of tropical agriculture—agronomy for sustainable development. SpringerLink, 5 July 2022. https://doi.org/10.1007/s13593-022-00799-z

Darré J-P (1996) L'invention des pratiques dans l'agriculture : vulgarisation et production locale de connaissance. Karthala, Paris

Darré JP (1994) Pairs et experts dans l'agriculture: dialogues et production de connaissance pour l'action. Érès Paris

De Silva S, Ranasinghe K (2017) Farmers' learning and dissemination of agricultural information through social networks: the case of rice farmers in Sri Lanka. Inf Dev 33(3):261–272

De Tourdonnet S, Brives H, Denis M, Omon B, Thomas F (2013) Accompagner le changement en agriculture : du non labour à l'agriculture de conservation. Agron, Environ Soc 2(3):19–27. Touzard

Deléage E (2012) Les paysans dans la modernité. Revue Française Soc-Écon 1(9):117–131. https://doi.org/10.3917/rfse.009.0117

Demeulenaere É, Goulet F (2012) Du singulier au collectif. Terrains Trav 1:121–138

Dumont B, Fort F, Petit M (2014) Collectives as change agents: the role of social networks in farmers' agroecological transition. Ecol Soc 19(3):16

FAO (2018) The 10 elements of agroecology. Guiding the transition to sustainable food and agricultural systems. https://openknowledge.fao.org/server/api/core/bitstreams/3d7778b3-8fba-4a32-8d13-f21dd5ef31cf/content

Gliessman SR (2015) Agroecology: the ecology of sustainable food systems, 3rd edn. CRC Press, Francis and Taylor Group, Boca Raton, London New York

Lamine C, Adam M, Magda D (2019) The role of networks in the diffusion of agroecological practices: evidence from France. Sustainability 11(18):5065

Meynard JM, Messéan A, Charlier A, Charrier F, Fares M, Le Bail M et al (2013) Freins et leviers à la diversification des cultures. Étude au niveau des exploitations agricoles et des filières. Synthèse du rapport d'étude. Inra, Paris

Nguyen G, Corso JPD, Kephaliacos C, Tavernier H (2013) Pratiques agricoles pour la réduction des produits phytosanitaires. le rôle de l'apprentissage collectif. Économie rurale. Agric, Alimentations, Territoires 333:105–121

Prost L, Coquil X, Fiorelli JL (2019) Social networks and resilience of rural communities to food insecurity: a case study in Northern Mali. Sustainability 11(7):2122

RIAM (2018) Cahier Des Charges «Productions Végétales» Du SPG du RIAM

Scherr SJ, McNeely JA (2008) Biodiversity conservation and agricultural sustainability: towards a new paradigm of "ecoagriculture" landscapes. Philos Trans R Soc Lond Ser b, Biol Sci 363:477–494

Snapp SS, Fisher M, Millar D (2010) Farming systems research and extension. Springer, Dordrecht

Touzard J-M, Fournier S (2014) La complexité des systèmes alimentaires: un atout pour la sécurité alimentaire ? VertigO 14(1). https://doi.org/10.4000/vertigo.14840

Vergote M-H, Tanguy C, Garcia E (2019) Les collectifs producteurs de savoirs pour l'action: retour d'expériences sur la production de légumineuses en Bourgogne-Franche-Comté. Innov Agronomiques 2019(74):105–120. https://doi.org/10.15454/vrxqlw.hal-02342392

Wenger E (1998) Communities of practice: learning, meaning, and identity. Cambridge University Press

Wezel A, Bellon S, Doré T, Francis, C, Vallod D, David C (2009) Agroecology as a science, a movement and a practice. A review. Agron Sustain Dev 29:503–515. https://doi.org/10.1051/agro/2009004

https://siredd.environnement.gov.ma/Rabat-Sale-Kenitra/Home/Region

https://www.hcp.ma/Recensement-population-RGPH-2014_a2941.html

Mr. Mouad Chari is an agricultural engineer from the National School of Agriculture (ENA) of Meknès (Morocco). After working on research and development projects, he joined the Moroccan Ministry of Agriculture.

Prof. Dr. Larbi Aziz is a doctor in environmental science and management (socio-economics of the environment) from the University of Liège (Belgium) and an agricultural engineer from the National School of Agriculture (ENA) of Meknès (Morocco). He is a professor in the Rural Development Engineering Department at ENA of Meknes (Morocco), and coordinator of the "Rural Societies and Territorial Development" research team. His research focuses on agricultural and rural development, in particular the social management of natural resources in the context of global and climatic change

Postface

As we conclude the collection, "Food Systems and Biodiversity in the Context of Environmental and Climate Risks: Dynamics and Evolving Solutions", we do so against a backdrop of unprecedented climate challenges. Between 2023 and 2024, numerous studies showed that global carbon sinks, as discussed within the context of tropical forest ecosystems by Sahoo, Dash, Presty, and Sharma, had temporarily collapsed, absorbing little to no carbon. This revelation is a sharp reminder of the ecological limitations we are facing and the critical need for solutions, such as the ones proposed in this book.

The urgency of responses is not limited to carbon sinks; each chapter in this book has demonstrated that we are in a state of environmental crisis and that now is the time to take meaningful steps toward conservation and sustainability. The proposed solutions are wide-ranging, from the use of saline-tolerant crops in agroforestry systems to adapting indigenous bamboo drip irrigation systems to enhance water use efficiency in water-scarce regions. Despite the range of proposed solutions, successful implementation rests on addressing three factors: the unequal distribution of climate change impacts, the role of economic and policy frameworks, and the need for locally tailored solutions.

First, one of the key takeaways of this volume is the disproportionate impact of climate change on the Global South. These communities are already grappling with rising sea levels, irregular weather patterns, and threats of ecosystem collapse, while also confronting existing vulnerabilities in economies and food systems. Numerous chapters in this work illustrate not just the dangers of dramatic climate change events but even the expected status quo for these areas. For example, Ali, Khan, and Ahmed's discussion on desert locusts reveals how even slight temperature increases raise the likelihood of desert locust outbreaks, devastating crops in regions already experiencing food shortages. While many of the proposed solutions to this and similar challenges are low cost, such as introducing solar dryers in agricultural supply chains, the authors emphasize that communities often require external support to adopt these

M. Behnassi et al. (eds.), *Food Systems and Biodiversity in the Context of Environmental and Climate Risks*, https://doi.org/10.1007/978-3-031-89167-0

initiatives. This points to the critical role of the international community in facilitating technology and knowledge transfers and strengthening the adaptive capacity of vulnerable communities and ecosystems.

The second point emphasized in this collection is the need for strong economic and policy frameworks to drive sustainable solutions. As discussed by Sahoo et al. in their research on tropical forests, the authors advocate for stronger governance and international collaboration. Cohesive policies and enforcement are vital for protecting vital ecosystem services. At the same time, as Mrabet and Behnassi emphasize, in other areas, lowering economic and policy barriers can aid vulnerable farmers in the Global South in accessing new market opportunities. For example, a streamlined process for organic and other certifications can make high-demand markets, such as in the U.S. and the EU, more accessible for smallholder farmers while also incentivizing the adoption of sustainable practices. In both cases, policies must balance economic growth with conservation, as Asmiya et al. argue in their study of blue economy practices. For many communities, immediate economic needs can outweigh the long-term benefits of conservation. Policies that incorporate both livelihoods and climate adaptation are critical to creating attractive paths for communities in the Global South, ensuring environmental protection does not come at the expense of their survival.

Lastly, multiple studies in the collection reinforce the need for tailored adaptations to local contexts. Khan et al.'s discussion of agricultural adaptation in the northern and southern regions of Khyber Pakhtunkhwa demonstrates that similar socioeconomic conditions do not always lead to identical responses from local actors. A similar case study on agroforestry in Cameroon by Alemagi et al. demonstrates how difficulties in resource availability, market access, and community practices can impede the potential of implementing broadly advantageous practices, such as sustainable intensification. These examples demonstrate that climate solutions are not one size fits all and must be tailored to local contexts in order to be effective.

For this reason, this volume highlights the critical roles of local and indigenous knowledge in sustainable development. Smallholder and indigenous communities have shown to have unique expertise in land and water management, crop diversity, and natural resource conservation. In his exploration of the Zabo farming system of Nagaland, Swami discovered the ways in which indigenous communities have combined traditional practices of agroecology with ingenious technological methods to overcome resource constraints and climate change. Indigenous knowledge, cultivated over generations, needs to be centered into the decision-making processes for any genuine attempt at protecting biodiversity and food systems. For this reason, this final condition emphasizes the need for solutions that respect cultural contexts and invest in community innovation.

Throughout this collection, we have witnessed the deeply interconnected nature of food systems, biodiversity, and climate resilience. Each chapter provided concrete strategies to support vulnerable communities in taking agency over their food security while adapting to climate pressures. In concluding this collection, we emphasize the importance of international solidarity, comprehensive and adaptive policy, and local innovation in developing resilient food systems that can protect biodiversity and withstand the challenges of a changing climate.

Aisha Baig
Mohamed Behnassi
Mirza Barjees Baig

The manufacturer's authorised representative in the EU is Springer
Nature Customer Service Centre GmbH, Europaplatz 3, 69115 Heidelberg,
Germany. If you have any concerns regarding our products, please
contact ProductSafety@springernature.com

Printed and bound by CPI Group (UK) Ltd, Croydon, CR0 4YY
30/09/2026
02225588-0018